FUNDAMENTALS OF
NUCLEAR SCIENCE AND ENGINEERING

SECOND EDITION

FUNDAMENTALS OF
NUCLEAR SCIENCE AND ENGINEERING

SECOND EDITION

J. KENNETH SHULTIS
RICHARD E. FAW

CRC Press
Taylor & Francis Group
Boca Raton   London   New York

CRC Press is an imprint of the
Taylor & Francis Group, an informa business

CRC Press
Taylor & Francis Group
6000 Broken Sound Parkway NW, Suite 300
Boca Raton, FL 33487-2742

© 2008 by Taylor & Francis Group, LLC
CRC Press is an imprint of Taylor & Francis Group, an Informa business

No claim to original U.S. Government works
Printed in the United States of America on acid-free paper
10 9 8 7 6 5 4

International Standard Book Number-13: 978-1-4200-5135-3 (Hardcover)

This book contains information obtained from authentic and highly regarded sources. Reprinted material is quoted with permission, and sources are indicated. A wide variety of references are listed. Reasonable efforts have been made to publish reliable data and information, but the author and the publisher cannot assume responsibility for the validity of all materials or for the consequences of their use.

No part of this book may be reprinted, reproduced, transmitted, or utilized in any form by any electronic, mechanical, or other means, now known or hereafter invented, including photocopying, microfilming, and recording, or in any information storage or retrieval system, without written permission from the publishers.

For permission to photocopy or use material electronically from this work, please access www.copyright.com (http://www.copyright.com/) or contact the Copyright Clearance Center, Inc. (CCC) 222 Rosewood Drive, Danvers, MA 01923, 978-750-8400. CCC is a not-for-profit organization that provides licenses and registration for a variety of users. For organizations that have been granted a photocopy license by the CCC, a separate system of payment has been arranged.

Trademark Notice: Product or corporate names may be trademarks or registered trademarks, and are used only for identification and explanation without intent to infringe.

Library of Congress Cataloging-in-Publication Data

Shultis, J. Kenneth.
    Fundamentals of nuclear science and engineering / J. Kenneth Shultis, Richard E. Faw. -- 2nd ed.
      p. cm.
    Includes bibliographical references and index.
    ISBN 978-1-4200-5135-3 (hardbook : alk. paper)
    1. Nuclear engineering. I. Faw, Richard E. II. Title.

TK9145.S474 2007
621.48--dc22   2007023262

Visit the Taylor & Francis Web site at
http://www.taylorandfrancis.com

and the CRC Press Web site at
http://www.crcpress.com

Contents

Preface

In the few years since publication of the first edition of this book, there have been significant advances in nuclear science and engineering. As this new edition is being published, economic considerations as well as concerns over security of energy supply and carbon emissions have led to a renewed interest in nuclear power. New generations of power plants are in various stages of design and construction. New plants are about to be constructed in the United States after many years without any new plants, and nuclear power continues to grow in importance in the Far East and in Europe. The new plants generally take advantage of standardization for economic benefits and passive cooling systems for greater safety. Also in recent years there have been major changes in the industrial organization for nuclear plant construction and operation. Both the technical and the organizational changes are addressed in Chapter 11 of the second edition of this book.

These past few years have also seen substantial changes in nuclear instrumentation and in medical applications of nuclear technology. Many advances in instrumentation have been made in response to needs for screening of packages and shipping containers in search of radioactive materials and special nuclear material. These and other advances are described in a totally rewritten Chapter 8 on the detection and measurement of radiation. We welcome the contribution of this chapter by our colleague Professor Douglas McGregor.

Medical applications are taken up in a revised Chapter 14. Cancer treatment by radiation therapy has become safer and more efficient as a result of a continuing series of advances in the design and operation of linear-accelerator based therapy units and vastly improved collimation systems. Treatment planning based on Monte Carlo dose computation and three-dimensional analysis and display allows greater control of dose uniformity within target structures and minimization of doses to surrounding structures in the body.

Since the first edition of this book, our ability to understand and quantify the health risks of exposure to ionizing radiation has increased substantially because of refinements in the radiation dosimetry associated with nuclear weapons, as reported by the joint U.S.-Japanese Radiation Effects Research Foundation in Hiroshima. That work led to new cancer-risk estimates released by the National Research Council Committee on the Biological Effects of Ionizing Radiation. Furthermore, new risk estimates for hereditary effects have been released by the United Nations Scientific Committee on the Effects of Atomic Radiation. These recent advances are addressed in a major revision of Chapter 9.

In this second edition, a new section has been added to Chapter 12 about fusion power and the methods currently being pursued to demonstrate its feasibility. Currently, major experiments are underway to build both magnetically and inertially confined devices that, for the first time, will be net energy producers.

Particle accelerators are machines that have found application in a wide variety of disciplines. They are used, for example, to explore the fundamental structure of matter, to produce a variety of radionuclides that can be produced in no other way, to treat various medical conditions, and even to determine the authenticity of ancient art objects. In Chapter 13, an extensive section about particle accelerators has been added in this new edition.

As authors of this book, and throughout our lives as students and teachers, we have benefited profoundly from the logic, the clarity of thought, and the elegance of expression of the authors of the texts we studied and used in the classroom. It is inevitable that, consciously or unconsciously, we have emulated the methods of these authors. We acknowledge this with especial gratitude to Irving Kaplan, John Lamarsh, M.M. El-Wakil, and G. Robert Keepin.

In the second edition of this book, we have corrected errors and ambiguities found in the first edition. We hope that we have not introduced any new impediments to learning. Since the publication of the first edition, we have worked with the publisher in issuing a solution manual available to educators. We have increased the number of examples and problems in the second edition and plan to issue a new edition of the solution manual. We would value receiving additional problems and solutions as well as corrections or suggestions from users of the book. The authors may be reached by e-mail at `jks@ksu.edu` and `faw@mne.ksu.edu` and via the internet at `http://www.mne.ksu.edu/~jks`.

J. Kenneth Shultis and Richard E. Faw
Manhattan, Kansas

Preface to the First Edition

Nuclear engineering and the technology developed by this discipline began and reached an amazing level of maturity within the past 60 years. Although nuclear and atomic radiation had been used during the first half of the twentieth century, mainly for medical purposes, nuclear technology as a distinct engineering discipline began after World War II with the first efforts at harnessing nuclear energy for electrical power production and propulsion of ships. During the second half of the twentieth century, many innovative uses of nuclear radiation were introduced in the physical and life sciences, in industry and agriculture, and in space exploration.

The purpose of this book is two-fold as is apparent from the table of contents. The first half of the book is intended to serve as a review of the important results of "modern" physics and as an introduction to the basic nuclear science needed by a student embarking on the study of nuclear engineering and technology. Later in this book, we introduce the theory of nuclear reactors and its applications for electrical power production and propulsion. We also survey many other applications of nuclear technology encountered in space research, industry, and medicine.

The subjects presented in this book were conceived and developed by others. Our role is that of reporters who have taught nuclear engineering for more years than we care to admit. Our teaching and research have benefited from the efforts of many people. The host of researchers and technicians who have brought nuclear technology to its present level of maturity are too many to credit here. Only their important results are presented in this book. For their efforts, which have greatly benefited all nuclear engineers, not least ourselves, we extend our deepest appreciation. As university professors we have enjoyed learning of the work of our colleagues. We hope our present and future students also will appreciate these past accomplishments and will build on them to achieve even more useful applications of nuclear technology. We believe the uses of nuclear science and engineering will continue to play an important role in the betterment of human life.

At a more practical level, this book evolved from an effort at introducing a nuclear engineering option into a much larger mechanical engineering program at Kansas State University. This book was designed to serve both as an introduction to the students in the nuclear engineering option and as a text for other engineering students who want to obtain an overview of nuclear science and engineering. We believe that all modern engineering students need to understand the basic aspects of nuclear science and engineering such as radioactivity and radiation doses and their hazards.

Many people have contributed to this book. First and foremost we thank our colleagues Dean Eckhoff and Fred Merklin, whose initial collection of notes for an introductory course in nuclear engineering motivated our present book intended for a larger purpose and audience. We thank Professor Gale Simons, who helped prepare an early draft of the chapter on radiation detection. Finally, many revisions have been made in response to comments and suggestions made by our students on whom we have experimented with earlier versions of the manuscript. Finally, the camera-ready copy given the publisher has been prepared by us using LaTeX, and, thus, we must accept responsibility for all errors, typographical and other, that appear in this book.

J. Kenneth Shultis and Richard E. Faw

Chapter 1

Fundamental Concepts

The last half of the twentieth century was a time in which tremendous advances in science and technology revolutionized our entire way of life. Many new technologies were invented and developed in this time period from basic laboratory research to widespread commercial application. Communication technology, genetic engineering, personal computers, medical diagnostics and therapy, bioengineering, and material sciences are just a few areas that were greatly affected.

Nuclear science and engineering is another technology that has been transformed in less than fifty years from laboratory research into practical applications encountered in almost all aspects of our modern technological society. Nuclear power, from the first experimental reactor built in 1942, has become an important source of electrical power in many countries. Nuclear technology is widely used in medical imaging, diagnostics, and therapy. Agriculture and many other industries make wide use of radioisotopes and other radiation sources. Finally, nuclear applications are found in a wide range of research endeavors such as archaeology, biology, physics, chemistry, cosmology, and, of course, engineering.

The discipline of nuclear science and engineering is concerned with quantifying how various types of radiation interact with matter and how these interactions affect matter. In this book, we will describe sources of radiation, radiation interactions, and the results of such interactions. As the word "nuclear" suggests, we will address phenomena at a microscopic level, involving individual atoms and their constituent nuclei and electrons. The radiation we are concerned with is generally very penetrating and arises from physical processes at the atomic level.

However, before we begin our exploration of the atomic world, it is necessary to introduce some basic fundamental atomic concepts, properties, nomenclature, and units used to quantify the phenomena we will encounter. Such is the purpose of this introductory chapter.

1.1 Modern Units

With only a few exceptions, units used in nuclear science and engineering are those defined by the SI system of metric units. This system is known as the "International System of Units" with the abbreviation SI taken from the French "Le Système International d'Unités." In this system, there are four categories of units: (1) base units of which there are seven, (2) derived units which are combinations of the base units, (3) supplementary units, and (4) temporary units which are in widespread

1

Table 1.1. The SI system of units and their four categories.

Base SI units:

Physical quantity	Unit name	Symbol
length	meter	m
mass	kilogram	kg
time	second	s
electric current	ampere	A
thermodynamic temperature	kelvin	K
luminous intensity	candela	cd
quantity of substance	mole	mol

Examples of Derived SI units:

Physical quantity	Unit name	Symbol	Formula
force	newton	N	$kg\ m\ s^{-2}$
work, energy, quantity of heat	joule	J	$N\ m$
power	watt	W	$J\ s^{-1}$
electric charge	coulomb	C	$A\ s$
electric potential difference	volt	V	$W\ A^{-1}$
electric resistance	ohm	Ω	$V\ A^{-1}$
magnetic flux	weber	Wb	$V\ s$
magnetic flux density	tesla	T	$Wb\ m^{-2}$
frequency	hertz	Hz	s^{-1}
radioactive decay rate	becquerel	Bq	s^{-1}
pressure	pascal	Pa	$N\ m^{-2}$
velocity			$m\ s^{-1}$
mass density			$kg\ m^{-3}$
area			m^2
volume			m^3
molar energy			$J\ mol^{-1}$
electric charge density			$C\ m^{-3}$

Supplementary Units:

Physical quantity	Unit name	Symbol
plane angle	radian	rad
solid angle	steradian	sr

Temporary Units:

Physical quantity	Unit name	Symbol	Value in SI unit
length	nautical mile		1852 m
velocity	knot		$1852/3600\ m\ s^{-1}$
length	angström	Å	$0.1\ nm = 10^{-10}\ m$
area	hectare	ha	$1\ hm^2 = 10^4\ m^2$
pressure	bar	bar	0.1 MPa
pressure	standard atmosphere	atm	0.101325 MPa
area	barn	b	$10^{-24}\ cm^2$
radioactive activity	curie	Ci	$3.7 \times 10^{10}\ Bq$
radiation exposure	röentgen	R	$2.58 \times 10^{-4}\ C\ kg^{-1}$
absorbed radiation dose	gray	Gy	$1\ J\ kg^{-1}$
radiation dose equivalent	sievert	Sv	

Source: NBS Special Publication 330, National Bureau of Standards, Washington, DC, 1977.

use for special applications. These units are shown in Table 1.1. To accommodate very small and large quantities, the SI units and their symbols are scaled by using the SI prefixes given in Table 1.2.

There are several units outside the SI which are in wide use. These include the time units day (d), hour (h), and minute (min); the liter (L or ℓ); plane angle degree (°), minute ('), and second ("); and, of great use in nuclear and atomic physics, the electron volt (eV) and the atomic mass unit (u). Conversion factors to convert some non-SI units to their SI equivalent are given in Table 1.3.

Finally it should be noted that correct use of SI units requires some "grammar" on how to properly combine different units and the prefixes. A summary of the SI grammar is presented in Table 1.4.

Table 1.2. SI prefixes.

Factor	Prefix	Symbol
10^{24}	yotta	Y
10^{21}	zetta	Z
10^{18}	exa	E
10^{15}	peta	P
10^{12}	tera	T
10^{9}	giga	G
10^{6}	mega	M
10^{3}	kilo	k
10^{2}	hecto	h
10^{1}	deca	da
10^{-1}	deci	d
10^{-2}	centi	c
10^{-3}	milli	m
10^{-6}	micro	μ
10^{-9}	nano	n
10^{-12}	pico	p
10^{-15}	femto	f
10^{-18}	atto	a
10^{-21}	zepto	z
10^{-24}	yocto	y

Table 1.3. Conversion factors.

Property	Unit	SI equivalent
Length	in	2.54×10^{-2} m[a]
	ft	3.048×10^{-1} m[a]
	mile (int'l)	$1.609\,344 \times 10^{3}$ m[a]
Area	in^2	6.4516×10^{-4} m^{2a}
	ft^2	$9.290\,304 \times 10^{-2}$ m^{2a}
	acre	$4.046\,873 \times 10^{3}$ m^2
	square mile (int'l)	$2.589\,988 \times 10^{6}$ m^2
	hectare	1×10^{4} m^2
Volume	oz (U.S. liquid)	$2.957\,353 \times 10^{-5}$ m^3
	in^3	$1.638\,706 \times 10^{-5}$ m^3
	gallon (U.S.)	$3.785\,412 \times 10^{-3}$ m^3
	ft^3	$2.831\,685 \times 10^{-2}$ m^3
Mass	oz (avdp.)	$2.834\,952 \times 10^{-2}$ kg
	lb	$4.535\,924 \times 10^{-1}$ kg
	ton (short)	$9.071\,847 \times 10^{2}$ kg
Force	kg_f	$9.806\,650$ N [a]
	lb_f	$4.448\,222$ N
	ton	$8.896\,444 \times 10^{3}$ N
Pressure	lb_f/in^2 (psi)	$6.894\,757 \times 10^{3}$ Pa
	lb_f/ft^2	$4.788\,026 \times 10^{1}$ Pa
	atm (standard)	$1.013\,250 \times 10^{5}$ Pa[a]
	in. H_2O (@ 4 °C)	$2.490\,82 \times 10^{2}$ Pa
	in. Hg (@ 0 °C)	$3.386\,39 \times 10^{3}$ Pa
	mm Hg (@ 0 °C)	$1.333\,22 \times 10^{2}$ Pa
	bar	1×10^{5} Pa[a]
Energy	eV	$1.602\,19 \times 10^{-19}$ J
	cal	4.184 J[a]
	Btu	$1.054\,350 \times 10^{3}$ J
	kWh	3.6×10^{6} J[a]
	MWd	8.64×10^{10} J[a]

[a]Exact converson factor.

Source: *Standards for Metric Practice*, ANSI/ASTM E380-76, American National Standards Institute, New York, 1976.

Table 1.4. Summary of SI grammar.

Grammar	Comments
capitalization	A unit name is never capitalized even if it is a person's name. Thus curie, not Curie. However, the symbol or abbreviation of a unit named after a person is capitalized. Thus Sv, not sv.
space	Use 58 m, not 58m .
plural	A symbol is never pluralized. Thus 8 N, not 8 Ns or 8 N^s.
raised dots	Sometimes a raised dot is used when combining units such as N·m^2·s; however, a single space between unit symbols is preferred as in N m^2 s.
solidus	For simple unit combinations use g/cm^3 or g cm^{-3}. However, for more complex expressions, N m^{-2} s^{-1} is much clearer than N/m^2/s.
mixing units/names	Never mix unit names and symbols. Thus kg/s, not kg/second or kilogram/s.
prefix	Never use double prefixes such as $\mu\mu$g; use pg. Also put prefixes in the numerator. Thus km/s, not m/ms.
double vowels	When spelling out prefixes with names that begin with a vowel, supress the ending vowel on the prefix. Thus megohm and kilohm, not megaohm and kiloohm.
hyphens	Do not put hyphens between unit names. Thus newton meter, not newton-meter. Also never use a hyphen with a prefix. Hence, write microgram not micro-gram.
numbers	For numbers less than one, use 0.532 not .532. Use prefixes to avoid large numbers; thus 12.345 kg, not 12345 g. For numbers with more than 5 adjacent numerals, spaces are often used to group numerals into triplets; thus 123 456 789.123 456 33, not 123456789.12345633.

1.1.1 Special Nuclear Units

When treating atomic and nuclear phenomena, physical quantities such as energies and masses are extremely small in SI units, and special units are almost always used. Two such units are of particular importance.

The Electron Volt

The energy released or absorbed in a chemical reaction (arising from changes in electron bonds in the affected molecules) is typically of the order of 10^{-19} J. It is much more convenient to use a special energy unit called the *electron volt*. By definition, the electron volt is the kinetic energy gained by an electron (mass m_e and charge $-e$) that is accelerated through a potential difference ΔV of one volt $= 1$ W/A $= 1$ (J s^{-1})/(C s^{-1}) $= 1$ J/C. The work done by the electric field is $-e\Delta V = (1.60217646 \times 10^{-19}$ C)(1 J/C) $= 1.60217646 \times 10^{-19}$ J $\equiv 1$ eV. Thus

$$1 \text{ eV} = 1.602\,176\,46 \times 10^{-19} \text{ J}.$$

If the electron (mass m_e) starts at rest, then the kinetic energy T of the electron after being accelerated through a potential of 1 V must equal the work done on the electron, i.e.,

$$T \equiv \frac{1}{2}m_e v^2 = -e\Delta V = 1 \text{ eV}. \tag{1.1}$$

The speed of the electron is thus $v = \sqrt{2T/m_e} \simeq 5.93 \times 10^5$ m/s, fast by our everyday experience but slow compared to the speed of light ($c \simeq 3 \times 10^8$ m/s).

The Atomic Mass Unit

Because the mass of an atom is so much less than 1 kg, a mass unit more appropriate to measuring the mass of atoms has been defined independent of the SI kilogram mass standard (a platinum cylinder in Paris). The *atomic mass unit* (abbreviated as amu, or just u) is *defined* to be 1/12 the mass of a neutral ground-state atom of ^{12}C. Equivalently, the mass of N_a ^{12}C atoms (Avogadro's number = 1 mole) is 0.012 kg. Thus, 1 amu equals $(1/12)(0.012 \text{ kg}/N_a) = 1.660\,538\,7 \times 10^{-27}$ kg.

1.1.2 Physical Constants

Although science depends on a vast number of empirically measured constants to make quantitative predictions, there are some very fundamental constants which specify the scale and physics of our universe. These *physical constants*, such as the speed of light in vacuum c, the mass of the neutron m_n, Avogadro's number N_a, etc., are indeed true constants of our physical world, and can be viewed as auxiliary units. Thus, we can measure speed as a fraction of the speed of light or mass as a multiple of the neutron mass. Some of the important physical constants, which we use extensively, are given in Table 1.5.

1.2 The Atom

Crucial to an understanding of nuclear technology is the concept that all matter is composed of many small discrete units of mass called *atoms*. Atoms, while often viewed as the fundamental constituents of matter, are themselves composed of other particles. A simplistic view of an atom is a very small dense *nucleus*, composed of protons and neutrons (collectively called *nucleons*), that is surrounded by a swarm of negatively-charged electrons equal in number to the number of positively-charged protons in the nucleus. In later chapters, more detailed models of the atom are introduced.

It is often said that atoms are so small that they cannot be seen. Certainly, they cannot with the naked human eye or even with the best light microscope. However, so-called tunneling electron microscopes can produce electrical signals, which, when plotted, can produce images of individual atoms. In fact, the same instrument can also move individual atoms. An example is shown in Fig. 1.1. In this figure, iron atoms (the dark circular dots) on a copper surface are shown being moved to form a ring which causes electrons inside the ring and on the copper surface to form standing waves. This and other pictures of atoms can be found on the web at http://www.almaden.ibm.com/vis/stm/gallery.html.

Table 1.5. Values of some important physical constants as internationally recommended in 2005.

Constant	Symbol	Value
Speed of light (in vacuum)	c	$2.997\,924\,58 \times 10^{8}$ m s^{-1}
Electron charge	e	$1.602\,176\,53 \times 10^{-19}$ C
Atomic mass unit	u	$1.660\,538\,9 \times 10^{-27}$ kg ($931.494\,043$ MeV/c^2)
Electron rest mass	m_e	$9.109\,382\,6 \times 10^{-31}$ kg ($0.510\,998\,92$ MeV/c^2) ($5.485\,799\,09 \times 10^{-4}$ u)
Proton rest mass	m_p	$1.672\,621\,71 \times 10^{-27}$ kg ($938.272\,03$ MeV/c^2) ($1.007\,276\,466\,9$ u)
Neutron rest mass	m_n	$1.674\,927\,3 \times 10^{-27}$ kg ($939.565\,36$ MeV/c^2) ($1.008\,664\,915\,6$ u)
Planck's constant	h	$6.626\,069\,3 \times 10^{-34}$ J s $4.135\,667\,4 \times 10^{-15}$ eV s
Avogadro's constant	N_a	$6.022\,141\,5 \times 10^{23}$ mol^{-1}
Boltzmann constant	k	$1.380\,650\,5 \times 10^{-23}$ J K^{-1} ($8.617\,343 \times 10^{-5}$ eV K^{-1})
Ideal gas constant (STP)	R	$8.314\,472$ J mol^{-1} K^{-1}
Electric constant	ϵ_o	$8.854\,187\,817 \times 10^{-12}$ F m^{-1}

Source: P.J. Mohr and B.N. Taylor, "CODATA Recommended Values of the Fundamental Physical Constants," *Rev. Modern Physics*, **77**, 1, 2005.

1.2.1   The Fundamental Constituents of Ordinary Matter

Throughout this book, atoms, neutrons, and protons are treated as the only "fundamental" entities of interest. Indeed, for the energies involved in practical nuclear and atomic interactions, this view is well justified. However, at kinetic energies produced in particle accelerators or found in cosmic rays, the microscopic structure of matter is found to be much more complicated. Although this fine structure is not of practical importance in our study of nuclear science and engineering, a brief overview of the constituents of ordinary matter is presented here.

Although neutrons and protons are often considered "fundamental" particles, we now know that they are composed of other smaller particles called *quarks* held together by the exchange of massless particles called *gluons*, which are the carriers of the nuclear force. The protons and neutrons are bound together in a nucleus by the residual interactions of the gluons in the neutrons and protons.

From studies of nuclear interactions produced by energetic cosmic rays and by particle accelerators (see Section 13.6), it is believed that there are six different types or "flavors" of quarks with the fanciful names up (u), down (d), strange (s), charm (c), top (t), and bottom (b). For each quark there exists an antiquark, denoted by $\bar{u}$, $\bar{d}$, etc. Each quark has a charge that is a fraction of the electron charge, namely the u, d, c, s, t, and b quarks have charge $2/3, -1/3, 2/3, -1/3, 2/3,$

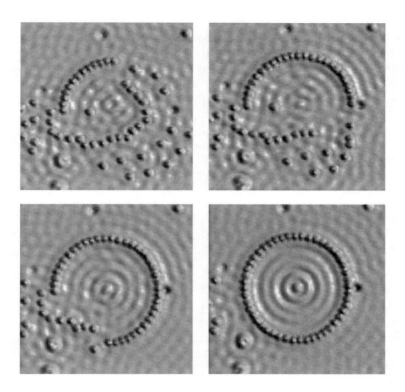

Figure 1.1. Pictures of iron atoms on a copper surface being moved to form a ring inside of which surface copper electrons are confined and form standing waves. Source: IBM Corp.

and $-1/3$, respectively. An antiquark has the same mass and angular momentum, but opposite charge, as its quark.

With the quarks and antiquarks two types of composite particles are formed. *Baryons* are composed of 3 quarks; for example, the neutron is a (ddu) composite (with a net charge of $-1/3 - 1/3 + 2/3 = 0$) and the proton is a (uud) composite (with a net charge of $2/3 + 2/3 - 1/3 = 1$). *Mesons* are composed of a quark-antiquark pair; for example, a π^+ meson is a $(u,\bar{d})$ pair and a π^- meson is a $(\bar{u},d)$ pair.

In addition to the quarks, there is another class of fundamental particles, called *leptons* consisting of the electron, muon, and tau particle, each with an associated neutrino. Each of the leptons also has an antiparticle. The quarks and leptons are exceedingly small with sizes less than about 10^{-19} m.

"Ordinary" matter is mostly composed of neutrons, protons, electrons, and neutrinos, each of which has an angular momentum of $\pm h/(2\pi)$ (or a "spin" of $1/2$) and, as a result, are called *fermions*. Most of the possible composite baryons and meson are extremely unstable, with the heavier quarks decaying rapidly to less massive ones by the forces mitigated by *bosons*, such as the photon or $W^\pm$ boson, all of which have spin $= 1$. For example, a free neutron can decay to a proton when a d quark changes to an u quark by emitting a W^- boson, which almost instantly decays to an electron (e^-) and an electron antineutrino $(\bar{\nu}_e)$. The known "fundamental" particles with their mass and charge are shown in Table 1.6. Whether there are additional fundamental particles or even whether these particles are themselves composite particles made from even more complex structures are unknown and are presently topics of great interest to high-energy physicists.

Table 1.6. The fundamental particles that make up "ordinary" matter. Each fermion has a corresponding antiparticle with the same mass but of opposite charge. Data from Yao et al. (Particle Data Group) *J. Phys G*, **33**, 1, 2006.

Fermions (spin = 1/2)					
quarks			**leptons**		
flavor	mass $(\mathrm{GeV/c^2})$	charge (e)	flavor	mass $(\mathrm{GeV/c^2})$	charge (e)
u	0.0015–0.0030	2/3	ν_e	$< 2 \times 10^{-9}$	0
d	0.003–0.007	$-1/3$	e^-	0.00051100	-1
c	1.25	2/3	ν_μ	< 0.00019	0
s	0.095 ± 0.025	$-1/3$	μ^-	0.10566	-1
t	176	2/3	ν_τ	< 0.018	0
b	4.70	$-1/3$	τ^-	1.7769	-1

Interaction Bosons (spin = 1)			
symbol	force	mass $(\mathrm{GeV/c^2})$	charge (e)
gluon g	strong	0	0
γ	electromagnetic	0	0
W^-	weak	80.4	-1
W^+	weak	80.4	1
Z^0	weak	91.187	0

However, in our study of nuclear science and engineering, we can view the electron, neutron, and proton as fundamental indivisible particles, because the composite nature of nucleons becomes apparent only under extreme conditions, such as those encountered during the first second after the creation of the universe (the "big bang") or in high-energy particle accelerators. We will not deal with such gigantic energies. Rather, the energy of radiation we consider in this book is sufficient only to rearrange or remove the electrons in an atom or the neutrons and protons in a nucleus.

1.2.2 Dark Matter and Energy

In our universe all *ordinary* matter is composed of electrons, neutrinos, and up and down quarks which are bound together in protons and neutrons. However, over the past several decades, it has become apparent that the gravitational forces from ordinary matter are insufficient to explain (1) the motion of stars near the center of galaxies, (2) the filigreed distribution of galaxies in the universe, and (3) the accelerating expansion of the universe itself. To explain these observations, astrophysicists have come to accept that there must be vast amounts of *dark matter* to provide extra gravitational forces on stars in a galaxy and between galaxies. Further, the amount of dark matter must be about five times that of ordinary matter. Although dark matter can exert gravitational forces, it does not interact with light and, hence, is invisible to us. What the fundamental particles of dark matter are unknown and is the subject of much current speculation.

Astrophysicists have also had to postulate the existence of something called *dark energy* that must pervade the universe in order to explain the observed accelerated expansion of our universe. Again, what constitutes dark energy is unknown, but the amount of dark energy represents about 70% of everything in the universe!

1.2.3   Atomic and Nuclear Nomenclature

The identity of an atom is uniquely specified by the number of neutrons N and protons Z in its nucleus. For an electrically neutral atom, the number of electrons equals the number of protons Z, which is called the *atomic number*. All atoms of the same *element* have the same atomic number. Thus, all oxygen atoms have 8 protons in the nucleus while all uranium atoms have 92 protons.

However, atoms of the same element may have different numbers of neutrons in the nucleus. Atoms of the same element, but with different numbers of neutrons, are called *isotopes*. The symbol used to denote a particular isotope is

$$\boxed{{}_{Z}^{A}\mathrm{X}}$$

where X is the chemical symbol and $A \equiv Z + N$, which is called the *mass number*. For example, two uranium isotopes, which will be discussed extensively later, are ${}_{92}^{235}\mathrm{U}$ and ${}_{92}^{238}\mathrm{U}$. The use of both Z and X is redundant because one specifies the other. Consequently, the subscript Z is often omitted, so that we may write, for example, simply ${}^{235}\mathrm{U}$ and ${}^{238}\mathrm{U}$.[1]

Because isotopes of the same element have the same number and arrangement of electrons around the nucleus, the chemical properties of such isotopes are nearly identical. Only for the lightest isotopes (e.g., ${}^{1}\mathrm{H}$, deuterium ${}^{2}\mathrm{H}$, and tritium ${}^{3}\mathrm{H}$) are small differences noted. For example, light water ${}^{1}\mathrm{H}_2\mathrm{O}$ freezes at 0 °C while heavy water ${}^{2}\mathrm{H}_2\mathrm{O}$ (or $\mathrm{D}_2\mathrm{O}$ since deuterium is often given the chemical symbol D) freezes at 3.82 °C.

A discussion of different isotopes and elements often involves the following basic nuclear jargon.

nuclide: a term used to refer to a particular atom or nucleus with a specific neutron number N and atomic (proton) number Z. Nuclides are either *stable* (i.e., unchanging in time unless perturbed) or radioactive (i.e., they spontaneously change to another nuclide with a different Z and/or N by emitting one or more particles). Such radioactive nuclides are termed *radionuclides*.

---

[1]To avoid superscripts, which were hard to make on old-fashioned typewriters, the simpler form U-235 and U-238 was often employed. However, with modern word processing, this form should no longer be used.

isobar: nuclides with the same mass number $A = N + Z$ but with different number of neutrons N and protons Z. Nuclides in the same isobar have nearly equal masses. For example, isotopes which have nearly the same isobaric mass of 14 u include $^{14}_{5}B$, $^{14}_{6}C$, $^{14}_{7}N$, and $^{14}_{8}O$.

isotone: nuclides with the same number of neutrons N but different number of protons Z. For example, nuclides in the isotone with 8 neutrons include $^{13}_{5}B$, $^{14}_{6}C$, $^{15}_{7}N$, and $^{16}_{8}O$.

isomer: the same nuclide (same Z and N) in which the nucleus is in different long-lived excited states. For example, an isomer of ^{99}Te is ^{99m}Te where the m denotes the longest-lived excited state (i.e., a state in which the nucleons in the nucleus are not in the lowest energy state).

1.2.4   Atomic and Molecular Weights

The *atomic weight* $\mathcal{A}$ of an atom is the ratio of the atom's mass to that of one-twelfth of a neutral atom of ^{12}C in its ground state. Similarly, the *molecular weight* of a molecule is the ratio of its mass to that of one-twelfth of an atom of ^{12}C. As ratios, the atomic and molecular weights are dimensionless numbers.

Closely related to the concept of atomic weight is the *atomic mass unit*, which we introduced in Section 1.1.1 as a special mass unit. Recall that the atomic mass unit is defined such that the mass of a ^{12}C atom is 12 u. It then follows that the mass M of an atom measured in atomic mass units numerically equals the atom's atomic weight $\mathcal{A}$. From Table 1.5 we see 1 u $\simeq 1.6605 \times 10^{-27}$ kg. A detailed listing of the atomic masses of the known nuclides is given in Appendix B. From this appendix, we see that the atomic mass (in u) and, hence, the atomic weight of a nuclide almost equals (within less than one percent) the atomic mass number A of the nuclide. Thus for approximate calculations, we can usually assume $\mathcal{A} \simeq A$.

Most naturally occurring elements are composed of two or more isotopes. The *isotopic abundance* γ_i of the i-th isotope in a given element is the fraction of the atoms in the element that are that isotope. Isotopic abundances are usually expressed in atom percent and are given in Appendix Table A.4. For a specified element, the *elemental atomic weight* is the weighted average of the atomic weights of all naturally occurring isotopes of the element, weighted by the isotopic abundance of each isotope, i.e.,

$$\mathcal{A} = \sum_i \frac{\gamma_i(\%)}{100} \mathcal{A}_i, \tag{1.2}$$

where the summation is over all the isotopic species comprising the element. Elemental atomic weights are listed in Appendix Tables A.2 and A.3.

1.2.5   Avogadro's Number

Avogadro's constant is the key to the atomic world since it relates the number of microscopic entities in a sample to a macroscopic measure of the sample. Specifically, Avogadro's constant $N_a \simeq 6.022 \times 10^{23}$ equals the number of atoms in 12 grams of ^{12}C. Few fundamental constants need be memorized, but an approximate value of Avogadro's constant should be.

Example 1.1: What is the atomic weight of boron? From Table A.4 we find that naturally occurring boron consists of two stable isotopes ^{10}B and ^{11}B with isotopic abundances of 19.9 and 80.1 atom-percent, respectively. From Appendix B the atomic weight of ^{10}B and ^{11}B are found to be 10.012937 and 11.009306, respectively. Then from Eq. (1.2) we find

$$\mathcal{A}_B = (\gamma_{10}\mathcal{A}_{10} + \gamma_{11}\mathcal{A}_{11})/100$$

$$= (0.199 \times 10.012937) + (0.801 \times 11.009306) = 10.81103.$$

This value agrees with the tabulated value $\mathcal{A}_B = 10.811$ as listed in Tables A.2 and A.3.

The importance of Avogadro's constant lies in the concept of the mole. A *mole* (abbreviated mol) of a substance is defined to contain as many "elementary particles" as there are atoms in 12 g of ^{12}C. In older texts, the mole was often called a "gram-mole" but is now called simply a mole. The "elementary particles" can refer to any identifiable unit that can be unambiguously counted. We can, for example, speak of a mole of stars, persons, molecules, or atoms.

Because the atomic weight of ^{12}C is defined to be 12 and because 12 g of ^{12}C is defined to contain 1 mol of atoms, it follows that the mass *in grams* of any atomic species that numerically equals the dimensionless atomic weight of the species must also contain 1 mole of atoms. The mass in grams of a substance that equals the dimensionless atomic or molecular weight is sometimes called the *gram atomic weight* or *gram molecular weight*. Thus, one gram atomic or molecular weight of any substance represents one mole of the substance and contains as many atoms or molecules as there are atoms in one mole of ^{12}C, namely, N_a atoms or molecules. That one mole of any substance contains N_a entities is known as *Avogadro's law* and is the fundamental principle that relates the microscopic world to the everyday macroscopic world. Table 1.7 illustrates how Avogadro's number N_a, together with the atomic weight $\mathcal{A}$, is the key to transforming between macroscopic masses and the number of atoms.

Table 1.7. Use of Avogadro's constant N_a (atoms/mol) and the atomic weight $\mathcal{A}$ (g/mol) to move between the atomic and macroscopic worlds.

Macroscopic World	Transformation $\overset{\longleftarrow}{\longrightarrow}$			Microscopic World
m (g)	$\overset{\div \mathcal{A}}{\longrightarrow}$	$\dfrac{m}{\mathcal{A}}$ (mols)	$\overset{\times N_a}{\longrightarrow}$	$\dfrac{m}{\mathcal{A}}N_a$ (atoms)
$m = \dfrac{N}{N_a}\mathcal{A}$ (g)	$\overset{\times \mathcal{A}}{\longleftarrow}$	$\dfrac{N}{N_a}$ (mols)	$\overset{\div N_a}{\longleftarrow}$	N (atoms)
$\rho \left(\dfrac{\text{g}}{\text{cm}^3}\right)$	$\overset{\div \mathcal{A}}{\longrightarrow}$	$\dfrac{\rho}{\mathcal{A}} \left(\dfrac{\text{mol}}{\text{cm}^3}\right)$	$\overset{\times N_a}{\longrightarrow}$	$\dfrac{\rho N_a}{\mathcal{A}} \left(\dfrac{\text{atoms}}{\text{cm}^3}\right)$
$m = \dfrac{\mathcal{A}}{N_a} \left(\dfrac{\text{g}}{\text{atom}}\right)$	$\overset{\times \mathcal{A}}{\longleftarrow}$	$\dfrac{1}{N_a}$ (mol)	$\overset{\div N_a}{\longleftarrow}$	1 (atom)

Example 1.2: How many atoms of ^{10}B are there in 5 grams of boron? From Table A.3, the atomic weight of elemental boron $\mathcal{A}_B = 10.811$. The 5-g sample of boron equals $m/\mathcal{A}_B$ moles of boron, and since each mole contains N_a atoms, the number of boron atoms is

$$N_B = \frac{mN_a}{\mathcal{A}_B} = \frac{(5 \text{ g})(0.6022 \times 10^{24} \text{ atoms/mol})}{(10.811 \text{ g/mol})} = 2.785 \times 10^{23} \text{ atoms.}$$

From Table A.4, the isotopic abundance of ^{10}B in elemental boron is found to be 19.9%. The number N_{10} of ^{10}B atoms in the sample is, therefore, $N_{10} = (0.199)(2.785 \times 10^{23}) = 5.542 \times 10^{22}$ atoms.

1.2.6 Mass of an Atom

With Avogadro's number many basic properties of atoms can be inferred. For example, the mass of an individual atom can be found. Since a mole of a group of identical atoms (with a mass of $\mathcal{A}$ grams) contains N_a atoms, the mass of an individual atom is

$$M \text{ (g/atom)} = \mathcal{A}/N_a \simeq A/N_a. \tag{1.3}$$

The approximation of $\mathcal{A}$ by A is usually quite acceptable for all but the most precise calculations. This approximation will be used often throughout this book.

In Appendix B, a comprehensive listing is provided for the masses of all the known atomic species. As will soon become apparent, atomic masses are central to quantifying the energetics of various nuclear reactions.

Example 1.3: Estimate the mass of an atom of ^{238}U. From Eq. (1.3) we find

$$M(^{238}\text{U}) \simeq \frac{238 \text{ (g/mol)}}{6.022 \times 10^{23} \text{ atoms/mol}} = 3.952 \times 10^{-22} \text{ g/atom.}$$

From Appendix B, the mass of ^{238}U is found to be 238.050782 u which numerically equals its gram atomic weight $\mathcal{A}$. A more precise value for the mass of an atom of ^{238}U is, therefore,

$$M(^{238}\text{U}) = \frac{238.050782 \text{ (g/mol)}}{6.022142 \times 10^{23} \text{ atoms/mol}} = 3.952925 \times 10^{-22} \text{ g/atom.}$$

Notice that approximating $\mathcal{A}$ by A leads to a very small and usually negligible error.

1.2.7 Atomic Number Density

In many calculations, we will need to know the number of atoms in 1 cm^3 of a substance. Again, Avogadro's number is the key to finding the atom density. For a homogeneous substance of a single species and with mass density ρ g/cm^3, 1 cm^3

contains $\rho/\mathcal{A}$ moles of the substance and $\rho N_a/\mathcal{A}$ atoms. The atom density N is thus

$$\boxed{N \text{ (atoms/cm}^3) = \frac{\rho}{\mathcal{A}} N_a.} \tag{1.4}$$

To find the atom density N_i of isotope i of an element with atom density N, simply multiply N by the fractional isotopic abundance $\gamma_i/100$ for the isotope, i.e., $N_i = (\gamma_i/100)N$.

Equation 1.4 also applies to substances composed of identical molecules. In this case, N is the molecular density and $\mathcal{A}$ the gram molecular weight. The number of atoms of a particular type, per unit volume, is found by multiplying the molecular density by the number of the same atoms per molecule. This is illustrated in the following example.

Example 1.4: What is the hydrogen atom density in water? The molecular weight of water $\mathcal{A}_{\mathrm{H_2O}} = 2\mathcal{A}_H + \mathcal{A}_O \simeq 2A_H + A_O = 18$. The molecular density of H_2O is thus

$$N(\mathrm{H_2O}) = \frac{\rho^{\mathrm{H_2O}} N_a}{\mathcal{A}_{\mathrm{H_2O}}} = \frac{(1 \text{ g cm}^{-3}) \times (6.022 \times 10^{23} \text{ molecules/mol})}{18 \text{ g/mol}}$$

$$= 3.35 \times 10^{22} \text{ molecules/cm}^3.$$

The hydrogen density $N(\mathrm{H}) = 2\,N(\mathrm{H_2O}) = 2(3.35 \times 10^{22}) = 6.69 \times 10^{22}$ atoms/cm^3.

The composition of a mixture such as concrete is often specified by the mass fraction w_i of each constituent. If the mixture has a mass density ρ, the mass density of the ith constituent is $\rho_i = w_i\rho$. The density N_i of the ith component is thus

$$N_i = \frac{\rho_i N_a}{\mathcal{A}_i} = \frac{w_i\rho N_a}{\mathcal{A}_i}. \tag{1.5}$$

If the composition of a substance is specified by a chemical formula, such as X_nY_m, the molecular weight of the mixture is $\mathcal{A} = n\mathcal{A}_X + m\mathcal{A}_Y$ and the mass fraction of component X is

$$w_X = \frac{n\mathcal{A}_X}{n\mathcal{A}_X + m\mathcal{A}_Y}. \tag{1.6}$$

Finally, as a general rule of thumb, it should be remembered that atom densities in solids and liquids are usually between 10^{21} and 10^{23} atoms cm^{-3}. Gases at standard temperature and pressure are typically less by a factor of 1000.

1.2.8 Size of an Atom

For a substance with an atom density of N atoms/cm^3, each atom has an associated volume of $V = 1/N$ cm^3. If this volume is considered a cube, the cube's width is $V^{1/3}$. For ^{238}U, the cubical size of an atom is thus $1/N^{1/3} = 2.7 \times 10^{-8}$ cm.

Measurements of the size of atoms reveal a diffuse electron cloud about the nucleus. Although there is no sharp edge to an atom, an effective radius can be defined such that outside this radius, an electron is very unlikely to be found. Almost all atoms with $Z > 10$ have radii between 1×10^{-8} and 2×10^{-8} cm. As Z increases, i.e., as more electrons and protons are added, the size of the electron cloud changes little, but simply becomes more dense. Hydrogen, the lightest element, is also the smallest with a radius of about 0.5×10^{-8} cm.

1.2.9   Atomic and Isotopic Abundances

During the first few minutes after the big bang only the lightest elements (hydrogen, helium, and lithium) were created. All the others were created inside stars either during their normal aging process or during supernova explosions. In both processes, nuclei are combined or fused to form heavier nuclei. Our earth with all the naturally occurring elements was formed from debris of dead stars. The abundances of the elements for our solar system is a consequence of the history of stellar formation and death in our corner of the universe. Elemental abundances are listed in Table A.3. For a given element, the different stable isotopes also have a natural relative abundance unique to our solar system. These isotopic abundances are listed in Table A.4.

1.2.10   Nuclear Dimensions

Size of a Nucleus

If each proton and neutron in the nucleus has the same volume, the volume of a nucleus should be proportional to A. This has been confirmed by many measurements that have explored the shape and size of nuclei. Nuclei, to a first approximation, are spherical or very slightly ellipsoidal with a somewhat diffuse surface. In particular, it is found that an effective spherical nuclear radius is

$$R = R_o A^{1/3}, \qquad \text{with } R_o \simeq 1.25 \times 10^{-13} \text{ cm.} \tag{1.7}$$

The associated volume is

$$V_{\text{nucleus}} = \frac{4}{3}\pi R^3 \simeq 7.25 \times 10^{-39} A \quad \text{cm}^3. \tag{1.8}$$

Since the atomic radius of about 2×10^{-8} cm is 10^5 times greater than the nuclear radius, the nucleus occupies only about 10^{-15} of the volume of an atom. If an atom were to be scaled to the size of a large concert hall, then the nucleus would be the size of a very small gnat!

Nuclear Density

Since the mass of a nucleon (neutron or proton) is much greater than the mass of electrons in an atom ($m_n \simeq 1837 \, m_e$), the mass density of a nucleus is

$$\rho_{\text{nucleus}} = \frac{m_{\text{nucleus}}}{V_{\text{nucleus}}} = \frac{A/N_a}{(4/3)\pi R^3} = 2.4 \times 10^{14} \text{ g/cm}^3.$$

This is the density of the earth if it were compressed to a ball 200 m in radius.

1.3 Chart of the Nuclides

The number of known different atoms, each with a distinct combination of Z and A, is large, numbering over 3200 nuclides. Of these, 266 are stable (i.e., non-radioactive) and are found in nature. There are also 65 long-lived radioisotopes found in nature. The remaining nuclides have been made by humans and are radioactive with lifetimes much shorter than the age of the solar system. The lightest atom $(A = 1)$ is ordinary hydrogen $^1_1$H, while the mass of the heaviest is continually increasing as heavier and heavier nuclides are produced in nuclear research laboratories. One of the heaviest $(A = 269)$ is meitnerium $^{269}_{109}$Mt.

A very compact way to portray this panoply of atoms and some of their properties is known as the *Chart of the Nuclides*. This chart is a two-dimensional matrix of squares (one for each known nuclide) arranged by atomic number Z (y-axis) versus neutron number N (x-axis). Each square contains information about the nuclide. The type and amount of information provided for each nuclide is limited only by the physical size of the chart. Several versions of the chart are available on the internet (see web addresses given in the next section and in Appendix A).

Perhaps, the most detailed Chart of the Nuclides is that provided by General Electric Co. (GE). This chart (like many other information resources) is not available on the web; rather, it can be purchased from GE ($15 for students) and is highly recommended as a basic data resource for any nuclear analysis. It is available as a 32 inch × 55 inch chart or as a 64-page book. Information for ordering this chart can be found on the web at `http://www.chartofthenuclides.com`.

1.3.1 Other Sources of Atomic/Nuclear Information

A vast amount of atomic and nuclear data is available on the world-wide web. However, it often takes considerable effort to find exactly what you need. The sites listed below contain many links to data sources, and you should explore these to become familiar with them and what data can be obtained through them.

The following site, in particular, has a large number of links to the major nuclear and atomic data repositories in the world.

`http://www.nndc.bnl.gov/usndp/usndp-subject.html`

The following sites offer large compilations of fundamental nuclear and atomic data as well as links to other data sites.

`http://www.nndc.bnl.gov/`
`http://physics.nist.gov/PhysRefData/contents.html`
`http://isotopes.lbl.gov/`
`http://wwwndc.tokai-sc.jaea.go.jp/index.html`
`http://wwwndc.tokai-sc.jaea.go.jp/nucldata/index.html`
`http://nucleardata.nuclear.lu.se/nucleardata/toi/`

The sites below contain much information about nuclear technology and other related topics. Many are home pages for various governmental agencies and some are sites offering useful links, software, reports, and other pertinent information.

```
http://physics.nist.gov/
http://www.nist.gov/
http://www.nrc.gov/
http://www.doe.gov/
http://www.epa.gov/oar/
http://www.hpa.org.uk/radiation/
http://www-rsicc.ornl.gov/
http://www.iaea.or.at/
http://www.nea.fr/
```

PROBLEMS

1. Both the hertz and the curie have dimensions of s^{-1}. Explain the difference between these two units.

2. Advantages of SI units are apparent when one is presented with units of *barrels*, *ounces*, *tons*, and many others.

   (a) Compare the British and U.S. units for the gallon and barrel (liquid and dry measure) in SI units of liters (L).

   (b) Compare the *long ton*, *short ton*, and *metric ton* in SI units of kg.

3. Compare the U.S. and British units of *ounce* (fluid), (apoth), (troy), and (avdp).

4. Explain the SI errors (if any) in and give the correct equivalent units for the following units: (a) m-grams/pL, (b) megaohms/nm, (c) N·m/s/s, (d) gram cm/(s^{-1}/mL), and (e) Bq/milli-Curie.

5. Consider H_2, D_2, and H_2O, treated as ideal gases at pressures of 1 atm and temperatures of $293.2°K$. What are the molecular and mass densities of each.

6. In vacuum, how far does light move in 1 ps?

7. In a medical test for a certain molecule, the concentration in the blood is reported as 123 mcg/dL. What is the concentration in proper SI notation?

8. How many neutrons and protons are there in each of the following nuclides: (a) ^{10}B, (b) ^{24}Na, (c) ^{59}Co, (d) ^{208}Pb, and (e) ^{235}U?

9. What are the molecular weights of (a) H_2 gas, (b) H_2O, and (c) HDO?

10. What is the mass in kg of a molecule of uranyl sulfate UO_2SO_4?

11. Show by argument that the reciprocal of Avogadro's constant is the gram equivalent of 1 atomic mass unit.

12. Prior to 1961, the physical standard for atomic masses was 1/16 the mass of the $^{16}_8O$ atom. The new standard is 1/12 the mass of the $^{12}_6C$ atom. The change led to advantages in mass spectrometry. Determine the conversion factor needed to convert from old to new atomic mass units. How did this change affect the value of the Avogadro constant?

13. How many atoms of ^{234}U are there in 1 kg of natural uranium?

14. How many atoms of deuterium are there in 2 kg of water?

15. Estimate the number of atoms in a 3000 pound automobile. State any assumptions you make.

16. Dry air at normal temperature and pressure has a mass density of 0.0012 g/cm^3 with a mass fraction of oxygen of 0.23. What is the atom density (atom/cm^3) of ^{18}O?

17. A reactor is fueled with 4 kg uranium enriched to 20 atom-percent in ^{235}U. The remainder of the fuel is ^{238}U. The fuel has a mass density of 19.2 g/cm^3. (a) What is the mass of ^{235}U in the reactor? (b) What are the atom densities of ^{235}U and ^{238}U in the fuel?

18. A sample of uranium is enriched to 3.2 atom-percent in ^{235}U with the remainder being ^{238}U. What is the enrichment of ^{235}U in weight-percent?

19. A crystal of NaI has a density of 2.17 g/cm^3. What is the atom density of sodium in the crystal?

20. A concrete with a density of 2.35 g/cm^3 has a hydrogen content of 0.0085 weight fraction. What is the atom density of hydrogen in the concrete?

21. How much larger in diameter is a uranium nucleus compared to an iron nucleus?

22. By inspecting the chart of the nuclides, determine which element has the most stable isotopes?

23. Find an internet site where the isotopic abundances of mercury may be found.

24. The earth has a radius of about 6.35×10^6 m and a mass of 5.98×10^{24} kg. What would be the radius if the earth had the same mass density as matter in a nucleus?

Chapter 2

Modern Physics Concepts

During the first three decades of the twentieth century, our understanding of the physical universe underwent tremendous changes. The classical physics of Newton and the other scientists of the eighteenth and nineteenth centuries was shown to be inadequate to describe completely our universe. The results of this revolution in physics are now called "modern" physics, although they are now almost a century old.

Three of these modern physical concepts are (1) Einstein's theory of special relativity, which extended Newtonian mechanics; (2) wave-particle duality, which says that both electromagnetic waves and atomic particles have dual wave and particle properties; and (3) quantum mechanics, which revealed that the microscopic atomic world is far different from our everyday macroscopic world. The results and insights provided by these three advances in physics are fundamental to an understanding of nuclear science and technology. This chapter is devoted to describing their basic ideas and results.

2.1 The Special Theory of Relativity

The classical laws of dynamics as developed by Newton were believed, for over 200 years, to describe all motion in nature. Students still spend considerable effort mastering the use of these laws of motion. For example, Newton's second law, in the form originally stated by Newton, says the rate of change of a body's momentum $\mathbf{p}$ equals the force $\mathbf{F}$ applied to it, i.e.,

$$\mathbf{F} = \frac{d\mathbf{p}}{dt} = \frac{d(m\mathbf{v})}{dt}. \tag{2.1}$$

For a constant mass m, as assumed by Newton, this equation immediately reduces to the modern form of the second law, $\mathbf{F} = m\mathbf{a}$, where $\mathbf{a} = d\mathbf{v}/dt$, the acceleration of the body.

In 1905 Einstein discovered an error in classical mechanics and also the necessary correction. In his theory of special relativity,[1] Einstein showed that Eq. (2.1) is still correct, but that the mass of a body is not constant, but increases with the body's speed v. The form $\mathbf{F} = m\mathbf{a}$ is thus incorrect. Specifically, Einstein showed that m

---

[1]In 1915 Einstein published the general theory of relativity, in which he generalized his special theory to include gravitation. We will not need this extension in our study of the microscopic world.

varies with the body's speed as

$$m = \frac{m_o}{\sqrt{1 - v^2/c^2}}, \tag{2.2}$$

where m_o is the body's "rest mass," i.e., the body's mass when it is at rest, and c is the speed of light ($\simeq 3 \times 10^8$ m/s). The validity of Einstein's correction was immediately confirmed by observing that the electron's mass did indeed increase as its speed increased in precisely the manner predicted by Eq. (2.2).

Most fundamental changes in physics arise in response to experimental results that reveal an old theory to be inadequate. However, Einstein's correction to the laws of motion was produced theoretically before being discovered experimentally. This is perhaps not too surprising since in our everyday world the difference between m and m_o is incredibly small. For example, a satellite in a circular earth orbit of 7100 km radius, moves with a speed of 7.5 km/s. As shown in Example 2.1, the mass correction factor $\sqrt{1 - v^2/c^2} = 1 - 0.31 \times 10^{-9}$, i.e., relativistic effects change the satellite's mass only in the ninth significant figure or by less than one part in a billion! Thus for practical engineering problems in our macroscopic world, relativistic effects can safely be ignored. However, at the atomic and nuclear level, these effects can be very important.

Example 2.1: What is the fractional increase in mass of a satellite traveling at a speed of 7.5 km/s? From Eq. (2.2) find the fractional mass increase to be

$$\frac{m - m_o}{m_o} = \left[\frac{1}{\sqrt{1 - v^2/c^2}} - 1 \right].$$

Here $(v/c)^2 = (7.5 \times 10^3 / 2.998 \times 10^8)^2 = 6.258 \times 10^{-10}$. With this value of v^2/c^2 most calculators will return a value of 0 for the fractional mass increase. Here's a trick for evaluating relativistically expressions for such small values of v^2/c^2. The expression $(1 + \epsilon)^n$ can be expanded in a Taylor series as

$$(1 + \epsilon)^n = 1 + n\epsilon + \frac{n(n+1)}{2!}\epsilon^2 + \frac{n(n-1)(n-2)}{3!}\epsilon^3 + \cdots \simeq 1 + n\epsilon \quad \text{for } |\epsilon| << 1.$$

Thus, with $\epsilon = -v^2/c^2$ and $n = -1/2$, we find

$$\left(1 - \frac{v^2}{c^2} \right)^{-1/2} \simeq 1 + \frac{1}{2}\frac{v^2}{c^2}$$

so that the fractional mass increase is

$$\frac{m - m_o}{m_o} \simeq \left[\frac{1}{2}\frac{v^2}{c^2} \right] = 3.13 \times 10^{-10}.$$

2.1.1 Principle of Relativity

The principle of relativity is older than Newton's laws of motion. In Newton's words (actually translated from Latin) "The motions of bodies included in a given space are the same amongst themselves, whether the space is at rest or moves uniformly forward in a straight line." This means that experiments made in a laboratory in uniform motion (e.g., in a nonaccelerating train) produce the same results as when the laboratory is at rest. Indeed, this principle of relativity is widely used to solve problems in mechanics by shifting to moving frames of reference to simplify the equations of motion.

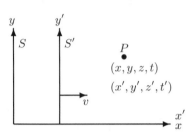

Figure 2.1. Two inertial coordinate systems.

The relativity principle is a simple intuitive and appealing idea. But do all the laws of physics indeed remain the same in all nonaccelerating (*inertial*) coordinate systems? Consider the two coordinate systems shown in Fig. 2.1. System S is at rest, while system S' is moving uniformly to the right with speed v. At $t = 0$, the origin of S' is at the origin of S. The coordinates of some point P are (x, y, z) in S and (x', y', z') in S'. Clearly, the primed and unprimed coordinates are related by

$$x' = x - vt; \quad y' = y; \quad z' = z; \quad \text{and} \quad t' = t. \tag{2.3}$$

If these coordinate transformations are substituted into Newton's laws of motion, we find they remain the same. For example, consider a force in the x-direction, F_x, acting on some mass m. Then the second law in the S' moving system is $F_x = m \, d^2x'/dt'^2$. Now transform this law to the stationary S system. We find (for v constant)

$$F_x = m \frac{d^2x'}{dt'^2} = m \frac{d^2(x - vt)}{d(t)^2} = m \frac{d^2x}{dt^2}.$$

Thus the second law has the same form in both systems. Since the laws of motion are the same in all inertial coordinate systems, it follows that it is impossible to tell, from results of mechanical experiments, whether or not the system is moving.

In the 1870s, Maxwell introduced his famous laws of electromagnetism. These laws explained all observed behavior of electricity, magnetism, and light in a uniform system. However, when Eqs. (2.3) are used to transform Maxwell's equations to another inertial system, they assume a different form. Thus from optical experiments in a moving system, one should be able to determine the speed of the system. For many years Maxwell's equations were thought to be somehow incorrect, but 20 years of research only continued to reconfirm them. Eventually, some scientists began to wonder if the problem lay in the Galilean transformation of Eqs. (2.3). Indeed, Lorentz observed in 1904 that if the transformation

$$x' = \frac{x - vt}{\sqrt{1 - v^2/c^2}}; \quad y' = y; \quad z' = z; \quad t' = \frac{t - vx/c^2}{\sqrt{1 - v^2/c^2}} \tag{2.4}$$

is used, Maxwell's equations become the same in all inertial coordinate systems. Poincaré, about this time, even conjectured that *all* laws of physics should remain unchanged under the peculiar looking Lorentz transformation. The Lorentz transformation is indeed strange, since it indicates that space and time are not independent quantities. Time in the S' system, as measured by an observer in the S system, is different from the time in the observer's system.

2.1.2   Results of the Special Theory of Relativity

It was Einstein who, in 1905, showed that the Lorentz transformation was indeed the correct transformation relating all inertial coordinate systems. He also showed how Newton's laws of motion must be modified to make them invariant under this transformation.

Einstein based his analysis on two postulates:

- The laws of physics are expressed by equations that have the same form in all coordinate systems moving at constant velocities relative to each other.

- The speed of light in free space is the same for all observers and is independent of the relative velocity between the source and the observer.

The first postulate is simply the principle of relativity, while the second states that we observe light to move with speed c even if the light source is moving with respect to us. From these postulates, Einstein demonstrated several amazing properties of our universe.

1. The laws of motion are correct, as stated by Newton, if the mass of an object is made a function of the object's speed v, i.e.,

$$m(v) = \frac{m_o}{\sqrt{1 - v^2/c^2}}. \tag{2.5}$$

   This result also shows that no material object can travel faster than the speed of light since the relativistic mass $m(v)$ must always be real. Further, an object with a rest mass ($m_o > 0$) cannot even travel at the speed of light; otherwise, its relativistic mass would become infinite and give it an infinite kinetic energy.

2. The length of a moving object in the direction of its motion appears smaller to an observer at rest, namely,

$$L = L_o\sqrt{1 - v^2/c^2}. \tag{2.6}$$

   where L_o is the "proper length" or length of the object when at rest.

3. The passage of time appears to slow in a system moving with respect to a stationary observer. The time t required for some physical phenomenon (e.g., the interval between two heart beats) in a moving inertial system appears to be longer (dilated) than the time t_o for the same phenomenon to occur in the stationary system. The relation between t and t_o is

$$t = \frac{t_o}{\sqrt{1 - v^2/c^2}}. \tag{2.7}$$

4. Perhaps the most famous result from special relativity is the demonstration of the equivalence of mass and energy by the well-known equation

$$E = mc^2. \tag{2.8}$$

This result says energy and mass can be converted to each other. Indeed, all changes in energy of a system result in a corresponding change in the mass of the system. This equivalence of mass and energy plays a critical role in the understanding of nuclear technology.

The first three of these results are derived in the Addendum 1 to this chapter. The last result, however, is so important that it is derived below.

Derivation of $E = mc^2$

Consider a particle with rest mass m_o initially at rest. At time $t = 0$, a force $\mathbf{F}$ begins to act on the particle accelerating the mass until at time t it has acquired a velocity $\mathbf{v}$ (see Fig. 2.2). From the conservation-of-energy principle, the work done on this particle as it moves along the path of length s must equal the kinetic energy T of the particle at the end of the path. The path along which the particle moves is arbitrary, depending on how $\mathbf{F}$ varies in time. The work done by $\mathbf{F}$ (a vector) on the particle as it moves through a displacement $d\mathbf{s}$ (also a vector) is $\mathbf{F} \cdot d\mathbf{s}$. The total work done on the particle over the whole path of length s is

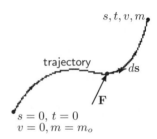

Figure 2.2. A force $\mathbf{F}$ accelerates a particle along a path of length s.

$$T = \int_0^s \mathbf{F} \cdot d\mathbf{s} = \int_0^s \frac{d(m\mathbf{v})}{dt} \cdot d\mathbf{s} = \int_0^t \frac{d(m\mathbf{v})}{dt} \cdot \frac{d\mathbf{s}}{dt} dt.$$

But, $d\mathbf{s}/dt = \mathbf{v}$, a vector parallel to $d(m\mathbf{v})/dt$; thus,

$$T = \int_0^t \frac{d(mv)}{dt} v\, dt = \int_0^{mv} v\, d(mv).$$

Substitution of Eq. (2.5) for m yields

$$T = m_o \int_0^{v/\sqrt{1-v^2/c^2}} v\, d\left(\frac{v}{\sqrt{1 - v^2/c^2}} \right)$$

$$= m_o \int_0^v v \left(\frac{1}{\sqrt{1 - v^2/c^2}} + \frac{v^2/c^2}{(1 - v^2/c^2)^{3/2}} \right) dv$$

$$= m_o \int_0^v \frac{v}{(1 - v^2/c^2)^{3/2}}\, dv$$

$$= m_o c^2 \left. \frac{1}{\sqrt{1 - v^2/c^2}} \right|_0^v$$

$$= \frac{m_o c^2}{\sqrt{1 - v^2/c^2}} - m_o c^2, \tag{2.9}$$

or finally

$$T = mc^2 - m_o c^2. \qquad (2.10)$$

Thus we see that the kinetic energy is associated with the increase in the mass of the particle.

Equivalently, we can write this result as $mc^2 = m_o c^2 + T$. We can interpret mc^2 as the particle's "total energy" E, which equals its rest-mass energy plus its kinetic energy. If the particle was also in some potential field, for example, an electric field, the total energy would also include the potential energy. Thus we have

$$E = mc^2. \qquad (2.11)$$

This well known equation is the cornerstone of nuclear energy analyses. It shows the equivalence of energy and mass. One can be converted into the other in precisely the amount specified by $E = mc^2$. When we later study various nuclear reactions, we will see many examples of energy being converted into mass and mass being converted into energy.

Example 2.2: What is the energy equivalent in MeV of the electron rest mass? From data in Table 1.5 and Eq. 1.1 we find

$$\begin{aligned}
E = m_e c^2 &= (9.109 \times 10^{-31} \text{ kg}) \times (2.998 \times 10^8 \text{ m/s})^2 \\
&\quad \times (1 \text{ J/(kg m}^2 \text{ s}^{-2})/(1.602 \times 10^{-13} \text{ J/MeV}) \\
&= 0.5110 \text{ MeV}
\end{aligned}$$

When dealing with masses on the atomic scale, it is often easier to use masses measured in atomic mass units (u) and the conversion factor of 931.49 MeV/u. With this important conversion factor we obtain

$$E = m_e c^2 = (5.486 \times 10^{-4} \text{ u}) \times (931.49 \text{ MeV/u}) = 0.5110 \text{ MeV}.$$

Reduction to Classical Mechanics

For slowly moving particles, that is, $v \ll c$, Eq. (2.10) yields the usual classical result. Since,

$$\frac{1}{\sqrt{1 - v^2/c^2}} \equiv (1 - v^2/c^2)^{-1/2} = 1 + \frac{v^2}{2c^2} + \frac{3v^4}{8c^4} + \cdots \simeq 1 + \frac{v^2}{2c^2}, \qquad (2.12)$$

the kinetic energy of a slowly moving particle is

$$T = m_o c^2 \left(\frac{1}{\sqrt{1 - v^2/c^2}} - 1 \right) = m_o c^2 \left(\left[1 + \frac{v^2}{2c^2} + \cdots \right] - 1 \right) \simeq \frac{1}{2} m_o v^2. \quad (2.13)$$

Thus the relativistic kinetic energy reduces to the classical expression for kinetic energy if $v \ll c$, a reassuring result since the validity of classical mechanics is well established in the macroscopic world.

Relation Between Kinetic Energy and Momentum

Both classically and relativistically, the momentum p of a particle is given by

$$p = mv. \tag{2.14}$$

In classical physics, a particle's kinetic energy T is given by

$$T = \frac{mv^2}{2} = \frac{p^2}{2m},$$

which yields

$$p = \sqrt{2mT}. \tag{2.15}$$

For relativistic particles, the relationship between momentum and kinetic energy is not as simple. Square Eq. (2.5) to obtain

$$m^2 \frac{c^2 - v^2}{c^2} = m_o^2,$$

or, upon rearrangement,

$$p^2 \equiv (mv)^2 = (mc)^2 - (m_o c)^2 = \frac{1}{c^2}[(mc^2)^2 - (m_o c^2)^2].$$

Then combine this result with Eq. (2.10) to obtain

$$p^2 = \frac{1}{c^2}\left[(T + m_o c^2)^2 - (m_o c^2)^2\right] = \frac{1}{c^2}\left[T^2 + 2Tm_o c^2\right]. \tag{2.16}$$

Thus for relativistic particles

$$\boxed{p = \frac{1}{c}\sqrt{T^2 + 2Tm_o c^2}.} \tag{2.17}$$

Particles

For most moving objects encountered in engineering analyses, the classical expression for kinetic energy can be used. Only if an object has a speed near c must we use relativistic expressions. From Eq. (2.10) one can readily calculate the kinetic energies required for a particle to have a given relativistic mass change. Listed in Table 2.1 for several important atomic particles are the rest mass energies and the kinetic energies required for a 0.1% mass change. At this threshold for relativistic effects, the particle's speed $v = 0.045c$ (see Problem 2).

2.2   Radiation as Waves and Particles

For many phenomena, radiant energy can be considered as electromagnetic waves. Indeed Maxwell's equations, which describe very accurately interactions of long wave-length radiation, readily yield a wave equation for the electric and magnetic fields of radiant energy. Phenomena such as diffraction, interference, and other related optical effects can be described only by a wave model for radiation.

Table 2.1. Rest mass energies and kinetic energies for a 0.1%
relativistic mass increase for four particles.

Particle	rest mass energy $m_o c^2$	kinetic energy for a 0.1% increase in mass
electron	0.511 MeV	511 eV $\simeq$ 0.5 keV
proton	938 MeV	938 keV $\simeq$ 1 MeV
neutron	940 MeV	940 keV $\simeq$ 1 MeV
α-particle	3751 MeV	3.8 MeV $\simeq$ 4 MeV

However, near the beginning of the twentieth century, several experiments involving light and x rays were performed that indicated that radiation also possessed particle-like properties. Today we understand through quantum theory that matter (e.g., electrons) and radiation (e.g., x rays) both have wave-like and particle properties. This dichotomy, known as the *wave-particle duality principle*, is a cornerstone of modern physics. For some phenomena, a wave description works best; for others, a particle model is appropriate. In this section, three pioneering experiments are reviewed that helped to establish the wave-particle nature of matter.

2.2.1 The Photoelectric Effect

In 1887, Hertz discovered that, when metal surfaces were irradiated with light, "electricity" was emitted. J.J. Thomson in 1898 showed that these emissions were electrons (thus the term *photoelectrons*). According to a classical (wave theory) description of light, the light energy was absorbed by the metal surface, and when sufficient energy was absorbed to free a bound electron, a photoelectron would "boil" off the surface. If light were truly a wave, we would expect the following observations:

- Photoelectrons should be produced by light of all frequencies.

- At low intensities, a time lag would be expected between the start of irradiation and the emission of a photoelectron since it takes time for the surface to absorb sufficient energy to eject an electron.

- As the light intensity (i.e., wave amplitude) increases, more energy is absorbed per unit time and, hence, the photoelectron emission rate should increase.

- The kinetic energy of the photoelectron should increase with the light intensity since more energy is absorbed by the surface.

However, experimental results differed dramatically with these results. It was observed:

- For each metal there is a minimum light frequency below which no photoelectrons are emitted no matter how high the intensity.

- There is no time lag between the start of irradiation and the emission of photoelectrons, no matter how low the intensity.

- The intensity of the light affects only the emission *rate* of photoelectrons.

- The kinetic energy of the photoelectron depends only on the frequency of the light and not on its intensity. The higher the frequency, the more energetic is the photoelectron.

In 1905 Einstein introduced a new light model which explained all these observations.[2] Einstein assumed that light energy consists of *photons* or "quanta of energy," each with an energy $E = h\nu$, where h is Planck's constant $(6.62 \times 10^{-34}$ J s) and ν is the light frequency. He further assumed that the energy associated with each photon interacts as a whole, i.e., either all the energy is absorbed by an atom or none is. With this "particle" model, the maximum kinetic energy of a photoelectron would be

$$E = h\nu - A, \tag{2.18}$$

where A is the amount of energy (the so-called *work function*) required to free an electron from the metal. Thus if $h\nu < A$, no photoelectrons are produced. Increasing the light intensity only increases the number of photons hitting the metal surface per unit time and, thus, the rate of photoelectron emission.

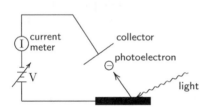

Figure 2.3. A schematic illustration of the experimental arrangement used to verify photoelectric effect.

Although Einstein was able to explain qualitatively the observed characteristics of the photoelectric effect, it was several years later before Einstein's prediction of the maximum energy of a photoelectron, Eq. (2.18), was verified quantitatively using the experiment shown schematically in Fig. 2.3. Photoelectrons emitted from freshly polished metallic surfaces were absorbed by a collector causing a current to flow between the collector and the irradiated metallic surface. As an increasing negative voltage was applied to the collector, fewer photoelectrons had sufficient kinetic energy to overcome this potential difference and the photoelectric current decreased to zero at a critical voltage V_o at which no photoelectrons had sufficient kinetic energy to overcome the opposing potential. At this voltage, the maximum kinetic energy of a photoelectron, Eq. (2.18), equals the potential energy $V_o e$ the photoelectron must overcome, i.e.,

$$V_o e = h\nu - A,$$

or

$$V_o = \frac{h\nu}{e} - \frac{A}{e}, \tag{2.19}$$

where e is the electron charge. In 1912 Hughes showed that, for a given metallic surface, V_o was a linear function of the light frequency ν. In 1916 Milliken, who had previously measured the electron charge e, verified that plots of V_o versus ν for different metallic surface had a slope of h/e, from which h could be evaluated.

---

[2]It is an interesting historical fact that Einstein received the Nobel prize for his photoelectric research and not for his theory of relativity, which he produced in the same year.

Milliken's value of h was in excellent agreement with the value determined from measurements of black-body radiation, in whose theoretical description Planck first introduced the constant h.

The prediction by Einstein and its subsequent experimental verification clearly demonstrated the quantum nature of radiant energy. Although the wave theory of light clearly explained diffraction and interference phenomena, scientists were forced to accept that the energy of electromagnetic radiation could somehow come together into individual quanta, which could enter an individual atom and be transferred to a single electron. This quantization occurs no matter how weak the radiant energy.

Example 2.3:   What is the maximum wavelength of light required to liberate photoelectrons from a metallic surface with a work function of 2.35 eV (the energy able to free a valence electron)? At the minimum frequency, a photon has just enough energy to free an electron. From Eq. (2.18) the minimum frequency to yield a photon with zero kinetic energy ($E = 0$) is

$$\nu_{\min} = A/h = 2.35 \text{ eV}/4.136 \times 10^{-15} \text{ eV/s} = 5.68 \times 10^{14} \text{ s}^{-1}.$$

The wavelength of such radiation is

$$\lambda_{\max} = c/\nu_{\min} = 2.998 \times 10^{8} \text{ m s}^{-1}/5.68 \times 10^{14} \text{ s}^{-1} = 5.28 \times 10^{-7} \text{ m}.$$

This corresponds to light with a wavelength of 528 nm which is in the green portion of the visible electromagnetic spectrum.

2.2.2   Compton Scattering

Other experimental observations showed that light, besides having quantized energy characteristics, must have another particle-like property, namely, momentum. According to the wave model of electromagnetic radiation, radiation should be scattered from an electron with no change in wavelength. However, in 1922 Compton observed that x rays scattered from electrons had a decrease in the wavelength $\Delta\lambda = \lambda' - \lambda$ proportional to $(1 - \cos\theta_s)$ where θ_s was the scattering angle (see Fig. 2.4). To explain this observation, it was necessary to treat x rays as particles with a linear momentum $p = h/\lambda$ and energy $E = h\nu = pc$.

In an x-ray scattering interaction, the energy and momentum before scattering must equal the energy and momentum after scattering. Conservation of linear momentum requires the initial momentum of the incident photon (the electron is assumed to be initially at rest) to equal the vector sum of the momenta of the scattered photon and the recoil electron. This requires the momentum vector triangle of Fig. 2.5 to be closed, i.e.,

$$\mathbf{P}_{\lambda} = \mathbf{P}_{\chi'} + \mathbf{P}_{e} \tag{2.20}$$

or from the law of cosines

$$p_e^2 = p_{\lambda}^2 + p_{\chi'}^2 - 2p_{\lambda}p_{\chi'}\cos\theta_s. \tag{2.21}$$

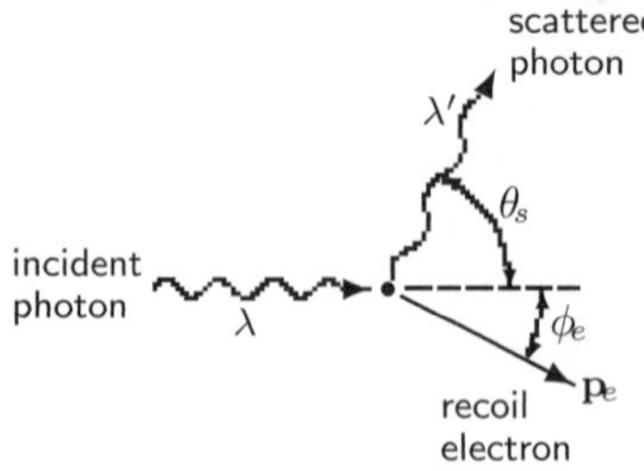

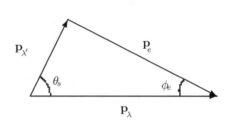

Figure 2.4. A photon with wavelength λ is scattered by an electron. After scattering, the photon has a longer wavelength λ' and the electron recoils with an energy T_e and momentum $\mathbf{p}_e$.

Figure 2.5. Conservation of momentum requires the initial momentum of the photon p_λ equal the vector sum of the momenta of the scattered photon and recoil electron.

The conservation of energy requires

$$p_\lambda c + m_e c^2 = p_{\lambda'} c + m c^2 \tag{2.22}$$

where m_e is the rest-mass of the electron before the collision when it has negligible kinetic energy, and m is its relativistic mass after scattering the photon. This result, combined with Eq. (2.16) (in which $m_e \equiv m_o$), can be rewritten as

$$p_\lambda + m_e c - p_{\lambda'} = \sqrt{p_e^2 + (m_e c)^2}. \tag{2.23}$$

Substitute for p_e from Eq. (2.21) into Eq. (2.23), square the result, and simplify to obtain

$$\frac{1}{p_{\lambda'}} - \frac{1}{p_\lambda} = \frac{1}{m_e c}(1 - \cos\theta_s). \tag{2.24}$$

Then since $\lambda = h/p$, this result gives the decrease in the scattered wavelength as

$$\lambda' - \lambda = \frac{h}{m_e c}(1 - \cos\theta_s), \tag{2.25}$$

where $h/(m_e c) = 2.431 \times 10^{-6}\mu$m. Thus, Compton was able to predict the wavelength change of scattered x rays by using a particle model for the x rays, a prediction which could not be obtained with a wave model.

This result can be expressed in terms of the incident and scattered photon energies, E and E', respectively. With the photon relations $\lambda = c/\nu$ and $E = h\nu$, Eq. (2.25) gives

$$\frac{1}{E'} - \frac{1}{E} = \frac{1}{m_e c^2}(1 - \cos\theta_s). \tag{2.26}$$

Example 2.4:   What is the recoil kinetic energy of the electron that scatters a 3-MeV photon by 45 degrees? In such a Compton scattering event, we first calculate the energy of the scattered photon. From Eq. (2.26) the energy E' of the scattered photon is found to be

$$E' = \left[\frac{1}{E} + \frac{1}{m_e c^2}[1 - \cos\theta_s]\right]^{-1}$$

$$= \left[\frac{1}{3 \text{ MeV}} + \frac{1}{0.511 \text{ MeV}}[1 - \cos(\pi/4)]\right]^{-1} = 1.10 \text{ MeV}.$$

Because energy is conserved, the kinetic energy T_e of the recoil electron must equal the energy lost by the photon, i.e., $T_e = E - E' = 3 - 1.10 = 1.90$ MeV.

2.2.3   Electromagnetic Radiation: Wave-Particle Duality

Electromagnetic radiation assumes many forms encompassing radio waves, microwaves, visible light, X rays, and gamma rays. Many properties are described by a wave model in which the wave travels at the speed of light c and has a wavelength λ and frequency ν, which are related by the wave speed formula

$$c = \lambda\nu. \tag{2.27}$$

The wave properties account for many phenomena involving light such as diffraction and interference effects.

However, as Einstein and Compton showed, electromagnetic radiation also has particle-like properties, namely, the light energy being carried by discrete quanta or packets of energy called *photons*. Each photon has an energy $E = h\nu$ and interacts with matter (atoms) in particle-like interactions (e.g., in the photoelectric interactions described above).

Thus, light has both wave-like and particle-like properties. The properties or model we use depend on the wavelength of the radiation being considered. If, for example, the electromagnetic radiation is visible or infrared, radar or radio, with wavelengths upwards of $\sim 10^{-6}$ m and thus much greater than atom dimensions, the wave model is usually most useful. However, if the electromagnetic radiation consists of ultraviolet, x rays or gamma rays, with wavelengths $\lesssim 10^{-8}$ m or less, the corpuscular or photon model is usually used. This is the model we will use in our study of nuclear science and technology, which deals primarily with penetrating short-wavelength electromagnetic radiation.

Photon Properties

Some particles must always be treated relativistically. For example, photons, by definition, travel with the speed of light c. From Eq. (2.5), one might think that photons have an infinite relativistic mass, and hence, from Eq. (2.17), infinite momentum. This is obviously not true since objects, when irradiated with light, are not observed to jump violently. This apparent paradox can easily be resolved if we insist that the rest mass of the photon be exactly zero, although its relativistic mass is finite. In fact, the total energy of a photon, $E = h\nu$, is due strictly to its motion. Equation (2.17) immediately gives the momentum of a photon (with $m_o \equiv 0$) as

$$p = \frac{E}{c} = \frac{h\nu}{c} = \frac{h}{\lambda}. \tag{2.28}$$

From Eq. (2.10), the photon's relativistic mass is

$$mc^2 = E = h\nu,$$

or

$$m = \frac{h\nu}{c^2}. \tag{2.29}$$

2.2.4 Electron Scattering

In 1924 de Broglie postulated that, since light had particle properties, then for symmetry (physicists love symmetry!), particles should have wave properties. Because photons had a discrete energy $E = h\nu$ and momentum $p = h/\lambda$, de Broglie suggested that a particle, because of its momentum, should have an associated wavelength $\lambda = h/p$.

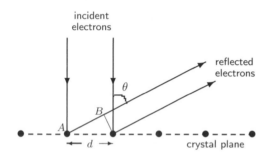
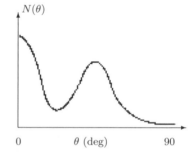

Figure 2.6. Electrons scattering from atoms on a crystalline plane, interfere constructively if the distance AB is a multiple of the electron's de Broglie wavelength.

Figure 2.7. Observed number of electrons $N(\theta)$ scattered into a fixed cone or directions about an angle θ by the atoms in a nickel crystal.

Davisson and Germer in 1927 confirmed that electrons did indeed behave like waves with de Broglie's predicted wavelength. In their experiment, shown schematically in Fig. 2.6, Davisson and Germer illuminated the surface of a Ni crystal by a perpendicular beam of 54-eV electrons and measured the number of electrons $N(\theta)$ reflected at different angles θ from the incident beam. According to the particle model, electrons should be scattered by individual atoms isotropically and $N(\theta)$ should exhibit no structure. However, $N(\theta)$ was observed to have a peak near 50° (see Fig. 2.7). This observation could only be explained by recognizing the peak as a constructive interference peak — a wave phenomenon. Specifically, two reflected electron waves are in phase (constructively interfere) if the difference in their path lengths AB in Fig. 2.6 is an integral number of wavelengths, i.e., if $d \sin \theta = n\lambda$, $n = 1, 2, \ldots$ where d is the distance between atoms of the crystal. This experiment and many similar ones clearly demonstrated that electrons (and other particles such as atoms) have wave-like properties.

2.2.5   Wave-Particle Duality

The fact that particles can behave like waves and that electromagnetic waves can behave like particles seems like a paradox. What really is a photon or an electron? Are they waves or particles? The answer is that entities in nature are more complex than we are used to thinking, and they have, simultaneously, both particle and wave properties. Which properties dominate depends on the object's energy and mass. In Fig. 2.8, the de Broglie wavelength

$$\lambda = \frac{h}{p} = \frac{hc}{\sqrt{T^2 + 2Tm_oc^2}} \tag{2.30}$$

is shown for several objects as the kinetic energy T increases. For a classical object ($T^2 << 2Tm_oc^c$), the wavelength is given by $\lambda = h/\sqrt{2m_oT}$. However, as the object's speed increases, its behavior eventually becomes relativistic ($T >> 2m_oc^2$) and the wavelength varies as $\lambda = h/T$, the same as that for a photon. When the wavelength of an object is much less than atomic dimensions ($\sim 10^{-10}$ m), it behaves more like a classical particle than a wave. However, for objects with longer wavelengths, wave properties tend to be more apparent than particle properties.

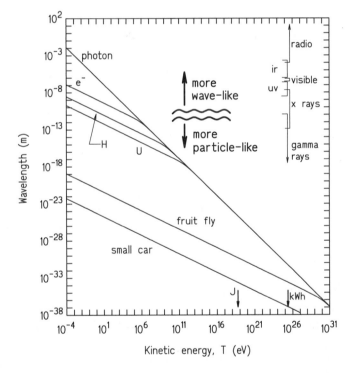

Figure 2.8. The de Broglie wavelength for several objects as their kinetic energy varies. Shown are the electron (e^-), a hydrogen atom (H), an atom of ^{238}U (U), a 1-mg fruit fly, and a 2000-lb car. For wavelengths larger than atomic sizes, the objects' wave-like behaviors dominate, while for smaller wavelengths, the objects behave like particles. On the right, regions of the electromagnetic spectrum are indicated.

Example 2.5: What is the de Broglie wavelength of a neutron with kinetic energy T in eV? We saw earlier that a neutron with less than about 10 MeV of kinetic energy can be treated classically, i.e., its momentum is given by $p = \sqrt{2m_nT}$. Thus, the de Broglie wavelength is

$$\lambda = \frac{h}{p} = \frac{h}{\sqrt{2m_nT}}.$$

Since T is in eV, we express m_n and h in eV units. From Table 1.5 we find $h = 4.136 \times 10^{-15}$ eV s and $m_n = 939.6 \times 10^6$ eV/c^2 = 939.6×10^6 eV/$(2.998 \times 10^8$ m/s$)^2$ = 1.045×10^{-8} eV m^{-2} s^2. Thus

$$\lambda = \frac{4.136 \times 10^{-15} \text{ eV s}}{\sqrt{2 \times 1.045 \times 10^{-8} T \text{ eV}^2 \text{ m}^{-2} \text{ s}^2}} = \frac{2.860 \times 10^{-11}}{\sqrt{T}} \text{ m}.$$

For very low energy neutrons, e.g., 10^{-6} eV with a wavelength of 2.86×10^{-8} m, the "size" of the neutron is comparable to the distances between atoms in a molecule, and such neutrons can scatter from several adjacent atoms simultaneously and create a neutron diffraction pattern from which the geometric structure of molecules and crystals can be determined. By contrast, neutrons with energies of 1 MeV have a wavelength of 2.86×10^{-14} m, comparable to the size of a nucleus. Thus, such neutrons can interact only with a single nucleus. At even higher energies, the wavelength becomes even smaller and a neutron begins to interact with individual nucleons inside a nucleus.

2.3 Quantum Mechanics

The demonstration that particles (point objects) also had wave properties led to another major advance of modern physics. Because a material object such as an electron has wave properties, it should obey some sort of wave equation. Indeed, Shrödinger in 1925 showed that atomic electrons could be well described as standing waves around the nucleus. Further, the electron associated with each wave could have only a discrete energy. The branch of physics devoted to this wave description of particles is called *quantum mechanics* or *wave mechanics*.

2.3.1 Schrödinger's Wave Equation

To illustrate Schrödinger's wave equation, we begin with an analogy to the standing waves produced by a plucked string anchored at both ends. The wave equation that describes the displacement $\Psi(x,t)$ as a function of position x from one end of the string, which has length L, and at time t is

$$\frac{\partial^2 \Psi(x,t)}{\partial x^2} = \frac{1}{u^2}\frac{\partial^2 \Psi(x,t)}{\partial t^2}. \tag{2.31}$$

Here u is the wave speed. There are infinitely many discrete solutions to this homogeneous partial differential equation, subject to the boundary condition $\Psi(0,t) =$

$\Psi(L, t) = 0$, namely,

$$\Psi(x,t) = A \sin\left(\frac{n\pi x}{L}\right) \sin\left(\frac{n\pi ut}{L}\right), \quad n = 1, 2, 3 \ldots \qquad (2.32)$$

That this is the general solution can be verified by substitution of Eq. (2.32) into Eq. (2.31). The *fundamental* solution ($n = 1$) and the first two *harmonics* ($n = 2$ and $n = 3$) are shown in Fig. 2.9. The frequencies ν of the solutions are also discrete. The time for one cycle $t_c = 1/\nu$ such that $n\pi ut_c/L = 2\pi$; thus

$$\nu = \frac{nu}{2L}, \quad n = 1, 2, 3, \ldots.$$

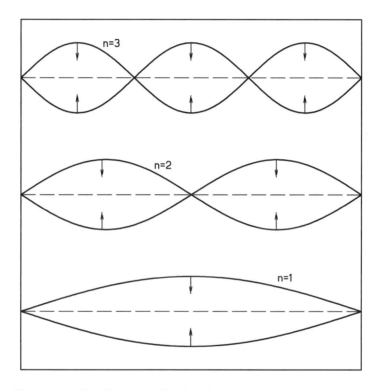

Figure 2.9. Standing wave solutions of a vibrating string. The solution corresponding to $n = 1$ is called the fundamental mode.

Notice that the solution of the wave equation Eq. (2.32) is *separable*, i.e., it has the form $\Psi(x,t) = \psi(x)T(t) = \psi(x)\sin(2\pi\nu t)$. Substitution of this separable form into Eq. (2.31) yields

$$\frac{d^2\psi(x)}{dx^2} + \frac{4\pi^2\nu^2}{u^2}\psi(x) = 0$$

or, since $u = \lambda\nu$,

$$\frac{d^2\psi(x)}{dx^2} + \frac{4\pi^2}{\lambda^2}\psi(x) = 0. \qquad (2.33)$$

To generalization to three-dimensions, the operator $d^2/dx^2 \to \partial^2/\partial x^2 + \partial^2/\partial y^2 + \partial^2/\partial z^2 \equiv \nabla^2$ gives

$$\nabla^2 \psi(x,y,z) + \frac{4\pi^2}{\lambda^2} \psi(x,y,z) = 0. \tag{2.34}$$

Now we apply this wave equation to an electron bound to an atomic nucleus. The nucleus produces an electric field or electric force on the electron $V(x,y,z)$. The electron with (rest) mass m has a total energy E, kinetic energy T, and a potential energy V such that $T = E - V$. The wavelength of the electron is $\lambda = h/p = h/\sqrt{2mT} = h/\sqrt{2m(E-V)}$ (assuming the electron is nonrelativistic). Substitution for λ into Eq. (2.34) gives

$$-\frac{h^2}{8\pi^2 m} \nabla^2 \psi(x,y,z) + V(x,y,z)\psi(x,y,z) = E\psi(x,y,z). \tag{2.35}$$

This equation is known as the steady-state *Schrödinger's wave equation*, and is the fundamental equation of quantum mechanics. This is a homogeneous equation in which everything on the left-hand side is known (except, of course, $\psi(x,y,z)$, but in which the electron energy E on the right is not known. Such an equation generally has only the trivial null solution ($\psi = 0$); however, non-trivial solutions can be found if E has very precise and discrete values[3] $E = E_n, n = 0,1,2,\ldots$. This equation then says that an electron around a nucleus can have only very discrete values of $E = E_n$, a fact well verified by experiment. Moreover, the wave solution of $\psi_n(x,y,z)$ associated with a given energy level E_n describes the amplitude of the electron wave. The interpretation of ψ_n is discussed below.

In Addendum 2, example solutions of Schrödinger's wave equation are presented for the interested reader.

2.3.2   The Wave Function

The non-trivial solution $\psi_n(x,y,z)$ of Eq. (2.35) when $E = E_n$ (an eigenvalue of the equation) is called a *wave function*. In general, this is a complex quantity which extends over all space, and may be thought of as the relative amplitude of a wave associated with the particle described by Eq. (2.35). Further, because Eq. (2.35) is a homogeneous equation, then, if ψ' is a solution, so is $\psi = A\psi'$, where A is an arbitrary constant. It is usual to choose A so that the integral of $\psi\psi^* = |\psi|^2$ over all space equals unity, i.e.,[4]

$$\iiint \psi(x,y,z)\psi^*(x,y,z)\,dV = 1. \tag{2.36}$$

Just as the square of the amplitude of a classical wave defines the intensity of the wave, the square of the amplitude of the wave function $|\psi|^2$ gives the probability of

---

[3]Mathematicians call such an equation (subject to appropriate boundary conditions) an *eigenvalue* problem in which E_n is called the *eigenvalue* and the corresponding solution $\psi_n(x,y,z)$ the *eigenfunction*.

[4]Here ψ^* denotes the complex conjugate of ψ.

finding the particle at any position in space. Thus, the probability that the particle is in some small volume dV around the point (x, y, z) is

$$\text{Prob} = |\psi(x, y, z)|^2 \, dV = \psi(x, y, z)\psi^*(x, y, z) \, dV$$

We see from this interpretation of ψ that the normalization condition of Eq. (2.36) requires that the particle be somewhere in space.

2.3.3 The Uncertainty Principle

With quantum/wave mechanics, we see it is no longer possible to say that a particle is at a particular location; rather, we can say only that the particle has a probability $\psi\psi^* \, dV$ of being in a volume dV. It is possible to construct solutions to the wave equation such that $\psi\psi^*$ is negligibly small except in a very small region of space. Such a wave function thus localizes the particle to the very small region of space. However, such localized *wave packets* spread out very quickly so that the subsequent path and momentum of the particle is known only within very broad limits. This idea of there being uncertainty in a particle's path and its speed or momentum was first considered by Heisenberg in 1927.

If one attempts to measure both a particle's position along the x-axis and its momentum, there will be an uncertainty Δx in the measured position and an uncertainty Δp in the momentum. Heisenberg's uncertainty principle says there is a limit to how small these uncertainties can be, namely,

$$\Delta x \, \Delta p \geq \frac{h}{2\pi}. \tag{2.37}$$

This limitation is a direct consequence of the wave properties of a particle. The uncertainty principle can be derived rigorously from Schrödinger's wave equation; however, a more phenomenological approach is to consider an attempt to measure the location of an electron with very high accuracy. Conceptually, one could use an idealized microscope, which can focus very short wavelength light to resolve points that are about 10^{-11} m apart. To "see" the electron, a photon must scatter from it and enter the microscope. The more accurately the position is to be determined (i.e., the smaller Δx), the smaller must be the light's wavelength (and the greater the photon's energy and momentum). Consequently, the greater is the uncertainty in the electron's momentum Δp since a higher energy photon, upon rebounding from the electron, will change the electron's momentum even more. *By observing a system, the system is necessarily altered.*

The limitation on the accuracies with which both position and momentum (speed) can be known is an important consideration only for systems of atomic dimensions. For example, to locate a mass of 1 g to within 0.1 mm, the minimum uncertainty in the mass's speed, as specified by Eq. (2.37), is about 10^{-26} m/s, far smaller than errors introduced by practical instrumentation. However, at the atomic and nuclear levels, the uncertainty principle provides a very severe restriction on how position and speed of a particle are fundamentally intertwined.

There is a second uncertainty principle (also by Heisenberg) relating the uncertainty ΔE in a particle's energy E and the uncertainty in the time Δt at which the particle had the energy, namely,

$$\Delta E \, \Delta t \geq \frac{h}{2\pi}. \tag{2.38}$$

This restriction on the accuracy of energy and time measurements is a consequence of the time-dependent form of Schrödinger's wave equation (not presented here), and is of practical importance only in the atomic world. In the atomic and subatomic world involving transitions between different energy states, energy need not be rigorously conserved during very short time intervals Δt, provided the amount of energy violation ΔE is limited to $\Delta E \simeq h/(2\pi\Delta t)$. This uncertainty principle is an important relation used to estimate the lifetimes of excited nuclear states.

2.3.4   Success of Quantum Mechanics

Quantum mechanics has been an extremely powerful tool for describing the energy levels and the distributions of atomic electrons around a nucleus. Each energy level and configuration is uniquely defined by four quantum numbers: n the principal quantum number, ℓ the orbital angular momentum quantum number, m_ℓ the z-component of the angular momentum, and $m_s = \pm 1/2$, the electron spin number. These numbers arise naturally from the analytical solution of the wave equation (as modified by Dirac to include special relativity effects) and thus avoid the *ad hoc* introduction of orbital quantum numbers required in earlier atomic models.

Inside the nucleus, quantum mechanics is also thought to govern. However, the nuclear forces holding the neutrons and protons together are much more complicated than the electromagnetic forces binding electrons to the nucleus. Consequently, much work continues in the application of quantum mechanics (and its more general successor quantum electrodynamics) to predicting energy and configuration states of nucleons. Nonetheless, the fact that electronic energy levels of an atom and nuclear excited states are discrete with very specific configurations is a key concept in modern physics. Moreover, when one state changes spontaneously to another state, energy is emitted or absorbed in specific discrete amounts.

2.4   Addendum 1: Derivation of Some Special Relativity Results

In this addendum, the relativistic effects for time dilation, length contraction, and mass increase are derived.

2.4.1   Time Dilation

Consider a timing device that emits a pulse of light from a source and then records the time the light takes to travel a distance d to a detector. In a stationary frame of reference (left-hand figure of Fig. 2.10), the travel time is denoted by t_o (the *proper time*) and the separation distance is given by $d = ct_o$.

Now observe the timing device as it moves to the right at a steady speed v relative to a stationary observer (right-hand figure of Fig. 2.10). During the time t it takes the pulse to appear to travel from the source to the detector, the detector has moved a distance vt to the right. From Einstein's first postulate (c is the same to all observers), the total distance the photon appears to travel is ct. The distance between source and detector is still $d = ct_o$. From the right-hand figure, these three

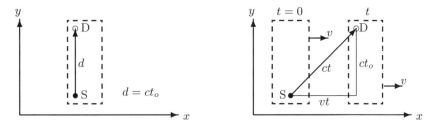

Figure 2.10. A light pulse leaves a source (S) and travels to a detector (D) in a stationary system (left) and in a system moving to the right with a uniform speed v (right).

distances are seen to be related by

$$c^2t^2 = v^2t^2 + c^2t_o^2,$$

from which it follows

$$t = \frac{t_o}{\sqrt{1 - v^2/c^2}}. \tag{2.39}$$

Thus we see that the travel time for the pulse of light *appears* to lengthen or dilate as the timing device moves faster with respect to a stationary observer. In other words, a clock in a moving frame of reference (relative to some observer) appears to run more slowly. This effect (like all other special relativity effects) is reciprocal. To the moving clock, the stationary observer's watch appears to be running more slowly.

2.4.2 Length Contraction

Consider a rod of length L_o when stationary (its *proper length*). If this rod is allowed to move to the right with constant speed v (see left figure in Fig. 2.11), we can measure its apparent length by the time t_o it takes to move past some stationary reference pointer. This is a proper time, since the transit time has been made at the same position in the observer's frame of reference. Thus the rod length appears to be

$$L = vt_o. \tag{2.40}$$

Now consider the same measurement made by an observer moving with the rod (see the right-hand figure in Fig. 2.11). In this frame of reference, the reference

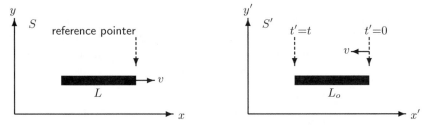

Figure 2.11. Left figure: a rod moves past a stationary measurement pointer at a steady speed v. Right figure: in the system moving with the rod, the pointer appears to be moving backward with speed v.

pointer appears to be moving to the left with speed v. The rod has length L_o to this observer, so the time it takes for the reference pointer to move along the rod is $t = L_o/v$. This is a dilated time since the reference pointer is moving with respect to the observer. Substitution of Eq. (2.39), gives $L_o/v = t = t_o/\sqrt{1 - v^2/c^2}$, or

$$t_o = \frac{L_o}{v}\sqrt{1 - v^2/c^2}.$$

Finally, substitution of this result into Eq. (2.40) shows the proper length L_o is related to the rod's apparent length L by

$$L = L_o\sqrt{1 - v^2/c^2}. \tag{2.41}$$

Thus, we see that the width of an object appears to decrease or contract as it moves at a uniform speed past an observer.

2.4.3 Mass Increase

By considering an elastic collision between two bodies, we can infer that the mass of a body varies with its speed. Suppose there are two experimenters, one at rest in system S and the other at rest in system S', which is moving with speed v relative to S along the x-direction. Both experimenters, as they pass, launch identical elastic spheres with speed u (as judged by each experimenter) in a direction perpendicular to the x-axis (again, as judged by each), so that a head-on collision occurs midway between them. Each sphere has rest mass m_o. For a launch speed $u \ll c$, then each experimenter's sphere has mass m_o in his frame of reference. Fig. 2.12 illustrates the trajectories of the spheres that each experimenter observes.

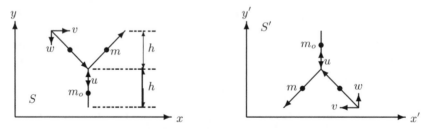

Figure 2.12. The trajectories of two colliding elastic spheres as seen by the experimenter in the stationary S system (left-hand figure), and by the experimenter in the S' system (right-hand figure).

To the experimenter in S (left-hand figure of Fig. 2.12), the time for his sphere to travel to the collision point is $t_o = h/u$. However, his colleague's sphere appears to take longer to reach the collision, namely, the dilated time $t = t_o/\sqrt{1 - v^2/c^2}$. Thus to the S experimenter, the sphere from S' appears to be moving in the negative y-direction with a speed

$$w = \frac{h}{t} = \frac{h}{t_o}\sqrt{1 - v^2/c^2} = u\sqrt{1 - v^2/c^2}, \tag{2.42}$$

where we have used Eq. (2.39).

By Einstein's second postulate (physical laws are the same in all inertial frames of reference), momentum must be conserved in the collision of the spheres. To conserve momentum in this experiment, we require that the total momentum of the two spheres in the y-direction be the same before and after the collision, i.e.,

$$m_o u - mw = -m_o u + mw,$$

or using Eq. (2.42)

$$m_o u = mw = mu\sqrt{1 - v^2/c^2}.$$

This reduces to

$$m = \frac{m_o}{\sqrt{1 - v^2/c^2}}. \tag{2.43}$$

2.5 Addendum 2: Solutions to Schrödinger's Wave Equation

2.5.1 The Particle in a Box

The solution of the Schrödinger wave equation is generally very difficult and usually requires approximations or sophisticated numerical techniques. However, there are several simple problems for which analytical solutions can be obtained. These exact solutions can be used to show clearly the differences between quantum and classical mechanics. In addition, many of these simple problems can be used to model important phenomena such as electron conduction in metals and radioactive alpha decay.

One such problem is the description of the motion of a free particle of mass m confined in a "one-dimensional" box. Within the box, $0 < x < a$, the particle is free to move. However, it is not allowed past the walls at $x = 0$ and $x = a$ and, if the particle reaches a wall, it is reflected elastically back into the interior of the box.

In a classical description, the particle can move back and forth between the bounding walls with any kinetic energy E or speed v. Moreover, the probability of finding the particle in any differential width dx between the walls is the same for all dx.

In a quantum mechanical description of the particle, given by the Schrödinger wave equation, the potential energy or force on the particle $V(x) = 0$ inside the box, and infinite outside. Since the potential energy is infinite outside the box, the particle cannot exist outside the box, and the wave functions $\psi(x)$ must vanish. The Schrödinger equation, Eq. (2.35), for the particle is,

$$\frac{d^2\psi}{dx^2} + \frac{8\pi^2 mE}{h^2}\psi = 0, \qquad 0 < x < a. \tag{2.44}$$

If we let $k^2 \equiv 8\pi^2 mE/h^2$, the general solution of this equation is

$$\psi(x) = A\sin kx + B\cos kx, \tag{2.45}$$

where A and B are arbitrary constants. This solution must also satisfy the boundary conditions, namely, $\psi(0) = 0$ and $\psi(a) = 0$. These boundary conditions, as shown below, severely restrict the allowed values of not only A and B but also of k.

Application of the boundary condition $\psi(0) = 0$ at the left wall to Eq. (2.45) yields

$$\psi(0) = 0 = A\sin(0) + B\cos(0) = B \tag{2.46}$$

which forces us to set $B = 0$. Thus, the general solution reduces to $\psi(x) = A \sin kx$. Application of the boundary condition $\psi(a) = 0$ at the right wall produces the restriction that

$$\psi(a) = A \sin ka = 0. \tag{2.47}$$

Clearly, we could satisfy this restriction by taking $A = 0$. But this would say that the wave function $\psi(x) = 0$ everywhere in the box. This trivial null solution is unrealistic since it implies the particle cannot be in the box. However, we can satisfy Eq. (2.47) with $A \neq 0$ if we require $\sin ka = 0$. Since the sine function vanishes at all multiples of π, we see that k must be chosen as

$$k = \frac{n\pi}{a}, \quad n = 1, 2, 3, \ldots. \tag{2.48}$$

These discrete or quantized values of k are denoted by k_n.

Thus, the only allowed solutions of the Schrödinger wave equation are the functions

$$\psi_n(x) = A \sin \frac{n\pi x}{a}, \quad n = 1, 2, 3, \ldots \tag{2.49}$$

where A is any non-zero constant. Moreover, the particle's energy, $E = h^2 k/(8\pi^2 m)$, can have only discrete values given by

$$E_n = \frac{k_n h^2}{8\pi^2 m} = \frac{n^2 h^2}{8ma^2}, \quad n = 1, 2, 3, \ldots. \tag{2.50}$$

These discrete values E_n are sometimes called the *eigenvalues* of Eq. (2.44) and n is a *quantum number*. Thus, in a quantum mechanical description of the particle, the particle's kinetic energy (and hence its speed and momentum) must have very discrete or quantized values. By contrast, a classical description allows the particle to have any kinetic energy or speed.

The wave function associated with the particle with energy E_n (the so-called *eigenfunction* of Eq. (2.44)) is given by Eq. (2.49). The constant A in Eq. (2.49) is chosen to normalize the square of the wave function to unity, i.e.,

$$\int_0^a |\psi_n(x)|^2 dx = A^2 \int_0^a \sin^2 \frac{n\pi x}{a} dx = 1. \tag{2.51}$$

To satisfy this normalization, $A = \sqrt{2/a}$ and the normalized wave functions become

$$\psi_n(x) = \sqrt{\frac{2}{a}} \sin \frac{n\pi x}{a}, \quad n = 1, 2, 3, \ldots \tag{2.52}$$

This normalization condition of Eq. (2.51) is required to give a probability of unity that the particle is somewhere in the box (see Section 2.3.2).

The probability that a particle, in a given energy state E_n, occupies any given volume element dx of the one-dimensional box is given by the quantum mechanical probability distribution, namely,

$$|\psi(x)|^2 dx = \frac{2}{a} \sin^2 \left(\frac{n\pi x}{a} \right) dx, \tag{2.53}$$

which is a wave-like function and is plotted in Fig. 2.13 for the quantum state $n = 3$. Because the square of the wave function is the probability of occupancy in a unit volume element about x, we see that the probability of finding the particle at any particular location varies with the location. This probability distribution $|\psi_3(x)|^2$ vanishes not only at the wall ($x = 0$ and $x = a$), but also at $x = a/3$ and $x = 2a/3$. By contrast it has maxima at $x = a/6$, $x = a/2$, and $x = 5a/6$. This quantum mechanical result for where the particles is likely to be in the box is significantly different than the classical result for which the probability of occupancy is the same for any volume element dx. The differences are shown in Fig. 2.13.

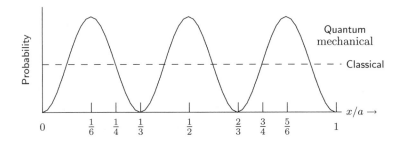

Figure 2.13. The probability distribution for finding a particle in quantum state $n = 3$ with energy E_3 at points in a one-dimensional box.

The quantum mechanical standing-wave distribution for the location of an electron trapped in a box can be seen in the lower-right figure of Fig. 1.1. Here the circle of iron atoms acts like the walls of our box, and the valence electrons, normally free to move around the surface of the copper substrate, are now confined. A standing wave for the location of these trapped electrons is clearly seen. This probability wave is completely analogous to the 1-dimensional analysis above.

If the particle is confined to a three-dimensional box with edges of length a, b, c, the normalized wave functions for the stationary states and the values for the energy levels are found to be

$$\psi_{n_1,n_2,n_3}(x,y,z) = 2\sqrt{\frac{2}{abc}} \sin\left(\frac{n_1\pi x}{a}\right) \sin\left(\frac{n_2\pi y}{b}\right) \sin\left(\frac{n_3\pi z}{c}\right) \qquad (2.54)$$

and

$$E_{n_1,n_2,n_3} = \frac{h^2}{8m}\left[\frac{n_1^2}{a^2} + \frac{n_2^2}{b^2} + \frac{n_3^2}{c^2}\right], \qquad (2.55)$$

where $n_1, n_2,$ and n_3 denote a set of positive integers.

2.5.2 The Hydrogen Atom

The hydrogen atom, the simplest atom, can be considered as a system of two interacting point charges, the proton (nucleus) and an electron. The electrostatic attraction between the electron and proton is described by Coulomb's law. The potential energy of a bound electron is given by,

$$V(r) = -\frac{e^2}{r},\tag{2.56}$$

where r is the distance between the electron and the proton.

The three-dimensional Schrödinger wave equation in spherical coordinates for the electron bound to the proton is

$$\frac{1}{r^2}\frac{\partial}{\partial r}\left(r^2\frac{\partial\psi}{\partial r}\right) + \frac{1}{r^2\sin\theta}\frac{\partial}{\partial\theta}\left(\sin\theta\frac{\partial\psi}{\partial\theta}\right) + \frac{1}{r^2\sin^2\theta}\frac{\partial^2\psi}{\partial\phi^2} + \frac{8\pi^2\mu}{h^2}\left[E - V(r)\right]\psi = 0,\tag{2.57}$$

where $\psi = \psi(r,\theta,\phi)$ and μ is the electron mass (more correctly the reduced mass of the system).

To solve this partial differential equation for the wave function $\psi(r,\theta,\phi)$, we first replace it by three equivalent ordinary differential equations involving functions of only a single independent variable. To this end, we use the "separation of variables" method, and seek a solution of the form

$$\psi(r,\theta,\phi) = R(r)\Theta(\theta)\Phi(\phi).\tag{2.58}$$

Substitute this form into Eq. (2.57) and multiply the result by $r^2\sin\theta/(R\Theta\Psi)$ to obtain

$$\frac{\sin^2\theta}{R(r)}\frac{d}{dr}\left(r^2\frac{dR(r)}{dr}\right) + \frac{1}{\Phi(\phi)}\frac{d^2\Phi(\phi)}{d\phi^2} + \frac{\sin\theta}{\Theta(\theta)}\frac{d^2\Theta(\theta)}{d\theta^2} + \frac{8\pi^2\mu}{h^2}r^2\sin^2\theta\left[E - V(r)\right] = 0.\tag{2.59}$$

The second term is only a function of ϕ while the other terms are independent of ϕ. This term, therefore, must equal a constant, $-m^2$ say. Thus,

$$\frac{d^2\Phi(\phi)}{d\phi^2} = -m^2\Phi(\phi).\tag{2.60}$$

Equation 2.59, upon division by $\sin^2\theta$ and rearrangement becomes

$$\frac{1}{R(r)}\frac{d}{dr}\left(r^2\frac{dR(r)}{dr}\right) + \frac{8\pi^2\mu r^2}{h^2}\left[E - V(r)\right] = -\frac{1}{\sin\theta\Theta(\theta)}\frac{d^2\Theta(\theta)}{d\theta^2} + \frac{m^2}{\sin^2\theta} = 0.\tag{2.61}$$

Since the terms on the left are functions only of r and the terms on the right are functions only of θ, both sides of this equation must be equal to the same constant, β say. Thus we obtain two ordinary differential equations, one for $R(r)$ and one for $\Theta(\theta)$, namely,

$$\frac{1}{\sin\theta}\frac{d^2\Theta(\theta)}{d\theta^2} - \frac{m^2}{\sin^2\theta}\Theta(\theta) + \beta\Theta(\theta) = 0,\tag{2.62}$$

and

$$\frac{1}{r^2}\frac{d}{dr}\left[r^2\frac{dR(r)}{dr}\right] + -\frac{\beta}{r^2}R(r) + \frac{8\pi^2\mu}{h^2}\left[E - V(r)\right]R(r) = 0.\tag{2.63}$$

We now have three ordinary, homogeneous, differential equations [Eq. (2.60), Eq. (2.62), and Eq. (2.63)], whose solutions, when combined, given the entire wave function $\psi(r,\theta,\phi) = R(r)\Theta(\theta)\Phi(\phi)$. Each of these equations is an eigenvalue problem which yields a "quantum number." These quantum numbers are used to describe the possible electron configurations in a hydrogen atom. The solutions also yield relationships between the quantum numbers. We omit the solution details—see any book on quantum mechanics for the explicit solutions. We restrict our discussion to the essential features that arise from each equation.

The most general solution of Eq. (2.60) is

$$\Phi(\phi) = A \sin m\phi + B \cos m\phi, \qquad (2.64)$$

where A and B are arbitrary. However, we require $\Phi(0) = \Phi(2\pi)$ since $\phi = 0$ and $\phi = 2\pi$ are the same azimuthal angle. This boundary condition then requires m to be an integer, i.e., $m = 0, \pm 1, \pm 2, \ldots$. For each *azimuthal* quantum number m, the corresponding solution is denoted by $\Phi_m(\phi)$.

Equation (2.62) in θ has normalizable solutions only if the separation constant has the form $\beta = \ell(\ell + 1)$ where ℓ is a positive integer or zero, and is called the *angular momentum* quantum number. Moreover, the azimuthal quantum number m must be restricted to $2\ell + 1$ integer values, namely, $m = 0, \pm 1, \pm 2, \ldots, \pm \ell$. The corresponding solutions of Eq. (2.62) are denoted by $\Theta_{\ell m}(\theta)$ and are known to mathematicians as the associate Legendre functions of the first kind; however, these details need not concern us here.

Finally, the solution of Eq. (2.63) for the radial component of the wave function, with $\beta = \ell(\ell+1)$ and $V(r) = -e^2/r$, has normalizable solutions only if the electron's energy has the (eigen)value

$$E_n = -\frac{2\pi^2 \mu e^2}{h^2 n^2}, \quad \text{where } n = 1, 2, 3, \ldots \qquad (2.65)$$

The integer n is called the *principal* quantum number. Moreover, to obtain a solution, the angular quantum number ℓ must be no greater than $n - 1$, i.e., $\ell = 0, 1, 2, \ldots, (n-1)$. The corresponding radial solution is denoted by $R_{n\ell}(r)$ and mathematically is related to the associated Laguerre function.

Thus, the wave functions for the electron bound in the hydrogen atom have only very discrete forms $\psi_{n\ell m}(r,\theta,\phi) = R_{n\ell}(r)\Theta_{\ell m}(\theta)\Phi_m(\phi)$. For the hydrogen atom, the energy of the electron is given by Eq. (2.65) and is independent of the angular moment or azimuthal quantum numbers m and ℓ (this is not true for multielectron atoms).

Special Notation for Electron States

A widely used, but strange, notation has been of long standing in describing the n and ℓ quantum numbers of particular electron states. The letters s, p, d, f, g, h, and i are used to denote values of the angular momentum quantum number ℓ of 0, 1, 2, 3, 4, 5, and 6, respectively. The value of n is then used as a prefix to the angular moment letter. Thus a bound electron designated as $5f$ refers to an electron with $n = 5$ and $\ell = 3$.

Examples of Wave Functions for Hydrogen

In Fig. 2.14, density plots of $|\psi_{n\ell m}|^2 = \psi_{n\ell m}^* \psi_{n\ell m}$ are shown for different electron states in the hydrogen atom. These plots are slices through the three-dimensional

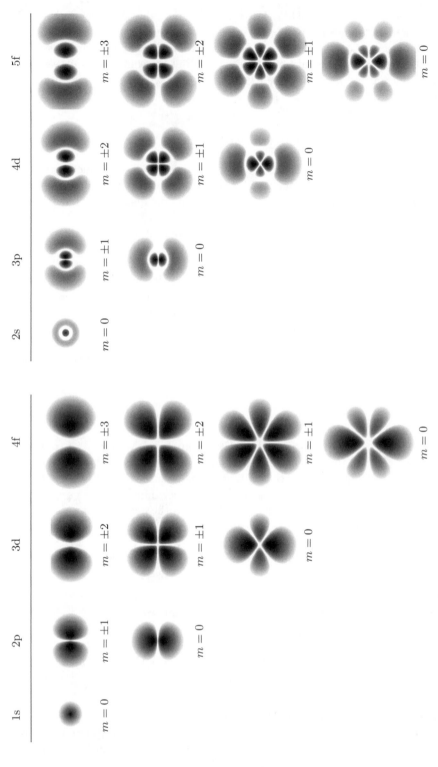

Figure 2.14. Density plots of $\psi^*\psi$ for different electron eigenstates in the hydrogen atom. Plots are sectional views of the probability density in a plane containing the polar axis, which is vertical and in the plane of the paper. Scale of the right group is about 15% bigger than that of the left group. After: R.B. Leighton, *Principles of Modern Physics*, McGraw Hill, New York, 1959.

$|\psi|^2$ in a plane perpendicular to the x-axis and through the atom's center. Since $|\psi|^2$ is the probability of finding the electron in a unit volume, the density plots directly show the regions where the electron is most likely to be found. The s states ($\ell = 0$) are spherically symmetric about the nucleus, while all the others have azimuthal and/or polar angle dependence.

Electron Energy Levels in Hydrogen

For hydrogen, the energy of the bound electron is a function of n only (see Eq. (2.65)). Thus, for $n > 1$, the quantum numbers ℓ and m may take various values without changing the electron binding energy, i.e., the allowed electron configurations fall into sets in which all members of the set have the same energy. Thus, in Fig. 2.14, an electron has the same energy in any of the four 4f electron states or the three 4d states, even though the distribution of the electron wave function around the nucleus is quite different for each state. Such states with the same electron binding energy are said to be *degenerate.*

The Spin Quantum Number

It was found, first from experiment and later by theory, that each quantum state (specified by values for n, ℓ, and m) can accommodate an electron in either of two spin orientations. The electron, like the proton and neutron, has an inherent angular momentum with a value of $\frac{1}{2}h/(2\pi)$. In a bound state (defined by n, ℓ, and m) the electron can have its spin "up" or "down" with respect to the z-axis used to define angular momentum. Thus, a fourth quantum number $m_s = \pm\frac{1}{2}$ is needed to unambiguously define each possible electron configuration in an atom.

Dirac showed in 1928 that when the Schrödinger wave equation is rewritten to include relativistic effects, the spin quantum number m_s is inherent in the solution along with the quantum numbers n, ℓ, and m, which were also inherent in the wave function solution of the non-relativistic Schrödinger's wave equation.

2.5.3   Energy Levels for Multielectron Atoms

The solution of the Schrödinger's equation for the hydrogen atom can be obtained analytically, but the solution for a multielectron system cannot. This difficulty arises because of the need to add a repulsive component to the potential energy term to account for the interactions among the electrons. To obtain a solution for a multielectron atom, numerical approximation techniques must be used together with high speed computers. Such a discussion is far beyond the scope of this text.

There is, however, one important result for multielectron atoms that should be described here. The energies of the electronic levels in atoms with more than one electron are functions of both n and ℓ. The energy degeneracy in the angular momentum quantum number disappears. Thus states with the same n but different ℓ values have slightly different energies, and there is a significant reordering of the electron energy levels compared to those in the hydrogen atom.

Electron energy levels with the same value of n and ℓ but different values of m are still degenerate in energy; however, this degeneracy is removed in the presence of a strong external magnetic field (the Zeeman effect). An electron moving about a nucleus creates a magnetic dipole whose strength of interaction with an external magnetic field varies with the quantum numbers ℓ and m.

Table 2.2. Electron shell arrangement for the lightest elements.

Element	Z	shell and electron designation							
		K	L		M			N	
		$1s$	$2s$	$2p$	$3s$	$3p$	$3d$	$4s$	$4p$
H	1	1							
He	2	2							
Li	3	2	1						
Be	4	2	2						
B	5	2	2	1					
C	6	2	2	2					
N	7	2	2	3					
O	8	2	2	4					
F	9	2	2	5					
Ne	10	2	2	6					
Na	11	neon configuration			1				
Mg	12				2				
Al	13				2	1			
Si	14				2	2			
P	15				2	3			
S	16				2	4			
Cl	17				2	5			
Ar	18				2	6			
K	19	argon configuration						1	
Ca	20							2	
Sc	21						1	2	
Ti	22						2	2	
V	23						3	2	
Cr	24						5	1	
Mn	25						5	2	
Fe	26						6	2	
Co	27						7	2	
Ni	28						8	2	
Cu	29						10	1	
Zn	30						10	2	
Ga	31						10	2	1
Ge	32						10	2	2
As	33						10	2	3
Se	34						10	2	4
Br	35						10	2	5
Kr	36						10	2	6

Source: H.D. Bush, *Atomic and Nuclear Physics*, Prentice Hall, Englewood Cliffs, NJ, 1962.

The Electronic Structure of Atoms

Each electron in an atom can be characterized by its four quantum numbers n, ℓ, m, and m_s. Further, according to Pauli's exclusion principle (1925), no two electrons in an atom can have the same quantum numbers. An assignment of a set of four quantum numbers to each electron in an atom, no sets being alike in all four numbers, then defines a quantum state for an atom as a whole. For ground-state atoms, the electrons are in the lowest energy electron states. For an atom in an excited state, one or more electrons are in electron states with energies higher than some vacant states. Electrons in excited states generally drop very rapidly ($\sim 10^{-7}$ s) into vacant lower energy states. During these spontaneous transitions, the difference in energy levels between the two states must be emitted as a photon (*fluorescence* or *x rays*) or be absorbed by other electrons in the atom (thereby causing them to change their energy states).

Crucial to the chemistry of atoms is the arrangement of atomic electrons into various electron *shells*. All electrons with the same n number constitute an *electron shell*. For $n = 1, 2, \ldots, 7$, the shells are designated $K, L, M, \ldots, Q$.

Consider electrons with $n = 1$ (K shell). Since $\ell = 0$ (s state electrons), then $m = 0$ and $m_s = \pm\frac{1}{2}$. Hence, there are only 2 $1s$ electrons, written as $1s^2$, in the K shell. In the L shell ($n = 2$), $\ell = 0$ or 1. For $\ell = 0$, there are two $2s$ electrons (denoted by $2s^2$), and for $\ell = 1$ ($m = -1, 0, 1$), there are six $2p$ electrons (denoted by $2p^6$). Thus in the L shell there are a total of eight electrons ($2s^2 2p^6$). Electrons with the same value of ℓ (and n) are referred as a *subshell*. For a given subshell, there are $(2\ell + 1)$ m values, each with two m_s values, giving a total of $2(2\ell + 1)$ electrons per subshell, and $2n^2$ electrons per shell. A shell or subshell containing the maximum number of electrons is said to be *closed*.

The Period Table of Elements (see Appendix A.2) can be described in terms of the possible number of electrons in the various subshells. The number of electrons in an atom equals its atomic number Z and determines its position in the Periodic Table. The chemical properties are determined by the number and arrangement of the electrons. Each element in the table is formed by adding one electron to that of the preceding element in the Periodic Table in such a way that the electron is most tightly bound to the atom. The arrangement of the electrons for the elements with electrons in only the first four shells is shown in Table 2.2.

BIBLIOGRAPHY

BUSH, H.D., *Atomic and Nuclear Physics*, Prentice Hall, Engelwood Cliffs, NJ, 1962.

EVANS, R.D., *The Atomic Nucleus*, McGraw-Hill, New York, 1955; republished by Krieger Publishing Co., Melborne, FL, 1982.

FRENCH, A.P., *Principles of Modern Physics*, Wiley, New York, 1962.

GLASSTONE, S., *Sourcebook on Atomic Energy*, Van Nostrand Reinhold, New York, 1967.

KAPLAN, I., *Nuclear Physics*, Addison-Wesley, Reading, MA, 1963.

LEIGHTON, R.B., *Principles of Modern Physics*, McGraw-Hill, New York, 1959.

MAYO, R.M., *Nuclear Concepts for Engineers*, American Nuclear Society, La Grange Park, IL, 1998.

SHULTIS, J.K. AND R.E. FAW, *Radiation Shielding*, American Nuclear Society, La Grange Park, IL, 2000.

PROBLEMS

1. An accelerator increases the total energy of electrons uniformly to 10 GeV over a 3000 m path. That means that at 30 m, 300 m, and 3000 m, the kinetic energy is 10^8, 10^9, and 10^{10} eV, respectively. At each of these distances, compute the velocity, relative to light (v/c), and the mass in atomic mass units.

2. Consider a fast moving particle whose relativistic mass m is 100ϵ percent greater than its rest mass m_o, i.e., $m = m_o(1 + \epsilon)$. (a) Show that the particle's speed v, relative to that of light, is

$$\frac{v}{c} = \sqrt{1 - \frac{1}{(1 + \epsilon)^2}}.$$

(b) For $v/c << 1$, show that this exact result reduces to $v/c \simeq \sqrt{2\epsilon}$.

3. In fission reactors, one deals with neutrons having kinetic energies as high as 10 MeV. How much error is incurred in computing the speed of 10-MeV neutrons by using the classical expression rather than the relativistic expression for kinetic energy?

4. What speed (m s^{-1}) and kinetic energy (MeV) would a neutron have if its relativistic mass were 10% greater than its rest mass?

5. In the Relativistic Heavy Ion Collider, nuclei of gold are accelerated to speeds of 99.95% the speed of light. These nuclei are almost spherical when at rest; however, as they move past the experimenters they appear considerably flattened in the direction of motion because of relativistic effects. Calculate the apparent diameter of such a gold nucleus in its direction of motion relative to that perpendicular to the motion.

6. Muons are subatomic particles that have the negative charge of an electron but are 206.77 times more massive. They are produced high in the atmosphere by cosmic rays colliding with nuclei of oxygen or nitrogen, and muons are the dominant cosmic-ray contribution to background radiation at the earth's surface. A muon, however, rapidly decays into an energetic electron, existing, from its point of view, for only 2.20 μs, on the average. Cosmic-ray generated muons typically have speeds of about $0.998c$ and thus should travel only a few hundred meters in air before decaying. Yet muons travel through several kilometers of air to reach the earth's surface. Using the results of special relativity explain how this is possible. HINT: consider the atmospheric travel distance as it appears to a muon, and the muon lifetime as it appears to an observer on the earth's surface.

7. A 1-MeV gamma ray loses 200 keV in a Compton scatter. Calculate the scattering angle.

8. At what energy (in MeV) can a photon lose at most one-half of its energy in Compton scattering?

9. Derive for the Compton scattering process the recoil electron energy T as a function of the incident photon energy E and the electron angle of scattering ϕ_e. Show that ϕ_e is never greater than $\pi/2$ radians.

10. A 1 MeV photon is Compton scattered at an angle of 55 degrees. Calculate (a) the energy of the scattered photon, (b) the change in wavelength, and (c) the recoil energy of the electron.

11. Show that the de Broglie wavelength of a particle with kinetic energy T can be written as

$$\lambda = \frac{h}{\sqrt{m_o}} \frac{1}{\sqrt{T}} \left[1 + \frac{m}{m_o} \right]^{-1/2}$$

where m_o is the particles's rest mass and m is its relativistic mass.

12. Apply the result of the previous problem to an electron. (a) Show that when the electron's kinetic energy is expressed in units of eV, its de Broglie wavelength can be written as

$$\lambda = \frac{17.35 \times 10^{-8}}{\sqrt{T}} \left[1 + \frac{m}{m_o} \right]^{-1/2} \text{ cm.}$$

(b) For non-relativistic electrons, i.e., $m \simeq m_o$, show that this result reduces to

$$\lambda = \frac{12.27 \times 10^{-8}}{\sqrt{T}} \text{ cm.}$$

(c) For very relativistic electrons, i.e., $m >> m_o$, show that the de Broglie wavelength is given by

$$\lambda = \frac{17.35 \times 10^{-8}}{\sqrt{T}} \sqrt{\frac{m_o}{m}} \text{ cm.}$$

13. What are the wavelengths of electrons with kinetic energies of (a) 10 eV, (b) 1000 eV, and (c) 10^7 eV?

14. What is the de Broglie wavelength of a water molecule moving at a speed of 2400 m/s? What is the wavelength of a 3-g bullet moving at 400 m/s?

15. If a neutron is confined somewhere inside a nucleus of characteristic dimension $\Delta x \simeq 10^{-14}$ m, what is the uncertainty in its momentum Δp? For a neutron with momentum equal to Δp, what is its total energy and its kinetic energy in MeV? Verify that classical expressions for momentum and kinetic energy may be used.

16. Repeat the previous problem for an electron trapped in the nucleus. HINT: relativistic expressions for momentum and kinetic energy must be used.

Chapter 3

Atomic/Nuclear Models

The concept of the atom is ancient. The Greek philosophers Leucippus and his pupil Democritus in the 5th century BC conjectured that all matter was composed of indivisible particles or *atoms* (lit. "not to be cut"). Unfortunately, Aristotle, whose ideas were more influential far into the Middle Ages, favored the "fire, air, earth, and water" theory of Empedocles, and the atom theory wasn't revived until a few hundred years ago.

The modern concept of the atom had its origin in the observations of chemical properties made by 18th and 19th century alchemists. In 1808, Dalton introduced his atomic hypothesis that stated: (1) each element consists of a large number of identical particles (called atoms) that cannot be subdivided and that preserve their identity in chemical reactions; (2) a compound's mass equals the sum of the masses of the constituent atoms; and (3) chemical compounds are formed by the combination of atoms of individual elements in simple proportions (e.g., 1:1, 1:2, etc.). This atomic hypothesis explained chemical reactions and the distinct ratios in which elements combined to form compounds. However, Dalton made no statement about the structure of an atom.

At the beginning of the twentieth century, a wealth of experimental evidence allowed scientists to develop ever more refined models of the atom, until our present model of the atom was essentially established by the 1940s. The structure of the nucleus of an atom has now also been well developed qualitatively and is supported by a wealth of nuclear data. However, work still continues on developing more refined mathematical models to quantify the properties of nuclei and atoms.

In this chapter, a brief historical summary of the development of atomic and nuclear models is presented. The emphasis of the presentation is on the novel ideas that were developed, and little discussion is devoted to the many experiments that provided the essential data for model development. Those interested in more detail, especially about the seminal experiments, should refer to any modern physics text.

3.1 Development of the Modern Atom Model

3.1.1 Discovery of Radioactivity

In 1896, Becquerel discovered that uranium salts emitted rays similar to x rays in that they also fogged photographic plates. Becquerel's discovery was followed by isolation of two other radioactive elements, radium and polonium, by the Curies in 1898. The radiation emission rate of radium was found to be more than a million times that of uranium, for the same mass.

Experiments in magnetic fields showed that three types of radiation could be emitted from naturally occurring radioactive materials. The identification of these radiations was made by the experimental arrangement shown in Fig. 3.1. Radioactive material was placed in a lead enclosure and the emitted radiation collimated in the upward direction by passing through the two collimating slits. The entire chamber was evacuated and a magnetic field was applied perpendicularly and directed outwardly from the plane of the page. With this arrangement, Becquerel found three distinct spots at which radiation struck the photographic plate used as a detector. The three different types of radiation, whose precise nature was then unknown, were called *alpha*, *beta*, and *gamma* rays.

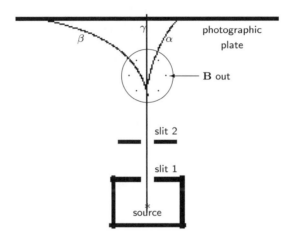

Figure 3.1. Deflection of α, β, and γ rays by a magnetic field out of and perpendicular to the page.

The beta rays, which were deflected to the left, were obviously negatively charged particles, and were later found to be the same as the "cathode" rays seen in gas discharge tubes. These rays were identified by J.J. Thomson in 1898 as electrons. The gamma rays were unaffected by the magnetic field and hence had to be uncharged. Today, we know that gamma rays are high frequency electromagnetic radiation whose energy is carried by particles called photons. The alpha particles, being deflected to the right, had positive charge. They were deflected far less than were the beta rays, an indication that the alpha particles have a charge-to-mass ratio e/m far less than that of beta particles. Either the positive charge of the alpha particle was far less than the negative charge of beta particles and/or the alpha particle's mass was far greater than that of a beta particle.

A quantitative analysis of the deflection of alpha particles showed that their speeds were of the order of 10^7 m/s. The charge-to-mass ratio was found to be 4.82×10^7 C kg^{-1}. By contrast, the charge-to-mass ratio for the hydrogen ion is twice as large, namely, 9.59×10^7 C kg^{-1}. Thus, if the alpha particle had the same charge as the hydrogen ion, its mass would have to be twice that of the hydrogen ion. If the alpha particle were doubly charged, its mass would be four times as large and would correspond to that of the helium atom.

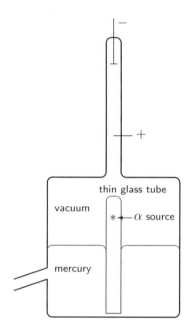

Figure 3.2. Experimental arrangement used to identify the identity of α particles.

That an alpha particle is an ionized helium atom was demonstrated by Rutherford who used the experimental arrangement of Fig. 3.2. The alpha particles from the radioactive source penetrate the thin-walled glass tube and are collected in the surrounding evacuated chamber. After slowing, the α particles capture ambient electrons to form neutral helium atoms. The accumulated helium gas is then compressed so that an electrical discharge will occur when a high voltage is applied between the electrodes. The emission spectrum from the excited gas atoms was found to have the same wavelengths as that produced by an ordinary helium-filled discharge tube. Therefore, the alpha particle must be a helium ion, and, since the mass of the alpha particle is four times the mass of hydrogen, the charge on the alpha particle is twice the charge on a hydrogen ion.

3.1.2   Thomson's Atomic Model: The Plum Pudding Model

The discovery of radioactivity by Becquerel in 1896 and Thomson's discovery of the electron in 1898 provided a basis for the first theories of atomic structure. In radioactive decay, atoms are transformed into different atoms by emitting positively charged or negatively charged particles. This led to the view that atoms are composed of positive and negative charges. If correct, the total negative charge in an atom must be an integral multiple of the electronic charge and, since atoms are electrically neutral, the positive and negative charges must be numerically equal.

The emission of electrons from atoms under widely varying conditions was convincing evidence that electrons exist as such inside atoms. The first theories of atomic structure were based on the idea that atoms were composed of electrons and positive charges. There was no particular assumption concerning the nature of the positive charges because the properties of the positive charges from radioactive decay and from gas discharge tubes did not have the e/m consistency that was shown by the negative charges (electrons).

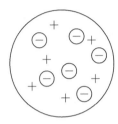

Figure 3.3. Thomson's plum-pudding model of the atom.

At the time of Thomson's research on atoms, there was no information about the way that the positive and negative charges were distributed in the atom. Thomson proposed a model, simple but fairly accurate, considering the lack of information about atoms at that time. He assumed that an atom consisted of a sphere of positive charge of uniform density, throughout which was distributed an equal and opposite charge in the form of electrons. The atom was like a "plum pudding," with the negative charges dispersed like raisins or plums in a dough of positive electricity (see Fig. 3.3). The diameter of the sphere was of the order of 10^{-10} m, the magnitude found for the size of the atom.

This model explained three important experimental observations: (1) an ion is just an atom from which electrons have been lost, (2) the charge on a singly ionized atom equals the negative of the charge of an electron, and (3) the number of electrons in an atom approximately equals one-half of the atomic weight of the atom, i.e., if the atom's mass doubles the number of electrons double.

Also it was known that the mass of the electron was known to be about one eighteen hundredth the mass of the hydrogen atom. Therefore, the total mass of the electrons in an atom is a very small part of the total mass of the atom, and, hence, practically all of the mass of an atom is associated with the positive charge of the "pudding."

3.1.3 The Rutherford Atomic Model

At the beginning of the 20th century, Geiger and Marsden used alpha particles from a radioactive source to irradiate a thin gold foil. According to the Thomson plum-pudding model, for a gold foil 4×10^{-5} cm in thickness, the probability an alpha particle scatters at least by an angle ϕ was calculated to be $\exp(-\phi/\phi_m)$, where $\phi_m \simeq 1°$. Thus, the probability that an alpha particle scatters by more than $90°$ should be about 10^{-40}. In fact, it was observed that one alpha particle in 8000 was scattered by more than $90°$!

To explain these observations, Rutherford in 1911 concluded that the positively charged mass of an atom must be concentrated within a sphere of radius about 10^{-12} cm. The electrons, therefore, must revolve about a massive, small, positively-charged nucleus in orbits with diameters of the size of atoms, namely, about 10^{-8} cm. With such a model of the nucleus, the predicted deflection of alpha particles by the gold foil fit the experimental data very well.

3.1.4   The Bohr Atomic Model

The Rutherford model of an atom was quickly found to have serious deficiencies. In particular, it violated classical laws of electromagnetism. According to classical theory, an accelerating charge (the electrons in circular orbits around a nucleus) should radiate away their kinetic energy within about 10^{-9} s and spiral into the nucleus.[1]

But obviously atoms do not collapse. Were they to do so, the electromagnetic radiation emitted by the collapsing electrons should be continuous in frequency, since as the electrons collapse, they should spiral ever faster around the nucleus and hence experience increasing central accelerations. When atoms are excited by an electrical discharge, for example, atoms are observed to emit light not in a continuous wavelength spectrum but with very discrete wavelengths characteristic of the element. For example, part of the spectrum of light emitted by hydrogen is shown in Fig. 3.4.

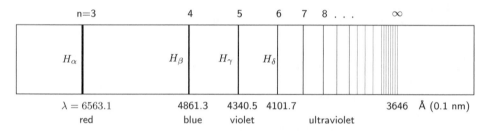

Figure 3.4. Diagram of the lines of the Balmer series of atomic hydrogen. After Kaplan [1963].

It was found empirically, the wavelength of light emitted by hydrogen (excited by electrical discharges) could be described very accurately by the simple equation

$$\frac{1}{\lambda} = R_H \left[\frac{1}{n_o^2} - \frac{1}{n^2} \right] \tag{3.1}$$

where n_o, n are positive integers with $n > n_o$ and R_H is the Rydberg constant found empirically to have the value $R_H = 10\,967\,758$ m^{-1}. The integer n_o defines a series of spectral lines discovered by different researchers identified in Table 3.1.

The observed discrete-wavelength nature of light emitted by excited atoms was in direct conflict with Rutherford's model of the atom. In a series of papers published between 1913 and 1915, Bohr developed an atomic model which predicted very closely the observed spectral measurements in hydrogen. Bohr visualized an atom much like the Rutherford model with the electrons in orbits around a small dense

---

[1] This radiation produced by an accelerated electric charge explains how radio waves are generated. Electrons are accelerated back and forth along a wire (antenna) by an alternating potential with a well defined frequency, namely that of the radio waves that are emitted from the wire.

Table 3.1. Various hydrogen spectral series.

n_o	Series name	Spectrum location
1	Lyman	ultraviolet
2	Balmer	visible and near ultraviolet
3	Paschen	infrared
4	Brackett	infrared
5	Pfund	infrared

central nucleus. However, his model included several non-classical constraints. Bohr postulated:

1. An electron moves in a circular orbit about the nucleus obeying the laws of classical mechanics.

2. Instead of an infinity of orbits, only those orbits whose angular momentum L is an integral multiple of $h/2\pi$ are allowed.

3. Electrons radiate energy only when moving from one allowed orbit to another. The energy $E = h\nu$ of the emitted radiation is the difference between characteristic energies associated with the two orbits. These characteristic energies are determined by a balance of centripetal and Coulombic forces.

Consider an electron of mass m_e moving with speed v in a circular orbit of radius r about a central nucleus with charge Ze. From postulate (1), the centripetal force on the electron must equal the Coulombic attractive force, i.e.,

$$\frac{m_e v^2}{r} = \frac{Ze^2}{4\pi\epsilon_o r^2}, \tag{3.2}$$

where ϵ_o is the *permittivity* of free space and e is the charge of the electron. From postulate (2)

$$L \equiv m_e vr = n\frac{h}{2\pi}, \quad n = 1, 2, 3, \ldots. \tag{3.3}$$

Solution of these relations for r and v yields

$$v_n = \frac{Ze^2}{2\epsilon_o nh} \quad \text{and} \quad r_n = \frac{n^2 h^2 \epsilon_o}{\pi m_e Ze^2}, \quad n = 1, 2, 3, \ldots \tag{3.4}$$

For $n = 1$ and $Z = 1$, these expressions yield $r_1 = 5.29 \times 10^{-11}$ m and $v_1 = 2.2 \times 10^6$ m/s $\ll c$, indicating that our use of non-relativistic mechanics is justified.

The electron has both potential energy[2] $V_n = -Ze^2/(4\pi\epsilon_o r_n)$ and kinetic energy $T_n = \frac{1}{2}m_e v_n^2$ or, from Eq. (3.2), $T_n = \frac{1}{2}[Ze^2/(4\pi\epsilon_o r_n)]$. Thus, the electron's total energy $E_n = T_n + V_n$ or, with Eq. (3.4),

$$E_n = T_n + V_n = \frac{1}{2}\left(\frac{Ze^2}{4\pi\epsilon_o r_n}\right) - \left(\frac{Ze^2}{4\pi\epsilon_o r_n}\right) = -\frac{1}{2}\frac{Ze^2}{4\pi\epsilon_o r_n} = -\frac{m_e(Ze^2)^2}{8\epsilon_o^2 n^2 h^2}. \tag{3.5}$$

---

[2]The potential energy is negative because of the attractive Coulombic force, with energy required to extract the electron from the atom. The zero reference for potential energy is associated with infinite orbital radius.

For the ground state of hydrogen ($n = 1$ and $Z = 1$), this energy is $E_1 = -13.606$ eV, which is the experimentally measured ionization energy for hydrogen.

Finally, by postulate (3), the frequency of radiation emitted as an electron moves from an orbit with "quantum number" n to an orbit denoted by $n_o < n$ is given by

$$h\nu_{n\rightarrow n_o} = E_{n_o} - E_n = \frac{m_e Z^2 e^4}{8\epsilon_o^2 h^2}\left[\frac{1}{n_o^2} - \frac{1}{n^2}\right]. \tag{3.6}$$

Since $1/\lambda = \nu/c$, the wavelength of the emitted light is thus

$$\frac{1}{\lambda_{n\rightarrow n_o}} = \frac{m_e Z^2 e^4}{8\epsilon_o^2 ch^3}\left[\frac{1}{n_o^2} - \frac{1}{n^2}\right] = R_\infty\left[\frac{1}{n_o^2} - \frac{1}{n^2}\right], \quad n > n_o. \tag{3.7}$$

This result has the same form as the empirical Eq. (3.1). However, in our analysis we have implicitly assumed the proton is infinitely heavy compared to the orbiting electron by requiring it to move in a circular orbit about a stationary proton. The Rydberg constant in the above result (with $Z = 1$) is thus denoted by R_∞, and is slightly different from R_H for the hydrogen atom. In reality, the proton and electron revolve about each other. To account for this, the electron mass in Eq. (3.7) should be replaced by the electron's *reduced* mass $\mu_e = m_e m_p/(m_e + m_p)$ [Kaplan 1963]. This is a small correction since $\mu_e = 0.999445568\, m_e$, nevertheless an important one. With this correction, $R_H = \mu_e e^4/(8\epsilon_o^2 ch^3) = 10\,967\,758$ m^{-1}, which agrees with the observed value to eight significant figures!

The innermost orbits of the electron in the hydrogen atom are shown schematically in Fig. 3.5 to scale. The electronic transitions that give rise to the various emission series of spectral lines are also shown. The Bohr model, with its excellent predictive ability, *quantizes* the allowed electron orbits. Each allowed orbit is defined by the *quantum number* n introduced through Bohr's second postulate, a postulate whose justification is that the resulting predicted emission spectrum for hydrogen is in amazing agreement with the observation.

Example 3.1: What is the energy (in eV) required to remove the electron in the ground state from singly ionized helium? For the helium nucleus $Z = 2$ and the energy of the ground state ($n=1$) is, from Eq. (3.5),

$$E_1 = -\frac{m_e(2e^2)^2}{8\epsilon_o^2 h^2}.$$

From Table 1.5, we find $h = 6.626 \times 10^{-34}$ J s, $m_e = 9.109 \times 10^{-31}$ kg, $e = 1.6022\times 10^{-19}$ C, and $\epsilon_o = 8.854\times 10^{-12}$ F m^{-1} ($= $ C^2 J^{-1} m^{-1}). Substitution of these values into the above expression for E_1 yields

$$E_1 = -8.720 \times 10^{-18}\left(\frac{\text{kg}}{\text{m}^{-2}\,\text{s}^2}\right) = -8.720 \times 10^{-18}\,\text{J}$$

$$= (-8.720 \times 10^{-18}\,\text{J})/(1.6022 \times 10^{-19}\,\text{J/eV}) = -54.43\text{ eV}.$$

Thus, it takes 54.43 eV of energy to remove the electron from singly ionized helium.

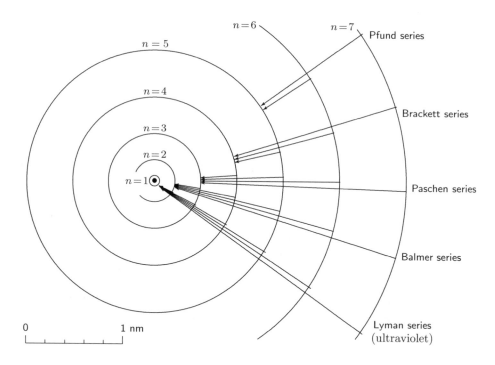

Figure 3.5. Bohr orbits for the hydrogen atom showing the deexcitation transitions that are responsible for the various spectral series observed experimentally. After Kaplan [1963].

3.1.5 Extension of the Bohr Theory: Elliptic Orbits

The Bohr theory was very successful in predicting with great accuracy the wavelengths of the spectral lines in hydrogen and singly ionized helium. Shortly after Bohr published his model of the atom, there was a significant improvement in spectroscopic resolution. Refined spectroscopic analysis showed that the spectral lines were not simple, but consisted of a number of lines very close together, a so-called *fine structure*. In terms of energy levels, the existence of a fine structure means that instead of a single electron energy level for each quantum number n, there are actually a number of energy levels lying close to one another. Sommerfeld partly succeeded in explaining some of the lines in hydrogen and singly ionized helium by postulating elliptic orbits as well as circular orbits (see Fig. 3.6). These elliptical orbits required the introduction of another quantum number ℓ to describe the angular momentum of orbits with varying eccentricities.

Sommerfeld showed that, in the case of a one-electron atom, the fine structure could be partially explained. When the predictions of Sommerfeld's model were compared to experimental results on the resolution of the Balmer lines for He$^+$, the theory predicted more lines than were observed in the fine structure. Agreement between theory and experiment required the new quantum number ℓ to be constrained by an empirical or *ad hoc* selection rule which limited the number of allowed elliptical orbits and the transitions between the orbits.

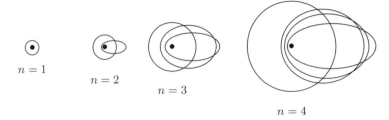

$n = 1$

$n = 2$

$n = 3$

$n = 4$

Figure 3.6. Relative positions and dimensions of the elliptical orbits corresponding to the first four values of the quantum number n in Sommerfeld's modification for the hydrogen atom. After Kaplan [1963].

For two-electron atoms, difficulties arose in the further refinement of Bohr's theory. To account for the observed splitting of the spectral lines in an applied magnetic field, it was necessary to introduce a third quantum number m, related to the orientation of the elliptical orbits in space. In the presence of a magnetic field, there is a component of the angular momentum in the direction of the field. This modification again led to the prediction of more spectral lines than were experimentally observed, and a second set of selection rules had to be empirically introduced to limit the number of lines.

Further difficulties arose when the model was applied to the spectrum of more complicated atoms. In the spectra of more complicated atoms, multiplet structure is observed. Multiplets differ from fine structure lines in that lines of a multiplet can be widely separated. The lines of sodium with wavelengths of 589.593 nm and 588.996 nm are an example of a doublet. Triplets are observed in the spectrum of magnesium.

These difficulties (and others) could not be resolved by further changes of the Bohr theory. It became apparent that these difficulties were intrinsic to the model. An entirely different approach was necessary to solve the problem of the structure of the atom and to avoid the need to introduce empirical selection rules.

3.1.6 The Quantum Mechanical Model of the Atom

From the *ad hoc* nature of quantum numbers and their selection rules used in refinements of the basic Bohr model of the atom, it was apparent an entirely different approach was needed to explain the details of atomic spectra. This new approach was introduced in 1925 by Schrödinger who brought wave or quantum mechanics to the world. Here, only a brief summary is given. Schrödinger's new theory (or model) showed that there were indeed three quantum numbers (n, ℓ, m) analogous to the three used in the refined Bohr models. Further, the values and constraints on these quantum numbers arose from the theory naturally—no *ad hoc* selection rules were needed. The solution of the wave equation for the hydrogen atom is discussed in Addendum 2 of Chapter 2.

To explain the multiple fine-line structure in the observed optical spectra from atoms and the splitting of some of these lines in a strong magnetic field (the *anomalous Zeeman effect*), the need for a fourth quantum number m_s was needed. This quantum number accounted for the inherent angular momentum of the electron equal to $\pm h/2\pi$. In 1928, Dirac showed that this fourth quantum number also arises from the wave equation if it is corrected for relativistic effects.

In the quantum mechanical model of the atom, the electrons are no longer point particles revolving around the central nucleus in orbits. Rather, each electron is visualized as a standing wave around the nucleus. The amplitude of this wave (called a *wave function*) at any particular position gives the probability that the electron is in that region of space. Example plots of these wave functions for the hydrogen atom are given in Fig. 2.14. Each electron wave function has a well-defined energy which is specified uniquely by the four quantum numbers defining the wave function.

3.2 Models of the Nucleus

Models of a nucleus tend to be relatively crude compared to those used to describe atomic electrons. There are two reasons for this lack of detail. First, the nuclear forces are not completely understood. And then, even with approximate nuclear force models, the calculation of the mutual interactions between all the nucleons in the nucleus is a formidable computational task. Nevertheless, a few very simple models, which allow interpretation and interpolation of much nuclear data, have been developed.

We begin our discussion on nuclear models by reviewing some fundamental properties of the nucleus and some early views about the nucleus. Then we review some systematics of the known nuclides and what these observations imply about the arrangements of neutrons and protons in the nucleus. Finally, the liquid drop and nuclear shell models are introduced.

3.2.1 Fundamental Properties of the Nucleus

Early research into the structure of atoms revealed several important properties of the nucleus. Experiments by J.J. Thomson in 1913 and Aston in 1919 showed that the masses of atoms were very nearly whole numbers (the *atomic mass numbers*), based on a mass scale in which elemental oxygen was given a mass of 16.[3] Thomson showed in 1898 that atoms contain electrons whose mass is considerably less than the mass of the atom. For example, an atom of ^{12}C has a mass of 12 u (by definition) while the mass of its six electrons is only 0.00329 u.

In Section 3.1.3 we discussed how, in 1911, Rutherford demonstrated from his alpha-scattering experiments that the positive charge and most of the mass of an atom were contained in a nucleus much smaller ($\simeq 10^{-14}$ m) than the size of the atom ($\simeq 10^{-8}$ m). More accurate measurements of the nuclear size were later made by scattering energetic beams of electrons from various nuclides to determine the density of protons (or positive charge) inside a nucleus. Since nuclei have dimensions of the order of several fm ("fermis"), the de Broglie wavelength of the incident electrons needs to be no more than a few fm. For example, if $\lambda = 2$ fm, then the kinetic energy of the electrons must be $T = \sqrt{p^2c^2 + m_ec^2} \simeq 620$ MeV.

---

[3]This oxygen standard used to define the atomic mass unit was replaced over thirty years ago by the ^{12}C standard that defines an atom of ^{12}C to have a mass of exactly 12 amu. The difference between the two standards, while small, is important when interpreting historical data.

From such electron scattering experiments, later confirmed by experiments measuring x-ray emission following muon capture by a nucleus, the density of protons inside spherical nuclei[4] was found to be well described by the simple formula

$$\rho_p(r) = \frac{\rho_p^o}{1 + \exp\left[(r - R)/a\right]} \quad (\text{protons/fm}^3), \tag{3.8}$$

where r is the distance from the center of the nucleus, R is the "radius" of the nucleus (defined as the distance at which the proton density falls to one-half of its central value), and a is the "surface thickness" over which the proton density becomes negligibly small. This distribution is shown in Fig. 3.7. The quantity ρ_p^o is obtained from the normalization requirement that the total number of protons must equal the atomic number, i.e.,

$$\iiint \rho_p(r)\, dV = 4\pi \int_0^\infty r^2 \rho_p(r)\, dr = Z. \tag{3.9}$$

Table 3.2. Nuclear density parameters for three spherical nuclei. From Cottingham and Greenwood [1986] using data from Barrett and Jackson [1977].

Nuclide	R (fm)	a (fm)	ρ_o (fm^{-3})	$R/A^{1/3}$
^{16}O	2.61	0.513	0.156	1.036
^{109}Ag	5.33	0.523	0.157	1.116
^{208}Pb	6.65	0.526	0.159	1.122

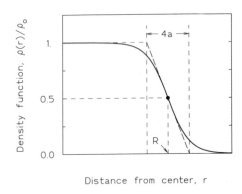

Figure 3.7. Distribution function used to describe the proton density inside the nucleus.

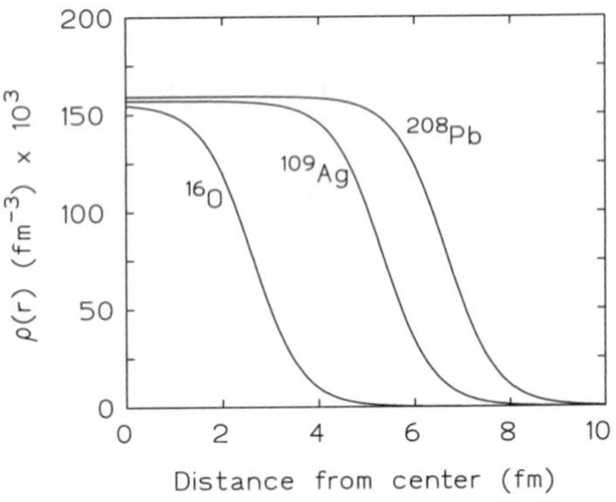

Figure 3.8. Nucleon number density distributions for three nuclides.

If it is assumed that the density ratio of neutrons $\rho_n(r)$ to protons $\rho_p(r)$ at all points in the nucleus is $N/Z = (A - Z)/Z$, the nucleon density is

$$\rho(r) = \rho_p(r) + \rho_n(r) = \left[1 + \frac{A - Z}{Z}\right] \rho_p(r) = (A/Z)\rho_p(r) \quad \text{nucleons/fm}^3. \tag{3.10}$$

---

[4]Not all nuclei are spherically symmetric, but we restrict our discussion to those nuclei that are symmetric or nearly so.

Thus, the nucleon density inside a nucleus can be described by

$$\rho(r) = \frac{\rho_o}{1 + \exp{[(r-R)/a]}} \quad (\text{nucleons/fm}^3), \qquad (3.11)$$

where $\rho_o = (A/Z)\rho_p^o$. Typical values for the density parameters are given in Table 3.2, and nucleon density plots for three representative nuclei are shown in Fig. 3.8. From these and similar results, it is found that the density at the nucleus center is about

$$\rho_o = 0.16 \times 10^{45} \quad \text{nucleons/m}^{-3}. \qquad (3.12)$$

Also from Table 3.2, the radius R of the nucleus is seen to be proportional to $A^{1/3}$, i.e.,

$$R = 1.1 A^{1/3} \times 10^{-15} \quad \text{m}. \qquad (3.13)$$

This $A^{1/3}$ variation of R suggests each nucleon in the nucleus has the same volume since the total volume of the nucleus $(4/3)\pi R^3$ is then proportional to A, the number of nucleons. Because a nucleon has a mass of about 1.67×10^{-24} g, the mass density inside a nucleus is about $(0.16 \times 10^{39} \text{ nucleons/cm}^{-3})(1.67 \times 10^{-24} \text{ g/nucleon}) = 2.7 \times 10^{14}$ g/cm^{-3}.

Any model of the nucleus and the forces that hold it together must explain the small nuclear volume and the corresponding high density. In addition, a nuclear model needs to explain which combinations of neutrons and protons produce stable nuclei and which combinations lead to unstable, or radioactive, nuclei.

3.2.2 The Proton-Electron Model

The masses of all nuclei, when measured in atomic mass units, are observed to be almost whole numbers. Because the hydrogen nucleus has the smallest mass (1 amu), a simple nuclear model was to assume that all heavier nuclei were composed of multiples of the hydrogen nucleus, namely protons.[5] This simple nuclear model neatly explains why nuclei have masses that are nearly an integer.

However, to give the correct nuclear charge with this model, it was also necessary to assume that some electrons (of negligible mass) were also included in the nucleus to cancel some of the positive charge of the protons. For example, sodium has an atomic mass number of 23 and an atomic charge of 11. Thus, the nucleus of a sodium atom must contain 23 protons to give it the correct mass and 12 electrons to give it the correct charge. In addition, there are 11 electrons surrounding the nucleus to produce a neutral sodium atom. Thus in this proton-electron model, an atom with atomic mass A and atomic number Z would have a nucleus containing A protons and $(A - Z)$ electrons with Z orbital electrons surrounding the nucleus. The atomic mass is determined by the number of protons, since the electrons make a negligible contribution.

However, there are two serious difficulties with this proton-electron model of the nucleus. The first problem is that predicted angular momentum (or *spin*) of the nuclei did not always agree with experiment. Both the proton and electron have an

---

[5]This conjecture is an extension of Prout's 1816 hypothesis which proposed that, since most atomic weights are nearly whole numbers and since the atomic weight of hydrogen is nearly one, then heavier atoms are composed of multiple hydrogen atoms.

inherent angular momentum (spin) equal to $\frac{1}{2}(h/2\pi)$. The spin of the nucleus must then be some combination of the spins of the nuclear constituents, e.g., (in units of $h/2\pi$) spin $= \frac{1}{2} + \frac{1}{2} + \frac{1}{2} - \frac{1}{2} + \frac{1}{2} - \frac{1}{2} \dots$. Beryllium ($A = 9, Z = 4$) should contain an even number of particles in the nucleus (9 protons and 5 electrons), and thus should have a spin which is a combination of 14 factors of $\frac{1}{2}$ (added and subtracted in some manner). Such a combination must necessarily give a spin that is an integer (or zero). But experimentally, the spin is found to be a half-integer. Similarly, nitrogen ($A = 14, Z = 7$) should have 21 particles in its nucleus (14 protons and 7 electrons). Its spin must, therefore, be a half-integer multiple of $h/2\pi$. Experimentally, nitrogen nuclei have integer spin.

The second problem with the proton-electron model, derives from the uncertainty principle. If an electron is confined to the nucleus of diameter about 10^{-14} m, its uncertainty in position is about $\Delta x \simeq 10^{-14}$ m. From Heisenberg's uncertainty principle, the uncertainty in the electron's momentum Δp is given by

$$\Delta p\, \Delta x \geq \frac{h}{2\pi}. \tag{3.14}$$

From this we find that the minimum uncertainty in momentum of the electron is about $\Delta p = 1.1 \times 10^{-20}$ J m^{-1} s. From Eq. (2.16), the electron's total energy $E = T + m_o c^2 = \sqrt{p^2 c^2 + m_o^2 c^4}$, which yields for $p = \Delta p$, $E \simeq T = 20$ MeV, an energy far greater than its rest-mass energy (0.51 MeV).[6] Electrons (beta particles) emitted by atoms seldom have energies above a few MeV. Clearly, there is some fundamental problem with having electrons confined to a nucleus.

From these difficulties, it was apparent that the proton-electron model of the nucleus was fundamentally flawed.

3.2.3   The Proton-Neutron Model

In 1932 Chadwick discovered the neutron, a chargeless particle with a mass just slightly greater than that of the proton ($m_n = 1.008665$ u versus $m_p = 1.007276$ u). Our current view that every nucleus is composed of only protons and neutrons was first suggested by Heisenberg in 1932. By this model, a nucleus with a mass number A contains Z protons and $N = A - Z$ neutrons.

This neutron-proton model avoids the failures of the proton-electron model. The empirical rule connecting mass number and nuclear angular momentum can be interpreted by showing that the neutron, as well as the proton, has half-integral spin. The spin of a nucleus containing A neutrons and protons must, therefore, be an integral or half-integral multiple of $h/(2\pi)$, according to whether A is even or odd. This has been well verified by experiment.

The neutron-proton model is also consistent with radioactivity. As previously discussed, there are reasons why electrons cannot exist in the nucleus, and hence it must be concluded that in beta-minus decay, the beta particle (electron) is created at the instant of emission. Such a decay is regarded as the conversion within the nucleus of a neutron into a proton, thereby transforming the nucleus into one with

---

[6]The same calculation applied to the proton shows that, because of the proton's much higher mass, most of the energy of a free proton confined to a nucleus is rest-mass energy (931 MeV), the kinetic energy being less than 1 MeV.

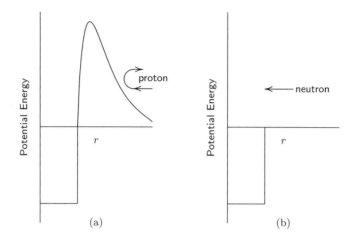

Figure 3.9. The potential energies of a proton (a) and a neutron (b) as a function of distance from the center of a nucleus.

one more proton and one less neutron. Similarly in alpha decay, two bound neutrons and protons are ejected from the nucleus.

A major difficulty with the proton-neutron model is how the protons and neutrons in the nucleus are held together. Because the protons are so close to each other, electromagnetic repulsive forces are very large and try to separate the protons. In the proton-electron model, the electrons with their negative charge tend to mitigate this repulsive force. To hold the neutrons and protons together in the proton-neutron model, there must be a very strong short-range force, the so-called *nuclear force*, that attracts nucleons to each other. A major challenge in nuclear physics is the understanding and quantification of the nature of the nuclear force which holds the nucleus together.

In a multi-proton nucleus, there must be nuclear attractive forces strong enough to overcome the repulsive forces among the protons. These attractive forces occur between a proton and neutron, between two neutrons, and between two protons. Further, they must be very strong at distances less than or equal to the nuclear radius. Outside the nucleus, however, they decrease radially very rapidly, and the Coulombic repulsive forces, responsible for the scattering of Rutherford's alpha particles, dominate. The magnitude of these forces is such that the work required to separate a nucleus into its constituent protons and neutrons (the *binding energy* of the nucleus) is orders of magnitude larger than the work required to remove an outer electron from an atom. The latter is usually a few electron volts; removing a nucleon from a nucleus requires energies typically a million times greater.

In Fig. 3.9(a), the potential energy of a proton is shown as a function of its distance from the center of a nucleus. The potential energy is zero at very large separation distances; but as the proton approaches the nucleus, the Coulombic repulsive force increases the proton's potential energy as it approaches the nucleus. When the proton reaches the surface of the nucleus, however, the nuclear force attracts and tries to bind the proton to the nucleus creating a negative potential energy. If a neutron approaches a nucleus [see Fig. 3.9(b)], there is no electrostatic repulsive force, and the potential energy of the neutron remains zero until it reaches the surface of the nucleus at which point it experiences the attractive nuclear force and becomes bound to the nucleus.

3.2.4 Stability of Nuclei

The neutron and proton numbers of the 3200 known nuclides are shown in the mini chart of the nuclides of Fig. 3.10. The 266 stable nuclides are shown as solid squares, and the radioactive nuclides are represented by a cross. The light stable nuclides are seen to have almost equal neutron and proton numbers, N and Z. However, as the mass of the nucleus increases, there have to be more neutrons than protons to produce a stable nucleus. With an increasing number of protons, the long-range electromagnetic repulsive forces trying to tear the nucleus apart increase dramatically. Although there are attractive nuclear forces between the protons, more neutrons are needed to provide additional attractive nuclear force to bind all the nucleons together. But because the nuclear force is short ranged (the order of the distance between a few nucleons), eventually when Z exceeds 83 (bismuth), extra neutrons can no longer provide sufficient extra nuclear force to produce a stable nucleus. All atoms with $Z > 83$ are thus radioactive.

Careful examination of the stable nuclides reveals several important features about the combinations of Z and N that produce stability in a nucleus.

1. In Figs. 3.11 and 3.12, the number of stable isotopes is plotted versus proton number Z and neutron number N, respectively. We see from these plots that there are many more stable isotopes with even N and/or Z (thick lines) than there are with odd N and/or Z (thin lines). In particular, there are 159 nuclides with even Z and even N; 53 nuclides with even Z and odd N; 50 nuclides with odd Z and even N; and, only 4 nuclides with both odd Z and odd N. It is apparent from these observations that stability is increased when neutrons and protons are both paired.

2. The nuclei of radioactive atoms undergo spontaneous changes whereby the number of neutrons and/or protons change. Such changes produce nuclei that are closer to the region of stability in Fig. 3.10. One might think that a nucleus with too many neutrons or protons for stability would simply expel neutrons or protons, perhaps one by one, in order to reach a stable configuration. However, neutron or proton emission in radioactive decay is rarely observed. Moreover, heavy radioactive atoms ($Z > 83$) often emit a small nucleus composed of 2 neutrons and 2 protons, called an *alpha particle* (which is the nucleus of ^{4}He). This observation indicates that, in a heavy nucleus, the neutrons and protons tend to group themselves into subunits of 2 neutrons and 2 protons. This alpha-particle substructure is also observed for light nuclei. In the next chapter, we will see that nuclei composed of multiples of alpha particles (e.g., ^{8}Be, ^{12}C, ^{16}O, ^{20}Ne) are extremely stable as indicated by the inordinately high energy required to break them apart.

3. From Figs. 3.11 and 3.12, it is seen that when either Z or N equals 8, 20, 50, 82, or 126, there are relatively greater numbers of stable nuclides. These numbers

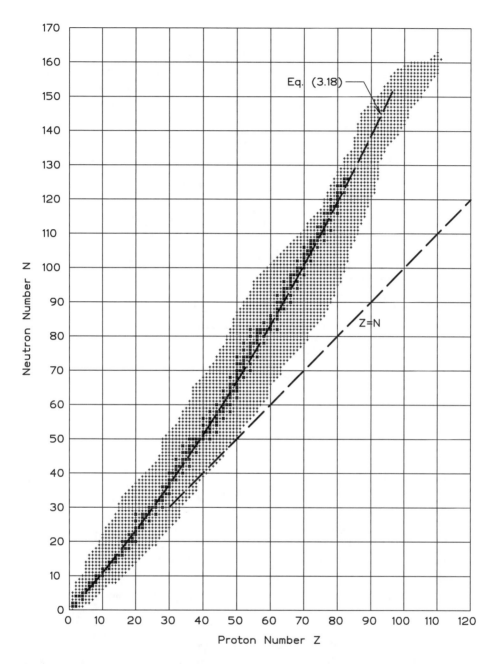

Figure 3.10. The known nuclides. Stable nuclides are indicated by solid squares and radioactive nuclides by +. The line of equal neutrons and protons is shown by the heavy dashed line.

are commonly called *magic numbers*. Other nuclear properties are also found to change abruptly when N or Z becomes magic. For example, $^{16}_{8}O$, $^{40}_{20}Ca$, $^{118}_{50}Sn$, $^{208}_{82}Pb$, and several other isotopes with magic N or Z are observed to have abnormally high natural abundances. Similarly, the readiness with which a neutron is absorbed by a nucleus with a magic number of neutrons is much less than for isotopes with one more or one less neutron. These observations suggest that neutrons and protons form into groups or shells, analogous to electron shells, which become closed when Z or N reaches a magic number.

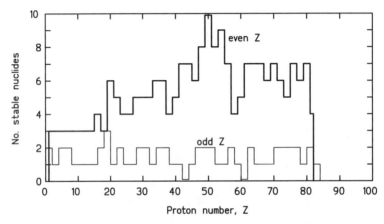

Figure 3.11. The number of stable isotopes for each element or proton number Z.

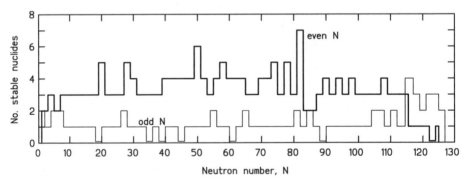

Figure 3.12. The number of stable isotopes as a function of neutron number N.

3.2.5   The Liquid Drop Model of the Nucleus

In Section 3.2.1, we saw that the density of matter inside a nucleus is nearly the same for all nuclides and that the volume of the nucleus is proportional to the number of nucleons A. These observations suggest that the nuclear forces holding the nucleus together are "saturated," i.e., a nucleon interacts only with its immediate neighbors. This is similar to a liquid drop in which each molecule interacts only with its nearest neighbors. Moreover, the near constancy of the density of nuclear matter is analogous to an incompressible liquid. To a first approximation, the mass of a nucleus (liquid droplet) equals the sum of the masses of the constituent nucleons (molecules). Thus, the nuclear mass $m(_Z^A X)$ of chemical element X whose nucleus is composed of Z protons and $N = A - Z$ neutrons is approximately

$$m(_Z^A X) = Zm_p + (A - Z)m_n \tag{3.15}$$

where m_p and m_n are the proton and neutron mass, respectively.

But the actual mass is somewhat less since it always requires energy to break up a nucleus into its individual neutrons and protons. Equivalently, when Z free protons and N neutrons come together to form a nucleus, energy is emitted. This *binding energy* BE comes from the conversion of a small fraction of the neutron and proton masses into energy (as prescribed by $E = mc^2$). Thus, the mass equivalent of the binding energy, BE/c^2, must be subtracted from the right-hand side of Eq. (3.15) to account for the conversion of a small fraction of the mass into energy when a nucleus is created. Several slightly different formulations are available. The one we present below is taken from Wapstra [1958].

One insightful way to make such binding energy corrections is to treat a nucleus as if it were a drop of nuclear liquid. Hence the name *liquid drop model*. As more nucleons bind to form a nucleus, more mass is converted into binding energy. Since each nucleon interacts only with its nearest neighbors, the volume binding energy component BE_v should be proportional to the volume of the nucleus or to the number of nucleons, i.e., $BE_v = a_v A$, where a_v is a constant. This volume binding energy, however, tends to overestimate the total binding energy and several binding energy corrections are needed. Phenomena leading to such binding energy corrections include the following.

1. Nucleons near the surface of the nucleus are not completely surrounded by other nucleons and, hence, are not as tightly bound as interior nucleons. This reduction in the binding energy of nucleons near the surface is analogous to surface tension effects in a liquid drop in which molecules near the surface are not as tightly bound as interior molecules. The surface tension effect decreases the volume binding energy and is proportional to the surface area of the nucleus, i.e., proportional to R^2 or $A^{2/3}$. Thus, a surface binding energy component, which has the form $BE_s = -a_s A^{2/3}$ where a_s is some positive constant, must be included.

2. The protons inside a nucleus produce Coulombic electrostatic repulsive forces among themselves. This repulsion decreases the stability of the nucleus, i.e., it reduces the nuclear binding energy by the negative potential energy of the mutual proton repulsion. This potential energy is proportional to $Z(Z - 1)/R$ or to $Z(Z - 1)/A^{1/3} \simeq Z^2/A^{1/3}$. Hence, a Coulombic binding energy correction, of the form $BE_c = -a_c Z^2/A^{1/3}$ with a_c a positive constant, is needed.

3. As we have observed, stable nuclides of small mass number A have equal numbers of protons and neutrons. We thus expect, neglecting Coulombic repulsion between the protons, that a departure from an equality, or symmetry,

in neutron and proton numbers tends to reduce nuclear stability. The greater the asymmetry $(N - Z) = (A - 2Z)$, the more the binding energy is decreased. Further, this asymmetry effect is greater for smaller nuclides. Thus, we need an asymmetry binding energy correction, which empirically has the form $BE_a = -a_a(A - 2Z)^2/A$, where a_a is a positive constant.

4. We have also observed that stable nuclides with even neutron and proton numbers are more abundant than stable even-odd or odd-even nuclides. Very few odd-odd stable nuclides exist. Thus, pairing both neutrons and protons produces more stable nuclides and increases the nuclear binding energy. By contrast, unpaired neutrons and unpaired protons decrease stability and the binding energy. To account for this effect, an empirical binding energy correction of the form $BE_p = -a_p/\sqrt{A}$ is used, where a_p is positive for even-even nuclei, negative for odd-odd nuclei, and zero for even-odd or odd-even nuclides.

Finally, the sum of these binding energy components $BE = BE_v + BE_s + BE_c + BE_a + BE_p$, converted to a mass equivalent by dividing by c^2, is subtracted from Eq. (3.15). With this refinement, the mass of a nucleus with mass number A and atomic number Z is

$$m(^A_Z X) = Z m_p + (A - Z)m_n$$
$$- \frac{1}{c^2}\left\{ a_v A - a_s A^{2/3} - a_c \frac{Z^2}{A^{1/3}} - a_a \frac{(A - 2Z)^2}{A} - \frac{a_p}{\sqrt{A}} \right\}. \quad (3.16)$$

Values for the empirical constants a_v, a_s, a_c, a_a, and a_p are as follows [Wapstra, 1958]:

$$a_v = 15.835 \text{ MeV}, \quad a_s = 18.33 \text{ MeV}, \quad a_c = 0.714 \text{ MeV}, \quad a_a = 23.20 \text{ MeV},$$

and

$$a_p = \begin{cases} +11.2 \text{ MeV} & \text{for odd } N \text{ and odd } Z \\ 0 & \text{for odd } N, \text{ even } Z \text{ or for even } N, \text{ odd } Z \\ -11.2 \text{ MeV} & \text{for even } N \text{ and even } Z \end{cases}$$

Line of Stability
The liquid drop model says nothing about the internal structure of the nucleus. However, it is very useful for predicting the variation of nuclear mass with changing A and Z number. In particular, it is able to predict those heavy nuclides that are easily fissioned, radioactive decay modes, energetics of unstable nuclides, and, of course, masses of isotopes not yet measured.

To illustrate the use of the liquid drop model, we use it to predict the variation of N and Z that produces the most stable nuclides. In the condensed chart of the nuclides shown in Fig. 3.10, it is seen that to produce stable nuclides for all but the lightest nuclides, the number of neutrons must exceed slightly the number of protons. This asymmetry effect becomes more pronounced as A increases.

For a given mass number A, the most stable nuclide of the isobar is the one with the smallest mass. To find the Z number that produces the smallest mass, differentiate Eq. (3.16) with respect to Z and set the result to zero. Thus,

$$\left(\frac{\partial m(^A_Z X)}{\partial Z}\right)_A = m_p - m_n + \frac{1}{c^2}\left\{2a_c\frac{Z}{A^{1/3}} - 4a_a\frac{(A-2Z)}{A}\right\} = 0. \qquad (3.17)$$

Solving for Z (at constant A) yields

$$Z(A) = \left(\frac{A}{2}\right)\frac{1 + (m_n - m_p)c^2/(4a_a)}{1 + a_c A^{2/3}/(4a_a)}. \qquad (3.18)$$

The value of Z versus $N = A - Z$ obtained from this result is shown in Fig. 3.10 and is seen to be in excellent agreement with the observed nuclide stability trend.

Atomic Masses

By adding Z electrons to a nucleus of $^A_Z X$, a neutral atom of $^A_Z X$ is formed. To obtain a formula for the *atomic* mass of a nuclide, simply add conceptually Z electron masses to both sides of Eq. (3.16) to form the mass of an atom of $^A_Z X$ on the left and the mass of Z hydrogen atoms $^1_1 H$ on the right. Although a very small amount of mass is lost as the electrons bind to the $^A_Z X$ nucleus or to the protons, these mass losses not only tend to cancel but, as we will see in the next chapter, are negligible compared to the nuclear binding energies (the former being of the order of eVs and the later millions of times greater). Thus, the mass of an atom of $^A_Z X$ is

$$M(^A_Z X) = ZM(^1_1 H) + (A - Z)m_n$$
$$- \frac{1}{c^2}\left\{a_v A - a_s A^{2/3} - a_c\frac{Z^2}{A^{1/3}} - a_a\frac{(A-2Z)^2}{A} - \frac{a_p}{\sqrt{A}}\right\}. \qquad (3.19)$$

Example 3.2:   Estimate the mass of an atom of $^{70}_{31}$Ga using the liquid drop model. If we neglect the binding energy of the electrons to the nucleus, we find that the atomic mass is

$$M(^{70}_{31}\text{Ga}) \simeq m(^{70}_{31}\text{Ga}) + 31m_e$$
$$= [31m_p + (70 - 31)m_n - \text{BE}(^{70}_{31}\text{Ga})/c^2] + 31m_e$$

where the nuclear binding energy is, from Eq. (3.16),

$$\text{BE}(^{70}_{31}\text{Ga})/c^2 = \left\{a_v 70 - a_s 70^{2/3} - a_c\frac{31^2}{70^{1/3}} - a_a\frac{(71-62)^2}{A} - \frac{a_p}{\sqrt{70}}\right\}\frac{1}{931.5 \text{ MeV/u}}$$
$$= 0.65280 \text{ u}.$$

Then, evaluating the expression for the atomic mass, we find

$$M(^{70}_{31}\text{Ga}) = [31m_p + 39m_n - \text{BE}(^{70}_{31}\text{Ga})/c^2] + 31m_e$$
$$= [31(1.0072765) + 39(1.0086649) - 0.65280 + 31(0.00054858)] \text{ u}$$
$$= 69.9277 \text{ u}.$$

The tabulated value from Appendix B is $M(^{70}_{31}\text{Ga}) = 69.9260$ u.

3.2.6   The Nuclear Shell Model

Although the liquid drop model of the nucleus has proved to be quite successful for predicting subtle variations in the mass of nuclides with slightly different mass and atomic numbers, it avoids any mention of the internal arrangement of the nucleons in the nucleus. Yet, there are hints of such an underlying structure. We have observed that there is an abnormally high number of stable nuclides whose proton and/or neutron numbers equal the *magic numbers*

$$2, 8, 14, 20, 28, 50, 82, 126.$$

Further evidence for such magic numbers is provided by the very high binding energy of nuclei with both Z and N being magic, and the abnormally high or low alpha and beta particle energies emitted by radioactive nuclei according to whether the daughter or parent nucleus has a magic number of neutrons. Similarly, nuclides with a magic number of neutrons are observed to have a relatively low probability of absorbing an extra neutron.

To explain such nuclear systematics and to provide some insight into the internal structure of the nucleus, a *shell model* of the nucleus has been developed. This model uses Shrödinger's wave equation or quantum mechanics to describe the energetics of the nucleons in a nucleus in a manner analogous to that used to describe the discrete energy states of electrons about a nucleus. This model assumes

1. Each nucleon moves independently in the nucleus uninfluenced by the motion of the other nucleons.

2. Each nucleon moves in a potential well which is constant from the center of the nucleus to its edge where it increases rapidly by several tens of MeV.

When the model's quantum-mechanical wave equation is solved numerically (not an easy task), the nucleons are found to distribute themselves into a number of energy levels. There is a set of energy levels for protons and an independent set of levels for neutrons. Filled shells are indicated by large gaps between adjacent energy levels and are computed to occur at the experimentally observed values of 2, 8, 14, 20, 28, 50, 82, and 126 neutrons or protons. Such closed shells are analogous to the closed shells of orbital electrons. However, to obtain such results that predict the magic numbers, it is necessary to make a number of ad hoc assumptions about the angular momenta of the energy states of nuclei.

The shell model has been particularly useful in predicting several properties of the nucleus, including (1) the total angular momentum of a nucleus, (2) characteristics of isomeric transitions, which are governed by large changes in nuclear angular momentum, (3) the characteristics of beta decay and gamma decay, and (4) the magnetic moments of nuclei.

3.2.7   Other Nuclear Models

The liquid drop model and the shell model are in sharp contrast, the former based on the strong interactions of a nucleon with its nearest neighbors and the latter based on the independent motion of each nucleon uninfluenced by other nucleons. A unified or *collective* model has been proposed in which an individual nucleon interacts with a "core" of other nucleons which is capable of complex deformations and oscillations. Although the shell characteristics are preserved with this model, the closed shells are distorted by the other nucleons. Such distortions produce regularly spaced excited energy levels above the ground state, levels which have been observed experimentally.

The development of this and other sophisticated nuclear models is a major research area in contemporary nuclear physics. For the most part, however, the nature of how nucleons interact inside a nucleus need not concern us. The simple neutron-proton nuclear model is sufficient for our study of nuclear energy.

BIBLIOGRAPHY

BUSH, H.D., *Atomic and Nuclear Physics*, Prentice Hall, Engelwoods Cliffs, NJ, 1962.

BARRETT, R.C. AND D.F. JACKSON, *Nuclear Sizes and Structures*, Clarendon, Oxford, 1977.

COTTINGHAM, W.N. AND D.A. GREENWOOD, *An Introduction to Nuclear Physics*, Cambridge Univ. Press, Cambridge, UK, 1986.

FRENCH, A.P., *Principles of Modern Physics*, Wiley, New York, 1962.

KAPLAN, I., *Nuclear Physics*, Addison-Wesley, Reading, MA, 1963.

WAPSTRA, A. H., " Atomic Masses of Nuclei," in *Handbuch der Physik*, S. Flügge, Ed., Vol. 38, Springer-Verlag, Berlin, 1958.

PROBLEMS

1. Estimate the wavelengths of the first three spectral lines in the Lyman spectral series for hydrogen. What energies (eV) do photons with these wavelengths have?

2. Consider an electron in the first Bohr orbit of a hydrogen atom. (a) What is the radius (in meters) of this orbit? (b) What is the total energy (in eV) of the electron in this orbit? (c) How much energy is required to ionize a hydrogen atom when the electron is in the ground state?

3. What photon energy (eV) is required to excite the hydrogen electron in the innermost (ground state) Bohr orbit to the first excited orbit?

4. Based on the nucleon distribution of Eq. (3.11), by what fraction does the density of the nucleus decrease between $r = R - 2a$ and $r = R + 2a$?

5. Using the liquid drop model, tabulate the nuclear binding energy and the various contributions to the binding energy for the nuclei ^{40}Ca and ^{208}Pb.

6. From the difference in mass of a hydrogen atom (Appendix B) to the mass of a proton and an electron (Table 1.5), estimate the binding energy of the electron in the hydrogen atom. Compare this to the ionization energy of the ground state electron as calculated by the Bohr model. What fraction of the total mass is lost as the electron binds to the proton?

7. Using the liquid drop model, plot on the same graph, as a function of A, in units of MeV/nucleon (a) the bulk or volume binding energy per nucleon, (b) the negative of the surface binding energy per nucleon, (c) the negative of the asymmetry contribution per nucleon, (d) the negative of the Coulombic contribution per nucleon, and (e) the total binding energy per nucleon ignoring the pairing term. For a given A value, use Z determined from Eq. (3.18) for the most stable member of the isobar.

8. In radioactive beta decay, the number of nucleons A remains constant although the individual number of neutrons and protons change. Members of such a beta-decay chain are isobars with nearly equal masses. Using the atomic mass data in Appendix B, plot the mass difference $[70 - {}^A_Z X]$ (in u) of the nuclei versus Z for the isobar chain ${}^{70}_{36}Kr$, ${}^{70}_{35}Br$, ${}^{70}_{34}Se$, ${}^{70}_{33}As$, ${}^{70}_{32}Ge$, ${}^{70}_{31}Ga$, ${}^{70}_{30}Zn$, ${}^{70}_{29}Cu$, ${}^{70}_{28}Ni$, and ${}^{70}_{27}Co$. Compare the position of maximum nuclear stability with that predicted by Eq. (3.18).

Chapter 4

Nuclear Energetics

In reactions (nuclear, atomic, mechanical) in which some quantity is changed into some other quantity (e.g., two cars into a mass of tangled metal), energy is usually emitted (*exothermic* reaction) or absorbed (*endothermic* reaction). This energy, according to Einstein's Special Theory of Relativity comes from a change in the rest mass of the reactants. This change in mass ΔM is related to the reaction energy ΔE by the famous relation

$$\Delta E = \Delta M \, c^2. \tag{4.1}$$

Here $\Delta M \equiv M_{\text{initial}} - M_{\text{final}}$, i.e., the loss of rest mass between the initial mass of the reactants and the mass of the final products. If mass is lost in the reaction $M_{\text{initial}} > M_{\text{final}}$, then $\Delta E > 0$ and the reaction is exothermic. If mass is gained in the reaction $M_{\text{final}} > M_{\text{initial}}$, then $\Delta E < 0$ and the reaction is endothermic.

Any reaction in which reactants $A, B \ldots$ form the products $C, D \ldots$ can be written symbolically as

$$A + B + \cdots \longrightarrow C + D + \cdots. \tag{4.2}$$

If M_i represents the mass of the ith component, the mass change in the reaction is

$$\Delta M = (M_A + M_B + \cdots) - (M_C + M_D + \cdots). \tag{4.3}$$

The energy *emitted* is then given by Eq. (4.1).

In principle, the energy emitted (or absorbed) in *any* reaction could be computed from Eq. (4.1). However, only for nuclear reactions is it feasible to calculate the ΔE from ΔM. To illustrate, consider a familiar exothermic chemical reaction involving the rearrangement of atomic electrons:

$$C + O_2 \longrightarrow CO_2, \quad \Delta H_{293K} = -94.05 \text{ kcal/mol}. \tag{4.4}$$

This says

$$1 \text{ mol C} + 1 \text{ mol O}_2 \longrightarrow 1 \text{ mol CO}_2 + 94\,050 \text{ cal}. \tag{4.5}$$

Since one mole contains $N_a = 6.022 \times 10^{23}$ entities, the creation of one molecule of CO_2 from one atom of C and one molecule O_2 liberates $\Delta E = -\Delta H_{293K}/N_a = 1.56 \times 10^{-19}$ cal $= 4.08$ eV. The energy emitted in this chemical reaction is typical of almost all chemical reactions, i.e., a few eV per molecule.

How much of a mass change occurs in the formation of a CO_2 molecule? From the energy release of $\Delta E = 4.08$ eV

$$\Delta M = 4.08 \left(\frac{\text{eV}}{CO_2 \text{ molec.}}\right) \times \left[931 \left(\frac{\text{MeV}}{\text{amu}}\right) \times 10^6 \left(\frac{\text{eV}}{\text{MeV}}\right)\right]^{-1}$$

$$= 4.4 \times 10^{-9} \left(\frac{\text{amu}}{CO_2 \text{ molec.}}\right).$$

One molecule of CO_2 has a mass of about 44 amu and, thus, about $100(4.4 \times 10^{-9}/44) = 10^{-8}\%$ of the reactant mass is converted into energy. To measure such a mass change would require the atomic and molecular masses to be correct to about 10 significant figures, far more accuracy than present technology allows.

For more macroscopic reactions, such as the deformation of a bullet penetrating a steel plate, even greater, and unrealistic, accuracy in the reactant masses would be needed in order to use Eq. (4.1) to calculate the energy released. However, for reactions in which nuclei are altered, typically a million times more energy is absorbed (or emitted) compared to a chemical reaction. Thus, for nuclear reactions, nuclear masses correct to only 4 or 5 significant figures (well within reach of current technology) are needed.

4.1 Binding Energy

Binding energy, in its most general meaning, is the energy required to disassemble a whole into its separate parts. A typical reaction is one in which two or more entities $A, B, \ldots$ come together to form a single product C, i.e.,

$$A + B + \cdots \longrightarrow C. \tag{4.6}$$

A bound system C has a lower potential energy than its constituent parts. This is what keeps the system together. The usual convention is that this corresponds to a positive binding energy, denoting an exothermic reaction.

As in any reaction, the BE arises because of a change in the mass of the reactants. The *mass defect* ΔM is defined as the difference between the sum of initial masses and the sum of the final masses. Here

$$\Delta M = [\text{mass } A + \text{mass } B + \cdots] - \text{mass } C. \tag{4.7}$$

The binding energy can then be computed as $\text{BE} = \Delta M c^2$, provided the masses of the reactants are known with sufficient accuracy.

A more precise, if more restrictive, meaning of binding energy is the nuclear binding energy. Then, for the reaction in Eq. 4.6 the energy emitted is given by the nuclear binding energy of the product C less the sum of the binding energies of reactants A and B.

4.1.1 Nuclear and Atomic Masses

First some notation: The rest mass of an atom is denoted by $M(^A_Z X)$ while that of its nucleus is denoted by $m(^A_Z X)$. A few frequently used symbols are also used for convenience: the neutron rest mass $m_n \equiv m(^1_0 n)$, the proton rest mass $m_p \equiv m(^1_1 H)$, and the electron rest mass $m_e \equiv m(_{-1}^{0}e)$. NOTE: the data in Appendix B are for the *atomic* rest masses $M(_{Z}^{A}X)$ of all the known isotopes, not the nuclear masses.

An atom of $_{Z}^{A}X$ is formed by combining Z electrons with its nucleus to form the neutral atom. As these electrons bind to the nucleus, energy is emitted equal to the binding energy BE_{Ze} of the electrons. To remove all the electrons from the atom would thus require an ionization energy of BE_{Ze}. This electron binding energy comes from a decrease in the mass of the atom compared to the sum of nuclear and electron masses, by an amount BE_{Ze}/c^2. Thus, atomic and nuclear masses are related by

$$M(_{Z}^{A}X) = m(_{Z}^{A}X) + Zm_e - \frac{BE_{Ze}}{c^2}. \tag{4.8}$$

The binding energy term in this relation is often neglected since it is always very small compared to the other terms. For example, it requires 13.6 eV to ionize the hydrogen atom. This electron binding energy represents a mass change of $BE_{1e}/c^2 = 13.6 \text{ (eV)}/9.315 \times 10^8 \text{ (eV/u)} = 1.4 \times 10^{-8}$ u. This mass change is negligible compared to the mass of the hydrogen atom and nucleus (each about 1 u) and even the electron (about 5.5×10^{-4} u).

4.1.2 Binding Energy of the Nucleus

To examine the energies involved with nuclear forces in the nucleus, consider the binding energy of a nucleus composed of Z protons and $N = A - Z$ neutrons. The formation of such a nucleus from its constituents is described by the reaction

$$Z \text{ protons } + (A - Z) \text{ neutrons } \longrightarrow \text{nucleus}(_{Z}^{A}X) + BE. \tag{4.9}$$

The binding energy is determined from the change of mass between the left- and right-hand sides of the reaction, i.e.,

$$\text{Mass Defect} = BE/c^2 = Z\,m_p + (A - Z)m_n - m(_{Z}^{A}X) \tag{4.10}$$

where $m(_{Z}^{A}X)$ is the *nuclear mass*. However, masses of nuclei are not available; only atomic masses are known with great accuracy (see Appendix B). To express Eq. (4.10) in terms of atomic masses, use Eq. (4.8) to obtain

$$\frac{BE}{c^2} = Z\left[M(_{1}^{1}H) - m_e + \frac{BE_{1e}}{c^2}\right] + (A - Z)m_n - \left[M(_{Z}^{A}X) - Zm_e + \frac{BE_{Ze}}{c^2}\right]$$

$$= ZM(_{1}^{1}H) + (A - Z)m_n - M(_{Z}^{A}X) + \frac{1}{c^2}[ZBE_{1e} - BE_{Ze}]. \tag{4.11}$$

The last term is the equivalent mass difference between the binding energies of Z hydrogen electrons and the Z electrons to the nucleus of interest. These electron binding energies are generally not known. However, the term involving electron binding energies can be ignored for two reasons: (1) the two electron binding energies tend to cancel, and (2) electron binding energies are millions of times less than the nuclear binding energy. Thus, neglect of the last term gives

$$BE(_{Z}^{A}X) = [ZM(_{1}^{1}H) + (A - Z)m_n - M(_{Z}^{A}X)]c^2. \tag{4.12}$$

Example 4.1: What is the binding energy of the nucleus in $_2^4$He? From the atomic mass data in Appendix B, the mass defect is

$$\text{Mass Defect} = \text{BE}/c^2 = 2M(_1^1\text{H}) + 2m_n - M(_2^4\text{He})$$

$$= 2(1.0078250) + 2(1.0086649) - 4.0026032 = 0.0303766 \text{ u.}$$

Thus,

$$\text{BE}(_2^4\text{He}) = \text{Mass Defect (u)} \times 931.5 \text{ (MeV/u)} = 28.30 \text{ MeV.}$$

Notice the manner in which mass is converted to equivalent energy. Simply multiply the mass deficit in u by the conversion factor 931.5 MeV/u. This is far easier than multiplying a mass deficit in kg by c^2 and going through the necessary units conversion.

4.1.3 Average Nuclear Binding Energies

The larger the BE of a nucleus, the more energy is needed to break it up into its constituent neutrons and protons. In Fig. 4.1, the average nuclear binding energy per nucleon (total BE divided by the number of nucleons A) is shown for all the naturally occurring isotopes. The value of BE/A is a stability measure of the nucleus; the larger BE/A, the more energy is needed, per nucleon, to tear apart the nucleus. From Fig. 4.1, we see that there is a broad maximum in the BE/A curve around $A \simeq 60$ (e.g., Cr, Mn, Fe isotopes).

From the expanded portion of the BE/A curve (left-hand part of Fig. 4.1), we see that several light nuclei have particularly high average BE/A values. These are isotopes whose nuclei are multiples of the $_2^4$He nucleus (i.e., an alpha particle). This observation implies that inside a nucleus, the nucleons tend to form alpha-particle groupings.

The BE/A versus A curve immediately suggests two ways to extract energy from the nucleus.

Fusion

If two light nuclei ($A \leq 25$) are joined (or fused) to form a single heavier nucleus, the average BE per nucleon will increase. This increase in BE/A times the number of nucleons A then equals the energy that must be emitted. For example, consider the fusion reaction in which two deuterium nuclei are combined to form a helium nucleus, namely,

$$_1^2\text{H} + _1^2\text{H} \longrightarrow _2^4\text{He.}$$

The energy released can be computed in several equivalent ways.

Method 1 (from the change in BE): From Eq. (4.12), one finds that BE($_1^2$H) = 2.225 MeV and BE($_2^4$He) = 28.296 MeV. Thus, the increase in the binding energy of the 4 nucleons involved is $28.296 - 2(2.225) = 23.85$ MeV. This is the energy E_{fusion} emitted in this particular fusion reaction.

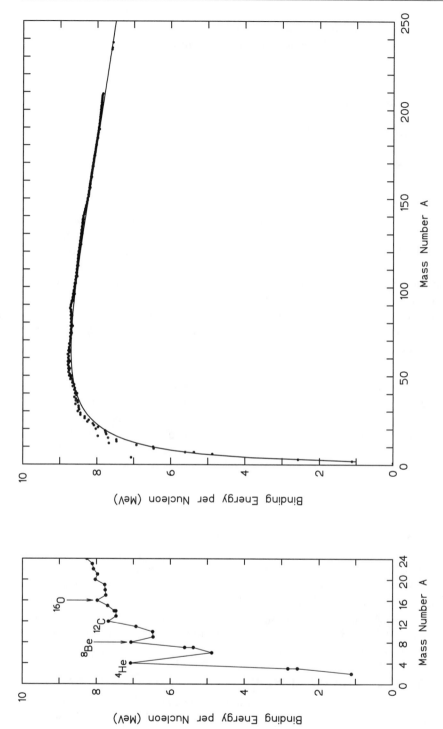

Figure 4.1. The average binding energy BE/A of the nucleons in a nucleus versus the atomic mass number A for the naturally occurring nuclides (and ^{8}Be). The left-hand figure shows the detailed variation of the average binding energy for the lightest nuclei, while the right-hand figure shows the overall variation. The smooth curve in the right-hand figure is the prediction obtained with the liquid drop model discussed in Section 3.2.5.

Method 2 (from the change in mass): The fusion energy released E_{fusion} must equal the energy equivalent of the decrease in mass caused by the formation of ^4_2He.

$$E_{\text{fusion}} = [2m(^2_1\text{H}) - m(^4_2\text{He})] \text{ (u)} \times 931.5 \text{ (MeV/u)}.$$

$$= [2(M(^2_1\text{H}) - m_e + \text{BE}_{1e}/c^2) - (M(^4_2\text{He}) - 2m_e + \text{BE}_{2e}/c^2)] \times 931.5.$$

$$\simeq [2M(^2_1\text{H}) - M(^4_2\text{He})] \times 931.5.$$

$$= [2(2.014102) - 4.002603] \times 931.5.$$

$$= 23.85 \text{ MeV}.$$

This release of energy through the fusion of two light nuclei is the mechanism responsible for energy generation in stars. Moreover, all the light elements above lithium and below about iron have been formed by fusion reactions during the normal lifetime of stars. In each exothermic fusion reaction, heavier products are formed that move up the BE/A curve. The heavier elements above iron have also been formed in stars, but only during the final moments of their cataclysmic death as an exploding nova. The tremendous energy released during a nova explosion allows intermediate mass nuclides to fuse (an endothermic reaction) to create the elements beyond iron.

Fission

An alternative way of extracting energy from a nucleus is to start at the upper end of the BE/A curve with a very heavy nucleus, such as $^{235}_{92}\text{U}$, and split (or *fission*) it into two lighter nuclei. To obtain an approximate idea of how much energy is released in a fission event, we see from Fig. 4.1 that if a nucleus of mass number 235 is split into two nuclei with $A \simeq 117$, the BE per nucleon increases from about 7.7 to 8.5 MeV/nucleon. Thus, the total fission energy released is about $235 \times (8.5 - 7.7) \simeq 210$ MeV.

Chemical versus Nuclear Reactions

The formation of a single molecule of CO_2 from carbon burning in oxygen releases 4.08 eV, typical of chemical reactions. By contrast, a nuclear fusion and fission event releases 5 million to 50 million times this energy, respectively. This huge concentration of nuclear energy gives nuclear power plants a major advantage over fossil-fuel power plants. For the same power capacity, a nuclear power plant consumes only about one millionth the mass of fuel that a fossil-fuel plant consumes. For example, a 1000 MW(e) coal-fired power plant consumes 11×10^6 kg of coal daily. By contrast, a nuclear power plant of the same capacity consumes only 3.8 kg of ^{235}U daily.

4.2 Nucleon Separation Energy

Closely related to the concept of nuclear binding energy is the energy required to remove a single nucleon from a nucleus. Consider the addition of a single neutron to form the nucleus of ^A_ZX, i.e.,

$$^{A-1}_Z\text{X} + ^1_0\text{n} \longrightarrow ^A_Z\text{X}.$$

The energy released in this reaction, $S_n({}_Z^A X)$, is the energy required to remove (or separate) a single neutron from the nucleus ${}_Z^A X$. This energy is analogous to the ionization energy required to remove an outer shell electron from an atom. The neutron separation energy equals the energy equivalent of the decrease in mass for the reaction, namely,

$$S_n({}_Z^A X) = [m({}^{A-1}_Z X) + m_n - m({}_Z^A X)]c^2 \simeq [M({}^{A-1}_Z X) + m_n - M({}_Z^A X)]c^2. \quad (4.13)$$

This separation energy can be expressed in terms of nuclear binding energies. Substitution of Eq. (4.12) into this expression yields

$$S_n({}_Z^A X) = BE({}_Z^A X) - BE({}^{A-1}_Z X). \quad (4.14)$$

A similar expression is obtained for the energy required to remove a single proton from the nucleus of ${}_Z^A Y$, or equivalently, the energy released when a proton is absorbed by a nucleus ${}^{A-1}_{Z-1} X$, i.e.,

$${}^{A-1}_{Z-1}X + {}_1^1p \longrightarrow {}_Z^A Y.$$

The energy equivalent of the mass decrease of this reaction is

$$S_p({}_Z^A Y) = [m({}^{A-1}_{Z-1}X) + m_p - m({}_Z^A Y)]c^2$$

$$\simeq [M({}^{A-1}_{Z-1}X) + M({}_1^1H) - M({}_Z^A Y)]c^2 = BE({}_Z^A Y) - BE({}^{A-1}_{Z-1}X). \quad (4.15)$$

The variation of S_n and S_p with Z and N provides information about nuclear structure. For example, nuclides with an even number of neutrons or protons have high values of S_n and S_p, respectively, indicating that stability is increased when neutrons or protons are "paired" inside the nucleus.

Example 4.2: What is the binding energy of the last neutron in ${}_8^{16}O$? This is the energy released in the reaction

$${}_8^{15}O + {}_0^1n \longrightarrow {}_8^{16}O.$$

Then, from Eq. (4.13),

$$S_n = [m({}_8^{15}O) + m_n - m({}_8^{16}O)]c^2$$

$$\simeq [M({}_8^{15}O) + m_n - M({}_8^{16}O)]c^2$$

$$= [15.0030654 + 1.00866492 - 15.9949146]\ u \times 931.5\ MeV/u$$

$$= 15.66\ MeV.$$

This is an exceptionally high value for a single nucleon. The average binding energy per nucleon for ${}_8^{16}O$ is 7.98 MeV. This large result for S_n indicates that ${}_8^{16}O$ is very stable compared to ${}_8^{15}O$.

4.3 Nuclear Reactions

Nuclear reactions play a very important role in nuclear science and engineering, since it is through such reactions that various types of radiation are produced or detected, or information about the internal structure of a nucleus is gained. There are two main categories of nuclear reactions.

In the first category, the initial reactant X is a single atom or nucleus that spontaneously changes by emitting one or more particles, i.e.,

$$X \longrightarrow Y_1 + Y_2 + \cdots.$$

Such a reaction is called *radioactive decay*. As we have seen from the Chart of the Nuclides, the vast majority of known nuclides are radioactive. We will examine in detail in the next chapter the different types of radioactive decay and how the number of radioactive atoms vary with time. For the present, we simply note that radioactive decay is a particular type of nuclear reaction that, for it to occur spontaneously, must necessarily be exothermic; i.e., mass must decrease in the decay process and energy must be emitted, usually in the form of the kinetic energy of the reaction products.

In the second broad category of nuclear reactions are *binary reactions* in which two nuclear particles (nucleons, nuclei or photons) interact to form different nuclear particles. The most common types of such nuclear reactions are those in which some nucleon or nucleus x moves with some kinetic energy and strikes and interacts with a nucleus X to form a pair of product nuclei y and Y, i.e.,[1]

$$x + X \longrightarrow Y + y.$$

As a shorthand notation, such a reaction is often written as $X(x,y)Y$ where x and y are usually the lightest of the reaction pairs.

4.4 Examples of Binary Nuclear Reactions

For every nuclear reaction, we can write a reaction equation. These reaction equations must be balanced, just as chemical reactions must be. Charge (the number of protons) and mass number (the number of nucleons) must be conserved. The number of protons and the number of neutrons must be the same before and after the reaction. We shall illustrate this with some examples of typical nuclear reactions.

(α, p) **reaction:** The first nuclear reaction was reported by Rutherford. He bombarded nitrogen in air with alpha particles (helium nuclei) and observed the production of protons (hydrogen nuclei),

$$\ce{^4_2He} + \ce{^{14}_7N} \longrightarrow \ce{^{17}_8O} + \ce{^1_1H} \qquad \text{or} \qquad \ce{^{14}_7N}(\alpha, p)\ce{^{17}_8O}.$$

The product of this reaction is $^{17}_8O$ and there are nine protons and nine neutrons on both sides of the equation, so the equation is balanced.

---

[1] Although there are a few important nuclear reactions (e.g., fission) in which an interacting pair produces more than two products, we restrict our attention, for the present, to those producing only a pair of product nuclei.

(α, n) **reaction:** In 1932, Chadwick discovered the neutron by bombarding beryllium with alpha particles to produce neutrons from the reaction

$$_2^4\text{He} + _4^9\text{Be} \longrightarrow _6^{12}\text{C} + _0^1\text{n} \qquad \text{or} \qquad _4^9\text{Be}(\alpha, n)_6^{12}\text{C}.$$

(γ, n) **reaction:** Energetic photons (gamma rays) can also interact with a nucleus. For example, neutrons can be produced by irradiating deuterium with sufficiently energetic photons according to the reaction

$$\gamma + _1^2\text{H} \longrightarrow _1^1\text{H} + _0^1\text{n} \qquad \text{or} \qquad _1^2\text{H}(\gamma, n)_1^1\text{H} \qquad \text{or} \qquad _1^2\text{H}(\gamma, p)_0^1\text{n}.$$

(p, γ) **reaction:** Protons can cause nuclear reactions such as the radiative capture of a proton by ^7Li, namely,

$$_1^1\text{H} + _3^7\text{Li} \longrightarrow _4^8\text{Be} + \gamma \qquad \text{or} \qquad _3^7\text{Li}(p, \gamma)_4^8\text{Be}.$$

The product nucleus $_4^8\text{Be}$ is not bound and breaks up (radioactively decays) almost immediately into two alpha particles, i.e.,

$$_4^8\text{Be} \longrightarrow _2^4\text{He} + _2^4\text{He}.$$

$(\gamma, \alpha n)$ **reaction:** As an example of a reaction in which more than two products are produced, a high-energy photon can cause ^{17}O to split into ^{12}C, an α particle, and a neutron through the reaction

$$\gamma + _8^{17}\text{O} \longrightarrow _6^{12}\text{C} + _2^4\text{He} + _0^1\text{n} \qquad \text{or} \qquad _8^{17}\text{O}(\gamma, n\alpha)_6^{12}\text{C}.$$

(n, p) **reaction:** Fast neutrons can cause a variety of nuclear reactions. For example, in a reactor core, fast neutrons can interact with ^{16}O to produce ^{16}N, which radioactively decays (half life of 7.12 s) with the emission of a 6.13-MeV (69%) or a 7.11-MeV (5%) photon. The radionuclide ^{16}N is produced by the reaction

$$_0^1\text{n} + _8^{16}\text{O} \longrightarrow _7^{16}\text{N} + _1^1\text{p} \qquad \text{or} \qquad _8^{16}\text{O}(n, p)_7^{16}\text{N}.$$

4.4.1 Multiple Reaction Outcomes

In general, more than one outcome can arise when a particle x reacts with a nucleus X. Let us consider a number of possible results when a neutron of moderately high energy (a few MeV) reacts with a $_{16}^{32}\text{S}$ nucleus. In the first place, there may be no reaction at all with the neutron being scattered elastically, i.e., the nucleus which scatters the neutron is left unaltered internally. This elastic scattering reaction is written

$$_0^1\text{n} + _{16}^{32}\text{S} \rightarrow _{16}^{32}\text{S} + _0^1\text{n}.$$

This reaction is written as (n, n). An inelastic scattering reaction may occur, in which the $_{16}^{32}\text{S}$ is left in a nuclear excited state, namely,

$$_0^1\text{n} + _{16}^{32}\text{S} \rightarrow (_{16}^{32}\text{S})^* + _0^1\text{n}'.$$

This is an (n, n') reaction. The asterisk means that the $^{32}_{16}S$ nucleus is left in an excited state[2] and the prime on the n' means that the neutron has less energy than it would have if it had been elastically scattered. The incident neutron can be absorbed and a proton ejected,

$$^{1}_{0}n + ^{32}_{16}S \rightarrow ^{32}_{15}P + ^{1}_{1}H,$$

which is called an (n, p) reaction. The neutron may also be simply captured resulting in the emission of a gamma ray from the product nucleus

$$^{1}_{0}n + ^{32}_{16}S \rightarrow ^{33}_{16}S + \gamma,$$

which is called a radiative capture reaction or simply an (n, γ) reaction in which the binding energy of the neutron to the $^{32}_{16}S$ nucleus is emitted as a photon called a *gamma ray*.

Elastic scattering and radiative capture are possible at all values of the incident neutron energy. However, the neutron's incident kinetic energy must exceed certain threshold energies to make the other reactions possible. If we use neutrons of sufficient energy, all of the four possibilities, as well as others, can occur when we bombard $^{32}_{16}S$ with neutrons.

4.5   Q-Value for a Reaction

In any nuclear reaction, energy must be conserved, i.e., the total energy including rest-mass energy of the initial particles must equal the total energy of the final products, i.e.,

$$\sum_i [E_i + m_i c^2] = \sum_i [E_i' + m_i' c^2] \tag{4.16}$$

where E_i (E_i') is the kinetic energy of the ith initial (final) particle with a rest mass m_i (m_i').

Any change in the total kinetic energy of particles before and after the reaction must be accompanied by an equivalent change in the total rest mass of the particles before and after the reaction. To quantify this change in kinetic energy or rest-mass change in a reaction, a so-called Q value is defined as

$$\boxed{Q = (\text{KE of final particles}) - (\text{KE of initial particles}).} \tag{4.17}$$

Thus, the Q value quantifies the amount of kinetic energy gained in a reaction. Equivalently, from Eq. (4.16), this gain in kinetic energy must come from a decrease in the rest mass, i.e.,

$$\boxed{Q = (\text{rest mass of initial particles})c^2 - (\text{rest mass of final particles})c^2.}$$

$$\tag{4.18}$$

---

[2]An atom whose nucleus is in an excited state has a mass greater than the ground-state atomic mass listed in Appendix B. The mass difference is just the mass equivalent of the nuclear excitation energy.

The Q value of a nuclear reaction may be either positive or negative. If the rest masses of the reactants exceed the rest masses of the products, the Q value of the reaction is positive with the decrease in rest mass being converted into a gain in kinetic energy. Such a reaction is exothermic.

Conversely, if Q is negative, the reaction is endothermic. For this case, kinetic energy of the initial particles is converted into rest-mass energy of the reaction products. The kinetic energy decrease equals the rest-mass energy increase. Such reactions cannot occur unless the colliding particles have at least a certain amount of kinetic energy.

4.5.1 Binary Reactions

For the binary reaction $x + X \rightarrow Y + y$, the Q value is given by

$$Q = (E_y + E_Y) - (E_x + E_X) = [(m_x + m_X) - (m_y + m_Y)]c^2. \qquad (4.19)$$

For most binary nuclear reactions, the number of protons is conserved so that the same number of electron masses may be added to both sides of the reactions and, neglecting differences in electron binding energies, the Q-value can be written in terms of atomic masses as

$$Q = (E_y + E_Y) - (E_x + E_X) = [(M_x + M_X) - (M_y + M_Y)]c^2. \qquad (4.20)$$

4.5.2 Radioactive Decay Reactions

For a radioactive decay reaction $X \rightarrow Y + y$, there is no particle x and the nucleus of X generally is at rest so $E_X = 0$. In this case

$$Q = (E_y + E_Y) = [m_X - (m_y + m_Y)]c^2 > 0. \qquad (4.21)$$

Thus radioactive decay is always exothermic and the mass of the parent nucleus must always be greater than the sum of the product masses. In some types of radioactive decay (e.g., beta decay and electron capture), the number of protons is *not* conserved and care must be exercised in expressing the nuclear masses in terms of atomic masses. This nuance is illustrated in detail in Section 4.6.1 and in the next chapter.

4.6 Conservation of Charge and the Calculation of Q-Values

In all nuclear reactions, total charge must be conserved. Sometimes this is not clear when writing the reaction and subtle errors can be made when calculating the Q-value from Eq. (4.18). Consider, for example, the reaction $^{16}_{8}\text{O}(n,p)^{16}_{7}\text{N}$ or

$$^{1}_{0}\text{n} + ^{16}_{8}\text{O} \longrightarrow ^{16}_{7}\text{N} + ^{1}_{1}\text{p}. \qquad (4.22)$$

One might be tempted to calculate the Q-value as

$$Q = \{m_n + M(^{16}_{8}\text{O}) - M(^{16}_{7}\text{N}) - m_p\}c^2.$$

However, this is incorrect since the number of electrons is not conserved on both sides of the reaction. On the left side there are 8 electrons around the $^{16}_{8}\text{O}$ nucleus, while on the right there are only 7 electrons in the product atom $^{16}_{7}$N. In reality, when the proton is ejected from the $^{16}_{8}$O nucleus by the incident neutron, an orbital electron is also lost from the resulting $^{16}_{7}$N nucleus. Thus, the reaction of Eq. (4.22), to conserve charge, should be written as

$$^{1}_{0}\text{n} + ^{16}_{8}\text{O} \longrightarrow ^{16}_{7}\text{N} + ^{0}_{-1}\text{e} + ^{1}_{1}\text{p}. \tag{4.23}$$

The released electron can now be conceptually combined with the proton to form a neutral atom of $^{1}_{1}$H, the electron binding energy to the proton being negligible compared to the reaction energy. Thus, the reaction can be approximated by

$$^{1}_{0}\text{n} + ^{16}_{8}\text{O} \longrightarrow ^{16}_{7}\text{N} + ^{1}_{1}\text{H} \tag{4.24}$$

and its Q-value is calculated as

$$Q = \{m_n + M(^{16}_{8}\text{O}) - M(^{16}_{7}\text{N}) - M(^{1}_{1}\text{H})\}c^2. \tag{4.25}$$

In any nuclear reaction, *in which the numbers of neutrons and protons are conserved*, the Q-value is calculated by replacing any charged particle by its neutral-atom counterpart. Two examples are shown below.

Example 4.3:   Calculate the Q value for the exothermic reaction $^{9}_{4}$Be $(\alpha, n)^{12}_{6}$C and for the endothermic reaction $^{16}_{8}$O$(n, \alpha)^{13}_{6}$C. In terms of neutral atoms, these reactions may be written as $^{9}_{4}$Be $+ ^{4}_{2}$He $\rightarrow ^{12}_{6}$C $+ ^{1}_{0}$n and $^{16}_{8}$O $+ ^{1}_{0}$n $\rightarrow ^{13}_{6}$C $+ ^{4}_{2}$He. The Q-values calculations are shown in detail in the tables below using the atomic masses tabulated in Appendix B.

$$^{9}_{4}\text{Be} + ^{4}_{2}\text{He} \rightarrow ^{12}_{6}\text{C} + ^{1}_{0}\text{n}$$

		Atomic Mass (u)
Reactants	$^{9}_{4}$Be	9.012182
	$^{4}_{2}$He	4.002603
	sum	13.014785
Products	$^{12}_{6}$C	12.000000
	$^{1}_{0}$n	1.008664
	sum	13.008664
Difference of sums		0.006121 u
$\times$931.5 MeV/u $= $ **5.702 MeV**		

$$^{16}_{8}\text{O} + ^{1}_{0}\text{n} \rightarrow ^{13}_{6}\text{C} + ^{4}_{2}\text{He}$$

		Atomic Mass (u)
Reactants	$^{16}_{8}$O	15.994915
	$^{1}_{0}$n	1.008664
	sum	17.003579
Products	$^{13}_{6}$C	13.003354
	$^{4}_{2}$He	4.002603
	sum	17.005957
Difference of sums		-0.002378 u
$\times$931.5 MeV/u $= $ **-2.215 MeV**		

4.6.1   Special Case for Changes in the Proton Number

The above procedure for calculating Q-values in terms of neutral-atom masses, *cannot* be used when the number of protons and neutrons are not conserved, and a more careful analysis is required. Such reactions involve the so-called weak force that is responsible for changing neutrons into protons or vice versa. These reactions are recognized by the presence of a neutrino ν (or antineutrino $\bar{\nu}$) as one of the reactants or products. In such reactions, the conceptual addition of electrons to both sides of the reaction to form neutral atoms often results in the appearance or disappearance of an extra free electron. We illustrate this by the following example.

Example 4.4: What is the Q-value of the reaction in which two protons fuse to form a deuteron (the nucleus of deuterium ^2_1H). Specifically, this reaction in which the number of protons and neutrons changes is

$$^1_1\text{p} + ^1_1\text{p} \longrightarrow ^2_1\text{d} + ^0_{+1}\text{e} + \nu, \tag{4.26}$$

where $^0_{+1}\text{e}$ is a positron (antielectron). This particular reaction is a key reaction occuring inside stars that begins the process of fusing light nuclei to release fusion energy. To calculate the Q-value of this reaction in terms of neutral atomic masses, conceptually add two electrons to both sides of the reaction and, neglecting electron binding energies, replace a proton plus an electron by a ^1_1H atom and the deuteron and electron by ^2_1H. Thus, the equivalent reaction is

$$^1_1\text{H} + ^1_1\text{H} \longrightarrow ^2_1\text{H} + ^0_{-1}\text{e} + ^0_{+1}\text{e} + \nu. \tag{4.27}$$

The Q-value of this reaction is then calculated as follows:

$$\begin{aligned}
Q &= \{2M(^1_1\text{H}) - M(^2_1\text{H}) - 2m_e - m_\nu\}c^2 \\
&= \{2 \times 1.00782503 - 2.01410178 - 2 \times 0.00054858\}(\text{u}) \times 931.5~(\text{MeV/u}) \\
&= 0.420~\text{MeV},
\end{aligned}$$

where we have assumed the rest mass of the neutrino is negligibly small (if not actually zero).

4.7 Q-Value for Reactions Producing Excited Nuclei

In many nuclear reactions, one of the product nuclei is left in an excited state, which subsequently decays by the emission of one or more gamma photons as the nucleus reverts to its ground state. In calculating the Q-value for such reactions, it must be remembered that the mass of the excited nucleus is greater than that of the corresponding ground state nucleus by the amount E^*/c^2, where E^* is the excitation energy. The mass of a neutral atom with its nucleus in an excited state is thus

$$M(^A_Z\text{X}^*) = M(^A_Z\text{X}) + E^*/c^2 \tag{4.28}$$

where $M(^A_Z\text{X})$ is the mass of ground state atoms given in Appendix B. An example of a Q-value calculation for a reaction leaving a product nucleus in an excited state is shown below.

Example 4.5:   What is the Q-value of the reaction $^{10}\text{B}(n, \alpha)^7\text{Li}^*$ in which the $^7\text{Li}^*$ nucleus is left in an excited state 0.48-MeV above its ground state? In terms of neutral atomic masses, the Q-value is

$$Q = [m_n + M(^{10}_5\text{B}) - M(^4_2\text{He}) - M(^7_3\text{Li}^*)]c^2.$$

The mass of the lithium nucleus with an excited nucleus $M(^7_3\text{Li}^*)$ is greater than that for the ground state atom $M(^7_3\text{Li})$, whose mass is tabulated in Appendix B, by an amount 0.48 (MeV)/931.5 (MeV/u), i.e., $M(^7_3\text{Li}^*) = M(^7_3\text{Li})+0.48$ MeV/c^2. Hence,

$$\begin{aligned}
Q &= [m_n + M(^{10}_5\text{B}) - M(^4_2\text{He}) - M(^7_3\text{Li})]c^2 - 0.48\text{MeV} \\
&= [1.0086649 + 10.0129370 - 4.0026032 - 7.0160040]\,\text{u} \times 931.5\ \text{MeV/u} \\
&\quad -0.48\ \text{MeV} \\
&= 2.310\ \text{MeV}.
\end{aligned}$$

BIBLIOGRAPHY

EVANS, R.D., *The Atomic Nucleus*, McGraw-Hill, New York, 1955; republished by Krieger Publishing Co., Melbourne, FL, 1982.

KAPLAN, I., *Nuclear Physics*, Addison-Wesley, Reading, MA, 1963.

MAYO, R.M., *Nuclear Concepts for Engineers*, American Nuclear Society, La Grange Park, IL, 1998.

SATCHLER, G.R., (Ed.), *Introduction to Nuclear Reactions*, 2nd ed., Oxford University Press, Oxford, 1990.

SHULTIS, J.K. AND R.E. FAW, *Radiation Shielding*, American Nuclear Society, La Grange Park, IL, 2000.

PROBLEMS

1. Complete the following nuclear reactions based on the conservation of nucleons:

   (a) $^{238}_{92}\text{U} + ^1_0\text{n} \longrightarrow (?)$

   (b) $^{14}_7\text{N} + ^1_0\text{n} \longrightarrow (?) + ^1_1\text{H}$

   (c) $^{226}_{88}\text{Ra} \longrightarrow (?) + ^4_2\text{He}$

   (d) $(?) \longrightarrow ^{230}_{90}\text{Th} + ^4_2\text{He}$

2. Determine the binding energy (in MeV) per nucleon for the nuclides: (a) $^{16}_8\text{O}$, (b) $^{17}_8\text{O}$, (c) $^{56}_{26}\text{Fe}$, and (d) $^{235}_{92}\text{U}$.

3. Calculate the binding energy per nucleon and the neutron separation energy for $^{16}_8\text{O}$ and $^{17}_8\text{O}$.

4. Verify Eq. (4.15) on the basis of the definition of the binding energy.

5. A nuclear scientist attempts to perform experiments on the stable nuclide $^{56}_{26}$Fe. Determine the energy (in MeV) the scientist will need to

    1. remove a single neutron.

    2. remove a single proton.

    3. completely dismantle the nucleus into its individual nucleons.

    4. fission it symmetrically into two identical lighter nuclides $^{28}_{13}$Al.

6. Write formulas for the Q-values of the reactions shown in Section 4.4. With these formulas, evaluate the Q-values.

7. What is the Q-value (in MeV) for each of the following possible nuclear reactions? Which are exothermic and which are endothermic?

$$^1_1\text{p} + {}^9_4\text{Be} \longrightarrow \begin{cases} {}^{10}_5\text{B} + \gamma \\ {}^9_5\text{B} + {}^1_0\text{n} \\ {}^9_4\text{Be} + {}^1_1\text{p} \\ {}^8_4\text{Be} + {}^2_1\text{H} \\ {}^7_4\text{Be} + {}^3_1\text{H} \\ {}^6_3\text{Li} + {}^4_2\text{He} \end{cases}$$

8. Neutron irradiation of ^{6}Li can produce the following reactions.

$$^1_0\text{n} + {}^6_3\text{Li} \longrightarrow \begin{cases} {}^7_3\text{Li} + \gamma \\ {}^6_3\text{Li} + {}^1_0\text{n} \\ {}^6_2\text{He} + {}^1_1\text{p} \\ {}^5_2\text{He} + {}^2_1\text{H} \\ {}^3_1\text{H} + {}^4_2\text{He} \end{cases}$$

What is the Q-value (in MeV) for each reaction?

9. What is the net energy released (in MeV) for each of the following fusion reactions? (a) $^2_1\text{H} + {}^2_1\text{H} \longrightarrow {}^3_2\text{He} + {}^1_0\text{n}$ and (b) $^2_1\text{H} + {}^3_1\text{H} \longrightarrow {}^4_2\text{He} + {}^1_0\text{n}$.

10. Calculate the Q-values for the following two beta radioactive decays. (a) $^{22}_{11}\text{Na} \longrightarrow {}^{22}_{10}\text{Ne} + {}^0_{+1}\text{e} + \nu$ and (b) $^{38}_{17}\text{Cl} \longrightarrow {}^{38}_{18}\text{Ar} + {}^0_{-1}\text{e} + \bar{\nu}$.

11. Reactions employed in cyclotron production of radionuclides for PET scanning are listed in Table 14.3. Select at least one reaction and compute the Q-value for the reaction.

Chapter 5

Radioactivity

5.1 Overview

Radioactive nuclei and their radiations have properties that are the basis of many of the ideas and techniques of atomic and nuclear physics. We have seen that the emission of alpha and beta particles has led to the concept that atoms are composed of smaller fundamental units. The scattering of alpha particles led to the idea of the nucleus, which is fundamental to the models used in atomic physics. The discovery of isotopes resulted from the analysis of the chemical relationships between the various radioactive elements. The bombardment of the nucleus by alpha particles caused the disintegration of nuclei and led to the discovery of the neutron and the present model for the composition of the nucleus.

The discovery of artificial, or induced, radioactivity started a new line of nuclear research and hundreds of artificial nuclei have been produced by many different nuclear reactions. The investigation of the emitted radiations from radionuclides has shown the existence of nuclear energy levels similar to the electronic energy levels. The identification and the classification of these levels are important sources of information about the structure of the nucleus.

A number of radioactive nuclides occur naturally on the earth. One of the most important is $^{40}_{19}$K, which has an isotopic abundance of 0.0118% and a half-life of 1.28×10^9 y. Potassium is an essential element needed by plants and animals, and is an important source of human internal and external radiation exposure. Other naturally occurring radionuclides are of cosmogenic origin. Tritium ($^3_1$H) and $^{14}_6$C are produced by cosmic ray interactions in the upper atmosphere, and also can cause measurable human exposures. $^{14}_6$C (half life 5730 y), which is the result of a neutron reaction with $^{14}_7$N in the atmosphere, is incorporated into plants by photosynthesis. By measuring the decay of ^{14}C in ancient plant material, the age of the material can be determined.

Other sources of terrestrial radiation are uranium, thorium, and their radioactive progeny. All elements with Z > 83 are radioactive. Uranium and thorium decay into daughter radionuclides, forming a series (or chain) of radionuclides that ends with a stable isotope of lead or bismuth. There are four naturally-occurring radioactive decay series; these will be discussed later in this chapter.

88

Table 5.1. Summary of important types of radioactive decay. The parent atom is denoted as P and the product or daughter atom by D.

Decay Type	Reaction	Description
gamma (γ)	$^A_Z P^* \longrightarrow \, ^A_Z P + \gamma$	An excited nucleus decays to its ground state by the emission of a gamma photon.
alpha (α)	$^A_Z P \longrightarrow \, ^{A-4}_{Z-2} D + \alpha$	An α particle is emitted leaving the daughter with 2 fewer neutrons and 2 fewer protons than the parent.
negatron (β^-)	$^A_Z P \longrightarrow \, ^A_{Z+1} D + \beta^- + \overline{\nu}$	A neutron in the nucleus changes to a proton. An electron (β^-) and an anti-neutrino ($\overline{\nu}$) are emitted.
positron (β^+)	$^A_Z P \longrightarrow \, ^A_{Z-1} D + \beta^+ + \nu$	A proton in the nucleus changes into a neutron. A positron (β^+) and a neutrino (ν) are emitted.
electron capture (EC)	$^A_Z P + e^- \longrightarrow \, ^A_{Z-1} D^* + \nu$	An orbital electron is absorbed by the nucleus, converts a nuclear proton into a neutron and a neutrino (ν), and, generally, leaves the nucleus in an excited state.
proton (p)	$^A_Z P \longrightarrow \, ^{A-1}_{Z-1} D + p$	A nuclear proton is ejected from the nucleus.
neutron (n)	$^A_Z P \longrightarrow \, ^{A-1}_{Z} D + n$	A nuclear neutron is ejected from the nucleus.
internal conversion (IC)	$^A_Z P^* \longrightarrow [^A_Z P]^+ + e^-$	The excitation energy of a nucleus is used to eject an orbital electron (usually a K-shell) electron.

In all nuclear interactions, including radioactive decay, there are several quantities that are always conserved or unchanged by the nuclear transmutation. The most important of these conservation laws include:

- Conservation of charge, i.e., the number of elementary positive and negative charges in the reactants must be the same as in the products.

- Conservation of the number of nucleons, i.e., A is always constant. With the exception of EC and $\beta^\pm$ radioactive decay, in which a neutron (proton) transmutes into a proton (neutron), the number of protons and neutrons is also generally conserved.

- Conservation of mass/energy (total energy). Although, neither rest mass nor kinetic energy is generally conserved, the total (rest-mass energy equivalent plus kinetic energy) is conserved.

- Conservation of linear momentum. This quantity must be conserved in all inertial frames of reference.

- Conservation of angular momentum. The total angular momentum (or the *spin*) of the reacting particles must always be conserved.

5.2   Types of Radioactive Decay

There are several types of spontaneous changes (or *transmutations*) that can occur in radioactive nuclides. In each transmutation, the nucleus of the parent atom $_Z^A$P is altered in some manner and one or more particles are emitted. If the number of protons in the nucleus is changed, then the number of orbital electrons in the daughter atom D must subsequently also be changed, either by releasing an electron to or absorbing an electron from the ambient medium. The most common types of radioactive decay are summarized in Table 5.1.

5.3   Radioactive Decay Diagrams

The quantification of the decay of a radionuclide involves much empirical data such as the types of decay it can undergo, the probability of each possible decay mode, the half-life of the radionuclide, the daughters produced, the nuclear excitation produced in the daughters, the lifetimes of long-lived excited daughters, and the energies and frequencies of secondary radiation produced when excited daughters decay to less excited states. One convenient method for presenting such data is through a radioactive decay diagram. In Fig. 5.1 the hypothetical decay of a radionuclide $_Z^A$X is shown.

In a radioactive decay diagram, also sometimes called a *nuclear-level diagram*, the y-axis represents the masses (or energy equivalents) of the various excited states of the daughters, which are denoted by horizontal lines. Ground states of both parent and daughter are shown by heavier horizontal lines. The x-axis denotes the number of protons in the various nuclei. Thus, when the decay produces a decrease in the atomic number (e.g., alpha, positron, or EC decay), the daughter is shown to the left of the parent. When there is an increase in the atomic number (e.g., β^- decay), the daughter is shown to the right of the parent.

Because all radioactive decays are exoergic, the mass of the daughter states must always be less than the mass of the parent. Thus, the various decay modes (shown by descending arrows) always proceed from the parent nuclide at the top of the diagram to daughter states below the parent, i.e., to states with less mass.

The vertical energy/mass axis is on a relative scale for each daughter with zero energy being the ground state of each daughter. The energy of each excited state is written on or adjacent to the horizontal line representing the state. For an excited state with energy E_i above the ground state, the excited nucleus has an excess mass $\Delta m_i = E_i/c^2$ greater than the mass of the nucleus in its ground state. Also often given is the Q-value for the decay to the ground state. The Q-value for decay to an excited state is the Q-value to the ground state less the nuclear excitation energy.

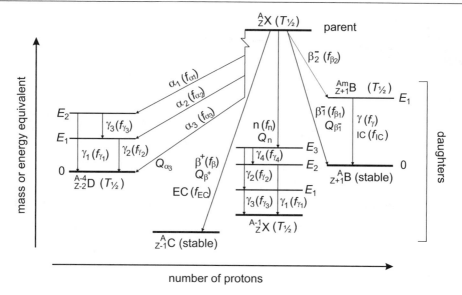

Figure 5.1. A radiactive decay diagram showing four decay modes of the hypothetical parent nuclide ^A_ZX in which four different daughter nuclides are produced.

In the decay diagram of Fig. 5.1, the parent nucleus ^A_ZX decays to four different daughters: by β^- decay to $_{Z+1}^A$B, by neutron decay to $^{A-1}_Z$X, by positron decay or electron capture to $_{Z-1}^A$C, and by α decay to $^{A-4}_{Z-2}$D. For the alpha decay, note the break in the vertical line descending from the parent to indicate the parent and daughters are on different energy/mass scales. Without such a break, the alpha decay daughter would be far (almost 4 GeV) below the the other decay daughters. Each decay arrow to a ground or excited state is labeled by the type of decay, e.g., α_1, and the decay probability or *frequency*, e.g., $f(\alpha_1)$. The frequency f_i, usually expressed in percent, gives the probability of that decay mode happening *per decay of the parent*. Thus, all the decay frequencies should sum to 100%.

Excited states of the daughter nucleus usually decay very rapidly to lower excited states or to the ground state either by internal conversion, whereby the excitation energy is transferred to an orbital electron, or by the emission of a photon called a *gamma ray*.[1] The gamma ray energy is essentially equal to the difference between the two energy states, because the kinetic energy of the recoiling atom is negligible (see Section 5.4.1 below). Thus, the unique energies of gamma rays emitted in radioactive decay can be found readily from the energy levels in the decay diagram.

Occasionally, an excited state can persist for an appreciable time. Such excited states are called *metastable* and labeled as such with its half-live (see Section 5.5.3). For example, in Fig. 5.1, the first excited state of the daughter $_{Z+1}^A$B is denoted as metastable by $_{Z+1}^{Am}$B. This metastable state then decays to the ground state by emission of a gamma ray or by internal conversion with the frequencies f_γ and f_{IC}, respectively.

---

[1] Any photon emitted from a nucleus is termed a *gamma* photon. Photons emitted when atomic electrons change their energy state are called *x rays*. Although x rays generally have lower energies than gamma photons, there are many exceptions.

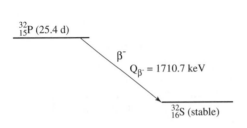

Figure 5.2. Radioactive decay diagram for the decay of ^{32}P.

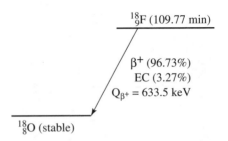

Figure 5.3. Radioactive decay diagram for the decay of ^{18}F.

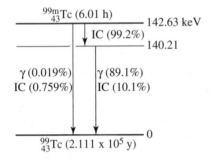

Figure 5.4. Radioactive decay diagram for the decay of ^{99m}Tc. Not shown is the beta decay to $^{99}_{44}$Ru with a frequency of 0.04%.

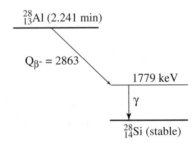

Figure 5.5. Radioactive decay diagram for the decay of ^{28}Al.

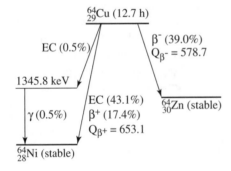

Figure 5.6. Radioactive decay diagram for the decay of ^{64}Cu.

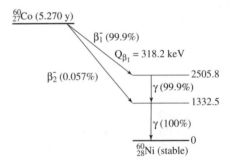

Figure 5.7. Radioactive decay diagram for the decay of ^{60}Co.

Some examples of beta decay and nuclear-level diagrams are shown, in simplified form, in Figs. 5.2 through 5.13. For many radionuclides, the diagrams can be much more complex with up to dozens of known excited states that result in a cascade of gamma rays each with a unique energy. The 3,100 page *Table of Isotopes* [Firestone and Shirley 1996] contains nuclear level diagrams for most of the known nuclides. Examples are also taken from Weber et al. [1989] and ICRP Publication 38 [1983].

5.4 Energetics of Radioactive Decay

In this section, the energies involved in the various types of radioactive decay are examined. Of particular interest are the energies of the particles emitted in the decay process.

5.4.1 Gamma Decay

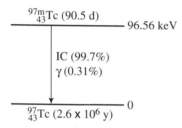

Figure 5.8. Radioactive decay diagram for the decay of ^{97m}Te.

In many nuclear reactions, a nuclide is produced whose nucleus is left in an excited state. These excited nuclei usually decay very rapidly within 10^{-9} s to the ground state by emitting the excitation energy in the form of a gamma ray. However, a few excited nuclei remain in an excited state for a much longer time before they decay by gamma emission. These long-lived excited nuclei are called *metastable* nuclei or *isomers*. When such an excited nuclide decays by gamma ray emission, the decay is called an *isomeric transition*. A simple example is shown in Fig. 5.8. The first excited state of $^{97m}_{43}$Tc is a metastable state with a half-life of 90.5 d. It decays to the ground state with the emission of a 96.5 keV gamma photon.

The gamma-decay reaction of an excited isotope of element P can be written as

$$\gamma \text{ decay:} \qquad ^{A}_{Z}\text{P}^* \rightarrow {^{A}_{Z}}\text{P} + \gamma \,. \tag{5.1}$$

Energy conservation for this nuclear reaction requires

$$M(^{A}_{Z}\text{P}^*)c^2 \equiv M(^{A}_{Z}\text{P})c^2 + E^* = M(^{A}_{Z}\text{P})c^2 + E_P + E_\gamma \qquad \text{or} \qquad E^* = E_P + E_\gamma, \tag{5.2}$$

where E_γ is the energy of the gamma photon, E^* is the excitation energy (above the ground state) of the initial parent nucleus, and E_P is the recoil kinetic energy of the resulting ground-state nuclide. If the initial nuclide is at rest, the Q-value of this reaction is simply the sum of the kinetic energies of the products, which, with Eq. (5.2), yields

$$Q_\gamma = E_P + E_\gamma = E^*. \tag{5.3}$$

Linear momentum must also be conserved.[2] Again with the parent at rest before the decay (i.e., with zero initial linear momentum), the gamma photon and recoil nucleus must move in opposite directions and have equal magnitudes of linear

---

[2]Throughout this section, because of the comparatively large mass and low energy of the recoil atom, momentum of the recoil is computed using laws of classical mechanics.

momentum. Since the photon has momentum $p_\gamma = E_\gamma/c$ and the recoil nuclide has momentum $M_P v_P = \sqrt{2M_P E_P}$, conservation of momentum requires

$$E_\gamma/c = \sqrt{2M_P E_P} \quad \text{or} \quad E_P = \frac{E_\gamma^2}{2M_P c^2} \tag{5.4}$$

where $M_P \equiv M(^A_Z P)$. Substitution of this result for E_P into Eq. (5.3) yields

$$E_\gamma = Q_\gamma \left[1 + \frac{E_\gamma}{2M_P c^2}\right]^{-1} \simeq Q_\gamma = E^*. \tag{5.5}$$

The approximation in this result follows from the fact that E_γ is at most 10–20 MeV, while $2M_P c^2 \geq 4000$ MeV. Thus, in gamma decay, the kinetic energy of the recoil nucleus is negligible compared to the energy of the gamma photon and $E_\gamma \simeq Q_\gamma = E^*$.

5.4.2 Alpha-Particle Decay

From our prior discussion of nuclear structure, we noted that nucleons tend to group themselves into subunits of $^4_2 He$ nuclei, often called alpha particles. Thus, it is not surprising that, for proton-rich heavy nuclei, a possible mode of decay to a more stable state is by alpha-particle emission.

In alpha decay, the nucleus of the parent atom $^A_Z P$ emits an alpha particle. The resulting nucleus of the daughter atom $^{A-4}_{Z-2} D$ then has two fewer neutrons and two fewer protons. Initially, the daughter still has Z electrons, two too many, and thus is a doubly negative ion $[^{A-4}_{Z-2} D]^{2-}$, but these extra electrons quickly break away from the atom, leaving it in a neutral state. The fast moving doubly charged alpha particle quickly loses its kinetic energy by ionizing and exciting atoms along its path of travel and acquires two orbital electrons to become a neutral $^4_2 He$ atom. Since the atomic number of the daughter is different from that of the parent, the daughter is a different chemical element. The alpha decay reaction is thus represented by

$$\alpha \text{ decay:} \qquad ^A_Z P \longrightarrow [^{A-4}_{Z-2} D]^{2-} + ^4_2\alpha \longrightarrow ^{A-4}_{Z-2} D + ^4_2 He. \tag{5.6}$$

Decay Energy

Recall from the previous chapter, the Q-value of a nuclear reaction is defined as the decrease in the rest mass energy (or increase in kinetic energy) of the product nuclei (see Eq. (4.18)). For radioactive decay, the Q-value is sometimes called the *disintegration energy*. For alpha decay we have

$$Q_\alpha/c^2 = M(^A_Z P) - [M([^{A-4}_{Z-2} D]^{2-}) + m(^4_2\alpha)]$$

$$\simeq M(^A_Z P) - [M(^{A-4}_{Z-2} D) + 2m_e + m(^4_2\alpha)]$$

$$\simeq M(^A_Z P) - [M(^{A-4}_{Z-2} D) + M(^4_2 He)]. \tag{5.7}$$

In this reduction to atomic masses, the binding energies of the two electrons in the daughter ion and in the He atom have been neglected since these are small (several eV) compared to the Q-value (several MeV). For alpha decay to occur, Q_α must be positive, or, equivalently, $M(^A_Z P) > M(^{A-4}_{Z-2} D) + M(^4_2 He)$.

Kinetic Energies of the Products

The disintegration energy Q_α equals the kinetic energy of the decay products. How this energy is divided between the daughter atom and the α particle is determined from the conservation of momentum. The momentum of the parent nucleus was zero before the decay, and thus, from the conservation of linear momentum, the total momentum of the products must also be zero. The alpha particle and the daughter nucleus must, therefore, leave the reaction site in opposite directions with equal magnitudes of their linear momentum to ensure the vector sum of their momenta is zero.

If we assume neither product particle is relativistic,[3] conservation of energy requires

$$Q_\alpha = E_D + E_\alpha = \frac{1}{2}M_D v_D^2 + \frac{1}{2}M_\alpha v_\alpha^2, \tag{5.8}$$

and conservation of linear momentum requires

$$M_D v_D = M_\alpha v_\alpha, \tag{5.9}$$

where $M_D \equiv M(_{Z-2}^{A-4}D)$ and $M_\alpha \equiv M(_2^4He)$. These two equations in the two unknowns v_D and v_α can be solved to obtain the kinetic energies of the products. Solve Eq. (5.9) for v_D and substitute the result into Eq. (5.8) to obtain

$$Q_\alpha = \frac{1}{2}\frac{M_\alpha^2}{M_D}v_\alpha^2 + \frac{1}{2}M_\alpha v_\alpha^2 = \frac{1}{2}M_\alpha v_\alpha^2\left[\frac{M_\alpha}{M_D}+1\right]. \tag{5.10}$$

Hence, we find that the kinetic energy of the alpha particle is

$$E_\alpha = Q_\alpha\left[\frac{M_D}{M_D+M_\alpha}\right] \simeq Q_\alpha\left[\frac{A_D}{A_D+A_\alpha}\right]. \tag{5.11}$$

Notice that, in alpha decay, the alpha particle is emitted with a well defined energy. The recoiling nucleus carries off the remainder of the available kinetic energy. From Eq. (5.8) we see $E_D = Q_\alpha - E_\alpha$, so that from the above result

$$E_D = Q_\alpha\left[\frac{M_\alpha}{M_D+M_\alpha}\right] \simeq Q_\alpha\left[\frac{A_\alpha}{A_D+A_\alpha}\right]. \tag{5.12}$$

Alpha Decay Diagrams

The above calculation assumes that the alpha decay proceeds from the ground state of the parent nucleus to the ground state of the daughter nucleus. As discussed earlier, sometimes the daughter nucleus is left in an excited nuclear state (which ultimately relaxes to the ground state by the emission of one or more gamma rays). In these cases, the Q_α-value of the decay is reduced by the excitation energy of the excited state. Often a radionuclide that decays by alpha emission is observed to emit alpha particles with several discrete energies. This is an indication that the daughter nucleus is left in excited states with nuclear masses greater than the ground state mass by the mass equivalent of the excitation energy.

---

[3]Here a non-relativistic analysis is justified since the energy liberated in alpha decay is less than 10 MeV, whereas the rest mass energy of an alpha particle is about 3700 MeV.

A simple case is shown in Fig. 5.9 for the decay of $^{226}_{88}$Ra. This nuclide decays to the first excited state of $^{222}_{86}$Rn 5.55% of the time and to the ground state 94.45% of the time. Two alpha particles with kinetic energies of $E_{\alpha1} = 4.783$ MeV and $E_{\alpha2} = 4.601$ MeV are thus observed. Also a 0.186-MeV gamma ray is observed, with an energy equal to the energy difference between the two alpha particles or the two lowest nuclear states in $^{222}_{88}$Rn.

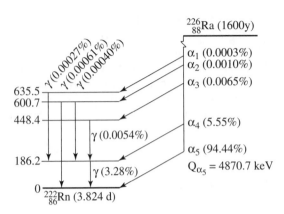

Figure 5.9. Energy levels for α decay of ^{226}Ra.

Example 5.1: What is the initial kinetic energy of the alpha particle produced in the radioactive decay $^{226}_{88}$Ra $\rightarrow$ $^{222}_{86}$Rn $+ ^4_2$He. The Q_α value in mass units (i.e., the mass defect) is, from Eq. (5.7),

$$Q_\alpha = [M(^{226}_{88}\text{Ra}) - M(^{222}_{86}\text{Rn}) - M(^4_2\text{He})]c^2$$
$$= [226.025402 - 222.017571 - 4.00260325 = 0.005228] \text{ u} \times 931.5 \text{ MeV/u}$$
$$= 4.870 \text{ MeV}.$$

The kinetic energy of the alpha particle from Eq. (5.11) is

$$E_\alpha = Q_\alpha \left[\frac{A_D}{A_D + A_\alpha}\right] = 4.870 \left[\frac{222}{222 + 4}\right] = 4.784 \text{ MeV}.$$

The remainder of the Q_α energy is the kinetic energy of the product nucleus, $^{222}_{86}$Rn, namely, 4.870 MeV $-$ 4.783 MeV $=$ 0.087 MeV.

5.4.3   Beta-Particle Decay

Many neutron-rich radioactive nuclides decay by changing a neutron in the parent (P) nucleus into a proton and emitting an energetic electron. Many different names are applied to this decay process: electron decay, beta minus decay, negatron decay, negative electron decay, negative beta decay, or simply beta decay. The ejected electron is called a *beta particle* denoted by β^-. The daughter atom, with one more proton in the nucleus, initially lacks one orbital electron, and thus is a single charged positive ion, denoted by $[_{Z-1}^A\text{D}]^+$. However, the daughter quickly acquires an extra orbital electron from the surrounding medium. The general β^- decay reaction is thus written as

$$\beta^- \text{ decay:} \qquad ^A_Z\text{P} \longrightarrow [^A_{Z+1}\text{D}]^+ + \,^{\,0}_{-1}\text{e} + \overline{\nu}_e. \qquad\qquad (5.13)$$

Here $\bar{\nu}$ is an antineutrino, a chargeless particle with very little, if any, rest mass.[4] That a third product particle is involved with β^- decay is implied from the observed energy and momentum of the emitted β^- particle. If the decay products were only the daughter nucleus and the β^- particle, then, as in α decay, conservation of energy and linear momentum would require that the decay energy be shared in very definite proportions between them. However, β^- particles are observed to be emitted with a continuous distribution of energies that has a well defined maximum energy (see Fig. 5.10). Rather than abandon the laws of conservation of energy and momentum, Pauli suggested in 1933 that at least three particles must be produced in a β^- decay. The vector sum of the linear momenta of three products can be zero without any unique division of the decay energy among them. In 1934 Fermi used Pauli's suggestion of a third neutral particle to produce a beta-decay theory which explained well the observed beta-particle energy distributions. This mysterious third particle, which Fermi named the neutrino (lit. "little neutral one"), has since been verified experimentally and today it is extensively studied by physicists trying to develop fundamental theories of our universe. The maximum energy of the β^- spectrum corresponds to a case in which the neutrino obtains zero kinetic energy, and the decay energy is divided between the daughter nucleus and the β^- particle.

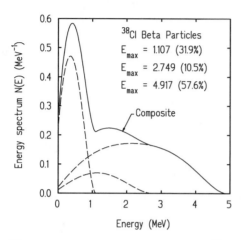

Figure 5.10. Energy spectra of principle ^{38}Cl β^- particles. From Shultis and Faw [2000].

Decay Energy

The beta decay energy is readily obtained from the Q-value of the decay reaction. Specifically,

$$Q_{\beta^-}/c^2 = M(^A_Z\mathrm{P}) - [M([_{Z+1}^A\mathrm{D}]^+) + m_{\beta^-} + m_{\bar{\nu}_e}]$$

$$\simeq M(^A_Z\mathrm{P}) - [\{M(_{Z+1}^A\mathrm{D}) - m_e\} + m_{\beta^-} + m_{\bar{\nu}_e}]$$

$$= M(^A_Z\mathrm{P}) - M(_{Z+1}^A\mathrm{D}). \tag{5.14}$$

For β^- decay to occur spontaneously, Q_{β^-} must be positive or, equivalently, the mass of the parent atom must exceed that of the daughter atom, i.e., $M(^A_Z\mathrm{P}) > M(_{Z+1}^A\mathrm{D})$.

---

[4]The mass of the emitted neutrino is essentially that of the electron neutrino ν_e. The mass of ν_e is currently a subject of great interest in the high-energy physics community. Recent experiments show that, if it has any mass, its mass is exceedingly small and is less than a few eV mass equivalent (see Table 1.6). For the energetics of radioactive beta decay, the neutrino rest mass is completely negligible and can be taken as zero.

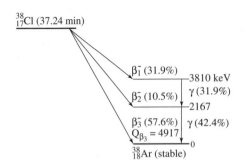

Figure 5.11. Radioactive decay diagram for the decay of ^{38}Cl. Three distinct groups of β^- particles are emitted.

Often in β^- decay, the nucleus of the daughter is left in an excited state. For example, ^{38}Cl decays both to the ground state of the daughter ^{38}Ar as well as to two excited states (see Fig. 5.11). The resulting β^- energy spectrum (Fig. 5.10) is a composite of the β^- particles emitted in the transition to each energy level of the daughter. For a decay to an energy level E^* above the ground level, the mass of the daughter atom in Eq. (5.14) $M(_{Z+1}^{A}\mathrm{D})$ must be replaced by the mass of the *excited* daughter $M(_{Z+1}^{A}\mathrm{D}^*) \simeq M(_{Z+1}^{A}\mathrm{D}) + E^*/c^2$. Thus, Q_{β^-} for β^- decay to an excited level with energy E^* above ground level in the daughter is

$$Q_{\beta^-}/c^2 = M(_{Z}^{A}\mathrm{P}) - M(_{Z+1}^{A}\mathrm{D}) - E^*/c^2. \qquad (5.15)$$

Because the kinetic energy of the parent nucleus is zero, the Q_{β^-} decay energy must be divided among the kinetic energies of the products. The maximum kinetic energy of the β^- particle occurs when the antineutrino obtains negligible energy. In this case, since the mass of the β^- particle is much less than that of the daughter nucleus, $Q = E_D + E_{\beta^-} \simeq E_{\beta^-}$ or

$$(E_{\beta^-})_{\max} \simeq Q_{\beta^-}. \qquad (5.16)$$

5.4.4   Positron Decay

Nuclei that have too many protons for stability often decay by changing a proton into a neutron. In this decay mechanism, an anti-electron or *positron* β^+ or $_{+1}^{0}\mathrm{e}$, and a neutrino ν are emitted. The daughter atom, with one less proton in the nucleus, initially has one too many orbital electrons, and thus is a negative ion, denoted by $[_{Z-1}^{A}\mathrm{D}]^-$. However, the daughter quickly releases the extra orbital electron to the surrounding medium and becomes a neutral atom. The β^+ decay reaction is written as

$$\boxed{\beta^+ \text{ decay:} \qquad _{Z}^{A}\mathrm{P} \longrightarrow [_{Z-1}^{A}\mathrm{D}]^- + \ _{+1}^{0}\mathrm{e} + \nu_e.} \qquad (5.17)$$

The positron has the same physical properties as an electron, except that it has one unit of positive charge. The positron β^+ is the antiparticle of the electron. The neutrino ν is required (as in β^- decay) to conserve energy and linear momentum since the β^+ particle is observed to be emitted with a continuous spectrum of energies up to some maximum value $(E_{\beta^+})_{\max}$. The neutrino ν in Eq. (5.17) is the antiparticle of the antineutrino $\bar{\nu}$ produced in beta-minus decay.

Decay Energy

The decay energy is readily obtained from the Q-value of the decay reaction. Specifically,

$$Q_{\beta^+}/c^2 = M(_{Z}^{A}\mathrm{P}) - [M([_{Z-1}^{A}\mathrm{D}]^-) + m(_{+1}^{0}\mathrm{e}) + m_{\nu_e}]$$

$$\simeq M(_Z^A\text{P}) - [\{M(_{Z-1}^{\ A}\text{D}) + m_e\} + m_{\beta^+} + m_{\nu_e}]$$

$$= M(_Z^A\text{P}) - M(_{Z-1}^{\ A}\text{D}) - 2m_e \tag{5.18}$$

where the binding energy of the electron to the daughter ion and the neutrino mass have been neglected. If the daughter nucleus is left in an excited state, the excitation energy E^* must also be included in the Q_{β^+} calculation, namely,

$$Q_{\beta^+}/c^2 = M(_Z^A\text{P}) - M(_{Z-1}^{\ A}\text{D}) - 2m_e - E^*/c^2. \tag{5.19}$$

Thus, for β^+ decay to occur spontaneously, Q_{β^+} must be positive, i.e., $M(_Z^A\text{P}) > M(_{Z-1}^{\ A}\text{D}) + 2m_e + E^*/c^2$.

The maximum energy of the emitted positron occurs when the neutrino acquires negligible kinetic energy, so that the Q_{β^-} energy is shared by the daughter atom and the positron. Because the daughter atom is so much more massive than the positron (by factors of thousands), almost all the Q_{β^-} energy is transferred as kinetic energy to the positron. Thus

$$(E_{\beta^+})_{\max} \simeq Q_{\beta^+}. \tag{5.20}$$

An example of a radionuclide that decays by positron emission is $_{11}^{22}\text{Na}$. The decay reaction is

$$_{11}^{22}\text{Na} \longrightarrow {}_{10}^{22}\text{Ne} + \beta^+ + \nu_e,$$

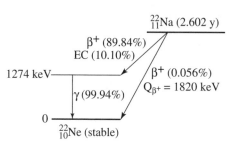

Figure 5.12. Radioactive decay diagram for positron emission from ^{22}Na.

and the level diagram for this decay is shown to the left. Notice that the daughter is almost always left in its first excited state. This state decays, with a mean lifetime of 3.63 ps, by emitting a 1.274-MeV gamma ray.

The emitted positron loses its kinetic energy by ionizing and exciting atomic electrons as it moves through the surrounding medium. Eventually, it captures an ambient electron, forming for a brief instant a pseudo-atom called *positronium* before they annihilate each other. Their entire rest mass energy $2m_ec^2$ is converted into photon energy (the kinetic energy at the time of annihilation usually being negligible). Before the annihilation, there is zero linear momentum, and there must be no net momentum remaining; thus, two photons traveling in opposite directions must be created, each with energy $E = m_ec^2 = 0.511$ MeV.

5.4.5 Electron Capture

In the quantum mechanical model of the atom, the orbital electrons have a finite (but small) probability of spending some time inside the nucleus, the innermost K-shell electrons having the greatest probability. It is possible for an orbital electron, while inside the nucleus, to be captured by a proton, which is thus transformed into a neutron. Conceptually we can visualize this transformation of the proton as $p + {}_{-1}^{\ 0}e \rightarrow n + \nu$, where the neutrino is again needed to conserve energy and momentum. The general electron capture (EC) decay reaction is written as

$$\boxed{\text{EC decay:} \qquad {}_{Z}^{A}\text{P} \longrightarrow {}_{Z-1}^{A}\text{D}^* + \nu_e.} \qquad (5.21)$$

where the daughter is generally left in an excited nuclear state with energy E^* above ground level. Unlike in most other types of radioactive decay, no charged particles are emitted. The only *nuclear* radiations emitted are gamma photons produced when the excited nucleus of the daughter relaxes to its ground state. As the outer electrons cascade down in energy to fill the inner shell vacancy, x rays and Auger electrons are also emitted.

Decay Energy

The decay energy is readily obtained from the Q-value of the decay reaction. If we assume the daughter nucleus is left in its ground state

$$Q_{EC}/c^2 = M({}_{Z}^{A}\text{P}) - [M({}_{Z-1}^{A}\text{D}) + m_{\nu_e}]$$

$$\simeq M({}_{Z}^{A}\text{P}) - M({}_{Z-1}^{A}\text{D}). \qquad (5.22)$$

If the daughter nucleus is left in an excited state, the excitation energy E^* must also be included in the Q_{EC} calculation, namely,

$$Q_{EC}/c^2 = M({}_{Z}^{A}\text{P}) - M({}_{Z-1}^{A}\text{D}) - E^*/c^2. \qquad (5.23)$$

Thus, for EC decay to occur spontaneously, Q_{EC} must be positive, i.e., $M({}_{Z}^{A}\text{P}) > M({}_{Z-1}^{A}\text{D}) + E^*/c^2$.

Notice that both β^+ and EC decay produce the same daughter nuclide. In fact, if the mass of the parent is sufficiently large compared to the daughter, both decay modes can occur for the same radionuclide. From Eq. (5.19) and Eq. (5.23), we see that if the parent's atomic mass is not at least two electron masses greater than the daughter's mass, Q_{β^+} is negative and β^+ decay cannot occur. However, Q_{EC} is positive as long as the parent's mass is even slightly greater than that of the daughter, and EC can still occur.

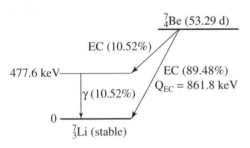

Figure 5.13. Radioactive decay diagram for electron capture in ^{7}Be.

An example of a radionuclide that decays by electron capture is ${}_{4}^{7}\text{Be}$. The decay reaction is

$${}_{4}^{7}\text{Be} \longrightarrow {}_{3}^{7}\text{Li} + \nu_e.$$

The level diagram for this decay is shown in Fig. 5.13. Notice that in this example, the Q_{EC} of 862 keV is less than $2m_ec^2 = 1022$ keV so that there can be no competing β^+ decay. Finally, in an EC transition, an orbital electron (usually from an inner shell) disappears leaving an inner electron vacancy. The remaining atomic electrons cascade to lower orbital energy levels to fill the vacancy, usually emitting x rays as they become more tightly bound. The energy change in an electronic transition, instead of being emitted as an x ray, may also be transferred to an outer orbital electron ejecting it from the atom. These ejected electrons are called *Auger* electrons, named after their discoverer, and appear in any process that leaves a vacancy in an inner electron shell.

5.4.6 Neutron Decay

A few neutron-rich nuclides decay by emitting a neutron producing a different isotope of the same parent element. Generally, the daughter nucleus is left in an excited state which subsequently emits gamma photons as it returns to its ground state. This decay reaction is

$$n \text{ decay:} \qquad {}^{A}_{Z}\text{P} \longrightarrow {}^{A-1}_{Z}\text{P}^* + {}^{1}_{0}\text{n}. \qquad (5.24)$$

An example of such a neutron decay reaction is ${}^{138}_{54}\text{Xe} \rightarrow {}^{137}_{54}\text{Xe} + n$. Although, neutron decay is rare, it plays a very important role in nuclear reactors. A small fraction of the radioactive atoms produced by fission reactions decay by neutron emission at times up to minutes after the fission event in which they were created. These neutrons contribute to the nuclear chain reaction and thus effectively slow it down, making it possible to control nuclear reactors.

The Q-value for such a decay is

$$Q_n/c^2 = M({}^{A}_{Z}\text{P}) - [M({}^{A-1}_{Z}\text{P}^*) + m_n]$$

$$= M({}^{A}_{Z}\text{P}) - M({}^{A-1}_{Z}\text{D}) - m_n - E^*/c^2 \qquad (5.25)$$

where E^* is the initial excitation energy of the daughter nucleus. Thus, for neutron decay to occur to even the ground state of the daughter, $M({}^{A}_{Z}\text{P}) > M({}^{A-1}_{Z}\text{D}) + m_n$.

5.4.7 Proton Decay

A few proton-rich radionuclides decay by emission of a proton. In such decays, the daughter atom has an extra electron (i.e., it is a singly charged negative ion). This extra electron is subsequently ejected from the atom's electron cloud to the surroundings and the daughter returns to an electrically neutral atom. The proton decay reaction is thus

$$p \text{ decay:} \qquad {}^{A}_{Z}\text{P} \longrightarrow [{}^{A-1}_{Z-1}\text{D}^*]^- + {}^{1}_{1}\text{p}. \qquad (5.26)$$

In this reaction P and D refer to atoms of the parent and daughter. Thus, the Q-value for this reaction is

$$Q_p/c^2 = M({}^{A}_{Z}\text{P}) - [M([{}^{A-1}_{Z-1}\text{D}^*]^-) + m_p]$$

$$\simeq M({}^{A}_{Z}\text{P}) - [\{M({}^{A-1}_{Z-1}\text{D}^*) + m_e\} + m_p]$$

$$\simeq M({}^{A}_{Z}\text{P}) - [\{M({}^{A-1}_{Z-1}\text{D}) + E^*/c^2 + m_e\} + m_p]$$

$$\simeq M({}^{A}_{Z}\text{P}) - M({}^{A-1}_{Z-1}\text{D}) - M({}^{1}_{1}\text{H}) - E^*/c^2. \qquad (5.27)$$

Thus, for proton decay to occur and leave the daughter in the ground state ($E^* = 0$), it is necessary that $M(_Z^A\text{P}) > M(_{Z-1}^{A-1}\text{D}) - M(_1^1\text{H})$.

5.4.8   Internal Conversion

Often the daughter nucleus is left in an excited state, which decays (usually within about 10^{-9} s) to the ground state by the emission of one or more gamma photons. However, the excitation may also be transferred to an atomic electron (usually a K-shell electron) causing it to be ejected from the atom leaving the nucleus in the ground state but the atom singly ionized with an inner electron-shell vacancy. Symbolically,

$$\text{IC decay:} \qquad _Z^A\text{P}^* \longrightarrow [_Z^A\text{P}]^+ + _{-1}^0\text{e}. \tag{5.28}$$

The inner electrons are very tightly bound to the nucleus with large binding energies BE_e^K for K-shell electrons in heavy atoms. The amount of kinetic energy shared by the recoil ion and the ejected electron should take this into account. The Q value for the IC decay is calculated as follows:

$$\begin{aligned}
Q_{\text{IC}}/c^2 &= M(_Z^A\text{P}^*) - [M([_Z^A\text{P}]^+) + m_e] \\
&\simeq \left\{ M(_Z^A\text{P}) + E^*/c^2 \right\} - [\left\{ M(_Z^A\text{P}) - m_e + \text{BE}_e^K/c^2 \right\} + m_e] \\
&= [E^* - \text{BE}_e^K]/c^2. \tag{5.29}
\end{aligned}$$

This decay energy is divided between the ejected electron and the daughter ion. To conserve the zero initial linear momentum, the daughter and IC electron must divide the decay energy as

$$E_e = \left(\frac{M(_Z^A\text{P})}{M(_Z^A\text{P}) + m_e} \right) [E^* - \text{BE}_e^K] \simeq E^* - \text{BE}_e^K \tag{5.30}$$

and

$$E_D = \left(\frac{m_e}{M(_Z^A\text{P}) + m_e} \right) [E^* - \text{BE}_e^K] \simeq 0. \tag{5.31}$$

Besides the monoenergetic IC conversion electron, other radiation is also emitted in the IC process. As the outer electrons cascade down in energy to fill the inner-shell vacancy, x rays and Auger electrons are also emitted.

5.5   Characteristics of Radioactive Decay

All radioactive decays, whatever the particles that are emitted or the rates at which they occur, are described by a single law: the radioactive decay law. The probability that an unstable parent nucleus will decay spontaneously into one or more particles of lower mass/energy is independent of the past history of the nucleus and is the same for all radionuclides of the same type.

5.5.1 The Decay Constant

Radioactive decay is a statistical (random or stochastic) process. There is no way of predicting whether or not a single nucleus will undergo decay in a given time period; however, we can predict the expected or average decay behavior of a very large number of identical radionuclides. Consider a sample containing a very large number N of the same radionuclide. In a small time interval Δt, ΔN of the atoms undergo radioactive decay.

The probability that any one of the radionuclides in the sample decays in Δt is thus $\Delta N/N$. Clearly, as Δt becomes smaller, so will the probability of decay per atom $\Delta N/N$. If one were to perform such a measurement of the decay probability $\Delta N/N$ for different time intervals Δt and plot the ratio, we would obtain results such as those shown in Fig. 5.14. The decay probability per unit time for a time interval Δt would vary smoothly for large Δt. But as Δt becomes smaller and smaller, the statistical fluctuations in the decay rate of the atoms in the sample would become apparent and the measured decay probability per unit time would have larger and larger statistical fluctuations.

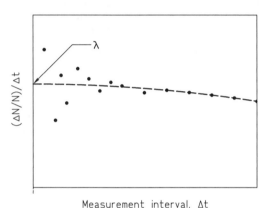

Figure 5.14. Measured decay probabilities $\Delta N/N$ divided by the measurement interval Δt.

If the experiment were repeated a large number of times, and the results of each experiment were averaged, the statistical fluctuations at small values of Δt would decrease and approach the dashed line in Fig. 5.14. This dashed line can also be obtained by extrapolating the decay probability per unit time from those values that do not have appreciable statistical fluctuations. This extrapolation averages the stochastic nature of the decay process for small time intervals. The statistically averaged decay probability per unit time, in the limit of infinitely small Δt, approaches a constant λ, i.e., we define

$$\lambda \equiv \lim_{\Delta t \to 0} \frac{(\Delta N/N)}{\Delta t}. \tag{5.32}$$

Each radionuclide has its own characteristic *decay constant* λ which, for the above definition, is the probability a radionuclide decays in a unit time for an infinitesimal time interval. The smaller λ, the more slowly the radionuclides decays. For stable nuclides, $\lambda = 0$. The decay constant for a radionuclide is independent of most experimental parameters such as temperature and pressure, since it depends only on the nuclear forces inside the nucleus.

5.5.2 Exponential Decay

Consider a sample composed of a large number of identical radionuclides with decay constant λ. With a large number of radionuclides ($N >>> 1$), one can use continuous mathematics to describe an inherently discrete process. In other words, $N(t)$

is interpreted as the average or expected number of radionuclides in the sample at time t, a continuous quantity. Then, the probability any one radionuclide decays in an interval dt is λdt, s the expected number of decays in the sample that occur in dt at time t is $\lambda\, dt\, N(t)$. This must equal the decrease $-dN$ in the number of radionuclides in the sample, i.e.,

$$-dN = \lambda N(t)dt$$

or

$$\frac{dN(t)}{dt} = -\lambda N(t). \tag{5.33}$$

The solution of this differential equation is

$$\boxed{N(t) = N_o e^{-\lambda t},} \tag{5.34}$$

where N_o is the number of radionuclides in the sample at $t = 0$. This exponential decay of a radioactive sample is known as the *radioactive decay law*. Such an exponential variation with time not only applies to radionuclides, but to any process governed by a constant rate of change, such as decay of excited electron states of an atoms, the rate of growth of money earning compound interest, and the growth of human populations.

5.5.3 The Half-Life

Any dynamic process governed by exponential decay (or growth) has a remarkable property. The time it takes for it to decay to one-half of (or to grow to twice) the initial value, $T_{1/2}$, is a constant called the *half-life*. From Eq. (5.34)

$$N(T_{1/2}) \equiv \frac{N_o}{2} = N_o e^{-\lambda T_{1/2}}. \tag{5.35}$$

Solving for $T_{1/2}$ yields

$$\boxed{T_{1/2} = \frac{\ln 2}{\lambda} \simeq \frac{0.693}{\lambda}.} \tag{5.36}$$

Notice that the half-life is independent of time t. Thus, after n half-lives, the initial number of radionuclides has decreased by a multiplicative factor of $1/2^n$, i.e.,

$$N(nT_{1/2}) = \frac{1}{2^n} N_o. \tag{5.37}$$

The number of half-lives n needed for a radioactive sample to decay to a fraction ϵ of its initial value is found from

$$\epsilon \equiv \frac{N(nT_{1/2})}{N_o} = \frac{1}{2^n}$$

which, upon solving for n, yields

$$n = -\frac{\ln \epsilon}{\ln 2} \simeq -1.44 \ln \epsilon. \tag{5.38}$$

Alternatively, the radioactive decay law of Eq. (5.34) can be expressed in terms of the half-life as

$$N(t) = N_o \left(\frac{1}{2}\right)^{t/T_{1/2}}. \tag{5.39}$$

5.5.4 Decay Probability for a Finite Time Interval

From the exponential decay law, we can determine some useful probabilities and averages. If we have N_o identical radionuclides at $t = 0$, we expect to have $N_o e^{-\lambda t}$ atoms at a later time t. Thus, the probability $\overline{P}$ that any one of the atoms *does not* decay in a time interval t is

$$\overline{P}(t) = \frac{N(t)}{N(0)} = e^{-\lambda t}. \tag{5.40}$$

The probability P that a radionuclide *does* decay in a time interval t is

$$P(t) = 1 - \overline{P}(t) = 1 - e^{-\lambda t}. \tag{5.41}$$

As the time interval becomes very small, i.e., $t \to \Delta t << 1$, we see

$$P(\Delta t) = 1 - e^{-\lambda \Delta t} = 1 - \left[1 - \lambda \Delta t + \frac{1}{2!}(\lambda \Delta t)^2 - + \ldots\right] \simeq \lambda \Delta t. \tag{5.42}$$

This approximation is in agreement with our earlier interpretation of the decay constant λ as being the decay probability per infinitesimal time interval.

From these results, we can obtain the probability distribution function for when a radionuclide decays. Specifically, let $p(t)dt$ be the probability a radionuclide, which exists at time $t = 0$, decays in the time interval between t and $t + dt$. Clearly,

$$p(t)dt = \{\text{prob. it doesn't decay in } (0, t)\}$$
$$\times \{\text{prob. it decays in the next } dt \text{ time interval}\}$$

$$= \{\overline{P}(t)\} \{P(dt)\} = \{e^{-\lambda t}\} \{\lambda \, dt\} = \lambda e^{-\lambda t} dt. \tag{5.43}$$

5.5.5 Mean Lifetime

In a radioactive sample, radionuclides decay at all times. From Eq. (5.34) we see that an infinite time is required for all the radioactive atoms to decay. However, as time increases, fewer and fewer atoms decay. We can calculate the average lifetime of a radionuclide by using the decay probability distribution $p(t)dt$ of Eq. (5.43). The average or mean lifetime T_{av} of a radionuclide is thus

$$T_{av} = \int_0^\infty t \, p(t) \, dt = \int_0^\infty t \lambda e^{-\lambda t} dt = -te^{-\lambda t} \Big|_0^\infty + \int_0^\infty e^{-\lambda t} dt = \frac{1}{\lambda}. \tag{5.44}$$

5.5.6 Activity

For detection and safety purposes, we are not really interested in the number of radioactive atoms in a sample; rather we are interested in the number of decays or transmutations per unit of time that occur within the sample. This decay rate, or *activity* $A(t)$, of a sample is given by[5]

$$A(t) \equiv -\frac{dN(t)}{dt} = \lambda N(t) = \lambda N_o e^{-\lambda t} = A_o e^{-\lambda t}, \tag{5.45}$$

where A_o is the activity at $t = 0$. Since the number of radionuclides in a sample decreases exponentially, the activity also decreases exponentially.

The SI unit used for activity is the becquerel (Bq) and is defined as one transformation per second. An older unit of activity, and one that is still sometimes encountered, is the curie (Ci) defined as 3.7×10^{10} Bq. One Ci is the approximate activity of one gram of $^{226}_{88}$Ra (radium). To obtain 1 Ci of tritium ($T_{1/2} = 12.6$ y, $\lambda = 1.74 \times 10^{-9}$ s^{-1}) we need 1.06×10^{-4} g of tritium. By contrast, one Ci of ^{238}U ($\lambda = 4.88 \times 10^{-18}$ s^{-1}) is 3.00×10^6 g.

In many instances, the activity of a radioactive sample is normalized to the mass or volume of the sample, e.g., curies/liter or Bq/g. This normalized activity is called the *specific activity*, which we denoted by $\widehat{A}$. Many safety limits and regulations are based on the specific activity concept.

However, it is important to be very clear when using specific activities to specify clearly to what exactly the activity is normalized. For example, a particular radionuclide dissolved in water could have its specific activity specified as Bq per g of solution. Alternatively, the same radionuclide could be normalized per gram of radionuclide. In this later case, a sample containing only $N(t)$ atoms of the radionuclide would have a mass $m(t) = N(t)M/N_a$, where M is the atomic weight (g/mol) of the radionuclide. Thus the specific activity, on a per unit mass of the radionuclide, is

$$\widehat{A}(t) = \frac{A(t)}{m(t)} = \frac{\lambda N(t)}{N(t)M/N_a} = \frac{\lambda N_a}{M} = \text{ constant.} \tag{5.46}$$

The specific activity of tritium ($\lambda = 1.79 \times 10^{-9}$ s^{-1}) is thus found to be 9.71×10^3 curies per gram of tritium. By contrast, ^{238}U ($\lambda = 4.88 \times 10^{-18}$ s^{-1}) has a specific activity of 3.34×10^{-7} curies per gram of ^{238}U.

5.5.7   Half-Life Measurement

Of great practical importance is the determination of a radionuclide's half-life $T_{1/2}$, or, equivalently, its decay constant λ. If the decay of the radionuclide's activity $A(t) \equiv \lambda N(t) = A(0)e^{-\lambda t}$ is plotted on a linear plot as, in Fig. 5.15, the exponential decay nature is very apparent. However, it is difficult to extract the decay constant from such a plot. On the other hand, if a semilog plot is used, i.e., plot $\log A(t) = \log A(0) - \lambda t$ versus t, the result is a straight line with a slope $-\lambda$ (see Fig. 5.16). Fitting a straight line to data, which always have measurement errors, is much easier than fitting an exponential curve to the data in Fig. 5.15.

When performing such measurements, we don't even need to measure the exact activity $A(t)$ of the radioactive source but only some quantity proportional to it

---

[5]Do not confuse the symbol A used here for activity with the same symbol used for the atomic mass number.

(such as the count rate obtained from a radiation detector). A semilog plot of $KA(t)$ (K being some proportionality constant) also has a slope of $-\lambda$.

The above approach is effective for determining the decay constant and the half-life of a radionuclide that decays appreciably during the time available for measurements, it cannot, however, be used for very long lived radionuclides, such as $^{248}_{92}U$ (half-life of 4.47×10^9 y). For such a radionuclide, determination of its half-life requires an absolute measurement of a sample's activity $A_o \equiv \lambda N_o \simeq$ constant and a careful measurement of the mass m of the radioactive atoms in the sample. The number of radionuclides is calculated as $N_o = mN_a/\mathcal{A}$, where $\mathcal{A}$ is the atomic weight of the radionuclide. The decay constant is then obtained as $\lambda = A_o/N_o = A_o\mathcal{A}/(mN_a)$.

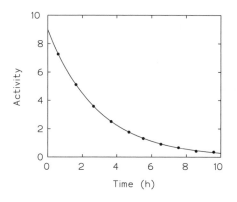

Figure 5.15. The activity of a radioactive sample with a half-life of two hours. At any time on the exponential curve, the activity is one-half of the activity two hours earlier.

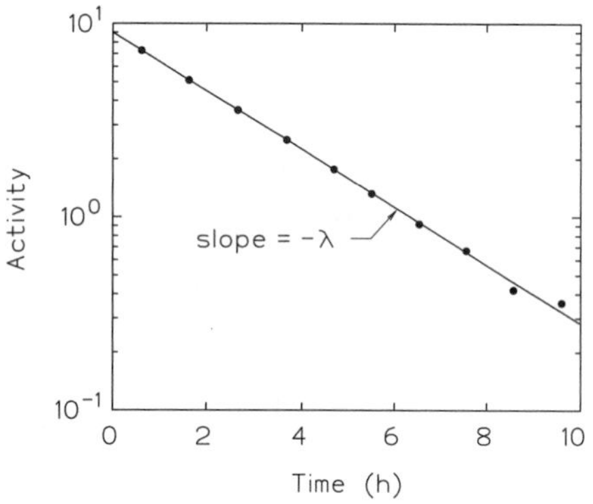

Figure 5.16. Semilog plot of the decay of the sample's activity. The decay curve is a straight line with a slope of $-\lambda$, from which the half-life $T_{1/2} = \ln 2/\lambda$ can be calculated.

5.5.8 Decay by Competing Processes

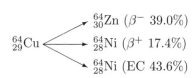

Figure 5.17. $^{62}_{29}Cu$ has three radioactive decay modes.

Some radionuclides decay by more than one process. For example, $^{64}_{29}Cu$ decays by β^+ emission 17.4% of the time, by β^- emission 39.0% of the time, and by electron capture 43.6% of the time. Each decay mode is characterized by its own decay constant λ_i. For the present example of $^{64}_{29}Cu$, the decay constants for the three decay modes are $\lambda_{\beta+} = 0.009497$ h^{-1}, $\lambda_{\beta-} = 0.02129$ h^{-1}, and $\lambda_{EC} = 0.02380$ h^{-1}.

To find the effective decay constant when the decay process has n competing decay modes, write the differential equation that models the rate of decay. Denote the decay constant for the ith mode by λ_i. Thus, the rate of decay of the parent radionuclide is given by,

$$\frac{dN(t)}{dt} = -\lambda_1 N(t) - \lambda_2 N(t) - \cdots - \lambda_n N(t) = -\sum_{i=1}^{n} \lambda_i N(t) \equiv -\lambda N(t). \quad (5.47)$$

where λ is the overall decay constant, namely,

$$\lambda \equiv \sum_{i=1}^{n} \lambda_i. \tag{5.48}$$

The probability f_i that the nuclide will decay by the ith mode is

$$f_i = \frac{\text{decay rate by } i\text{th mode}}{\text{decay rate by all modes}} = \frac{\lambda_i}{\lambda}. \tag{5.49}$$

See Example 5.2 for the calculation of the decay probabilities shown in Fig. 5.17.

Example 5.2: What is the probability ^{64}Cu decays by positron emission? The decay constants for the three decay modes of this radioisotope are $\lambda_{\beta+} = 0.009497$ h^{-1}, $\lambda_{\beta-} = 0.02129$ h^{-1}, and $\lambda_{\text{EC}} = 0.02380$ h^{-1}. The overall decay constant is

$$\lambda = \lambda_{\beta+} + \lambda_{\beta-} + \lambda_{\text{EC}} = 0.009497 + 0.02129 + 0.02380 = 0.05459 \text{ h}^{-1}.$$

The probability that an atom of ^{64}Cu eventually decays by positron emission is

$$\text{probability of } \beta^+ \text{ decay} = \lambda_{\beta+}/\lambda = 0.009497/0.05459 = 0.174,$$

a value in agreement with the branching probabilities shown in Fig. 5.17.

5.6 Decay Dynamics

The transient behavior of the number of atoms of a particular radionuclide in a sample depends on the nuclide's rates of production and decay, the initial values of it and its parents, and the rate at which it escapes from the sample. In this section, several common decay transients are discussed.

5.6.1 Decay with Production

In many cases the decay of radionuclides is accompanied by the creation of new ones, either from the decay of a parent or from production by nuclear reactions such as cosmic ray interactions in the atmosphere or from neutron interactions in a nuclear reactor. If $Q(t)$ is the rate at which the radionuclide of interest is being created, the rate of change of the number of radionuclides is

$$\frac{dN(t)}{dt} = -\text{rate of decay} + \text{rate of production} \tag{5.50}$$

or

$$\frac{dN(t)}{dt} = -\lambda N(t) + Q(t). \tag{5.51}$$

The most general solution of this differential equation is

$$\boxed{N(t) = N_o e^{-\lambda t} + \int_0^t dt' \, Q(t') e^{-\lambda(t-t')}} \tag{5.52}$$

where again N_o is the number of radionuclides at $t = 0$.

For the special case that $Q(t) = Q_o$ (a constant production rate), the integral in Eq. (5.52) can be evaluated analytically to give

$$N(t) = N_o e^{-\lambda t} + \frac{Q_o}{\lambda}[1 - e^{-\lambda t}]. \tag{5.53}$$

As $t \to \infty$ we see $N(t) \to N_e = Q_o/\lambda$. This constant equilibrium value N_e can be obtained more directly from Eq. (5.51). Upon setting the left-hand side of Eq. (5.51) to zero (the equilibrium condition), we see $0 = -\lambda N_e + Q_o$, or $N_e = Q_o/\lambda$.

Example 5.3: How long after a sample is placed in a reactor is it before the sample activity reaches 75% of the maximum activity? Assume the production of a single radionuclide species at a constant rate of Q_o s^{-1} and that there initially are no radionuclides in the sample material. Multiplication of Eq. (5.53) by λ and setting $A(0) = 0$ gives the sample activity

$$A(t) = Q_o[1 - \exp(-\lambda t)].$$

The maximum activity $A_{\max}$ is obtained as $t \to \infty$, namely, $A_{\max} = Q_o$, i.e., the sample activity when the production rate equals the decay rate. The time to reach 75% of this maximum activity is obtained from the above result with $A(t) = 0.75 Q_o$, i.e.,

$$0.75 Q_o = Q_o[1 - \exp(-\lambda t)].$$

Solving for t, we obtain

$$t = \frac{1}{\lambda} \ln[1 - 0.75] \simeq \frac{1.39}{\lambda}.$$

5.6.2 Three Component Decay Chains

Often a radionuclide decays to another radionuclide which in turn decays to yet another. The chain continues until a stable nuclide is reached. For simplicity, we first consider a three component chain. Such three component chains are quite common and an example is

$$^{90}_{38}\text{Sr} \xrightarrow{T_{1/2}=29.1\text{y}} {}^{90}_{39}\text{Y} \xrightarrow{T_{1/2}=64\text{h}} {}^{90}_{40}\text{Zn}(\text{ stable}),$$

in which both radionuclides decay by β^- emission. Such three-member decay chains can be written schematically as

$$X_1 \xrightarrow{\lambda_1} X_2 \xrightarrow{\lambda_2} X_3(\text{ stable}).$$

At $t = 0$ the number of atoms of each type in the sample under consideration is denoted by $N_i(0)$, $i = 1, 2, 3$. The differential decay equations for each species are (assuming no loss from or production in the sample)

$$\frac{dN_1(t)}{dt} = -\lambda_1 N_1(t). \tag{5.54}$$

$$\frac{dN_2(t)}{dt} = -\lambda_2 N_2(t) + \lambda_1 N_1(t).$$ (5.55)

$$\frac{dN_3(t)}{dt} = \lambda_2 N_2(t).$$ (5.56)

The solution of Eq. (5.54) is just the exponential decay law of Eq. (5.34), i.e.,

$$N_1(t) = N_1(0)e^{-\lambda_1 t}.$$ (5.57)

The number of first daughter atoms $N_2(t)$ is obtained from Eq. (5.52) with the production term $Q(t) = \lambda_1 N_1(t)$. The result is

$$N_2(t) = N_2(0)e^{-\lambda_2 t} + \frac{\lambda_1 N_1(0)}{\lambda_2 - \lambda_1}[e^{-\lambda_1 t} - e^{-\lambda_2 t}].$$ (5.58)

The number of second daughter (granddaughter) atoms is obtained by integrating Eq. (5.56). Thus,

$$N_3(t) = N_3(0) + \lambda_2 \int_0^t dt'\, N_2(t')$$

$$= N_3(0) + \lambda_2 N_2(0) \int_0^t dt'\, e^{-\lambda_2 t'} + \frac{N_1(0)\lambda_1 \lambda_2}{\lambda_2 - \lambda_1} \int_0^t dt'\, [e^{-\lambda_1 t'} - e^{-\lambda_2 t'}]$$

$$= N_3(0) + N_2(0)[1 - e^{-\lambda_2 t}] + \frac{N_1(0)}{\lambda_2 - \lambda_1}[\lambda_2(1 - e^{-\lambda_1 t}) - \lambda_1(1 - e^{-\lambda_2 t})].$$ (5.59)

Activity of the First Daughter

The activity of a parent radionuclide sample $A_1(t) = \lambda N_1(t)$ decays exponentially from its initial activity $A_1(0)$, i.e.,

$$A_1(t) = A_1(0)e^{-\lambda_1 t}, \qquad t > 0.$$ (5.60)

The activity of the radioactive daughter is found by multiplying Eq. (5.58) by λ_2 to give

$$A_2(t) = \lambda_2 N_2(t) = A_2(0)e^{-\lambda_2 t} + A_1(0)\frac{\lambda_2}{\lambda_2 - \lambda_1}[e^{-\lambda_1 t} - e^{-\lambda_2 t}].$$ (5.61)

The rate of decay of the daughter activity depends on how much larger or smaller its half-life is to that of the parent.

Daughter Decays Faster than the Parent

Equation (5.61) can be written as

$$A_2(t) = A_2(0)e^{-\lambda_2 t} + A_1(0)\frac{\lambda_2}{\lambda_2 - \lambda_1}e^{-\lambda_1 t}[1 - e^{-(\lambda_2 - \lambda_1)t}].$$ (5.62)

The initial activity $A_2(0)$ decays away quickly and because $\lambda_2 - \lambda_1 > 0$, then, as t becomes large, the asymptotic activity of the daughter is

$$A_2(t) \longrightarrow A_1(0)\frac{\lambda_2}{\lambda_2 - \lambda_1}e^{-\lambda_1 t},$$ (5.63)

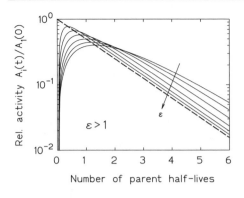
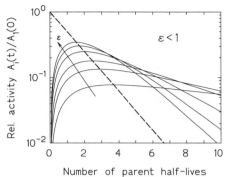

Figure 5.18. Activity of the first daughter with a half-life less than that of the parent, i.e., the daughter's decay constant $\lambda_2 = \epsilon\lambda_1, \epsilon > 1$. The six displayed daughter transients are for $\epsilon = 1.2, 1.5, 2, 3, 5,$ and 10. The heavy-dashed line is the parent's activity.

Figure 5.19. Activity of the first daughter with a half-life greater than that of the parent, i.e., the daughter's decay constant $\lambda_2 = \epsilon\lambda_1, \epsilon < 1$. The six displayed daughter transients are for $\epsilon = 0.9, 0.7, 0.5, 0.3, 0.2,$ and 0.1. The heavy-dashed line is the parent's activity.

i.e., the daughter decays asymptotically at almost the same rate as its parent. The buildup to this asymptotic behavior is shown in Fig. 5.18 for the case $A_2(0) = 0$. From this figure, we see that the asymptotic decay rate of the daughter, for $\lambda_1 < \lambda_2$, is never faster than the parent's decay rate. This is reasonable because the daughter cannot decay faster than the rate at which it is being created (the parent's activity). The asymptotic behavior, in which the daughter's decay rate is limited by the decay rate of the parent, is called *transient equilibrium*.

In the extreme case that the daughter decays much more rapidly than the parent (i.e., $\lambda_1 << \lambda_2$), Eq. (5.62) reduces to

$$A_2(t) \longrightarrow A_1(0)\frac{\lambda_2}{\lambda_2 - \lambda_1}e^{-\lambda_1 t} \simeq A_1(0)e^{-\lambda_1 t}. \tag{5.64}$$

The activity of the daughter approaches that of the parent. This extreme case is known as *secular equilibrium*, and is characteristic of natural decay chains each headed by a very long lived parent. Such decay chains are discussed in the next section.

Daughter Decays Slower than the Parent

Equation 5.61 can also be rewritten as

$$A_2(t) = A_2(0)e^{-\lambda_2 t} + A_1(0)\frac{\lambda_2}{\lambda_1 - \lambda_2}e^{-\lambda_2 t}[1 - e^{-(\lambda_1 - \lambda_2)t}]. \tag{5.65}$$

For large values of t with $\lambda_1 - \lambda_2 > 0$, we see

$$A_2(t) \simeq \left[A_2(0) + A_1(0)\frac{\lambda_2}{\lambda_1 - \lambda_2}\right]e^{-\lambda_2 t}, \tag{5.66}$$

i.e., the daughter decays in accordance with its normal decay rate. Activity transients for $\lambda_1 = \epsilon\lambda_2$ ($\epsilon > 1$) are shown in Fig. 5.19 for the case $A_2(0) = 0$.

Example 5.4:   A radioactive source is prepared by chemically separating ^{90}Sr from other elements. Initially, the source contains only ^{90}Sr ($T_{1/2} = 29.12$ y), but this radionuclide decays to a radioactive daughter ^{90}Y ($T_{1/2} = 64.0$ h), which, after some time, reaches secular equilibrium with its parent. What is the time after the source is created that the activity of the daughter ^{90}Y is within 5% of that of the parent?

First, consider the general case. The activity of the parent decays exponentially as $A_1(t) = A_1(0)\exp(-\lambda_1 t)$ and the activity of the daughter (obtained by multiplying Eq. (5.58) by λ_2) is

$$A_2(t) = A_2(0)e^{-\lambda_2 t} + \frac{A_1(0)\lambda_2}{\lambda_2 - \lambda_1}\left[e^{-\lambda_1 t} - e^{-\lambda_2 t}\right].$$

Here the sample initially contains no daughter nuclides, i.e., $A_2(0) = 0$, and we seek the time t such that

$$\frac{A_1(t) - A_2(t)}{A_1(t)} \simeq 0.05 = 1 - \frac{A_1(0)}{A_1(0)e^{-\lambda_1 t}}\frac{\lambda_2}{\lambda_2 - \lambda_1}\left[e^{-\lambda_1 t} - e^{-\lambda_2 t}\right]$$

$$= 1 - \frac{\lambda_2}{\lambda_2 - \lambda_1}\left[1 - e^{-(\lambda_2 - \lambda_1)t}\right].$$

Solving for t yields

$$\boxed{t = -\frac{1}{\lambda_2 - \lambda_1}\ln\left[1 - (1 - 0.05)\frac{\lambda_2 - \lambda_1}{\lambda_2}\right]}$$

For the present problem, $\lambda_1 = \ln 2/T_{1/2}^{\text{Sr}} = 2.715 \times 10^{-6}$ h^{-1} and $\lambda_2 = \ln 2/T_{1/2}^{\text{Y}} = 1.083 \times 10^{-2}$ h^{-1}. Because $\lambda_1 << \lambda_2$, the above general result reduces to

$$t = -\frac{1}{\lambda_2}\ln(1 - 0.95) \simeq \frac{2.996}{\lambda_2} = 276.6 \text{ h} = 11.52 \text{ d}.$$

In a radionuclide generator, a long-lived radioisotope decays to a short-lived daughter that is useful for some application. When the daughter is needed, it is extracted from the generator. The first such generator contained ^{226}Ra ($T_{1/2} = 1599$ y) from which ^{222}Rn ($T_{1/2} = 3.82$ d) was extracted. Today, the most widely used generator is one containing ^{99}Mo ($T_{1/2} = 65.5$ h) from which its daughter ^{99m}Tc ($T_{1/2} = 6.01$ h) is obtained by passing a saline solution through an aluminum oxide column containing the ^{99}Mo. In jargon, the ^{99m}Tc is said to be "milked" from the ^{99}Mo "cow." Other radionuclide generators include ^{81}Rb/^{81m}Kr, ^{82}Sr/^{82}Rb, ^{87}Y/^{87m}Sr, ^{113}Sn/^{113m}In, and ^{191}Os/^{191m}Ir.

Daughter is Stable

The limiting case of a daughter decaying much more slowly than the parent is a stable daughter ($\lambda_2 \to 0$). The number of stable daughter nuclei present after time t is the number of initial daughter atoms $N_2(0)$ plus the number of daughter atoms created by the decay of the parent up to time t (which equals $N_1(0)$ less the number of parent atoms remaining, namely, $N_1(0)e^{-\lambda t}$). Thus, the buildup of the daughter atoms is

$$N_2(t) = N_2(0) + N_1(0)\left(1 - e^{-\lambda t}\right). \tag{5.67}$$

This same result comes directly from Eq. (5.58) when $\lambda_2 \to 0$.

5.6.3 General Decay Chain

The general decay chain can be visualized as

$$\mathrm{X}_1 \xrightarrow{\lambda_1} \mathrm{X}_2 \xrightarrow{\lambda_2} \mathrm{X}_3 \xrightarrow{\lambda_3} \cdots \mathrm{X}_i \xrightarrow{\lambda_i} \cdots \xrightarrow{\lambda_{n-1}} \mathrm{X}_n \text{ (stable)}$$

The decay and buildup equations for each member of the decay chain are

$$\frac{dN_1(t)}{dt} = -\lambda_1 N_1(t)$$

$$\frac{dN_2(t)}{dt} = \lambda_1 N_1(t) - \lambda_2 N_2(t)$$

$$\frac{dN_3(t)}{dt} = \lambda_2 N_2(t) - \lambda_3 N_3(t)$$

$$\vdots \qquad \vdots$$

$$\frac{dN_{n-1}(t)}{dt} = \lambda_{n-2} N_{n-2}(t) - \lambda_{n-1} N_{n-1}(t)$$

$$\frac{dN_n(t)}{dt} = \lambda_{n-1} N_{n-1}(t). \tag{5.68}$$

For the case when only radionuclides of the parent X_1 are initially present, the initial conditions are $N_1(0) \neq 0$ and $N_i(0) = 0$, $i > 1$. The solution of these so-called Bateman equations with these initial conditions is [Mayo 1998]

$$A_j(t) = \lambda_j N_j(t) = N_1(0)(C_1 e^{-\lambda_1 t} + C_2 e^{-\lambda_2 t} + \cdots + C_j e^{-\lambda_j t}) = N_1(0) \sum_{m=1}^{j} C_m e^{-\lambda_m t}. \tag{5.69}$$

The coefficients C_m are

$$C_m = \frac{\prod_{i=1}^{j} \lambda_j}{\prod_{\substack{i=1 \\ i \neq m}}^{j} (\lambda_i - \lambda_m)} = \frac{\lambda_1 \lambda_2 \lambda_3 \cdots \lambda_j}{(\lambda_1 - \lambda_m)(\lambda_2 - \lambda_m) \cdots (\lambda_j - \lambda_m)}, \tag{5.70}$$

where the $i = m$ term is excluded from the denominator.

Example 5.5: What is the activity of the first daughter in a multicomponent decay chain? From Eq. (5.70) for $j = 2$

$$C_1 = \frac{\lambda_1 \lambda_2}{\lambda_2 - \lambda_1} \quad \text{and} \quad C_2 = \frac{\lambda_1 \lambda_2}{\lambda_1 - \lambda_2}.$$

Substitution of these coefficients into Eq. (5.69) then gives

$$A_2(t) = N_1(0) \left[\frac{\lambda_1 \lambda_2}{\lambda_2 - \lambda_1} e^{-\lambda_1} + \frac{\lambda_1 \lambda_2}{\lambda_1 - \lambda_2} e^{-\lambda_1} \right] = A_1(0) \frac{\lambda_2 [e^{-\lambda_1 t} - e^{-\lambda_2 t}]}{\lambda_2 - \lambda_1}.$$

This result agrees, as it should, with Eq. (5.61).

5.7 Naturally Occurring Radionuclides

Radionuclides existing on earth arose from two sources. First, earth has always been bombarded by cosmic rays coming from the sun and from intergalactic space. For the most part, cosmic rays consist of protons and alpha particles, with heavier nuclei contributing less than 1% of the incident nucleons.[6] These cosmic rays interact with atoms in the atmosphere and produce a variety of light radionuclides.

The second source of naturally occurring radionuclides is the residual radioactivity in the matter from which the earth was formed. These heritage or primordial radionuclides were formed in the stars from whose matter the solar system was formed. Most of these radionuclides have since decayed away during the 4 billion years since the earth was formed.

5.7.1 Cosmogenic Radionuclides

Cosmic rays interact with constituents of the atmosphere, sea, or earth, but mostly with the atmosphere, leading directly to radioactive products. Capture of secondary neutrons produced in primary interactions of cosmic rays leads to many more. Only those produced from interactions in the atmosphere lead to significant radiation exposure to humans.

The most prominent of the cosmogenic radionuclides are tritium ^{3}H and ^{14}C. The ^{3}H (symbol T) is produced mainly from the ^{14}N$(n,\text{T})^{12}$C and ^{16}O$(n,\text{T})^{14}$N reactions. Tritium has a half-life of 12.3 years, and, upon decay, emits one β^- particle with a maximum energy of 18.6 (average energy 5.7) keV. Tritium exists in nature almost exclusively as HTO.

The nuclide ^{14}C is produced mainly from the ^{14}N$(n,p)^{14}$C reaction. It exists in the atmosphere as CO_2, but the main reservoirs are the oceans. ^{14}C has a half life of 5730 years and decays by β^- emission with a maximum energy of 157 keV and average energy of 49.5 keV.

Over the past century, combustion of fossil fuels with the emission of CO_2 without any ^{14}C has diluted the cosmogenic content of ^{14}C in the environment. Moreover, since the use of nuclear weapons in World War II, artificial introduction of

---

[6]Electrons, primarily from the sun, also are part of the cosmic rays incident on the earth. But they are deflected by earth's magnetosphere and do not produce cosmogenic radionuclides.

^{3}H and ^{14}C (and other radionuclides) by human activity has been significant, especially from atmospheric nuclear-weapons tests. Consequently, these isotopes no longer exist in natural equilibria in the environment.

5.7.2 Singly Occurring Primordial Radionuclides

Of the many radionuclide species present when the solar system was formed about 5 billion years ago, some 17 very long-lived radionuclides still exist as singly occurring or isolated radionuclides, that is, as radionuclides not belonging to a radioactive decay chain. These radionuclides are listed in Table 5.2, and are seen to all have half-lives greater than the age of the solar system. Of these radionuclides, the most significant (from a human exposure perspective) are ^{40}K and ^{87}Rb since they are inherently part of our body tissue.

Table 5.2. The 17 isolated primordial radionuclides. Data taken from GE-NE [1996].

Radionuclide & the Decay Modes		Half-life (years)	% El. Abund.	Radionuclide & the Decay Modes		Half-life (years)	% El. Abund.
$^{40}_{19}$K	β^- EC β^+	1.27×10^9	0.0117	$^{50}_{23}$V	β^- EC	1.4×10^{17}	0.250
$^{87}_{37}$Rb	β^-	4.88×10^{10}	27.84	$^{113}_{48}$Cd	β^-	9×10^{15}	12.22
$^{115}_{49}$In	β^-	4.4×10^{14}	95.71	$^{123}_{52}$Te	EC	$> 1.3 \times 10^{13}$	0.908
$^{138}_{57}$La	EC β^-	1.05×10^{11}	0.090	$^{144}_{60}$Nd	α	2.38×10^{15}	23.80
$^{147}_{62}$Sm	α	1.06×10^{11}	15.0	$^{148}_{62}$Sm	α	7×10^{15}	11.3
$^{152}_{64}$Gd	α	1.1×10^{14}	0.20	$^{176}_{71}$Lu	β^-	3.78×10^{10}	2.59
$^{174}_{72}$Hf	α	2.0×10^{15}	0.162	$^{180}_{73}$Ta	EC β^+	$> 1.2 \times 10^{15}$	0.012
$^{187}_{75}$Re	β^-	4.3×10^{10}	62.60	$^{186}_{76}$Os	α	2×10^{15}	1.58
$^{190}_{78}$Pt	α	6.5×10^{11}	0.01				

5.7.3 Decay Series of Primordial Origin

Each naturally occurring radioactive nuclide with $Z > 83$ is a member of one of three long decay chains, or radioactive series, stretching through the upper part of the Chart of the Nuclides. These radionuclides decay by α or β^- emission and they have the property that the number of nucleons (mass number) A for each member of a given decay series can be expressed as $4n + i$, where n is an integer and i is a constant (0, 2, or 3) for each series. The three naturally occurring series are named the thorium ($4n$), uranium ($4n + 2$), and actinium ($4n + 3$) series, named after the radionuclide at, or near, the head of the series. The head of each series has a half-life much greater than any of its daughters.

There is no naturally occurring series represented by $4n + 1$. This series was recreated after $^{241}_{94}$Pu was made in nuclear reactors. This series does not occur naturally since the half-life of the longest lived member of the series, $^{237}_{93}$Np, is only 2.14×10^6 y, much shorter than the lifetime of the earth. Hence, any members of this series that were in the original material of the solar system have long since decayed away.

The decay chains for the three naturally occurring series are shown in Figs. 5.20 and 5.21. In all of these decay chains, a few members decay by both α and β^- decay causing the decay chain to branch. Nevertheless, each decay chain ends in the same stable isotope.

5.7.4 Secular Equilibrium

For a radioactive decay series, such as the natural decay series, in which the parent is long-lived compared to the daughters, an equilibrium exists in samples that have been undisturbed for a very long period of time. The rate of decay of the parent is negligibly slow, that is, $A_1(t) = \lambda_1 N(t) \simeq A_o$ (a constant) since the number of atoms of the parent does not change in time intervals comparable to the half-lives of the daughters. Thus, $dN_1(t)/dt \simeq 0$.

The number of first daughter atoms can decay away no faster than they are formed at the constant rate A_o, i.e., $\lambda_2 N_2(t) \simeq \lambda_1 N_1 = A_o$. Thus, $dN_2(t)/dt \simeq 0$. Similarly, the second daughter can decay away no faster than the rate it is formed by the decay of the first daughter (at the constant rate A_o, so that $dN_3(t)/dt \simeq 0$. This pseudo equilibrium, called *secular equilibrium*, continues down the decay chain for each daughter.

Under secular equilibrium, we thus have for each radioactive member $dN_1/dt \simeq dN_2/dt \simeq \cdots dN_{n-1}/dt \simeq 0$. By setting each of the derivatives to zero in the decay equations, given by Eq. (5.68), we obtain

$$\lambda_1 N_1 = \lambda_2 N_2 = \cdots = \lambda_{n-1} N_{n-1} \tag{5.71}$$

or, equivalently,

$$A_o = A_1 = A_2 = \cdots = A_{n-1}. \tag{5.72}$$

Thus, under secular equilibrium, each member of the decay chain has the same activity.

However, this very useful result applies only in samples that have been undisturbed for periods greater than several half-lives of the longest lived daughter. The daughters in samples of ^{238}U obtained by extraction from ore, are not in secular equilibrium.

Application of Secular Equilibrium

The secular equilibrium condition can be used to determine the decay constant of long-lived nuclides. For example, a uranium ore sample contains the daughter $^{226}_{88}$Ra, which has a half life of 1620 y. From chemical measurements, it is found that the ratio of the number of atoms of ^{226}Ra, N_{226}, to the number of atoms of ^{238}U, N_{238} is $N_{226}/N_{238} = 1/2.3 \times 10^6$. If ^{238}U and ^{226}Ra are assumed to be in secular equilibrium, the half life of ^{238}U can be determined from

$$\lambda_{238} N_{238} = \lambda_{226} N_{226}.$$

From this and the relationship $\lambda = \ln 2/T_{1/2}$, we have the following,

$$\frac{N_{238}}{T_{238}} = \frac{N_{226}}{T_{226}}.$$

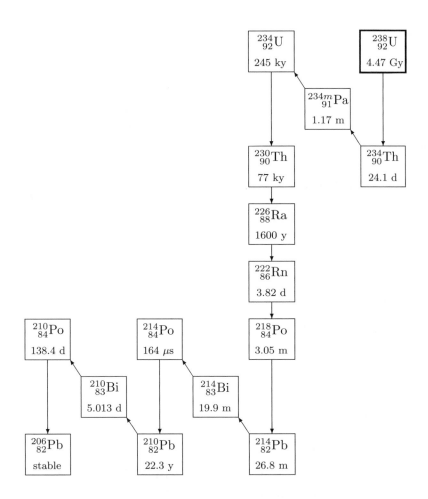

Figure 5.20. The $^{238}_{92}$U ($4n + 2$) natural decay series. Alpha decay is depicted by downward arrows and β^- decay by arrows upward and to the left. Not shown are (1) the isomeric transition to $^{234}_{91}$Pa (0.16%) followed by beta decay to ^{234}U, (2) beta decay of ^{218}Po to ^{218}At (0.020%) followed by alpha decay to ^{214}Bi, (3) alpha decay of ^{214}Bi to ^{210}Tl (0.0210%) followed by beta decay to ^{210}Pb, and (4) alpha decay of ^{210}Bi to ^{206}Tl (0.000132%) followed by beta decay to ^{206}Pb. After Faw and Shultis [1999].

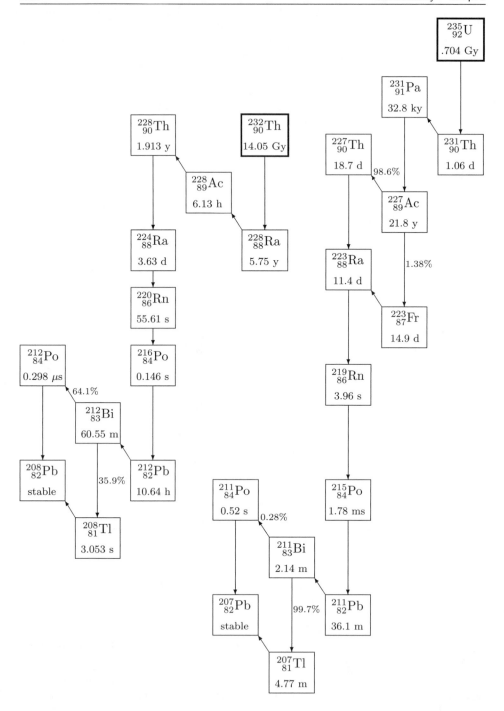

Figure 5.21. The $^{232}_{90}$Th ($4n$) natural decay series (left) and the $^{235}_{92}$U ($4n + 3$) natural decay series (right). Alpha decay is depicted by downward arrows and β^- decay by arrows upward and to the left. Not shown in the $^{235}_{92}$U series on the right are two very minor side chains: (1) alpha decay of ^{223}Fr (0.004%) to ^{85}At which beta decays to ^{223}Ra (3%) or alpha decays to ^{215}Bi which beta decays to ^{215}Po, and (2) the beta decay of ^{215}At ($< 0.001\%$) followed by alpha decay to ^{211}Bi. After Faw and Shultis [1999].

Thus, the half life for ^{238}U is

$$T_{238} = T_{226} \frac{N_{238}}{N_{226}} = (1.62 \times 10^3)(2.3 \times 10^6) = 4.5 \times 10^9 \text{y}.$$

5.8 Radiodating

A very important application of radioactive decay is the dating of geological and archaeological specimens. Measurements of daughter and parent concentrations in a specimen allow us to determine the sample's age because the decay rates of the radionuclides in the sample serve as nuclear clocks running at a constant rate. Several variations of radiodating have been developed and are now used routinely.

5.8.1 Measuring the Decay of a Parent

If, at the time a sample was created, a known number of radioactive atoms of the same type, $N(0)$, was incorporated, the sample age t is readily found from the remaining number of these atoms, $N(t)$, and the radioactive decay law $N(t) = N(0)\exp(-\lambda t)$, namely,

$$t = -\frac{1}{\lambda} \ln\left(\frac{N(t)}{N(0)}\right).$$

Unfortunately, we never know $N(0)$. However, sometimes the initial ratio $N(0)/N_s$ of the radionuclide and some stable isotope of the same element can be estimated with reliability. This ratio also decays with the same radioactive decay law as the radionuclide. Thus,

$$t = -\frac{1}{\lambda} \ln\left(\frac{N(t)/N_s}{N(0)/N_s}\right). \tag{5.73}$$

C-14 Dating

The most common dating method using this approach is ^{14}C dating. The radionuclide ^{14}C has a half-life of 5730 y and is introduced into the environment by cosmic-ray ^{14}N$(n,p)^{14}$C interactions in the atmosphere and removed by radioactive decay, leading if unperturbed to an equilibrium ratio of ^{14}C to all carbon atoms in the environment of about $N_{14}/N_C = 1.23 \times 10^{-12}$. It is usually easier to measure the specific activity of ^{14}C in a sample, i.e., A_{14} per gram of carbon. This specific activity is proportional to the N_{14}/N_C ratio, since

$$\frac{A_{14}}{\text{g(C)}} = \left(\frac{N_{14}}{N_C}\right)\frac{\lambda_{14}\,N_a}{12} = 0.237\,\frac{\text{Bq}}{\text{g(C)}} = 6.4\,\frac{\text{pCi}}{\text{g(C)}}.$$

All isotopes of carbon are incorporated by a living (biological) organism, either through ingestion or photosynthesis, in the same proportion that exists in its environment. Once the entity dies, the $N_{14}(t)/N_C$ ratio decreases as the ^{14}C atoms decay. Thus, a carbonaceous archaeological artifact, such as an ancient wooden axe handle or a mummy, had an initial $A_{14}(0)/\text{g(C)}$ ratio of about 6.4 pCi/g. From a measurement of the present $N_{14}(t)/N_C$ ratio, the age of the artifact can be determined from Eq. (5.73), namely,

$$t = -\frac{1}{\lambda} \ln\left(\frac{N_{14}(t)/N_C}{N_{14}(0)/N_C}\right) = -\frac{1}{\lambda} \ln\left(\frac{A_{14}(t)/\text{g(C)}}{A_{14}(0)/\text{g(C)}}\right). \tag{5.74}$$

This time should be thought of as an uncorrected "radiocarbon age." In fact, the production of ^{14}C in the atmosphere has varied because of fluctuations in solar activity and in the earth's magnetic field. In recent times, combustion of fossil fuels, absent of ^{14}C because of their age, has diluted atmospheric concentrations. Atmospheric nuclear weapon tests have artificially added to atmospheric concentrations. Thus, it has been necessary to develop calibration data relating "true" to "radiocarbon" age, with efforts now under way to extend the calibration to ages dating to 50,000 years [Balter 2006].

Example 5.6: What is the age of an archaeological sample of charcoal from an ancient fire that has a $A_{14}(t)/g(C)$ ratio of 1.2 pCi/g of carbon? The uncorrected radiocarbon age of the wood, from Eq. (5.74),

$$t = -\frac{1}{\lambda} \ln\left(\frac{1.2 \times 10^{-12}}{6.4 \times 10^{-12}}\right) = -\frac{5730}{\ln 2} \ln\left(\frac{1.2}{6.4}\right) \simeq 13,800 \text{ y.}$$

5.8.2   Measuring the Buildup of a Stable Daughter

Consider a sample containing an initial number of identical parent radioactive atoms $N_1(0)$, each of which decays (either directly or through a series of comparatively short-lived intermediate radionuclides) to N_2 stable daughter atoms. For simplicity, assume there are initially no daughter nuclides in the sample. If we further assume that there is no loss (or gain) of the parent N_1 and daughter N_2 atoms from the sample since its formation, the number of these atoms in the sample at time t is then

$$N_1(t) = N_1(0)e^{-\lambda t} \qquad \text{and} \qquad N_2(t) = N_1(0)[1 - e^{-\lambda t}]. \tag{5.75}$$

Division of the second equation by the first and solution of the result for the sample age t yields

$$t = \frac{1}{\lambda_1} \ln\left(1 + \frac{N_2(t)}{N_1(t)}\right). \tag{5.76}$$

The atom ratio $N_2(t)/N_1(t)$ in the sample is the same as the concentration ratio and can readily be found from mass spectroscopy or chemical analysis.

Example 5.7: Long-lived ^{232}Th $(T_{1/2} = 14.05 \times 10^9$ y) decays through a series of much shorter lived daughters to the stable isotope ^{208}Pb. The number of atoms of ^{208}Pb in a geological rock sample, assuming no initial inventory of ^{208}Pb in the sample, equals the number of initial ^{232}Th atoms that have decayed since the rock was formed. The number of decayed ^{232}Th atoms in the form of intermediate daughters, which are in secular equilibrium and have not yet reached ^{208}Pb, is negligibly small.

What is the age of a rock sample that is found to have $m_{Th} = 1.37$ g of ^{232}Th and $m_{Pb} = 0.31$ g of ^{208}Pb? The corresponding atom ratio of these two isotopes is thus

$$\frac{N_{Pb}(t)}{N_{Th}(t)} = \frac{m_{Pb}N_a/A_{Pb}}{m_{Th}N_a/A_{Th}} \simeq \frac{m_{Pb}}{m_{Th}}\frac{A_{Th}}{A_{Pb}}.$$

If there is no ^{208}Pb initially in the rock, then from Eq. (5.76) the rock's age is

$$t = \frac{1}{\ln 2/T_{1/2}} \ln\left(1 + \frac{m_{Pb}}{m_{Th}}\frac{A_{Th}}{A_{Pb}}\right)$$

$$= \frac{1}{\ln 2/14.05 \times 10^9 \text{ y}} \ln\left(1 + \frac{0.31}{1.37}\frac{232}{208}\right) = 4.56 \times 10^9 \text{ y}.$$

Stable Daughter with Initial Concentration

Often, however, the initial value $N_2(0)$ of the stable daughter is not zero, and a slightly more refined analysis must be used. In this case, the number of atoms of a radioactive parent and a stable daughter in a sample at age t is

$$N_1(t) = N_1(0)e^{-\lambda t} \qquad \text{and} \qquad N_2(t) = N_2(0) + N_1(0)[1 - e^{-\lambda t}]. \qquad (5.77)$$

We now have two equations and three unknowns ($N_1(0)$, $N_2(0)$, and, of course, t). We need one additional relationship.

Suppose there exists in a sample N_2' atoms of another stable isotope of the same element as the daughter, one that is not formed as a product of another naturally occurring decay chain. The ratio $R = N_2/N_2'$ is known from the relative abundances of the stable isotopes of the element in question and is a result of the mix of isotopes in the primordial matter from which the solar system was formed. In samples, without any of the radioactive parent, this ratio remains constant, i.e., $N_2(t)/N_2'(t) = N_2(0)/N_2'(0) = R_o$. Samples in which R is observed to be larger than R_o must have been enriched as a result of the decay of parent radionuclides.

If we assume a sample has experienced no loss (or gain) of parent and daughter nuclides since its formation, then we may assume $N_2'(t) = N_2'(0)$. Division of Eq. (5.77) by $N_2'(0) = N_2'(t)$ yields

$$\frac{N_1(t)}{N_2'(t)} = \frac{N_1(0)}{N_2'(0)}e^{-\lambda t} \qquad \text{and} \qquad R(t) = R_o + \frac{N_1(0)}{N_2'(0)}[1 - e^{-\lambda t}].$$

Substitution for $N_1(0)/N_2(0)$ from the first equation into the second equation gives

$$R(t) = R_o + \frac{N_1(t)}{N_2'(t)}e^{\lambda t}[1 - e^{-\lambda t}] = R_o + \frac{N_1(t)}{N_2'(t)}[e^{\lambda t} - 1],$$

from which the sample age is found, namely,

$$t = \frac{1}{\lambda} \ln\left\{1 + \frac{N_2'(t)}{N_1(t)}[R(t) - R_o]\right\}. \qquad (5.78)$$

Thus, the age of the sample can be determined by three atom ratios (easier to obtain than absolute atom densities): $N_2'(t)/N_1(t)$ and $N_2(t)/N_2'(t)$ (obtained from measurements) and the primordial relative isotopic ratio $R_o = N_2(0)/N_2'(0)$.

Example 5.8: Some geological samples contain the long-lived radionuclide ^{87}Rb (half-life 4.88×10^{10} y) which decays to stable ^{87}Sr. Strontium has another stable isotope ^{86}Sr. In samples without ^{87}Rb, the normal atomic ^{87}Sr to ^{86}Sr ratio is $R_o = 7.00/9.86 = 0.710$ (see data in Table A.4). What is the age of a rock that has an atomic ^{87}Sr to ^{86}Sr ratio $R(t) = 0.80$, and an atomic ^{87}Rb to ^{86}Sr ratio of 1.48? The age of this rock is estimated from Eq. (5.78) as

$$t = \frac{1}{\ln 2/T_{1/2}} \ln\left\{ 1 + \frac{N_{86\text{Sr}}(t)}{N_{97\text{Rb}}(t)}[R(t) - R_o] \right\}$$

$$= \frac{1}{\ln 2/4.88 \times 10^{10} \text{ y}} \ln\left\{ 1 + \frac{1}{1.48}[0.80 - 0.710] \right\} = 4.16 \times 10^9 \text{ y}.$$

5.9   Radioactive Decay Data

The *Chart of the Nuclides* conveniently provides the half-life, decay mode, and the energy of the principal secondary radiation for every known radionuclide. Table A.4 in Appendix A lists the half-life, decay modes with their branching ratios of all radionuclides with half-lifes greater than 1 hour.

Often, however, more detailed information is needed. In Appendix D, a compilation is given of the energies and frequencies of the principal particles and photons emitted by selected radionuclides which are of most concern for reactor operation and environmental radiological assessments. Also given is the total energy emitted per decay through particles and photons, data useful, for example, in determining the thermal power generated in a sample of a particular radionuclide. Appendix D also gives references where more extensive compilations can be found.

BIBLIOGRAPHY

BALTER, M., "Radiocarbon Dating's Final Frontier," *Science* **313**: 1560-1563 (2006).

EISENBUD, M. AND T.F. GESELL, *Environmental Radioactivity*, 4th ed., Academic Press, New York, 1987.

EVANS, R.D., *The Atomic Nucleus*, McGraw-Hill, New York, 1955; republished by Krieger Publishing Co., Melbourne, FL, 1982.

FAW, R.E., AND J.K. SHULTIS, *Radiological Assessment: Sources and Doses*, American Nuclear Society, La Grange Park, IL, 1999.

FIRESTONE, R.B. AND V.S. SHIRLEY, Eds., *Table of Isotopes*, Vols. 1 and 2, 8th ed., Wiley, New York, 1996.

GE-NE (GENERAL ELECTRIC NUCLEAR ENERGY), *Nuclides and Isotopes: Chart of the Nuclides*, 15th ed., prepared by J.R. Parrington, H.D. Knox, S.L. Breneman, E.M. Baum, and F. Feiner, GE Nuclear Energy, San Jose, CA, 1996.

ICRP, *Radionuclide Transformations*, Publication 38, International Commission on Radiological Protection, *Annals of the ICRP* **11-13**: 1983.

KAPLAN, I., *Nuclear Physics*, Addison-Wesley, Reading, MA, 1963.

L'ANNUNZIATA, M.F. (Ed.), *Handbook of Radioactive Analysis*, Academic Press, New York, 1998.

MAYO, R.M., *Nuclear Concepts for Engineers*, American Nuclear Society, La Grange Park, IL, 1998.

ROMER, A. (Ed.), *Radioactivity and the Discovery of Isotopes*, Dover Publ., New York, 1970.

SHULTIS, J.K. AND R.E. FAW, *Radiation Shielding*, American Nuclear Society, La Grange Park, IL, 2000.

WEBER, D.A., K.F. ECKERMAN, L.T. DILLMAN, AND J.C. RYMAN, *MIRD: Radionuclide Data and Decay Schemes*, Society of Nuclear Medicine, New York, 1989.

PROBLEMS

1. Consider a stationary nucleus of mass m_n in an excited state with energy E^* above the ground state. When this nucleus decays to the ground state by gamma decay, the emitted photon has an energy E_γ. (a) By considering the conservation of both energy and momentum of the decay reaction explain why $E_\gamma < E^*$. (b) Show that the two energies are related by

$$E_\gamma = m_n c^2 \left\{ \sqrt{1 + \frac{2E^*}{m_n c^2}} - 1 \right\} \simeq E^* \left[1 - \frac{E^*}{2 m_n c^2} \right].$$

(c) Use an explicit example to verify that the difference between E^* and E_γ is for all practical purposes negligible.

2. The radioisotope ^{224}Ra decays by α emission primarily to the ground state of ^{220}Rn (94% probability) and to the first excited state 0.241 MeV above the ground state (5.5% probability). What are the energies of the two associated α particles?

3. The radionuclide ^{41}Ar decays by β^- emission to an excited level of ^{41}K that is 1.293 MeV above the ground state. What is the maximum kinetic energy of the emitted β^- particle?

4. As shown in Fig. 5.6, ^{64}Cu decays by several mechanisms. (a) List the energy and frequency (number per decay) of all photons emitted from a sample of material containing ^{64}Cu. (b) List the maximum energy and frequency of all electrons and positrons emitted by this sample.

5. What is the value of the decay constant and the mean lifetime of ^{40}K (half-life 1.29 Gy)?

6. From the energy level diagram of Fig. 5.12, what are the decay constants for electron capture and positron decay of ^{22}Na? What is the total decay constant?

7. The activity of a radioisotope is found to decrease by 30% in one week. What are the values of its (a) decay constant, (b) half-life, and (c) mean life?

8. The isotope ^{132}I decays by β^- emission to ^{132}Xe with a half-life of 2.3 h. (a) How long will it take for 7/8 of the original number of ^{132}I nuclides to decay? (b) How long will it take for a sample of ^{132}I to lose 95% of its activity?

9. How many grams of ^{32}P are there in a 5 mCi source?

10. How many atoms are there in a 1.20 MBq source of (a) ^{24}Na and (b) ^{238}U?

11. A very old specimen of wood contained 10^{12} atoms of ^{14}C in 1986. (a) How many ^{14}C atoms did it contain in 9474 B.C.? (b) How many ^{14}C atoms did it contain in 1986 B.C.?

12. A 6.2 mg sample of ^{90}Sr (half-life 29.12 y) is in secular equilibrium with its daughter ^{90}Y (half-life 64.0 h). (a) How many Bq of ^{90}Sr are present? (b) How many Bq of ^{90}Y are present? (c) What is the mass of ^{90}Y present? (d) What will the activity of ^{90}Y be after 100 y?

13. A sample contains 1.0 GBq of ^{90}Sr and 0.62 GBq of ^{90}Y. What will be the activity of each nuclide (a) 10 days later and (b) 29.12 years later?

14. Consider the following β^- decay chain with the half-lives indicated,

$$^{210}\text{Pb} \xrightarrow[22\ y]{} {}^{210}\text{Bi} \xrightarrow[5.0\ d]{} {}^{210}\text{Po}.$$

A sample contains 30 MBq of ^{210}Pb and 15 MBq of ^{210}Bi at time zero. (a) Calculate the activity of ^{210}Bi at time $t = 10$ d. (b) If the sample were originally pure ^{210}Pb, how old would it have been at time $t = 0$?

15. A 40-mg sample of pure ^{226}Ra is encapsulated. (a) How long will it take for the activity of ^{222}Rn to build up to 10 mCi? (b) What will be the activity of ^{222}Rn after 2 years? (c) What will be the activity of ^{222}Rn after 1000 y?

16. Tritium (symbol T) is produced in the upper atmosphere by neutron cosmic rays primarily through the ^{14}N(n,T)^{12}C and ^{16}O(n,T)^{14}N reactions. Tritium has a half-life of 12.3 years and decays by emitting a β^- particle with a maximum energy of 18.6 keV. Tritium exists in nature almost exclusively as HTO, and in the continental surface waters and the human body, it has an atomic T:H ratio of 3.3×10^{-18}. Since the human body is about 10% hydrogen by weight, estimate the tritium activity (Bq) in a 100-kg person.

17. The global inventory of ^{14}C is about 8.5 EBq. If all that inventory is a result of cosmic ray interactions in the atmosphere, how many kilograms of ^{14}C are produced each year in the atmosphere?

18. The average mass of potassium in the human body is about 140 g. From the abundance and half-life of ^{40}K (see Table 5.2), estimate the average activity (Bq) of ^{40}K in the body.

19. The naturally occurring radionuclides ^{222}Rn and ^{220}Rn decay by emitting α particles which have a range of only a few centimeters in air. Yet there is considerable concern about the presence of these radioisotopes in interior spaces. Explain how these radioisotopes enter homes and workspaces and why they are considered hazardous to the residents.

20. Charcoal found in a deep layer of sediment in a cave is found to have an atomic ^{14}C/^{12}C ratio only 30% that of a charcoal sample from a higher level with a known age of 1850 y. What is the (radiocarbon) age of the deeper layer?

21. ^{238}Pu, an alpha-particle emitter, has been used as a thermal power source, as described in Table 12.2. What is the energy recoverable as heat per decay?

22. ^{90}Sr, in secular equilibrium with its daughter ^{90}Y, both beta-particle emitters, has been used as a thermal power source, as described in Table 12.2. What is the energy recoverable as heat per decay of ^{90}Sr?

23. The text discusses ^{14}C dating of biogenic materials, useful for ages up to about 50,000 years. Another dating method, useful for dating rock of volcanic origin of ages 100,000 to 4 billion years, is the ^{40}K–^{40}Ar method. The former nuclide decays to the latter in 10.7% of its decays. Argon is expelled from molten rock but, after solidification, remains in the rock as a stable nuclide along with the slowly decaying ^{40}K. In a rock specimen under analysis, the rock is melted and the ^{40}Ar is collected and measured using mass spectrometry. The ^{40}K in the rock is measured using flame photometry or atomic absorption spectrometry. What is the atomic ratio of ^{40}Ar:^{40}K at rock ages of 10^5, 10^7, and 10^9 years.

Chapter 6

Binary Nuclear Reactions

By far the most important type of nuclear reaction is that in which two nuclei interact and produce one or more products. Such reactions involving two reactants are termed *binary* reactions. All naturally occurring nuclides with more than a few protons were created by binary nuclear reactions inside stars. Such nuclides are essential for life, and the energy that nurtures it likewise derives from energy released by binary nuclear reactions in stars.

Binary nuclear reactions are shown schematically as

$$A + B \longrightarrow C + D + E + \cdots$$

In all such reactions involving only the nuclear force, several quantities are conserved: (1) total energy (rest mass energy plus kinetic and potential energies), (2) linear momentum, (3) angular momentum (or *spin*), (4) charge, (5) number of protons,[1] (6) number of neutrons,[1] and several other quantities (such as the parity of the quantum mechanical wave function, which we need not consider).

Many binary nuclear reactions are caused by a nucleon or light nuclide x striking a heavier nucleus X at rest in our frame of reference. Although several products may result, very often only two products y (the lighter of the two) and Y are produced. Such binary, two-product reactions are written as

$$x + X \longrightarrow y + Y \qquad \text{or, more compactly, as} \qquad X(x, y)Y. \qquad (6.1)$$

In this chapter, the consequences of conservation of energy and momentum in such two-product binary reactions are examined. The kinetic energies of the products and their directions of travel with respect to the incident particle are of particular interest. Such considerations are usually referred to as the *kinematics* of the reaction.

---

[1]The number of protons and neutrons is *not* necessarily conserved in reactions involving the so-called *weak force*, which is responsible for beta and electron-capture radioactive decays.

6.1 Types of Binary Reactions

For a given pair of reactants, many possible sets of products are possible. The probability of obtaining a specific reaction outcome generally depends strongly on the kinetic energy of the incident reactant and the mechanism by which the products are produced.

Reactions Mechanisms

Nucleons or light nuclei, as projectiles with kinetic energies greater than about 40 MeV, have de Broglie wavelengths that are comparable to the size of the nucleons in a target nucleus. Such projectiles, consequently, are likely to interact with one or, at most, a few neighboring nucleons in the target nucleus. The remainder of the nucleons in the target play little part in the interaction. These *direct interactions* usually occur near the surface of the target nucleus and are, thus, sometimes called *peripheral processes*.

Incident projectiles with less than a few MeV of kinetic energy have de Broglie wavelengths much larger than the nucleons in the target nucleus. Consequently, the projectile is more likely to interact with the nucleus as whole, i.e., with all the nucleons simultaneously, forming initially a highly excited *compound nucleus*. This compound nucleus decays within about 10^{-14} s of its formation into one of several different sets of possible reaction products. The concept of the compound nucleus, introduced by Bohr in 1936, is discussed in detail below.

6.1.1 The Compound Nucleus

As explained above, incident particles with kinetic energies less than a few MeV are more likely to interact with the whole target nucleus than with any individual constituent nucleon. Upon absorption, the incident particle's kinetic energy is absorbed and transferred to the nucleus as a whole, creating a single excited nucleus called a *compound nucleus*. This excited nucleus typically has a lifetime much greater than the *nuclear lifetime*, the time it would take the incident particle to pass through the target nucleus (between 10^{-21} and 10^{-17} s). Within 10^{-15} to 10^{-13} seconds, the compound nucleus decays through the emission of one or more particles. The excited states of the compound nucleus are called *virtual states* or *virtual levels* since they decay by particle emission. This is to distinguish them from *bound states* of excited nuclei which decay only by γ-ray emission.

Reactions involving a compound nucleus are, therefore, two-step processes. First a relatively long-lived excited compound nucleus is formed, which subsequently decays into two or more products. We can visualize such a reaction as

$$x + X \longrightarrow (x + X)^* \longrightarrow y + Y. \tag{6.2}$$

Because the lifetime of the compound nucleus is much longer than the nuclear lifetime, the compound nucleus "forgets" how it was formed or the direction of the incident particle. Upon disintegration of the compound nucleus, the reaction products are emitted, with equal probability, in all directions (from the point of view of the compound nucleus). Moreover, the modes of decay of the compound nucleus are independent of how it was formed. Indeed, this has been well confirmed experimentally. For example, the compound nucleus ^{64}Zn* can be formed in two ways and can decay in the following three ways [Mayo 1998]:

$$\left.\begin{array}{c} p + {}^{63}\text{Cu} \\ \alpha + {}^{60}\text{Ni} \end{array}\right\} \longrightarrow {}^{64}\text{Zn}^* \longrightarrow \left\{\begin{array}{l} {}^{63}\text{Zn} + n \\ {}^{62}\text{Cu} + n + p \\ {}^{62}\text{Zn} + 2n \end{array}\right. .$$

The different ways the compound nucleus disintegrates into different reaction products are called *exit channels*. The probability of observing a particular exit channel is independent of how the compound nucleus was formed.

Reaction Nomenclature

There is some standard nomenclature used to classify different types of binary reactions based on the identity of the projectile and light product. Here we summarize the most frequently encountered types of reactions.

transfer reactions: These are direct reactions in which one or two nucleons are transferred between the projectile and light product. Examples are (α, d) or (d, n) reactions.

scattering reactions: When the projectile and light product are the same, the projectile is said to have scattered. If the target nucleus is left in the ground state, the scattering is *elastic*, whereas, if some of the incident kinetic energy is used to leave the target nucleus in an excited state, the scattering is called *inelastic*. In inelastic scattering, the excited heavy product usually decays rapidly with the emission of a gamma photon. Elastic scattering reactions are denoted by (x, x) and inelastic scattering reactions by (x, x').

knockout reactions: In certain direct interactions, the initial projectile is emitted accompanied by one or more nucleons from the target nucleus. Examples are $(n, 2n)$, (n, np), and $(n, 3n)$ reactions.

capture reactions: Sometimes the incident projectile is absorbed by the target leaving only a product nucleus. These product nuclei are usually left in an excited state and rapidly decay by emitting one or more gamma photons. The ground-state product nucleus may or may not be radioactive. The (n, γ) reaction is one of the most important capture reactions, since this reaction allows us to produce radioisotopes from stable target nuclei by bombarding them with neutrons inside a nuclear reactor.

nuclear photoeffect: If the incident projectile is a gamma photon, it can liberate a nucleon from the target. For example, (γ, n) reactions are a means of producing neutrons for laboratory use.

6.2   Kinematics of Binary Two-Product Nuclear Reactions

A very common type of nuclear reaction is one in which two nuclei react to form two product nuclei. Such a reaction may be written as

$${}^{A_1}_{Z_1}X_1 + {}^{A_2}_{Z_2}X_2 \longrightarrow {}^{A_3}_{Z_3}X_3 + {}^{A_4}_{Z_4}X_4, \quad \text{or, more compactly, as} \quad x + X \longrightarrow Y + y. \quad (6.3)$$

Conservation of the number of protons and neutrons requires that $Z_1 + Z_2 = Z_3 + Z_4$ and $A_1 + A_2 = A_3 + A_4$.

In these two-product reactions, the constraints of energy and momentum conservation require that the two products divide between them, in a unique manner, the initial kinetic energy and the reaction energy from any mass changes.

6.2.1 Energy/Mass Conservation

The energy released in the binary reaction of Eq. (6.3) is the Q-value of the reaction. This energy is manifested as a net gain in kinetic energy of the products, or, equivalently, a net decrease in the rest masses, i.e.,

$$Q = E_y + E_Y - E_x - E_X = (m_x + m_X - m_y - m_Y)c^2. \tag{6.4}$$

Because the number of protons is usually conserved in this binary reaction,[2] the nuclear masses m_i in this expression can be replaced by the corresponding atomic masses M_i of the particles. The electron masses on both sides of the reaction cancel, and the small difference in electron binding energies is negligible. Thus, the Q-value may also be written

$$Q = (M_x + M_X - M_y - M_Y)c^2. \tag{6.5}$$

In many binary nuclear reactions, the target nucleus is stationary in the laboratory coordinate system, i.e., $E_X = 0$. If the target nucleus is not at rest, we can always perform our analyses in a coordinate system moving with the target since the laws of physics are the same in all inertial frames of reference. Thus, with $E_X = 0$, the Q-value is

$$Q = E_y + E_Y - E_x = (m_x + m_X - m_y - m_Y)c^2 = (M_x + M_X - M_y - M_Y)c^2. \tag{6.6}$$

If $Q > 0$, the reaction is exoergic (exothermic) and there is a net increase in the kinetic energy of the products accompanied by a net decrease in the rest mass of the products. By contrast, if $Q < 0$, the reaction is endoergic (endothermic), and energy is absorbed to increase the rest mass of the products. The special case of $Q = 0$ occurs only for elastic scattering, i.e., when $m_x = m_y$ and $m_X = m_Y$. In inelastic scattering, $m_x = m_y$ but $m_Y > m_X$, since Y is an excited configuration of the target nucleus X, and hence $Q < 0$.

If one of the reactants or products is a photon, its rest mass is zero and its kinetic energy is $E = h\nu = pc$, where p is the photon's momentum.

6.2.2 Conservation of Energy and Linear Momentum

Consider the binary reaction shown in Fig. 6.1 in which the target nucleus is at rest. The products y and Y move away from the collision site at angles θ_y and θ_Y, respectively, with respect to the incident direction of x.

From the constraints imposed by conservation of total energy and linear momentum, the energy and direction of a product nuclei can be determined in terms of those of the reactants. If E_i denotes the kinetic energy of the ith nucleus and Q is the Q-value of the reaction, conservation of energy requires (with $E_X = 0$)

$$E_x = E_y + E_Y - Q. \tag{6.7}$$

---

[2]The exceptions being electron capture and similar reactions involving the weak force.

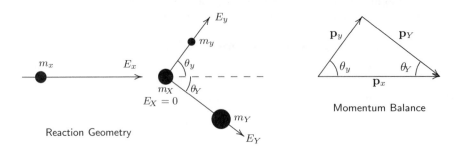

Figure 6.1. The geometry for a binary two-product nuclear reaction in the laboratory frame of reference in which the target nucleus is stationary (left-hand figure). Conservation of linear momentum requires that the momentum vectors of the three moving particles be closed (right-hand figure).

Conservation of momentum requires $\mathbf{p}_x = \mathbf{p}_y + \mathbf{p}_Y$ where $\mathbf{p}_i$ is the momentum vector for the ith nucleus. For linear momentum to be conserved, these three vectors must lie in the same plane, and the vector triangle formed by them (see Fig. 6.1) must be closed. Thus, by the law of cosines,

$$p_Y^2 = p_x^2 + p_y^2 - 2p_x p_y \cos\theta_y. \tag{6.8}$$

For non-relativistic particles,[3] $p_i = \sqrt{2m_i E_i}$. Then from Eq. (6.8)

$$2m_Y E_Y = 2m_x E_x + 2m_y E_y - 2\sqrt{4m_x m_y E_x E_y}\cos\theta_y. \tag{6.9}$$

Substitution from Eq. (6.7) for E_Y and rearrangement yields

$$(m_Y + m_y)E_y - 2\sqrt{m_x m_y E_x}\,\omega_y \sqrt{E_y} + E_x(m_x - m_Y) - m_Y Q = 0 \tag{6.10}$$

where $\omega_y \equiv \cos\theta_y$. This quadratic equation for $\sqrt{E_y}$ has the following general solution:

$$\sqrt{E_y} = \sqrt{\frac{m_x m_y E_x}{(m_y + m_Y)^2}}\,\omega_y$$
$$\pm \sqrt{\frac{m_x m_y E_x}{(m_y + m_Y)^2}\,\omega_y^2 + \left[\frac{m_Y - m_x}{(m_y + m_Y)}E_x + \frac{m_Y Q}{(m_y + m_Y)}\right]}. \tag{6.11}$$

This solution has the form $\sqrt{E_y} = a \pm \sqrt{a^2 + b}$, and, depending on the magnitude and sign of $a^2 + b$, there can be zero, one, or even two acceptable values for $\sqrt{E_y}$.

---

[3]For the more general relativistic treatment, see Shultis and Faw [2000].

For a physically realistic solution, $\sqrt{E_y}$ must be real and positive. If the right-hand side of Eq. (6.11) is negative or complex ($a^2 + b < 0$), then the solution is not physically possible and the reaction cannot occur. The physical factors which can contribute toward making a particular emission angle θ_y for m_y energetically impossible are (1) a negative Q-value, (2) a heavy projectile so that $m_Y - m_x < 0$, and (3) a large scattering angle θ_y so that $\omega_y \equiv \cos\theta_y < 0$.

Exoergic Reactions ($Q > 0$)

For $Q > 0$ and with $m_Y > m_x$, the term in square brackets in Eq. (6.11) is always positive and only the positive sign of the $\pm$ pair is meaningful. Thus, $E_y = (a + \sqrt{a^2 + b})^2$ is the only solution.

As the bombarding energy E_x approaches zero we see $\sqrt{E_y} \to \sqrt{b}$ or

$$E_y \longrightarrow \frac{m_Y}{m_y + m_Y} Q, \qquad Q > 0. \tag{6.12}$$

In this case of zero linear momentum, the Q-value is simply the sum of the kinetic energies of the products, which must be apportioned according the mass fractions of the products to preserve zero linear momentum. Notice, that the kinetic energy E_y is the same for all angles θ_y since conservation of energy and momentum require

$$Q = E_y + E_Y \qquad \text{and} \qquad \theta_y + \theta_Y = \pi.$$

Endoergic Reactions ($Q < 0$)

For $Q < 0$ and with $m_Y > m_x$, the argument of the square-root term in Eq. (6.11) is negative if E_x is too small, and the reaction is thus not possible. However as E_x increases, $a^2 + b$ becomes positive and two positive values of $\sqrt{E_y}$ can be obtained for forward angles $\omega_y > 0$. These *double values* of E_y correspond to two groups of particles y that are emitted in the forward and backward directions from the perspective of the compound nucleus[4] in the center-of-mass coordinate system (i.e., that of the compound nucleus' perspective). The backward directed group is, however, carried forward in the laboratory coordinate system whenever the velocity of the center of mass, V_c, is greater than that of particle y in that system (see Fig. 6.2). However, as E_x increases further only the $E_y = (a + \sqrt{a^2 + b})^2$ becomes physically, admissible.

For endoergic reactions, we thus expect the incident projectile to have to supply a minimum amount of kinetic energy before the reaction can occur, i.e., before physically meaningful values of E_y can be obtained from Eq. (6.11). This reaction threshold behavior is discussed next.

---

[4]This coordinate system, in which the compound nucleus is at rest and in which there is no net linear momentum before or after the reaction, is often referred to as the *center-of-mass* coordinate system. Kinematic relations similar those of Section 6.2.2 can be derived in this alternative coordinate system and are sometimes used since, from the point of view of the compound nucleus, the reaction products are emitted (nearly) isotropically, because during the relatively long lifetime of the compound nucleus, all information about the initial direction of the reactants is lost. The velocity vectors **v** in the laboratory coordinate system (in which the target is stationary and the compound nucleus moves) can be obtained by adding to the corresponding velocity vectors **v**′ in the center-of-mass coordinate system the velocity vector of the center of mass, $\mathbf{V}_c = m_x v_x / (m_x + m_X)$.

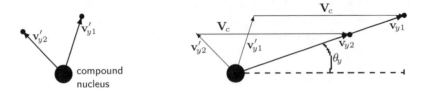

Figure 6.2. In the left figure, two particle y are emitted with the same speed v_y' from the compound nucleus with respect to the compound nucleus (which emits particles equally in all directions). By adding the velocity vector of the compound nucleus, V_c, the speeds are transformed in the right-hand figure to the laboratory coordinate system (target at rest). If the speed of the compound nucleus is greater than v_y', both example particles will appear to be moving along the same direction θ_y, but with different speeds in the laboratory coordinate system.

6.3   Reaction Threshold Energy

For reactions with $Q < 0$ (or even for reactions with $Q > 0$ when $m_Y < m_x$), the incident particle must supply a certain minimum amount of energy E_x before the reaction can occur. This minimum incident energy is termed the reaction *threshold energy*. Two types of reaction threshold energies are encountered. These are discussed below.

6.3.1   Kinematic Threshold

For endoergic reactions, the reaction is possible only if E_x is greater than some threshold energy, above which the argument of the square-root term in Eq. (6.11) is positive. For the argument of the square-root term in Eq. (6.11) to be non-negative, it is necessary that

$$
\begin{aligned}
(E_x)_{\theta_y} &\geq -\frac{m_Y(m_y + m_Y)Q}{(m_y + m_Y)(m_Y - m_x) + m_x m_y \cos^2 \theta_y} \\
&= -\frac{(m_y + m_Y)Q}{m_Y + m_y - m_x - (m_x m_y / m_Y)\sin^2 \theta_y}.
\end{aligned}
\tag{6.13}
$$

For this expression to have the smallest value, the denominator must be as large as possible, i.e., θ_y must go to zero. Thus, the *kinematic* reaction threshold energy is

$$
E_x^{th} = -\frac{m_y + m_Y}{m_y + m_Y - m_x}Q.
\tag{6.14}
$$

Since $m_i \gg Q/c^2$ for most nuclear reactions, $m_y + m_Y \simeq m_x + m_X$ and $m_y + m_Y - m_x \simeq m_X$ so that Eq. (6.14) simplifies to

$$
\boxed{E_x^{th} \simeq -\left(1 + \frac{m_x}{m_X}\right)Q.}
\tag{6.15}
$$

At the threshold for endoergic reactions, the product particle y is emitted from the compound nucleus with negligible speed (see Fig. 6.2). However, in the laboratory system, y has the speed of the compound nucleus and is moving with $\theta_y = 0$. At bombarding energies E_x just slightly greater than E_x^{th}, the product nucleus m_y will appear at a forward angle θ_y with two distinct energies E_y given by Eq. (6.11) with the $\pm$ pair. However, for larger values of incident energy E_x, the square-root term in Eq. (6.11) exceeds the $\sqrt{m_x m_y E_x} \omega_y$ term (i.e., $\sqrt{a^2 + b} > a$), and only the $+$ sign gives a physically realistic value for $\sqrt{E_y}$.

6.3.2 Coulomb Barrier Threshold

If the incident projectile x is a neutron or gamma photon, it can reach the nucleus of the target without hindrance. The minimum energy needed to initiate a given nuclear reaction is, thus, that specified by the threshold energy of the previous section. For exoergic reaction ($Q > 0$), neutrons or photons need negligible incident kinetic energy to cause the reaction. For endoergic reactions ($Q < 0$), the minimum energy of the incident neutron or photon, needed to conserve both energy and linear momentum, is given by Eq. (6.15).

However, if the incident projectile is a nucleus, it necessarily has a positive charge. As it approaches the target nucleus, Coulombic nuclear forces repel the projectile, and, if the projectile does not have sufficient momentum, it cannot reach the target nucleus. Only if the projectile reaches the surface of the target nucleus can nuclear forces cause the two particles to interact and produce a nuclear reaction. Thus, even for a reaction with a positive Q-value, if a positively charged incident projectile does not have enough kinetic energy to reach the target nucleus, the reaction cannot occur even though energy and momentum constraints for the reaction can be satisfied. No matter what the magnitude or sign of the Q-value, a charged incident projectile must come into close proximity of the target nucleus for the nuclear forces to interact and cause the nuclear reaction.

The Coulombic repulsive forces between the incident projectile (charge $Z_x e$) and the target nucleus (charge $Z_X e$), when separated by a distance r, is

$$F_C = \frac{Z_x Z_X e^2}{4\pi\epsilon_o r^2} \tag{6.16}$$

where ϵ_o is the permittivity of free space. Consider an incident particle (charge Z_x) starting infinitely far from the target nucleus and moving directly towards it. The work done by the bombarding particle against the electric field of the target nucleus, by the time it is a distance b from the target, is

$$W_C = -\int_\infty^b \mathbf{F_c} \cdot d\mathbf{r} = -\frac{Z_x Z_X e^2}{4\pi\epsilon_o} \int_\infty^b \frac{dr}{r^2} = \frac{Z_x Z_X e^2}{4\pi\epsilon_o b}. \tag{6.17}$$

This work comes at the expense of a reduction in the kinetic energy of the incident particle by an amount W_C. The incident projectile must possess kinetic energy equal to W_C to get its center within a distance b of the center of the target nucleus. For a nuclear reaction to occur, b must be such that the surfaces of the projectile and target nucleus come into close proximity.

Nuclear surfaces are not sharply defined; and quantum mechanics allows particles that do not "touch" to still interact (the "tunneling effect") [Eisberg and Resnick 1985]. Moreover, the nuclear and electrostatic forces overlap. However, we may assume, to a first approximation, that $b = R_x + R_X$, where R_x and R_X are the radii of the incident projectile and the target nucleus, respectively. From Eq. (1.7) we obtain

$$b = R_x + R_X = R_o(A_x^{1/3} + A_X^{1/3}). \qquad (6.18)$$

Thus, for an incident particle with a positively charged nucleus to interact with the target nucleus, it must have a kinetic energy to penetrate the Coulomb barrier of about (after substituting appropriate values for the constants) [Mayo 1998]

$$E_x^C \simeq W_c = 1.2\frac{Z_x Z_X}{A_x^{1/3} + A_X^{1/3}} \quad \text{(MeV)}. \qquad (6.19)$$

This threshold energy required to penetrate the Coulombic barrier is not lost from the reaction. As the incident particle works against the Coulomb field of the target nucleus, the target nucleus recoils and gains the kinetic energy lost by the incident projectile. The momentum of the initial projectile must equal that of the temporary compound nucleus, i.e., $m_x v_x = M_{cn} V_{cn}$, where M_{cn} and V_{cn} and the mass and speed of the compound nucleus. Hence, the kinetic energy of the compound nucleus is $E_{cn} = E_x^C(m_x/M_{cn})$. The remainder of E_x^C is used to excite the compound nucleus. Upon the decay of the compound nucleus into the reaction products, the entire E_x^C is recovered in the mass and kinetic energy of the reaction products.

6.3.3   Overall Threshold Energy

For a chargeless incident particle (e.g., neutron or photon), no Coulomb barrier has to be overcome before the reaction takes place. For exoergic reactions ($Q > 0$), there is no threshold. For endoergic reactions $Q < 0$, the only threshold is the kinematic threshold of Eq. (6.15).

However, for incident particles with a charged nucleus, the Coulomb barrier always has to be surmounted. For exoergic reactions ($Q > 0$), there is no kinematic threshold, only the Coulomb barrier threshold E_x^C, given by Eq. (6.19). For endoergic reactions ($Q < 0$), there are both kinematic and Coulomb thresholds. For this case, the minimum energy $(E_x)_{\min}$ needed for the reaction to occur is

$$(E_x^{th})_{\min} = \max(E_x^C, E_x^{th}). \qquad (6.20)$$

Example 6.1:   The compound nucleus ^{15}N* is produced by many nuclear reactions including ^{13}C(d,t)^{12}C, ^{14}C(p,n)^{14}N, and ^{14}N(n,α)^{11}B. What is the minimum kinetic energy of the incident deuteron, proton, or neutron for each of these reactions to occur?

The first step is to calculate the Q-value for each reaction as discussed in Section 4.5.1. The kinematic threshold E_x^{th} and the Coulombic threshold E_x^C are found by the methods described in this section. The minimum kinetic energy of the products is $\min(E_y + E_Y) = Q + (E_x^{th})_{\min}$. The calculations are summarized below.

Reaction Path	Q-value (MeV)	E_x^C (MeV)	E_x^{th} (MeV)	Reaction Condition	$\min(E_y + E_Y)$ (MeV)
$^{13}\text{C}(d,t)^{12}\text{C}$	1.311	1.994	0	$E_x > E_x^C$	3.305
$^{14}\text{C}(p,n)^{14}\text{N}$	-0.6259	2.111	0.6706	$E_x > E_x^C$	1.485
$^{14}\text{N}(n,\alpha)^{11}\text{B}$	-0.1582	0	0.1695	$E_x > E_x^{th}$	0.0113

6.4 Applications of Binary Kinematics

The results presented so far in this chapter apply to any binary, two-product nuclear reaction. We now apply these general results to three example binary reactions.

6.4.1 A Neutron Detection Reaction

Most radiation is detected by measuring the ionization it creates as it passes through some material. However, neutrons without any charge cannot directly ionize matter as they pass through it. Neutrons must first cause a nuclear reaction whose charged products are then detected from the ionization the products create. One such reaction, widely used to detect neutrons, is the $^3\text{He}(n,p)^3\text{H}$ reaction. This is an exoergic reaction with $Q = 0.764$ MeV. Because the neutron can approach the nucleus of ^3He without Coulombic repulsion, there is no threshold energy. For incident neutrons with negligible kinetic energy, the resulting proton has kinetic energy $E_p = 0.573$ MeV (from Eq. (6.12)) and the recoil tritium nucleus 0.191 MeV

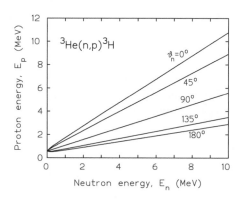

Figure 6.3. Proton energy versus incident neutron energy for the $^3\text{He}(n,p)^3\text{H}$ reaction.

$(Q - E_p)$. As the energy of the incident neutron increases, the product proton energy E_p also increases in accordance with Eq. (6.11). The variation of E_p with the incident neutron energy E_n for several emission angles is shown in Fig. 6.3. No double-energy region or threshold energy exists for this reaction. By measuring the proton's kinetic energy at a specific scattering angle, the kinetic energy of the incident neutron can be inferred. In the limit of very high incident neutron energy ($E_n >> Q$), E_p is directly proportional to E_n, as can be seen from the figure and from Eq. (6.11).

6.4.2 A Neutron Production Reaction

Neutrons can be produced by many reactions. One such reaction is the $^7\text{Li}(p,n)^7\text{Be}$ reaction. This is an endoergic reaction with $Q = -1.644$ MeV, and hence has a small double-energy region slightly above the threshold energy for E_p. From Eqs. (6.15) and (6.19), the kinematic and Coulombic-barrier threshold energies are, respectively, 1.875 and 1.236 MeV. The reaction threshold is, thus, determined by the kinematic constraint. The neutron energy versus the incident proton energy is shown in Figs. 6.4 and 6.5.

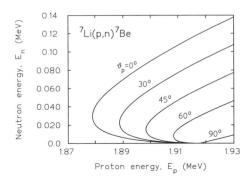

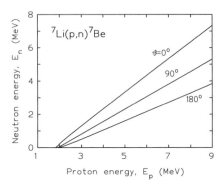

Figure 6.4. Neutron energy versus incident proton energy for the ^{7}Li$(p, n)^7$Be reaction in the double energy region.

Figure 6.5. Neutron energy versus incident proton energy for the ^{7}Li$(p, n)^7$Be reaction above the double energy region.

6.4.3   Heavy Particle Scattering from an Electron

As heavy charged particles such as alpha particles pass through matter, they interact through the Coulombic force, mostly with the electrons of the medium since they occupy most of the matter's volume. For heavy charged particles with MeV kinetic energies, the much smaller binding energy of an electron to the nucleus is negligible. Thus, the electrons, with which an incident alpha particle interacts, can be considered as "free" electrons at rest.

To analyze this scattering reaction, identify particles X and y in Eq. (6.1) as the electron, so $m_X = m_y = m_e$, $E_y = E_e$ (the recoil electron energy), $m_x = m_Y = M$ (the mass of the heavy particle), and $E_x = E_M$ (the kinetic energy of the incident heavy particle). For this scattering process, there is no change in the rest masses of the reactants, i.e., $Q = 0$. With the substitution of these variables into Eq. (6.11), we obtain

$$\sqrt{E_e} = \frac{2}{M + m_e}\sqrt{Mm_eE_M}\cos\theta_e.$$

Squaring this result and using $M >> m_e$, we find that the recoil energy of the electron is

$$E_e = 4\frac{m_e}{M}E_M\cos^2\theta_e. \tag{6.21}$$

The maximum electron recoil energy and the maximum kinetic energy loss by the incident heavy particle, occurs for $\cos^2\theta_e = 1$ (i.e., for $\theta_e = 0$). Thus, the maximum energy of the recoil electron is

$$(E_e)_{\max} = 4\frac{m_e}{M}E_M. \tag{6.22}$$

Example 6.2: What is the maximum energy loss $(\Delta E_\alpha)_{\text{max}}$ for a 4-MeV alpha particle scattering from an electron? From Eq. (6.22), we have

$$(\Delta E_\alpha)_{\text{max}} = (E_e)_{\text{max}} = 4\frac{m_e}{M}E_M = 4\frac{0.0005486 \text{ u}}{4.003 \text{ u}}(4 \text{ MeV}) = 2.20 \text{ keV}.$$

This is an energy sufficient to free most electrons from their atoms and create an ion-electron pair. Most collisions transfer less energy from the alpha particle, and, hence, we see that tens of thousands of ionization and excitation interactions are required for an alpha particle (or any other heavy charged particle with several MeV of kinetic energy) to slow down and become part of the ambient medium.

6.5 Reactions Involving Neutrons

Of special interest to nuclear engineers are nuclear reactions caused by incident neutrons or ones that produce neutrons. In this section, several of the most important neutron reactions are described.

6.5.1 Neutron Scattering

When a neutron scatters from a nucleus, the nucleus is either left in the ground state (*elastic scattering*) or in an excited state (*inelastic scattering*). Such scattering is shown schematically below.

$$n + X \longrightarrow \begin{cases} X + n & \text{elastic scattering} \\ \\ X^* + n' & \text{inelastic scattering} \end{cases} \tag{6.23}$$

The scattered neutron has less kinetic energy than the incident neutron by the amount of recoil and excitation energy (if any) of the scattering nucleus.

The energy of the scattered neutron can be obtained directly from the general kinematics result for binary reactions given by Eq. (6.11). For neutron scattering from some nucleus, particles x and y are identified as the neutron so that $m_x = m_y \equiv m_n$ (the neutron mass) and $\theta_y \equiv \theta_s$ (the neutron scattering angle). Neutron energies before and after scattering are denoted as $E_x \equiv E$ and $E_y \equiv E'$. Particles X and Y are the same nucleus, although particle Y may be an excited state of the scattering nucleus if the scattering is inelastic (i.e., if $Q < 0$). However, $m_X \simeq m_Y \equiv M$ since $Mc^2 >> |Q|$. With this change of notation, Eq. (6.11) becomes

$$\sqrt{E'} = \frac{1}{M + m_n}\left\{\sqrt{m_n^2 E}\cos\theta_s \pm \sqrt{E(M^2 + m_n^2\cos^2\theta_s - 1) + M(M + m_n)Q}\right\} \tag{6.24}$$

or, equivalently,

$$E' = \frac{1}{(A + 1)^2}\left\{\sqrt{E}\cos\theta_s \pm \sqrt{E(A^2 - 1 + \cos^2\theta_s) + A(A + 1)Q}\right\}^2, \tag{6.25}$$

where $A \equiv M/m_n \simeq$ the atomic mass number of the scattering nucleus. For elastic scattering ($Q = 0$), only the $+$ sign in this equation is physically meaningful. Rearrangement of this equation allows the scattering angle to be expressed in terms of the initial and final neutron kinetic energies, namely

$$\cos\theta_s = \frac{1}{2}\left[(A+1)\sqrt{\frac{E'}{E}} - (A-1)\sqrt{\frac{E}{E'}} - \frac{QA}{\sqrt{EE'}}\right]. \tag{6.26}$$

Example 6.3: What is the minimum and maximum energy of an elastically scattered neutron? Equation (6.25) for elastic scattering ($Q = 0$) becomes

$$E' = \frac{1}{(A+1)^2}\left\{\sqrt{E}\cos\theta_s + \sqrt{E(A^2-1+\cos^2\theta_s)}\right\}^2.$$

The minimum energy of the scattered neutron occurs when the scattered neutron rebounds directly backward from its initial direction, i.e., the scattering angle $\theta_s = \pi$ so that $\cos\theta_s = -1$. The above result gives

$$E'_{\min} = \frac{\left\{\sqrt{E}(-1) + \sqrt{E(A^2-1+1)}\right\}^2}{(A+1)^2} = \frac{\left\{(A-1)\sqrt{E}\right\}^2}{(A+1)^2} \equiv \alpha E,$$

where $\alpha \equiv (A-1)^2/(A+1)^2$. The maximum energy of the scattered neutron occurs for a glancing collision, i.e., for $\theta_s = 0$ so that $\cos\theta_s = 1$. Thus,

$$E'_{\max} = \frac{\left\{\sqrt{E}(+1) + \sqrt{E(A^2-1+1)}\right\}^2}{(A+1)^2} = \frac{\left\{(A+1)\sqrt{E}\right\}^2}{(A+1)^2} = E.$$

These results hold for $A > 1$. For scattering from hydrogen $A = 1$, we must be more careful. Conservation of energy and momentum gave us Eq. (6.10) which we then solved for $\sqrt{E'}$. For elastic scattering from hydrogen, $m_x = m_y = m_X = m_Y \simeq 1$ (u), so that this quadratic equation for $\sqrt{E'}$ reduces to

$$2E' - 2\sqrt{E}\cos\theta_s\sqrt{E'} + 0 = 0.$$

From this we find $\sqrt{E'} = \sqrt{E}\cos\theta_s$. Since $\sqrt{E'}$ must be non-negative, the only physically allowed scattering angles are $0 \leq \theta_s \leq \pi/2$, i.e., there can be no backscatter—a result well known to every pool player who scatters a ball from a ball of equal mass. The minimum energy of the scattered neutron now occurs for $\theta_s = \pi/2$ for which $E'_{min} = 0$. The maximum energy of the scattered neutron occurs, as in the general case, for $\theta_s = 0$ which gives $E'_{\max} = E$.

Average Neutron Energy Loss

For elastic scattering ($Q = 0$), the minimum and maximum energy of the scattered neutron is found from Eq. (6.25) for maximum and minimum scattering angles $[(\theta_s)_{\max} = \pi$ and $(\theta_s)_{\min} = 0]$, respectively. The result is

$$E'_{\min} = \alpha E \qquad \text{and} \qquad E'_{\max} = E, \tag{6.27}$$

where $\alpha \equiv (A-1)^2/(A+1)^2$. The corresponding neutron energy loss (equal to the average kinetic energy of the recoiling nucleus) is, for isotropic scattering,

$$(\Delta E)_{\text{av}} \equiv E - E'_{\text{av}} = E - \frac{1}{2}(E + \alpha E) = \frac{1}{2}(1 - \alpha)E. \qquad (6.28)$$

Notice, that as the neutron energy decreases, the average energy loss also decreases.

Average Logarithmic Energy Loss

Because neutrons in a reactor can have a wide range of energies, ranging from a few milli to several mega electron volts, a logarithmic energy scale is often used when plotting the energy spectrum of the neutrons. One can also compute the average logarithmic energy loss per elastic scatter, i.e., the average value of $(\ln E - \ln E') = \log(E/E')$. For isotropic scattering with respect to the compound nucleus, it can be shown [Mayo 1998]

$$\overline{\ln\left(\frac{E}{E'}\right)} = 1 + \frac{\alpha}{1-\alpha} \ln \alpha \equiv \xi. \qquad (6.29)$$

Notice that this result is independent of the initial neutron energy. Thus, on a logarithmic energy scale, a neutron loses the same amount of logarithmic energy per elastic scatter, regardless of its initial energy.

From Eq. (6.29) we can calculate the average number of scatters required to bring a neutron of initial energy E_1 to a lower energy E_2. Because the change in the logarithm of the energy is $\ln(E_1/E_2)$, and because ξ is the average loss in the logarithm of the energy per scatter, the number of scatters n needed, (on the average), is

$$n = \frac{1}{\xi} \ln\left(\frac{E_1}{E_2}\right). \qquad (6.30)$$

Since the smaller the A number the larger is ξ, material with small A numbers are more effective at slowing down neutrons than materials with large A numbers. The effectiveness of several elements in slowing a 2-MeV neutron to thermal energies (0.025 eV) is shown in Table 6.1.

Table 6.1. Slowing of neutron by various materials. Here n is the number of elastic scatters to slow, on the average, a neutron from 2 MeV to 0.025 eV.

Material	A	α	ξ	n
H	1	0	1	18.2
H_2O	1 & 16	—	0.920	19.8
D	2	0.111	0.725	25.1
D_2O	2 & 16	—	0.509	35.7
He	4	0.360	0.425	42.8
Be	9	0.640	0.207	88.1
C	12	0.716	0.158	115
^{238}U	238	0.983	0.0084	2172

Thermal Neutrons

As fast neutrons slow down, they eventually come into *thermal equilibrium* with the thermal motion of the atoms in the medium through which they are moving. The atoms of the medium are moving with a distribution of speeds, known as the *Maxwellian* distribution. In such a thermal equilibrium, a neutron is as likely to gain kinetic energy upon scattering from a rapidly moving nucleus as it is to lose energy upon scattering from a slowly moving nucleus. Such neutrons are referred to as *thermal neutrons*. At room temperature, 293 K, the most probable kinetic energy of thermal neutrons is 0.025 eV corresponding to a neutron speed of about 2200 m/s. This is a minuscule energy compared to the energy released in nuclear reactions induced by such slow moving neutrons.

6.5.2 Neutron Capture Reactions

The ultimate fate of a free neutron is to be absorbed by a nucleus, and thereby sometimes convert a stable nucleus into a radioactive one. Such transformations may be intentional (such as in the production of useful radioisotopes) or an undesirable by-product of a neutron field. The capture of a neutron by a nucleus usually leaves the compound nucleus in a highly excited state (recall the average binding energy of a nucleon is about 7 to 8 MeV). The excited nucleus usually decays rapidly (within about 10^{-13} s) by emitting one or more *capture* gamma rays. These energetic capture gamma photons may present a significant radiation hazard if produced in regions near humans. Such (n, γ) reactions may be depicted as

$$n + {}^{A}X \longrightarrow [{}^{A+1}X]^* \longrightarrow {}^{A+1}X + \gamma. \tag{6.31}$$

To shield against neutrons, we often allow the neutron to first slow down by passing through a material composed of elements with small A (e.g., a hydrogen rich material such as water or paraffin) and then pass into a material that readily absorbs slow moving (thermal) neutrons. One such material is boron whose natural isotope ^{10}B has a large propensity to absorb neutrons with kinetic energies less than 1 eV. This neutron absorption reaction is

$$^{10}_{5}B + n \longrightarrow [^{11}_{5}B]^* \longrightarrow {}^{7}_{3}Li + \alpha + \gamma.$$

6.5.3 Fission Reactions

We saw in Chapter 4, in the discussion of binding energies of nuclei, that nuclear energy can be obtained by splitting a very heavy nucleus (e.g., ^{235}U) into two lighter nuclei. This fission reaction is an important and practical source of energy. A large fraction of the world's electrical energy is currently generated from energy liberated through the fission process.

The trick to generating fission energy is to get a heavy nucleus to fission and release the fission energy. Some very heavy nuclides actually undergo radioactive decay by spontaneously fissioning into two lighter nuclides. For example, ^{252}Cf has a half-life of 2.638 y and usually decays by alpha particle emission. However, in 3.09% of the decays, ^{252}Cf spontaneously fissions. Such spontaneously fissioning nuclides are rare and usually decay more likely by alpha decay. Some properties of spontaneously fission radionuclides are summarized in Table 6.2.

Table 6.2. Nuclides which spontaneously fission. All also decay by alpha emission, which is usually the only other decay mode. However, nuclides ^{241}Pu, ^{250}Cm, and ^{249}Bk also undergo beta decay with probabilities of 99.99755%, 14%, and 99.99856%, respectively.

Nuclide	Half-life	Fission prob. per decay (%)	Neutrons per fission	α per fission	Neutrons per (g s)
^{233}U	1.59×10^5 y	1.3×10^{-10}	1.76	7.6×10^{11}	8.6×10^{-4}
^{235}U	7.04×10^8 y	2.0×10^{-7}	1.86	5.0×10^8	3.0×10^{-4}
^{238}U	4.47×10^9 y	5.4×10^{-5}	2.07	1.9×10^6	0.0136
^{237}Np	2.14×10^6 y	2.1×10^{-12}	2.05	4.7×10^{11}	1.1×10^{-4}
^{236}Pu	2.85 y	8.1×10^{-8}	2.23	1.2×10^9	3.6×10^4
^{238}Pu	87.7 y	1.8×10^{-7}	2.28	5.4×10^8	2.7×10^3
^{239}Pu	2.41×10^4 y	4.4×10^{-10}	2.16	2.3×10^{11}	2.2×10^{-2}
^{240}Pu	6569 y	5.0×10^{-6}	2.21	2.0×10^7	920
^{241}Pu	14.35 y	5.7×10^{-13}	2.25	4.3×10^9	0.05
^{242}Pu	3.76×10^5 y	5.5×10^{-4}	2.24	1.8×10^5	1.8×10^3
^{244}Pu	8.26×10^7 y	0.125	2.28	8.0×10^2	1.9×10^3
^{241}Am	433.6 y	4.1×10^{-10}	3.22	2.4×10^{11}	1.18
^{242}Cm	163 d	6.8×10^{-6}	2.70	1.5×10^7	2.3×10^7
^{244}Cm	18.11 y	1.3×10^{-4}	2.77	7.5×10^5	1.1×10^7
^{246}Cm	4730 y	0.0261	2.86	3.8×10^3	8.5×10^6
^{248}Cm	3.39×10^5 y	8.26	3.14	11	4.1×10^{12}
^{250}Cm	6900 y	61.0	3.31	0.40	1.6×10^{10}
^{249}Bk	320 d	4.7×10^{-8}	3.67	3.1×10^4	1.1×10^5
^{246}Cf	35.7 h	2.0×10^{-4}	2.83	5.0×10^5	7.5×10^{10}
^{248}Cf	333.5 d	2.9×10^{-3}	3.00	3.5×10^4	5.1×10^9
^{249}Cf	350.6 y	5.2×10^{-7}	3.20	1.9×10^8	2.5×10^3
^{250}Cf	13.08 y	0.077	3.49	1.3×10^3	1.1×10^{10}
^{252}Cf	2.638 y	3.09	3.73	31	2.3×10^{12}
^{254}Cf	60.5 d	99.69	3.89	0.0031	1.2×10^{15}
^{253}Es	20.47 d	8.7×10^{-6}	3.70	1.2×10^7	3.0×10^8
^{254}Fm	3.24 h	0.053	4.00	1.9×10^3	3.0×10^{14}

Source: Dillman [1980], Kocher [1981], and Reilly et al. [1991]].

Generally, however, it is necessary to hit a heavy nucleus with a subatomic particle, such as a neutron, to produce a compound nucleus in an extremely highly excited state. The unstable compound nucleus may decay by a variety of methods including the fission process. For example, if ^{235}U is bombarded with neutrons, some possible reactions are

$$
{}_{0}^{1}\text{n} + {}_{92}^{235}\text{U} \longrightarrow {}_{92}^{236}\text{U}^* \longrightarrow
\begin{cases}
{}_{92}^{235}\text{U} + {}_{0}^{1}\text{n} & \text{elastic scatter} \\
{}_{92}^{235}\text{U} + {}_{0}^{1}\text{n}' + \gamma & \text{inelastic scatter} \\
{}_{92}^{236}\text{U} + \gamma & \text{radiative capture} \\
Y_H + Y_L + y_1 + y_2 + \cdots & \text{fission}
\end{cases}
$$

where, in the fission reaction, Y_H and Y_L are the heavy and light fission products, respectively, and the y_i are other lighter subatomic particles such as neutrons and beta particles. Direct reactions such as $(n, 2n)$, (n, α), and (n, p) proceed without the formation of the compound nucleus. These reactions occur only for incident neutron energies of more than several MeV, and, hence, are not important in fission reactors where few neutrons have energies above 5 MeV.

The crucial aspect of fission reactions induced by neutrons is that some of the subatomic particles emitted are neutrons, usually more than one. These fission neutrons can then be used to cause other uranium nuclei to fission and produce more neutrons. In this way, a self-sustaining fission chain reaction can be established to release fission energy at a constant rate. This is the principle upon which fission reactors are based. The maintenance and control of such a chain reaction is discussed in detail in a later chapter.

Any very heavy nucleus can be made to fission accompanied by a net release of energy if hit sufficiently hard by some incident particle. However, very few nuclides can fission by the absorption of a neutron with negligible incident kinetic energy.[5] The binding energy of this neutron in the compound nucleus provides sufficient excitation energy that the compound nucleus can deform sufficiently to split apart. Nuclides, such as ^{235}U, ^{233}U, and ^{239}Pu, that fission upon the absorption of a slow moving neutron, are called *fissile* nuclei and play an important role in present-day nuclear reactors. Nuclides that fission only when struck with a neutron with one or more MeV of kinetic energy, such as ^{238}U and ^{240}Pu, are said to be *fissionable*, i.e., they undergo fast fission.

Some nuclides, which are themselves not fissile, can be converted into fissile nuclides upon the absorption of a slow moving neutron. Such *fertile* nuclides can be converted into useful nuclear fuel for fission reactors in which most of the neutrons are slow moving. The two most important of these fissile breeding reactions are

$$^{232}_{90}\text{Th} + ^{1}_{0}\text{n} \quad \longrightarrow \quad ^{233}_{90}\text{Th} \quad \xrightarrow[22\text{ m}]{\beta^-} \quad ^{233}_{91}\text{Pa} \quad \xrightarrow[27\text{ d}]{\beta^-} \quad ^{233}_{92}\text{U}$$

$$^{238}_{92}\text{U} + ^{1}_{0}\text{n} \quad \longrightarrow \quad ^{239}_{92}\text{U} \quad \xrightarrow[24\text{ m}]{\beta^-} \quad ^{239}_{93}\text{Np} \quad \xrightarrow[56\text{ h}]{\beta^-} \quad ^{239}_{94}\text{Pu}.$$

6.6 Characteristics of the Fission Reaction

In this section, several important properties of the fission reaction, which are of great importance in designing fission power reactors, are discussed.

The Fission Process

When an incident subatomic projectile is absorbed by a heavy nucleus, the resulting compound nucleus is produced in a very excited state. The excitation energy E^* of the compound nucleus equals the binding energy of the incident particle to it plus the kinetic energy of the incident projectile, less a negligible recoil energy of

---

[5]Such neutrons typically are *thermal neutrons*, neutrons which have lost almost all of their kinetic energy and are moving with speeds comparable to the speed of the thermal motion of the ambient atoms. At room temperature, thermal neutrons have an average kinetic energy of only about 0.025 eV. Neutrons with energies just above thermal energies are call *epithermal*, and those with energies above several tens of eV are called *fast* neutrons.

the compound nucleus. The excited compound nucleus is very unstable, with a lifetime of about 10^{-14} s. The "nuclear fluid" of such an excited nucleus undergoes large oscillations and deformations in shape. If the compound nucleus is sufficiently excited, it may, during one of its oscillations in shape, deform into an elongated or dumbbell configuration in which the two ends Coulombically repel each other and the nuclear forces, being very short ranged, are no longer able to hold the two ends together. The two ends then separate (*scission*) within $\sim 10^{-20}$ s into two nuclear pieces, repelling each other with such tremendous Coulombic force that many of the orbital electrons are left behind. Two highly charged *fission fragments* are thus created.[6] For example, in neutron-induced fission of ^{235}U, the reaction sequence, immediately following scission of the compound nucleus, is

$$\isotope{1}{0}{n} + \isotope{235}{92}{U} \longrightarrow \isotope{236}{92}{U}^* \longrightarrow Y_H^{+n} + Y_L^{+m} + (n+m)e^-, \qquad (6.32)$$

where the Y_H and Y_L indicate, respectively, the heavy and light primary fission products, and the ionic charge n and m on the fission fragments is about 20.

The primary fission fragments are produced in such highly excited nuclear states that neutrons can "boil" off them. Anywhere from 0 to about 8 neutrons, denoted by ν_p, evaporate from the primary fission fragments within about 10^{-17} s of the scission or splitting apart of the compound nucleus. These neutrons are called *prompt* fission neutrons, hence the subscript p in ν_p, in order to distinguish them from *delayed* neutrons emitted at later times during radioactive decay of the fission products.

Following prompt neutron emission, the fission fragments are still in excited states, but with excitation energies insufficient to cause particle emission. They promptly decay to lower energy levels only by the emission of *prompt* gamma rays γ_p. This emission is usually completed within about 2×10^{-14} s after the emission of the prompt neutrons.

The highly charged fission fragments pass through the surrounding medium causing millions of Coulombic ionization and excitation interacts with the electrons of the medium. As the fission fragments slow, they gradually acquire electrons reducing their ionization charge, until, by the time they are stopped, they have become electrically neutral atoms. This transfer of the fission fragments' kinetic energy to the ambient medium takes about 10^{-12} s. Thus, after about 10^{-12} s following scission, the fission reaction may be written

$$\isotope{1}{0}{n} + \isotope{235}{92}{U} \longrightarrow \isotope{236}{92}{U}^* \longrightarrow Y_H + Y_L + \nu_p(\isotope{1}{0}{n}) + \gamma_p. \qquad (6.33)$$

In the fission process, the number of neutrons and protons must be conserved. For neutron-induced fission of $^{235}_{92}$U this requires

$$\begin{aligned} A_L + A_H + \nu_p &= 236 \\ N_L + N_H + \nu_p &= 144 \\ Z_L + Z_H &= 92 \end{aligned} \qquad (6.34)$$

---

[6]Sometimes three fission fragments are formed in *ternary* fission with the third being a small nucleus. Alpha particles are created in about 0.2% of the fissions, and nuclei of ^{2}H, ^{3}H, and others up to about ^{10}B with much less frequency. Such ternary fission, though rare, accounts in part for buildup of tritium in water-cooled nuclear reactors.

After prompt neutron and gamma ray emission, the fission fragments are termed *fission products*. The fission products are still radioactively unstable since they still have too many neutrons compared to protons to form stable atoms. They thus form the start of decay chains whose members radioactively decay, usually by isobaric β^- decay, (constant A) until a stable end-nuclide is reached. The half-lives of the fission product daughters range from fractions of a second to many thousands of years. The proper management of these long-lived fission products is a major challenge to proponents of nuclear fission power.

6.6.1   Fission Products

Several hundred different nuclides can be produced by a fission event or by the subsequent decay of the fission products. Understanding how the radioactive fission products accumulate and decay in time is important in the management of nuclear waste. Also since the kinetic energy of the fission fragments represents the bulk of the released fission energy, the energetics of these products is also very important. In this section these and other related properties are discussed.

Fission Product Decay Chains

The neutron to proton ratio in ^{236}U is $144/92 = 1.57$, and the fission fragments must initially have this same ratio. However, stable nuclides of the mass of the fission fragments have much lower ratios, typically between 1.2 to 1.4. Thus, the initial fission products are neutron rich and begin to decrease their neutron to proton ratio through β^- decay.[7] Consequently, a fission product chain is created, whereby a series of isobaric (constant A) β^- decays occur until a stable nuclide is obtained. Schematically,

$$\ _Z^A X \xrightarrow{\beta^-} \ _{Z+1}^A X \xrightarrow{\beta^-} \ _{Z+2}^A X \xrightarrow{\beta^-} \cdots \xrightarrow{\beta^-} \ _{Z+n}^A X \text{ (stable)}.$$

Decay chains may be short or long and vary tremendously in how long it takes to reach the stable end nuclide. A few important examples of decay fission product chains are presented below.

It is through the following fission product chain

$$^{140}_{54}\text{Xe} \xrightarrow[16\text{ s}]{\beta^-} \ ^{140}_{55}\text{Cs} \xrightarrow[66\text{ s}]{\beta^-} \ ^{140}_{56}\text{Ba} \xrightarrow[12.8\text{ d}]{\beta^-} \ ^{140}_{57}\text{La} \xrightarrow[40\text{ h}]{\beta^-} \ ^{140}_{58}\text{Ce} \text{ (stable)} \tag{6.35}$$

that fission was discovered. Otto Hahn, working with Fritz Strassman and Lise Meitner, observed in 1939 that the elements La and Ba were produced in an initially pure uranium sample bombarded by neutrons and thus had to be produced by the splitting of the uranium atom.

The short fission product chain

$$^{147}_{60}\text{Nd} \xrightarrow[11\text{ d}]{\beta^-} \ ^{147}_{61}\text{Pm} \xrightarrow[2.6\text{ y}]{\beta^-} \ ^{147}_{62}\text{Sm} \ (T_{1/2} = 10^{11}\text{ y}). \tag{6.36}$$

---

[7]Very rarely, a fission product decays by neutron emission. Although very important for reactor control, decay by neutron emission is unlikely and is usually ignored in assessing fission product inventories.

produces an isotope of the element promethium, which has no stable isotopes, and which had been poorly identified before the discovery of fission.

Another long fission product chain,

$$\ce{^{99}_{38}Sr} \xrightarrow[0.27\ \text{s}]{\beta^-} \ce{^{99}_{39}Y} \xrightarrow[1.5\ \text{s}]{\beta^-} \ce{^{99}_{40}Zr} \xrightarrow[2.2\ \text{s}]{\beta^-} \ce{^{99}_{41}Nb} \xrightarrow[15\ \text{s}]{\beta^-} \ce{^{99}_{42}Mo} \xrightarrow[2.75\ \text{d}]{\beta^-}$$

$$\ce{^{99m}_{43}Tc^*} \xrightarrow[6.01\ \text{h}]{\gamma} \ce{^{99}_{43}Tc} \xrightarrow[0.21\ \text{My}]{\beta^-} \ce{^{99}_{44}Ru}\ \text{(stable)},\quad (6.37)$$

contains the longest lived isotope of technetium, all of which are radioactive and do not occur naturally on earth. This element was first discovered through this fission product chain. Also of interest, the metastable ^{99m}Tc is the radioisotope most often used today in medical diagnoses.

Finally, a decay chain of great importance to nuclear reactor operations is

$$\ce{^{135}_{51}Sb} \xrightarrow[1.7\ \text{s}]{\beta^-} \ce{^{135}_{52}Te} \xrightarrow[19\ \text{s}]{\beta^-} \ce{^{135}_{53}I} \xrightarrow[6.57\ \text{h}]{\beta^-} \ce{^{135}_{54}Xe} \xrightarrow[9.10\ \text{h}]{\beta^-} \ce{^{135}_{55}Cs} \xrightarrow[2.6\ \text{My}]{\beta^-} \ce{^{135}_{56}Ba}\ \text{(stable)}.\quad (6.38)$$

The nuclide $^{135}_{54}$Xe has such a huge propensity (the largest of all nuclides) to absorb slow moving neutrons, that a very little buildup of this nuclide in a reactor core can absorb enough neutrons to stop the self-sustaining chain reaction. This ^{135}Xe effect is discussed at length in Chapter 10.

Mass Distribution of Fission Products

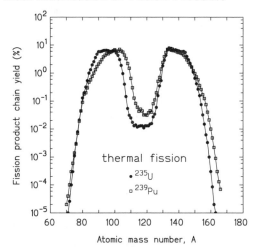

Figure 6.6. Fission product yields for the thermal fission of ^{235}U and ^{239}Pu. Data source: GE-NE [1996]

The two fission fragments can have mass numbers ranging from about 70 to about 170. Thus, about 100 different fission product decay chains, each with a constant A, are formed. However, not all fission fragment masses are equally likely. Likewise, not all nuclides with the same A number are produced equally.

The probability a fission fragment is a nuclide with mass number A, i.e., it is a member of the decay chain of mass number A, is called the *fission chain yield* and is denoted by $y(A)$. The fission chain yield for the thermal fission of ^{235}U and ^{239}Pu are shown in Fig. 6.6. Both fissile isotopes have very similar yield curves, that for ^{239}Pu being shifted slightly towards higher masses. Notice also that there is an asymmetry in the mass of the fission fragments. It is much more likely to obtain a fission fragment with mass numbers of 94 and 140 (obtained in about 6.5% of all fissions) than it is to obtain a symmetric splitting $A \simeq 118$ (obtained in only 0.01% of all fissions of ^{235}U).

It should be noted, that the mass of one fission fragment determines that of the other. For thermal fission of ^{235}U, if the light fragment has mass A_L, conservation of nucleons requires the other fragment to have mass $A_H = 236 - A_L - \nu_p$. Thus, the probability of producing a fragment of a given mass is closely correlated with the mass of the other fragment.

We see that thermal fission is an asymmetric process. However, as the energy of the incident neutron increases, the compound nucleus is produced in a higher excited state. The resulting fission chain curve shows a smaller asymmetry. For thermal fission, the peak maximum to valley minimum is about $6.5\%/0.01\% \simeq 650$. For 14-MeV neutrons this asymmetry ratio is about 100, and for 90-MeV neutrons, there is only a single central peak.

Many members of a decay chain can be produced as fission fragments. The probability of obtaining a fission fragment with mass number A and with Z protons is the *fission yield* $y(A, Z)$. The fission chain yield through member k is simply the cumulative yield through that member, i.e.,

$$y_k(A) = \sum_{j \leq k} y(A, Z_j). \qquad (6.39)$$

Initial Energy of Fission Fragments

Following the scission of the compound nucleus, as in the reaction of Eq. (6.32), the reaction's Q-value plus the incident neutrons's kinetic energy, if any, must equal the kinetic energy of the two highly excited fission fragments Y_L and Y_H. For a fission caused by the absorption of an incident neutron with negligible kinetic energy, such as a *thermal neutron*, conservation of momentum requires

$$m_L v_L = m_H v_H \quad \text{or} \quad \frac{v_L}{v_H} = \frac{m_H}{m_L}, \qquad (6.40)$$

where m_i and v_i are the masses and speeds of the two fragments. The kinetic energies E_i of the two fragments are then in the ratio (using the above result),

$$\frac{E_L}{E_H} = \frac{m_L v_L^2/2}{m_H v_H^2/2} = \frac{m_H}{m_L}. \qquad (6.41)$$

Thus, a fragment's initial kinetic energy is inversely proportional to its mass, so that the lighter fragment has a greater kinetic energy than does the heavy fragment.

Because the mass distribution of the fission fragments is bimodal (two peaks), we expect the initial kinetic energy of the fragments to also be bimodal. In Fig. 6.7, the kinetic energy distribution of the fission fragments produced in the thermal fission of ^{235}U is shown. This energy is rapidly transferred to the ambient medium as thermal energy as the fragments slow and become incorporated into the medium.

6.6.2   Neutron Emission in Fission

The neutrons emitted by a fission reaction are of great importance for the practical release of nuclear power. With these fission neutrons, other fission reactions can be caused, and, if properly arranged, a self sustaining chain reaction can be established.

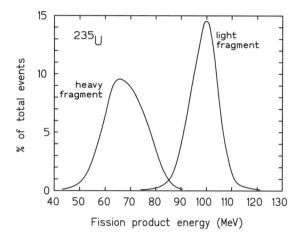

Figure 6.7. Energy distribution of fission products produced by the thermal fission of ^{235}U. The mean energy of the light fragment is 99.2 MeV and that of the heavy fragment is 68.1 MeV. After Keepin [1965].

Example 6.4: Consider the following fission reaction:

$$^{235}_{92}\text{U} + ^{1}_{0}\text{n} \longrightarrow ^{140}_{54}\text{Xe}^* + ^{96}_{38}\text{Sr}^* \xrightarrow[\text{1 s}]{\text{after}} {}^{139}_{54}\text{Xe} + ^{95}_{38}\text{Sr} + 2(^{1}_{0}\text{n}) + 7(\gamma),$$

where the 2 prompt fission neutrons have a total kinetic energy of 5.2 MeV and the prompt gamma rays have a total energy of 6.7 MeV. What is the prompt energy released for this fission event and what is the initial kinetic energy of the $^{140}_{54}\text{Xe}^*$ fission fragment?

The prompt energy released E_p is obtained from the mass deficit of this reaction as

$$E_p = \left[M(^{235}_{92}\text{U}) + m_n - M(^{139}_{54}\text{Xe}) - M(^{95}_{38}\text{Sr}) - 2m_n \right] c^2$$
$$= [235.043923 + 1.008665 - 138.918787 - 94.919358 -$$
$$2(1.008665)] \, \text{u} \times 931.5 \, \text{MeV/u}$$
$$= 183.6 \, \text{MeV}.$$

The total kinetic energy E_{ff} of the $^{140}_{54}\text{Xe}^*$ and $^{96}_{38}\text{Sr}^*$ fission fragments equals E_p less the kinetic energy of the prompt neutrons and gamma rays since their energy comes from the excitation energy of the fission fragments. Thus, $E_{ff} = 183.6 - 5.2 - 6.7 = 171.7$ MeV $= E_H + E_L$, where E_H and E_L are the kinetic energies of the heavy and light fission fragment, respectively. From Eq. (6.41)

$$E_H = E_L \frac{m_L}{m_H} = (E_{ff} - E_H) \frac{m_L}{m_H} = E_{ff} \frac{m_L}{m_L + m_H} = 69.8 \, \text{MeV}.$$

Prompt and Delayed Neutrons

Almost all of the neutrons produced from a fission event are emitted within 10^{-14} s of the fission event. The number ν_p of these prompt neutrons can vary between 0 and about 8 depending on the specific pair of fission fragments created. The average number $\overline{\nu}_p$ of prompt fission neutrons produced by thermal fission, i.e., averaged over all possible pairs of fission fragments, is typically about 2.5, although its precise value depends on the fissile nuclide. As the energy of the incident neutron increases, the compound nucleus and the resulting fission fragments are left in higher excited states, and more prompt neutrons are produced.

A small fraction, always less than 1% for thermal fission, of fission neutrons are emitted as *delayed neutrons*, which are produced by the neutron decay of fission products at times up to many seconds or even minutes after the fission event. The average number $\overline{\nu}_d$ of delayed neutrons emitted per fission depends on the fissioning nucleus and the energy of the neutron causing the fission.

The average number of prompt plus delayed fission neutrons per fission is then $\overline{\nu} = \overline{\nu}_p + \overline{\nu}_d$. The fraction of these neutrons that are emitted by the fission products as delayed neutrons is called the *delayed neutron fraction* $\beta \equiv \overline{\nu}_d/\overline{\nu}$.[8]

Much research has been performed to determine the exact fission neutron yield for many fissionable nuclides because of the importance this parameter plays in the theory of nuclear reactors. In Table 6.3 the average yield of fission neutrons, $\overline{\nu}$, is given for five important fissionable nuclides for fission caused by a thermal neutron and by a fast fission neutron (average energy 2 MeV).

Table 6.3. Total fission neutrons per fission, $\overline{\nu}$, and the delayed neutron fraction $\beta \equiv \overline{\nu}_d/\overline{\nu}$ for several important nuclides. Results for fast fission are for a fission neutron energy spectrum.

Nuclide	Fast Fission		Thermal Fission	
	$\overline{\nu}$	β	$\overline{\nu}$	β
^{235}U	2.57	0.0064	2.43	0.0065
^{233}U	2.62	0.0026	2.48	0.0026
^{239}Pu	3.09	0.0020	2.87	0.0021
^{241}Pu	–	–	3.14	0.0049
^{238}U	2.79	0.0148	–	–
^{232}Th	2.44	0.0203	–	–
^{240}Pu	3.3	0.0026	–	–

Source: Keepin [1965].

Energies of Fission Neutrons

The energy spectrum of fission neutron has been investigated extensively, particularly for the important isotope ^{235}U. All fissionable nuclides produce a distribution of prompt fission-neutron energies which goes to zero at low and high energies and reaches a maximum at about 0.7 MeV. The energies of the prompt fission neutrons are described by a function $\chi(E)$ defined such that the fraction of prompt neutrons emitted with kinetic energies in dE about E is $\chi(E)\,dE$. This prompt *fission-neutron energy spectrum* $\chi(E)$ is shown in Fig. 6.8 for three fissioning nuclides. Spectra for other fissionable isotopes are very similar to these results. The average energy of prompt fission neutrons is about 2 MeV.

---

[8]The symbol ν is overly used in nuclear technology. It is used for photon frequency, the neutrino, and for the delayed neutron fraction. However, the context is usually sufficient to indicate its meaning.

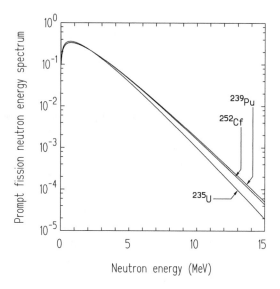

Figure 6.8. Energy distribution of the prompt fission neutrons produced by the thermal fission of ^{235}U and ^{239}Pu. The results for ^{252}Cf are for spontaneous fission.

The fraction of prompt fission neutrons emitted per unit energy about E, $\chi(E)$, can be described quite accurately by the following modified two-parameter Maxwellian distribution.

$$\chi(E) = \frac{e^{-(E+E_w)/T_w}}{\sqrt{\pi E_w T_w}} \sinh\sqrt{\frac{4E_w E}{T_w^2}}. \tag{6.42}$$

This Watt distribution can be written more compactly as

$$\chi(E) = ae^{-E/b}\sinh\sqrt{cE}, \tag{6.43}$$

where $a \equiv e^{-E_w/T_w}/\sqrt{\pi E_w T_w}$, $b \equiv T_w$, and $c \equiv 4E_w/T_w^2$. Values of parameters for several fissionable nuclides are given in Table 6.4

Table 6.4. Parameters for the Watt approximation to the prompt fission-neutron energy distribution for several fissionable nuclides. From Shultis and Faw [2000].

Nuclide	Type of Fission	Eq. (4.1)		Eq. (4.2)		
		E_w	T_w	a	b	c
^{233}U	thermal	0.3870	1.108	0.6077	1.1080	1.2608
^{235}U	thermal	0.4340	1.035	0.5535	1.0347	1.6214
^{239}Pu	thermal	0.4130	1.159	0.5710	1.1593	1.2292
^{232}Th	fast (2 MeV)	0.4305	0.971	0.5601	0.9711	1.8262
^{238}U	fast (2 MeV)	0.4159	1.027	0.5759	1.0269	1.5776
^{252}Cf	spontaneous	0.359	1.175	0.6400	1.1750	1.0401

Delayed neutrons, being emitted by less excited fission products, have considerably less kinetic energy, typically only several hundred keV. Their energy spectrum is quite complex because it is a superposition of the neutron energy spectra from each of the several dozen fission products which can decay by neutron emission.

6.6.3   Energy Released in Fission

The magnitude of the energy ultimately released per fission is about 200 MeV. The energy released in other nuclear events is of the order of several MeV. The amount of energy released per fission can be estimated by considering the binding energy per nucleon, shown in Fig. 4.1. The value of the average binding energy per nucleon has a broad maximum of about 8.4 MeV in the mass number range from 80 to 150, and all the fission products are in this range. The average binding energy per nucleon in the vicinity of uranium is about 7.5 MeV. The difference of the average binding energy is 0.9 MeV between the compound nucleus, $^{235}_{92}$U, and the fission products and the excess, 0.9 MeV, is released in fission. The total amount of energy released is approximately equal to the product of the number of nucleons (236) and the excess binding energy per nucleon (0.9 MeV), or about 200 MeV.

The fission energy released has two distinct time scales. Within the first 10^{-12} s the majority of the fission energy is released. Subsequently, the initial fission products decay (usually by β^- decay) until stable nuclides are reached. In both cases the total energy released per fission can be calculated from the nuclear masses of the reactants and the masses of the products. We have seen that the fission products for ^{235}U with the highest yields have mass numbers near $A = 95$ and $A = 139$. An example of the "prompt" energy release from fission is given in Example 6.5, and, after the fission product chains have decayed to their final stable end-chain nuclei, is given in Example 6.5.

Example 6.5:   What is the energy released by the decay of the initial $^{139}_{54}$Xe and $^{95}_{38}$Sr fission products of Example 6.4 as they decay to the stable end points of their decay chains? From the Chart of the Nuclides, we find that after 3 β^- decays $^{139}_{54}$Xe reaches the stable nuclide $^{139}_{57}$La and after 4 β^- decays $^{95}_{38}$Sr reaches stable $^{95}_{42}$Mo. The delayed energy release reaction is, therefore,

$$^{139}_{54}\text{Xe} + ^{95}_{38}\text{Sr} \longrightarrow ^{139}_{57}\text{La} + ^{95}_{42}\text{Mo} + 7(^{0}_{-1}\beta) + 7(\bar{\nu}).$$

Although seven β^- particles are emitted by the decay chains, seven ambient electrons have been absorbed by the decaying fission products to make each member of the decay chains electrically neutral. The delayed energy release reaction should be more properly written as

$$^{139}_{54}\text{Xe} + ^{95}_{38}\text{Sr} + 7(^{0}_{-1}\text{e}) \longrightarrow ^{139}_{57}\text{La} + ^{95}_{42}\text{Mo} + 7(^{0}_{-1}\beta) + 7(\bar{\nu}).$$

In this way we see the seven electron masses on both sides of this reaction cancel so the delayed emitted energy is

$$E_d = [M(^{139}_{54}\text{Xe}) + M(^{95}_{38}\text{Sr}) - M(^{139}_{57}\text{La}) - M(^{95}_{42}\text{Mo})]c^2$$
$$= [138.918787 + 94.919358 - 138.906348 - 94.905842] \text{ u} \times 931.5 \text{ MeV/u}$$
$$= 24.2 \text{ MeV}.$$

Because the fissioning of the compound nucleus can produce many different pairs of fission fragments, each leading to different fission product decay chains, there is a variation among different fission events in the amount of fission energy released. However, since huge numbers of fissions occur in a nuclear reactor, we are interested in the amount of fission energy per fission, averaged over all possible fission outcomes. The average energy release in the thermal fission of ^{235}U is shown in Table 6.5. Some of the delayed energy emission is not recoverable either because the carrier particle does not interact with the surrounding medium (neutrinos) or because it is emitted at long times after the fission products are removed from a reactor core (delayed betas and gamma rays). It is the recoverable energy that is converted into thermal energy in a reactor core, which is subsequently used to produce electrical energy.

Table 6.5. The average energy produced in the fission of ^{235}U by thermal neutrons. In a reactor core, some fission neutrons are radiatively captured producing gamma rays whose energy is absorbed. The amount of capture gamma-ray energy produced depends on the core material and design. None of the neutrino energy is recoverable.

	Energy from Fission (MeV)	Recoverable in Core (MeV)
Prompt:		
kinetic energy of the fission fragments	168	168
kinetic energy of prompt fission neutrons	5	5
fission γ-rays	7	7
γ rays from neutron capture	—	3–9
Delayed:		
fission product β-decay energy	8	8
fission product γ-decay energy	7	7
neutrino kinetic energy	12	0
Total energy (MeV)	207	198–204

Energy from Fission Products

Most fission products decay within a few years, but many others have much longer half-lives. The decay heat produced by such slowly decaying fission products is of concern to those who have to deal with spent nuclear fuel. The average power from delayed gamma photons and beta particles that are emitted by the fission products produced by a single fission of ^{235}U at time 0 can be estimated from the empirical formulas

$$F_\gamma(t) = 1.4\, t^{-1.2} \quad \text{MeV s}^{-1} \text{ fission}^{-1} \tag{6.44}$$

$$F_\beta(t) = 1.26\, t^{-1.2} \quad \text{MeV s}^{-1} \text{ fission}^{-1} \tag{6.45}$$

where t is the time (in seconds) after the fission with $10 \text{ s} < t < 10^5$ s. The results of more detailed calculations are shown in Fig. 6.9.

Useful Fission Energy Conversion Factors

Quite often we need to convert energy units useful at the microscopic level (e.g., MeV) to macroscopic units (e.g., MWh). This is particularly true for nuclear power

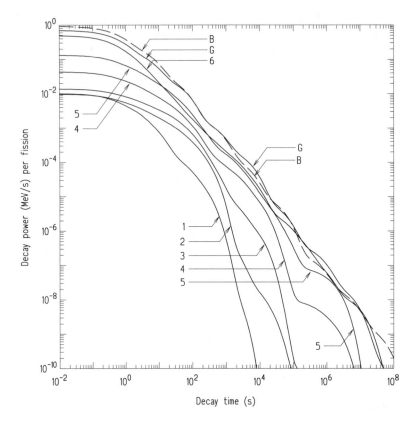

Figure 6.9. Total gamma-ray (G) and beta-particle (B) energy emission rates as a function of time after the thermal fission of ^{235}U. The curves identified by the numbers 1 to 6 are gamma emission rates for photons in the energy ranges 5–7.5, 4–5, 3–4, 2–3, 1–2, and 0–1 MeV, respectively. Source: Based on data from LaBauve et al. [1980].

where a vast number of fissions must occur to release a macroscopically useful amount of energy. If we assume 200 MeV of recoverable energy is obtained per fission, then

$$1 \text{ W} = 1(\text{J/s}) \times \frac{1}{1.602 \times 10^{-13} \text{ (J/MeV)}} \times \frac{1}{200 \text{ (MeV/fission)}}$$

$$= 3.1 \times 10^{10} \text{ (fissions/s)}.$$

Since A/N_a is the mass (g) per atom, the rate at which ^{235}U must be fissioned to generate a power of 1 MW is

$$1 \text{ MW} = 3.1 \times 10^{16} \text{ (fissions/s)} \times 86400 \text{ (s/d)} \times \frac{235}{6.023 \times 10^{23}} \left(\frac{\text{g-fissioned}}{\text{atom-fissioned}} \right)$$

$$= 1.05 \text{ (g/d)}.$$

In other words, the thermal fission of 1 g of ^{235}U yields about 1 MWd of thermal energy. However, a thermal neutron absorbed by ^{235}U leads to fission only 85% of the time; the other 15% of absorptions lead to (n, γ) reactions. Thus, the *consumption* rate of ^{235}U equals (the fission rate)/0.85, or

$$1 \text{ MWd} = 1.24 \text{ grams of } ^{235}\text{U consumed.}$$

6.7 Fusion Reactions

An alternative to the fission of heavy nuclides for extracting nuclear energy is the fusion of light nuclides. Some possible fusion reactions involving the lightest nuclides are listed below.

$$_1^2\text{D} + {}_1^2\text{D} \longrightarrow {}_1^3\text{T} + {}_1^1\text{H}, \qquad Q = 4.03 \text{ MeV}$$

$$_1^2\text{D} + {}_1^2\text{D} \longrightarrow {}_2^3\text{He} + {}_0^1\text{n}, \qquad Q = 3.27 \text{ MeV}$$

$$_1^2\text{D} + {}_1^3\text{T} \longrightarrow {}_2^4\text{He} + {}_0^1\text{n}, \qquad Q = 17.59 \text{ MeV}$$

$$_1^2\text{D} + {}_2^3\text{He} \longrightarrow {}_2^4\text{He} + {}_1^1\text{H}, \qquad Q = 18.35 \text{ MeV}$$

$$_1^3\text{T} + {}_1^3\text{T} \longrightarrow {}_2^4\text{He} + 2{}_0^1\text{n}, \qquad Q = 11.33 \text{ MeV}$$

$$_1^1\text{H} + {}_3^6\text{Li} \longrightarrow {}_2^4\text{He} + {}_2^3\text{He}, \qquad Q = 4.02 \text{ MeV}$$

$$_1^1\text{H} + {}_5^{11}\text{B} \longrightarrow 3({}_2^4\text{He}), \qquad Q = 8.08 \text{ MeV}$$

Although these fusion reactions are exoergic, they have threshold energies because of the repulsive Coulomb forces between the two reactants. Consequently, for these reactions to occur, it is necessary that the reactants have sufficient incident kinetic energy to overcome the Coulomb barrier and allow the two reacting nuclides to reach each other, thereby enabling the nuclear forces in each particle to interact and precipitate the fusion reaction. The kinetic energy needed is typically a few keV to several hundred keV. The needed incident kinetic energy could be provided by a particle accelerator to give one of the reactants sufficient kinetic energy and to direct it towards a target composed of the second reactant. Indeed, this is done in many laboratories in which fusion reactions are studied. However, the energy needed to operate the accelerator far exceeds the energy released by the subsequent fusion reactions, and this approach is not a practical source of fusion energy.

6.7.1 Thermonuclear Fusion

The reactants can also be given sufficient kinetic energy to interact if they are raised to high temperature T so that their thermal motion provides the necessary kinetic energy (average kinetic energy $E_{av} = 3kT/2$, where k is Boltzmann's constant). Fusion reactions induced by the thermal motion of the reactants are often called *thermonuclear* reactions. However, the temperatures needed are very high, typically 10–300×10^6 K. At these temperatures, the reactants exist as a *plasma*, a state of matter in which the electrons are stripped from the nuclei so that the hot plasma consists of electrons and positively charged nuclei. The challenge to produce fusion energy in useful quantities is to confine a high temperature plasma of light elements for sufficiently long times and at sufficiently high nuclide densities to allow the fusion reactions to produce more energy than is required to produce the plasma. Three approaches are available.

Gravitational Confinement

One plasma confinement approach, the basis of life itself, uses gravity to confine a plasma of light nuclei. Such is the method used by stars where, deep in their interiors, extreme temperatures and pressures are produced. For example, at our sun's core, the temperature is about 15 MK and the pressure is 4×10^{16} Pa ($\simeq 400$ billion earth atmospheres). Obviously, this is not a very useful approach to use with earth-bound energy-generating systems. The fusion energy production in stars is discussed in detail in Section 6.7.2.

Magnetic Confinement

The second confinement approach, is to use magnetic fields to confine the plasma. Since a plasma is composed of fast moving charged particles, their motion can be affected by external magnetic fields. Currently, several multinational efforts are underway to design fusion plasma devices with clever arrangements of electromagnetic fields to heat and confine a plasma. The hot plasma fluid in these devices is generally very tenuous ($\sim 10^{15}$ particles per cm^3) and is notoriously unstable, exhibiting many modes of breaking up and allowing the plasma to come into contact with the walls of the reaction chamber where it is immediately chilled to normal temperatures and any possibility of thermonuclear reactions ceases. These instabilities increase dramatically as the density and temperature of the plasma increase.

Moreover, the magnetic pressure needed to confine a plasma is enormous. For example, to confine a plasma with an average thermal energy of 10 keV per particle and a particle density equal to atmospheric density ($\sim 10^{19}$ cm^{-3}) requires pressures exceeding 10^5 atm [Mayo 1998]. The external magnetic field coils and their support structure must be designed to withstand such pressures.

Finally, the reaction chamber must be surrounded with some system to cool the chamber walls, which are heated by the charged fusion products, and to stop and absorb the more penetrating fusion neutrons. The heat generated by the fusion products must then be converted into electrical energy before these magnetic confinement devices can be considered a practical source of nuclear energy.

All current experimental research into fusion power is based on the D-T fusion reaction, which requires the lowest plasma temperature to overcome the Coulomb barrier problem. This reaction is

$$^2_1\text{D} + {}^3_1\text{T} \longrightarrow {}^4_2\text{He} \ (3.54 \text{ MeV}) + {}^1_0\text{n} \ (14.05 \text{ MeV}). \tag{6.46}$$

The radioactive tritium ($T_{1/2} = 12.3$ y) needed for this reaction can be generated by having the fusion neutron produced in this reaction interact with lithium through the following reactions:

$$^1_0\text{n} + {}^6_3\text{Li} \longrightarrow {}^4_2\text{He} + {}^3_1\text{T} \qquad Q = 4.78 \text{ MeV},$$

$$^1_0\text{n} + {}^7_3\text{Li} \longrightarrow {}^4_2\text{He} + {}^3_1\text{T} + {}^1_0\text{n} \qquad Q = 4.78 \text{ MeV}.$$

Although great advances have been made in magnetic confinement, no system using this approach has yet to produce more energy than that used to operate the system. Yet the enormous amount of energy that would be made available if practical fusion devices could be fabricated motivates the current research efforts.

Inertial Confinement

The third approach to extract useful fusion energy from light nuclei is to first place the reactants in a small pellet at normal temperatures and pressures. The pellet is then rapidly heated into a plasma and compressed to a very high density by bombarding it from several directions instantly and simultaneously with enormously powerful pulses of laser light. During the brief instant, in which the pellet implodes and before it rebounds and blows apart, the resulting high-density, high-temperature plasma will produce many fusion reactions. In this *inertial confinement* approach, the challenge is to have the fusion reactions produce more energy than is needed to produce the laser light. So far, *break-even* has yet to be achieved.

This same inertial confinement principle is the basis of thermonuclear weapons. In these weapons, a small fission explosion is used to rapidly heat and compress a deuterium-tritium mixture, thereby producing a vast number of D-T fusion reactions in the small fraction of a second before the weapon is blown apart and the reactions cease.

Example 6.6: What is the maximum energy released by the fusion of deuterium to form $^4_2$He? When deuterons fuse, $^4_2$He is not produced directly. Rather, the two reactions $^2_1$D $+ ^2_1$D $\longrightarrow ^3_1$T $+ ^1_1$H and $^2_1$D $+ ^2_1$D $\longrightarrow ^3_2$He $+ ^1_0$n occur with almost equal probabilities. However, the products of each of these reactions can fuse with each other or another deuteron as shown below.

$$^2_1\text{D} + ^2_1\text{D} \longrightarrow ^3_1\text{T} + ^1_1\text{H} + 4.03 \text{ MeV} \qquad ^2_1\text{D} + ^2_1\text{D} \longrightarrow ^3_2\text{He} + ^1_0\text{n} + 3.27 \text{ MeV}$$

$$^2_1\text{D} + ^3_1\text{T} \longrightarrow ^4_2\text{He} + ^1_0\text{n} + 17.59 \text{ MeV} \qquad ^2_1\text{D} + ^3_2\text{He} \longrightarrow ^4_2\text{He} + ^1_1\text{H} + 18.35 \text{ MeV}$$

$$^1_0\text{n} + ^1_1\text{H} \longrightarrow ^2_1\text{D} + 2.22 \text{ MeV} \qquad ^1_0\text{n} + ^1_1\text{H} \longrightarrow ^2_1\text{D} + 2.22 \text{ MeV}$$

Summation of reactions in each path give the same result, namely,

$$^2_1\text{D} + ^2_1\text{D} \longrightarrow ^4_2\text{He} + 23.84 \text{ MeV}.$$

The Potential of Fusion Energy

Although nature has yet to produce the only practical fusion energy device (stars), the potential energy yield by an earth-bound fusion system is enormous. For example [Mayo 1998], deuterium has an abundance of 0.015% of all hydrogen isotopes. Thus, in the 1.4×10^{18} m^3 of earth's water, there are about 1.4×10^{43} atoms of deuterium. If all this deuterium could be fused in d–d reactions (11.9 MeV/d, as calculated in Example 6.6), about 1.8×10^{31} J of energy could be obtained. This amount of energy source could provide our energy needs (at the world's current rate of use) for about 50 billion years.

This enormous potential for fusion energy, provides the current incentive for the large research effort into how to extract useful amounts of energy from fusion reactions.

6.7.2 Energy Production in Stars

Fusion energy, either directly or indirectly, provides all the energy used on earth. Even fission energy requires the stars to produce heavy nuclides like uranium. In a star such as our sun, light nuclei are fused to form heavier nuclei with the release of several MeV of energy per fusion reaction. Our sun has produced radiant energy at a rate of about 4×10^{26} W for the last few billions years. All this power arises from fusion reactions that occur in the hot, dense, energy-producing core of the sun. This central core is about 7,000 km in radius or about 0.1% of the sun's total volume. The remainder of the sun is a "thermal blanket" through which the fusion energy must pass by conductive, convective, and radiative energy transfer processes before it reaches the surface of the sun where it is radiated into space, a small fraction of which subsequently reaches earth and makes life possible.

Stars are composed mostly of primordial matter created in the first few minutes after the "big bang." This primordial matter is over 90% hydrogen, a few percent He, and less than 1% of heavier elements. In stars, the hydrogen is "burned" into helium, which, as the star ages, is converted into ever heavier nuclides up to $A \simeq 56$ (iron and nickel), after which fusion reactions are not exothermic. The understanding of how stars produce their energy through fusion reactions and how elements heavier than iron are created in stars has been one of the major scientific achievements of the 20th century. The following description of stellar evolution and the associated nuclear reactions is adapted from Mayo [1998] and Cottingham and Greenwood [1986].

Young Stars: Helium Production

The birth of a star begins with gravity (and possibly shock waves from nearby supernovas) pulling together interstellar gas, composed primarily of hydrogen with small amounts of helium and trace amounts of heavier nuclides (often called *metals* in astrophysical jargon), to form a dense cloud of matter. As gravity pulls the atoms together, gravitational potential energy is converted into thermal energy, thereby heating the collapsing gas cloud. At some point, the temperature and nuclide density at the center of the collapsing matter become sufficiently high that fusion reactions begin to occur (i.e., the star *ignites*). Eventually, the pressure of the hot gas heated by thermonuclear reactions balances the gravitational forces and a (temporary) equilibrium is established as the hydrogen is converted into helium. This state, which our sun is now in, can exist for several billion years.

During this initial life of a star, hydrogen is fused to produce ^{4}He. The principal sequence of fusion reactions that transforms hydrogen ions into helium nuclei begins with

$$^1_1\text{H} + ^1_1\text{H} \longrightarrow ^2_1\text{D} + \beta^+ + \nu, \qquad Q = 0.42 \text{ MeV}.$$

In the sun, mass $= 2.0 \times 10^{30}$ kg, the central core, in which these reactions occur, has a temperature of about 15 MK and a hydrogen nuclide density N_p of about 10^{26} cm^{-3}. This proton-proton fusion reaction involves the weak force instead of the nuclear force and, consequently, it occurs very slowly with only one proton out of 10^{18} fusing each second. This is such an improbable reaction that it has never been measured in a laboratory. Nevertheless, the high proton density in the central core leads to a proton-proton fusion reaction rate R_{pp} of about 10^8 cm^{-3} s^{-1}. The average lifetime of a proton in the core is thus about $N_p/R_{pp} \simeq 3 \times 10^{10}$ y. This slow p-p reaction controls the rate at which a new star consumes its hydrogen fuel, and is sometimes referred to as the "bottleneck" in the transformation of ^{1}H to ^{4}He.

The deuterium product of the above reaction very rapidly absorbs a proton through the reaction[9]

$$\,^2_1\text{D} + \,^1_1\text{H} \longrightarrow \,^3_2\text{He} + \gamma, \qquad Q = 5.49 \text{ MeV}.$$

The ^{3}He product does not transform by absorbing a proton, since the product ^{4}Li rapidly breaks up into ^{3}He and a proton. Rather the ^{3}He migrates through the core until it encounters another ^{3}He nucleus, whereupon

$$\,^3_2\text{He} + \,^3_2\text{He} \longrightarrow \,^4_2\text{He} + 2\,^1_1\text{H} + \gamma, \qquad Q = 12.86 \text{ MeV}.$$

The net result of the above three-step process is the transformation of four protons into ^{4}He, i.e.,

$$\boxed{4(\,^1_1\text{H}) \longrightarrow \,^4_2\text{He} + 2\gamma + 2\beta^+ + 2\nu, \qquad Q = 26.72 \text{ MeV}.} \tag{6.47}$$

Minor Variations: I. The transmutation of ^{3}He into ^{4}He in the third step above can occasionally proceed along another path:

$$\,^3_2\text{He} + \,^4_2\text{He} \longrightarrow \,^7_4\text{Be} + \gamma.$$

The $^7_4$Be then goes to ^{4}He through one of the following two sets of reactions:

$$\,^7_4\text{Be} + \,^0_{-1}\text{e} \longrightarrow \,^7_3\text{Li} + \nu \qquad\qquad \,^7_4\text{Be} + \,^1_1\text{H} \longrightarrow \,^8_5\text{B} + \gamma$$
$$\,^7_3\text{Li} + \,^1_1\text{H} \longrightarrow 2(\,^4_2\text{He}) \qquad\qquad\quad \,^8_5\text{B} \longrightarrow \,^8_4\text{Be} + \beta^+ + \nu$$
$$\,^8_4\text{Be} \longrightarrow 2(\,^4_2\text{He})$$

Both branches lead to ^{4}He with the same net energy released, per ^{4}He nucleus formed.

Minor Variations: II. In stars with some heavier nuclides, ^{12}C can act as a catalyst to transform ^{1}H into ^{4}He through the "carbon" or "CNO" cycle. This sequence of reactions is

$$\,^1_1\text{H} + \,^{12}_6\text{C} \longrightarrow \,^{13}_7\text{N} + \gamma$$
$$\,^{13}_7\text{N} \longrightarrow \,^{13}_6\text{C} + \beta^+ + \nu$$
$$\,^1_1\text{H} + \,^{13}_6\text{C} \longrightarrow \,^{14}_7\text{N} + \gamma$$
$$\,^1_1\text{H} + \,^{14}_7\text{N} \longrightarrow \,^{15}_8\text{O} + \gamma$$
$$\,^{15}_8\text{O} \longrightarrow \,^{15}_7\text{N} + \beta^+ + \nu$$
$$\,^1_1\text{H} + \,^{15}_7\text{N} \longrightarrow \,^{12}_6\text{C} + \,^4_2\text{He}$$

---

[9]The $^2_2$D $+$ $^2_2$D $\longrightarrow$ $^4_2$He reaction is extremely unlikely since the density of D is exceedingly low. Moreover, the large Q-value for this reaction (28.5 MeV) is well above the neutron or proton separation energies for ^{4}He, thereby making it much more likely that ^{3}He or ^{3}T would be produced.

The overall result of this CNO reaction sequence is

$$4(^1_1\text{H}) \longrightarrow {}^4_2\text{He} + 3\gamma + 2\beta^+ + 2\nu, \qquad Q = 26.72 \text{ MeV}. \qquad (6.48)$$

This CNO reaction is usually a minor contributor to a star's energy production, representing only about 1% of the total in a young star.

Mature Stars

Once the hydrogen in a star's core is nearly consumed, leaving mostly ^{4}He, the helium burning phase of a star's life begins. With dwindling energy production by hydrogen fusion, the gas pressure in the core decreases and gravity causes the core to contract. As the core shrinks, the temperature increases as gravitational energy is converted into thermal energy. Eventually, the temperature increases to the point where the increased thermal motion of the He nuclei overcomes the Coulombic repulsive forces between them and helium fusion begins. The outward gas pressure produced by this new fusion energy source establishes a new balance with the inward gravitation force during this helium burning phase of a star's life. The new higher core temperature heats the gas surrounding the core and causes the outer gas layer to expand and increase the star's size. As the star expands, the surface temperature decreases and the star becomes a *red giant*. Although the surface temperature decreases, the greater surface area allows more radiant energy to escape into space as needed to balance the higher energy production in the core.

The first step to forming elements heavier than ^{4}He is not, as might be expected, the fusion of two ^{4}He nuclei to form ^{8}Be. The average binding energy per nucleon in ^{8}Be is less than that for ^{4}He (see Fig. 4.1), and ^{8}Be in the ground state is very unstable and disintegrates rapidly (half-life 7×10^{-17} s) into two alpha particles. The path by which stars converted ^{4}He into heavier elements was, for a long time, a mystery. However, it was eventually discovered that the fusion of two ^{4}He nuclei could produce ^{8}Be in an excited resonance state which had a sufficiently long (although still short) lifetime that, in the dense hot ^{4}He rich core, there is enough time for a third ^{4}He nucleus to be absorbed by an excited ^{8}Be complex to form ^{12}C. In effect, the path to heavier nuclei begins with the ternary reaction

$$3(^4_2\text{He}) \longrightarrow {}^{12}_6\text{C}, \qquad Q = 7.27 \text{ MeV}.$$

With the production of ^{12}C, heavier elements are formed through fusion reactions with ^{4}He.

$$
\begin{aligned}
^4_2\text{He} + {}^{12}_6\text{C} &\longrightarrow {}^{16}_8\text{O} + \gamma, & Q &= 7.16 \text{ MeV} \\
^4_2\text{He} + {}^{16}_8\text{O} &\longrightarrow {}^{20}_{10}\text{Ne} + \gamma, & Q &= 4.73 \text{ MeV} \\
^4_2\text{He} + {}^{20}_{10}\text{Ne} &\longrightarrow {}^{24}_{12}\text{Mg} + \gamma, & Q &= 9.31 \text{ MeV}
\end{aligned}
$$

When the ^{4}He fuel is exhausted, the star again shrinks and raises its core temperature by the conversion of gravitational energy into thermal energy. When the core becomes sufficiently hot, the heavier elements begin to fuse, forming even heavier elements and generating additional pressure to offset the increased gravitational forces. Reactions during this phase include, for example,

$$
{}^{12}_{6}\text{C} + {}^{12}_{6}\text{C} \longrightarrow \begin{cases} {}^{20}_{10}\text{Ne} + {}^{4}_{2}\text{He}, & Q = 4.62 \text{ MeV} \\ {}^{23}_{11}\text{Na} + {}^{1}_{1}\text{H}, & Q = 2.24 \text{ MeV} \end{cases}
$$

and

$$
{}^{16}_{8}\text{O} + {}^{16}_{8}\text{O} \longrightarrow \begin{cases} {}^{28}_{14}\text{Si} + {}^{4}_{2}\text{He}, & Q = 9.59 \text{ MeV} \\ {}^{31}_{15}\text{P} + {}^{1}_{1}\text{H}, & Q = 7.68 \text{ MeV} \end{cases}
$$

As a star ages it generates heavier nuclei and the star shrinks to create ever increasing core temperatures to sustain the fusion of the heavier nuclei. The details of energy production in stars fusing elements heavier than helium are quite complex, depending, for example, on the star's mass, rotation rate, and initial concentrations of trace elements. Often an aging star acquires an onion-like structure. After the helium burning phase it has an oxygen-neon core, surrounded by a layer of carbon, then one of helium with the whole enveloped by a thick skin of unfused hydrogen. As the temperature of the core increases, heavier elements are created. The neon combines to form magnesium, which in turn combines to form silicon, and then, in turn, to iron. In some aging stars, hydrogen from the outer layer is periodically pulled towards the core causing periodic increases in power production which, in turn, causes the star's brightness to oscillate. Sometimes the momentary increased power production can be so high that material from the outer layer can be expelled into space.

With an iron core, the star reaches the end of the fusion chain. Such an old star has several concentric shells in each of which a different fuel is being consumed. The central temperature reaches several billion degrees. The time a star spends fusing helium and the other heavier elements is very short compared to the very long time it spends as a young star fusing hydrogen. The slide into old age is very rapid.

Death of a Star

The production of fusion energy ceases in a star when most of the nuclides in the core have been transformed into nuclides near the peak of the BE/A versus A curve (see Fig. 4.1), i.e., near $A \simeq 56$. At this point, energy production begins to dwindle, and there is no more possible fusion energy production to resist further gravitational collapse. The result is a very unstable star. What happens next, depends on the star's mass. A less massive star, like our sun, slowly shrinks, converting more gravitational energy to thermal energy, thereby increasing its temperature to become a *white dwarf*. Shrinkage is arrested when the electrons in the core cannot be forced closer together by the gravitational forces. The star continues to radiate away its thermal energy and eventually becomes a *brown dwarf*, the evolutionary end of the star.

More massive stars, however, undergo a much more dramatic finale. When the radiation and gas pressures can no longer counteract the gravitational forces, the star, within a fraction of a second, implodes generating intense pressure shock waves that produce immense temperatures. In one final burst of energy production from the rapid fusion of unburned fuel in the star's outer layers, the star is blown apart and becomes a *supernova* whose radiation output can exceed that of an entire galaxy. The inward moving shock waves produce such large pressures that the electrons in the core plasma are forced to combine with the protons in nuclei allowing the core to shrink even more. These electron-capture reactions are endothermic and tend to cool the core which, in turn, causes more gravitational collapse and the absorption of even more electrons. During this transformation of protons to neutrons, a huge burst of neutrinos is emitted (one for every p-n transformation). These neutrino bursts from collapsing stars have been observed on earth.

What often remains of the core after this catastrophic collapse is a small ($\simeq$ 40 km diameter), dense (comparable to that of nuclei), rapidly rotating (thousands of rpm), neutron star. However, if the mass of the original star is sufficiently great, the residual core is so small and massive that space-time around the core is pinched off from the rest of the universe and a *black hole* is formed.

6.7.3 Nucleogenesis

Nuclear reactions in stars are responsible for the creation of all the naturally occurring nuclides, save for the very lightest. Our present expanding universe was created, by current understanding, in a "big bang" event about 15 to 20 billion years ago. This big bang created a small, intensely hot, dense, and rapidly expanding universe filled initially with only radiant energy.

As this infant universe expanded and cooled, some radiation was converted into quarks and other elementary particles. After about 1/100 s, the temperature had dropped to about 10^{11} K and the elementary particles had further condensed or decayed into neutrons, protons, and electrons (and their antiparticles) with a huge residual of photons. After about 3 min, neutrons and protons had combined to form nuclei of ^{4}He, with a p:He-nuclei ratio of 100:6, and trace amounts of D, Li, Be, and a few other light nuclei. Finally, after about 100,000 y, the universe had cooled to about 3000 K and the electrons had combined with nuclei to form neutral atoms of H, He, and trace amounts of a few light atoms. All heavier nuclides have been created subsequently by nuclear reactions in stars. During the explosive deaths of early stars, newly created heavy elements were expelled into the interstellar gas from which our solar system and sun later condensed. Thus, the heavy atoms found in our solar system reflect the history of stellar evolution in our corner of the universe.

How the heavy elements are created in stars is still not completely understood. This uncertainty arises, in part, because of our incomplete understanding of the details of stellar evolution. Nevertheless, the basic processes are clear. We have seen in the preceding section how hydrogen is transformed into the light elements up to iron through the fusion reactions responsible for a star's energy production. However, because iron and nickel are at the peak of the binding energy curve, fusion reactions cannot produce heavier elements.

During the final silicon burning phase of a star's life, in which iron is produced, the temperature of the core is roughly 10^{10} K ($kT \simeq 1$ MeV). Some of the photons of the associated thermal radiation have such high energies that they interact with nuclei causing neutrons, protons, alpha particles, etc. to be ejected. The neutrons and protons are, in turn, absorbed by other nuclei. In particular, neutron absorption by iron produces heavier elements. As more neutrons are absorbed, the nuclides become β^- unstable and decay to elements with an increased number of protons, i.e., heavier elements. This mechanism of producing heavier elements through neutron absorption and subsequent beta decay, however, is unable to produce some of the proton-rich stable heavy nuclides. Interestingly, abundances of such proton-rich nuclides are observed to be much smaller than their stable neutron-rich isotopes.

Two basic time scales appear to be involved with this mechanism of neutron absorption followed by beta-decay to produce heavier nuclides. If the neutron capture is slow compared to the beta decay (the slow or *s-process*), the transformation path on the chart of the nuclides follows closely the bottom of the beta-stability valley just beneath the line of stable nuclides. By contrast, if neutron capture is rapid compared to beta decay (the rapid or *r-process*), a highly neutron-rich nuclide is produced, which subsequently cascades down through a chain of β^- decays, to a stable nuclide with a much higher proton number than the nuclide absorbing the first neutron. From the observed nuclear abundances, especially near magic-number nuclei and heavy nuclei, it is apparent that both the r- and s-processes were important in creating the heavy nuclides found in our solar system.

The s-process most likely occurs during the He-burning phase of a star. In this phase, there is a small population of free neutrons produced by several reactions such as the $^{13}C(\alpha, n)^{16}O$ reaction. To produce elements heavier than iron during this phase, there must be some iron, generated by earlier stars, present in the core. The r-process, on the other hand, occurs at the end of a star's life during the supernova explosion. Close to the core (and future neutron star), an enormous neutron population exists that can produce very neutron-rich nuclides through multiple neutron absorptions in a very short time.

BIBLIOGRAPHY

COTTINGHAM, W.N. AND D.A. GEENWOOD, *An Introduction to Nuclear Physics*, Cambridge Univ. Press, Cambridge, 1986.

EVANS, R.D., *The Atomic Nucleus*, McGraw-Hill, New York, 1955; republished by Krieger Publishing Co., Melbourne, FL, 1982.

DILLMAN, L.T., *EDISTR—A Computer Program to Obtain a Nuclear Data Base for Nuclear Dosimetry*, Report ORNL/TM-6689, Oak Ridge National Laboratory, Oak Ridge, TN, 1980.

EISBERG, R. AND R. RESNICK, *Quantum Physics of Atoms, Molecules, Solids, Nuclei, and Particles*, John Wiley & Sons, NY, 1985.

GE-NE (General Electric Nuclear Energy), *Nuclides and Isotopes: Chart of the Nuclides*, 15th ed., prepared by J.R. Parrington, H.D. Knox, S.L. Breneman, E.M. Baum, and F. Feiner, GE Nuclear Energy, San Jose, CA, 1996.

HEPPENHEIMER, T.A., *The Man-Made Sun: The Quest for Fusion Power*, Little, Brown & Co., Boston, 1984.

HERMAN, R., *The Search for Endless Energy*, Cambridge University Press, New York, 1990.

KEEPIN, G.R., *Physics of Nuclear Kinetics*, Addison-Wesley, Reading, MA, 1965.

KOCHER, D.C., *Radioactive Decay Data Tables*, DOE/TIC-11026, Technical Information Center, U.S. Department of Energy, Washington, DC, 1981.

LABAUVE, R.J., T.R. ENGLAND, D.C. GEORGE, AND C.W. MAYNARD, "Fission Product Analytic Impulse Source Functions," *Nucl. Technol.*, **56**, 322–339 (1982).

LAMARSH, J.R., *Nuclear Reactor Theory*, Addison-Wesley, Reading, MA, 1966.

MAYO, R.M., *Nuclear Concepts for Engineers*, American Nuclear Society, La Grange Park, IL, 1998.

REILLY, D., N. ENSSLIN, H. SMITH, JR. AND S. KREINER, *Passive Nondestructive Assay of Nuclear Materials*, NUREG/CR-5550, U.S. Nuclear Regulatory Commission, Washington, DC, 1991.

SHULTIS, J.K. AND R.E. FAW, *Radiation Shielding*, American Nuclear Society, La Grange Park, IL, 2000.

PROBLEMS

1. A 2-MeV neutron is scattered elastically by ^{12}C through an angle of 45 degrees. What is the scattered neutron's energy?

2. The first nuclear excited state of $^{12}_{6}$C is 4.439 MeV above the ground state. (a) What is the Q-value for neutron inelastic scattering that leaves ^{12}C in this excited state? (b) What is the threshold energy for this scattering reaction? (c) What is the kinetic energy of an 8-MeV neutron scattered inelastically from this level at 45 degrees?

3. Derive Eq. (6.21) from Eq. (6.11).

4. For each of the following possible reactions, all of which create the compound nucleus ^{7}Li,

$$^{1}\text{n} + {}^{6}\text{Li} \longrightarrow {}^{7}\text{Li}^{*} \longrightarrow \begin{cases} ^{7}\text{Li} + \gamma \\ ^{6}\text{Li} + n \\ ^{6}\text{He} + p \\ ^{5}\text{He} + d \\ ^{3}\text{H} + \alpha \end{cases}$$

calculate (a) the Q-value, (b) the kinematic threshold energy, and (c) the minimum kinetic energy of the products. Summarize your calculations in a table.

5. For each of the following possible reactions, all of which create the compound nucleus ^{10}B,

$$^{1}\text{p} + {}^{9}\text{Be} \longrightarrow {}^{10}\text{B}^{*} \longrightarrow \begin{cases} ^{10}\text{B} + \gamma \\ ^{6}\text{Li} + \alpha \\ ^{8}\text{Be} + d \\ ^{9}\text{B} + n \\ ^{5}\text{Li} + \alpha + n \end{cases}$$

calculate (a) the Q-value, (b) the kinematic threshold energy of the proton, (c) the threshold energy of the proton for the reaction, and (d) the minimum kinetic energy of the products. Summarize your calculations in a table.

6. Consider the following reactions caused by tritons, nuclei of ^{3}H, interacting with ^{16}O to produce the compound nucleus ^{19}F

$$^{3}\text{H} + {}^{16}\text{O} \longrightarrow {}^{19}\text{F}^{*} \longrightarrow \begin{cases} ^{18}\text{F} + n \\ ^{17}\text{O} + d \\ ^{18}\text{O} + p \\ ^{16}\text{N} + {}^{3}\text{He} \end{cases}$$

For each of these reactions calculate (a) the Q-value, (b) the kinematic threshold energy of the triton, (c) the threshold energy of the triton for the reaction, and (d) the minimum kinetic energy of the products. Summarize your calculations in a table.

7. An important radionuclide produced in water-cooled nuclear reactors is ^{16}N which has a half-life of 7.13 s and emits very energetic gamma rays of 6.1 and 7.1 MeV. This nuclide is produced by the endoergic reaction $^{16}O(n,p)^{16}$N. What is the minimum energy of the neutron needed to produce ^{16}N?

8. The isotope ^{18}F is a radionuclide used in medical diagnoses of tumors and, although usually produced by the $^{18}O(p,n)^{18}$F reaction, it can also be produced by irradiating lithium carbonate (Li_2CO_3) with neutrons. The neutrons interact with ^{6}Li to produce tritons (nuclei of ^{3}H), which, in turn, interact with the oxygen to produce ^{18}F. (a) What are the two nuclear reactions? (b) Calculate the Q-value for each reaction. (c) Calculate the threshold energy for each reaction. (d) Can thermal neutrons be used to create ^{18}F?

9. Consider the following two neutron-producing reactions caused by incident 5.5-MeV α particles.

$$\alpha + {}^7\text{Li} \longrightarrow {}^{10}\text{B} + \text{n}$$

$$\alpha + {}^9\text{Be} \longrightarrow {}^{12}\text{C} + \text{n}$$

(a) What is the Q-value of each reaction? (b) What are the kinetic energies of neutrons emitted at angles of 0, 30, 45, 90, and 180 degrees.

10. Show that for inelastic neutron scattering, the minimum and maximum energies of the scattered neutron are

$$E'_{\min} = E \left[\frac{A\sqrt{1+\Delta} - 1}{A+1} \right]^2 \qquad \text{and} \qquad E'_{\max} = E \left[\frac{A\sqrt{1+\Delta} + 1}{A+1} \right]^2$$

11. How many elastic scatters, on the average, are required to slow a 1-MeV neutron to below 1 eV in (a) ^{16}O and in (b) ^{56}Fe?

12. How many neutrons per second are emitted spontaneously from a 1 mg sample of ^{252}Cf?

13. In a particular neutron-induced fission of ^{235}U, 4 prompt neutrons are produced and one fission fragment is ^{121}Ag. (a) What is the other fission fragment? (b) How much energy is liberated promptly (i.e., before the fission fragments begin to decay)? (c) If the total initial kinetic energy of the fission fragments is 150 MeV, what is the initial kinetic energy of each? (d) What is the total kinetic energy shared by the four prompt neutrons.

14. Consider the following fission reaction

$$_0^1 n + {}_{92}^{235}U \longrightarrow {}_{36}^{90}Kr + {}_{56}^{142}Ba + 4(_0^1 n) + 6\gamma$$

where ^{90}Kr and ^{142}Ba are the initial fission fragments. (a) What is the fission product chain created by each of these fission fragments? (b) What is the equivalent fission reaction taken to the stable end fission products? (c) How much energy is liberated promptly? (d) What is the total energy eventually emitted?

15. A 10-g sample of ^{235}U is placed in a nuclear reactor where it generates 100 W of thermal fission energy. (a) What is the fission rate (fission/s) in the sample? (b) After one year in the core, estimate the number of atoms of $_{43}^{99}Tc$ in the sample produced through the decay chain shown in Eq. (6.37). Notice that all fission products above $_{43}^{99}Tc$ in the decay chain have half-lives much shorter than 1 year; hence, all of these fission products can be assumed to decay to $_{43}^{99}Tc$ immediately.

16. (a) How much ^{235}U is consumed per year (in g/y) to produce enough electricity to continuously run a 100-W light bulb? (b) How much coal (in g/y) would be needed (coal has a heat content of about 12 GJ/ton)? Assume a conversion efficiency of 33% of thermal energy into electrical energy.

17. An accident in a fuel reprocessing plant, caused by improper mixing of ^{235}U, produced a burst of fission energy liberating energy equivalent to the detonation of 7 kg of TNT (4.2 GJ/ton = 4.6 kJ/g). About 80% of the fission products were retained in the building. (a) How many fissions occurred? (b) Three months after the accident, what is the rate (W) at which energy is released by all the fission products left in the building?

18. In an 8-ounce glass of water, estimate the available D-D fusion energy. For how long could this energy provide the energy needs of a house with an average power consumption of 10 kW?

19. The sun currently is still in its hydrogen burning phase, converting hydrogen into helium through the net reaction of Eq. (6.47). It produces power at the rate of about 4×10^{26} W. (a) What is the rate (in kg/s) at which mass is being converted to energy? (b) How many ^4He nuclei are produced per second? (c) What is the radiant energy flux (W cm^{-2}) incident on the earth? The average distance of the earth from the sun is 1.5×10^{11} m.

20. Explain how electricity generated from hydroelectric power, wind turbines, coal-fired power plants, and nuclear power plants are all indirect manifestations of fusion energy generated in stars.

21. Consider the reaction $_8^{18}O(_1^1 p, _0^1 n)_9^{18}F$ mentioned in problem 8. This is one of the reactions employed in cyclotron production of radionuclides for PET scanning. Calculate (a) the Q-value, (b) the kinematic threshold energy of the proton, (c) the threshold energy of the proton for the reaction, and (d) the minimum kinetic energy of the products.

Chapter 7

Radiation Interactions with Matter

We live in an environment awash in radiation. Radiation emitted by radioactive nuclides, both inside and outside our bodies, interacts with our tissues. Electromagnetic radiation of all wavelengths, including radio waves, microwaves, radar, and light, of both manmade and natural origins, constantly, bombard us. Photons are far more abundant than matter in our universe; for every nucleon there are about 10^9 photons. Cosmic rays and the subatomic debris they create during interactions in the atmosphere also impinge on us. Neutrinos from fusion reactions in stars pervade the universe in such numbers that billions per second pass through every square centimeter of our skin. Most of this radiation such as neutrinos and radio waves, fortunately, passes harmlessly through us. Other radiation such as light and longer wavelength electromagnetic radiation usually interact harmlessly with our tissues. However, shorter wavelength electromagnetic radiation (ultraviolet light, x rays, and gamma rays) and charged particles produced by nuclear reactions can cause various degrees of damage to our cells.

For radiation to produce biological damage, it must first interact with the tissue and ionize cellular atoms, which, in turn, alter molecular bonds and change the chemistry of the cells. Likewise, for radiation to produce damage in structural and electrical materials, it must cause interactions that disrupt crystalline and molecular bonds. Such radiation must be capable of creating ion-electron pairs and is termed *ionizing* radiation. Ionizing radiation is further subdivided into two classes: *directly ionizing* radiation whose interactions produce ionization and excitation of the medium, and *indirectly ionizing* radiation that cannot ionize atoms but can cause interactions whose charged products, known as *secondary* radiation, are directly ionizing. Fast moving charged particles, such as alpha particles, beta particles, and fission fragments, can directly ionize matter. Neutral particles, such as photons and neutrons, cannot interact Coulombically with the electrons of the matter through which they pass; rather, they cause interactions that transfer some of their incident kinetic energy to charged secondary particles.

165

In this chapter, we present how these two types of ionizing radiation interact with matter. Particular emphasis is given to how these radiations are attenuated as they pass through a medium, and to quantify the rate at which they interact and transfer energy to the medium.

7.1   Attenuation of Neutral Particle Beams

The interaction of a photon or neutron with constituents of matter is dominated by short-range forces. Consequently, unlike charged particles, neutral particles move in straight lines through a medium, punctuated by occasional "point" interactions, in which the neutral particle may be absorbed or scattered or cause some other type of reaction. The interactions are stochastic in nature, i.e., the travel distance between interactions with the medium can be predicted only in some average or expected sense.

The interaction of a given type of neutral radiation with matter may be classified according to the type of interaction and the matter with which the interaction takes place. The interaction may take place with an electron, and in many cases the electron behaves as though it were free. Similarly, the interaction may take place with an atomic nucleus, which in many cases behaves as though it were not bound in a molecule or crystal lattice. However, in some cases, particularly for radiation particles of comparatively low energy, molecular or lattice binding must be taken into account.

The interaction may be a scattering of the incident radiation accompanied by a change in its energy. A scattering interaction may be elastic or inelastic. Consider, for example, the interaction of a gamma photon with an electron in what is called Compton scattering. In the sense that the interaction is with the entire atom within which the electron is bound, the interaction must be considered as inelastic, since some of the incident photon's energy must compensate for the binding energy of the electron in the atom. However, in most practical cases, electron binding energies are orders of magnitude lower than gamma-photon energies, and the interaction may be treated as a purely elastic scattering of the photon by a free electron. Neutron scattering by an atomic nucleus may be elastic, in which case the incident neutron's kinetic energy is shared by that of the scattered neutron and that of the recoil nucleus, or it may be inelastic, in which case, some of the incident neutron's kinetic energy is transformed to internal energy of the nucleus and thence to a gamma ray emitted from the excited nucleus. It is important to note that, for both elastic and inelastic scattering, unique relationships between energy exchanges and angles of scattering arise from conservation of energy and linear momentum.

Other types of interactions are absorptive in nature. The identity of the incident particle is lost, and total relativistic momentum and energy are conserved, some of the energy appearing as nuclear excitation energy, some as translational, vibrational, and rotational energy. The ultimate result may be the emission of particulate radiation, as occurs in the photoelectric effect and in neutron radiative capture.

The discussion in this section of how a beam of radiation is attenuated as it passes through matter applies equally to both neutrons and photons. In later sections, descriptions specific to each type of neutral radiation and the particular radiation-medium interactions involved are given. We begin by introducing the concept of the interaction coefficient to describe how readily radiation particles interact with matter, and then use it to quantify the attenuation of a beam of neutral particles passing through some material.

7.1.1 The Linear Interaction Coefficient

The interaction of radiation with matter is always statistical in nature, and, therefore, must be described in probabilistic terms. Consider a particle traversing a homogeneous material and let $P_i(\Delta x)$ denote the probability that this particle, while traveling a distance Δx in the material, causes a reaction of type i (e.g., it is scattered). It is found empirically that the probability per unit distance traveled, $P_i(\Delta x)/\Delta x$, approaches a constant as Δx becomes very small, i.e.,

$$\mu_i \equiv \lim_{\Delta x \to 0} \frac{P_i(\Delta x)}{\Delta x}, \tag{7.1}$$

where the limiting process is performed in the same average statistical limit as was used in the definition of the radioactive decay constant λ discussed in Section 5.5.1. The quantity μ_i is a property of the material for a given incident particle and interaction. In the limit of small path lengths, μ_i is seen to be the probability, per unit differential path length of travel, that a particle undergoes an ith type of interaction.

That μ_i is constant for a given material and for a given type of interaction implies that the probability of interaction, per unit differential path length, is independent of the path length traveled prior to the interaction. In this book, when the interaction coefficient is referred to as the *probability per unit path length* of an interaction, it is understood that this is true only in the limit of very small path lengths.

The constant μ_i is called the *linear coefficient* for reaction i. For each type of reaction, there is a corresponding linear coefficient. For, example, μ_a is the *linear absorption coefficient*, μ_s the *linear scattering coefficient*, and so on. Although this nomenclature is widely used to describe photon interactions, μ_i is often referred to as the *macroscopic cross section* for reactions of type i, and is usually given the symbol Σ_i when describing neutron interactions. In this section, we will use the photon jargon, although the present discussion applies equally to neutrons.

The probability, per unit path length, that a neutral particle undergoes some sort of reaction, μ_t, is the sum of the probabilities, per unit path length of travel, for each type of possible reaction, i.e.,

$$\mu_t(E) = \sum_i \mu_i(E). \tag{7.2}$$

Since these coefficients generally depend on the particle's kinetic energy E, this dependence has been shown explicitly. The total interaction probability per unit path length, μ_t, is fundamental in describing how indirectly-ionizing radiation interacts with matter and is usually called the *linear attenuation coefficient*. It is perhaps more appropriate to use the words *total linear interaction coefficient* since many interactions do not "attenuate" the particle in the sense of an absorption interaction.

7.1.2 Attenuation of Uncollided Radiation

Consider a plane parallel beam of neutral particles of intensity I_o particles cm^{-2}s^{-1} normally incident on the surface of a thick slab. (see Fig. 7.1). As the particles pass

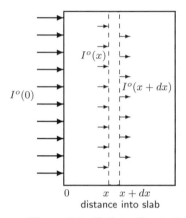

into the slab, some interact with the slab material. Of interest in many situations, is the intensity $I^o(x)$ of *uncollided* particles at depth x into the slab. At some distance x into the slab, some uncollided particles undergo interactions for the first time as they cross the next Δx of distance, thereby reducing the uncollided beam intensity at x, $I^o(x)$, to some smaller value $I^o(x + \Delta x)$ at $x + \Delta x$. The probability an uncollided particle interacts as it crosses Δx is

$$P(\Delta x) = \frac{I^o(x) - I^o(x + \Delta x)}{I^o(x)}.$$

Figure 7.1. Uniform illumination of a slab by radiation.

In the limit as $\Delta x \to 0$, we have from Eq. (7.1)

$$\mu_t = \lim_{\Delta x \to 0} \frac{P(\Delta x)}{\Delta x} = \lim_{\Delta x \to 0} \frac{I^o(x) - I^o(x + \Delta x)}{\Delta x} \frac{1}{I^o(x)} \equiv -\frac{dI^o(x)}{dx} \frac{1}{I^o(x)}$$

or

$$\frac{dI^o(x)}{dx} = -\mu_t I^o(x). \tag{7.3}$$

This has the same form as the radioactive decay Eq. (5.33). The solution for the uncollided intensity is

$$\boxed{I^o(x) = I^o(0)e^{-\mu_t x}.} \tag{7.4}$$

Uncollided indirectly-ionizing radiation is thus exponentially attenuated as it passes through a medium.

From this result, the interaction probability $P(x)$ that a particle interacts somewhere along a path of length x is

$$P(x) = 1 - \frac{I^o(x)}{I^o(0)} = 1 - e^{-\mu_t x}. \tag{7.5}$$

The probability $\overline{P}(x)$ that a particle does *not* interact while traveling a distance x is

$$\overline{P}(x) = 1 - P(x) = e^{-\mu_t x}. \tag{7.6}$$

As $x \to dx$, we find $P(dx) \to \mu_t dx$, which is in agreement with the definition of μ_t.

7.1.3 Average Travel Distance Before an Interaction

From the above results, the probability distribution for how far a neutral particle travels before interacting can be derived. Let $p(x)dx$ be the probability that a particle interacts for the first time between x and $x + dx$. Then

$$p(x)dx = \{\text{prob. particle travels a distance } x \text{ without interaction}\} \times$$
$$\{\text{prob. it interacts in the next } dx\}$$

$$= \{\overline{P}(x)\}\{P(dx)\} = \{e^{-\mu_t x}\}\{\mu_t dx\} = \mu_t e^{-\mu_t x} dx. \qquad (7.7)$$

Note that $\int_0^\infty p(x)\,dx = 1$, as is required for a proper probability distribution function.

This probability distribution can be used to find the average distance $\overline{x}$ traveled by a neutral particle to the site of its first interaction, namely, the average distance such a particle travels before it interacts. The average of x is

$$\overline{x} = \int_0^\infty x\,p(x)\,dx = \mu_t \int_0^\infty x\,e^{-\mu_t x}\,dx = \frac{1}{\mu_t}. \qquad (7.8)$$

This average travel distance before an interaction, $1/\mu_t$, is called the *mean-free-path length*.

The total linear attenuation coefficient μ_t can be interpreted, equivalently, as (1) the probability, per unit differential path length of travel, that a particle interacts, or (2) the inverse of the average distance traveled before interacting with the medium.

Note the analogy to radioactive decay, where the decay constant λ is the inverse of the mean lifetime of a radionuclide.

7.1.4   Half-Thickness

To shield against photons or neutrons, a convenient concept is that of the *half-thickness*, $x_{1/2}$, namely, the thickness of a medium required for half of the incident radiation to undergo an interaction. For the uncollided beam intensity to be reduced to one-half of its initial value, we have

$$\frac{1}{2} = \frac{I^o(x_{1/2})}{I^o(0)} = e^{-\mu_t x_{1/2}}$$

from which we find

$$x_{1/2} = \frac{\ln 2}{\mu_t}. \qquad (7.9)$$

Again, note the similarity to the half-life of radioactive decay.

Example 7.1:   What is the thickness of a water shield and of a lead shield needed to reduce a normally incident beam of 1-MeV photons to one-tenth of the incident intensity? We denote this thickness by $x_{1/10}$ and, by a similar analysis to that in Section 7.1.4, we find

$$x_{1/10} = \ln 10/\mu.$$

In Appendix C we find for water $\mu(1\ \text{MeV}) = 0.07066\ \text{cm}^{-1}$ and for lead $\mu(E = 1\ \text{MeV}) = 0.7721\ \text{cm}^{-1}$. With these values for μ and the above formula, we find for water $x_{1/10} = 32.59\ \text{cm} = 12.8\ \text{in}$ and for lead $x_{1/10} = 2.98\ \text{cm} = 1.17\ \text{in}$.

7.1.5 Scattered Radiation

The above discussion is concerned with uncollided neutral-particle radiation. However, once a particle collides, it is not necessarily absorbed. Some of the particles, upon interacting with the medium, scatter and change their energy and direction of travel. Thus, at any distance x into the attenuating slab, the population of radiation particles consists of both uncollided and scattered (collided) particles. Calculating the number of the particles that have scattered one or more times is difficult, requiring the use of *transport theory*. This theory keeps track of all radiation particles, collided and uncollided; however, it generally requires complex calculations and its discussion is well beyond the scope of this text.

However, in some situations, particularly for photon radiation, the total radiation field (collided plus uncollided) can be obtained from the uncollided intensity as

$$I(x) = B(x)I^o(x),$$

where $B(x)$ is an appropriate *buildup factor*, determined by more elaborate radiation transport calculations and whose values are tabulated or reduced to approximate formulas. In this manner, the total radiation field can be obtained from a simple (exponential attenuation) calculation of the uncollided radiation field.[1]

7.1.6 Microscopic Cross Sections

The linear coefficient $\mu_i(E)$ depends on the type and energy E of the incident particle, the type of interaction i, and the composition and density of the interacting medium. One of the more important quantities that determines μ is the density of target atoms or electrons in the material. It seems reasonable to expect that μ_i should be proportional to the "target" atom density N in the material, that is,

$$\mu_i = \sigma_i N = \sigma_i \frac{\rho N_a}{A}, \tag{7.10}$$

where σ_i is a constant of proportionality independent of N. Here ρ is the mass density of the medium, N_a is Avogadro's number (mol^{-1}), and A is the atomic weight of the medium.

The proportionality constant σ_i is called the *microscopic cross section* for reaction i, and is seen to have dimensions of area. It is often interpreted as being the effective cross-sectional area presented by the target atom to the incident particle for a given interaction. Indeed, in many cases σ_i has dimensions comparable to those expected from the physical size of the nucleus. However, this simplistic interpretation of the microscopic cross section, although conceptually easy to grasp, leads to philosophical difficulties when it is observed that σ_i generally varies with the energy of the incident particle and, for a crystalline material, the particle direction. The view that σ is the interaction probability per unit differential path length, normalized to one target atom per unit volume, avoids such conceptual difficulties while emphasizing the statistical nature of the interaction process. Cross sections

---

[1] Most often, the buildup factor is applied not to the number of particles but to the "dose" induced by the particles.

are usually expressed in units of cm^2. A widely used special unit is the *barn*, equal to 10^{-24} cm^2.

Data on cross sections and linear interaction coefficients, especially for photons, are frequently expressed as the ratio of μ_i to the density ρ, called the *mass interaction coefficient* for reaction i. Upon division of Eq. (7.10) by ρ, we have

$$\frac{\mu_i}{\rho} = \frac{\sigma_i N}{\rho} = \frac{N_a}{A}\sigma_i. \tag{7.11}$$

From this result, we see μ_i/ρ is an intrinsic property of the interacting medium— independent of its density. This method of data presentation is used much more for photons than for neutrons, in part because, for a wide variety of materials and a wide range of photon energies, μ_i/ρ is only weakly dependent on the nature of the interacting medium.

For compounds or homogeneous mixtures, the linear and mass interaction coefficients for interactions of type i are, respectively,

$$\mu_i = \sum_j \mu_i^j = \sum_j N^j \sigma_i^j \tag{7.12}$$

and

$$\frac{\mu_i}{\rho} = \sum_j w_j \left(\frac{\mu_i}{\rho}\right)^j, \tag{7.13}$$

in which the superscript j refers to the jth component of the material, the subscript i to the type of interaction, and w_j is the weight fraction of component j. In Eq. (7.12), the atomic density N^j and the linear interaction coefficient μ_i^j are values for the jth material *after* mixing.

Example 7.2: What is the total interaction coefficient ($\mu_t \equiv \mu$) for a 1-MeV photon in a medium composed of an intimate mixture of iron and lead with equal proportions by weight? From Eq. (7.13), we have

$$(\mu/\rho)^{\mathrm{mix}} = w_{\mathrm{Fe}}(\mu/\rho)^{\mathrm{Fe}} + w_{\mathrm{Pb}}(\mu/\rho)^{\mathrm{Pb}}.$$

For the shield mixture $w_{\mathrm{Fe}} = w_{\mathrm{Pb}} = 0.5$ and from the μ/ρ values in Appendix C, we obtain

$$(\mu/\rho)^{\mathrm{mix}} = 0.5(0.05951) + 0.5(0.06803) = 0.06377 \ cm^2/g.$$

If there is no volume change on mixing, not necessarily true, then 2 grams of the mixture contains a gram each of iron and lead. Then 2 grams of the mixture has a volume of $(1 \ g/\rho_{\mathrm{Fe}}) + (1 \ g/\rho_{\mathrm{Pb}}) = (1/7.784) + (1/11.35) = 0.2151 \ cm^3$. The density of the mixture is, therefore, $\rho^{\mathrm{mix}} = 2 \ g/0.2151 \ cm^3 = 9.298 \ g/cm^3$. The total interaction coefficient for a 1-MeV photon in the mixture is

$$\mu^{\mathrm{mix}} = \rho^{\mathrm{mix}}(\mu/\rho)^{\mathrm{mix}} = 9.298 \times 0.06377 = 0.5929 \ cm^{-1}.$$

Example 7.3: What is the absorption coefficient (macroscopic absorption cross section) for thermal neutrons in water? From Eq. (7.12) there are, in principle, contributions from all stable isotopes of hydrogen and oxygen, namely,

$$\mu_a^{H_2O} \equiv \Sigma_a^{H_2O} = N^1\sigma_a^1 + N^2\sigma_a^2 + N^{16}\sigma_a^{16} + N^{17}\sigma_a^{17} + N^{18}\sigma_a^{18}$$
$$= f^1 N^H \sigma_a^1 + f^2 N^H \sigma_a^2 + f^{16} N^O \sigma_a^{16} + f^{17} N^O \sigma_a^{17} + f^{18} N^O \sigma_a^{18}.$$

Here the superscripts 1, 2, 16, 17, and 18 refer to ^{1}H, ^{2}H, ^{16}O, ^{17}O, and ^{18}O, respectively, and f^i is the isotopic abundance of the ith isotope. The atomic densities $N^H = 2N^{H_2O}$, $N^O = N^{H_2O}$ so that

$$\mu_a^{H_2O} \equiv \Sigma_a^{H_2O} = N^{H_2O}\left[2f^1\sigma_a^1 + 2f^2\sigma_a^2 + f^{16}\sigma_a^{16} + f^{17}\sigma_a^{17} + f^{18}\sigma_a^{18}\right].$$

The molecular density of water is $N^{H_2O} = (\rho^{H_2O} N_a)/\mathcal{A}^{H_2O} = (1)(6.022 \times 10^{23})/$ 18.0153 $= 0.03343 \times 10^{24}$ molecules/cm^3. It is convenient to express atom densities in units of 10^{24} cm^{-3} because σ has units of 10^{-24} cm^2 and the exponents, thus, cancel when multiplying atom densities by microscopic cross sections. From the data in Table C.1, this expression is evaluated as

$$\mu_a^{H_2O} \equiv \Sigma_a^{H_2O} = 0.03343[2(0.99985)(0.333) + 2(0.00015)(0.000506)$$
$$+ (0.99756)(0.000190) + (0.00039)(0.239) + (0.00205)(0.000160)]$$
$$= 0.0223 \text{ cm}^{-1}.$$

Note that for this particular case, only the contribution of ^{1}H is significant, i.e.,

$$\Sigma_a^{H_2O} \simeq N^{H_2O}\left[2f^1\sigma_a^1\right] = 0.0223 \text{ cm}^{-1}.$$

In general, isotopes with both small abundances and small cross sections compared to other isotopes in a mixture or compound, can be ignored.

7.2   Calculation of Radiation Interaction Rates

To quantify the number of interactions caused by neutral particles as they pass through matter, we often need to estimate the number of a specific type of interaction that occurs, in a unit time and in a unit volume, at some point in the matter. This density of reactions per unit time is called the *reaction rate density*, denoted by R_i. This quantity depends on (1) the amount or strength of the radiation at the point of interest, and (2) the readiness with which the radiation can cause the interaction of interest.

7.2.1   Flux Density

To quantify the "strength" of a radiation field, we can use several measures. For simplicity, we assume initially all the radiation particles move with the same speed v and that the radiation field is constant in time.

For a parallel beam of radiation, we could use the *intensity* of the beam as a measure of the radiation field in the beam. The intensity of a beam is the flow (number per unit time) of radiation particles that cross a unit area perpendicular to the beam direction, and has dimensions, for example, of $(\text{cm}^{-2}\ \text{s}^{-1})$.

However, in most radiation fields, the particles are moving in many different (if not all) directions, and the concept of beam intensity loses its usefulness. One possible measure of such a multidirectional radiation field is the density of radiation particles $n(\mathbf{r})$ at position $\mathbf{r}$, i.e., the number of radiation particles in a differential unit volume at $\mathbf{r}$.

Another measure, and one which is very useful for calculating reaction rates, is the particle *flux density*, defined as

$$\phi(\mathbf{r}) \equiv v n(\mathbf{r}). \tag{7.14}$$

Because v is the distance one particle travels in a unit time and, because $n(\mathbf{r})$ is the number of particles in a unit volume, we see that the product $v n(\mathbf{r})$ is the distance traveled by all the particles in a unit volume at $\mathbf{r}$ in a unit time. In other words, the flux density is the total particle track length per unit volume and per unit time. The units of flux density are, for example, $v n$ $(\text{cm s}^{-1})\ \text{cm}^{-3} = \text{cm}^{-2}\ \text{s}^{-1}$.

7.2.2 Reaction-Rate Density

From the concept of flux density and cross sections, we can now calculate the density of interactions. Specifically, let $\widehat{R}_i(\mathbf{r})$ be the number of ith type interactions per unit time that occur in a unit volume at $\mathbf{r}$. Thus,

$\widehat{R}_i(\mathbf{r}) = \{\text{total path-length traveled by the radiation in one cm}^3 \text{ in one second}\}/$

$\qquad \{\text{average distance radiation must travel for an } i\text{th type interaction}$

$\quad = \{\phi(\mathbf{r})\}/\{1/\mu_i\}$

or

$$\widehat{R}_i(\mathbf{r}) = \mu_i\, \phi(\mathbf{r}). \tag{7.15}$$

For, neutron interactions, Σ_i is usually used instead of μ_i, so the density of ith type of neutron interaction per unit time is

$$\widehat{R}_i(\mathbf{r}) = \Sigma_i\, \phi(\mathbf{r}). \tag{7.16}$$

This simple expression for reaction-rate densities is a key equation for many nuclear engineering calculations.

7.2.3 Generalization to Energy- and Time-Dependent Situations

Sometimes, the radiation field at a particular location varies in time. For example, an individual might walk past a radioactive source, which emits gamma photons at a constant rate. The flux density of photons in the individual's body, however, varies in time, reaching a maximum when the person is closest to the source and decreasing as the person moves away from the source.

More frequently, the radiation field is composed of particles of many energies. For example, radioactive sources often emit photons with several discrete energies, and these photons as they scatter in the surround medium can assume a continuous distribution of lower energies. Similarly, fission neutrons are born with a continuous distribution of high energies, and, as they slow down by scattering, a continuous range of lower energies results. Thus, in general, the density of radiation particles at some position $\mathbf{r}$ is a function of energy E and time t. We define the particle density $n(\mathbf{r}, E, t)$ such that $n(\mathbf{r}, E, t)dV\, dE$ is the expected number of particles in differential volume element dV about $\mathbf{r}$ with energies in dE about E at time t. The corresponding energy- and time-dependent flux density is

$$\phi(\mathbf{r}, E, t) \equiv v(E)\, n(\mathbf{r}, E, t). \tag{7.17}$$

This flux density, by extension of the arguments in the previous section, is interpreted such that $\phi(\mathbf{r}, E, t)dE$ is the total path length traveled, in a unit time at t, by all the particles that are in a unit volume at $\mathbf{r}$ and that have energies in dE about E. The mean-free-path length for the ith interaction for such particles is $1/\mu_i(\mathbf{r}, E, t)$, where the interaction coefficient, generally, may vary with position, the incident particle energy, and even time if the medium's composition changes. Then the expected number of ith type interactions occurring in a unit volume at $\mathbf{r}$, in a unit time at time t, caused by particles with energies in dE about E is

$$\boxed{\widehat{R}_i(\mathbf{r}, E, t)\, dE = \mu_i(\mathbf{r}, E, t)\, \phi(\mathbf{r}, E, t)\, dE.} \tag{7.18}$$

From this interaction rate density, all sorts of useful information can be calculated about the effects of the radiation. For example, the total number of fissions that occur in some volume V inside a reactor core between times t_1 and t_2 by neutrons of all energies is

$$\text{Number of fissions in } (t_1, t_2) = \int_{t_1}^{t_2} dt \iiint_V dV \int_0^{E_{\max}} dE\, \Sigma_f(\mathbf{r}, E, t)\phi(\mathbf{r}, E, t). \tag{7.19}$$

7.2.4 Radiation Fluence

In most interaction rate calculations, the interaction properties of the medium do not change appreciably in time. For this usual situation, the example of Eq. (7.19) reduces to

$$\text{Number of fissions in } (t_1, t_2) = \iiint_V dV \int_0^{E_{\max}} dE\, \Sigma_f(\mathbf{r}, E)\, \Phi(\mathbf{r}, E), \tag{7.20}$$

where the *fluence* of radiation between t_1 and t_2 is the time-integrated flux density, namely

$$\boxed{\Phi(\mathbf{r}, E) \equiv \int_{t_1}^{t_2} dt\, \phi(\mathbf{r}, E, t).} \tag{7.21}$$

For a steady-state radiation field, the fluence is simply $\Phi(\mathbf{r}, E) = (t_2 - t_1)\phi(\mathbf{r}, E)$

Often the time interval over which the flux density is integrated to obtain the fluence is implicitly assumed to be over all prior time. Thus, the fluence at time t is

$$\Phi(\mathbf{r}, E, t) \equiv \int_{-\infty}^{t} dt' \, \phi(\mathbf{r}, E, t'). \tag{7.22}$$

The rate of change of the fluence with time, or the fluence rate, is simply the flux density. The fluence is thus used to quantify the cumulative effect of radiation interactions, while the flux density is used to quantify the rate of interactions.

7.2.5 Uncollided Flux Density from an Isotropic Point Source

From the above discussion, we see that to quantify the number of radiation-induced interactions, we need to know (1) the flux density $\phi(\mathbf{r}, E, t)$ of the radiation field, and (2) the interaction coefficient (macroscopic cross section) for the material of interest. The interaction coefficients are readily found from Eq. (7.10) with the use of tabulated microscopic cross sections σ_i (the microscopic cross sections for photons and neutrons are discussed at length in the following Sections). Calculation of the flux density $\phi(\mathbf{r}, E, t)$ is more difficult and, generally, requires particle transport calculations, a subject well beyond the scope of this book.

Fortunately, in many practical situations, the flux density can be approximated by the flux density of uncollided source particles, a quantity which is relatively easy to obtain. For example, flux density of particles in a plane parallel beam is simply the beam intensity. Then for the slab problem considered in Section 7.1.2, we obtain from Eq. (7.4)

$$\phi^o(x) = \phi(0)e^{-\mu_t x}.$$

The most basic source of radiation is a "point source," a mathematical idealization used to represent sources that are very small compared to other problem dimensions. For example, a small sample of a radionuclide can often be treated as if it were a point source emitting S_p radiation particles per unit time. Even spatially large sources can be viewed as a set of point-like sources by conceptually dividing the large source volume into a set of contiguous small volumes, each of which acts like a point source. Most radiation (such as radiation from radioactive decay or neutrons from a fission) is also emitted *isotropically*, i.e., the radiation is emitted in all directions with equal probability. Consequently, the point isotropic source of radiation is of fundamental importance in quantifying how radiation spreads out from its source. The uncollided flux density arising from an isotropic point source is derived below for several frequently encountered situations.

Point Source in Vacuum

Consider a point isotropic source that emits S_p particles per unit time, all with energy E, into an infinite vacuum as in Fig. 7.2(a). All particles move radially outward without interaction, and because of the source isotropy, each unit area on an imaginary spherical shell of radius r has the same number of particles crossing it, namely, $S_p/(4\pi r^2)$. It then follows from the definition of the flux density that the flux density ϕ^o of uncollided particles at a distance r from the source is

$$\phi^o(r) = \frac{S_p}{4\pi r^2}. \tag{7.23}$$

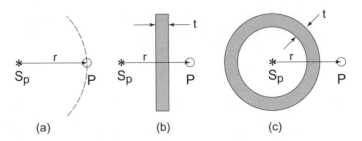

Figure 7.2. Point isotropic source (a) in a vacuum, (b) with a slab shield, and (c) with a spherical-shell shield. Point P is the location of the point detector.

If, at distance r, we place a small homogeneous mass, such as a small radiation detector, with a volume ΔV_d, the interaction rate R_d in the mass obtained from Eq. (7.15) is

$$R^o(r) = \mu_d(E)\Delta V_d \frac{S_p}{4\pi r^2}, \tag{7.24}$$

where $\mu_d(E)$ is the linear interaction coefficient for the reaction of interest in the detector material.

Notice that the flux density and interaction rate R decrease as $1/r^2$ as the distance from the source is increased. This decreasing dose with increasing distance is sometimes referred to as *geometric attenuation*.

Point Source in a Homogeneous Attenuating Medium

Now consider the same point monoenergetic isotropic source embedded in an infinite homogeneous medium characterized by a total interaction coefficient μ. As the source particles stream radially outward, some interact before they reach the imaginary sphere of radius r and do not contribute to the uncollided fluence. The number of source particles that travel at least a distance r without interaction is $S_p e^{-\mu r}$, so that the uncollided flux density is

$$\phi^o(r) = \frac{S_p}{4\pi r^2} e^{-\mu r}. \tag{7.25}$$

The term $e^{-\mu r}$ is referred to as the *material attenuation* term to distinguish it from the $1/r^2$ geometric attenuation term.

Point Source with a Shield

Now suppose that the only attenuating material separating the source and the detector is a slab of material with attenuation coefficient μ and thickness t as shown in Fig. 7.2(b). In this case, the probability that a source particle reaches the detector without interaction is $e^{-\mu t}$, so that the uncollided flux density is

$$\phi^o(r) = \frac{S_p}{4\pi r^2} e^{-\mu t}. \tag{7.26}$$

This same result holds if the attenuating medium has any shape [e.g., a spherical shell of thickness t as shown in Fig. 7.2(c)] provided that a ray drawn from the source to the detector passes through a thickness t of the attenuating material.

If the interposing shield is composed of a series of different materials such that an uncollided particle must penetrate a thickness t_i of a material with attenuation coefficient μ_i before reaching the detector, the uncollided flux density is

$$\phi^o(r) = \frac{S_p}{4\pi r^2} \exp(-\textstyle\sum_i \mu_i t_i). \tag{7.27}$$

Here $\sum_i \mu_i t_i$ is the total number of mean-free-path lengths of attenuating material that an uncollided particle must traverse without interaction, and $\exp(-\sum_i \mu_i t_i)$ is the probability that a source particle traverses this number of mean-free-path lengths without interaction.

Point Source in a Heterogeneous Medium
The foregoing treatment for a series of different attenuating materials can be extended to an arbitrary heterogeneous medium in which the interaction coefficient $\mu(\mathbf{r})$ is a function of position in the medium. In this case, the probability that a source particle reaches the detector a distance r away is $e^{-\ell}$, where ℓ is the total number of mean-free-path lengths of material that a particle must traverse without interaction to reach the detector. This dimensionless distance can be written formally as

$$\ell = \int_0^r ds\, \mu(s), \tag{7.28}$$

where the integration is along the ray from source to detector.[2] The uncollided flux density then is given by

$$\phi^o(r) = \frac{S_p}{4\pi r^2} \exp\left[-\int_0^r ds\, \mu(s)\right]. \tag{7.29}$$

Polyenergetic Point Source
So far it has been assumed that the source emits particles of a single energy. If the source emits particles at several discrete energies, the reaction rate in a small detector mass is found simply by summing the reaction rates for each monoenergetic group of source particles. Thus if a fraction f_i of the S_p source particles are emitted with energy E_i, the total interaction rate caused by uncollided particles streaming through the small detector mass at distance r from the source is

$$R^o(r) = \sum_i \frac{S_p f_i \mu_d(E_i) \Delta V_d}{4\pi r^2} \exp\left[-\int_0^r ds\, \mu(s, E_i)\right]. \tag{7.30}$$

For a source emitting a continuum of particle energies such that $N(E)\, dE$ is the probability that a source particle is emitted with energy in dE about E, the total uncollided interaction rate is

$$R^o(r) = \int_0^\infty dE\, \frac{S_p N(E) \mu_d(E_i) \Delta V_d}{4\pi r^2} \exp\left[-\int_0^r ds\, \mu(s, E)\right]. \tag{7.31}$$

---

[2]The distance ℓ is sometimes referred to as the *optical depth* or *optical thickness*, in analogy to the equivalent quantity used to describe the exponential attenuation of light. This nomenclature is used here to distinguish it from the geometric distance r.

Example 7.4: A point source with an activity of 500 Ci emits 2-MeV photons with a frequency of 70% per decay. What is the flux density of 2-MeV photons 1 meter from the source? What thickness of iron needs to be placed between the source and detector to reduce the uncollided flux density to $\phi_{\max}^o = 2 \times 10^3$ cm^{-2} s^{-1}?

The strength S_p (photons/s) $=$ (500 Ci)(3.7 $\times$ 10^{10} Bq/Ci)(0.7 photons/decay) $=$ 1.295 $\times$ 10^{13} s^{-1}. Because the mean-free-path length ($= 1/\mu$) for a 2-MeV photon in air is 187 m, we can ignore air attenuation over a 1-m distance. The flux density of 2-MeV photons 1 meter from the source is then, from Eq. (7.23),

$$\phi_1^o = \frac{S_p}{4\pi r^2} = \frac{1.295 \times 10^{13}}{4\pi(100 \text{ cm})^2} = 1.031 \times 10^6 \text{ cm}^{-2} \text{ s}^{-1}.$$

The uncollided flux density with an iron shield of thickness t placed between the source and the 1-meter detection point is

$$\phi_2^o = \frac{S_p}{4\pi r^2} \exp[-\mu t] = \phi_1^o \exp[-\mu t].$$

Solving for the shield thickness gives

$$t = -\frac{1}{\mu} \ln\left(\frac{\phi_2^o}{\phi_1^0}\right).$$

From App. C, μ(2 MeV) for iron equals $(\mu/\rho)\rho =$ (0.04254 cm^2/g)(7.874 g/cm^3) $=$ 0.3222 cm^{-1}. With $\phi_2^0/\phi_1^0 = \phi_{\max}^o/\phi_1^o$, the thickness of the iron shield is

$$t = -\frac{1}{\mu} \ln\left(\frac{\phi_{\max}^o}{\phi_1^0}\right) = \frac{1}{0.3222 \text{ cm}^{-1}} \ln\left(\frac{2 \times 10^3}{1.031 \times 10^6}\right) = 19.4 \text{ cm}.$$

7.3 Photon Interactions

A photon is a quantum (particle) of electromagnetic radiation which travels with the speed of light $c = 2.9979 \times 10^8$ m/s, has zero rest mass, and an energy $E = h\nu$, where $h = 6.62608 \times 10^{-34}$ J s is Planck's constant, and ν is the photon frequency. Photon energies between 10 eV and 20 MeV are important in radiation shielding design and analysis. For this energy range, only the photoelectric effect, pair production, and Compton scattering mechanisms of interaction are significant. Of these three, the photoelectric effect predominates at the lower photon energies. Pair production is important only for higher-energy photons. Compton scattering predominates at intermediate energies. In rare instances one may need to account also for coherent scattering (scattering from multiple atomic electrons simultaneously). These three photon interaction mechanisms are discussed below.

7.3.1 Photoelectric Effect

In the photoelectric effect, a photon interacts with an entire atom, resulting in the emission of a photoelectron, usually from the K shell of the atom. Although the difference between the photon energy E and the electron binding energy E_b is distributed between the electron and the recoil atom, virtually all of that energy is carried as kinetic energy of the photoelectron because of the comparatively small electron mass. Thus, $E_e = E - E_b$.

K-shell binding energies E_k vary from 13.6 eV for hydrogen to 7.11 keV for iron, 88 keV for lead, and 116 keV for uranium. As the photon energy drops below E_k, the cross section drops discontinuously as E decreases further, the cross section increases until the first L edge is reached, at which energy the cross section drops again, then rises once more, and so on for the remaining edges. These "edges" for lead are readily apparent in Fig. 7.3. The cross section varies as E^{-n}, where $n \simeq 3$ for energies less than about 150 keV and $n \simeq 1$ for energies greater than about 5 MeV. The atomic photoelectric cross section varies as Z^m, where m varies from about 4 at $E = 100$ keV to 4.6 at $E = 3$ MeV. As a very crude approximation in the energy region for which the photoelectric effect is dominant,

$$\sigma_{ph}(E) \propto \frac{Z^4}{E^3}. \tag{7.32}$$

For light nuclei, K-shell electrons are responsible for almost all photoelectric interactions. For heavy nuclei, about 80% of photoelectric interactions result in the ejection of a K-shell electron. Consequently, the approximation is often made for heavy nuclei that the total photoelectric cross section is 1.25 times the cross section for K-shell electrons.

As the vacancy left by the photoelectron is filled by an electron from an outer shell, either fluorescence x rays or Auger electrons[3] may be emitted. The energies of the fluorescence x rays are unique for each element and, hence, are also called *charactistic x rays*. The probability of x-ray emission is given by the *fluorescent yield*. For the K shell, fluorescent yields vary from 0.005 for $Z = 8$ to 0.965 for $Z = 90$. Although x rays of various energies may be emitted, the approximation is often made that only one x ray or Auger electron is emitted, with energy equal to the binding energy of the photoelectron.

7.3.2 Compton Scattering

In the process known as Compton scattering, a photon is scattered by a free (unbound) electron. The kinematics of such a scattering process are a bit more complicated than the earlier discussion of binary nuclear reactions presented in Chapter 6, in which we used classical mechanics. However, in Section 2.2.2, the relativistic scattering of a photon, of initial energy E, from a free electron is analyzed. From Eq. (2.26), the energy E' of the scattered photon is

$$E' = \frac{E}{1 + (E/m_e c^2)(1 - \cos\theta_s)}, \tag{7.33}$$

---

[3]If an electron in an outer shell, say Y, makes a transition to a vacancy in an inner shell, say X, an x ray may be emitted with energy equal to the difference in binding energy between the two shells. Alternatively, an electron in some other shell, say Y′, which may be the same as Y, may be emitted with energy equal to the binding energy of the electron in shell X less the sum of the binding energies of electrons in shells Y and Y′. This electron is called an *Auger electron*.

where $\cos\theta_s$ is the photon scattering angle, and $m_e c^2$ is the rest-mass energy of the electron, 0.51100 MeV or 8.1871×10^{-14} J.

The total cross section, per atom with Z electrons, based on the free-electron approximation, is given by the well-known Klein-Nishina formula [Evans 1955]

$$\sigma_c(E) = \pi Z r_e^2 \lambda \left[(1 - 2\lambda - 2\lambda^2) \ln\left(1 + \frac{2}{\lambda}\right) + \frac{2(1 + 9\lambda + 8\lambda^2 + 2\lambda^3)}{(\lambda + 2)^2} \right]. \quad (7.34)$$

Here $\lambda \equiv m_e c^2 / E$, a dimensionless quantity, and r_e is the *classical electron radius*. The value of r_e is given by

$$r_e = \frac{e^2}{4\pi\epsilon_o m_e c^2}, \quad (7.35)$$

where e is the electronic charge, 1.6022×10^{-19} C, and ϵ_o is the permittivity of free space, 8.8542×10^{-14} F cm^{-1}. Thus, $r_e = 2.8179 \times 10^{-13}$ cm.

Incoherent Scattering Cross Sections for Bound Electrons

Equations (7.33) and (7.34) for scattering from free electrons break down when the kinetic energy of the recoil electron is comparable to its binding energy in the atom. More complicated semiempirical models must be used to evaluate the incoherent scattering cross section, denoted by σ_{inc}. An example of the difference between the incoherent scattering cross section from free electrons σ_c and from bound electrons σ_{inc} is shown in Fig. 7.3, where it is seen that the scattering cross section from bound electrons decreases at very low photon energies while the Compton cross section stays constant.

Thus, binding effects might be thought to be an important consideration for the attenuation of low-energy photons in media of high atomic number. For example, the binding energy of K-shell electrons in lead is 88 keV. However, at this energy, the cross section for the photoelectric interactions greatly exceeds the incoherent scattering cross section. Radiation attenuation in this energy region is dominated by photoelectric interactions and, in most attenuation calculations, the neglect of electron binding effects in incoherent scattering causes negligible error. Figure 7.3 shows the relative importance, in lead, of electron binding effects by comparing photoelectric cross sections with those for incoherent scattering from both free and bound electrons.

Coherent Scattering

In competition with the incoherent scattering of photons by individual electrons is coherent (Rayleigh) scattering by the electrons of an atom collectively. Since the recoil momentum in the Rayleigh interaction is taken up by the atom as a whole, the energy loss of the gamma photon is slight and the scattering angle small. For example, for 1-MeV photons scattering coherently from iron atoms, 75% of the photons are scattered within a cone of less than 4° half-angle [Hubbell 1969]. As is apparent from Fig. 7.3, coherent scattering cross sections may greatly exceed incoherent scattering cross sections, especially for low-energy photons and high-Z materials. However, because of the minimal effect on photon energy and direction, and because the coherent scattering cross section is so much less than the cross section for the photoelectric effect, it is common practice to ignore Rayleigh scattering in radiation shielding calculations, especially when electron binding effects mentioned in the preceding paragraph are ignored.

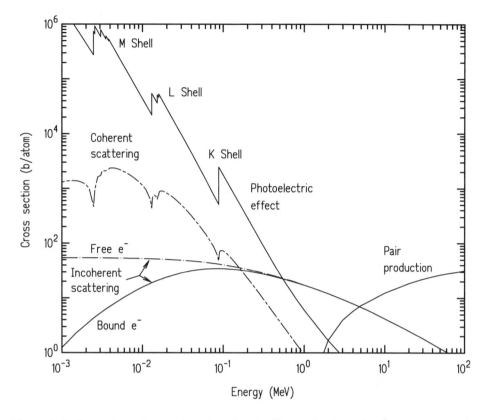

Figure 7.3. Comparison of scattering, photoelectric-effect, and pair-production cross sections for photon interactions in lead, based on data from the EPIC data library [Cullen 1994].

7.3.3   Pair Production

In this process, the incident photon with energy E_γ is completely absorbed and, in its place, appears a positron-electron pair. The interaction is induced by the strong electric field in the vicinity of the nucleus and has a photon threshold energy of $2m_e c^2$ ($= 1.02$ MeV). It is possible, but much less likely, that the phenomenon is induced by the electric field of an electron (triplet production), for which case the threshold energy is $4m_e c^2$. The discussion that follows is limited to the *nuclear* pair production process.

In this process, the nucleus acquires indeterminate momentum but negligible kinetic energy. Thus,

$$E_+ + E_- = E_\gamma - 2m_e c^2, \tag{7.36}$$

in which E_+ and E_- are the kinetic energies of the positron and electron, respectively. To a first approximation, the total atomic pair production cross section varies as Z^2. The cross section increases with photon energy, approaching a constant value at high energy. The electron and positron both have directions not far from the direction of the incident photon (although they are separated by π radians in azimuth about the photon direction). As an approximation, the positron-electron angles with respect to the incident photon direction are $m_e c^2/E_\gamma$ radians.

The fate of the positron is annihilation in an interaction with an ambient electron, generally after slowing to practically zero kinetic energy. The annihilation process results in the creation of two photons moving in opposite directions, each with energy $m_e c^2$.

7.3.4   Photon Attenuation Coefficients

The photon linear attenuation coefficient μ is, in the limit of small path lengths, the probability per unit distance of travel that a gamma photon undergoes any significant interaction. Thus, for a specified medium, the effective total attenuation coefficient is

$$\mu(E) \equiv N \left[\sigma_{ph}(E) + \sigma_{inc}(E) + \sigma_{pp}(E) \right]$$
$$\simeq N \left[\sigma_{ph}(E) + \sigma_c(E) + \sigma_{pp}(E) \right], \tag{7.37}$$

in which $N = \rho N_a / A$ is the atom density. Note that Rayleigh coherent scattering and other minor effects are specifically excluded from this definition.

More common in data presentation is the effective total *mass interaction coefficient*

$$\frac{\mu}{\rho} = \frac{N_a}{A} \left[\sigma_{ph}(E) + \sigma_c(E) + \sigma_{pp}(E) \right] = \frac{\mu_{ph}}{\rho} + \frac{\mu_c}{\rho} + \frac{\mu_{pp}}{\rho}, \tag{7.38}$$

which, from Eq. (7.11), is independent of the density of the medium. Effective total mass interaction coefficients for four common gamma-ray shielding materials are shown in Fig. 7.4, and the mass coefficients μ_i / ρ for various reactions are tabulated in Appendix C.

Correction for Secondary Radiation

When a photon interacts with matter, not all of the photon's energy is transferred to the medium locally. Secondary photons (fluorescence, Compton scattered photons, annihilation photons, and bremsstrahlung) are produced and continue to travel through the medium eventually depositing their energy at points far removed from the original interaction site. To allow easy calculations of energy deposition, various photon interaction coefficients are defined as the total coefficient μ / ρ times the fraction of photon energy that does not escape the interaction site as different types of secondary photons. Of considerable use is the *linear energy absorption coefficient* defined as

$$\mu_{en} \simeq \mu f, \tag{7.39}$$

where f is the fraction of the incident photon energy E that is transferred to secondary charged particles and subsequently transferred to the medium through ionization and excitation interactions. Other, photon coefficients, denoted by μ_a, $\mu_{tr'}$, and μ_{tr}, which account for fewer secondary photons escaping from the interaction site, are sometimes encountered. These derived coefficients are shown schematically in Fig. 7.5.

7.4   Neutron Interactions

The interaction processes of neutrons with matter are fundamentally different from those for the interactions of photons. Whereas photons interact, more often, with

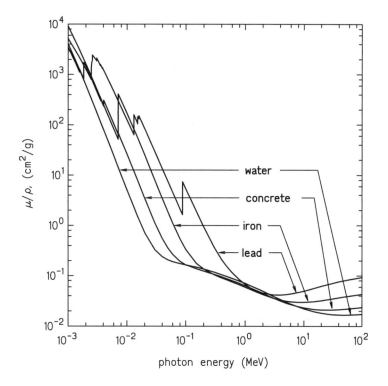

Figure 7.4. Comparison of the total mass interaction coefficients, less coherent scattering, for four important shielding materials.

the atomic electrons, neutrons interact essentially only with the atomic nucleus. The cross sections that describe the various neutron interactions are also very unlike those for photons. Neutron cross sections not only can vary rapidly with the incident neutron energy, but also can vary erratically from one element to another and even between isotopes of the same element. The description of the interaction of a neutron with a nucleus involves complex interactions between all the nucleons in the nucleus and the incident neutron, and, consequently, fundamental theories which can be used to predict neutron cross-section variations in any accurate way are still lacking. As a result, all cross-section data are empirical in nature, with little guidance available for interpolation between different energies or isotopes.

Over the years, many compilations of neutron cross sections have been generated, and the more extensive, such as the evaluated nuclear data files (ENDF) [Kinsey 1979; Rose and Dunford 1991], contain so much information that digital computers are used to process these cross-section libraries to extract cross sections or data for a particular neutron interaction. Even with the large amount of cross-section information available for neutrons, there are still energy regions and special interactions for which the cross sections are poorly known. For example, cross sections which exhibit sudden decreases as the neutron energy changes only slightly or cross sections for interactions which produce energetic secondary photons are of

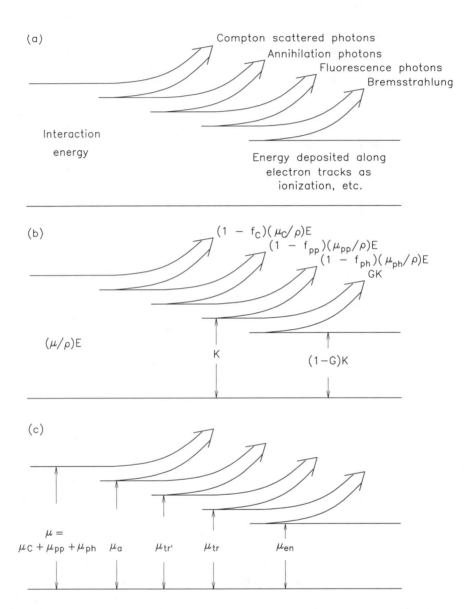

Figure 7.5. Relationships leading to definitions of various energy deposition coefficients for photons. (a) Energy deposition for photon energy involved in the interactions in an incremental volume of material. (b) Formulas for the energy per unit mass of the material in the incremental volume, corresponding to the various energy increments in (a). (c) Linear coefficients defined by their proportionality to the mass energy relationships in diagrams (a) and (b).

obvious concern in shielding problems and often still are not known to an accuracy sufficient to perform precise deep-penetration shielding calculations.

7.4.1 Classification of Types of Interactions

There are many possible neutron-nuclear interactions, only some of which are of concern in radiation-protection calculations. Ultra-high-energy interactions can produce many and varied secondary particles; however, the energies required for such reactions are usually well above the neutron energies commonly encountered, and, therefore, such interactions can be neglected here. Similarly, for low-energy neutrons, many complex neutron interactions are possible—Bragg scattering from crystal planes, phonon excitation in a crystal, coherent scattering from molecules, and so on—none of which is of particular importance in neutron shielding or dosimetry problems. As summarized in Table 7.1, the reactions of principal importance for radiation shielding applications are the absorption reactions and the high-energy scattering reactions with their various associated angular distributions.

Table 7.1. Principal nuclear data required for neutron shielding calculations.

High-energy interactions (1 eV $< E <$ 20 MeV)
    Elastic scattering cross sections
    Angular distribution of elastically scattered neutrons
    Inelastic scattering cross sections
    Angular distribution of inelastically scattered neutrons
    Gamma-photon yields from inelastic neutron scattering
    Resonance absorption cross sections

Low-energy interactions ($<$ 1 eV)
    Thermal-averaged absorption cross sections
    Yield of neutron-capture gamma photons
    Fission cross sections and associated gamma-photon and neutron yields

---

The total cross section, which is the sum of cross sections for all possible interactions, gives a measure of the probability that a neutron of a certain energy will interact in some manner with the medium. The components of the total cross section for absorption and scattering interactions are usually of primary concern for shielding analysis. Nonetheless, when the total cross section is large, the probability of some type of interaction is great and thus the total cross section, which is the most easily measured and the most widely reported, gives at least an indication of the neutron energy regions over which one must investigate the neutron interactions in greater detail.

The total cross sections, although they vary from nuclide to nuclide and with the incident neutron energy, have certain common features. For the sake of classification, the nuclides are usually divided into three broad categories: light nuclei, with mass number < 25; intermediate nuclei; and heavy nuclei, with mass number > 150. Example total cross sections for each category are shown in Figs. 7.6 through 7.9. Thermal cross sections for several important isotopes are given in Table C.1 of Appendix C.

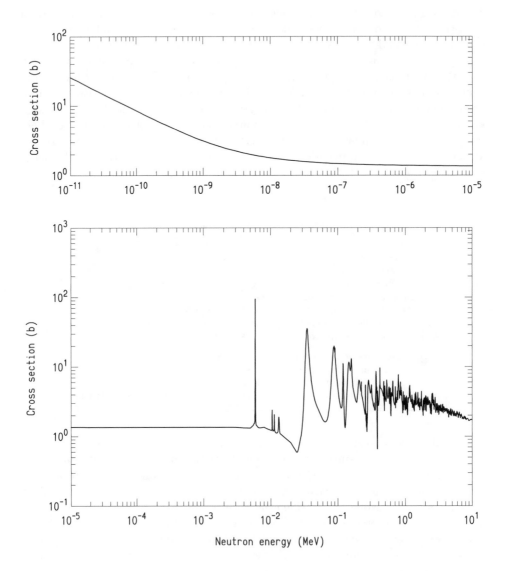

Figure 7.6. Total neutron cross section for aluminum computed using NJOY-processed ENDF/B (version V) data.

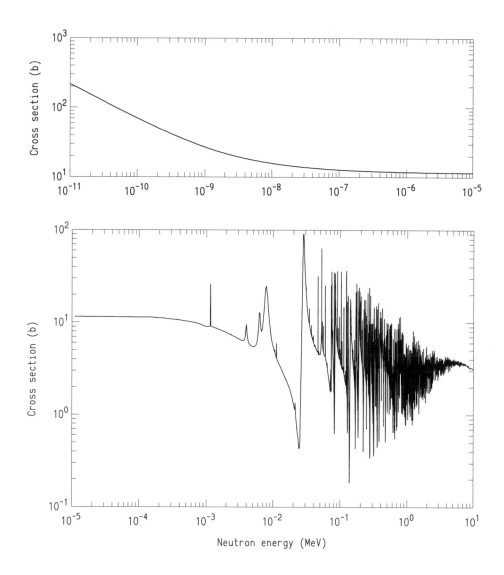

Figure 7.7. Total neutron cross section for iron computed using NJOY-processed ENDF/B (version V) data.

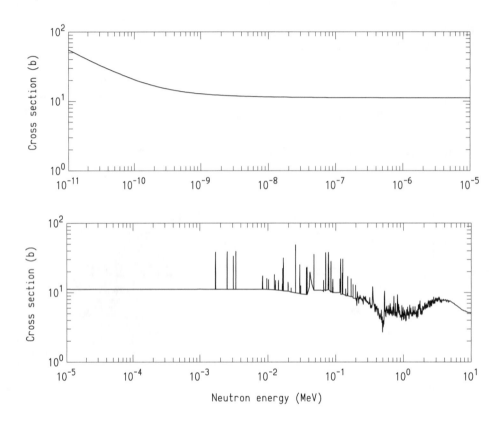

Figure 7.8. Total neutron cross section for lead computed using NJOY-processed ENDF/B (version V) data.

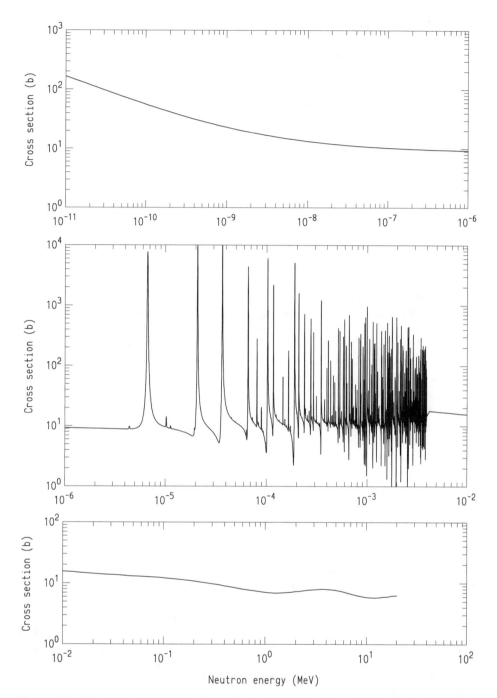

Figure 7.9. Total neutron cross section for uranium computed using NJOY-processed ENDF/B (version V) data. Above 4 keV, the resonances are no longer resolved and only the average cross-section behavior is shown.

For light nuclei and some magic number nuclei, the cross section at energies less than 1 keV often varies as

$$\sigma_t = \sigma_1 + \frac{\sigma_2}{\sqrt{E}}, \tag{7.40}$$

where E is the neutron energy, σ_1 and σ_2 are constants, and the two terms on the right-hand side represent the elastic scattering and the radiative capture (or absorption) reactions, respectively. For solids at energies less than about 0.01 eV, there may be *Bragg cutoffs* representing energies below which coherent scattering from the various crystalline planes of the material is no longer possible. These cutoffs are not shown in Figs. 7.6 through 7.9. For energies from about 0.1 eV to a few keV, the cross sections are usually slowly varying and featureless. At higher energies, resonances appear that are fairly wide (keV to MeV). Of all the nuclides, only hydrogen and its isotope deuterium exhibit no resonances. For both isotopes, the cross sections above 1 eV are almost constant up to the MeV region, above which they decrease with increasing energy.

For heavy nuclei, as illustrated in Fig. 7.9, the total cross section, unless masked by a low-energy resonance, exhibits a $1/\sqrt{E}$ behavior at low energies and usually a Bragg cutoff (not shown). The resonances appear at much lower energies than for the light nuclei, usually in the eV region, and have very narrow widths (1 eV or less) with large peak values. Above a few keV, the resonances are so close together and so narrow that they cannot be resolved, and the cross sections appear to be smooth except for a few broad resonances. Finally, the intermediate nuclei, as would be expected, are of intermediate character between the light and heavy nuclei, with resonances in the region from 100 eV to several keV. The resonances are neither as high nor as narrow as for the heavy nuclei.

For neutron shielding purposes, one is usually concerned with neutrons of energy 500 keV or greater. Neutron penetration studies are somewhat simplified in this energy region compared to the lower-energy region since the heavy nuclides have no resolved resonances and the cross sections vary smoothly with the neutron energy. Only for the light elements must the complicating resonances be considered. Further, the absorption cross sections for all nuclides are usually very small compared to those for other reactions at all energies except thermal energies. Over the fission-neutron energy spectrum, the (n, γ) reaction cross sections seldom exceed 200 mb for the heavy elements and, for the lighter elements, this cross section is considerably smaller. Only for thermal neutrons and a few isolated absorption resonances in the keV region for heavy elements is the (n, γ) reaction important. Some important thermal activation cross sections are given in Table C.2 of Appendix C.

In the high-energy region, by far the most important neutron interaction is the scattering process. Generally, elastic scattering is more important, although, when the neutron energy somewhat exceeds the energy level of the first excited state of the scattering nucleus, inelastic scattering becomes possible. For energies greater than about 8 MeV, multiple-particle reactions such as $(n, 2n)$ and $(n, n+p)$ become possible. In most materials of interest for shielding purposes, the thresholds for these reactions are sufficiently high and the cross sections sufficiently small that these neutron-producing reactions may be ignored compared to the inelastic scattering reactions. However, a few rare reactions producing secondary neutrons should be noted. The threshold for $(n, 2n)$ reactions are particularly low for D and Be (3.3

MeV and 1.84 MeV, respectively) and for these two nuclei there is no inelastic scattering competition. Also, fission induced by fast neutrons may release multiple neutrons. In most shielding situations, though, reactions which produce secondary neutrons are not encountered.

Finally, interactions such as (n, p) and (n, α), which produce charged particles, may be of importance when light elements are involved. In the MeV region, the (n, α) reaction cross sections for Be, N, and O are appreciable fractions of the total cross sections and may exceed the inelastic scattering contributions. This situation is probably true for most light elements, although only partial data are available. For heavy and intermediate nuclei, the charged-particle emission interactions are at most a few percent of the total inelastic interaction cross sections and, hence, are usually ignored.

Example 7.5: A ^{54}Mn sample with mass $m = 2$ g and density $\rho = 7.3$ g cm^{-3} is exposed for $\Delta t = 2$ minutes in a thermal neutron field with a flux density of 10^{13} cm^{-2} s^{-1}. What is the activity of the sample immediately after the exposure as a result of the radioactive ^{55}Mn ($T_{1/2} = 2.579$ h) produced by (n,γ) reactions in the sample?

Since the irradiation time is very small compared to the half-life of ^{55}Mn, we can neglect any radioactive decay. Then the number of ^{55}Mn atoms produced, N^{55}, equals the number of (n,γ) reactions that occur in the sample. The sample volume $\Delta V = m/\rho$, and from Eq. (7.24) it follows

$$N^{55} = \Delta V \mu_\gamma \phi \Delta t = \left(\frac{m}{\rho} \right) \left(\frac{\rho N_a}{A^{54}} \sigma_\gamma^{54} \right) \phi \Delta t = \frac{m N_a}{A^{54}} \sigma_\gamma^{54} \phi \Delta t.$$

From Table C.2, we find the thermal-neutron, microscopic, (n,γ) cross section for ^{54}Mn is $\sigma_\gamma^{54} = 13.3$ b. The decay constant for ^{55}Mn is $\lambda = \ln 2/T_{1/2} = \ln 2/(2.579 \text{ h} \times 3600 \text{ s/h}) = 7.466 \times 10^{-5}$ s^{-1}. Hence, the activity of the sample after irradiation is

$$\text{Activity} = \lambda N^{55} = \lambda \frac{m N_a}{A^{54}} \sigma_\gamma^{54} \phi \Delta t$$

$$= (7.466 \times 10^{-5} \text{ s}^{-1}) \frac{(2 \text{ g})(0.6022 \times 10^{24} \text{ atoms/mol})}{55 \text{ g/mol}} \times$$

$$(13.3 \times 10^{-24} \text{ cm}^2/\text{atom})(10^{13} \text{ cm}^{-2} \text{ s}^{-1})(120 \text{ s})$$

$$= 2.609 \times 10^{10} \text{ Bq}.$$

7.4.2 Fission Cross Sections

Of special interest in the design of nuclear reactors are the fission cross sections of heavy isotopes. *Fissile isotopes* are those which can undergo fission upon the absorption of a thermal neutron. Three important fissile isotopes are ^{233}U, ^{235}U, and ^{239}Pu. The total and fission cross sections for the most important uranium fissile isotope ^{235}U is shown in Fig. 7.10. Fission cross sections for other important nuclides are given in Table C.1 of Appendix C.

Fissionable isotopes are those which can be made to fission upon the absorption of a neutron with sufficiently high kinetic energy. Most heavy isotopes are thus fissionable; however, only a few will fission readily upon absorption of a neutron with energies around 1 MeV, comparable to the energies of fission neutrons. The fission cross sections of three fissionable isotopes are shown in Fig. 7.11.

7.5 Attenuation of Charged Particles

Understanding how charged particles interact with matter, the ranges of these directly ionizing particles, and the rates at which their energy is dissipated along their paths is important for several practical reasons. First, secondary charged particles resulting from neutron and photon interactions are responsible for the radiation effects of principal concern, namely, biological, chemical, and structural changes. Second, detection and measurement of photons or neutrons are almost always effected through interactions that produce secondary charged particles. Indeed, the roentgen unit of photon exposure (see Chapter 9) is defined in terms of ionization produced by secondary electrons. Third, a knowledge of the ranges of charged particles leads directly to the determination of shield thicknesses necessary to stop them or the extents of the regions in which they can cause biological damage.

7.5.1 Interaction Mechanisms

Charged particles such as beta particles, alpha particles, and fission fragments, are *directly ionizing* radiation and interact with the ambient medium primarily through the long-range electromagnetic force. Consequently, charged particles continuously interact with the electrons of the ambient atoms, interacting simultaneously with multiple electrons. These interactions primarily cause the atomic electrons of the atoms along the particle's path to jump into excited states or to be torn from the atoms creating electron-ion pairs. Thousands of such interactions are needed to transfer the particle's initial kinetic energy to the medium and slow the particle to thermal energies. With almost all their kinetic energy exhausted, the charged particles capture ambient electrons and become neutral atoms, or, in the case of beta particles, become part of the ambient electron field. Charged particles of a given initial energy have a maximum distance or *range* they can travel through a medium before they are stopped and become incorporated into the medium.

An indirectly ionizing photon or neutron, by contrast, travels in straight-line segments past enormous numbers of ambient atoms before interacting and altering its direction and energy. Unlike charged particles, neutrons and photons typically interact only a few to several tens of times before they are captured by the ambient medium. Also, indirectly ionizing particles have no definite range, as do charged particles, but rather are exponentially attenuated as they traverse the medium.

Although, excitation and ionization of the surrounding medium is the primary method by which a charged particle loses its energy, it may (very rarely) cause a nuclear reaction or transmutation, or scatter from an atomic nucleus (usually, only when the particle has slowed and is near the end of its path). Also, if the energy of the charged particle is sufficiently high and/or the mass the of particle is sufficiently small, the charged particles will lose energy through the emission of electromagnetic radiation, called *bremsstrahlung* (lit. "braking radiation"), as

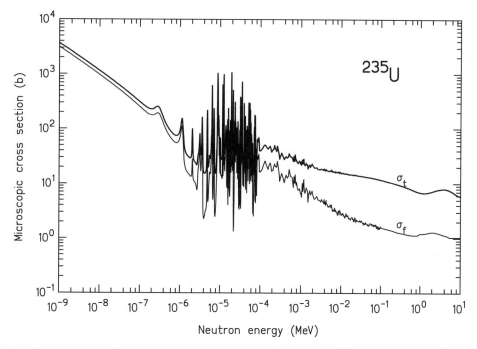

Figure 7.10. The total and fission cross section for ^{235}U based on NJOY-processed ENDF/B (version V) data.

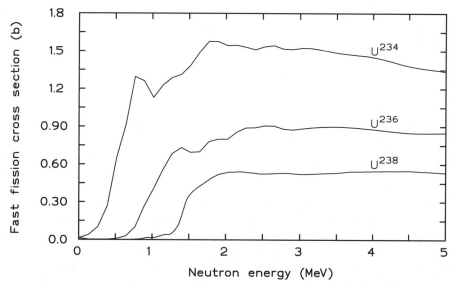

Figure 7.11. The fast fission cross section for three fissionable uranium isotopes based on NJOY-processed ENDF/B (version V) data.

they are deflected (centripetally accelerated) primarily by the atomic nuclei of the medium.

7.5.2   Particle Range

Energetic charged particles cause thousands of ionizations and excitations of the atoms along their path before they are slowed and become part of the ambient medium. Charged particles, unlike photons and neutrons, can travel only a certain maximum distance in matter, called the *range*, before they are stopped. The paths traveled by light particles, like electrons and positrons, are quite different from those of heavy charged particles such as protons and alpha particles. We considered each separately.

Heavy Charged Particles

As an example, consider a 4 MeV alpha particle. From Eq. (6.22), the maximum energy loss of such an alpha particle when it interacts with an electron is 2.2 keV. Most such alpha-electron interacts result in far less energy loss, although usually sufficient energy transfer to free the electron from its atom or at least to raise the atom to a high excitation state. Consequently, many thousands of these interactions are required for the alpha particle to lose its kinetic energy.

Moreover, the mass of the alpha particle being over 7000 times that of an electron, the alpha particle experiences negligible deflection from its path of travel. An important characteristic of such heavy charged particles, as they move through a medium, is that they travel in almost straight lines, unless they are deflected by a nucleus. Such nucleus-particle interactions are rare and usually occur only when the charged particles have lost most of their kinetic energy.

Because of the large number of interactions with electrons of the ambient medium, heavy charged particles, of the same mass and initial kinetic energy, travel in straight lines and penetrate almost the same distance into a medium before they are stopped. Because of the stochastic nature of the particle-medium interactions and because, near the end of their paths, some heavy particles experience large angle deflections from atomic nuclei, the penetration distances exhibit a small variation called *straggling*

Many measurements of the *range* of heavy particles have been made. In Fig. 7.12, the fraction of alpha particles, with the same initial energy, that travel various distances s into a medium is illustrated. We see from this figure, that the number of alpha particles penetrating a distance s into a medium remains constant until near the end of their range. The number penetrating beyond this maximum distance falls rapidly.

The distance a heavy charged particle travels in a medium before it is stopped is not a precise quantity because of the stochastic nature of the number and energy losses of all the particle-medium interactions, and because of possible large-angle deflections near the end of the path. From Fig. 7.12, we see that the range of a heavy charged particle is not an exact quantity. Several measures of the range can be defined. We can define $\overline{R}$ as the *mean range*, that is, the distance at which the intensity of heavy particles has been halved. This range is at the maximum of the differential energy curve or *straggling curve* $dI(s)/ds$, shown as the dotted curve in Fig. 7.12. Alternatively, the range can be taken as R_e, the *extrapolated range*, that

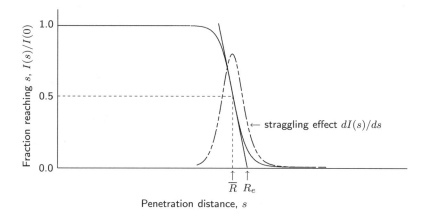

Figure 7.12. Fraction of α-particles that penetrate a distance s. The dotted line shows the derivative $dI(s)/ds$ and is termed the straggling curve, a result of the large-angle scattering near the end of the particles' paths and the statistical fluctuations in the number of interactions needed to absorb the particles' initial energy.

is, the value obtained by drawing the tangent to the curve at its inflection point and extrapolating the tangent until it crosses the s-axis. Finally, if we treat the slowing down of heavy charged particles as a continuous process, a unique range, called the *continuous slowing-down approximation* (CSDA), a unique CSDA range can be calculated. The CSDA range, denoted simply by R, is now the standard way of specifying the penetration depth of the heavy charged particle. Although slightly different, all three of these definitions of a heavy particle's range are, for practical purposes, nearly the same.

Electrons

Beta particles (electrons or positrons) also produce numerous ionizations and electronic excitations as they move through a medium. Most interactions are small deflections accompanied by small energy losses. However, beta particles, having the same mass as the electrons of the medium, can also undergo large-angle scatters producing secondary electrons with substantial recoil energy. These energetic secondary electrons (called *delta* rays) in turn pass through the medium causing additional ionization and excitation. Large-angle deflections from atomic nuclei can also occur with negligible energy loss. Consequently, the paths travelled by beta particles as they transfer their energy to the surrounding medium are far from straight. Such paths are shown in Figs. 7.13 and 7.14.

For such paths, we see that the distances beta particles travel into a medium vary tremendously. The range of beta particles is defined as the path length of the particles, i.e., the distance they travel along their twisted trajectories. This range is also the maximum distance a beta particle can penetrate, although most, because of the twisting of their paths, seldom penetrate this far (see Section 7.5.4).

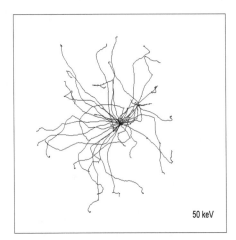

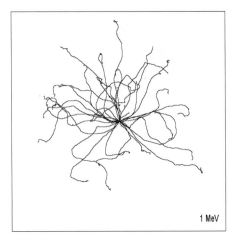

Figure 7.13. Tracks of 30 electrons from a 50-keV point isotropic source in water, shown as orthographic projections of tracks into a single plane. The box has dimensions $2r_o \times 2r_o$. Calculations performed using the EGS4 code, courtesy of Robert Stewart, Kansas State University.

Figure 7.14. Tracks of 30 electrons from a 1-MeV point isotropic source in water, shown as orthographic projections of tracks into a single plane. The box has dimensions $2r_o \times 2r_o$. Calculations performed using the EGS4 code, courtesy of Robert Stewart, Kansas State University.

7.5.3 Stopping Power

With initial kinetic energy E_o, a particle is slowed to kinetic energy $E(s)$, after traveling a distance s along its path, as a result of both Coulombic interactions with (atomic) electrons and radiation losses (bremsstrahlung). During deceleration, the stopping power, $(-dE/ds)$, generally increases until the energy of the particle is so low that charge neutralization or quantum effects bring about a reduction in stopping power. Only for particles of very low energy do collisions with atomic nuclei of the stopping medium become important.

Heavy charged particles (those with masses greater than or equal to the proton mass), with kinetic energies much less than their rest-mass energies, slow down almost entirely due to Coulombic interactions. A multitude of such interactions takes place—so many that the slowing down is virtually continuous and along a straight-line path. These interactions, taken individually, may range from ionization processes producing energetic recoil electrons (*delta rays*) to weak atomic or molecular excitation which may not result in ionization at all. The stopping power resulting from Coulombic interactions, $(-dE/ds)_{coll}$, is called *collisional stopping power*, or the *ionization stopping power*.

Another mechanism, which is important for electrons, is radiative energy loss, characterized by the *radiative stopping power* $(-dE/ds)_{rad}$. Also, a careful treatment of electron slowing down requires accounting for the possibility of delta-ray production and the concomitant deflection of the incident electron from its original direction. In this discussion, electron range will refer to the mean path length rather than the straight-line penetration distance.

Collisional Stopping Power

The collisional stopping power depends strongly on the charge number z and speed v of the particle. The speed v is commonly expressed in terms of β, the ratio of v to the speed of light c. If m is the rest mass of the particle and E its kinetic energy, then $\beta = v/c = \sqrt{E^2 + 2Emc^2}/(E + mc^2)$. The collisional stopping power also depends on the density ρ of the stopping medium, the ratio Z/A of the medium's atomic number to the atomic mass (atomic mass units), and the effective ionization potential I (typically a few tens of eV) of the medium. For protons with energies between about 2 and 10 MeV (or other heavy charged particles with comparable speeds) and for electrons of energies between 0.01 and 10 MeV, the collisional stopping power is [Evans 1955]

$$\left(-\frac{dE}{ds} \right)_{coll} = \rho \frac{Z}{A} z^2 f(I, \beta), \qquad (7.41)$$

where $f(I, \beta)$ is a complicated function different for heavy charged particles and for electrons. Selected collisional stopping-power data are presented in Fig. 7.16.

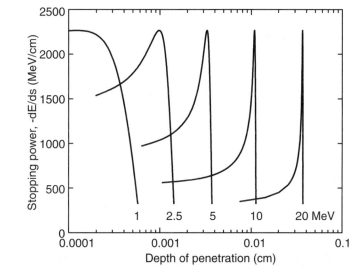

Figure 7.15. "Bragg curves" for alpha particles (helium nuclei) in liquid water: stopping power versus depth of penetration. Data obtained using the ASTAR program [Berger 1992].

The Bragg Curve

In reporting on his 1904 research on the ionization produced along the track of an alpha-particle, W.H. Bragg plotted a graph of the ion pairs produced per unit distance as a function of the distance penetrated by alpha particles in air. Known as the Bragg curve, this type of graph has been used countless times to illustrate the concepts of range and stopping power. Closely related to the Bragg curve is a graph of stopping power versus distance penetrated along the track of an alpha particle or other charged particle. This is done in Fig. 7.15 illustrating energy loss along

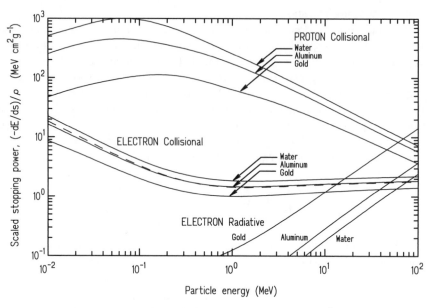

Figure 7.16. Stopping power $(-dE/ds)/\rho$ in mass units (MeV cm^2/g) for protons and electrons. The dashed line for aluminum is the collisional LET for positrons. Electron data from ICRU [1984] and proton data from Janni [1982].

the tracks of alpha particles in water. The stopping power is maximum for alpha particles of energy about 0.7 MeV. For alpha particles with initial energies above 0.7 MeV, stopping power increases along the particle path, reaches a maximum, and then decreases rapidly as the particle releases most of its kinetic energy near the end of its path.

Radiative Stopping Power for Electrons
A charged particle gives up its kinetic energy either by collisions with electrons along its path or by photon emission as it is deflected by the electric fields of nuclei. The photons produced by the deflection of the charged particle are called *bremsstrahlung*.

No single formula can adequately describe the radiative stopping power over a wide range of electron kinetic energies. Bremsstrahlung is emitted when the electron is deflected by the electromagnet field of both ambient electrons and ambient nuclei, the later usually being the dominant mechanism. The radiative stopping power may be written [ICRU 1984]

$$\left(-\frac{dE}{ds}\right)_{rad} = \left(\frac{\rho N_a}{A}\right)(E + m_e c^2)Z^2 F(E, Z), \tag{7.42}$$

where $F(E, Z)$ is a function strongly dependent on E and weakly dependent on Z. The function varies slowly with E for energies up to about 1 MeV, then increases as E increases, reaching another plateau for energies greater than about 100 MeV. Selected data for the radiative stopping power are presented in Fig. 7.16.

For an electron with kinetic energy $E \gg m_e c^2$, we see from Eq. (7.42) the radiative stopping power, $(-dE/ds)_{rad}$, varies as $Z^2 E$. The collisional stopping power for relativistic electrons ($\beta \simeq 1$), however, varies as Z (see Eq. (7.41)). The ratio of radiative to collision stopping powers is then proportional to ZE. From this we see bremsstrahlung production is most important for energetic electrons in a medium with large Z.

For a relativistic heavy charged particle of rest mass M, with $E >> Mc^2$, it can be shown that the ratio of radiative to ionization losses is approximately [Evans 1955]

$$\frac{(-dE/ds)_{rad}}{(-dE/ds)_{coll}} \simeq \frac{EZ}{700} \left(\frac{m_e}{M}\right)^2, \tag{7.43}$$

where E is in MeV. From this result it is seen that bremsstrahlung is more important for high-energy particles of small mass incident on high-Z material. In shielding situations, only electrons ($m_e/M = 1$) are ever of importance for their associated bremsstrahlung. All other charged particles are far too massive to produce significant amounts of bremsstrahlung. Bremsstrahlung from electrons is of particular radiological interest for devices that accelerate electrons, such as betatrons and x-ray tubes, or for situations involving radionuclides that emit only beta particles.

Example 7.6: At what energy does an electron moving through gold lose as much energy by bremsstrahlung as it does by ionizing and exciting gold atoms? From Eq. (7.43),

$$E \simeq \frac{700}{Z} \left(\frac{M}{m_e}\right)^2 \frac{(-dE/ds)_{rad}}{(-dE/ds)_{coll}}.$$

For gold $Z = 79$ and for equal energy loss by both mechanisms, we find for electrons $M = m_e$ that $E = 700/79 = 8.9$ MeV. This cross-over energy is in agreement with the data shown in Fig. 7.16.

7.5.4 Estimating Charged-Particle Ranges

It is common to neglect energy-loss fluctuations and assume that particles lose energy continuously along their tracks, with a mean energy loss per unit path length given by the total stopping power. Under this approximation [ICRU 1984], the range of a charged particle with initial kinetic energy E_o is given by

$$R = \int_0^R ds = \int_0^{E_o} \frac{dE}{(-dE/ds)_{tot}}. \tag{7.44}$$

where $(-dE/ds)_{tot} = (-dE/ds)_{coll} + (-dE/ds)_{rad}$, the total stopping power.

Evaluation of the integral is complicated by difficulties in formulating both the radiative stopping power and the collisional stopping power for low-energy particles, particularly electrons, and it is common to assume that the reciprocal of the stopping power is zero at zero energy and increases linearly to the known value at the least energy. The range defined in this way is identified as the CSDA range, i.e., the range in the *continuous slowing-down approximation*. Proton and electron CSDA ranges are presented in Fig. 7.17.

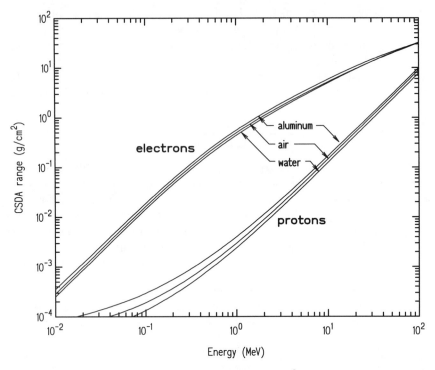

Figure 7.17. Range or path length ρR, in mass units (g/cm^2), in the continuous slowing-down approximation. Electron data from ICRU [1984] and proton data from Janni [1982].

It may be inferred from data given by Cross, Freedman, and Wong [1992] that for a beam of electrons, ranging in energy from 0.025 to 4 MeV, normally incident on water, about 80% of the electron energy is deposited within a depth of about 60% of the CSDA range, 90% of the energy within about 70% of the range, and 95% of the energy within about 80% of the range. For a point isotropic source of monoenergetic electrons of the same energies in water, 90% of the energy is deposited within about 80% of the range, and 95% of the energy within about 85% of the range. In either geometry, all of the energy is deposited within about 110% of the CSDA range. For protons in aluminum, for example, ρR is 0.26 mg/cm^2 at 100 keV. However, for such protons normally incident on a sheet of aluminum, the likely penetration depth,[4] in mass thickness, is only 0.22 mg/cm^2. For 10-keV protons incident on gold, the average penetration depth is only about 20% of the CSDA range.

For *heavy* charged particles, we may use Eq. (7.44) to obtain the dependence of the range on the particle's mass m and charge z. Because bremsstrahlung is seldom important for heavy charged particles with energies below several tens of MeV, we may take $(-dE/ds)_{tot} \simeq (-dE/ds)_{coll}$. From Eq. (7.41) we see the collisional

---

[4]If a particle starts its trajectory along the z-axis, the penetration depth is the final z-coordinate when the particle has come to rest.

stopping power is of the form $(-dE/ds)_{coll} = \rho(Z/A)z^2 f(v)$ where $f(v)$ is a function only of v or $\beta = v/c$. Then from Eq. (7.44), the mass thickness range, ρR is

$$\rho R = \rho \int_0^{E_o} \frac{dE}{(-dE/ds)_{coll}} = \rho \int_0^{E_o} \frac{1}{(-dE/ds)_{coll}} \frac{dE}{dv} \, dv. \qquad (7.45)$$

Substitution of $dE/dv = mv$ and $(-dE/ds)_{coll} = \rho(Z/A)z^2 f(v)$ then gives

$$\rho R = \left(\frac{A}{Z}\right) \frac{m}{z^2} \int_0^{v_o} \frac{v \, dv}{f(v)} = \left(\frac{A}{Z}\right) \frac{m}{z^2} g(v_o), \qquad (7.46)$$

where $g(v_o)$ is a function only of the initial speed v_o of the particle. From this result, three useful "rules" for relating different ranges follow.

1. The mass-thickness range ρR is independent of the density of the medium.

2. For particles of the *same initial speed* in the same medium, ρR is *approximately* proportional to m/z^2, where m and z are for the charged particle.

3. For particles of the *same initial speed*, in different media, ρR is *approximately* proportional to $(m/z^2)(Z/A)$, where Z is the atomic number of the stopping medium and A is its atomic weight.

Thus, from rule 2, in a given medium, a 4-MeV alpha particle has about the same range as a 1-MeV proton. However, these rules fail for particle energies less than about 1 MeV per atomic mass unit. For example, the CSDA range of a 0.4-MeV alpha particle in aluminum is about twice that of a 0.1-MeV proton. The application of these rules to find the range of other heavy charged particles using proton ranges is illustrated in Exercise 7.7.

Table 7.2. Constants for the empirical formula $y = a + bx + cx^2$ relating charged-particle energy and CSDA range, in which y is $\log_{10}$ of ρR (g/cm^2) and x is $\log_{10}$ of the initial particle energy E_o (MeV). Valid for energies between 0.01 and 100 MeV.

	Protons			Electrons		
Material	a	b	c	a	b	c
Aluminum	-2.3829	1.3494	0.19670	-0.27957	1.2492	-0.18247
Iron	-2.2262	1.2467	0.22281	-0.23199	1.2165	-0.19504
Gold	-1.8769	1.1664	0.20658	-0.13552	1.1292	-0.20889
Air	-2.5207	1.3729	0.21045	-0.33545	1.2615	-0.18124
Water	-2.5814	1.3767	0.20954	-0.38240	1.2799	-0.17378
Tissue[a]	-2.5839	1.3851	0.20710	-0.37829	1.2803	-0.17374
Bone[b]	-2.5154	1.3775	0.20466	-0.33563	1.2661	-0.17924

[a] Striated muscle (ICRU).
[b] Cortical bone (ICRP).
Source: Shultis and Faw [2000].

Approximate Formula for Proton and Electron Range

Very useful empirical methods are available for making approximate estimates of charged particle ranges. CSDA ranges for *electrons and protons* with initial energy E may be estimated using the formula

$$\rho R = 10^{a+bx+cx^2}, \tag{7.47}$$

where $x = \log_{10} E$. The empirical constants a, b, and c are given in Table 7.2.

Example 7.7: Estimate the range in water of a 6-MeV triton (nucleus of ^{3}H). Data for the empirical range formula of Eq. (7.47) are not provided for a triton. We must, therefore, estimate the triton's range from the observation that, for a given medium, the range R of heavy charged particles with the same speed is proportion to m/z^2, where m and z are the mass and charge of the particle.

Step 1: First find the kinetic energy of a proton with the same speed as the 6-MeV triton. Classical mechanics, appropriate for these energies, gives

$$\frac{E_p}{E_T} = \frac{(1/2)m_p v_p^2}{(1/2)m_T v_T^2} = \frac{m_p}{m_T} = \frac{1}{3}.$$

The proton energy with the same speed as a 6-MeV triton is, thus, $E_p = E_T/3 = 2$ MeV.

Step 2: Now we find the range in water of a 2-MeV proton using the empirical formula of Eq. (7.47) and the data in Table 7.2. With $x = \log_{10} 2 = 0.30103$

$$\rho R_p(2 \text{ MeV}) = 10^{-2.5814+1.3767x+0.20954x^2} = 0.00711 \text{ g/cm}^2.$$

Since $\rho = 1$ g/cm^3, $R_p(2$ MeV$) = 0.00711$ cm.

Step 3: From the range "rule" for the same medium, we have

$$\frac{R_T(6 \text{ MeV})}{R_p(2 \text{ MeV})} = \frac{m_T}{m_p} \frac{z_p^2}{z_T^2} = \frac{3}{1}\frac{1}{1} = 3.$$

The range of the triton is then $R_T(6$ MeV$) = 3R_p(2$ MeV$) = 0.021$ cm.

Range of Fission Fragments

The range of a *fission fragment* is difficult to calculate because the fission fragment's charge changes as it slows down. The lighter of the two fragments, having more initial kinetic energy, has a somewhat longer range. Figures 7.18 and 7.19 demonstrate the residual energy after a fission fragment passes through a mass thickness ρR. The x-axis intercept is the range of the fragment. An empirical formula with an accuracy of $\pm 10\%$ for the range (in units of mass thickness) of a fission product, with initial kinetic energy E (in MeV), is [Alexander and Gazdik 1960]

$$\rho R \text{ (mg/cm}^2) = C\, E^{2/3}, \qquad C = \begin{cases} 0.14, & \text{air} \\ 0.19, & \text{aluminum} \\ 0.50, & \text{gold} \end{cases} . \tag{7.48}$$

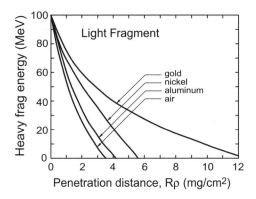
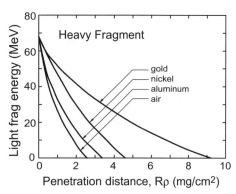

Figure 7.18. Energy of the median light fission product (initial energy 99.9 MeV) after penetrating a distance R into various materials. After Keepin [1965].

Figure 7.19. Energy of the median heavy fission product (initial energy 67.9 MeV) after penetrating a distance R into various materials. After Keepin [1965].

BIBLIOGRAPHY

ALEXANDER, J.M. AND M.F. GAZDIK, *Phys. Rev*, **120**, 874 (1960).

BARKAS, W.H., AND M.J. BERGER, *Tables of Energy Losses and Ranges of Heavy Charged Particles*, Report NASA SP-3013, National Aeronautics and Space Administration, Washington, DC, 1964.

BERGER, M.J., *ESTAR, PSTAR, and ASTAR: Computer Programs for Calculating Stopping-Power and Range Tables for Electrons, Protons and Helium Ions*, Report NISTIR 4999, National Institute of Standards and Technology, 1992.

CULLEN, D.E., *Program EPICSHOW: A Computer Program to Allow Viewing of the EPIC Data Libraries*, Report UCRL-ID-116819, Lawrence Livermore National Laboratory, Livermore, CA, 1994.

CROSS, W.G., N.O. FREEDMAN, AND P.Y. WONG, *Tables of Beta-Ray Dose Distributions in Water*, Report AECL-10521, Chalk River Laboratories, Atomic Energy of Canada, Ltd., Chalk River, Ontario, 1992.

EVANS, R.D., *The Atomic Nucleus*, McGraw-Hill, New York, 1955; republished by Krieger Publishing Co., Melbourne, FL, 1982.

EISBERG, R. AND R. RESNICK, *Quantum Physics of Atoms, Molecules, Solids, Nuclei, and Particles*, John Wiley & Sons, NY, 1985.

HUBBELL, J.H., *Photon Cross Sections, Attenuation Coefficients, and Energy Absorption Coefficients from 10 keV to 100 GeV*, Report NSRDS-NBS 29, National Bureau of Standards, Washington, DC, 1969.

ICRU, *Stopping Powers for Electrons and Positrons*, Report 37, International Commission on Radiation Units and Measurements, Washington, DC, 1984.

JANNI, J.F., "Proton Range-Energy Tables, 1 keV–10 GeV," *At. Data and Nucl. Data Tables*, **27**, 147–529 (1982).

KEEPIN, G.R., *Physics of Nuclear Kinetics*, Addison-Wesley, Reading, MA, 1965.

KINSEY, R., *ENDF/B-V Summary Documentation*, Report BNL-NCS-17541 (ENDF-201), 3rd ed., Brookhaven National Laboratory, Upton, NY, 1979.

LAMARSH, J.R., *Nuclear Reactor Theory*, Addison-Wesley, Reading, MA, 1966.

MAYO, R.M., *Nuclear Concepts for Engineers*, American Nuclear Society, La Grange Park, IL, 1998.

ROSE, P.F. AND C.L. DUNFORD (Eds.), *ENDF-102, Data Formats and Procedures for the Evaluated Nuclear Data File ENDF-6*, Report BNL-NCS 44945 (Rev.), Brookhaven National Laboratory, Upton, NY, 1991.

SHULTIS, J.K. AND R.E. FAW, *Radiation Shielding*, American Nuclear Society, La Grange Park, IL, 2000.

PROBLEMS

1. A broad beam of neutrons is normally incident on a homogeneous slab 6-cm thick. The intensity of neutrons transmitted through the slab without interactions is found to be 30% of the incident intensity. (a) What is the total interaction coefficient μ_t for the slab material? (b) What is the average distance a neutron travels in this material before undergoing an interaction?

2. With the data of Appendix C, calculate the half-thickness for 1-MeV photons in (a) water, (b) iron, and (c) lead.

3. Based on the interaction coefficients tabluated in Appendix C, plot the tenth-thickness (in centimeters) versus photon energy from 0.1 to 10 MeV for water, concrete, iron, and lead.

4. A material is found to have a tenth-thickness of 2.3 cm for 1.25-MeV gamma rays. (a) What is the linear attenuation coefficient for this material? (b) What is the half-thickness? (c) What is the mean-free-path length for 1.25-MeV photons in this material?

5. In natural uranium, 0.720% of the atoms are the isotope ^{235}U, 0.0055% are ^{234}U, and the remainder ^{238}U. From the data in Table C.1, what is the total linear interaction coefficient (macroscopic cross section) for a thermal neutron in natural uranium? What is the total macroscopic fission cross section for thermal neutrons?

6. Calculate the linear interaction coefficients in pure air at 20°C and 1 atm pressure for a 1-MeV photon and a thermal neutron (2200 m s^{-1}). Assume that air has the composition 75.3% nitrogen, 23.2% oxygen, and 1.4% argon *by mass*. Use the following data:

	Photon	Neutron
Element	μ/ρ (cm^2 g^{-1})	σ_{tot} (b)
Nitrogen	0.0636	11.9
Oxygen	0.0636	4.2
Argon	0.0574	2.2

7. At a particular position, the flux density of particles is 2×10^{12} cm^{-2} s^{-1}. (a) If the particles are photons, what is the density of photons at that position? (b) If the particles are thermal neutrons (2200 m/s), what is the density of neutrons?

8. A beam of 2-MeV photons with intensity 10^8 cm^{-2}s^{-1} irradiates a small sample of water. (a) How many photon-water interactions occur in one second in one cm^3 of the water? (b) How many positrons are produced per second in one cm^3 of the water?

9. A small homogeneous sample of mass m (g) with atomic mass A is irradiated uniformly by a constant flux density ϕ (cm^{-2} s^{-1}). If the total atomic cross section for the sample material with the irradiating particles is denoted by σ_t (cm^2), derive an expression for the fraction of the atoms in the sample that interact during a 1-h irradiation. State any assumptions made.

10. A 1-mCi source of ^{60}Co is placed in the center of a cylindrical water-filled tank with an inside diameter of 20 cm and depth of 100 cm. The tank is made of iron with a wall thickness of 1 cm. What is the uncollided flux density at the outer surface of the tank nearest the source?

11. What is the maximum possible kinetic energy (keV) of a Compton electron and the corresponding minimum energy of a scattered photon resulting from scattering of **(a)** a 100-keV photon, **(b)** a 1-MeV photon, and **(c)** a 10-MeV photon? Estimate for each case the range the electron would have in air of 1.2 mg/cm^3 density and in water of 1 g/cm^3 density.

12. From Fig. 7.7, the total microscopic cross section in iron for neutrons with energy of 27 keV is about 0.4 b, and for a neutron with an energy of 28 keV about 90 b. (a) Estimate the fraction of 27-keV neutrons that pass through a 10-cm thick slab without interaction. (b) What is this fraction for 28-keV neutrons?

13. When an electron moving through air has 5 MeV of energy, what is the ratio of the rates of energy loss by bremsstrahlung to that by collision? What is this ratio for lead?

14. About what thickness of aluminum is needed to stop a beam of (a) 2.5-MeV electrons, (b) 2.5-MeV protons, and (c) 10-MeV alpha particles? Hint: For parts (a) and (b), use Table 7.2 and compare your values to ranges shown in Fig. 7.17. For part (c), use the range interpolation rules on page 201.

15. Estimate the range of a 10-MeV tritium nucleus in air.

Chapter 8

Detection and Measurement of Radiation

Douglas S. McGregor
Kansas State University

Over 100 years ago, radiation detectors were used to discover x rays and natural radiation emissions, and those discoveries led to Wilhelm Roentgen receiving the first Nobel prize in physics in 1901, and Henri Becquerel sharing in the third physics prize in 1903. Roentgen's radiation detector was a plate coated with a scintillating platino-cyanide material. Becquerel's radiation detector was simply a photographic plate. Radiation detectors of a wide variety are used for detecting, measuring, characterizing, and classifying radiation emissions. The three main functions that characterize a radiation detector are (1) a radiation absorber, (2) an observable phenomenon from the interaction, and (3) a method to measure the observable. Regarding the first requirement, clearly the detector will not function if radiation simply passes through it without interacting in some capacity, so the device must be designed to somehow absorb the radiation of interest, whether that be neutrons, gamma rays, x rays, charged particles, or even cosmic rays and neutrinos. Not all materials work equally well for each of these different forms of ionizing radiation, hence, the choice of absorber is important. Second, after absorbing energy from a radiation interaction, the detector must yield some observable phenomenon to the user, otherwise, the interaction will go unnoticed. Third, whatever the phenomenon is, a method must exist to measure it quantitatively or qualitatively. Although these concepts are simple, their effective implementation can often be elusive.

A radiation detector may be composed of gaseous, liquid, or solid substances, or a combination of each, confined within a controlled region. Radiation detectors range in size from only a few microns to, in some cases, dimensions so large that they span acres of ground. The observable phenomena from radiation detectors include changes in temperature, scintillation light emission, changes in electrical conductivity, changes in color, and accumulation of a visible image. There are many types of radiation detectors, and because ionizing radiation, both natural and induced, comes in many forms, no single radiation detector is available that uniformly detects all forms of ionizing radiation equally well. Hence, the proper detector must be chosen to accomplish the desired radiation measurement.

There are detectors designed to simply notify the user that ionizing radiation is present, while other detectors are designed to inform the user of the radiation type and its energy. Some detectors are used to image where the radiation is emanating from, and some detectors are used to assist with medical imaging. Some detectors are small, can be worn on a person, and are used to accumulate radiation interactions over an extended period of time, thereby notifying the wearer of the radiation dose received. In fact, some radiation detectors commonly found in household environments are actually used as fire alarms. The list is so long, that it is quite possible that the reader has been using some sort of radiation detector without knowing it! In the present chapter, the basic operating principles behind radiation detector designs and their uses are explored. The most common types are introduced, and some of the more interesting exotic devices are briefly discussed.

8.1 Gas-Filled Detectors

In 1908, Ernest Rutherford and Hans Geiger constructed a device composed of a metallic cylinder with a thin wire arranged axially inside. The gas medium in the device was simply air. With the application of a voltage, alpha particles projected into the device produced sizable currents as measured with an electrometer. Rutherford and Geiger had devised the first radiation counter. They also noticed that the behavior of the detector changed with increasing voltage, mainly that alpha particles could be detected at much lower applied voltages than beta particles, a technique and application that later became known as *proportional* counting. Experiments conducted with the gas-filled detectors clearly showed distinctive regions of operation, as shown in Fig. 8.1.

The principle behind a gas-filled detector is quite simple. Radiation interactions in the gas or ejected particles from radiation interactions in the chamber walls cause the detector gas to become ionized, and a charge cloud composed of electrons and positive ions appears. A voltage placed across electrodes in the gas chamber causes the electrons and ions to drift apart, where electrons drift towards the anode and the positive ions drift towards the cathode. As the charged particles, or *charge carriers*, move through the chamber, they induce current to flow in a circuit externally connected to the chamber. This current, or change in current, can then be measured as an indication that a radiation interaction occurred in the chamber.

8.1.1 General Operation

Gas detectors can be operated in *pulse mode* or *current mode*. Pulse mode is generally used in low to moderate radiation fields. In such a case, a single radiation quantum, such as an alpha particle, beta particle, or gamma ray, interacts in the chamber volume, giving rise to an ionized cloud. The charge carriers drift apart and, as they induce current to flow to the device terminals, a charging circuit, usually consisting of a preamplifier and feedback loop, integrates the current and stores the charge, thereby producing a voltage potential. This voltage is measured as a *single* event, indicating that a single radiation quantum has been detected. The preamplifier circuit is subsequently discharged and reset, allowing the device to measure the next radiation interaction event. Hence, each voltage pulse from the detector indicates an individual radiation interaction event. Although extremely useful, there

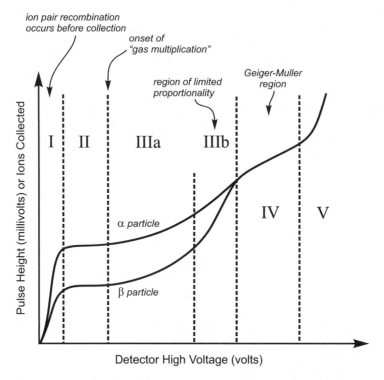

Figure 8.1. The observed output pulse height versus the applied high voltage for a gas-filled detector, showing the main regions: (I) recombination, (II) ion chamber, (III) proportional, (IV) Geiger-Müller, and (V) continuous discharge. This plot is often referred to as the *gas curve*.

are drawbacks to this method. Should another radiation interaction occur while the detector is integrating or discharging the current from a previous interaction event, the device may not, and usually does not, record the new interaction, a condition referred to as *pulse pile up*. The time duration in which a new pulse cannot be recorded is the detector *recovery time*, sometimes referred to as *dead time*. A pulse mode detector operated in low radiation fields has little problem with dead time count losses; however, a detector operated in high radiation fields may have significant dead time losses, thereby yielding an incorrect measurement of the radiation activity in the vicinity.

For high radiation fields, gas detectors are operated in current mode, in which the radiation induced current is measured on a current meter. Under such conditions, many interactions can occur in the device in short periods of time, and the current observed increases with total radiation exposure rate. Hence, current mode can be used to measure high radiation fields, with the magnitude of the current being a measure of the radiation induced ionization rate in the detector, thereby giving a measure of the radiation field in which the device is being operated. The disadvantage of current mode is that it does not identify individual radiation interactions.

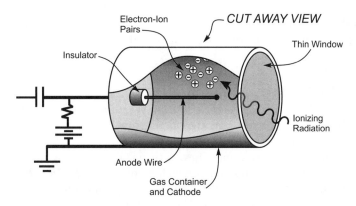

Figure 8.2. Schematic view of a coaxial gas detector, which is commonly used for Geiger-Müller tubes, and sometimes used for proportional counters. High voltage is applied to the central wire anode, while the outer cylinder wall, the cathode, is held at ground.

Figure 8.2 illustrates a gas-filled detector similar to that first explored by Geiger and Rutherford. The detector is exposed to directly ionizing radiation, which would include α-particles and β-particles. Either of these particles can cause ionization in the gas-filled device, thereby, producing *electron-ion pairs*. Hence, there are both an absorber and an observable, so that to produce a radiation detector only a method is needed to measure the amount of ionization. Suppose the device is connected to a simple electrometer so as to measure the current produced by the motion of the electron-ion pairs. Without an applied voltage, the electron-ion pairs diffuse randomly *in all directions* and eventually recombine. As a result, the net current from the electrometer is zero. Now apply a positive voltage to the thin wire of the device, or anode, so that the free electrons (negative charge) drift towards the anode and the free ions (positive charge) drift towards the detector wall. At low voltages, some measurable current is seen, yet considerable recombination still occurs, which is the *recombination region* identified as Region I in Fig. 8.1. As the voltage is increased, electron-ion pair separation becomes more efficient until practically no recombination occurs. Hence, the current measured is a measure of the total number of electron-ion pairs formed, which is Region II of Fig. 8.1, and is referred to as the *ionization chamber region*.

As the voltage is increased further, the electrons gain enough kinetic energy to create more electron-ion pairs through impact ionization. This provides a mechanism for signal gain, often referred to as *gas multiplication*. As a result, the observed current increases as the voltage increases, but is still proportional to the energy of the original radiation particle. This multiplication occurs in Region IIIa, the *proportional region*. Increasing the applied voltage further causes disproportional current increases to form, marked in Fig. 8.1 as Region IIIb, beyond which, in Region IV, *ALL* currents, regardless of origin, radiation species, or energies, are the same magnitude. Region IV is the *Geiger-Müller region*. Finally, excessive voltage drives the detector into Region V where the voltage causes sporadic arcing and other spontaneous electron emissions to occur, hence causing *continuous discharging* in the detector. Gas detectors should NOT be operated in the continuous discharge region. In the following subsections, detector operation in Regions II, III, and IV is described in more detail.

8.1.2   Ion Chambers

The simplest gas-filled detector is the *ion chamber*. There are many configurations of ion chambers, and they are operated in Region II of the gas curve shown in Fig. 8.1. The detection method is simple. Ionizing radiation, such as alpha or beta particles, or gamma or x rays, enter into a region filled with a gas such as Ar or air. The chamber has electrodes across which a voltage is applied. When radiation interactions occur in the gas, they cause the gas to become ionized, which produces electron-ion pairs relative in number to the radiation energy absorbed. The voltage applied across the electrodes causes the negative electrons to separate from the positive ions and drift across the chamber volume. Electrons drift towards the anode and positive ions drift towards the cathode, and their movement *induces* current to flow in the external circuit. Typically, this induced current is sensed by either directly measuring the current or by storing the charge in a capacitor and measuring the resulting voltage. The first case is referred to as *current mode* operation and the second case is *pulse mode* operation. Current mode operation is used in high radiation fields, and the magnitude of the current measured gives a measure of the intensity of the radiation field. Pulse mode is used for lower radiation fields, and allows for each individual radiation interaction in the chamber to be counted. Ion chambers come in many forms, and can be used for reactor power measurements, where the radiation field is very high, or as small personnel dosimeters, for use where radiation levels are typically low. Although simple in concept, two main problems occur in the ion chamber for pulse mode operation, those being (1) the signal measured is small, due to the fact that the current measured is only from the primary (or initial) electron-ion pairs excited by the radiation quantum and (2) the signal formation time can be long due to the slow motion of the heavy positive ions. Often, an RC circuit is connected to an ion chamber to reduce the time constant of the system and discharge the capacitor *before* all of the ions are collected, thereby reducing the time response.

Gamma-Ray Ion Chambers

Gamma-ray ion chambers operated in the current mode are very stable and have a long life. They can be fabricated in a variety of sizes and shapes. Large ion chambers are used as area monitors for ionizing radiation and high-pressure chambers offer a relatively high sensitivity, permitting measurement of exposure rates as low as 1 μR/h. Small chambers with low gas pressures can be operated in radiation fields with exposure rates as great as 10^7 R/h.

Air-filled ionization chambers vented to atmospheric pressure are often used to measure radiation exposure. After the formation of an electron-ion pair, the electrons and ions begin to move through the gas towards their collecting electrodes. However, the electronegative oxygen atoms quickly capture free electrons to produce heavy negative ions. Because of this attachment, most of the free electrons disappear and the chamber current is produced by the motion of negative and positive heavy ions. This is not a problem when a detector is operated in the current mode as long as the potential difference is sufficient to prevent significant recombination. However, such a chamber is not suited for pulse-mode operation, since the positive and negative ions move very slowly compared to electron speeds in the same electric field, and very broad and indistinct pulses are produced.

Neutron-Sensitive Ion Chambers

If an ion chamber is coated with a strongly-absorbing neutron-reactive material or filled with a neutron reactive gas, such that ionizing particles are released from the neutron reactions, it can be used as a neutron detector. Commonly used isotopes for neutron detectors are ^{3}He, ^{10}B, ^{6}Li, and ^{235}U. Neutron sensitive ion chambers are usually filled with ^{10}BF$_3$ or ^{3}He gas, or the inside walls of the chamber are coated with ^{10}B, ^{6}LiF, or ^{235}U. These gas-filled neutron detectors can be operated as ion chambers or proportional counters.

Ion chambers that use ^{235}U are often referred to as fission chambers, since it is the fission fragments from the ^{235}U that ionize the chamber gas. Fission chambers are often used where there is a mixed radiation field containing a large component of gamma rays. Fission fragments can deposit as much as 50 times the energy as gamma rays in a fission chamber. Hence, when operated in pulse mode, the voltage pulses formed by fission fragments are much larger than gamma-ray pulses, thereby, making it possible to discriminate between the two radiations. Due to problems with pulse pile up, ion chambers and fission chambers are generally not operated in pulse mode when in high radiation fields, although some special pulse mode designs incorporating ^{235}U are used for in-core nuclear reactor monitoring.

Compensated Ion Chambers

A form of an ion chamber commonly used for nuclear reactor control is the *compensated ion chamber*. Ion chambers, when operated in *current mode*, can be used in high radiation environments. If a gas-filled neutron detector is placed near a nuclear reactor, it responds to both neutrons and gamma rays. Yet, current mode operation does not allow for pulse height discrimination between neutron and gamma-ray interactions as does pulse mode operation. The compensated ion chamber design is used to distinguish between the two types of radiation.

Typically the chamber has two concentric electrodes and a central wire electrode. One concentric electrode is coated with a neutron sensitive material such as ^{235}U or a compound containing ^{10}B, as shown in Fig. 8.3. The ^{10}B (or ^{235}U) coated chamber is generally referred to as the "working chamber" and the uncoated chamber is referred to as the "compensating chamber." When exposed to a combined gamma ray and neutron source, the voltage potential for the working chamber causes current to flow in one direction. The voltage potential in the compensating chamber, sensitive only to gamma rays, causes current to flow in the opposite direction. The voltage potentials on the chambers can be adjusted so that the two gamma-ray currents exactly cancel. As a result, the compensating chamber cancels the current due to gamma rays, thereby, allowing the user to interpret the net current as the neutron induced current. Compensated ion chambers are widely used in nuclear reactors because of their ability to respond to neutron fields that vary up to ten orders of magnitude; i.e., these detectors have a very large "dynamic range."

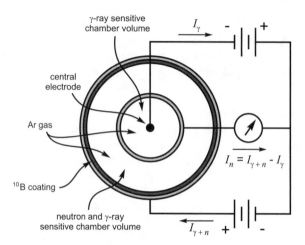

Figure 8.3. Cross section diagram of concentric compensated ion chamber. The configuration allows for both chambers to experience the same radiation field. Differences between the two chambers can be properly calibrated by adjusting the operating voltages.

Free Air Ion Chambers

A standard instrument used for radiation exposure measurements is the *free air ion chamber*. The radiation exposure unit, the roentgen (R), is defined as the quantity of x or gamma radiation that produces one statcoulomb of charge of either sign per cm^3 of air at 0 °C and 760 mm Hg. The SI exposure unit, the x unit, is defined as the amount of x or gamma radiation that produces 1 coulomb of charge per kg of air, and is equal to 3881 R. Hence, if the amount of ionization produced in the air can be measured, the radiation exposure field can be measured. This measurement is made with a free air ionization chamber. The chamber is constructed as shown in Fig. 8.4, where one electrode is segmented to form a guard ring structure. The chamber is shielded with lead and is filled only with air at ambient temperature. X or gamma rays enter through an aperture in the box and ionize the air. However, only the ionization formed in that region defined by the radiation aperture and the center electrode (of width L) is measured, thereby giving the air volume and the amount of ionization. As a result, the radiation exposure can be determined.

Smoke Detector Ionization Chambers

A typical form of smoke detector commonly found in a household environment is actually a small free air ionization chamber with an embedded ^{241}Am alpha particle emitting radiation source. It can detect particles of smoke that are too small to be visible. The alpha particles produce ionization in the tiny ion chamber, which consists of an air-filled space between two electrodes, and permits a small, constant current to flow between the electrodes. Smoke entering the chamber absorbs and neutralizes the alpha particles, which reduces the alpha particle induced ionization and current flow, thereby setting off the alarm. Hot air entering the chamber can change the rate of ionization, which also alters the current and sets off an alarm.

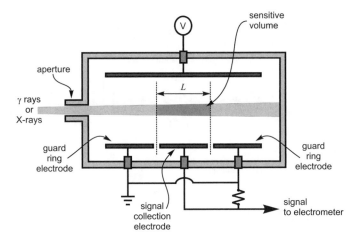

Figure 8.4. Free air ionization chamber configuration.

8.1.3 Proportional Counters

Observe in Fig. 8.1 that Region III is separated into subregions, namely, Region IIIa (proportional) and Region IIIb (limited proportionality). Proportional counters are operated in region IIIa of the gas curve, in which an electronic pulse produced by ions moving through the detector is proportional to the original energy absorbed in the detector by a quantum of radiation, be they charged particles, neutrons, gamma rays, or x rays. Although the gas-flow proportional counter was invented in 1943 by John Simpson, the actual effect of pulse height proportionality was known from those initial experiments conducted by Rutherford and Geiger with their gas-filled chambers. Ar is the most commonly used gas in a proportional counter, although there are many other gases that can be used, which include ^{3}He, Xe, and ^{10}BF$_3$.

Let us understand just exactly how the proportional counter operates. As with the ion chamber, a quantum of radiation can interact in the device's volume, either with the gas or with the chamber walls. If, for instance, a gamma ray interacts with the chamber wall, an energetic electron can be ejected into the gas volume, which then produces a cloud of electron-ion pairs. If the gamma ray interacts directly with the gas, then the primary energetic electron again produces a cloud of electron-ion pairs. In either case, a cloud of electron-ion pairs is formed in which the total number of ion pairs produced is proportional to the radiation energy deposited in the detector. Hence, by measuring the number of ion pairs formed, the energy deposited in the gas volume by the interacting radiation quantum can be determined. This measurement can be performed by applying a voltage across the detector and measuring the current produced as the electrons and ions drift through the chamber volume. Yet, as explained with the ion chamber, such a current can be minuscule and hard to measure.

At high enough voltages, electrons can gain enough kinetic energy to cause more ionization and excitation in the gas, an effect called *impact ionization*. These newly liberated electrons gain enough energy from the electric field to cause even more ionization. The process continues until the electrons are collected at the anode.

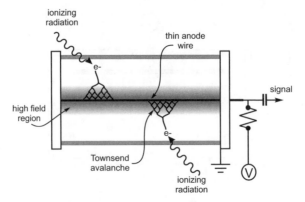

Figure 8.5. With a high electric field near the anode of a gas-filled detector, signal gain is realized through impact or Townsend avalanching, often referred to as "gas multiplication."

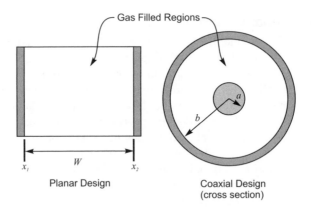

Figure 8.6. Planar and coaxial geometries are often used for gas-filled radiation detectors.

The entire process of generating the impact ionization cloud is called a *Townsend avalanche*, or sometimes *gas multiplication*, as illustrated in Fig. 8.5. There is a critical electric field E_A at which gas multiplication begins and below which the electrons do not gain sufficient energy to cause impact ionization. This threshold electric field defines the difference between Region II and Region III in the gas curve.

Parallel plate detector configurations may work for ion chambers, but are seldom used for proportional counters. A preferred geometry is a coaxial configuration, as depicted in Figs. 8.2 and 8.6. To see why, compare the difference in electric fields between coaxial and parallel plate geometries, as shown below.

Consider the parallel plate detector configuration shown in Fig. 8.6. If the voltage is V_o at $x = x_1$ and zero (grounded) at $x = x_2$, then it can be shown that the electric field is

$$E(x) = \frac{V_o}{x_2 - x_1} = \frac{V_o}{W},$$ (8.1)

where W is the width between the parallel contacts. Notice that the electric field for the planar configuration is constant, hence, a relatively large voltage is required to reach the critical avalanching field E_A.

Now consider the coaxial case also shown in Fig. 8.6. It can be shown that, for a voltage V_o applied to the inner anode with the outer surface at ground potential, the electric field at radial distance r is

$$E(r) = \frac{V_o}{r \ln(b/a)},$$ (8.2)

where a is the radius of the inner anode and b is the radius of the cathode shell wall. Unlike the planar case, the electric field is not constant for the coaxial case, and the highest electric field occurs at $r = a$.

Suppose the distance between b and a in the cylindrical case is the same as the distance between x_2 and x_1 in the planar, i.e., $b - a = x_2 - x_1 = W$. Now assume that highest value of the electric field in both cases just reaches the critical electric field E_A such that

$$E_A = \frac{V_o^{cylindrical}}{a \ln(b/a)} = \frac{V_o^{planar}}{W},$$ (8.3)

which, upon rearrangement, yields

$$\frac{V_o^{planar}}{V_o^{cylindrical}} = \frac{W}{a \ln(b/a)}.$$ (8.4)

If $a \ll b$, then $W = b - a \approx b$, so that the above result becomes

$$\frac{V_o^{planar}}{V_o^{cylindrical}} \approx \frac{b/a}{\ln(b/a)} > 1.$$ (8.5)

Because $a \ll b$, for similar chamber dimensions, it is seen that the voltage needed to reach E_A for the planar device is always greater than that needed for the cylindrical device!

Atomic electrons elevated in energy through impact ionization can also generate additional free electrons. The excited atoms de-excite by the emission of ultraviolet (UV) light which, in turn, can remove loosely bound electrons from other atoms through the process known as photoionization. Such electrons from photoionization can cause problems. To see this, let δ be the probability that a secondary electron produces a tertiary electron as a result of UV photoionization. If f is the gas multiplication from the initial avalanche, the overall multiplication from successive avalanches caused by the UV produced photoionization electrons is

$$M = f + \delta f^2 + \delta^2 f^3 + ... + \delta^{n-1} f^n = \sum_{i=1}^{n} \delta^{i-1} f^i$$ (8.6)

where i represents the consecutive avalanche waves (first, second, third, and so on) up to the final avalanche n. The quantity δf is strongly dependant upon the applied operating voltage. If $\delta f < 1$, the series in Eq. (8.6) reduces to

$$M = \frac{f}{1 - \delta f}. \qquad (8.7)$$

If, however, $\delta f > 1$, the avalanching process becomes uncontrollable and the detector develops a self-sustaining discharge. This may occur when too high a voltage is applied (as in Region V of Fig. 8.1). Continuous waves of avalanches can occur if UV light released by the excited electrons ionize too many Ar atoms, and if the Ar atoms, when arriving at the cathode wall, strike with enough kinetic energy to cause the ejection of more electrons, as depicted in Fig. 8.7a. To prevent continuous waves of avalanching from occurring in the chamber after a radiation interaction, a *quenching gas* is added to the gas mixture, typically a polyorganic molecule. A common proportional counter gas is P-10, which is a mixture of 90% Ar and 10% methane (the quenching gas). When an ionizing particle enters the detector, it ionizes both the Ar and the quenching gas. However, as the Ar gas ions drift through the chamber, they transfer their charge to the quench gas molecules, which then continue to drift and carry the positive charge to the cathode wall. When a quench gas is struck by a UV photon or strikes the cathode wall, it dissociates by releasing a hydrogen atom rather than ejecting an electron, as shown in Fig. 8.7b. As a result, the quench gas prevents continuous waves of avalanches.

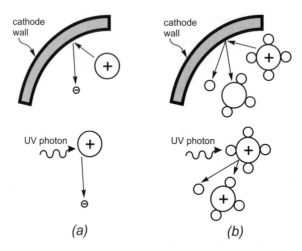

Figure 8.7. A quench gas is used to prevent continuous avalanches in the proportional counter. When an argon ion strikes the cathode wall or absorbs excited UV photons, an electron may be ejected that can start another avalanche, as depicted in *(a)*. The quench gas, usually an organic molecule, breaks apart when it strikes the cathode wall or when it absorbs a UV photon, hence does not release an electron that can start a new avalanche, as depicted in *(b)*.

Multiwire Proportional Counter

Multiwire proportional counters, developed in 1968 by Charpak, are similar to single wire devices, except that they use a criss-cross array of wires. Typically there are two planar arrays of parallel cathode wires with the arrays positioned orthogonal to each other. One might consider one set of wires parallel to the x direction and the other set parallel to the y direction. In between the two cathode wire array planes is a parallel planar array of anode wires, which are typically arranged at a 45° angle to the cathode wires (see Fig. 8.8). As with the simple proportional counter, ionizing radiation produces primary electron-ion pairs in the detector gas. Electrons travel towards the nearest anode wires in the array, which then produce a Townsend avalanche of electron-ion pairs. The cloud of positive ions separate and travel towards the nearest cathode wires in the planes on both sides of the anodes. Hence, the position of the event is determined by which cathode wires deliver a signal on the x-y plane. Overall, the multiwire proportional counter can provide both energy information and position information of the ionizing event. Charpak was awarded the 1992 Nobel Prize in Physics for his invention of the multiwire proportional chamber.

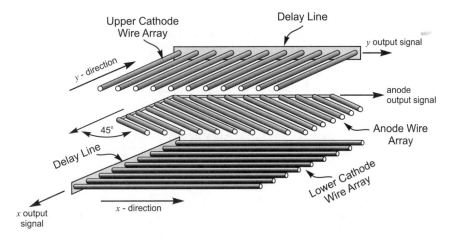

Figure 8.8. A multiwire gas filled proportional counter is composed of parallel layers of wire arrays. Shown is a system with three parallel wire arrays, in which the upper and lower arrays are cathode wires arranged orthogonally. The middle anode array is arranged at a 45° angle to the cathode arrays.

Gas Electron Multiplier (GEM)

The gas electron multiplier was introduced by Sauli in 1997, and is a unique method to produce high gain from a gas-filled detector. A thin insulating film between 50 to 100 microns thick, typically a Kapton film, is stretched on a frame and coated on both sides with a conductive metal (typically Cu). Afterwards, tiny holes approximately 100 microns in diameter are etched through the metal and insulating film (Fig. 8.9). The GEM is placed in a gas filled chamber between cathode and anode plates, usually closer to the anode. When a voltage potential is applied across the GEM, a high electric field is formed in each of the holes, high enough to cause ion-electron pair avalanching, much like a proportional counter (see Fig. 8.10). Typically, the GEM is held at a potential that is positive with respect to the cathode plate and negative to the anode plate.

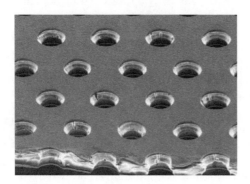

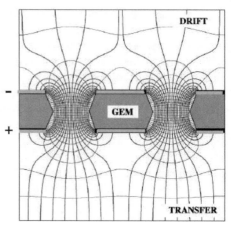

Figure 8.9. A scanning electron microscope photograph of a GEM section, showing the Kapton film, the Cu conductive surfaces, and the holes. The holes in the present photograph are approximately 75 microns in diameter. Courtesy F. Sauli.

Figure 8.10. The electric fields around and in holes of a GEM detector. Courtesy F. Sauli.

When an ionizing particle enters the gas detector between the cathode and GEM, primary electron-ion pairs are created as in a typical ion chamber or proportional counter. The electrons drift towards the GEM and the positive ions drift towards the cathode. When electrons reach the GEM, they are funneled into the tiny holes where avalanching multiplication occurs in the gas. Hence, a much larger cloud of electrons emerges from the GEM with gains up to 200. These electrons continue to drift towards the anode where they induce an output signal. By using several GEMs in stages, much higher gains can be achieved, often exceeding 1000. Position sensitive GEMs can be manufactured by using segmented anodes. Various additional coatings on top of the actual GEM conductor coatings can make the GEM more sensitive to gamma rays and neutrons. For instance, a GEM coated with Gd, ^{10}B, or ^{6}LiF is sensitive to neutrons, while a GEM coated with CsI or NaI has enhanced sensitivity to gamma rays.

Neutron-Sensitive Proportional Counters

As with the ion chamber, proportional counters that are either coated with a strongly absorbing neutron reactive material or are filled with a neutron reactive gas can be used as neutron detectors. The most commonly used materials for proportional counter neutron detectors are the gases ^{3}He and ^{10}BF$_3$, and the solid ^{10}B. Although neutron sensitive, neither ^{10}BF$_3$ nor ^{3}He are ideal proportional gases, but they perform adequately well. Because the device operates in proportional mode, a low resolution spectrum associated with the reaction product energies of the ^{10}B(n,α)^{7}Li reactions or the ^{3}He(n,p)^{3}H reactions can be identified, depending on the gas used in the counter. The neutron detection efficiency can be increased by increasing the gas pressure of the counter, hence providing more neutron absorber. Typical pressures range from 1 atm to 10 atm. Electron and ion velocities decrease inversely proportional to gas pressure: consequently, increasing the gas pressure in the tube causes the counter dead time to increase. Gas-filled tubes come in a variety of sizes, ranging from small chambers only a few cm long and one cm in diameter to large chambers several feet long and several inches in diameter.

A better proportional gas such as P-10 may be used in the chamber if, instead of filling the chamber with a neutron reactive gas, the walls are coated with ^{10}B. Unfortunately, the spectral features from such a device are harder to interpret due to interference from background gamma rays, and the total neutron detection efficiency is limited by the thinness of the optimum ^{10}B absorber coating, typically only 2 to 3 microns thick. The detectors can be made more efficient by increasing the diameter, or by inserting additional ^{10}B-coated plates in the chamber.

8.1.4   Geiger-Müller Counters

Although Hans Geiger originally created the gas-filled detector in 1908 (with Ernest Rutherford), the device used today is based on an improved version that his first PhD student, Walther Müller, constructed in 1928. Hence, the proper name for the device is the "Geiger-Müller" counter. The original "Geiger" counter was sensitive to alpha particles, but not so much to other forms of ionizing radiation. Müller's improvements included the implementation of vacuum tube technology, which allowed for the device to be formed into a compact and portable tube sensitive to alpha, beta, and gamma radiation. In 1947, Sidney Liebson further improved the device by substituting a halogen as the quenching gas, which allowed the detector to operate at lower applied voltages while lasting a significantly longer time. Geiger counters are typically arranged in a coaxial configuration, in which a thin anode wire is projected inside a tube that serves as the cathode. A high voltage is applied to the central anode wire, while the cathode is held at ground, as shown in Fig. 8.2.

Geiger-Müller counters are operated in Region IV of the gas counter curve. The device depends upon gas multiplication as a signal amplification mechanism, much like the proportional counter, however a single important difference is that, at any specific applied voltage, *ALL* output pulses from a Geiger-Müller counter are of the same magnitude *regardless of the ionizing radiation energy or type*. Hence, Geiger-Müller counters do not intrinsically possess the ability to discern between alpha, beta, or gamma radiation, nor can they distinguish between different energies of these radiations. Let us understand why this is.

When an ionizing particle enters a Geiger-Müller counter, the counting gas becomes ionized creating a small cloud of electron-ion pairs (depicted in Fig. 8.11(1)). Because a high voltage is applied to the anode, the device operates in region IV of the gas curve. The electrons drift rapidly to the anode while the ions slowly drift towards the cathode, as shown in Fig. 8.11(2). When the electrons enter into the high electric field near the anode above the critical field E_A needed to produce avalanche ionization, they gain enough kinetic energy to produce more electron-ion pairs through impact ionization, and a large and dense cloud of electron-ion pairs is formed. In addition, impact ionizations excite electrons in some gas atoms which emit UV photons when they de-excite and produce more ionization through photoionization. This large accumulation of positive ions near the anode affects the electric field and reduces its strength. These processes are depicted in Fig. 8.11 (3) and (4). There is a point at which the large accumulation of space charge around the anode increases so much that the electric field is reduced below the critical field strength E_A needed to sustain avalanching; hence, impact ionization ceases, as

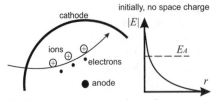

1. Primary event creates ion pairs.

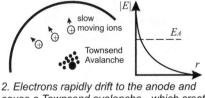

2. Electrons rapidly drift to the anode and cause a Townsend avalanche - which creates a tremendous number of ion pairs.

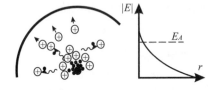

3. UV light from excited atoms in the avalanche excite more ion pairs.

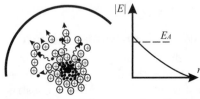

4. Waves of avalanches occur from the ion pairs excited by released UV light. Positive space charge begins to build up around the anode.

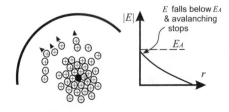

5. Positive space charge builds up around the anode to the point that the electric field is reduced below the critical value for avalanching. The avalanching ceases.

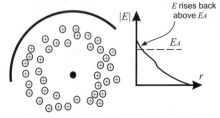

6. The space charge drifts away from the anode towards the cathode (wall). The electric field recovers such that another Geiger discharge can occur.

Figure 8.11. Geiger-Müller tube cross section depicting the progression of the Geiger discharge. The avalanching continues until the space charge accumulated around the anode wire decreases the electric field below the avalanche threshold causing the progression to cease.

shown in Fig. 8.11(5). The positive ions drift to the cathode, which produces the output pulse for the detector. As they move towards the cathode, the electric field near the anode recovers to full strength once again, and the detector is now set to detect the next radiation interaction event, as depicted by Fig. 8.11(6).

A few matters should be noted: (a) the electric field in the detector increases with an increase in applied voltage; (b) the Geiger-Müller discharge ceases when the electric field is reduced below E_A at the anode and, therefore, the positive ion accumulation density must increase with applied operating voltage to stop the avalanche; (c) to prevent more electrons from being ejected when the ions strike the cathode, a quenching gas must be used just as with the proportional counter; and (d) the entire Geiger discharge process is slower than that of a proportional counter, mainly because of the time required to produce the dense cloud of positive ions. Hence, the size of the output pulse is determined by how much space charge must accumulate to reduce the electric field below E_A and not the energy deposited within the detector! As a result, the pulse height for various energies of α-particles,

Figure 8.12. Two commercial hand held portable gas-filled detectors. The detector on the left is an ion chamber manufactured by Eberline and the detector on the right is a Geiger-Müller counter manufactured by Ludlum.

β-particles, and γ-rays are all the same, within statistical variation, and the output pulse height is predetermined by the applied operating voltage. Dead times for Geiger-Müller counters can be on the order of 10 times longer than those of proportional counters of similar size. Lastly, because Geiger-Müller counters are typically closed tubes, the quenching gas inside can be exhausted over time if traditional organic molecules such as the methane component of P-10 gas are used. Instead, Geiger-Müller counters use halogens for a quenching gas, in which the diatomic molecules dissociate when they strike the cathode. Halogens, unlike methane, can heal themselves by recombining into diatomic molecules, thereby extending the life of the gas in the detector. A portable Geiger-Müller counter and a portable ion chamber are shown in Fig 8.12.

8.2 Scintillation Detectors

Scintillating radiation detectors can rightfully be recognized as the first type of radiation detector, considering that it was a scintillating barium platinocyanide plate that Roentgen used to discover x rays in 1895. However, the emission of light was basically a qualitative measure of radiation interactions and not so much a quantitative measure. The fluorescence of the material indicated the presence of penetrating ionizing radiation, yet the actual intensity of radiation was not easily gauged.

Scintillators are generally separated into two classes, those being inorganic and organic. The method by which either produces scintillation light is physically different; hence, the distinction. Inorganic scintillators can be found as crystalline, polycrystalline, or microcrystalline materials. Organic scintillators come in many forms, including crystalline materials, plastics, liquids, and even gases.

The scintillation principle is quite simple. Radiation interactions occurring in a scintillator cause either the atomic or molecular structure in the scintillator to become excited such that electrons are increased in potential energy. These excited electrons then de-excite, during which some radiant light energy is emitted. These light emissions can then be detected with light sensitive instrumentation.

Although simple in concept, there are certain attributes that a material must possess to be a successful scintillator. First, the material must be capable of absorbing the radiation energy of interest. Second, the energy released from the scintillator must be largely radiative such that a large percentage of energy is converted to photons. Third, the light release must be spontaneous such that the scintillator *fluoresces* rather than *phosphoresces*. Fourth, the scintillator must be transparent to its own scintillation light. Finally, fifth, the light must be of a wavelength that can be detected with conventional light detecting systems, such as photomultiplier tubes or semiconductor photodiodes. There are many materials that scintillate, but only a select few that have all of the necessary properties listed above. In fact, many scintillators were discovered and used in a limited sense for the first 50 years after Roentgen's discovery of x rays.

Although scintillators were well known as radiation sensitive devices, they had low light yields. Further, light detection devices, needed to detect the scintillations, were inadequate at the time. As a result, gas filled detectors dominated the detector industry prior to the 1940s. It was in 1941 that matters began to change. In 1941, RCA introduced the first commercial *photo-multiplier tube* (PMT), a highly light-sensitive electron-amplification vacuum tube that could detect tiny amounts of visible light. It was this invention of the PMT that allowed scintillating materials to become practical radiation detectors. In 1948, Robert Hofstadter discovered NaI(Tl), probably the most important scintillator in use over the last 60 years. The first practical scintillation counter and spectrometer was demonstrated when a NaI(Tl) crystal and a PMT were coupled together. Properties of commonly used scintillator materials are listed in Table 8.1.

A typical scintillation spectrometer consists of a scintillating material hermetically sealed in an internally light-reflecting canister. Typical canisters are cylindrical, with one end of the cylinder being an optically transparent window with all remaining surfaces being Lambertian reflectors.[1] The optically transparent window is coupled to a light collection device, such as a PMT, with an optical compound. The optical compound helps match the indices of refraction between the scintillation canister and the light collection device so as to reduce reflective losses. The PMT provides a voltage output that is linear with respect to the light emitted from the scintillator. Hence, the voltage output "spectrum" is a linear indication of the radiation energy spectrum deposited in the detector. It is typical for commercial vendors to provide the scintillation canister and the PMT as one complete unit, although they can be acquired separately.

---

[1] A Lambertian surface, named after Johann Heinrich Lambert, has luminance independent of the angle of view. Typically these reflectors are white and *not* glossy or mirror-like.

Table 8.1. Common scintillator materials and properties.

Inorganic Scintillators			
Scintillator	Wavelength of Maximum Emission (nm)	Decay Time (ns)	Light Yield (photons per MeV)	Relative PMT Response Compared to NaI(Tl)
NaI(Tl)	415	230	38000	1.00
CsI(Na)	420	680, 3340	39000	1.10
CsI(Tl)	540	460, 4180	65000	0.49
LiI(Eu)	470	1400	11000	0.23
BGO	480	300	8200	0.13
$CaF_2(Eu)$	435	900	24000	0.50
GSO(Ce)	440	56, 400	9000	0.20
YAP(Ce)	370	27	18000	0.45
YAG(Ce)	550	88, 302	17000	0.50
LSO(Ce)	420	47	25000	0.75
$LaCl_3(Ce)$	350	28	49000	0.70–0.90
$LaBr_3(Ce)$	380	16	63000	1.30

Organic Scintillators			
Scintillator	Wavelength of Maximum Emission (nm)	Decay Time (ns)	Light Yield Compared to Anthracene (%)	Special Conditions and Uses
Crystalline				
Anthracene*	447	30	100	
Stilbene	410	4.5	50	
Plastic				
BC-400	423	2.4	65	general purpose
EJ-410	450	200		ZnS(Ag) embedded, fast n
EJ-204 or BC-404	408	1.8	68	large area
EJ-232 or BC-422	370	1.4	55	fast timing
EJ-240 or BC-444	435	285	41	phoswich detectors
EJ-252 or BC-470	423	2.4	46	dosimetry
Liquid				
EJ-301 or BC-501A	425	3.2	78	γ, fast n
EJ-305 or BC-505	425	2.7	80	γ, fast n
EJ-331 or BC-521	424	3.6	68	Gd-loaded n-spec, neutrino
EJ-339 or BC-523A	425	3.7	65	^{10}B-loaded n-spec

*For comparison, the NaI(Tl) light yield is 2.3 times that of anthracene.

8.2.1   Inorganic Scintillators

Inorganic scintillators depend primarily on the crystalline energy band structure of the material for the scintillation mechanism. In Fig. 8.13, an energy band diagram is shown for a typical inorganic scintillator. A lower energy band, referred to as the valence energy band, has a reservoir of electrons. It is this band of electrons that participates in the binding of atoms. The next higher band is commonly referred to as the conduction band, which for inorganic scintillators is usually devoid of electrons. Between the two bands is a forbidden region where electrons are not allowed to exist, typically referred to as the energy band gap.

If a radiation energy quantum, such as a gamma ray or charged particle, interacts in the scintillation material, it can excite numerous electrons from the valence band

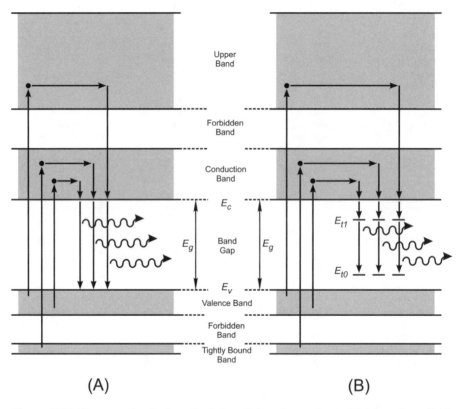

(A)   **(B)**

Figure 8.13. Shown are two basic methods by which an inorganic scintillator produces light. The intrinsic case (A) has no added activator dopants. Absorbed radiation energy excites electrons from the valence and tightly bound bands up into the higher energy conduction bands. These electrons quickly de-excite to the lowest conduction band edge E_C. As they drop back to the valence band, they release light photons. Some intrinsic scintillators emit light through optical transitions from ionized elemental constituents, which can be of lower energy than the band gap. The extrinsic case (B) has activator dopants that produce energy levels in the band gap. Absorbed radiation energy excites electrons from the valence and tightly bound bands up into the higher conduction bands. These electrons quickly de-excite to the lowest conduction band edge E_C as before. However, many drop into the upper activator site energy state E_{t1}. As they then drop to the activator ground state E_{t0}, they release light photons of lower energy than the scintillator band gap.

and the tightly bound bands up into the conduction bands (see Fig. 8.13A). These electrons rapidly lose energy and fall to the conduction band edge E_C. As they de-excite and drop back into the valence band, they can lose energy through light emissions. Unfortunately, since the radiated energy of the photons is equivalent to the band gap energy, these same photons can be reabsorbed in the scintillator and again excite electrons into the conduction band. Hence, the scintillator usually is opaque to its own light emissions. There are exceptions in which intrinsic scintillators work well. For example, bismuth germanate (BGO) releases light through optical transitions of Bi^{+3} ions, which release light that is lower in energy than the band gap, hence, is relatively transparent to its own light emissions.

However, if an *impurity* or *dopant* is added to the crystal, it can produce allowed states in the band gap, as depicted in Fig. 8.13B. Such a scintillator is referred as being *activated*. In the best of cases, the impurity atoms are uniformly distributed throughout the scintillator. When electrons are excited by a radiation event, they migrate through the crystal and many subsequently drop into the excited state of the impurity atom. Upon de-excitation, a photon is produced equal in energy to the difference between the impurity atom's excited and ground states. Hence, it is unlikely to be reabsorbed by the scintillator material. Careful selection of the proper impurity *dopant* can allow for the light emission wavelength to be tailored specifically to match the sensitivity of the light collection device.

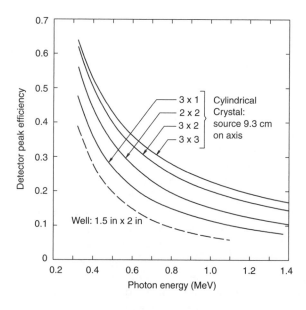

Figure 8.14. Intrinsic peak efficiency for NaI(Tl) detectors. This efficiency is the probability that all of the incident photon's energy is absorbed in the scintillator.

NaI(Tl) Scintillation Detectors

The most widely used inorganic scintillator today is NaI(Tl), meaning that the scintillator is the salt NaI that has been activated with the dopant Tl. NaI(Tl) yields approximately 38000 photons per MeV of energy absorbed in the crystal. Light emitted from NaI(Tl) has a continuous wavelength spectrum, with the most probable emission at 415 nm that matches well to typical commercial photomultiplier tubes. The intensity of light emitted at time t after radiation is absorbed in the scintillator is $I(t) = I(0) \exp(-t/\tau)$, where τ is the *decay time*. The decay time for NaI(Tl) is 230 ns, similar to that of most other inorganic, but long compared to almost all organic scintillators. Yet, it is the availability of large sizes and the relative linear response to gamma rays that makes NaI(Tl) so important. Many different sizes are available, ranging in size from cylinders that are only 0.5 inch diameter to almost a meter in diameter. Yet, the most preferred geometry remains the 3×3 inch right circular cylinder. It is the most characterized NaI(Tl) detector size with extensive efficiency data in the literature. Further, it is the standard by which all other inorganic scintillators are measured.

Because of its high efficiency for electromagnetic radiation (see Fig. 8.14), NaI(Tl) is widely used to measure x rays and gamma rays. X-ray detectors with a thin entrance window containing a very thin NaI(Tl) detector are often used to measure the intensity and/or spectrum of low energy electromagnetic radiation. NaI(Tl) detectors do not require cooling during operation and can be used in a great variety of applications. The bare NaI(Tl) crystal is hygroscopic and fragile. However, when properly packaged, field applications are possible since they can operate over a long time period in warm and humid environments, resist a reasonable level of mechanical shock, and are resistant to radiation damage. For any application requiring a detector with a high gamma-ray efficiency and a modest resolution, the NaI(Tl) detector is clearly a good choice. A gamma-ray spectrum from a 2×2 right circular cylindrical NaI(Tl) detector is shown in Fig. 8.15.

Other Inorganic Scintillation Detectors

Since the discovery of NaI(Tl) in 1948, the search has continued for a better scintillator for higher energy resolution gamma-ray spectroscopy. There have been some limited successes, which include those scintillators listed in Table 8.1. For instance, CsI(Na) is similar in performance to NaI(Tl), but has a longer decay time. CsI(Tl) has much higher light output than NaI(Tl), but the emission spectrum has a maximum at 560 nm, a wavelength that does not couple well to PMTs. However, CsI(Tl) has been coupled to Si photodiode sensors quite successfully. Bismuth germanate (BGO) has lower light output, but is much denser and a better absorber of gamma rays. As a result, BGO is used for medical imaging systems to reduce the overall radiation dose that a patient receives during the imaging procedure. LiI(Eu) is a scintillator that is primarily used for neutron detection, relying upon the ^{6}Li content in the crystal. In recent years, LaBr$_3$, a new scintillator with exceptional properties for gamma-ray spectroscopy, has become available. LaBr$_3$ has much higher light yield and a much shorter decay constant than NaI(Tl). Further, it is composed of higher Z elements, hence, is a better gamma-ray absorber than NaI(Tl). However, it is extremely hygroscopic and fragile; hence, it is difficult to produce and handle. Although it has recently become commercially available, it is presently 40 times more

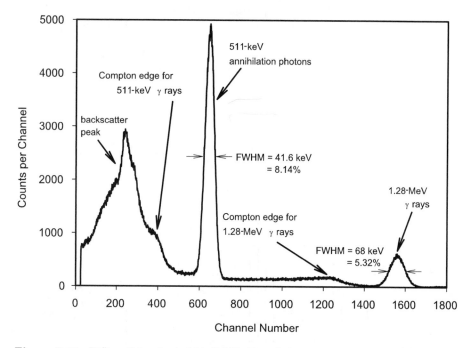

Figure 8.15. Differential pulse height distribution of the gamma rays emitted by the radioactive decay of ^{22}Na as measured by a NaI(Tl) scintillation detector. ^{22}Na is a positron and gamma-ray emitter, hence, shown are the 511 keV full-energy peaks from positron annihilation and the 1.28 MeV γ ray emissions. In addition to the two full-energy peaks, several other features are also apparent. The end points of continua for Compton scattered photons are shown, referred to as "Compton edges," one for the 511 keV annihilation photons and one for the 1.28 MeV γ rays. The backscatter peak arises from ^{22}Na emissions scattering in the material surrounding the detector, and then entering and interacting in the NaI detector.

expensive than NaI(Tl) because of production and fabrication problems. Overall, there are numerous inorganic scintillators available for special radiation detection purposes.

8.2.2 Organic Scintillators

Organic scintillators depend primarily on the molecular structure of the material for the scintillation mechanism. In Fig. 8.16, an energy diagram is shown that is typical of an organic scintillator. An independent molecule of organic scintillation material can have an electron excited through the π states up from the ground state into an excited singlet state, of which there are many levels.

There are many vibrational levels associated with the ground states, typically denoted by S_{0x} where x refers to one of the vibrational sub-states. There are also numerous excited singlet states as well as excited triplet states associated with the carbon π bonds. Electrons that gain energy rise to one of the excited vibrational states and generally fall rapidly to the lowest S_{10} state, which then de-excite through two possible channels. If the electrons de-excite directly from the S_{10} state

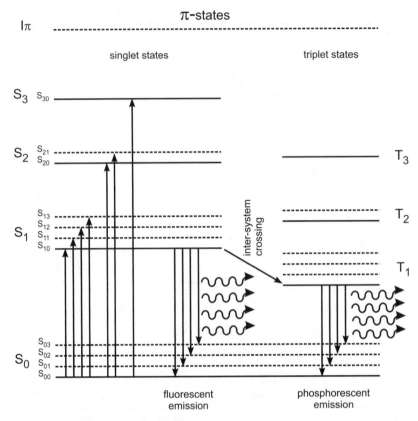

Figure 8.16. Shown are two basic methods by which an organic scintillator produces light. π electrons in the organic molecule are excited into upper vibrational states from a radiation event and rapidly de-excite to the lowest S_{10} state. Electrons that then de-excite directly to the S_{0x} states contribute to scintillation fluorescence. Those electrons that transfer to the triplet states fall to the T_{10} state, and gradually de-excite to the S_{0x} states, a process known as phosphorescence. After Birks [1964].

to one of the S_{0x} states, the light emission is rapid and is referred to as scintillation fluorescence. Decay times for fluorescence are typically only a few nanoseconds, and fluorescent emission can be easily linked to individual radiation events. However, if the electrons de-excite by crossing to the triplet states T_{1x} and then fall to one of the S_{0x} states, the light emission is slow and is referred to as scintillation phosphorescence. This second light producing mechanism is undesirable because phosphorescent emission is slow and continues to produce afterglow for extended periods of time and, hence, cannot be directly linked to individual radiation events, especially in high-radiation fields. Regardless, a main point to notice is that organic scintillators depend on the organic structure, often a benzene ring structure, and do not need activator dopants for the scintillation mechanism. Hence, they also do not need to be crystalline or polycrystalline in structure. As a result, organic scintillators can be formed as solids, liquids, gases, and plastics. Some common organic scintillators are listed in Table 8.1.

Organic scintillators are comprised mostly of hydrogen and carbon, both of which are poor absorbers of gamma rays. They are also notoriously non-linear in light output for heavy ion radiation. However, they are much more linear in response to electrons and beta particles, and the low atomic numbers for the constituents tend to make light ion backscattering almost negligible. Hence, organic scintillators are commonly used for beta particle and electron detection.

Another use for hydrogen rich detector materials is fast neutron detection. As discussed in Chapter 6, energetic neutrons lose more energy by scattering from low A materials than high A materials. Fast neutrons scatter off of the hydrogen and carbon in the organic scintillator, producing recoil hydrogen and carbon atoms that then slow down by causing ionization and excitation of other molecules in the organic scintillator. Since organic scintillators depend on molecular structure for light emission, other materials can be mixed in the scintillator without destroying the scintillation process. For instance ^{10}B, ^{6}LiF, or Gd can be mixed into an organic solution or a plastic to make them more sensitive to neutrons. Likewise, heavy metal particles, such as Pb, can be mixed into the organic or plastic to make them more sensitive to gamma rays. There is of course a limit to the amount of absorber material that can be added since the scintillator transparency reduces with increasing foreign material.

Overall, organic scintillators provide an excellent option when a larger less expensive detector is needed. Although not well suited for use in gamma-ray spectroscopy, they are quite useful as beta particle and fast neutron detectors. Moreover, since these detectors are composed of hydrogen, carbon, and oxygen with an average density of 1.032 g cm^{-3}, they make near "tissue equivalent" detectors, which are good for dosimetry measurements.

8.2.3 Light Collection

Although a scintillator produces a light signal when a radiation particle interacts in the scintillator, the light must be converted to an electronic signal if it is to be recorded. Below, three devices for doing such a conversion are discussed.

Photomultiplier Tubes

Secondary electron emission was discovered in 1902 by Austin and Starke when they noticed that exposing metal surfaces to cathode rays (electrons) caused the emission of more electrons than were incident. In 1919, Slepian proposed, and patented, the concept of using secondary electron emission as an amplification device. Some attempts were made to use the process with vacuum tube technology, but it was not until 1941 that the RCA Company released the first amplifier tube that used secondary electron emission, referred to as photomultiplier tube (PMT). The RCA Type 931 was originally used as a signal amplifier for electronics, and was used for radar jamming technology during the second world war.

Many years later, in 1947, Broser and Kallman coupled a PMT to the organic scintillator naphthalene and produced the first scintillation detector system. In 1948, Hofstadter discovered the inorganic scintillator NaI(Tl), which when coupled with a PMT produced the first practical solid-state gamma-ray spectrometer.

The photomultiplier is a simple device to understand. The basic PMT (see Fig. 8.17) has a photocathode that is located to absorb light emissions from a light source such as a scintillating material. When photons of light strike the coating on the photocathode, they excite electrons that can diffuse to the surface facing the vacuum of the tube. A fraction of these excited electrons then escape the surface and leap into the vacuum tube. A voltage applied to the tube guides the liberated electrons to an adjacent electrode called a *dynode*. As an electron approaches the dynode, it gains velocity and energy from the applied voltage and electric field. Hence, when it strikes the dynode, it again causes more electrons to become liberated into the tube. These newly liberated electrons are then guided to the next dynode where more electrons are liberated and so on. As a result, the total number of electrons released is a function of the number of dynodes in the PMT and the photoefficiency of the photocathode and the dynodes.

The total charge released in the PMT is

$$Q = qN_0G^n \tag{8.8}$$

where q is the charge of an electron, N_0 is the initial number of electrons released at the photocathode, G is the number of electrons released per dynode per incident electron (the gain), and n is the total number of dynodes in the PMT. For instance suppose that a PMT has 10 dynodes each operated with a gain of 4. An event that initially releases 1000 electrons (N_0) causes over 10^9 electrons to emerge from the PMT! Figure 8.18 shows the components of a NaI(Tl) detector along with the complete detector.

The photomultiplier tube is an important tool in radiation detection because it is the device that allowed scintillation materials to be used as practical detectors. The PMT can take a minute amount of light produced in a scintillator from a single radiation absorption and turn it into a large electrical signal. It is this electrical

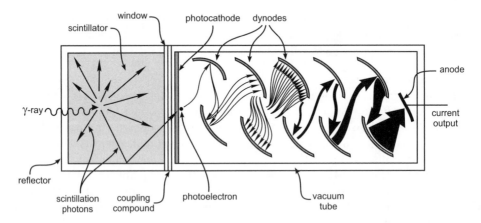

Figure 8.17. The basic mechanism of a photomultiplier tube (PMT). An absorbed γ ray causes the emission of numerous light photons which can strike the photocathode. A scintillation photon that strikes the photocathode excites a photoelectron. The photoelectron is accelerated and guided to the first dynode by an electric field, where it strikes the dynode and ejects more electrons. These electrons are accelerated to the next dynode and excite more electrons. The process continues through the dynode chain until the cascade of electrons is collected at the anode whose output current is used to produce a voltage pulse.

signal, typically converted to a voltage pulse, that is measured.

PMTs are stable and electronically quiet (low noise). Modern varieties have exceptional photocathode and dynode efficiencies, often referred to as *quantum efficiency*, with gains that can exceed 30. However there is a drawback. PMT materials used as photocathodes are generally fabricated from alkaline metals, which are most sensitive to light ˙in the 350 to 450 nm range. Scintillators emitting light outside this range can still be used under some circumstances, although their effectiveness can be severely compromised.

Figure 8.18. An encapsulated NaI(Tl) crystal in a reflective canister is shown in the upper left and a photomultiplier tube is shown in the upper right. The two components are coupled together to form a scintillation counter such as the one shown below them. The NaI(Tl) scintillation spectrometer is composed of a 2×2 NaI(Tl) crystal coupled to a PMT, both of which are sealed in the container together. The system preamplifier is attached to the back (right side) of the PMT.

Microchannel Plates

Microchannel plates are an alternative method of amplifying signals from a scintillator. Microchannel plates are glass tubes whose insides are coated with secondary electron emissive materials. A voltage is applied across the tube length which causes electrons to cascade down the tube. Every time an electron strikes the tube wall, more electrons are emitted, much like with dynodes in a PMT. Hence, a single electron can cause a cascade that can eventually produce 10^6 electrons emitted from the other end of the tube. Typically, hundreds of these microchannels are bonded together to form a plate of channels running in parallel. The microchannel plate can be fastened to a scintillator to operate in a similar fashion as a PMT. Light photons entering the microchannel plate cause the ejection of primary photoelectrons, which cascade down the microchannels to liberate millions more electrons. The main advantage to a microchannel plate is its compact size, in which a microchannel plate only one inch thick can produce a signal of similar strength as a PMT. The main problem with microchannel plates is the signal produced per monoenergetic radiation event is statistically much noisier than that produced by a PMT, hence the energy resolution for spectroscopy is typically worse than that provided by a PMT.

Photodiodes

Photodiodes are actually semiconductor devices formed into a *pn* or *pin* junction diode. When photons strike the semiconductor, usually Si or GaAs based materials, electrons are excited. A voltage bias across the diode causes the electrons to drift across the device and induce charge much as occurs in a gas-filled ion chamber. The quantum efficiency of the semiconductor diode varies with the configuration and packaging of the diode. For instance, various different commercial Si photodiodes have peak efficiencies at wavelengths ranging between 700 to 1000 nm. Regardless, they are typically more sensitive to longer wavelengths than commercial PMTs. As a result, CsI(Tl) emissions match better to Si photodiodes than PMTs. Photodiodes operate with low voltage, are small, rugged and relatively inexpensive, hence, offer a compact method of sensing light emissions from scintillators. However, they typically do not couple well to light emissions near the 400 nm range (blue-green) and have low gain, if any at all. Consequently, the signals from photodiodes need more amplification than signals from PMTs, and scintillator/photodiode systems generally do not have an energy resolution as good as that of scintillator/PMT systems.

8.3   Semiconductor Detectors

Shortly after the invention of the Geiger gas-filled detector, there were attempts to make a similar device using a solid state material rather than a gas. Unfortunately, the materials chosen were strong insulators (glass, mica, quartz, etc.), hence, responses to intense radiation were highly speculative. It was not until 1945 that Van Heerden witnessed electronic pulses induced by alpha particles in a tiny sample of the semiconductor AgCl chilled to low temperature. Initial improvements with semiconductor detectors were few, and scintillators continued to dominate the field of radiation spectroscopy. Early problems with semiconductors included impurity contamination and crystalline defects in the materials, both of which act to degrade the performance for radiation spectroscopy. However, as these problems were remedied by the mid-1960s, semiconductor detectors became the most sought after devices for serious radiation spectroscopy applications, and today constitute a third major branch of radiation detectors, along with gas-filled and scintillation detectors.

Semiconductor detectors did not become practical until Pell introduced the method of Li drifting in 1960. This process consists of applying a paste or film of Li on a surface of Si or Ge and warming the crystal. A large electric field applied to the crystal drives the highly-mobile positively-charged Li ions deep into the semiconductor. When an individual Li ion comes into close proximity with a negatively charged defect, it ceases moving and effectively neutralizes or *compensates* the detrimental effect of this undesirable defect. The Li drifting process basically negated the adverse effects of impurities in Ge and Si. This process, along with improved crystal growing methods, allowed semiconductor detectors to become practical radiation spectrometers.

The operation of a semiconductor detector combines the concepts of the charge excitation method in a crystalline inorganic scintillator and the charge collection method of a gas-filled ion chamber. As seen in Fig. 8.19, gamma rays or charged particles that are absorbed in the semiconductor excite electrons from the valence and tightly bound energy bands up into the numerous conduction bands. The empty states left behind by the negative electrons act as positively charged particles, called *holes*. The excited electrons rapidly de-excite to the conduction band edge E_C. Likewise, as electrons high in the valence band fall to lower empty states in the valence and tightly bound bands, they act as holes moving up towards the valence band edge E_V.

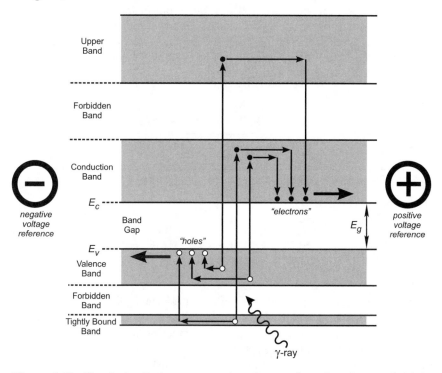

Figure 8.19. Absorbed radiation energy excites electrons from the valence and tightly bound bands up into the higher conduction bands, in a manner similar to that in a crystalline inorganic scintillator. The empty states or *holes* left behind behave as positive charges. The electrons quickly de-excite to the lowest conduction band edge E_C and the holes rapidly de-excite to the top of the valence band E_V. A voltage applied to the detector causes the electron and hole charge carriers to drift to the device's contacts as electron-ion pairs drift to the electrodes in a gas-filled ion chamber.

A single major difference between a semiconductor and almost all scintillators is that the mobility of charge carriers in semiconductors is high enough to allow for conduction, whereas scintillation materials are mostly insulating materials that do not conduct. As a result, a voltage can be applied across a semiconductor material to cause the negative electrons and positive holes, commonly referred to as *electron-hole pairs*, to drift in opposite directions, much like the electron-ion pairs in a gas-filled ion chamber. In fact, at one time semiconductor detectors were referred to as "solid-state ion chambers." As these charges drift across the semiconductor, they induce a current to flow in an external circuit which can be measured as a current or stored across a capacitor to form a voltage.

Semiconductors are far more desirable for energy spectroscopy than gas-filled detectors or scintillation detectors because they are capable of much higher energy resolution. The observed improvement is largely due to the better statistics regarding the number of charges produced by a radiation interaction. Typically, it only takes 3 to 5 eV to produce an electron-hole pair in a semiconductor. By comparison, it takes between 25 and 40 eV to produce an electron-ion pair in a gas-filled detector and between 100 eV and 1 keV to produce a single photoelectron ejection from the PMT photocathode in a scintillation/PMT detector (primarily due to light reflections and poor quantum efficiency at the photocathode). Thus, a semiconductor produces more charge carriers from the primary ionization event and thus reduces the statistical fluctuation in the energy resolution.[2] Some common semiconductors and their properties are listed in Table 8.2.

Table 8.2. Common semiconductors and their properties.

Semiconductor	At. Number (Z)	Density (g cm^{-3})	Band Gap (eV)	Ionization Energy (eV/e-h pair)
Si	14	2.33	1.12	3.61
Ge	32	5.33	0.72	2.98
GaAs	31/33	5.32	1.42	4.2
CdTe	48/52	6.06	1.52	4.43
$Cd_{0.9}Zn_{0.1}Te$	48/30/52	6.0	1.60	5.0
HgI_2	80/53	6.4	2.13	4.3

Most semiconductor detectors are configured as either planar or coaxial devices, as shown in Fig. 8.20. Small semiconductor detectors are configured as planar devices and can be used for charged-particle detection and gamma-ray detection. Large semiconductor gamma ray spectrometers are usually configured in a coaxial form to reduce the capacitance of the detector (which can affect the overall energy resolution). There are three basic methods generally used to reduce leakage currents through semiconductor detectors. Most frequently, the semiconductors are formed into reverse biased *pn* or *pin* junction diodes, which is the case for Ge, Si, GaAs, and InP detectors. Alternatively, highly resistive semiconductors, such as CdTe, CdZnTe, HgI_2 need only have ohmic contacts since the bulk resistance of the material is high enough to prevent leakage currents. Finally, large detectors, such as high-purity Ge detectors and lithium drifted Si detectors are chilled with liquid nitrogen (LN2) or a mechanical refrigerator to reduce thermally generated leakage currents.

---

[2]Note that, although PMTs, proportional counters and Geiger-Müller counters all greatly multiply the number of electrons from the initial ionizing event, it is the primary number of electrons initially excited that determines the energy resolution and not the amplified number.

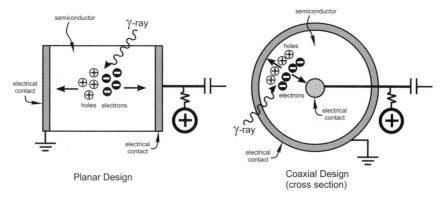

Figure 8.20. The most common designs for semiconductor detectors are the planar and coaxial configurations.

8.3.1 Ge Detectors

Although Li drifting made possible Ge-based semiconductor gamma-ray spectrometers, denoted as Ge(Li) detectors,[3] these detectors are not without problems. Li is highly mobile in Ge and must be locked into place by immediately freezing the Ge crystal with LN2 after the drifting process is finished. Further, if a Ge(Li) detector is ever allowed to warm, the Li would diffuse and redistribute, thereby, ruining the detector. As a result, Ge(Li) detectors have to be keep constantly at LN2 temperatures, a major inconvenience. Zone refinement of Ge materials now can remove the impurities from the material so that Li drifting is no longer necessary. Hence, Ge(Li) detectors have largely been replaced by high-purity Ge detectors, denoted as HPGe detectors. However, HPGe detectors must still be chilled with LN2 *when operated* in order to reduce excessive thermally generated leakage currents.

HPGe detectors have exceptional energy resolution compared to scintillation and gas-filled detectors. The dramatic difference in the energy resolution between NaI(Tl) and HPGe spectrometers is shown in Fig. 8.21, where there is a spectroscopic comparison of measurements made of a mixed ^{152}Eu, ^{154}Eu, and ^{155}Eu radiation source. HPGe detectors are now standard high resolution spectroscopy devices used in the laboratory. Their high energy resolution allows them to easily identify radioactive isotopes for a variety of applications including impurity analysis, composition analysis, and medical isotope characterization, to name a few. Portable devices with small LN2 dewars are also available for remote spectroscopy measurements, although the dewar capacity allows for only one day of operation. Hence, a source of LN2 must be available.

The gamma-ray absorption efficiency for Ge ($Z = 32$) is much less than the iodine ($Z = 53$) in NaI(Tl). Due to the higher atomic number and generally larger size, NaI(Tl) detectors often have higher detection efficiency for high energy gamma rays than do HPGe detectors (but much poorer energy resolution). When first introduced, a Ge detector's efficiency was usually compared to that of a 3-inch diameter × 3-inch long (3×3) right circular cylinder of NaI(Tl) at ^{60}Co gamma-

---

[3]Ge(Li) detectors are often pronounced as "jelly" detectors.

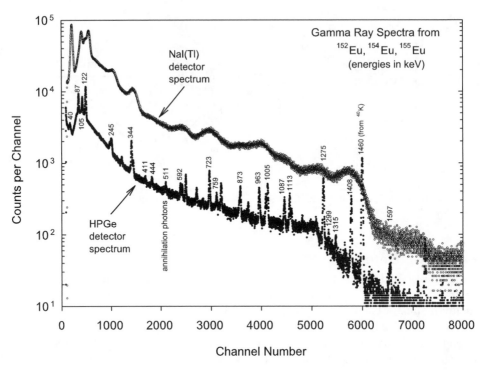

Figure 8.21. Comparison of the energy resolution between a NaI(Tl) and an HPGe detector. The gamma-ray source is a mixture of ^{152}Eu, ^{154}Eu, and ^{155}Eu.

ray energies. Even today, the efficiency of a Ge detector is quoted in terms of a 3×3 NaI(Tl) detector. For instance, a 60% efficient HPGe detector has 60% of the efficiency that a 3×3 NaI(Tl) detector has for ^{60}Co gamma rays. HPGe detectors are much more expensive than NaI(Tl) detectors, hence, are best used when gamma-ray energy resolution is most important for measurements. If efficiency is of greatest concern, it is often wiser to use a NaI(Tl) detector. Still, although very expensive, modern manufacturers do produce larger HPGe detectors with 200% efficiency (as compared to a 3×3 NaI(Tl) detector). Figure 8.22 shows a 40% efficient HPGe detector, pointing upwards, mounted onto an LN2 dewar.

8.3.2 Si Detectors

The problem with Li redistribution when the detector warms does not apply to Si; hence, Si(Li) detectors are still available. Since Si(Li) detectors have a much lower atomic number than HPGe, their relative efficiency, per unit thickness, is significantly lower for electromagnetic radiation. However, for x-ray or gamma-ray energies less than about 30 keV, commercially available Si(Li) detectors are thick enough to provide performance comparable to HPGe detectors. For example, a 3 to 5 mm thick Si(Li) detector with a thin entrance window has an efficiency of almost

Figure 8.22. Shown is a HPGe detector mounted onto an LN2 dewar. The HPGe detector is located between heavy concrete shielding to reduce the effects of background gamma rays on sensitive gamma-ray spectral measurements.

100% for 10-keV photons. Si(Li)[4] detectors are preferred over HPGe detectors for low energy x-ray measurements, primarily because of the lower energy x-ray escape peak features that appear in a Si(Li) detector spectrum as opposed to a HPGe detector spectrum. Further, background gamma rays tend to interact more strongly in HPGe detectors than in Si(Li) detectors, which also complicates the x-ray spectrum. Because a majority of the applications require a thin window, Si(Li) detectors are often manufactured with very thin beryllium windows. Typically, Si(Li) detectors are chilled with LN2 to reduce thermal leakage currents to allow optimum performance.

High-purity Si detectors, which do not incorporate Li drifting, are also available, but are significantly smaller than HPGe and Si(Li) detectors. Such devices are typically only a few hundred microns thick and are designed for charged particle spectroscopy. They range in diameter from one centimeter to several centimeters. The detectors are formed as diodes to reduce leakage currents, and use either a thin metal contact or a thin implanted dopant layer contact to produce a rectifying

---

[4]Si(Li) detectors are usually pronounced "silly" detectors.

diode. These devices are always operated in reverse bias to reduce leakage currents. Heavy charged particles, such as alpha particles, rapidly lose energy as they pass through a substance, including the detector contacts. Hence, to preserve the original energy of charged particles under investigation, the detector's contacts and implanted junctions are very thin, typically being only a few hundred nanometers thick to reduce energy loss in the contact layer. Further, the measurements are typically made in a vacuum chamber to reduce energy loss otherwise encountered by the alpha particles in air. Since the detectors are not very thick, they do not have much thermal charge carrier generation and, consequently, do not need to be cooled during operation.

8.3.3   Compound Semiconductor Detectors

Although HPGe and Si(Li) detectors have proven to be useful and important semiconductor detectors, the fact that they must be chilled with LN2 is a considerable inconvenience. Hence, much research has been devoted to the search for semiconductors that can be used at room temperature. The main requirement is that the band gap energy be greater than 1.4 eV, a requirement that seriously limits the attractiveness of many semiconductor materials. Further, the material must be composed of high atomic numbers for adequate gamma-ray absorption. As a result, there are only a few useful materials, all of which are compound semiconductors, meaning that they are composed of two or more elements. Hence, the issues regarding crystal growth defects and impurities become far more problematic. Still, there are several materials that show promise, three of which are briefly mentioned here.

HgI$_2$, CdTe, and CdZnTe Detectors

Mercuric iodide (HgI$_2$) has been studied since the early 1970s as a candidate gamma-ray spectrometer, and has been used for commercial x-ray spectrometry analysis tools. The high atomic numbers of Hg ($Z = 80$) and iodine ($Z = 53$) make it attractive as an efficient gamma-ray absorber, and its large band gap of 2.13 eV allows it to be used as a room temperature gamma-ray spectrometer. However, the bright red crystals are difficult to grow and manufacture into detectors. The voltage required to operate the devices is excessive, usually 1000 volts or more for a device only a few mm thick. HgI$_2$ detectors degrade over time, an effect referred to as *polarization*, which is another reason why they are not used widely.

Cadmium telluride (CdTe) has been studied since the late 1960's as a candidate gamma-ray spectrometer. It has relatively good gamma-ray absorption efficiency with Cd ($Z = 48$) and Te ($Z = 52$). The band gap of 1.52 eV allows CdTe to be operated at room temperature. Compared to HgI$_2$, the CdTe crystals are easier to grow and are not as fragile. Further, although still difficult to manufacture, CdTe detectors are easier to produce than HgI$_2$. There are commercial vendors of CdTe detectors, although the devices are relatively small, typically being only a few mm thick with area of only a few mm^2. CdTe detectors have been used for room-temperature-operated low-energy gamma-ray spectroscopy systems, and also for electronic personal dosimeters. Over time, CdTe detectors also suffer from polarization.

Cadmium zinc telluride (CdZnTe or CZT) has been studied as a gamma-ray spectrometer since 1990. By far, the most studied version of CZT has 5% Zn, 45%

Cd, and 50% Te molar concentrations, which yields a material with a band-gap energy of approximately 1.6 eV. CZT detectors offer an excellent option for low-energy x-ray spectroscopy when cooling is not possible. Although the detectors are quite small compared to HPGe and Si(Li) detectors, they are manufactured in sizes ranging from 0.1 cm^3 to 2.5 cm^3. Because of their small size they perform best at gamma-ray energies below 1.0 MeV. Various clever electrode designs have been incorporated into new CZT detectors to improve their energy resolution, and CZT has become the most used compound semiconductor for gamma-ray spectroscopy. Some detector cooling (near −30 °C), usually performed with miniature electronic Peltier coolers, improves their resolution, although excellent resolution can be achieved at room temperature. The average ionization energy is 5.0 eV per electron-hole pair, greater than that for Ge (2.98 eV) or Si (3.6 eV). Hence, the resolution of CZT detectors is not as good as HPGe or Si(Li) detectors, although it is much better than that of gas-filled and scintillation detectors. When LN2 chilling is not an option, CZT detectors are a good choice for radiation measurement applications requiring good energy resolution. Typically, CZT detectors do not show polarization effects.

8.4 Personal Dosimeters

An important application of radiation detection is to measure the radiation dose received by workers in a radiation environment. Here the most common personal dosimeters are discussed.

8.4.1 Photographic Film

Photographic film is one of the oldest radiation detection devices, having been used by Roentgen after his discovery of x rays in 1895. To this day, photographic film plays an important part in radiation detection. The *film badge* consists of a packet of photographic film sealed in a holder with various attenuating filters. Ionizing radiation darkens the film as in the production of an x-ray image. The filtration is designed to modify the degree of film darkening, as nearly as possible, to a known function of gamma-ray exposure, independent of the energy of the incident gamma rays. After the badge is carried by a radiation worker for a period of time, the film is processed, along with calibration films with the same emulsion batch exposed to known radiation doses. The worker's radiation dose for the period is assessed and ordinarily maintained in a lifetime record of exposure. In some cases, special attenuation filters are used to relate the darkened portions of the film to beta-particle or even neutron dose. Special badge holders in the form of rings or bracelets are used to monitor the radiation exposure of hands, wrists, and ankles.

8.4.2 Pocket Ion Chambers

A compact form of ion chamber closely related to an electrometer is the *pocket ion chamber*. These devices are routinely used to measure the radiation dose received by the wearer. The device consists of a tube approximately 10 cm long and 1 cm in diameter (see Fig. 8.23). Inside the tube are two a small metal-coated quartz fibers, each approximately 4 microns in diameter. One of the quartz fibers is stationary and the other is hinged, and both are inside an air cavity of the tube. The tube also has viewing optics such that when held up to a light, the observer can see the

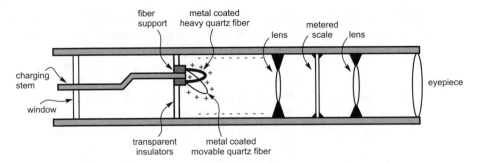

Figure 8.23. The basic components of a pocket ion dosimeter. The quartz fibers are both positively charged and become separated from the Coulombic force. Electrons excited by radiation interactions in the chamber are attracted to the quartz fibers and reduce their charge. As a result, the mobile fiber moves and its position is visually seen on the metered scale.

shadow of the hinged quartz fiber against a dose scale. The chamber is inserted into a power supply and charged so the two fibers, having like charges, are spread apart. This charged device can now be worn as a dosimeter. Electrons excited by radiation interactions in the air cavity are attracted to the quartz fibers and reduce the charge on the quartz fibers causing these fibers to move closer together. The change in location of the hinged fiber yields the dose to the wearer by determining the change in location of the fiber with respect to the metered display in the tube (the scale maximum is usually 200 mR).

8.4.3   TLDs and OSLs

Thermoluminescent dosimeters (TLDs) and optically stimulated luminescent (OSL) dosimeters are mostly used for radiation dosimetry measurements. Both a TLD and an OSL dosimeter are reusable, hence, can be worn as dosimeters, read out for dose, cleared and used again repeatedly unlike photographic film (which can only be used one time per dose assessment). However, unlike developed film, which can be archived, once a TLD or OSL dosimeter is read, the initial dose information is lost and cannot be reevaluated or archived. Hence, a dose from a used TLD or an OSL dosimeter can be measured only once before it is cleared, and the information from that measurement must be recorded and preserved.

TLDs consist of a small crystal, commonly lithium fluoride (LiF), calcium fluoride (CaF_2), or calcium sulphate ($CaSO_4$). Of the three, LiF is far more commonly used. LiF fluoride has a density similar to human tissue and so is more popular than other TLD materials for personal dosimetry. Radiation interactions in the TLD crystal cause electrons to become excited to higher energy states. These excited electrons can migrate to impurity sites where they become temporarily trapped in an upper energy state. Impurities commonly used are Mn (for CaF_2 and $CaSO_4$), Mg, Ti, Cu, and P (for LiF). The number density of electrons trapped in these impurity states is generally a linear function of the radiation energy absorbed. The dosimeter is checked periodically, between a few weeks to a month and, hence, does not yield an immediate indication that radiation is present. Rather, it is used to record the radiation dose accumulated by the wearer. The trapped electrons are later released by heating the TLD on a resistive filament. When the electrons are released, they drop back to their original ground state and release visible light. Typically, TLDs are gradually heated to a maximum temperature that ensures the complete release of all trapped electrons, and the light output is monitored during the process with a PMT. The light released, as a function of temperature, is referred to as the *glow curve* and it is unique to the type of TLD under investigation (LiF, CaF_2, $CaSO_4$). If the TLD is not read out within a reasonable amount of time (approximately 1–2 months), a substantial number of electrons can spontaneously drop back to the ground state, an effect referred to as *fading.* Under such a condition, information about the actual radiation dose received by the wearer is lost.[5]

TLDs can be designed for use in specific radiation fields. For instance, TLDs composed of natural Li, which contains 7.4% of the neutron sensitive isotope ^{6}Li, are sensitive to both neutrons and gamma rays. Neutron sensitive LiF TLDs are fabricated with enriched ^{6}Li, where the ^{6}Li is a strong absorber of neutrons that, upon absorbing a neutron, yields an immediate energetic reaction that is recorded by the TLD. Neutron insensitive TLDs can be made from LiF enriched with ^{7}Li, which is not sensitive to neutrons. The response to gamma rays is enhanced for CaF_2, as compared to LiF, and is used when sensitive gamma-ray measurements are important. TLDs have a large linear dose sensitivity range extending from 10^{-2} rad up to 400 rad, but become non-linear in response at higher doses. Also, as with photographic film, filters are often used in conjunction with TLDs to distinguish between gamma and beta dose.

OSL dosimeters have become popular over the last decade and have in many cases replaced TLDs. OSL dosimeters are usually carbon doped samples of Al_2O_3, and behave almost exactly the same as TLDs. Electrons are excited into the impurity traps (from the carbon component) where they stay lodged until the OSL dosimeter is read. However, instead of thermal excitation, trapped electrons in OSL dosimeters are dislodged with laser stimulation. The total light released by the OSL dosimeter during laser-irradiation is linearly related to the total dose accumulated by the dosimeter. OSL dosimeters are more sensitive than TLDs to low levels of radiation, being capable of recording doses below 10^{-3} rad.

8.5   Other Interesting Detectors

There are many novel detectors, some unique, that are built to reveal the paths of radiation particles, to detect radiation through new physical mechanisms, or to record exotic radiations such as neutrinos and mesons. The description of the many specialized detectors, particularly those used in high-energy physics research, would require a separate chapter. Here just a few of these unusual detectors are introduced.

---

[5]Of interest is the fact that thermoluminescence has been used for radiation dose reconstruction in Hiroshima and Nagasaki. The measurements are made on quartz, found in structural materials such as roof tiles, which is more resistant to fading than common TLD materials. These measurements continue to this day.

8.5.1   Cloud Chambers, Bubble Chambers, and Superheated Drop Detectors

Charles Wilson invented the cloud chamber in 1911, a device used to detect ionizing charged particles. In its most basic form, a cloud chamber is a sealed transparent container holding a supercooled, supersaturated vapor of water, alcohol, or aromatic hydrocarbon. Alpha particles or beta particles ionize the saturated vapor as they pass through it. The ionized vapor molecules form nucleation centers for condensation and a visible mist forms along the ionization track. Wilson shared the 1927 Nobel Prize in Physics for inventing the cloud chamber.

Donald Glaser invented the bubble chamber in 1952, a type of detector closely related to the cloud chamber. The bubble chamber is a vessel filled with a super-heated transparent liquid used to detect charged particles as they move through the detector. Charged particles deposit sufficient energy in the liquid so that it begins to boil along the ionization track, thereby, forming a visible string of bubbles. Glaser received the 1960 Nobel Prize in Physics for inventing the bubble chamber.

In 1979, Robert Apfel invented the superheated drop detector, which is similar to the cloud chamber and the bubble chamber. A fluid can be superheated to temperatures and pressures corresponding to the vapor region in the phase diagram. This metastable state is fragile and typically short-lived due to the high number of microscopic particles or gas pockets normally present at the container surfaces. However, a liquid may be kept in steady-state superheated conditions by fractionating it into droplets and dispersing them in an immiscible host fluid, for example, suspending freon droplets in a gel solution. This procedure creates perfectly smooth spherical interfaces, free of nucleating impurities or irregularities. Radiation particles, including neutrons, interacting in the fluid can disrupt this balance between the liquid and vapor phase of a superheated droplet, causing it to burst into a sizable bubble of vapor. Bubble formation is measured from the volume of vapor expelled or by detecting individual vaporizations acoustically, i.e., "listening" to the radiation interactions.

8.5.2   Cryogenic Detectors

Radiation detectors that are becoming increasingly more interesting are the cryogenic or low-temperature radiation detectors, of which there are many forms under investigation. One such device is the *microcalorimeter*. These unique devices sense the change in temperature of an absorber material caused by a single radiation quantum event, such as absorption of a gamma ray, charged particle or neutron. The device operates by attaching a small sample of thermally conductive material to a cryogenic refrigerator platform. The sample is cooled to milli-Kelvin temperature levels. Although such temperatures have been reached with liquid-helium refrigerators in the past, in recent times portable mechanical refrigerators, capable of chilling samples to mK levels, have become available. When a single quantum of radiation, such as a gamma ray or neutron, interacts in the chilled material, the temperature change ΔT in the material is measured and recorded. Because the temperature change is a linear function of the energy absorbed, a spectrum is accumulated as a function of temperature change. Extremely high resolution spectroscopy is possible, nearly 20 times better than achievable with Si(Li) detectors. Because the device must rapidly disperse the heat, the detector samples must be small. Consequently, cryogenic detectors work best for low-energy gamma rays and x rays.

Another unique cryogenic detector is the *Josephson tunnel junction* detector, which uses the low temperature superconducting properties of certain special materials. When a superconducting material is cooled below the critical temperature needed for superconductivity, electrons are formed into *Cooper pairs*. Although short lived, the Cooper pairs share in momentum such that any movement from one electron is matched in an equal but opposite manner by the other. As a result, the electrons do not scatter or interact with the host material (the superconductor) which consequently offers no electrical resistance. The detector is fabricated from a superconducting Josephson junction. A Josephson junction consists of two superconducting materials linked by a thin insulating barrier. When an x ray hits a Josephson junction, the Cooper pairs break up, and quasi-particles are created. These quasi-particles can tunnel through the thin insulating barrier of the Josephson junction, producing a short current pulse. By measuring the magnitude of the pulse, the energy of the x ray can be determined with energy resolution nearly 10 times better than achievable with Si(Li) detectors.

8.5.3   AMANDA and IceCube

There is an enormous detector array in the south polar region, the Antarctic Muon and Neutrino Detector Array (AMANDA) with another on the way, the IceCube array. AMANDA is a neutrino observatory located beneath the Amundsen-Scott South Pole Station. It consists of optical modules, each containing one photomultiplier tube (PMT), sunk in the Antarctic ice cap at depths between 1500 to 1900 meters. AMANDA is an array composed of 677 optical modules mounted on 19 separate strings spread out in a circle roughly 200 meters in diameter. Each string is placed in the ice by first drilling holes with a hot water hose, lowering the cable string and attached optical modules into the hole, and then allowing the water to refreeze around it.

AMANDA detects high energy neutrinos (> 50 GeV) which pass through the Earth from the northern hemisphere and then react just as they are leaving upwards through the Antarctic ice. A neutrino collides with nuclei of oxygen or hydrogen atoms contained in the surrounding water ice, producing a muon and a hadronic shower. The optical modules detect the Cherenkov radiation produced by these particles, and by analysis of the timing and location of when and where photons strike the PMTs, the direction of the original neutrino can be determined with a spatial resolution of approximately 2 degrees.

The AMANDA project is directed at neutrino astronomy and is to identify and characterize extra-solar sources of neutrinos. Compared to other large underground neutrino detectors consisting of large water tanks and hundreds of PMTs, AMANDA is capable of looking at higher energy neutrinos because it is not limited in volume. However, accuracy is compromised because the PMTs are spaced further apart and the conditions of the Antarctic ice are not controllable. Traditional neutrino observatories, such as the former Sudbury Neutrino Observatory (powered off on Nov. 28, 2006) or the Super-Kamiokande Neutrino Observatory, were designed to provide greater detail about neutrinos emanating from the Sun or generated in the Earth's atmosphere. However, higher neutrino energies in the spectrum should be dominated by neutrinos from sources external to our solar system. In 2005, after nine years of operation, AMANDA officially became part of its successor IceCube.

Similar to AMANDA, IceCube is being constructed in deep Antarctic ice by deploying thousands of spherical PMTs. The IceCube detector consists of a minimum of 4200 optical modules deployed on 70 vertical strings buried 1450 to 2450 meters under the surface of the ice. The sensors are deployed into the ice on strings each with sixty PMT modules. The IceTop detector array, which is a surface array above IceCube arranged to detect neutrino air showers, is composed of an additional 280 or more optical modules. The main goal of the IceCube experiment is to detect neutrinos in the energy range from 10^{11} eV to about 10^{21} eV.

8.6 Measurement Theory

No measurement can be exact. There always are errors and uncertainties. With radiation measurements, there is also the stochastic nature of radiation sources that further leads to uncertainties. In this section, such concerns are addressed.

8.6.1 Types of Measurement Uncertainties

The analysis of any type of experimental data always requires an assessment of the uncertainties associated with each measurement. Without such an uncertainty estimate, the data have very limited value. There are several types of uncertainties associated with any measurement. These include stochastic and sampling uncertainties and errors as well as systematic errors. For example, the decay of radioactive atoms occurs stochastically so that a measurement of the number of decays in a fixed time interval has an inherent stochastic uncertainty. Repeated measurements would give slightly different results. *Systematic* errors are introduced by some constant bias or error in the measuring system and are often very difficult to assess since they arise from biases unknown to the experimenter. *Sampling errors* arise from making measurements on a population different from the one desired. These biases too are hard to detect, let alone quantify.

Engineers and scientists must always be aware of the difference between *accuracy* and *precision*, even though popular usage often blurs or ignores the important distinction. Precision refers to the degree of measurement quantification as determined, for example, by the number of significant figures. Accuracy is a measure of how closely the measured value is to the true (and usually unknown) value. A very precise measurement may also be very inaccurate.

8.6.2 Uncertainty Assignment Based Upon Counting Statistics

Data measured with an ionizing-radiation detection system embody both random uncertainties and systematic errors. Uncertainty assignment requires knowledge of both. Even with a perfect measurement system capable of operating over a long time period without introducing significant systematic error, one must always consider the random nature of the data caused by the stochastic decay of radioactive nuclei. This section deals only with one component of the total uncertainty—the random or statistical uncertainty, which can be estimated using the binomial distribution since it is the basic statistical distribution describing the stochastic decay process of radionuclides. However, the binomial probability distribution function is difficult to apply to large numbers and sample sets. For samples where the probability of observing a discrete value is small ($\ll 1$) and the average value observed is large (typically > 20), the Gaussian distribution describes the probability of observing an event within some dx quite well, namely,

$$G(x)dx = \frac{1}{(\sqrt{2\pi})\sigma} \exp\left[-\frac{(x - \overline{x})^2}{2\sigma^2}\right] dx \qquad (8.9)$$

where the Gaussian $G(x)$ is the probability distribution function, $\overline{x}$ is the average value of the random variable x being observed, and σ^2 is the variance of the distribution. For a stochastic process, as with radiation counting, the Gaussian or normal distribution describes the probability distribution function well. A typical Gaussian distribution is shown in Fig. 8.24, where it has been normalized to unit height.

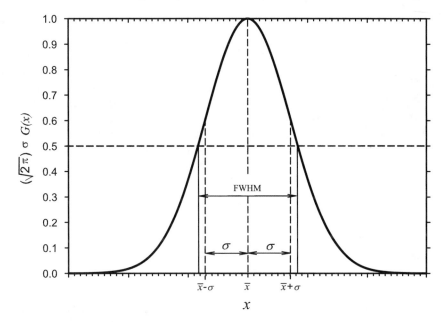

Figure 8.24. The Gaussian, or normal, distribution describes the expected probability distribution for radiation counting, where $\overline{x}$ is the average number of counts observed and σ is the standard deviation. Also, the full energy peaks of a pulse height spectrum, such as those shown in Fig. 8.15, are generally Gaussian in shape, where the common reference to energy resolution is the full width at half the maximum (FWHM) of the peak, equal to 2.355σ.

Data are routinely reported as $\overline{x} \pm \sigma_{\overline{x}}$ where $\overline{x}$ is the average number of measured counts during a specified time period and $\sigma_{\overline{x}}$ is the *standard deviation* of $\overline{x}$. When an error is reported as one standard deviation (or "one sigma"), it is called the *standard error*. For a single measurement x of a radioactive sample, it can be shown that the standard deviation, for x greater than about 20, is

$$\sigma = \sqrt{x} \qquad (8.10)$$

Thus, for a single count measurement, the mean value is estimated as x with an estimated standard deviation of $\sigma = \sqrt{x}$, so the reported value is $x \pm \sigma$. If a Gaussian

Table 8.3. Standard deviation of count data measured with a radiation detector operating in pulse mode.

Observed Total Counts	% Standard Deviation (1σ error)
100	10%
400	5%
1100	3%
2500	2%
10000	1%

distribution can approximate the data, then σ has a specific meaning, namely, there is a 68.3% probability that the next observed measurement will have a value that falls within the range $x \pm \sigma$. Table 8.3 shows the relationship between the number of radiation induced counts recorded and the percent standard deviation. For example, a standard error of one percent or less would require recording at least 10,000 counts.

Frequently, more than one count measurement is recorded. In this case, if it is assumed the radiation source has constant activity, estimates of the mean and standard deviation can be obtained directly from the N measured values, $x_1, x_2, \ldots, x_N$. The experimental mean is the average value of the counts,

$$\overline{x} = \frac{1}{N} \sum_{i=1}^{N} x_i. \tag{8.11}$$

Because $\sigma_i^2 = x_i$, the total variance for a series of measurements is

$$\sigma_{SUM}^2 = \sum_{i=1}^{N} x_i = \overline{x}N. \tag{8.12}$$

For a series of N like measurements, assuming that the activity of the radioactive source remains fairly constant during the measurement series, it can be shown with Eqs. (8.11) and (8.12) that the standard deviation of $\overline{x}$ is

$$\sigma_{\overline{x}} = \frac{\sqrt{\sigma_{SUM}^2}}{N} = \frac{\sqrt{\overline{x}N}}{N} = \sqrt{\frac{\overline{x}}{N}}. \tag{8.13}$$

For the case in which $N = 1$, Eq. (8.13) reduces to equation Eq. (8.10). Note, that for replicate measurements, the error for the average count observed is reduced by the square root of N. This assumes, of course, that the measurements are independent and that no changes occur in the radiation field or in the detector's response during the time that the N data points are being recorded. If another series of N measurements were to be taken, a new average value for $\overline{x}$ would be found, yet there would be a 68.3% chance that the new average would fall within the interval ranging from $\overline{x} - \sigma_{\overline{x}}$ and $\overline{x} + \sigma_{\overline{x}}$. The result of N measurements is

$$\overline{x} \pm \sigma_{\overline{x}} = \overline{x} \pm \sqrt{\frac{\overline{x}}{N}}. \tag{8.14}$$

Table 8.4. Probability interval relative to the number of standard deviations based on the Gaussian distribution.

Number of Standard Deviations ($k\sigma$)	Probability Event is Observed within $\pm k\sigma$
0.67σ	0.500
1.00σ	0.683
1.65σ	0.900
1.96σ	0.950
3.00σ	0.997

Uncertainties are not always reported as one standard deviation. Sometimes a larger uncertainty is reported in order to increase the probability that the mean is included within the range between $x - k\sigma$ and $x + k\sigma$. As shown in Table 8.4, if the error is reported as one standard deviation, the probability is 68.3%. However, for an error range of $x - 1.65\sigma$ to $x + 1.65\sigma$, the probability increases to 90%. This would be reported as the 90% *confidence interval*. However, the convention is to report errors as one standard deviation. For anything other than one standard deviation, one should specify the number of standard deviations or the probability interval.

Lastly, the shape of full energy peaks observed in a gamma-ray spectrum are typically Gaussian in shape, as can be seen by comparing the peaks in Fig. 8.15 to Fig. 8.24. The Gaussian gamma-ray peak shape arises from the statistical fluctuations in the number of free charge carriers excited per monoenergetic gamma-ray interaction event. Typically, the resolution of the gamma-ray peak from the detector is quoted as a function of the peak's *full width at half the maximum* value (FWHM), which for a Gaussian distribution is 2.355σ, where σ is in units of energy. In terms of percent, the *energy resolution* of the spectrometer is calculated by dividing 235.5σ by the gamma-ray energy.

8.6.3 Dead Time

All radiation detection systems operating in pulse mode have a limit on the maximum rate at which data can be recorded. The limiting component may be either the time response of the radiation detector, which is the case for the Geiger counter, or may be the resolving capability of the electronics. The true counting rate (n) for a detector with zero dead time losses is related to the recorded counting rate (m) for a detector with significant dead time losses by

$$n = \frac{m}{1 - m\tau}, \tag{8.15}$$

where τ is the dead time of the detector. Note that the term $m\tau$ is the fraction of the time that the detector is unable to respond to additional ionization in the active volume of the detector. When designing an experiment, it is advisable to keep these losses to a minimum. If possible, this means that $m\tau$ should be less than 0.05. For example, for a GM counter with a typical dead time of $\tau = 100$ μs, the maximum count rate would be 500 counts/s.

8.7   Detection Equipment

Instrumentation for the nuclear industry was standardized in 1969 according to the U.S. Atomic Energy Commission (now the Department of Energy) to what are referred to as *Nuclear Instrument Modules* (NIM). The standard defines the dimensions, standard connections, and power requirements for modular nuclear electronic components, as well as the bin into which these components are inserted. The NIM standard is a flexible and simple system that can incorporate many useful detection tools, such as amplifiers, analog to digital converters, counting and timing electronics, discriminators, and power supplies. The NIM standard provides a common footprint for electronic modules (amplifiers, ADCs, DACs, discriminators, etc.), which plug into a larger chassis (the NIM bin). The bin must supply 12 and 24 volts DC power to the modules via a backplane; the standard also specifies 6V DC and 220V or 110V AC pins, but not all NIM bins provide them. For more complex detection systems, as may be expected at large research facilities and accelerator centers, there is an international standard of modularized electronics, Computer Automated Measurement And Control, or CAMAC, a standard bus for data acquisition and control. CAMAC specifications were defined by the European Standards On Nuclear Electronics (ESONE) Committee. The CAMAC bus allows data exchange between plug-in modules (up to 24 in a single crate) and a crate controller, which then interfaces to a PC or to a VME-CAMAC interface. Typically, CAMAC components are significantly more costly than NIM components, and are best used when computer automation is necessary. Regardless of the standard, both NIM and CAMAC systems use basic components to operate radiation detectors. Figure 8.25 shows a block diagram of components commonly used in a radiation pulse-mode counting system, and Fig. 8.26 shows a NIM bin with common components described in the sections that follow.

8.7.1   Power Supply

High Voltage Power Supply

The high voltage (HV) power supply, as the name implies, supplies high voltage to the detector with either positive or negative polarity. Typically, the voltage is applied to the detector using a load resistor with the connection routed through the preamplifier unit. The HV power supply is used for low current devices, such as PMT/scintillator combinations or gas-filled detectors. A power supply usually has a meter to indicate the applied voltage. The meter reading is reasonably accurate since the detector resistance is typically much greater than the preamplifier load resistance. HV power supplies can be acquired with maximum allowable voltages exceeding 3.5 kV. Due to the high power requirement for PMTs and gas-filled devices, HV power supplies are plugged directly into a wall outlet, although many models have rails that can slip into a NIM bin for convenience.

High Voltage Bias Supply

For semiconductor detectors, the power supply is generally referred to as a "bias supply." As with the HV power supply, the bias supply provides a choice of positive or negative high voltage to the detector. It does not have a meter, but rather only a dial indicator for the total applied voltage to the system. Instead of a meter, the bias supply has current output jacks such that the operator can monitor the

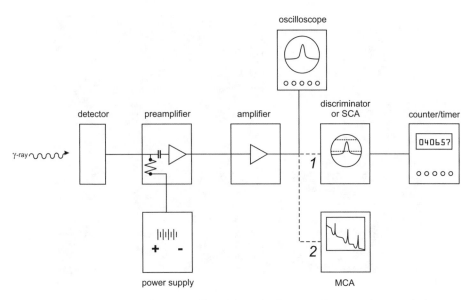

Figure 8.25. The basic components of a pulse-mode electronic detection system. A detector is powered by a bias supply or power supply, usually through the preamplifier. The induced electronic signal from the detector is shaped and amplified through the preamplifier and amplifier circuits. The signal can be observed and monitored with an oscilloscope. When used as a basic counter, the shaped signal pulse is routed through a discriminator or single channel analyzer (route 1) and then to a counter/timer. When used as a spectrometer, the shaped signal pulse is routed to a multichannel analyzer (route 2) where the pulses are discriminated and categorized according to pulse height.

reverse bias leakage current. There is actually good reason for the arrangement. Since semiconductor detectors have varying resistances, the actual voltage across the device is a function of the resistive divider as defined by the preamplifier load resistor. Hence, although a bias supply dial may indicate the total voltage, only a fraction of the voltage is dropped across the detector as defined by the ratio of the detector resistance divided by the summed resistance of the detector and the load resistor. The leakage current and detector resistance can be used to determine the voltage dropped across the detector. Bias supplies can be acquired with maximum allowable voltages exceeding 5 kV. Because of lower power detector requirements, most bias supplies are plugged directly into the NIM bus for power, unlike the HV power supplies.

8.7.2 Preamplifier

A preamplifier unit serves two basic functions, namely, (a) to provide low noise coupling of the detector to the string of amplifier and readout electronics, and (b) to produce a first stage of signal amplification. The preamplifier acts as a filter to reduce objectionable line noise that may alter the input signal. Preamplifers are designed to be either voltage sensitive, typically used for scintillation detectors and common gas-filled counters, or charge sensitive, typically used for semiconductor spectrometers and gas-filled proportional detectors.

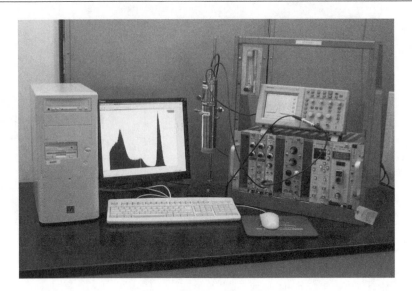

Figure 8.26. NIM components are installed into the NIM bin. Shown is a NIM bin with several nuclear instrument modules, including a power supply, a pulser, an amplifier, an SCA, and a counter/timer. The detector in the middle is a NaI(Tl) device with the preamplifier attached to the back. The system is also attached to an oscilloscope, shown on top of the NIM bin. The detector is connected to a computer-based MCA, which shows a spectrum of ^{137}Cs gamma-rays on the monitor. The ^{137}Cs gamma-ray check source is located directly beneath the NaI(Tl) detector.

8.7.3 Amplifier

The amplifier unit performs two main functions, namely, (a) pulse shaping and (b) amplification. A *tail pulse*, typically at the output of the preamplifier, has a fast rise time and a slower decay time as determined by the RC time constant of the detector/preamplifier combination. This same pulse is integrated and shaped by the amplifier to provide a Gaussian-like pulse that is far easier to manipulate with counting electronics further along in the system. The amplifier, depending on the design, may allow for the user to change rise times, decay times, and pulse widths. Further, the gain on an amplifier has course and fine adjustments, important features when calibrating the system. Lastly, the NIM standard allows for a 10 volt maximum output from the amplifier, meaning that any pulses larger than 10 volts are "clipped" at 10 volts. Hence, it is wise to adjust the gain so that the largest expected output pulses are less than 10 volts, otherwise, the clipped signal may not be recognized or counted.

8.7.4 Oscilloscope

Although an oscilloscope is not part of the NIM system, it is a fundamental and important piece of equipment that all NIM users should become accustomed to using. The oscilloscope allows the user to determine how the system is manipulating the electronic pulses at every stage of the system by simply hooking into that portion of the system with a "tee" connector. It can be used to measure pulse heights, pulse widths, repetition rates, pulse symmetry and variety of other values that allow for trouble-shooting and system calibration.

8.7.5   Discriminator/Single Channel Analyzer

The discriminator is used to reject pulses outside set boundaries, thereby passing only pulses of certain voltages that are of interest. The voltage pulse emerging from the amplifier is indicative of the amount of energy deposited within the detector. Typically the pulse amplitude is linear with energy in well-designed systems. Hence, the pulse height is a measure of the energy absorbed in the detector. The discriminator allows the user to select a voltage at a lower level discriminator (LLD) below which pulses are rejected and also an upper level discriminator (ULD) above which pulses are rejected. Often a "discriminator" unit only has an LLD (passing everything above the LLD), whereas the single channel analyzer (SCA) has both an LLD and a ULD. Hence, the discriminator or SCA allows the user to pick a voltage region (or energy region) of interest, and count only those pulses that fall within the region.

8.7.6   Counter/Timer

Voltage pulses that are allowed to pass through the discriminator or SCA are routed to a counter/timer. The function of the counter/timer is to record the number of radiation induced pulses within a predetermined amount of time. Hence, a count rate from a radiation source can be determined. Most counter/timers can be set to automatically count for a preset amount of time and stop so as to reveal the number of pulses or counts obtained within that time period. Conversely, they can be set to record a certain number of counts and then stop when that limit is obtained to reveal the time required for the measurement. Counter/timers are common instruments used for radiation measurement and dosimetry.

8.7.7   Multichannel Analyzer

A multichannel analyzer (MCA) separates by amplitude the voltage pulses exiting the amplifier unit into numerous bins. For instance, the NIM standard for pulse output mandates a 10 volt upper limit exiting the amplifier. Any pulses larger than 10 volts are truncated or "clipped." Hence, the amplifier gain is set to ensure that pulses exiting the amplifier are less than 10 volts. The MCA sorts the incoming pulses by amplitude into separate bins or *channels*, each bin being defined by an LLD and ULD. Hence, the pulses are categorized according to the channel into which they are counted. Suppose, using binary electronics, the 10 volts are divided into 1024 channels. Then each channel represents 0.977 volts, consecutively stacked in order. The *pulse height spectrum* that is formed from radiation induced pulses forms a histogram on the MCA output, usually a computer screen. The histogram represents the distribution of pulse amplitudes provided by the detector, with the pulse amplitudes being linearly related to the radiation energy absorbed for each event. Hence, the pulse height spectrum is a relative measure of the spectroscopic energy distribution absorbed in the detector. Overall, the MCA performs the function of numerous SCAs in series as well as the counter/timer. Modern MCAs can be set with over 16000 channels spread over 10 volts, a condition required for ultra high resolution radiation spectroscopy applications. Figure 8.15 shows an MCA differential pulse height spectrum of ^{22}Na as taken with a NaI(Tl) detector, and Fig. 8.21 shows MCA differential pulse height spectra from NaI(Tl) and HPGE detectors.

8.7.8   Pulser

Finally, a valuable piece of equipment that assists with calibration of the radiation counting system is the pulser. Pulsers provide accurate voltage test pulses into the radiation detection system. The pulses can be set to have different shapes (tail pulse being the most common), decay times, pulse frequencies, and amplitudes in order to calibrate the amplifier, MCA, SCA, or counter/timers. The test pulses can also be used to help determine system dead times and system electronic noise. Although not necessary for a radiation counting system, they are quite valuable when first configuring the system and calibrating it for radiation measurements. Further, it is typical that a pulser be operating during spectroscopic radiation measurements in order to have a comparison peak for system noise determination. Typically the pulser peak is set to an MCA channel that does not interfere with the accumulated radiation spectrum.

8.7.9   Other NIM Components

Although the main NIM (or CAMAC) components commonly used in a detecting system are listed above, there are many other NIM components that can be acquired for special counting purposes. Such units include timing modules, coincidence modules, delay lines, time-to-amplitude converters (TAC), and analog-to-digital converters. There are also many variations of those NIM components discussed in the preceding sections. The user should consult the specifications regarding the NIM components to determine which modules are best for the experiments or measurements to be made.

BIBLIOGRAPHY

A Handbook of Radioactivity Measurements Procedures, 2nd ed., NCRP Report No. 58, National Council on Radiation Protection and Measurements, Bethesda, 1985.

BERTOLINI, G. AND A. COCHE, *Semiconductor Detectors*, Wiley, New York, 1968.

BEVINGTON, P.R. AND D.K. ROBINSON, *Data Reduction and Error Analysis for the Physical Sciences*, 2nd ed., McGraw-Hill, New York, 1992.

BIRKS, J.B., *The Theory and Practice of Scintillation Counting*, Pergamon Press, Oxford, 1964.

BROWN, W.L., W.A. HIGINBOTHAM, G.L. MILLER, and R.L. CHASE, Eds., *Semiconductor Nuclear-Particle Detectors and Circuits*, Pub. 1593, National Academy of Sciences, Washington, 1969.

CEMBER, H., *Introduction to Health Physics*, 2nd ed., Pergamon Press, New York, 1983.

DEARNALEY, G. AND D.C. NORTHROP, *Semiconductor Counters for Nuclear Radiations*, 2nd ed., Wiley, New York, 1966.

DEME, S. *Semiconductor Detectors for Nuclear Radiation Measurement*, Wiley, New York, 1971.

EICHHOLZ, G.G. AND J.W. POSTON, *Principles of Radiation Detection*, Ann Arbor Science, Ann Arbor, 1979.

FEYNES, E. AND O. HAIMAN, *The Physical Principles of Nuclear Radiation Measurements*, Academic Press, New York, 1969.

GARDNER, R.P. AND R.L. ELY, Jr., *Radioisotope Measurement Applications in Engineering*, Reinhold, New York, 1967.

HORROCKS, D.L. AND C.T. PENG, *Organic Scintillators and Liquid Scintillation Counting*, Academic Press, New York, 1971.

KORFF, S.A., *Electron and Nuclear Counters*, Van Nostrand, New York, 1946.

KLEINKNECHT, K., *Detectors for Particle Radiation*, Cambridge University Press, U.K., 1999.

KNOLL, G.F., *Radiation Detection and Measurement*, 3rd ed., Wiley, New York, 2000.

LEROY, C. AND P.G. RANCOITA, *Principles of Radiation Interaction in Matter and Detection*, World Scientific, Singapore, 2004.

MCGREGOR, D.S. and H. HERMON, "Room Temperature Compound Semiconductor Radiation Detectors," *Nuclear Instruments and Methods* **A395**: 101–124 (1997).

O'KELLEY, G.D., *Detection and Measurement of Nuclear Radiation*, NAS-NS-3105, National Academy of Sciences, Washington, 1962.

Photomultiplier Handbook, Burle Technologies, Inc., Lancaster, 1980.

PELL, E.M., "Ion Drift in an n-p Junction," *J. Appl. Physics* **31**: 291–302 (1960).

POENARU, D.N. AND N. VILCOV, *Measurement of Nuclear Radiation with Semiconductor Detectors*, Chemical Publishing Company, New York, 1969.

PRICE, W.J., *Nuclear Radiation Detection*, 2nd. ed., McGraw-Hill, New York, 1964.

SAULI, F. AND A. SHARMA, "Micropattern Gaseous Detectors," *Annu. Rev. Nucl. Part. Sci.* **49**: 341–388 (1999).

SHARPE, J. *Nuclear Radiation Detectors*, 2nd. ed., Methuen, London, 1964.

SHAFROTH, S.M., Ed., *Scintillation Spectroscopy of Gamma Radiation*, Gordon and Breach, London, 1967.

Standard Nuclear Instrument Modules, TID-20893 (revision 2), U.S. Atomic Energy Commission, 1968. The new designation is DOE/ER-0457T introduced in 1990.

SZE, S.M., *Physics of Semiconductor Devices*, 2nd Ed., Wiley, New York, 1981.

TAIT, W.H., *Radiation Detection*, Butterworths, London, 1980.

TAYLOR, J.M., *Semiconductor Particle Detectors*, Butterworths, London, 1963.

TSOULFANIDIS, N., *Measurement and Detection of Radiation*, 2nd. Ed., Taylor and Francis, Washington, 1995.

PROBLEMS

1. What effect does each of the following changes have on the performance of a proportional counter? (a) The diameter of the anode wire is increased. (b) The pressure of the fill gas is increased. (c) The atomic number of the wall material is increased.

2. A given GM tube has a dead time of 0.25 ms. If the measured count rate is 900 counts per second, what would be the count rate if there were no dead time?

3. The decay constant for NaI(Tl) fluorescence radiation is about 230 ns. How long must one wait to collect 90% of the scintillation photons?

4. In anthracene, scintillation photons have a wavelength of 447 nm. If 1 MeV of energy is deposited in an anthracene crystal and 20,000 scintillation photons are produced, what is the scintillation efficiency?

5. Why is air often used as a tube fill gas in an ionization chamber, but not in a proportional counter?

6. If the energy resolution of a NaI(Tl) detector is 8%, what is the FWHM in terms of the energy of the full-energy peak for a ^{137}Cs source?

7. You take measurements, each over a 1 minute time interval, and record the following number of counts for each measurement: 1255, 1286, 1234, 1301, and 1221. What is the standard deviation for the average of the 1 minute counts? If you take one more count, by what factor should the standard deviation of the average change?

8. Consider inorganic and organic scintillators. Which type of scintillator would you use to detect γ rays? Fast neutrons? β particles? Explain your reasoning.

9. Relatively large HPGe and Si(Li) detectors are available. Why are HPGe detectors important and what are their advantages and disadvantages? Why are Si(Li) detectors important and what are their advantages and disadvantages?

Chapter 9

Radiation Doses and Hazard Assessment

In this chapter, we examine the biological risks associated with ionizing radiation and how they are quantified. That such radiation creates chemical free radicals and promotes oxidation-reduction reactions as it passes through biological tissue is well known. However, how these chemical processes affect the cell and produce subsequent detrimental effects to an organism is not easily determined. Much research has been directed towards understanding the hazards associated with ionizing radiation.

Consequences of exposure to ionizing radiation may be classified broadly as *hereditary effects* and *somatic effects*. Damage to the genetic material in germ cells, without effect on the individual exposed, may result in hereditary illness expressed in succeeding generations. Somatic effects are effects on the individual exposed and may be classified by the nature of the exposure, e.g., acute or chronic, and by the time scale of expression, e.g., short term or long term. The short-term acute effects on the gastrointestinal, respiratory, and hematological systems are described as the *acute radiation syndrome*.

The manner in which the hazards of human exposure to ionizing radiation are expressed depends on both the exposure and its duration. Acute, life-threatening exposure leads to *deterministic* consequences and requires a definite course of medical treatment. Illness is certain, with the scope and degree depending on the radiation dose and the physical condition of the individual exposed.

On the other hand, minor acute or chronic low-level exposure produces stochastic damage to cells and the subsequent manifestation of ill effects is likewise quantifiable only in a probabilistic sense. Hereditary illness may or may not result; cancer may or may not result. Only the probability of illness, not its severity, is dependent on the radiation dose. The consequences of such radiation exposures are, thus, *stochastic* as distinct from *deterministic*. Although the effects of low-level radiation exposures to a large number of individuals can be estimated, the effect to a single individual can be described only probabilistically.

9.1 Historical Roots

The culmination of decades of study of cathode rays and luminescence phenomena was Roentgen's 1895 discovery of x rays and Becquerel's 1896 discovery of what Pierre and Marie Curie in 1898 called *radioactivity*. The beginning of the twentieth century saw the discovery of gamma rays and the identification of the unique properties of alpha and beta particles. It saw the identification of first polonium, then radium and radon, among the radioactive decay products of uranium and thorium.

The importance of x rays in medical diagnosis was immediately apparent and, within months of their discovery, their bactericidal action and ability to destroy tumors were revealed. The high concentrations of radium and radon associated with the waters of many mineral spas led to the mistaken belief that these radionuclides possessed some subtle, broadly curative powers. The genuine effectiveness of radium and radon in treatment of certain tumors was also discovered and put to use in medical practice.

What were only later understood to be ill effects of exposure to ionizing radiation had been observed long before the discovery of radioactivity. Fatal lung disease, later diagnosed as cancer, was the fate of many miners exposed to the airborne daughter products of radon gas, itself a daughter product of the decay of uranium or thorium. Both uranium and thorium have long been used in commerce, thorium in gas-light mantles, and both in ceramics and glassware, with unknown health consequences. The medical quackery and commercial exploitation associated with the supposed curative powers of radium and radon may well have led to needless cancer suffering in later years.

Certain ill effects of radiation exposure such as skin burns were observed shortly after the discovery of x rays and radioactivity. Other effects such as cancer were not soon discovered because of the long latency period, often of many years duration, between radiation exposure and overt cancer expression.

Quantification of the degree of radiation exposure and, indeed, the standardization of x-ray equipment, were for many years major challenges in the evaluation of radiation risks and the establishment of standards for radiation protection. Fluorescence, darkening of photographic plates, and threshold erythema (skin reddening as though by first-degree burn) were among the measures early used to quantify radiation exposure. However, variability in equipment design and applied voltages greatly complicated dosimetry. What we now identify as the technical unit of exposure, measurable as ionization in air, was proposed in 1908 by Villard, but was not adopted for some years because of instrumentation difficulties. Until 1928, the threshold erythema dose (TED) was the primary measure of x-ray exposure. Unfortunately this measure depends on many variables such as exposure rate, site and area of exposure, age and complexion of the person exposed, and energy spectrum of the incident radiation.

The first organized efforts to promote radiation protection took place in Europe and America in the period 1913–1916 under the auspices of various national advisory committees. After the 1914–1918 world war, more detailed recommendations were made, such as those of the British X-ray and Radium Protection Committee in 1921. Notable events in the history of radiation protection were the 1925 London and 1928 Stockholm International Congresses on Radiology. The first led to the establishment of the International Commission on Radiological Units and Measurements (ICRU). The 1928 Congress led to the establishment of the International Commission on Radiological Protection (ICRP). Both these Commissions continue to foster the interchange of scientific and technical information and to provide scientific support and guidance in the establishment of standards for radiation protection.

In 1929, upon the recommendation of the ICRP, the Advisory Committee on X-ray and Radium Protection was formed in the United States under the auspices of the National Bureau of Standards (NBS). In 1946, the name of the Committee was changed to the National Committee on Radiation Protection. In 1964, upon receipt of a Congressional Charter, the name was changed to the National Council on Radiation Protection and Measurements (NCRP). It and its counterpart national advisory organizations maintain affiliations with the ICRP, and committees commonly overlap substantially in membership. A typical pattern of operation is the issuance of recommendations and guidance on radiation protection by both the NCRP and the ICRP, with those of the former being more closely related to national needs and institutional structure. The recommendations of these organizations are in no way mandatory upon government institutions charged with promulgation and enforcement of laws and regulations pertaining to radiation protection. In the United States, for example, the Environmental Protection Agency (EPA) has the responsibility for establishing radiation protection standards. Other federal agencies, such as the Nuclear Regulatory Commission (NRC), or state agencies have the responsibility for issuance and enforcement of laws and regulations.

9.2 Dosimetric Quantities

Consider the question of what properties of a radiation field are best correlated with its effect on matter, which may be a biological, chemical, or even mechanical effect. Dosimetric quantities are intended to provide, at a point or in a region of interest, a physical measure correlated with a radiation effect. The radiometric quantity called the fluence is not closely enough related to most radiation effects to be a useful determinant. If faced with exposure to a radiation field of fixed fluence, one would certainly care about the nature of the radiation or its energy. Energy fluence appears to be more closely correlated with radiation effect than is fluence alone, since the energy carried by a particle must have some correlation with the damage it can do to material such as biological matter. Even so, this choice is not entirely adequate—not even for particles of one fixed type. For example, one has good reason to believe that the effect of a thermal neutron, with energy about 0.025 eV, is much greater than one would expect based on its relatively small kinetic energy.

One must examine more deeply the mechanism of the effect of radiation on matter in order to determine what properties of the radiation are best correlated with its effects, especially its biological hazards. It is apparent that one should be concerned not so much with the passage of particles or energy through a region of material, but with the creation of certain physical effects in that material. These effects may be, or may result from, the deposition of energy, ionization of the medium, induction of atomic displacements, production of molecular changes, or other phenomena. Major efforts have been made to quantify these phenomena by measurement or calculation, the results to be used as indices of radiation damage. Historically, since medical radiologists were the first persons concerned with the effects of radiation on biological material, any such quantification is called a *dose* if accumulated over a period of time, or a *dose rate* if the effect per unit time is of interest.

A number of different physical phenomena may be involved and these may be quantified in a variety of ways. Thus, *dose* is not a precise term and is best used in a generic sense to relate to any actual or conceptual measure of physical phenomena involving the effect of radiation on a material. There are a few dosimetric quantities that have been precisely defined and that are particularly useful in radiological assessment. To understand these definitions, which follow, one must appreciate the several-stage process in the passage of energy from its original to its final location and form. In general, these stages are as follows for uncharged, indirectly ionizing radiation:

1. Uncharged *primary* radiation such as neutrons or photons interact with the nuclei or the electrons of the material through which they are passing.

2. As a result of the interactions, *secondary* charged particles are emitted from the atoms involved, and each of these particles starts out with kinetic energy related to the energy of the primary particle and the type of interaction that led to creation of the secondary particle.

3. The secondary charged particles lose energy while traversing the material either (a) through ionization and associated processes such as atomic and molecular excitation and molecular rearrangement, or (b) through emission of photons called *bremsstrahlung*. The progress of energy degradation need not be considered further at this point, except to say that the energy removed from the secondary charged particles by the process of ionization and associated mechanisms is distributed along the tracks of the charged particles and, for the most part, is degraded promptly into thermal energy of the medium.

4. The uncharged primary particles may produce additional uncharged particles through scattering or other processes. These uncharged particles also carry off part of the energy of the interaction, but this is not of immediate concern in the definitions which follow.

The definitions given in this section are for dose quantities. Dose rate quantities follow naturally. When it is necessary to make a distinction, these rate quantities are symbolized in standard fashion by placing a dot over the appropriate symbol (e.g., $\dot{D}$, $\dot{K}$, and $\dot{X}$).

9.2.1 Energy Imparted to the Medium

For a given volume of matter of mass m, the energy ϵ "imparted" in some time interval is the sum of the energies (excluding rest-mass energies) of all charged and uncharged ionizing particles entering the volume minus the sum of the energies (excluding rest-mass energies) of all charged and uncharged ionizing particles leaving the volume, further corrected by subtracting the energy equivalent of any increase in rest-mass energy of the material in the volume. Thus, the energy imparted is that which is involved in the ionization and excitation of atoms and molecules within the volume and the associated chemical changes. This energy is eventually degraded almost entirely into thermal energy.

9.2.2   Absorbed Dose

The *absorbed dose* is the quotient of the *mean* energy imparted $\Delta\bar{\epsilon}$ to matter of mass Δm, in the limit as the mass approaches zero [ICRU 1971],

$$D \equiv \lim_{\Delta m \to 0} \frac{\Delta\bar{\epsilon}}{\Delta m}. \tag{9.1}$$

Here $\bar{\epsilon}$ is the expected energy imparted to the medium averaged over all stochastic fluctuations. The absorbed dose is thus the average energy absorbed from the radiation field, per unit differential mass of the medium. The concept of absorbed dose is very useful in radiation protection. Energy imparted per unit mass in tissue is closely, but not perfectly, correlated with radiation hazard.

The standard unit of absorbed dose is the gray (Gy), 1 Gy being equal to an imparted energy of 1 joule per kilogram. A traditional unit for absorbed dose is the rad (no abbreviation), defined as 100 ergs per gram. Thus, 1 rad = 0.01 Gy.

9.2.3   Kerma

The absorbed dose is, in principle, a measurable quantity; but in many circumstances, it is difficult to calculate the absorbed dose from radiation fluence and material properties. A closely related deterministic quantity, used only in connection with indirectly ionizing (uncharged) radiation, is the *kerma*, an acronym for *k*inetic *e*nergy of *r*adiation absorbed per unit *m*ass. If E_{tr} is the sum of the initial kinetic energies of all the charged ionizing particles released by interaction of indirectly ionizing particles in matter of mass m, then

$$K \equiv \lim_{\Delta m \to 0} \frac{\Delta\overline{E}_{tr}}{\Delta m}. \tag{9.2}$$

Again the bar over the E_{tr} indicates the expected or stochastic average. That some of the initial kinetic energy may be transferred ultimately to bremsstrahlung, for example, is irrelevant.

The use of the kerma requires some knowledge of the material present in the incremental volume, possibly hypothetical, used as an idealized receptor of radiation. Thus, one may speak conceptually of tissue kerma in a concrete shield or in a vacuum, even though the incremental volume of tissue may not actually be present.

Absorbed dose and kerma are frequently almost equal in magnitude. Under a condition known as *charged-particle equilibrium*, they are equal. This equilibrium exists in a small incremental volume about the point of interest if, for every charged particle leaving the volume, another of the same type and with the same kinetic energy enters the volume traveling in the same direction. In many practical situations, this charge-particle equilibrium is closely achieved so that the kerma is a good approximation of the absorbed dose.

9.2.4   Calculating Kerma and Absorbed Doses

The calculation of the kerma (rate) is closely related to the reaction (rate) density introduced in Section 7.2.2. If, at some point of interest in a medium, the fluence of radiation with energy E is Φ, the kerma at that point is

$$K = \{\text{no. of interactions per unit mass}\} \times \{\text{energy imparted per interaction}\}$$

$$= \left\{ \frac{\mu(E)\,\Phi}{\rho} \right\} \times \{E\,f(E)\}. \tag{9.3}$$

Here $f(E)$ is the fraction of the fraction of the incident radiation particle's energy E that is transferred to secondary charged particles, and $\mu(E)/\rho$ is the mass interaction coefficient for the detector material. This result can be rearranged as

$$K = \left(\frac{f(E)\,\mu(E)}{\rho} \right) E\,\Phi. \tag{9.4}$$

The kerma *rate* $\dot{K}$ is obtained by replacing the fluence by the fluence rate or flux density ϕ. This result for the kerma applies equally well to neutrons and photons.

Photon Kerma and Absorbed Dose

The product $f\mu$, called μ_{tr}, is illustrated in Fig. 7.5. This coefficient accounts only for charged secondary particles and excludes the energy carried away from the interaction site by secondary photons (Compton scattered photons, annihilation photons, and fluorescence). Thus for photon kerma calculations,

$$K(E) = \left(\frac{\mu_{tr}(E)}{\rho} \right) E\,\Phi.$$

For the kerma in units of Gy, E in MeV, (μ_{tr}/ρ) in cm^2/g, and Φ in cm^{-2},

$$K\ (\text{Gy}) = \frac{\mu_{tr}}{\rho}\ \left(\frac{cm^2}{g} \right) \times E\ (\text{MeV}) \times 1.602 \times 10^{-13}\ (\text{J/MeV}) \times$$

$$\frac{1\ (\text{Gy})}{1\ (\text{J/kg})} \times 1000\ (\text{g/kg}) \times \Phi\ (cm^{-2})$$

or

$$K = 1.602 \times 10^{-10}\, E \left(\frac{\mu_{tr}(E)}{\rho} \right) \Phi. \tag{9.5}$$

If the secondary charged particles produce substantial bremsstrahlung, a significant portion of the charged-particles' kinetic energy is reradiated away as bremsstrahlung from the region of interest. Even under charged-particle equilibrium, the kerma may overpredict the absorbed dose. The production of bremsstrahlung can be taken into account by the substitution in Eq. (9.5) of μ_{en}/ρ for μ_{tr}/ρ (see Fig. 7.5). Then, under the assumptions of charged particle equilibrium and no local energy transfer from bremsstrahlung,

$$D = 1.602 \times 10^{-10}\, E \left(\frac{\mu_{en}(E)}{\rho} \right) \Phi. \tag{9.6}$$

Example 9.1: What are the iron kerma and absorbed dose rates from uncollided photons 1 meter from a point isotropic source emitting 10^{14} 5-MeV gamma rays per second into an infinite water medium?

From Appendix C, the total mass interaction coefficient for 5-MeV photons is found to be $(\mu/\rho)_{H_2O} = 0.03031$ cm^2/g or $\mu_{H_2O} = 0.03031$ cm^{-1}. The uncollided flux density 1 meter from the source is, from Eq. (7.25),

$$\phi^o = \frac{S_p}{4\pi r^2} \exp[-\mu r]$$

$$= \frac{10^{14} \text{ s}^{-1}}{4\pi(100 \text{ cm})^2} \exp[-(0.03031 \text{ cm}^{-1})(100 \text{ cm})] = 38.41 \text{ cm}^{-2}\text{s}^{-1}.$$

We find from Appendix C for 5-MeV photons in iron $(\mu_{tr}/\rho)_{Fe} = 0.02112$ cm^2/g and $(\mu_{en}/\rho)_{Fe} = 0.01983$ cm^2/g. Then, from Eq. (9.5), the iron kerma rate is

$$\dot{K}^o = 1.602 \times 10^{-10} E \left(\frac{\mu_{tr}}{\rho}\right)_{Fe} \phi^o = 6.50 \times 10^{-10} \text{ Gy/s} = 2.34 \ \mu\text{Gy/h}.$$

With the assumption of charged particle equilibrium, the absorbed dose rate in iron is obtained from Eq. (9.6) as

$$\dot{D}^o = 1.602 \times 10^{-10} E \left(\frac{\mu_{en}}{\rho}\right)_{Fe} \phi^o = 6.10 \times 10^{-10} \text{ Gy/s} = 2.20 \ \mu\text{Gy/h}.$$

Notice that the detection medium can be different from the medium through which the radiation is traveling. Also, even with the assumption of charged particle equilibrium, $\dot{K}$ is slightly larger than $\dot{D}$ since the bremsstrahlung energy emitted by secondary electrons is absorbed away from the point at which the absorbed dose is calculated; however, it is included (through the initial kinetic energy of the secondary electrons) in the concept of the kerma.

Kerma for Fast Neutrons

When fast neutrons pass through a medium, the primary mechanism for transferring the neutrons' kinetic energy to the medium is from neutron scattering interactions. In a neutron scatter, the scattering nucleus recoils through the medium creating ionization and excitation of the ambient atoms. For isotropic elastic scattering in the center-of-mass system of a neutron with initial energy E, the average neutron energy loss (and hence average energy of the recoil nucleus) is, from Eq. (6.28),

$$f_s(E) = \frac{1}{2}(1-\alpha)E = \frac{2A}{(A+1)^2}E \tag{9.7}$$

Thus, if only elastic scattering is of importance, the neutron kerma, from Eq. (9.4), is

$$\boxed{K = 1.602 \times 10^{-10} E \left(\frac{f_s(E)\,\mu_s(E)}{\rho}\right)\Phi.} \tag{9.8}$$

Here K has units of Gy, when E is in MeV, Φ is in cm^{-2}, the macroscopic cross section for elastic scattering μ_s is in cm^{-1}, and the medium's mass density ρ is in g/cm^3. For slow or thermal neutrons, calculation of the neutron kerma is more difficult since charged particles produced by nuclear reactions must be considered.

Example 9.2: What is the kerma rate in a small sample of water irradiated with a beam of 0.1-MeV neutrons with an intensity of $I = \phi = 10^{10}$ cm^{-2} s^{-1}? The microscopic scattering cross sections at 0.1 MeV are $\sigma_s^{\mathrm{H}} = 12.8$ b and $\sigma_s^{\mathrm{O}} = 3.5$ b. Assume isotropic neutron scattering in the center of mass.

The fraction f_s of the incident neutron energy E transferred to the recoiling scattering nucleus is given by Eq. (9.7), namely, $f_s = 2A/(A+1)^2$. For hydrogen ($A = 1$) $f_s^{\mathrm{H}} = 0.5$ and for oxygen ($A = 16$) $f_s^{\mathrm{O}} = 0.1107$. Thus, the mass scattering coefficient weighted by the fraction of energy transferred to recoil nuclei is

$$(f_s \mu_s/\rho)_{\mathrm{H_2O}} = [N^{\mathrm{H}} \sigma_s^{\mathrm{H}} f_s^{\mathrm{H}} + N^{\mathrm{O}} \sigma_s^{\mathrm{O}} f_s^{\mathrm{O}}]/\rho^{\mathrm{H_2O}} = (N/\rho)^{\mathrm{H_2O}} [2\sigma_s^{\mathrm{H}} f_s^{\mathrm{H}} + \sigma_s^{\mathrm{O}} f_s^{\mathrm{O}}]$$

$$= (N_a/A^{\mathrm{H_2O}}) [2\sigma_s^{\mathrm{H}} f_s^{\mathrm{H}} + \sigma_s^{\mathrm{O}} f_s^{\mathrm{O}}]$$

$$= (0.6022/18)[(2)(12.8)(0.5) + (3.5)(0.1107)] = 0.4412 \text{ cm}^2/\text{g}.$$

Then, from Eq. (9.8) with $E = 0.1$ MeV and $\phi = 10^{10}$ cm^{-1} s^{-1}, we calculate the kerma rate as

$$\dot{K} = 1.602 \times 10^{-10} E \left(\frac{f_s(E) \, \mu_s(E)}{\rho} \right)_{\mathrm{H_2O}} \phi = 0.071 \text{ Gy/s} = 254 \text{ Gy/h}.$$

9.2.5 Exposure

The quantity called *exposure*, with abbreviation X, is used traditionally to specify the radiation field of gamma or x-ray photons. It is applied only to photons. Exposure is defined as the absolute value of the ion charge of one sign produced anywhere in air by the complete stoppage of all negative and positive electrons, except those produced by bremsstrahlung, that are liberated in an incremental volume of air, per unit mass of air in that volume. The exposure is closely related to air kerma but differs in one important respect. The phenomenon measured by the interaction of the photons in the incremental volume of air is not the kinetic energy of the secondary electrons but the ionization caused by the further interaction of these secondary electrons with air. The SI unit of exposure is coulombs per kilogram. The traditional unit is the roentgen, abbreviated R, which is defined as precisely 2.58×10^{-4} coulomb of separated charge of either sign per kilogram of air in the incremental volume where the primary photon interactions occur.

Kerma in air and exposure are very closely related. A known proportion of the initial kinetic energy of secondary charged particles results in ionization of the air. The conversion factor, given the symbol W, fortunately is almost energy independent for any given material. For air, it is estimated to be 33.85 ± 0.15 electron volts of kinetic energy per ion pair [ICRU 1979]. The product of K and W, with appropriate unit conversions, is the exposure X. The product, however, must be reduced somewhat to account for the fact that some of the original energy of the secondary electrons may result in bremsstrahlung, not in ionization or excitation. Thus the interaction coefficient to be used is μ_{en}/ρ for air, which accounts for the fact that some of the charged particles' kinetic energy is taken away as bremsstrahlung and thus does not produce ionization locally.

For exposure in units of roentgen, E in MeV, (μ_{en}/ρ) for air in cm^2/g, and Φ in cm^{-2}

$$X \text{ (R)} = \frac{\mu_{en}}{\rho} \left(\frac{cm^2}{g}\right) \times 1000 \left(\frac{g}{kg}\right) \times E \text{ (MeV)} \times 10^6 \left(\frac{eV}{MeV}\right) \times \frac{1 \text{ ion-pair}}{33.85 \text{ eV}}$$

$$\times 1.602 \times 10^{-19} \left(\frac{C}{\text{ion-pair}}\right) \times \frac{1}{2.58 \times 10^{-4} \text{ (C/R)}} \Phi \text{ (cm}^{-2})$$

or

$$X = 1.835 \times 10^{-8} E \left(\frac{\mu_{en}(E)}{\rho}\right)_{air} \Phi. \tag{9.9}$$

Exposure is a kerma-like concept, and, consequently, there are difficulties in its measurement. The measurement must be such as to ensure that all the ionic charge resulting from interaction in the air within the sensitive volume of a detector is collected for measurement. However, it is likely that much of the ionization occurs outside the test volume and some ionization in the test volume takes place as a result of photon interactions taking place outside the volume. This makes charge collection complicated. This difficulty can be surmounted, at least for low-energy photons, if in the vicinity of the test volume there is *electronic equilibrium*.

The use of exposure as a measure of the photon field is sometimes criticized but has survived because it is a measurable quantity correlated reasonably well with biological hazard. This is because, per unit mass, air and tissue are similar in their interaction properties with photons. On the other hand, absorbed dose in tissue is more closely related to radiation hazard and is more fundamentally sound as a radiation quantity for assessing biological effects.

9.2.6   Relative Biological Effectiveness

If the energy imparted by ionizing radiation per unit mass of tissue were by itself an adequate measure of biological hazard, absorbed dose would be the best dosimetric quantity to use for radiation protection purposes. However, there are also other factors to consider that are related to the spatial distribution of radiation-induced ionization and excitation. The charged particles responsible for the ionization may themselves constitute the primary radiation, or they may arise secondarily from interactions of uncharged, indirectly ionizing primary radiation.

In dealing with the fundamental behavior of biological material or organisms subjected to radiation, one must account for variations in the sensitivity of the biological material to different types or energies of radiation. For this purpose, radiobiologists define a *relative biological effectiveness* (RBE) for each type and energy of radiation, and, indeed, for each biological effect or *endpoint*. The RBE is the ratio of the absorbed dose of a reference type of radiation (typically, 250-kVp x rays[1] or

---

[1] The designation kVp refers to the peak energy of x rays released when the generating electron beam is produced by a fluctuating voltage.

1.25 MeV ^{60}Co gamma rays) producing a certain kind and degree of biological effect to the absorbed dose of the radiation under consideration required to produce the same kind and degree of effect. RBE is normally determined experimentally and takes into account all factors affecting biological response to radiation in addition to absorbed dose.

9.2.7 Dose Equivalent

The RBE depends on many variables: the physical nature of the radiation field, the type of biological material, the particular biological response, the degree of response, the radiation dose, and the dose rate or dose fractionation. For this reason it is too complicated a concept to be applied in the routine practice of radiation protection or in the establishment of broadly applied standards and regulations. In 1964, therefore, a group of specialists established a related but more explicitly defined surrogate called the *quality factor QF*. Unlike the RBE, which has an objective definition in terms of a particular biological endpoint, QF is meant to apply generically to those endpoints of importance in low-level radiation exposure, namely, cancer and hereditary illness. The quality factor and the absorbed dose are both point functions, that is, deterministic quantities that may be evaluated at points in space. Their product

$$H \equiv QF \times D \tag{9.10}$$

is identified as the *dose equivalent H* and is recognized as an appropriate measure of radiation risk when applied in the context of establishing radiation protection guidelines and dose limits for population groups.

The standard unit of dose equivalent is the sievert (Sv) equal to the absorbed dose in Gy times the quality factor. A traditional unit for dose equivalent is the rem, based on the absorbed dose in rad, and 1 rem is equivalent to 0.01 Sv.

9.2.8 Quality Factor

Originally, methods for evaluation of quality factors were somewhat subjective in the values assigned even though the prescriptions were very precisely defined in terms of macroscopic physical characteristics of the radiation fields. Over the years, subjectivity has been reduced in degree, but not eliminated, and the prescriptions have become more highly reliant on very sophisticated microscopic descriptions of the fields. Although modern usage defines the value of the quality factor for a particular radiation in terms of its stopping power (or *linear energy transfer* LET) [ICRP 1991, NCRP 1993], typical and practical values are those given in Table 9.1. Quality factors used in U.S. federal regulations differ for neutrons, varying from 2 for low

Table 9.1. Values of the quality factor for different radiations. Source: ICRP [1991]; NCRP [1993].

Radiation	QF
X, γ, $\beta^{\pm}$, (all energies)	1
Neutrons < 10 keV	5
10–100 keV	10
0.1–2 MeV	20
2– 20 MeV	10
> 20 MeV	5
Protons (> 2 MeV) [ICRP]	5
Protons (> 2 MeV) [NCRP]	2
Alpha particles	20

energy neutrons to 11 for 0.5-1-MeV neutrons, and taken as 10 for neutrons of unknown energy.

Since the quality factor is defined for use in conventional and routine radiation protection practice, the dose equivalent is firmly established only for radiation doses less than annual limits. It is nevertheless common for "dose" to be expressed in sieverts in the context of very high doses. One school of thought is that such practice should be deplored and such exposures should be expressed in terms of quantities such as absorbed dose; the opposite school of thought is that the use of dose equivalent units is permitted provided the user applies, not the standard quality factor, but a relative biological effectiveness appropriate to the circumstances.

Example 9.3: What is the dose equivalent 15 meters from a point source that emitted 1 MeV photons isotropically into an infinite air medium for 5 minutes at a rate of 10^9 photons per second?

We can neglect air attenuation over a distance of 15 m so that the fluence 15 m from the source is

$$\Phi = \frac{S_p \Delta t}{4\pi r^2} = \frac{(10^9 \text{ s}^{-1})(600 \text{ s})}{4\pi(1500 \text{ cm})^2} = 2.122 \times 10^4 \text{ cm}^{-2}.$$

The dose equivalent $H = QF \times D$, where D is the absorbed dose at the point of interest. Implicit in the concept of the dose equivalent is that the energy absorbing medium is *tissue*. If (μ_{en}/ρ) data is not available for tissue, we can approximate tissue well by water. Finally, since the radiation is photons, the quality factor from Table 9.1 is $QF = 1$. Then with the data in Appendix C for water, the dose equivalent is

$$H = QF \times D \simeq QF\, 1.602 \times 10^{-10} E \left(\frac{\mu_{en}}{\rho}\right)^{H_2O} \Phi$$

$$= (1)(1.602 \times 10^{-10})(1)(0.03103)(2.122 \times 10^4) = 10.5 \; \mu Sv.$$

9.2.9 Effective Dose Equivalent

In a human, different organs have different radiological sensitivities, i.e., the same dose equivalent delivered to different organs results in different consequences. Moreover, a beam of radiation incident on a human body generally delivers different dose equivalents to the major body organs and tissues. Finally, ingested or inhaled sources of radiation usually produce different dose equivalents in the various body organs and tissues. To account for different organ sensitivities and the different doses received by the various organs, a special dose unit, the *effective dose equivalent H_E*, is used to describe better the hazard a human body experiences when placed in a radiation field. The effective dose equivalent is a weighted average of the dose equivalents received by the major body organs and tissue, namely,

$$H_E = \sum_T w_T \, \overline{QF_T} \, D_T \equiv \sum_T w_T H_T, \tag{9.11}$$

where, for organ/tissue T, w_T is the *tissue weighting factor* (see Table 9.2), D_T is the absorbed dose, $\overline{QF_T}$ is the quality factor, averaged over LET and position in the organ and weighted by dose as a function of LET and position in the organ. Here $H_T \equiv \overline{QF_T}\, D_T$ is the dose equivalent.

Table 9.2. Tissue weighting factors adopted by the ICRP [1977] for use in determining the effective dose equivalent.

Organ	w_T	Organ	w_T
Gonads	0.25	Thyroid	0.03
Breast	0.15	Bone surface	0.03
Red marrow	0.12	Remainder^a	0.30
Lung	0.12	TOTAL	1.00

^{a}A weight of 0.06 is applied to each of the five organs or tissues of the remainder receiving the highest dose equivalents, the components of the GI system being treated as separate organs.

Example 9.4:   Naturally occurring radionuclides in the human body deliver an annual dose to the various tissues and organs of the body as follows: lung 36 mrem, bone surfaces 110 mrem, red marrow 50 mrem, and all other soft tissues 36 mrem. What is the annual effective dose equivalent that a human receives?

With the organ weighting factors of Table 9.2, the calculations are summarized in the following table.

Organ T	H_T (mrem/y)	w_T	$H_T w_T$ (mrem/y)
Lung	36	0.12	4.32
Bone surfaces	110	0.03	3.30
Red marrow	50	0.12	6.00
Soft Tissues:			
Gonads	36	0.25	9.00
Breast	36	0.15	5.40
Thyroid	36	0.03	1.08
Remainder	36	0.30	10.80
Total: (effective dose equivalent)			39.90

9.2.10   Effective Dose

In 1991, the ICRP recommended replacement of the effective dose equivalent by the effective dose. This recommendation was endorsed in 1993 by the NCRP in the United States and has been adopted in the European Community. The effective dose $\mathcal{E}$ is defined as follows. Suppose that the body is irradiated externally by a mixture of particles of different type and different energy, the different radiations being identified by the subscript R. The effective dose may then be determined as

$$\mathcal{E} = \sum_T w_T \sum_R w_R\, D_{T,R} \equiv \sum_T w_T\, H_T, \qquad (9.12)$$

in which H_T is the *equivalent dose* in organ or tissue T, $D_{T,R}$ is the mean absorbed dose in organ or tissue T from radiation R, w_R is the quality factor, now called the *radiation weighting factor*, for radiation R, as determined from Table 9.1, and w_T is a tissue weight factor given in Table 9.3. Note that in this formulation, w_R is independent of the organ or tissue and w_T is independent of the radiation.

Table 9.3. Tissue weighting factors adopted by the ICRP [1991] for use in determining the effective dose.

0.01	0.05	0.12	0.20
Bone surface	Bladder	Bone marrow	Gonads
Skin	Breast	Colon	
	Liver	Lung	
	Esophagus	Stomach	
	Thyroid		
	Remainder[a]		

[a]The remainder is composed of the following additional organs and tissues: adrenals, brain, small intestine, large intestine, kidney, muscle, pancreas, spleen, thymus, uterus, and others selectively irradiated. The weight factor of 0.05 is applied to the average dose in the remainder tissues and organs except as follows. If one of the remainder tissues or organs receives an equivalent dose in excess of the highest in any of the 12 organs or tissues for which factors are specified, a weight factor of 0.025 should be applied to that tissue or organ and a weight factor of 0.025 to the average dose in the other remainder tissues or organs.

9.3   Natural Exposures for Humans

Life on earth is continually subjected to radiation of natural origin. Exposure is both external and internal, the former arising from cosmic radiation and radionuclides in the environment, the latter arising from radionuclides taken into the body by ingestion or inhalation. Natural sources are the major contributors to human radiation exposure. Study of these sources is important for several reasons. Natural exposure represents a reference against which exposure to man-made sources may be compared, not only for standards-setting purposes, but also in epidemiological studies of the consequences of man-made sources or even of unusually concentrated natural sources in certain areas.

Cosmic radiation at the earth's surface consists mainly of muons and electrons which are debris from cosmic ray showers caused by energetic galactic or intergalactic cosmic rays interacting with atoms in the high atmosphere. The intensity of cosmic radiation varies with atmospheric elevation and with latitude. The 11-year cyclic variation of solar-flare activity affects the earth's magnetic field which, in turn, modulates the intensity of cosmic radiation reaching the earth.

Cosmic radiation reaching the earth's atmosphere consists mainly of high energy hydrogen nuclei (protons). Cosmic rays interact primarily with constituents of

Table 9.4. Summary of the annual global-average effective dose equivalents from various sources of natural background radiation. Source: UN [2000] and NAS [2006].

	Annual average effective dose (mSv)	
Source	World average	Typical range
External exposure		
Cosmic rays		
High-LET component	0.1	
Low-LET component	0.3	
Subtotal cosmic rays	0.4	0.3–1.0
Terrestrial gamma rays (low-LET)	0.5	0.3–0.6
Subtotal external exposure	0.9	
Internal exposure		
Ingestion		
High-LET ingestion	0.1	
Low-LET ingestion	0.2	
Subtotal ingestion	0.3	0.2–0.8
Inhalation (high-LET)	1.2	0.2–10
Subtotal internal exposure	1.5	
Total	2.4	1–10

the atmosphere to produce showers of secondary particles including a number of "cosmogenic" radionuclides. Chief among these are ^{3}H and ^{14}C. These radionuclides are produced at relatively uniform rates and, except as augmented by man-made sources, exist in the biosphere in equilibrium, i.e., with equal production and decay rates.

Naturally existing radionuclides not of cosmic-ray origin and not members of decay chains must have half lives comparable to the several billion-year age of the earth (see Section 5.7.2. These radionuclides are few and only two, ^{40}K and ^{87}Rb, result in significant portions of the dose rate in humans due to natural sources of radiation.

There are also several decay chains of radionuclides which occur naturally and whose parent radionuclides necessarily have half lives comparable to the age of the earth. Identified by the name of the parent, the only series of significance to human exposure are those of ^{238}U and ^{232}Th (see Section 5.7.3). Within these chains or series are subgroups comprising the daughter products of ^{222}Rn and ^{220}Rn. Radon as a source leads to the greatest component of natural radiation exposure and may be responsible for a significant fraction of lung cancer mortality. Health risks from exposure to radon daughter products have been studied intensively and, in many countries, efforts are being taken to mitigate these risks.

Global average radiation doses from natural background sources are listed in Table 9.4. Note the wide range of doses around the world, reflecting variations in diet, elevation, latitude, and soil or rock composition. Note too that the major component of dose is attributed to inhalation sources of high-LET radiation. Inhaled radon and its daughters represent the highest exposure component, and, although this component can be reduced somewhat by better ventilation of domiciles to prevent radon buildup, it cannot be altogether eliminated. Other components such as the doses from the ^{40}K in our bodies cannot be reduced. For the U.S. population,

Table 9.5. Summary of the annual effective dose equivalents from various sources of natural background radiation in the United States. Source: NCRP [1987].

	Annual dose equivalent (mSv)				
Radiation source	Bronchial epithelium	Other soft tissues	Bone surfaces	Bone marrow	H_E
Cosmic radiation	0.27	0.27	0.27	0.27	0.27
Cosmogenic radionuclides	0.01	0.01	0.01	0.03	0.01
External terrestrial radionuclides	0.28	0.28	0.28	0.28	0.28
Inhaled radionuclides	24				2.00
Radionuclides in body	0.36	0.36	1.10	0.50	0.39
Totals (rounded)	25	0.90	1.7	1.1	3.0

the average annual radiation dose from natural sources is enumerated in Table 9.5 in terms of both source and tissue affected.

In addition to natural exposures, human activities have produced other sources of radiation exposure. Primary among these sources are exposures resulting from medical uses of radiation for diagnosis and therapy. Authors of the BEIR-VII Report [NAS 2006], based on data from the NCRP [1987], deduced that, for the U.S. population, doses from man-made radiation sources are 18% of the total, the remaining 82% being from natural sources. Of the man-made component, approximately 0.65 mSv annually, 58% is from medical x rays, 21% from nuclear medicine, 16% from consumer products, 1% from the nuclear fuel cycle, and 2% each from occupational exposure and fallout from atmospheric testing of nuclear weapons. The Health Physics Society[2] lists effective radiation doses from diagnostic x rays ranging from 0.02 mSv for a chest x ray to 1.2 mSv for an abdominal x ray, with most procedures, including mammography, resulting in about 0.7 mSv. CT exposures range from 1.5 to 10 mSv for most exposures, but even greater for angiography procedures. Effective doses from nuclear medicine procedures range from 1 to about 20 mSv. U.S. adults who lived through the years of atmospheric testing of nuclear weapons on average received the greatest dose from intake of ^{131}I—5 mGy to the thyroid from tests at the Nevada Test Site and 0.4 mGy from global fallout. However, for a U.S. child born in 1951, thyroid doses of 30 mGy (U.S.) and 2 mGy (global) were received [Simon, et al. 2006].

9.4 Health Effects from Large Acute Doses

There are two circumstances under which high doses of ionizing radiation may be received. The first is accidental, and likely to involve a single instance of short duration. The second is medical, and likely to involve doses fractionated into a series delivered daily for several weeks, and perhaps administered under conditions designed to modify response in certain organs and tissues. Oxygen, among a group of chemical agents known to modify dose response, promotes the action of radiation presumably through the enhanced production of oxidizing agents along the tracks

---

[2]Fact Sheet on Radiation Exposure from Medical Diagnostic Imaging Procedures, available from http://www.hps.org/.

of the radiation. This section deals only with single, acute exposure of all or part of the body. It does not address issues such as fractionation and effect modification which pertain largely to medical exposure.

9.4.1   Effects on Individual Cells

The likelihood that a cell will be killed or prevented from division as a result of a given radiation exposure or dose depends on many factors. Primary factors are the dose rate and the collisional stopping power (i.e., the linear energy transfer LET). Doses delivered at low dose rates allow the cell's natural repair mechanism to repair some of the damage so that the consequences are generally not as severe as if the dose were delivered at high dose rates. High LET radiation, like alpha particles, creates more ion-electron pairs closer together than does low LET radiation. Consequently, high LET radiation produces more damage to a cell it passes through than would, say, a photon.

Another important factor affecting the consequences of radiation on a cell is the state of the cell's life cycle at the time of exposure. Cell death is more likely if the cell is in the process of division than if it is in a quiescent state. Thus, radiation exposure results in a greater proportion of cell death in those organs and tissues with rapidly reproducing cells. Examples include the fetus, especially in the early stages of gestation, the bone marrow, and the intestinal lining. Whole-body absorbed doses of several Gy are life-threatening largely because of cell killing in the bone marrow and lining of the intestines. However, in these tissues and in most other tissues and organs of the body, there are ample reserves of cells, and absorbed doses of much less than one Gy are tolerable without significant short term effect. Similarly, radiation doses which would be fatal if delivered in minutes or hours may be tolerable if delivered over significantly longer periods of time. Age, general health, and nutritional status are also factors in the course of events following radiation exposure.

For those tissues of the body for which cell division is slow, absorbed doses which might be fatal if delivered to the whole body may be sustained with little or no effect. On the other hand, much higher absorbed doses may lead ultimately to such a high proportion of cell death that, because replacement is so slow, structural or functional impairment appears perhaps long after exposure and persists perhaps indefinitely.

9.4.2   Deterministic Effects in Organs and Tissues

This section deals only with deterministic somatic effects—effects in the person exposed, effects with well-defined patterns of expression and thresholds of dose below which the effects are not experienced, and effects for which the severity is a function of dose. The stochastic carcinogenic and genetic effects of radiation are addressed in later sections.

The risk, or probability of suffering a particular effect or degree of harm, as a function of radiation dose above a threshold dose D_{th}, can be expressed in terms of a 50th-percentile dose D_{50}, or median effective dose, which would lead to a specified effect or degree of harm in half the persons receiving that dose. The D_{50} dose depends, in general, on the rate at which the dose is received. For doses below D_{th}, no one experiences the effect.

Table 9.6. Median effective absorbed doses D_{50} and threshold doses D_{th} for exposure of different organs and tissues in the human adult to gamma photons at dose rates ≤ 0.06 Gy h^{-1}. Source: Scott and Hahn [1989].

Organ/Tissue	Endpoint	D_{50} (Gy)	$D_{\text{th}}(Gy)$
Skin	Erythema	6 ± 1	3 ± 1
	Moist desquamation	30 ± 6	10 ± 2
Ovary	Permanent ovulation supression	3 ± 1	0.6 ± 0.4
Testes	Sperm count supressed for 2 y	0.6 ± 0.1	0.3 ± 0.1
Eye lens	Cataract	3.1 ± 0.9	0.5 ± 0.5
Lung	Death[a]	70 ± 30	40 ± 20
GI system	Vomiting	2 ± 0.5	0.5
	Diarrhea	3 ± 0.8	1
	Death	15 ± 5	8
Bone marrow	Death	3.8 ± 0.6	1.8 ± 0.3

[a]Dose rate 0.5 Gy/h.

Effects of radiation on individual organs and tissues are described in the subsections that follow. Information is taken from the following sources: [ICRP 1984, 1991; Langham 1967; NCRP 1971; Pochin 1983; Upton and Kimball 1967; Vogel and Motulsky 1979; Wald 1967; UN 1988]. A summary is provided in Table 9.6

Skin

Erythema, equivalent to a first-degree burn, appears about two weeks after an acute absorbed dose of 2 to 3 Gy, although reddening and itching may occur shortly after exposure. Loss of hair may result beginning about two weeks after acute exposure to 2 Gy or more, the effect being complete for doses greater than about 5 Gy and permanent for doses greater than about 6 Gy.

Wet or dry desquamation (transepidermal injury), equivalent to a second-degree burn, may appear one or two weeks after an acute absorbed dose in excess of 10 to 15 Gy. The desquamation results from destruction of basal cells of the epidermis. Dry desquamation is experienced with doses between about 10 and 20 Gy. Greater doses lead to moist desquamation. Doses on the order of 50 Gy result in prompt and extremely painful injury (dermal necrosis) in the nature of scalding or chemical burn.

Repair of sublethal damage to surviving cells is nearly complete in one day, though cellular repopulation is much slower. The radiation dose for a given effect is approximately doubled if the exposure is divided into ten equal daily fractions.

The various layers of skin have differing sensitivity to radiation damage. In terms of both chronic low-level exposure (potentially leading to skin cancer), and transepidermal injury, the basal-cell layer of the epidermis, at a depth range from approximately 0.05 to 0.1 mm, is the most sensitive region because of the mitotic activity that takes place therein. For extremely high doses leading to ulceration, damage to the dermal region, at depth greater than 0.1 mm, is of greater significance than damage to the epidermis.

Lens of the Eye

The lens of the eye is particularly susceptible to ionizing radiation, as it is to microwave radiation. As little as 1 to 2 Gy absorbed dose from x rays results in detectable opacification of the lens. Effects are cumulative, resulting from the migration of dead cells to a central position at the back of the lens. Dose equivalents in excess of about 10 Sv lead to cataract formation causing detectable impairment of vision.

Blood Forming Tissues

Except for lymphocytes, the mature circulating blood cells are relatively insensitive to radiation. The precursor cells present in bone marrow and lymphoid tissue, being rapidly dividing, are, however, highly sensitive. Exposure of the whole body leads to rapid depletion of lymphocytes and a relatively slow decline in platelets, granulocytes, and the long-lived red blood cells. Depression of lymphocytes is central to the impairment of immune response resulting from whole-body exposure to high levels of ionizing radiation. Recovery of the blood-forming tissues occurs through multiplication of surviving precursor cells in the bone marrow, lymphoid organs, spleen, and thymus. Blood cell depression may be countered by transfusion. Recovery may be enhanced by bone-marrow transplant.

Gastrointestinal Tract

The rapidly dividing cells in the lining of the gastrointestinal tract suffer significant mitotic inhibition at doses on the order of 0.5 Gy for low-LET radiation. For absorbed doses in excess of about 5 Gy, the lining cannot be adequately supplied with new cells. Excessive fluid and salts enter the gastrointestinal system, while bacteria and toxic materials enter the blood stream. The consequences are diarrhea, dehydration, infection, and toxemia.

Urinary, Respiratory, and Neurovascular Systems

The urinary system is relatively radiation resistant. Absorbed doses to the kidney in excess of about 3 Gy of low-LET radiation leads to hypertension. Absorbed doses of greater than 5 to 10 Gy of low-LET radiation lead to atrophy of the kidney and potentially lethal renal failure.

The respiratory system as a whole is relatively radiation resistant as compared to the blood-forming organs and tissues. Very high absorbed doses delivered only to the thorax (as in radiation therapy) may lead to scarring of lung tissue and pulmonary blood vessels, leading to chronic pneumonia-like illness.

The nervous system, with little or no mitotic activity, is extremely radiation resistant. Acute whole-body doses in excess of 100 Gy lead, in hours or days, to death from cerebro-vascular injury. Sub-lethal exposure may lead to impairment or degeneration of the nervous system as a result of radiation damage to blood vessels of supporting tissues.

Ovaries and Testes

In certain stages of cellular life, germ cells and gametes are very susceptible to radiation damage. In other stages, the cells are relatively resistant. There are significant differences between the courses of radiation damage in the male and the female.

Production of ova is completed early during fetal development. Some 2 million ova are present in the ovaries at birth, but the number declines to less than about 10 thousand by age 40. The immature ova, or *oocytes*, begin meiotic division during fetal development. However, the process is suspended in each ovum until a few weeks before that ovum takes part in ovulation. That few-week period is the time of greatest sensitivity for both cell killing and mutation induction. Mature oocytes may be killed by dose equivalents of a few Sv, and complete sterilization requires dose equivalents of 2.5 to 6 Sv, the required dose decreasing with subject age.

The spermatogenesis process which has a duration of about 74 days, occurs in a series of stages. Spermatogonia, produced from spermatogonial stem cells are the most radiosensitive cells. Cytological effects are observable within a few hours after delivery of doses as low as 0.15 Gy. The spermatogonia divide mitotically to spermatocytes. In a process which requires about 46 days, spermatocytes divide by meiosis into spermatids which mature to spermatozoa. Cells in these latter three stages are unaffected by absorbed doses less than about 3 Gy. A 0.15 Gy absorbed dose causes reduction in the sperm count beginning some 46 days later. By 74 days, the count may be reduced by about 80 percent. A 1.0 Gy or greater absorbed dose may lead to complete loss of sperm after the 74-day period. At absorbed doses less than about 3 Gy, fertility is restored completely, but only after many months. Permanent sterility results from greater absorbed doses.

The Embryo

In the developing embryo, the loss of only a few cells may risk serious consequences more akin to severe hereditary illness than to cell killing in fully developed somatic tissues. The natural probability for mortality and induction of malformations, mental retardation, tumors, and leukemia during gestation is about 0.06. The combined radiation risk is about 0.2 per Sv delivered throughout gestation. The risk of childhood cancers and leukemia up to age 10 is about 0.02 per Sv. The risk of severe mental retardation is greatest during the 8th to 15th week of gestation, with a radiation risk of about 0.4 per Sv. The central nervous system is sensitive throughout gestation, but teratological (i.e., malformation) risks may be greater by a factor of 2 or 3 during the major organogenesis of the 2nd through 8th weeks.

9.4.3 Potentially Lethal Exposure to Low-LET Radiation

The question of what constitutes a *lethal dose* of radiation has of course received a great deal of study. There is no simple answer. Certainly the age and general health of the exposed person are key factors in the determination. So, too, are the availability and administration of specialized medical treatment. Inadequacies of dosimetry make interpretation of sparse human data difficult. Data from animal studies, when applied to human exposure, are subject to uncertain-

Table 9.7. Lethal doses of radiation.

Lethality	Mid-line absorbed dose (Gy)
$LD_{5/60}$	2.0–2.5
$LD_{10/60}$	2.5–3.0
$LD_{50/60}$	3.0–3.5
$LD_{90/60}$	3.5–4.5
$LD_{99/60}$	4.5–5.5

ties in extrapolation. Delay times in the response to radiation, and the statistical variability in response have led to expression of the lethal dose in the form, for example, $LD_{50/60}$, meaning the dose fatal to 50 percent of those exposed within 60 days. The *dose* itself requires careful interpretation. One way of defining the *dose* is the free-field exposure, in roentgen units, for gamma or x rays. A second is the average absorbed dose to the whole body. A third is the *mid-line* absorbed dose, i.e., the average absorbed dose near the abdomen of the body. For gamma rays and x rays, the mid-line dose, in units of rads, is about two-thirds the free-field exposure, in units of roentgens. Anno et al. [1989] suggestions for the lethal doses of ionizing radiation are given in Table 9.7. Representative effects of doses below the threshold for lethality are given in Table 9.8.

The lethal effects of radiation exposure may take several courses. For extremely high doses ($>$500 Gy), death is nearly instantaneous, resulting from enzyme inactivation or possibly from immediate effects on the electrical response of the heart [Kathren 1985]. Lesser but still fatal doses lead promptly to symptoms known collectively as the *prodromal syndrome*. The symptoms, which are expressed within a 48-hour period, are of two main types as summarized in Table 9.9.

Table 9.8. Effects of high, sublethal doses. Source: NCRP [1971], Anno et al. [1989].

Minimal dose detectable by chromosome analysis	0.05–0.25 Gy
Minimal dose detectable in groups by change in white-blood cell count	0.25–0.50 Gy
Minimal acute dose readily detectable in a specific individual	0.50–0.75 Gy
Mild effects only during first day post-exposure with slight depression of blood counts	0.50–1.00 Gy
Minimal acute dose to produce vomiting in 10 percent of exposed individuals	0.75–1.25 Gy
Nausea and vomiting in 20 to 70% of persons exposed fatigue and weakness in 30 to 60%; 20 to 35% drop in blood cell production due to loss of bone marrow stem cells	1.00–2.00 Gy
Acute dose likely to produce transient disability and clear hematological changes in a majority of individuals so exposed.	1.50–2.00 Gy

At doses from about 50 to 500 Gy, most symptoms appear within minutes of exposure, and death results from neurological or cardiovascular failure. At doses on the order of 50 Gy, death is likely to result from cardiac failure within a few days. At lower doses, down to about 6 Gy, death is more likely to result from damage to the gastrointestinal system with attendant dehydration and electrolyte loss.

At doses for which survivability is questionable, the course of radiation illness takes roughly the following pattern [Wald 1967]:

Table 9.9. Symptoms of high, sublethal doses. Source: Langham [1967].

Gastrointestinal	Neuromuscular
Anorexia	Fatigue
Nausea	Apathy
Vomiting	Sweating
Diarrhea	Fever
Intestinal cramps	Headache
Dalivation	Hypotension
Dehydration	Hypotensive shock

- **Prodromal Stage (0 to 48 hours postexposure)**: Symptoms are gastrointestinal and neuromuscular. Reduction in lymphocytes is apparent within 24 hours.

- **Latent Stage (2 to 3 weeks postexposure)**: Remission of prodromal symptoms occurs and there is a period of apparent well being.

- **Manifest Illness Stage (2 to 3 weeks to 6 to 8 weeks postexposure)**: Gastrointestinal and hematological symptoms are sharply defined. Blood elements except for red blood cells are depleted and secondary infection is a threat. Dehydration and loss of electrolytes are life threatening. Skin pigmentation changes and epilation takes place.

- **Recovery Stage (beyond 6 to 8 weeks postexposure)**: If the exposed person survives for six weeks, the outlook for permanent recovery is favorable.

9.5 Hereditary Effects

Inheritance of radiation-induced abnormalities was discovered by Hermann Muller in 1927 while studying high-dose irradiation of fruit flies. Since 1927, studies of plants and animals have revealed approximately linear relationships between mutation frequency and dose, for doses as low as 3 mSv. Very little information is available on mutational effects of radiation in humans. Indeed, the only unequivocal evidence relates to chromosomal rearrangement in spermatocytes. However, the evidence from animal studies has left no doubt that heritable mutational effects in the human are possible. Estimation of risks to human populations are based largely on extrapolation of studies of radiation effects in other mammals.

Beginning in the 1950s, radiation protection standards recognized the potential significance of radiation induced hereditary illness. Both the International Commission on Radiation Protection and the National Council on Radiation Protection and Measurements issued recommendations designed to minimize genetic risks in future generations, and which may be interpreted as limiting exposure to human populations to 1.7 mSv average dose equivalent per year in excess of medical and background exposure.

This section deals only with exposure to low-LET ionizing radiation, at dose rates comparable to those received from the natural radiation environment. It deals only with heritable genetic effects, and not with radiation induced mutations in somatic cells.

9.5.1   Classification of Genetic Effects

Table 9.10 reports estimates of the incidence of human hereditary or partially hereditary traits causing serious handicap at some time during life. Inheritance of a deleterious trait results from mutation(s) in one or both maternal and paternal lines of germ cells. By mutations, we mean either microscopically visible chromosome abnormality or submicroscopic disruption in the DNA making up the individual genes within the chromosomes. Mutations take place in both germ cells and somatic cells, but only mutations in germ cells are of concern here.

Regularly inherited traits are those whose inheritance follows Mendelian laws. These are autosomal dominant, X-linked, and recessive traits. Examples of autosomal dominant disorders, i.e., those which are expressed even when the person is heterozygous for that trait, are certain types of muscular dystrophy, retinoblastoma, Huntington's chorea, and various skeletal malformations. Examples of recessive disorders, i.e., those which are expressed only when the individual is homozygous for the trait, include Tay-Sachs disease, phenylketonuria, sickle-cell anemia, and cystic fibrosis. X-linked disorders, i.e., those traits identified with genes in the X chromosome of the X-Y pair and which are expressed mostly in males, include hemophilia, color-blindness, and one type of muscular dystrophy. In the X-Y chromosome pair, otherwise recessive genetic traits carried by the "stronger" maternal X chromosome are expressed as though the traits were dominant. Chromosome abnormalities are of two types, those involving changes in the numbers of chromosomes and those involving the structure of the chromosomes themselves. Down syndrome is an example of the former. With natural occurrence, numerical abnormalities are more common. Radiation-induced abnormalities are more frequently structural abnormalities.

There is a very broad category comprising what are variously called irregularly inherited traits, multifactorial diseases, or traits of complex etiology. This category includes abnormalities and diseases to which genetic mutations doubtlessly contribute but which have inheritances much more complex than result from chromosome abnormalities or mutations of single genes. They are exemplified by inherited predispositions for a wide variety of ailments and conditions. Included in Table 9.10 is a subgroup of irregularly inherited traits, identified as congenital abnormalities. These are well identified conditions such as spina bifida and cleft palate, with reasonably well known degrees of heritability. One or more other multifactorial disorders, including cancer, are thought to afflict nearly all persons sometime during life; however, the mutational components of these disorders are unknown even as to orders of magnitude [NAS 1990].

9.5.2   Summary of Risk Estimates

Table 9.10 summarizes the 2006 BEIR-VII and 2001 UNSCEAR genetic risk estimates. The results are for low-LET radiation (quality factor $QF = 1$); thus, the absorbed dose and dose equivalent are the same. These estimates are based on a population-average gonad absorbed dose 1 Gy (100 rad) to the reproductive population which produce one-million liveborne. Because of the linearity of the dose-effect

Table 9.10. Genetic risks from continuing exposure to low-LET, low-dose, or chronic radiation as estimated by the BEIR and UNSCEAR committees on the basis of a doubling-dose of 1 Gy. Source: NAS [2006] and UN [2001].

Type of Disorder	Per million progeny		
	Baseline frequency	Cases/Gy in first generation	Cases/Gy in second generation[a]
Mendelian autosomal			
Dominant and X-linked	16,500	750–1500	1300–2500
Recessive	7,500	0	0
Chromosomal	4,000	[b]	[b]
Multifactorial			
Chronic multifactorial	650,000	250–1200	250–1200
Congenital abnormalities	60,000	2000	2400–3000
Total	738,000	3000–4700	3930–6700
Total risk as percent of baseline		0.41–0.64	0.53–0.91

[a]Risk to the second generation includes that to the first except for congentital abnormalities for which it is assumed that between 20% and 50% of the abnormal progeny in the first generation may transmit the damage to the second post-radiation generation, the remainder causing lethality.

[b]Assumed to be included with Mendelian diseases and congenital abnormalities.

models used, these estimated hereditary risks are the same whether the gonad dose is received in a single occurrence or over the 30-year reproduction interval. The population for these results is assumed static in number, so that one million born into one generation replace one million in the parental generation.

Data in Table 9.10 give the expected number of genetic illness cases appearing in the first and second generations, each receiving radiation exposure. Except as indicated, cases in the second generation include the new cases from exposure of that generation plus cases resulting from exposure of the previous generation, e.g., 550 to 1000 cases of autosomal dominant and X-linked class and no cases of the autosomal recessive class. After many generations of continuous exposure, a new equilibrium would be approached, with cases in any one generation arising from exposure to that generation as well as all previous generations. However, authors of the BEIR-VII report resist projecting risks at a new equilibrium, arguing that risks to generations of children and grandchildren are of principal concern and that predictions over tens to hundreds of generations require unrealistic and untestable assumptions that demographics and health care possibilities remain constant. Understanding of Table 9.10 requires further exploration of two issues: basing estimates on *gonad dose*, and application of the *doubling dose*.

Gonad Dose versus Effective Dose

For many conditions of exposure, notably exposure of the entire body to broad beams of gamma or x rays, the "whole body" dose, the effective dose, and the gonad dose are essentially the same. For other conditions of exposure, expression of genetic risks in terms of population average or collective effective dose equivalent is subject to considerable uncertainty. The weight factor for the gonads in evaluating the effective dose is 0.2 (see Table 9.3). This means that for an effective dose of 1 Sv (1 Gy for low-LET radiation), the actual gonad dose could be as little as zero (if the gonads were not exposed) to as great as 5 Gy (if only the gonads were exposed). Thus, one could say that, per 10^4 person-Sv collective effective dose *to the reproductive population only* in one generation, one could expect, from the risk estimates of Table 9.10, an increase in the number of cases of hereditary illness in the first generation of

$$\frac{3000 \text{ to } 4700 \text{ cases}}{10^6 \text{ person-Sv}} \times (0 \text{ to } 5\times10^4 \text{ person-Sv}) \simeq 0 \text{ to } 235 \text{ cases.}$$

For a population with a mean life expectancy of 75 years and a reproductive period of 30 years, one could also say that, per 10^4 person-Sv collective effective dose *to the entire population* in one generation, one could expect an increase of

$$\frac{3000 \text{ to } 4700 \text{ cases}}{10^6 \text{ person-Sv}} \times \frac{30 \text{ years}}{75 \text{ years}} \times (0 \text{ to } 5\times10^4 \text{ person-Sv}) \simeq 0 \text{ to } 94 \text{ cases}$$

of hereditary illness in the first generation.

The Doubling Dose and Risk Estimates
In simplest terms, the *doubling dose DD* is that gonad dose to a parent generation leading to a first-generation increase in mutations equal to the baseline number of mutations, i.e., a doubling of the number of mutations. Human data are available for baseline mutations. Induced mutation rates in human populations are unknown, even for atomic bomb survivors, so it is necessary to rely upon data from laboratory tests on the mouse. In use conceptually since the 1950s, the doubling dose saw its initial use for risk estimates in the first BEIR report [NAS 1972]. Risk estimates, though, are expressed in terms of disease or disorder frequencies, not mutation frequencies, and must take into account that not all mutations lead to genetic disorders. Multifactorial disorders have causes other than mutational, requiring a *mutation component MC* correction factor defined as the relative increase in disease frequency per unit relative increase in mutation rate. Similarly, a *potential recoverability correction factor PRCF* accounts for the observation that there are fundamental differences between mutations that cause human disease and radiation-induced mutations studied in experimental systems such as the mouse. Thus, if P is the baseline frequency of disease, the risk per unit dose is given by

$$P \times (1/DD) \times MC \times PRCF. \tag{9.13}$$

The radiation-induced mutation rate in mice is $R_{mice}^{rad} \simeq 3.6\times10^{-6}$ per Gy. Spontaneous rates in humans are $R_{human} \simeq 2.95 \times 10^{-6}$ per gene locus per generation. Thus, $DD = R_{human}/R_{mice}^{rad} = 0.82$ Gy, which is rounded to 1 Gy.

From Table 9.10, the baseline risk P equals 16,500 per million progeny for dominant and X-linked disorders The latest estimates of the correction factors [NAS 2006] are as follows: $MC = 0.3$ and $PRCF = 0.15$ to 0.30 for dominant and X-linked disorders; and $MC = 0.02$ and $PRCF = 0.02$ to 0.09 for chronic diseases. It follows that the specific risk for dominant and X-linked disorders is

$$\text{Risk} = \frac{P \times MC \times PRCF}{DD} = \frac{16,500 \times 0.3 \times (0.15 \text{ to } 0.30)}{1 \text{ Gy}}$$

$$\simeq 750 \text{ to } 1500 \text{ per Gy per million progeny.} \tag{9.14}$$

9.6 Cancer Risks from Radiation Exposures

A large body of evidence leaves no doubt that ionizing radiation, when delivered in high doses, is one of the many causes of cancer in the human. Excess cancer risk cannot be observed at doses less than about 0.2 Gy and, therefore, risks for lower doses cannot be determined directly [UN 1988]. At high doses, in almost all body tissues and organs, radiation can produce cancers that are indistinguishable from those occurring naturally. Consequently, radiation induced cancer can be inferred only from a statistical excess above natural occurrence (see Table 9.11). There is a large variation in the sensitivity of tissues and organs to cancer induction by radiation.

For whole-body exposure to radiation, solid tumors are of greater numerical significance than leukemia. The excess risk of leukemia appears within a few years after radiation exposure and largely disappears within 30 years after exposure. By contrast, solid cancers, which occur primarily in the female breast, the thyroid, the lung, and some digestive organs, characteristically have long latent periods, seldom appearing before 10 years after radiation exposure and continuing to appear for 30 years or more. It is also apparent that age at exposure is a major factor in the risk of radiation induced cancer. Various host or environmental factors influence the incidence of radiation induced cancer. These may include hormonal influences, immunological status, and exposure to various oncogenic agents.

Just how cancer is induced is not understood fully; but it is clear that (1) there are no unique cancer types created solely by ionizing radiation, and (2) that induction is a multistep process involving initiation, promotion, and progression. The initiating event is undoubtedly disruption of the genetic coding within a cell nucleus. The mutation must not be so severe that the cell is unable to reproduce and it must somehow confer a selective advantage in multiplication over sister cells in the host organ or tissue. As Pochin [1983] observes, the transformation of a normal cell into one with the potential for abnormal multiplication does not necessarily promote the multiplication. There are many carcinogenic agents, some of which are effective in the initiation phase, some of which are effective in the promotion phase, and some of which, like ionizing radiation, are effective in both phases. There

Table 9.11. Annual cancer incidence and death rates per 100,000 population in the 2002 United States population. Source: HHS [2005].

Primary site	Incidence per 10^5 per year		Deaths per 10^5 per year	
	Males	Females	Males	Females
Leukemia	14.6	8.8	10.1	5.7
Lymphoma	25.1	17.9	10.2	6.5
Respiratory	94.7	55.8	76.2	42.2
Digestive	106.9	71.4	59.1	35.9
Breast		124.9		25.5
Genital	167.3	48.9	28.6	16.7
Urinary	56.3	19.1	13.9	5.3
Other	80.6	58.2	41.8	24.9
Total	545.5	405.0	239.9	162.7

is solid evidence that there is a latent period between radiation exposure and the onset of cancer. While it is not certain that the latent period may be explained simply in terms of initiation-promotion phenomena, the fact that there is a latent period is evidence of the multistep process of carcinogenesis. Tumor progression, known to be enhanced by radiation, denotes either conversion of a benign growth to a malignant growth or the attainment of increasingly malignant properties in an established cancer. In either case, rapid growth of a subpopulation of cancer cells overwhelms less active normal cells.

9.6.1   Estimating Radiogenic Cancer Risks

Our knowledge about radiation induced cancer is based on epidemiological studies of people who have received large radiation doses. These populations include atomic bomb survivors, radiation therapy patients, and people who have received large occupational doses. Some 91,000 survivors of the atomic weapon attacks on Hiroshima and Nagasaki and their offspring remain under continuing study, and much of our knowledge about radiation induced cancer derives from this group. Occupational groups include medical and industrial radiologists and technicians, women who ingested large amounts of radium while painting instrument dials during World War I, and miners exposed to high concentrations of radon and its daughter radionuclides. Finally, radiation therapy patients have provided much information on radiation carcinogenesis. These include many treated with x rays between 1930 and 1950 for severe spinal arthritis, Europeans given ^{226}Ra injections, and many women given radiation therapy for cervical cancer.

The Dose and Dose-Rate Effectiveness Factor

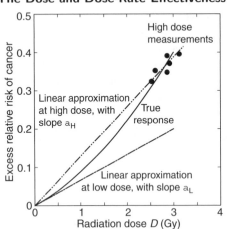

Figure 9.1. Basis for the dose and dose-rate effectiveness factor.

Assessment of cancer risks from radiation exposure is concerned primarily with exposures of population groups to low doses at low dose rates. As just indicated, however, there is little choice but to base risk estimates on consequences of exposures at high doses and dose rates. Furthermore, organizations such as the ICRP and the NCRP have endorsed, and government organizations have agreed to base risk estimates on a linear, no-threshold relationship between cancer risk and radiation dose. An exception applies to radiation induced leukemia, for which a quadratic, no-threshold relationship has been adopted. How does one reconcile low-dose risk estimates based on high-dose data? The answer is addressed in Fig. 9.1. Symbols display a limited base of data for high doses and dose rates. A central curved line displays what may be the true dose response (say a linear quadratic relationship). The upper straight line, with slope α_H is the linear, no-threshold approximation for the high-dose data. The lower straight line, with slope α_L is tangent to the true response curve in the limit of low dose and low dose rate, conditions allowing for partial repair of radiation damage. Risks at low doses and dose rates, if computed on the basis of α_H, need to be corrected by division by the ratio α_H/α_L defined as the $DDREF$, the *dose and dose rate correction factor*.

9.6.2 Dose-Response Models for Cancer

Evidence is clear that absorbed doses of ionizing radiation at levels of 1 Gy or greater may lead stochastically to abnormally high cancer incidence in exposed populations. However, there is no direct evidence that chronic exposure to low levels of ionizing radiation may likewise lead to abnormally high cancer incidence. Risk estimates for chronic, low-level exposure requires extrapolation of high-dose and high dose-rate response data to low doses. Methods used for extrapolation are often controversial, any one method being criticized by some as over-predictive and by others as under-predictive.

Models relating cancer incidence or mortality to radiation dose are required both for interpretation of study data using regression analysis and for risk estimation. Estimation models are usually applicable only to low-LET radiation doses (quality factor $QF = 1$). However, study data may be based in part on high-LET radiation ($QF > 5$), as is the case for the atomic-bomb survivors, some of whom received large neutron doses. Thus, data interpretation may require consideration of the relative biological effectiveness of the high-LET component. The doses, on which projections are based, are average whole-body doses that, for low-LET radiation, well represent dose equivalents.

Current risk estimates for cancer have as the basic elements dose responses that are functions of the cancer site, the age a_o (y) at exposure, the age a (y) at which the cancer is expressed or the age at death, and the sex s of the subject. The radiogenic cancer risk is expressed as

$$\text{risk} = R_o(s,a) \times EER(D,s,a_o,a). \qquad (9.15)$$

Here R_o is the natural cancer risk as a function of sex, site, and age at cancer expression, for both incidence and mortality, and EER is the *excess relative risk* function that is determined by fitting a model to observed radiogenic cancer incidence or mortaility for cancer at a particular site. For example, the excess relative risk for all solid cancer, except thyroid and non-melanoma skin cancer, is expressed as [NAS 2006]

$$ERR(D,s,e,a) = \beta_s D \exp[e*\gamma](a/60)^\eta, \qquad (9.16)$$

in which D is the dose in Sv. In this particular model, the empirical parameter $e* = (a_o - 30)/10$ for $a_o < 30$ and zero for $a_o \geq 30$. Parameters β, γ, and η depend on whether the estimate is for incidence or mortality. For example, for cancer incidence, $\gamma = -0.30$, $\eta = -1.4$, and $\beta_s = 0.33$ for males and 0.57 for females.

To use Eq. (9.15), the natural risk $R_o(s,a)$ for the type of cancer of concern must be known. Data for natural risk are available, but are too extensive to be presented here. They may be found in publications of the Centers for Disease Control [HHS 2005] and on line at `http://www.cdc.gov/cancer/ncpr/uscs` or `http://seer.cancer.gov/statistics`. For example, U.S. death rates per 100,000

U.S. males, for all races and all cancer sites combined increase from 2.1 for infants to 11.3 at age 30-34 to 520 at age 60-64. For females, corresponding rates are 1.6, 13.8 and 382.

Ethnic and environmental factors are not taken into account explicitly, but do affect the estimates. For example, risk estimates based on studies of Hiroshima and Nagasaki nuclear-weapon survivors are strictly applicable only to the Japanese population. Nevertheless, such estimates are widely applied to estimation of risks for other populations. For details of these risk estimation models, the interested reader is referred to reports by the National Academy of Sciences and the United Nations [NAS 1980, 1990, 2006; UN 2000]. In the section to follow, a few results based on these risk models are given.

It should be emphasized that, in examining these risks of cancer from radiation exposure, one should keep in mind the overall or natural risk of cancer. As indicated in Table 9.11, two persons per thousand in the United States die each year from cancer. As will be seen in the next section, the overall lifetime risk of cancer mortality is about one in five for males and about one in six for females.

9.6.3 Average Cancer Risks for Exposed Populations

The BEIR-VII Committee of the National Academy of Sciences [NAS 2006] made various estimates of the risk of excess cancer mortality resulting from low LET (gamma-ray) exposures. These risks are summarized in Table 9.12 by sex and by age at exposure. Though the data are for conditions of low dose and dose rate, they were generated in large part from cancer incidence and mortality experienced by survivors of atomic weapons at Hiroshima and Nagasaki. At lower doses and dose rates, risks are somewhat less since biological repair mechanisms can repair a greater fraction of the genetic damage produced by the radiation. This effect is accounted for in risk estimates for leukemia, which are based on a linear-quadratic dose-response model. For solid cancers, risks have been modified by application of a dose and dose-rate effectiveness factor, namely, by dividing high dose and dose-rate data by the DDREF value of 1.5.

The BEIR-VII Committee also calculated risks to the United States population under three (low-LET) exposure scenarios: (1) single exposure to 0.1 Gy, (2) continuous lifetime exposure to 1 mGy per year, and (3) exposure to 10 mGy per year from age 18 to age 65. Results are given in Table 9.13. The first scenario is representative of accidental exposure of a large population (the 1999 U.S. population), the second of chronic exposure, and the third of occupational exposure. For example, for leukemia mortality, with 95% confidence limits (not given in Table 9.13, for a single exposure of the U.S. population to 0.1 Gy, the risk per 100,000 is 70 (20 to 220) for the male and 50 (10 to 190) for the female. For nonleukemia mortality per 1000,000, fatalities are 410 (200 to 830) for the male and 610 (300 to 1200) for the female. For this case, the total low-dose cancer mortality risk for the U.S. population is $0.5 \times (480 + 660)/(0.1 \text{ Gy} \times 10^5) = 0.057$ per Gy, which can be rounded to 0.05 per Gy, or 5×10^{-4} per rem. This risk may be used as an overall cancer risk factor for environmental exposures, i.e., small exposures obtained at low dose rates.

The radiation cancer risks summarized in Tables 9.12 and 9.13 can used to estimate risks for a variety of exposures. See Example 9.5 for use of these data.

Table 9.12. Excess lifetime cancer incidence and mortality for the U.S. population by age at exposure for a whole-body dose of 0.1 Gy (10 rad) from low LET radiation to populations of 10^5 males or females. Source: based on [NAS 2006].

	Age at exposure (years)								
	0	10	20	30	40	50	60	70	80
	Females								
Incidence									
Leukemia	185	86	71	63	62	62	57	51	37
All solid	4592	2525	1575	1002	824	678	529	358	177
Mortality									
Leukemia	53	53	51	51	52	54	55	52	38
All solid	1717	1051	711	491	455	415	354	265	152
	Males								
Incidence									
Leukemia	237	120	96	84	84	84	82	73	48
All solid	2326	1325	881	602	564	507	407	270	126
Mortality									
Leukemia	71	71	67	64	67	71	73	69	51
All solid	1028	641	444	317	310	289	246	181	102

Table 9.13. Excess cancer incidence and mortality per 100,000 males and 100,000 females in the stationary U.S. population for three low-dose exposure scenarios. Source: based on [NAS 2006].

	Cases per 10^5		Deaths per 10^5	
Cancer Type	Males	Females	Males	Females
Single Exposure to 0.1 Gy (10 rad):				
Radiation Induced:				
leukemia	100	70	70	50
nonleukemia	800	1300	410	610
total	900	1370	480	660
Natural Expectation:				
leukemia	830	590	710	530
nonleukemia	45500	36900	22100	17500
total	46330	37490	22810	18030
Continuous Lifetime Exposure to 1 mGy (100 mrad) per year:				
Radiation Induced:				
leukemia	67	51	47	38
nonleukemia	554	968	285	459
total	621	1019	332	497
Continuous Exposure to 10 mGy (1 rad) per year from age 18 to 65:				
Radiation Induced:				
leukemia	360	270	290	220
nonleukemia	2699	4025	1410	2169
total	3059	4295	1700	2389

Example 9.5:   What are the probabilities a 30-y old male who receives an accidental gamma-ray exposure of 2 rad (0.02 Gy) will eventually die from (1) radiogenic leukemia and from (2) any other cancer caused by his exposure?

From Table 9.12, the leukemia risk for a 0.1 Gy exposure is $64/10^5$. Thus, for a 0.02 Gy exposure, the person's risk of dying from radiogenic leukemia is $(64/10^5) \times (0.02/0.1) = 0.000128$ or 1 chance in about 7800. Similarly, the probability he will die from another type of radiogenic cancer is $(317/10^5) \times (0.02/0.1) = 0.000634$ or 1 chance in about 1600.

9.6.4   Probability of Causation Calculations

As a requirement of the Orphan Drug Act (Public Law 97-414), the National Institutes of Health were assigned the responsibility of preparing *probability of causation* tables for use in adjudication of claims of radiation carcinogenesis, an actuarial method useful for summarizing the existing scientific evidence bearing on the likelihood that prior radiation exposure might be causally related to cancer occurrence under various circumstances

Tables were first released in 1985 and revised in 2003 [NIH]. An online program for computing probabilities is offered by the National Institute for Occupational Safety and Health (NIOSH). It is available at the link: http://www.cdc.gov/niosh/ocas/ocasirep.html. Legalities of implementation by the Department of Health and Human Services are contained in 42 CFR Part 81, "Guidelines for Determining the Probability of Causation Under the Energy Employees Occupational Illness Compensation Program Act of 2000."

Probability of causation (PC) also called assigned share (AS) for a specific radiation exposure at a certain age and diagnosis of cancer at a later age is computed as the ratio of the radiation risk to the sum of the baseline risk plus the radiation risk. In terms of excess risk relative to baseline (ERR),

$$PC = \frac{ERR}{1 + ERR}. \tag{9.17}$$

9.7   Radon and Lung Cancer Risks

A recently recognized radiation hazard to humans is that posed by naturally occurring radon. This radioactive noble gas diffuses into the atmosphere from rocks, soils, and building materials containing progenitor radionuclides. Each of the three naturally occurring primordial decay series (see Figs. 5.20 and 5.21) include different isotopes of radon.

Radon itself presents little radiation hazard on inhalation, and only minor hazard if ingested in aqueous solution. The principal hazard associated with radon is due to its short-lived daughter products. After a mixture of radon and daughters is inhaled, the radon, a noble gas, is exhaled. However, the daughters, either attached to aerosol particles or present in highly chemically reactive atomic or ionic states, are retained in the respiratory system, decaying one into the other and depositing their decay energies within the cells on the surfaces of the respiratory system. Daughters that decay by alpha emission are of greatest concern since they can produce very large localized doses to the basal cells of the bronchial and pulmonary epithelia. The importance of this localized lung exposure is evident in the natural background exposures listed in Table 9.5.

^{222}Rn and its daughters ordinarily present a greater hazard than ^{220}Rn and its daughters, largely because the much shorter half life of ^{220}Rn (55.6 s) makes its decay very likely before it can be released into the atmosphere. The air concentration of ^{219}Rn (3.96 s), a product of the ^{235}U decay series, is negligible. Globally, the mean annual effective dose equivalent due to ^{222}Rn daughters is about 1 mSv (100 mrem) while that due to ^{220}Rn daughters is estimated to be about 0.2 mSv (20 mrem) [UN 1982]. Consequently, this section deals primarily with hazards associated with ^{222}Rn and its short-lived daughters whose decay chain is shown below.

$$^{222}_{86}\text{Rn} \xrightarrow[3.82\text{ d}]{\alpha} {}^{218}_{84}\text{Po} \xrightarrow[3.05\text{ m}]{\alpha} {}^{214}_{82}\text{Pb} \xrightarrow[26.8\text{ m}]{\beta^-} {}^{214}_{83}\text{Bi} \xrightarrow[19.9\text{ m}]{\beta^-} {}^{214}_{84}\text{Po} \xrightarrow[164\mu s]{\alpha} {}^{210}_{82}\text{Pb} \ (22.2\text{ y}).$$

$$(9.18)$$

Decay products beyond ^{214}Po are of little consequence since ^{210}Pb has such a long half-life and is eliminated from lung tissues before it decays.

Equilibrium Equivalent Concentration

Of prime importance in evaluating doses to the lung is the *potential alpha energy concentration* of the radon daughters in the inhaled air. The potential alpha energy concentration (MeV m^{-3}) is the total alpha particle energy that would be emitted by all the daughters in a unit volume of air after they have all decayed to the end of the decay chain.

Consider the ^{222}Rn series of Eq. (9.18). Denote the activity concentrations (Bq m^{-3}) of the four daughters ^{218}Po, ^{214}Pb, ^{214}Bi, and ^{214}Po by C_i, $i = 1$ to 4 and their decay constants by λ_i (s^{-1}). ^{214}Po releases $E_4 = 7.687$ MeV of alpha particle energy per decay. It thus has a potential alpha particle energy concentration of $E_4 C_4/\lambda_4$ MeV m^{-3}. ^{214}Bi emits no alpha particles. However, once it decays to ^{214}Po, the alpha particle of the latter is released almost immediately. ^{214}Bi thus has a potential alpha particle energy concentration of $E_4 C_3/\lambda_3$ MeV m^{-3}. That for ^{214}Pb is similarly $E_4 C_2/\lambda_2$. ^{218}Po itself emits $E_1 = 6.003$ MeV of alpha particle energy per decay. Its potential alpha energy concentration is thus $(E_1 + E_4)C_1/\lambda_1$. Thus, the total potential alpha energy concentration is

$$E_{tot} = (E_1 + E_4)C_1/\lambda_1 + E_4(C_2/\lambda_2 + C_3/\lambda_3 + C_4/\lambda_4). \qquad (9.19)$$

It is this alpha-particle energy that is ultimately released in the tissues of the lung.

Because of the long lifetime of ^{222}Rn compared to those of its daughters in Eq. (9.18), the activity concentration of the daughters in the air should ideally equal that of ^{222}Rn if secular equilibrium were established (see Section 5.6.4). However, such equilibrium conditions are seldom realized.

Decay of ^{222}Rn and its daughters in the atmosphere leads first to individual unattached ions or neutral atoms. The ions or atoms may become attached to aerosol or dust particles, the attachment rate depending in a complex manner on the size distribution of the particles. Radioactive decay of an attached ion or atom, because of recoil, usually results in an unattached daughter ion or atom. Either attached or unattached species may be deposited (plate out) on surfaces, especially in indoor spaces, the rate depending on the surface to volume ratio of the space. Because of plate out, ^{222}Rn daughter products in the atmosphere are not likely to be in equilibrium with the parent.

However, if the daughters were in secular equilibrium with ^{222}Rn with an activity concentration C_o Bq m^{-3}, then $C_o = C_1 = C_2 = C_3 = C_4$ and the potential alpha particle energy concentration would be

$$E_{tot}^{equil} = C_o[(E_1 + E_4)/\lambda_1 + E_4(1/\lambda_2 + 1/\lambda_3 + 1/\lambda_4)], \qquad (9.20)$$

a value that is always greater than that in an actual air mixture since the disequilibrium decreases the activity concentration of the daughters.

To quantify the disequilibrium that usually exists between the daughters and ^{222}Rn, an *equilibrium factor* is defined as $F = E_{tot}/E_{tot}^{equil} < 1$. This equilibrium factor can then be used to define an *equilibrium equivalent concentration* EEC that is the activity concentration of radon in equilibrium with its short-lived daughters which has the same total potential alpha energy concentration as the actual nonequilibrium air mixture, i.e.,

$$\text{EEC} = F \times C_o. \qquad (9.21)$$

Thus, the ^{222}Rn activity concentration in a nonequilibrium atmosphere generally must be somewhat higher than that in an atmosphere containing daughters in equilibrium with the radon in order to achieve the same potential alpha energy concentration and to present the same hazard to humans.

9.7.1   Radon Activity Concentrations

Outdoor Concentrations

The exhalation rate of ^{222}Rn from rocks and soils is highly variable, ranging from 0.2 to 70 mBq m^{-2} s^{-1}. An area-weighted average for continental areas, exclusive of Antarctica and Greenland, is 16 mBq m^{-2} s^{-1} [UN 1988]. Exhalation from the surface of the sea is only about 1% of that from land areas. Rain, snow, and freezing decrease exhalation rates so that they are generally lower in winter than in summer. Barometric pressure and wind speed also affect exhalation rates, decreasing pressure or increasing wind speed causing the rate to increase.

Mean annual activity concentrations of ^{222}Rn above continental areas range from 1 to 10 Bq m^{-3}, with 5 Bq m^{-3} being typical [UN 1988]. Concentrations of ^{220}Rn daughters are typically 10% of those of ^{222}Rn daughters. Anomalously high levels often exist near coal-fired and geothermal power stations and near uranium-mine tailings.

Indoor Concentrations

Indoor activity concentrations of radon can be many times greater than outdoor levels. Sources include exhalation from soil, building materials, water, ventilation air, and natural gas if unvented (as used in cooking). Exhalation from soil can be a significant contribution if a building has cracks or other penetrations in the basement structure or if the building has unpaved and unventilated crawl spaces. Radon precursor concentrations in building materials are highly variable. Greater concentrations occur in phosphogypsum and in concrete based on fly ash or alum shale. Sealing concrete surfaces with materials such as epoxy-resin paints greatly reduces radon exhalation.

Surveys of indoor radon concentrations reveal a wide variation among houses. A survey in Canada found median and average concentrations of 7.4 and 17 Bq m^{-3} (equilibrium equivalent ^{222}Rn) with an equilibrium factor of 0.52 ± 0.12. Geographic variations were significant, but variations within cities were greater than variations between cities.

For the U.S. as a whole, a survey by the U.S. Environmental Protection Agency [EPA 1991] found an average radon concentration of 46 Bq m^{-3} (1.25 pCi/L) in the living spaces of U.S. houses. Only 6.3% of the houses had concentrations above 150 Bq m^{-3} (4 pCi/L). These concentrations are uncorrected for disequilibrium, and, in the absence of measured equilibrium factors, the EPA uses $F = 0.5$. There is little correlation between radon concentration and nearly all factors thought to be important in affecting radon levels—basement vs. crawl space vs. no space, integrity of the barrier between the ground and the house, windiness, draftiness, construction materials, use of natural gas, and, to some extent, even ventilation. Geographical variations are apparently so important as to overwhelm all other factors.

The U.S. Environmental Protection Agency [EPA 1986] has recommended that remedial action be taken when the EEC of ^{222}Rn in a home exceeds 150 Bq m^{-3} (4 pCi/L). Methods for radon remediation vary widely in absolute and marginal cost and effectiveness. They range from sealing cracks and covering earth spaces, through various schemes for increasing ventilation, to elaborate house pressurization or sub-foundation suction techniques.

9.7.2 Lung Cancer Risks

The assessment of lung cancer risks from breathing radon and its daughters is a two step process. First, for an atmosphere with a given annual average EEC, the annual dose equivalent to the various respiratory regions (naso-pharynx, trachea and bronchial tree, and the pulmonary region) must be determined. These are very complicated calculations with large uncertainties. The doses depend on many factors such as breathing rate, size and chemical form of aerosol particles to which radon daughters are attached, the ratio of attached to unattached daughters, the movement of the radionuclides through the body and eventual excretion, to name a few. Details of these calculations can be found in Faw and Shultis [1999].

Once the lung dose is determined, the risk of mortality from radiation-induced lung cancer is estimated based on results of epidemiological studies, particularly among underground miners. These risk estimates are generally based on a linear, no-threshold, dose-response relationship derived mostly from data obtained at relatively high radon exposures.

The details of the risk estimation methodology need not concern us here. Rather, we present some results of such studies that allow us to estimate lung cancer risks for a specified radon exposure. The unit of radon exposure used in these risk estimates is the annual average equilibrium equivalent activity concentration (Bq m^{-3}) that an individual breathes multiplied by the number of hours in the year (8766 h).[3]

---

[3]Many analyses of radon risk use an older unit called the *working level month* (WLM). The conversion relation is 1 WLM = 629,000 Bq h m^{-3} (EEC basis).

In Table 9.14, the probability of death from radon-induced lung cancer, per unit annual exposure, is given for males and females and for smokers and nonsmokers. In this table, the excess risk is the lifetime mortality probability from radon-induced cancer per unit of annual radon exposure expressed as MBq h m^{-3} (EEC basis). Of particular note in these results is that the radiogenic risk for smokers is ten times that for non-smokers!

In Table 9.15, the lifetime probability of dying from radon-induced lung cancer, per unit annual exposure, is given as a function of the duration of the exposure and age of first exposure. The results are based on the 1975 U.S. population and do not distinguish between smokers and nonsmokers.

Table 9.14. Radiogenic cancer risk for chronic lifetime exposure to ^{222}Rn daughter products for various populations. Risk data is derived from [NAS 1988] for annual exposures of 63 kBq h m^{-3}. Risks vary slightly for other exposures.

Population	Life expect. (years)	Excess risk per annual EEC MBq h m^{-3}
General		
Male	69.7	0.055
Female	76.4	0.022
Mixed	73.1	0.039
Nonsmokers		
Male	70.5	0.016
Female	76.7	0.0088
Smokers		
Male	69.0	0.16
Female	75.9	0.081

The results in Tables 9.14 and 9.15 allow us to estimate risks from radon inhalation once we know the radon exposure to individuals. However, the risk data in these tables are from different studies and sometimes yield somewhat different risk estimates. The following problem illustrates the use of these tables.

Table 9.15. Lifetime lung cancer risk per MBq h m^{-3} (EEC) annual exposure as a function of exposure duration and age at first exposure. Data derived from NCRP [1984].

Age at first exposure	Exposure duration (y)				
	1	5	10	30	Life
1	0.00010	0.00054	0.0012	0.0054	0.014
10	0.00014	0.00079	0.0017	0.0076	0.014
20	0.00021	0.0011	0.0024	0.0087	0.012
30	0.00029	0.0016	0.0033	0.0087	0.012
40	0.00033	0.0016	0.0032	0.0067	0.0072
50	0.00027	0.0013	0.0022	0.0040	0.0043
60	0.00021	0.00074	0.0014	0.0021	0.0021
70	0.00011	0.00045	0.00060	0.00060	0.00060

Example 9.6: Consider a static population of 10^6 people exposed to an average lifetime ^{222}Rn activity concentration of 20 Bq m^{-3}. According to NAS and NCRP risk models, how many annual lung-cancer fatalities are caused by this exposure to radon daughter products? What is the annual number of lung-cancer fatalities from all causes? Assume an equilibrium factor of $F = 0.5$.

The annual radon exposure (EEC equivalent) $= 0.5 \times 20$ Bq m^{-3} $\times 8766$ h/y $= 0.0876$ MBq h m^{-3} per year. Now, for this annual exposure, use the results of Tables 9.14 and 9.15 to estimate the annual radon-induced death rate.

NAS Estimate: From Table 9.14 we find, for a general mixed population, the individual probability of mortality from radiogenic lung cancer is 0.039 per MBq h m^{-3}. For the specified exposure to 10^6 people, there are $0.039 \times 0.0876 \times 10^6 = 3116$ radon-induced deaths in the population. For a steady-state population with an average life span of 73 y, the annual radon-related death rate is then $3116/73 = 47$ deaths per year.

NCRP Estimate: From Table 9.15 the individual probability of mortality from radiogenic lung cancer for a lifetime exposure is 0.014 per MBq h m^{-3}. For the given annual exposure of 0.0876 MBq h m^{-3} to the 10^6 people, the expected number of radon-induced deaths is $0.014 \times 0.0876 \times 10^6 = 1230$. This corresponds to an annual death rate of $1230/73 = 17$ deaths per year, about one-third the NAS estimate.

Finally, the annual natural mortality from respiratory cancer is, from Table 9.11, 71.9 per 10^5 for males and 25.2 per 10^5 for females. For equal numbers of males and females in our 10^6 population, the natural mortality rate from respiratory cancer is $[0.5 \times (71.9/10^5) + 0.5 \times (25.2/10^5)] \times 10^6 = 486$ deaths per year.

9.8 Radiation Protection Standards

It was recognized near the beginning of the twentieth century that standards were needed to protect workers and patients from the harmful consequences of radiation. Many sets of standards, based on different philosophies, have been proposed by several national and international standards groups. The earliest standards were based on the concept of *tolerable doses* below which no ill effects would occur. This was replaced in 1948 by the National Council on Radiation Protection and Measurements (NCRP) in the United States which introduced standards based on the idea of *permissible doses*, i.e., a dose of ionizing radiation which was not expected to cause appreciable body injury to any person during his or her lifetime.

9.8.1 Risk-Related Dose Limits

Today it is understood that low-level radiation exposure leads to stochastic hazards and that modern radiation standards should be based on probabilistic assessments of radiation hazards. This new line of thinking is exemplified by a 1972 report to the ICRP by the Task Group on Dose Limits. Key portions of the report are summarized below. It must be noted that the report is unpublished and not necessarily reflective of the official ICRP position. The tentative dose limits examined in the report were not based on explicit balancing of risks and benefits, then thought to be an unattainable ideal. Rather, they were based on the practical alternative of identifying acceptable limits of occupational radiation risk in comparison with risks in other occupations generally identified as having a high standard of safety and also having risks of environmental hazards generally accepted by the public in everyday life.

Linear, no-threshold dose-response relationships were assumed for carcinogenic and genetic effects, namely, a 1×10^{-4} probability per rem whole-body dose equivalent for malignant illness or a 4×10^{-5} probability per rem for hereditary illness within the first two generations of descendants [ICRP 1977]. For other radiation effects, absolute thresholds were assumed.

For occupational risks, it was observed that "occupations with a high standard of safety" are those in which the average annual death rate due to occupational hazards is no more than 100 per million workers. An acceptable risk was taken as 50 per million workers per year, or a 40-year occupational lifetime risk of 2 fatalities per 1000 workers, i.e., 0.002. It was also observed that in most installations in which radiation work is carried out, the average annual doses are about 10 percent of the doses of the most highly exposed individuals, with the distribution highly skewed toward the lower doses. To ensure an *average* lifetime risk limit of 0.002, an upper limit of 10 times this value was placed on the lifetime risk for any one individual. The annual whole-body dose-equivalent limit for stochastic effects was thus taken as $(10 \times 0.002)/(40 \times 0.0001) = 5$ rem (5 cSv) per year.

For members of the public, it was observed that risks readily accepted in everyday life, and of a nature that could be assumed to be understood by the public and not readily avoidable, are typically 5 deaths per year per million population, or a 70-year lifetime risk of about 4 per 10,000 population. It was observed that some individuals accept risks in everyday life an order of magnitude greater. Thus, the lifetime risk incurred by a well-selected and homogeneous small "critical group" could be considered acceptable if it does not exceed the higher lifetime risk of 4 per 1000. Setting this limit for the generality of *critical groups* would assure that the lifetime risk to the *average* member of the public would be much below 4 in 10000. The whole-body dose equivalent limit for stochastic risks to individual members of the public would thus be $0.004/(70 \times 0.0001) = 0.5$ rem (0.5 cSv) per year.

The concept of "risk-based" or "comparable-risk" dose limits provides the rational for the 1977 ICRP and the 1987 NCRP recommendations for radiation protection, and which serve as the basis for the U.S. radiation protection standards.

9.8.2   The 1987 NCRP Exposure Limits

The 1987 NCRP recommendations state that "The goal of radiation protection is to limit the probability of radiation induced diseases in persons exposed to radiation (somatic effects) and in their progeny (genetic effects) to a degree that is reasonable and acceptable in relation to the benefits from the activities that involve such exposure." Acceptability implies degrees of risk comparable to other risks accepted in the work place and by the general public in their everyday affairs. This comparable risk concept requires comparison of the *estimated* risk of radiogenic cancer with *measured* accidental death rates or general mortality rates. Taken by the NCRP as a reasonable basis for estimated risk is the cautious assumption that the dose-risk relationship is without a threshold and is strictly proportional (linear) throughout the range of dose equivalents and dose equivalent rates of importance in routine radiation protection. Specifically, the dose-risk relationship is taken as a lifetime risk of fatal cancer of 10^{-2} Sv^{-1} (10^{-4} rem^{-1}) for both sexes and for a normal population age distribution of 18 to 60 years. The genetic component of risk in the first two generations is taken to be 0.4×10^{-2} Sv^{-1} (0.4×10^{-4} rem^{-1}). In support

Table 9.16. The 1987 NCRP recommendations for exposure limits. Source: NCRP [1987].

Type of Dose		mSv	rem
Occupational exposures (annual):			
1. Limit for stochastic effects		50	5
2. Limit for non-stochastic effects:			
a. Lens of the eye		150	15
b. All other organs		500	50
3. Guidance: cumulative exposure	age (y) ×	10	1
Public exposures (annual):			
1. Continuous or frequency exposure		1	0.1
2. Infrequent exposure		5	0.5
3. Remedial action levels:		5	0.5
4. Lens of the eys, skin and extremities		50	5
Embryo-fetus exposure:			
1. Effective dose equivalent		5	0.5
2. Dose equivalent limit in a month		0.5	0.05
Negligible individual risk level (annual):			
1. Effective dose equivalent per source or practice		0.01	0.001

of the limits presented in Table 9.16, the following observations are relevant:

- The annual fatal accident rate per million workers ranges from 50 in trade occupations to 600 in mining and quarrying occupations, the all-industry average being 110, i.e., about 1 person per year in 10,000 workers.

- "Safe" industries are taken as those with an annual average fatal accident rate less than 1 in 10,000, i.e., with an annual average risk of less than 1×10^{-4}.

- Among radiation workers, the annual fatal accident rate (non-radiation) is about 0.25×10^{-4}. The annual effective dose equivalent to radiation workers is about 0.23 rem which, based on the estimated risk for radiogenic cancer, results in an annual risk of about 0.25×10^{-4}, comparable to the non-radiation fatal accident rate. Radiation workers thus have a total annual risk of fatality of about 0.5×10^{-4}, well within the range for "safe" industries.

- Overall average mortality risks from accident or disease for members of the public range from about 10^{-4} to 10^{-6} annually. Natural background radiation, resulting in an average dose of 0.1 rem annually, results in a risk of mortality of about 10^{-5} per year.

The NCRP recommendations of 1987 are summarized in Table 9.16. These standards are the basis for the current U.S. federal regulations on radiation exposures to workers and members of the public. U.S. regulations, are found in Title 10, Part 20 of the Code of Federal Regulations. For example, the annual limit for occupational exposure is the more restrictive of (1) 5 rem (50 mSv) for the total effective dose equivalent (TEDE), that is, the sum of the committed effective dose equivalent for ingested or inhaled radionuclides and the deep dose equivalent for external exposure, or (2) 50 rem (500 mSv) for the sum of the committed dose equivalent to any organ and the deep dose equivalent. The first of these limits (5 rem) is associated with stochastic effects, the second (50 rem) with mechanistic. The annual limit for any individual member of the public is a TEDE of 0.1 rem (1 mSv) exclusive of background and medical exposures.

BIBLIOGRAPHY

ANNO, G.H., S.J. BAUM, H.R. WITHERS, AND R.W. YOUNG, "Symptomatology of Acute Radiation Effects in Humans After Exposure to Doses of 0.5-30 Gy," *Health Physics*, **56**, 821-838 (1989).

EPA, *A Citizen's Guide to Radon*, Report OPA-86-004 of the U.S. Environmental Protection Agency and the Centers for Disease Control of the U.S. Department of Health and Human Services, U.S. Government Printing Office, 1986.

EPA, *National Residential Radon Survey*, U.S. Environmental Protection Agency, Office of Radiation Programs, Washington, D.C., 1991.

FAW, R.E. AND J.K. SHULTIS, *Radiological Assessment: Sources and Doses*, American Nuclear Society, La Grange Park, IL, 1999.

HHS, *U.S. Decennial Life Tables for 1979-81*, Publication PHS 85-1150-1, National Center for Health Statistics, Public Health Service, Department of Health and Human Services, Washington, D.C., 1985.

HHS, *U.S. Cancer Statistics, 2002 Incidence and Mortality*, U.S. Cancer Statistics Working Group, U.S. Department of Health and Human Services, Center for Disease Control and Prevention and National Cancer Institute, 2005.

ICRP, *Recommendations of the International Commission on Radiological Protection*, Publication 26, International Commission on Radiological Protection, *Annals of the ICRP*, Vol. 1, No. 3, Pergamon Press, Oxford, 1977.

ICRP, *Nonstochastic Effects of Ionizing Radiation*, Publication 41, International Commission on Radiological Protection, Pergamon Press, Oxford, 1984.

ICRP, *1990 Recommendations of the International Commission on Radiological Protection*, Publication 60, International Commission on Radiological Protection, *Annals of the ICRP* **2** No. 1-3 (1991).

ICRU, *Radiation Quantities and Units*, Report 19, International Commission on Radiation Units and Measurements, Washington, D.C., 1971.

ICRU, *Average Energy Required to Produce an Ion Pair*, Report 31, International Commission on Radiation Units and Measurements, Washington, D.C., 1979.

KATHREN, R.L., *Radiation Protection*, Medical Physics Handbook 16, Adam Hilger Ltd., Bristol, 1985.

LANGHAM, W.H., (Ed.), *Radiobiological Factors in Manned Space Flight*, Report of the Space Radiation Study Panel, National Academy of Sciences, National Research Council, Washington, D.C., 1967.

NAS, National Research Council, Advisory Committee on the Biological Effects of Ionizing Radiations. *The Effects on Populations of Exposure to Low Levels of Ionizing Radiation*, National Academy of Sciences, Washington, D.C., 1972. [The BEIR-I Report]

NAS, National Research Council, Advisory Committee on the Biological Effects of Ionizing Radiations. *The Effects on Populations of Exposure to Low Levels of Ionizing Radiation*, [The BEIR-III Report], Report of the BEIR Committee, National Academy of Sciences, Washington, D.C., 1980.

NAS, *Health Risks of Radon and Other Internally Deposited Alpha-Emitters*, Report of the BEIR Committee [The BEIR-IV Report], National Research Council, National Academy of Sciences, Washington, D.C., 1988.

NAS, National Research Council, Advisory Committee on the Biological Effects of Ionizing Radiations. *Health Effects of Exposure to Low Levels of Ionizing Radiation*, National Academy of Sciences, Washington, D.C., 1990. [The BEIR-V Report]

NAS, National Research Council, Advisory Committee on the Biological Effects of Ionizing Radiations. *Health Effects of Exposure to Low Levels of Ionizing Radiation*, National Academy of Sciences, Washington, D.C., 2006. [The BEIR-VII Report, Phase 2]

NCRP, *Basic Radiation Protection Criteria*, NCRP Report 39, National Council on Radiation Protection and Measurements, Washington, D.C., 1971.

NCRP, *Evaluation of Occupational and Environmental Exposures to Radon and Radon Daughters*, Report 78, National Council on Radiation Protection and Measurements, Washington, D.C., 1984.

NCRP, *Radiation Exposure of the U.S. Population from Consumer Products and Miscellaneous Sources*, Report 95, National Council on Radiation Protection and Measurements, Washington, D.C., 1987.

NCRP, *Recommendations on Limits for Exposure to Ionizing Radiation*, Report 116, National Council on Radiation Protection and Measurements, Bethesda, MD, 1993.

NIH, *NCI-CDC Report of the NCI-CDC Working Group to Revise the 1985 NIH Radioepidemiological Tables*, Publication 03-5387, National Institutes of Health, Bethesda, MD, 2003.

POCHIN, E., *Nuclear Radiation Risks and Benefits*, Clarendon Press, Oxford, 1983.

SCOTT, B.R. AND F.F. HAHN, "Early Occurring and Continuing Effects," in *Health Effects Models for Nuclear Power Plant Accident Consequence Analysis—Low LET Radiation, Part II: Scientific Bases for Health Effects Models*, Report NUREG/CR-4214, Rev. 1, Part II, U.S. Nuclear Regulatory Commission, 1989.

SIMON, S.L., A. BOUVILLE, AND C.E. LAND, "Fallout from Nuclear Weapons Tests and Cancer Risks," *American Scientist*, **94**, 48-57 (2006).

UN, *Ionizing Radiation Sources and Biological Effects*, United Nations Scientific Committee on the Effects of Atomic Radiation, United Nations, New York, 1982.

UN, *Sources, Effects and Risks of Ionizing Radiation*, United Nations Scientific Committee on the Effects of Atomic Radiation, United Nations, New York, 1988.

UN, *Sources, Effects and Risks of Ionizing Radiation*, United Nations Scientific Committee on the Effects of Atomic Radiation, United Nations, New York, 2000.

UN, *Hereditary Effects of Radiation*, United Nations Scientific Committee on the Effects of Atomic Radiation, United Nations, New York, 2001.

UPTON, A.C. AND R.F. KIMBALL, "Radiation Biology," in *Principles of Radiation Protection*, K.Z. Morgan and J.E. Turner (Eds), John Wiley & Sons, New York, 1967.

VOGEL, F. AND A.G. MOTULSKY, *Human Genetics*, Springer-Verlag, Berlin, 1979.

WALD, N., "Evaluation of Human Exposure Data," in *Principles of Radiation Protection*, K.Z. Morgan and J.E. Turner (Eds), John Wiley & Sons, New York, 1967.

PROBLEMS

1. In an infinite homogeneous medium containing a uniformly distributed radionuclide source emitting radiation energy at a rate of E MeV cm^{-3} s^{-1}, energy must be absorbed uniformly by the medium at the same rate. Consider an infinite air medium with a density of 0.0012 g/cm^3 containing tritium (half-life 12.33 y and emitting beta particles with an average energy of 5.37 MeV/decay) at a concentration of 2.3 pCi/L. What is the air-kerma rate (Gy/h)?

2. A ^{137}Cs source has an activity of 700 μCi. A gamma photon with energy 0.662 MeV is emitted with a frequency of 0.849 per decay. At a distance of 2 meters from the source, what is (a) the exposure rate, (b) the kerma rate in air, and (c) the dose equivalent rate?

3. Suppose the source in the previous problem is placed in a large tank of water. Considering only the uncollided photons, at 0.5 m from the source, what is (a) the exposure rate, (b) the kerma rate in air, and (c) the dose equivalent rate?

4. What is the gamma-ray absorbed dose rate (Gy/h) in an infinite air medium at a distance of 10 cm from a 1-mCi point source of (a) ^{13}N and (b) ^{43}K? Energies and frequencies of the emitted gamma rays from these radionuclides are given in Appendix D.

5. What are the absorbed dose rates (Gy/h) in air at a distance of 10 cm from the point sources of the previous problem if the sources are placed in an infinite iron medium?

6. A rule of thumb for exposure from point sources of photons in air at distances over which exponential attenuation is negligible is as follows:

$$\dot{X} = \frac{6CEN}{r^2},$$

where C is the source strength (Ci), E is the photon energy (MeV), N is the number of photons per disintegration, r is the distance in feet from the source, and $\dot{X}$ is the exposure rate (R h^{-1}).

1. Reexpress this rule in units of Bq for the source strength and meters for the distance.

2. Over what ranges of energies is this rule accurate within 20%?

7. A male worker at a nuclear facility receives an accidental whole-body exposure of 2.3 Gy. Describe what physical symptoms the individual is likely to have and when they occur.

8. A population of 500,000 around a nuclear facility receives an average whole-body dose of 0.5 rad as a result of an accidental release of radionuclides. (a) Estimate the collective gonad dose received. (b) Estimate how many children subsequently born would experience significant hereditary defects as a result of this exposure to their parents. (c) Estimate how many such defects would occur naturally in the absence of this accidental exposure. (d) Estimate the cancer mortality risk imposed on this population, both absolutely and relative to natural cancer mortality.

9. A static population of 900,000 people lives in an area with a background radiation level that gives each person an extra 125 mrem (whole-body) per year of low LET exposure over that received by people in other parts of the country. (a) Estimate the collective gonad dose to the reproductive population. (b) Estimate the number of radiation-induced cases of hereditary illness that occurs in each generation. (c) Compare this number to the natural incidence of genetic illness in this population.

10. A 35-year old female worker receives an x-ray exposure of 5.2 rad (whole-body) while carrying out emergency procedures in a nuclear accident. Discuss the health risks assumed by this worker as a result of the radiation exposure.

11. How many people in the U.S. might be expected to die each year as a result of cancer caused by natural background radiation (excluding radon lung exposures)? Assume an average whole-body exposure of 200 mrem and a population of 250 million. Compare this to the natural total death rate by cancer.

12. A radiology technician receives an average occupational dose of 0.3 rad per year over her professional lifetime. What cancer risk does she assume as a result of this activity?

13. A male reactor operator receives a whole body dose equivalent of 0.95 rem over a period of one week while working with gamma-ray sources. What is the probability he will (a) die of cancer, (b) die of cancer caused by the radiation exposure, (c) have a child with an hereditary illness due to all causes, and (d) have a child with an hereditary illness due to the radiation exposure?

14. An individual is exposed 75% of the time to radon with a physical concentration of 4.6 pCi/L and an equilibrium factor of $F = 0.6$. The remaining 25% of the time, the individual is exposed to radon at a concentration of 1.3 pCi/L and with an equilibrium factor of $F = 0.8$. What is the annual radon exposure (on an EEC basis) in MBq h m^{-3}?

15. If the individual in the previous problem is a nonsmoking male and receives the same annual radon exposure for his entire life, what is the probability he will die from lung cancer as a result of his radon exposure?

16. Consider a female exposed 75% of the time to radon with a ^{222}Rn EEC of 25 Bq m^{-3} and 25% of the time to a concentration of 5 Bq m^{-3} for her entire life. (a) What is the probability that she will die from radon-induced lung cancer if she is a nonsmoker? (b) What is the probability if she smokes?

17. Consider people who receive an annual radon exposure (EEC basis) of 0.11 MBq h m^{-3} for the first 30 years of their lives. At age 30, through radon reduction remediation actions, they decrease their annual exposure to 0.02 MBq h m^{-3} for the remainder of their lives. (a) What is their mortality risk for radon-induced lung cancer? (b) If they had not undertaken remedial measures, what would be their radon mortality risk?

18. Estimate how many radon-caused lung cancer deaths there are every year in the U.S. if everyone were exposed to the U.S. residential average radon concentration of 1.25 pCi/L 75% of the time. Assume an equilibrium factor of $F = 0.5$. Assume a population of 250 million. State and justify any other assumptions you make.

19. Why do the skin and extremities of the body have a higher annual dose limit than for other organs and tissues?

20. Describe in your own words the rationale for the NCRP [1987] limit of 5 rem a year whole-body exposure for occupational workers. Give arguments why you do or do not believe this limit to be reasonable.

Chapter 10

Principles of
Nuclear Reactors

Since December 2, 1942, when the first man-made nuclear reactor produced a self-sustaining chain reaction, several hundred different types of reactor systems have been constructed. Despite the many possible differences in design, there are a number of general features which all reactors have in common. The heart of every reactor is an *active core* in which the fission chain reaction is sustained. The active core contains (1) fissile *fuel* which through its fissioning is the main source of neutrons, (2) *moderator* material if the fission neutrons are to slow down, (3) *coolant* if the heat generated by the fissions is to be removed from the core, and (4) *structural* material which maintains the physical integrity of the core. Surrounding the active core is usually either a *reflector* whose purpose is to scatter neutrons back towards the core or a *blanket* region which captures neutrons leaking from the core to produce useful isotopes such as ^{60}Co or ^{239}Pu. The reactor core and reflector/blanket are, in turn, surrounded by a *shield* to minimize radiation reaching personnel and equipment near the reactor. Finally, all reactors must have a means of *control* to allow the chain reaction to be started up, maintained at some desired level, and safely shut down.

Reactors are broadly classified according to the energy of the neutrons that cause most of the fissions. In a *fast reactor*, the fast fission neutrons do not slow down very much before they are absorbed by the fuel and cause the production of a new generation of fission neutrons. By contrast, in a *thermal reactor* almost all fissions are caused by neutrons that have slowed down and are moving with speeds comparable to those of the atoms of the core material, i.e., the neutrons are in *thermal equilibrium* with the surrounding material.

In this chapter, the basic principles of nuclear reactors and fission chain reactions are discussed. Initially, we consider steady-state neutron populations in a reactor core, and seek methods to quantify the conditions necessary for a self-sustaining chain reaction with a constant neutron population and fission power release. In particular, we concentrate on "thermal" reactors, although the principles for fast reactors are quite similar. Later in the chapter, we consider the dynamics of reactors as the power increases or decreases in response to physical changes in the reactor as a result of externally applied changes or from feedback effects.

10.1   Neutron Moderation

In a thermal reactor, the fast fission neutrons lose their kinetic energy primarily through elastic scattering from moderator nuclei with small mass numbers. In our earlier discussion of the kinematics of neutron elastic scattering from a stationary nucleus,[1] we found that the energy of the scattered neutron is between $E_{\max} = E$ and $E_{\min} = \alpha E$, where E is the energy of the incident neutron and $\alpha \equiv (A - 1)^2/(A + 1)^2$. The number of scatters, on the average, required to bring a neutron with initial energy E_1 to a lower energy E_2 is given by Eq. (6.30). A summary of these important properties for several moderators is given in Table 6.1.

From the values in Table 6.1, it is seen that scattering nuclei with small A numbers cause greater average energy loss and thus thermalize fast neutrons with fewer scatters than do nuclei with large mass numbers. In addition to having a small A number, a good *moderator* should have a large scattering cross section Σ_s (to encourage scattering) and a small absorption cross section Σ_a (to avoid loss of neutrons before they can cause fissions). In all thermal reactors, a large portion of the core material is the moderator, usually light or heavy water, graphite, or beryllium. In addition to acting as a moderator, light water can also serve as the main coolant in a power reactor, and, for this reason, it is not surprising that light water reactors (LWRs) are the dominant type of power reactors in service today.

10.2   Thermal-Neutron Properties of Fuels

In thermal reactors, only fissile isotopes such as ^{233}U, ^{235}U, and ^{239}Pu can be used. By far the most widely used nuclear fuel is uranium dioxide, with the uranium enriched in ^{235}U from its natural 0.720 atom-% to several percent. Only some heavy-water and graphite-moderated reactors can use natural uranium as fuel. Most reactors use uranium that has been enriched, typically 2 to 3%, in ^{235}U. The fissile nuclide ^{239}Pu is created during the operation of a nuclear power reactor whose fuel contains ^{238}U (see Section 6.5.3), and at the end of fuel life (typically three years), almost half of the power is generated by the fission of ^{239}Pu. In some power reactors, ^{239}Pu is mixed with enriched uranium in the form of a "mixed oxide" fuel. Some important properties of nuclear fuels are presented in Table 10.1. The data lead to the following observations:

1. ^{233}U has the largest value of η, the number of fission neutrons produced per thermal neutron absorbed, and hence is the best prospect for a thermal *breeder reactor*, one in which more fissile fuel is produced by neutron absorptions than is consumed in the chain reaction. A breeder reactor needs an η of at least two, since one neutron is needed to sustain the chain reaction and one neutron must be absorbed in the fertile material to breed a new fissile fuel atom. Fertile materials are those such as ^{232}Th and ^{238}U that, upon thermal neutron absorption, may yield fissile materials (see Section 6.5.3).

2. Although the plutonium isotopes produce almost 3 fission neutrons per thermal fission, their relatively high radiative capture (n,γ) cross sections (indicated by the relatively large σ_γ/σ_f ratio) result in low values of η. However,

---

[1]The thermal motion of the nucleus is negligible compared to the speed of fast neutrons.

Table 10.1. Thermal-neutron properties of important fuel isotopes. Cross sections are at 0.0253 eV (2200 m/s).

Nuclide	cross section (b)			$\frac{\sigma_\gamma}{\sigma_f}$	ν	$\eta = \nu\frac{\sigma_f}{\sigma_a}$
	$\sigma_a = \sigma_\gamma + \sigma_f$	σ_γ	σ_f			
^{232}Th	5.13	5.13	−	−	−	−
^{233}U	575	46	529	0.087	2.49	2.29
^{235}U	687	99.3	587	0.169	2.42	2.07
^{238}U	2.73	2.73	−	−	−	−
^{239}Pu	1020	271	749	0.362	2.87	2.11
^{240}Pu	289.5	289.5	0.064	−	−	−
^{241}Pu	1378	363	1015	0.358	2.92	2.15

for fissions caused by neutrons with energies above several hundred keV, the η for both ^{239}Pu and ^{241}Pu is greater than 3. Thus, fast reactors using plutonium as fuel are attractive as breeder reactors.

3. The fertile isotopes ^{232}Th and ^{238}U have absorption cross sections of about 1% or less than those of their conversion fissile isotopes ^{233}U and ^{239}Pu.

4. The fertile isotope ^{240}Pu has a large capture cross section for the production of the fissile isotope ^{241}Pu.

5. Although not shown in this table, the fission and absorption cross sections for reactions of high energy fission neutrons with fissile isotopes are several hundred times less than for reactions with thermal neutrons (see, for example, Fig. 7.10).

10.3 The Neutron Life Cycle in a Thermal Reactor

In a thermal reactor, some of the fast neutrons ($\sim$ 2 MeV) born from fission slow down to thermal energies ($\sim$ 0.025 eV), are absorbed by the fuel, cause the fuel to fission, and thus produce a second generation of fast neutrons. The remainder of the fast neutrons suffer a variety of fates that do not result in fission.

Consider the life cycle shown below for one generation of neutrons in a thermal reactor. Here n fast neutrons are introduced into the core. These fast neutrons can cause a few fast fission events (thereby creating some second generation neutrons), can leak from the core, or can be absorbed while slowing down to thermal energies. Those neutrons that reach thermal energies begin to diffuse throughout the core. Some of these thermal neutrons may be lost by leaking from the core or by being absorbed by the "nonfuel." Most thermal neutrons, however, are absorbed in the fuel, but only a certain fraction of these absorptions cause the fuel to fission, thereby releasing ν fast second-generation neutrons per fission. Thus, at the end of the cycle, there is a new generation of n' fast neutrons which begin the cycle again. Clearly for the neutron population to remain constant cycle after cycle, n' must equal n, i.e., the chain reaction must be self-sustaining.

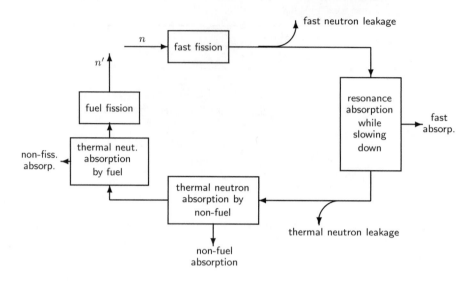

Figure 10.1. The neutron life cycle in a thermal reactor showing the major mechanisms for the loss and gain of neutrons. The n fast neutrons beginning the cycle produce n' second-generation fast neutrons which, in turn, begin their life cycle.

10.3.1 Quantification of the Neutron Cycle

To quantify the neutron life cycle in a thermal reactor, as shown in Fig. 10.1, we define the following six factors.

1. Fast fission factor ϵ: To account for fast fission, we define ϵ as the ratio of the total number of fast neutrons produced by both thermal and fast fission to the number produced by thermal fission alone. This quantity is quite difficult to calculate accurately. It is generally near unity, varying typically between 1.02 and 1.08 for reactors fueled with natural or slightly enriched uranium.

Fast fission is almost completely due to ^{238}U which has a fission threshold of about 1 MeV (see Fig. 7.11). Thus, in a core using highly enriched uranium, ϵ is very close to unity. Also, in a very dilute homogeneous mixture of uranium in a moderator, ϵ is very close to unity since a fast fission neutron is far more likely to first encounter a moderator atom and lose some energy than it is to encounter an atom of ^{238}U.

The fast fission factor is generally higher in a heterogeneous core, in which the fuel is lumped together, compared to a homogeneous core. This increase for a heterogeneous core is a result of the fast fission neutrons having a better chance of encountering ^{238}U atoms before they begin slowing down and hence a better chance to cause fast fission. Typically, ϵ in a heterogeneous core is 5–10% higher than in an equivalent homogeneous mixture.

Slowing down of fast neutron probability

2. Resonance escape probability p: To account for fast neutron absorption, we define p as the probability that a fast fission neutron slows to thermal energies without being absorbed. In uranium fueled reactors, most neutrons that are absorbed while slowing down are absorbed by ^{238}U which has very large absorption resonances in its cross section in the slowing down energy region. Calculation of p for a core is also a very difficult task because of the complicated variation of core-material cross section with neutron energy in the slowing down energy region. Many empirical results have been obtained and are often used in preliminary analyses (see Lamarsh [1966] and Lamarsh and Baratta [2001]).

For reactors fueled with fully enriched ^{235}U or ^{239}Pu, p is very close to unity. For homogeneous cores with uranium of only a few percent enrichment, p can vary between 0.9 and 0.6 depending on the fuel-to-moderator ratio. In heterogeneous cores, p is considerably higher than in a homogeneous core of the same fuel-to-moderator ratio, typically varying between 0.8 and 0.9. By lumping the fuel atoms together, the neutrons can slow down in the moderator away from the fuel and, consequently, are less likely to encounter a ^{238}U atom and be absorbed. By contrast, in a homogeneous core material, as neutrons slow past the ^{238}U resonance energies, they are more likely to be near a ^{238}U atom and, thus, to be absorbed.

High enrichment
$p \simeq 1$
low enrichment
$p < 0.9$
$p > 0.6$

Fuel vs non fuel absorption (thermal)

3. Thermal utilization f: Because not all thermal neutrons are absorbed by the fuel, we define f as the probability that, when a thermal neutron is absorbed, it is absorbed by the "fuel" (F) and not by the "nonfuel" (NF). Equivalently, f is the ratio of the average thermal neutron absorption rate in the fuel to the total thermal neutron absorption rate in the fuel and nonfuel. Mathematically,

$$f = \frac{\Sigma_a^F \phi^F V^F}{\Sigma_a^F \phi^F V^F + \Sigma_a^{NF} \phi^{NF} V^{NF}} = \frac{\Sigma_a^F}{\Sigma_a^F + \Sigma_a^{NF}(V^{NF}/V^F)(\phi^{NF}/\phi^F)}, \tag{10.1}$$

where ϕ^F and ϕ^{NF} are the average thermal flux densities in the fuel (of volume V^F) and nonfuel (with volume V^{NF}), respectively. For a homogeneous core $\phi^F = \phi^{NF}$ and $V^F = V^{NF}$, so that

$$f = \frac{\Sigma_a^F}{\Sigma_a^F + \Sigma_a^{NF}} = \frac{\sigma_a^F}{\sigma_a^F + \sigma_a^{NF}(N^{NF}/N^F)}, \tag{10.2}$$

where N^F/N^{NF} is the ratio of fuel-to-nonfuel atomic concentrations. The value of f can range from near zero for a very dilute fuel mixture to unity for a core composed only of fuel.

4. Thermal fission factor η: We define η as the number of fast fission neutrons produced per thermal neutron absorbed by the "fuel." Equivalently, η is the average number of neutrons per thermal fission (ν) times the probability a fission occurs when a thermal neutron is absorbed by the fuel, i.e.,

fuel fission

$$\eta = \nu \frac{\Sigma_f^F}{\Sigma_a^F}. \tag{10.3}$$

fuel absorption.

V = neutrons out (of fission)

$\eta = \eta_{th}$

Since non-fuel has a zero fission cross section, $\Sigma_f^F = \Sigma_f$, the total fission cross section of the core material. Notice this factor is a property of the fuel material alone and is unaffected by the type and amount of nonfuel material in the core.

This factor must be greater than unity if a self-sustaining chain reaction is to be realized. Values of η for isotopically pure fuels are given in Table 10.1. For fuel composed of a mixture of isotopes, we need to calculate the macroscopic cross sections for each isotope and add the results to get the total cross section (see Example 10.2).

$$\eta > 1 \quad \text{for chain reaction}$$

Example 10.1: What is the thermal utilization factor in a mixture of graphite and natural uranium with a carbon-to-uranium atomic ratio N^C/N^U of 450?

Natural uranium contains the three isotopes ^{234}U, ^{235}U, and ^{238}U with atomic abundances (see Appendix A.4) $f_{234} = 0.0055\%$, $f_{235} = 0.720\%$, $f_{238} = 99.2745\%$, respectively. From Appendix C.1 and Table 10.1, $\sigma_a^{234} = 103.4$ b, $\sigma_a^{235} = 687$ b, and $\sigma_a^{238} = 2.73$ b. Thus, the total thermal macroscopic cross section for the uranium in the carbon-uranium mixture is

$$\Sigma_a^U = [f_{234}\sigma_a^{234} + f_{235}\sigma_a^{235} + f_{238}\sigma_a^{238}]N^U = 7.662 N^U \text{ cm}^{-1}.$$

We can assume that graphite is ^{12}C since other naturally occurring carbon isotopes have very low abundances and cross sections. From Table C.2, $\sigma_a^C = 0.0034$ b so that

$$\Sigma_a^C = N^C \sigma_a^C = 0.0034 N^C.$$

Then from Eq. (10.2)

$$f = \frac{\Sigma_a^U}{\Sigma_a^U + \Sigma_a^C} = \frac{7.662 N^U}{7.662 N^U + 0.0034 N^C}$$

$$= \frac{7.662}{7.662 + 0.0034(N^C/N^U)} = \frac{7.662}{7.662 + 0.0034 \times 450} = 0.8336.$$

Example 10.2: What is the thermal fission factor η for uranium enriched to 2 atom-% in ^{235}U, the remainder being ^{238}U? From Eq. (10.3) and the data in Table 10.1 we have

$$\eta = \frac{\nu^{235}\Sigma_f^{235}}{\Sigma_a^F} = \frac{\nu^{235} N^{235} \sigma_f^{235}}{N^{235}\sigma_a^{235} + N^{238}\sigma_a^{238}} = \frac{\nu^{235}\sigma_f^{235}}{\sigma_a^{235} + \sigma_a^{238}(N^{238}/N^{235})}$$

$$= \frac{2.42 \times 587}{687 + 2.73(0.98/0.02)} = 1.73 .$$

Self explanatory

5. Thermal non-leakage probability P_{NL}^{th}: To account for thermal neutrons leaking from the core, we define P_{NL}^{th} as the probability a thermal neutron does not leak from the core before it is absorbed. It can be shown [Lamarsh and Baratta 2001] that P_{NL}^{th} can be estimated for a bare core from

$$P_{NL}^{th} = \frac{1}{1 + L^2 B_c^2}, \tag{10.4}$$

length
thermal diffusion core
$L^2 = \dfrac{D}{\Sigma_a}$ ← therm diff coef
← macro absorptn cross section

Table 10.2. Moderator properties for thermal (0.00253 eV) neutrons. L is the thermal diffusion length and τ is the Fermi age from fission to thermal energies. Source: [ANL 1963].

Moderator	Density ρ (g cm^{-3})	Σ_a (cm^{-1})	D (cm)	L^2 (cm^2)	L (cm)	τ (cm^2)
H_2O	1.00	0.0197	0.16	8.1	2.85	27
D_2O	1.10	2.9×10^{-5}	0.87	30,000	170	131
Be	1.85	1.04×10^{-3}	0.50	480	21	102
BeO	2.96	6.0×10^{-4}	0.47	790	28	100
C	1.60	2.4×10^{-4}	0.84	3500	59	368

where L is the *thermal diffusion length* and B_c^2 is the *critical buckling*. The evaluation of these two parameters is discussed next.

The square of the thermal diffusion length $L^2 \equiv D/\Sigma_a$, where D is the *thermal diffusion coefficient* and Σ_a is the thermal macroscopic absorption cross section of the core material. It can be shown that L is one-half of the average distance a thermal neutron diffuses from the point where it becomes thermal to the point at which it is absorbed in an infinite medium of core material. Values of D, Σ_a, and L^2 are given in Table 10.2 for several pure moderators. When material such as fissile fuel is added to a moderator to make core material, the value of L^2 decreases significantly while D remains relatively constant for dilute fuel-moderator mixtures. Thus, for a homogeneous mixture of fuel (F) and moderator (M),

$$L^2 \equiv \frac{D}{\Sigma_a} = \frac{D^M}{\Sigma_a^F + \Sigma_a^M} = \frac{D^M}{\Sigma_a^M} \frac{\Sigma_a^M}{\Sigma_a^F + \Sigma_a^M} = L_M^2 \left(1 - \frac{\Sigma_a^F}{\Sigma_a^F + \Sigma_a^M} \right) = L_M^2(1-f). \tag{10.5}$$

The critical buckling, B_c^2, discussed in Addendum 1, depends only on the geometry and size of the reactor core. Expressions for B_c^2 are given in Table 10.6 as B_g^2 for some simple core geometries. For example, a spherical core of radius R has a critical buckling of $B_c^2 = (\pi/R)^2$. Thus as the core increases in size, $B_c^2 \to 0$ and $P_{NL}^{th} \to 1$, i.e., there is no leakage.

6. Fast non-leakage probability P_{NL}^f: The leakage of fast neutrons from the core is accounted for by P_{NL}^f, defined as the probability a fast neutron does not leak from the core as it slows to thermal energies. This factor can be estimated for a bare core from [Lamarsh and Baratta 2001]

$$P_{NL}^f = e^{-B_c^2 \tau}, \tag{10.6}$$

where τ is the *Fermi age* to thermal energies and can be interpreted as one-sixth the mean squared distance between the point at which a fast fission neutron is born and begins to slow down and the point at which it reaches thermal energies. In Table 10.2 values of τ are given for pure moderators. In dilute fuel-moderator mixtures, τ changes little from its value for the pure moderator. Again we see from Eq. (10.6) that, as the core size increases, $B_c^2 \to 0$, and the non-leakage probability $P_{NL}^{th} \to 1$.

With the above six definitions, the number of neutrons at various stages in the life cycle can be calculated as shown in Fig. 10.2.

Example 10.3: What are the thermal and fast neutron non-leakage probabilities for a bare spherical reactor core of radius $R = 120$ cm composed of a homogeneous mixture of graphite and ^{235}U in an atomic ratio of 40,000 to 1? The geometric buckling for the spherical core is $B_c^2 = (\pi/R)^2 = (\pi/120)^2 = 6.85 \times 10^{-4}$ cm^{-2}. For such a dilute mixture of uranium in graphite, the Fermi age to thermal energies is the same as that in a pure graphite medium, namely, $\tau = 368$ cm^2. Thus, from Eq. (10.6), the fast-neutron nonleakage probability is

$$P_{NL}^f = \exp(-B_c^2 \tau) = \exp(-6.85 \times 10^{-4} \times 368) = 0.777.$$

From Example 10.4, the thermal utilization for this core mixture is $f = 0.835$, so that the square of the thermal diffusion length $L^2 = L_C^2(1-f) = 3500(1. - 0.835) = 578$. The thermal-neutron nonleakage probability, given by Eq. (10.2), is thus

$$P_{NL}^{th} = \frac{1}{1 + L^2 B_c^2} = \frac{1}{1 + (578)(6.85 \times 10^{-4})} = 0.716.$$

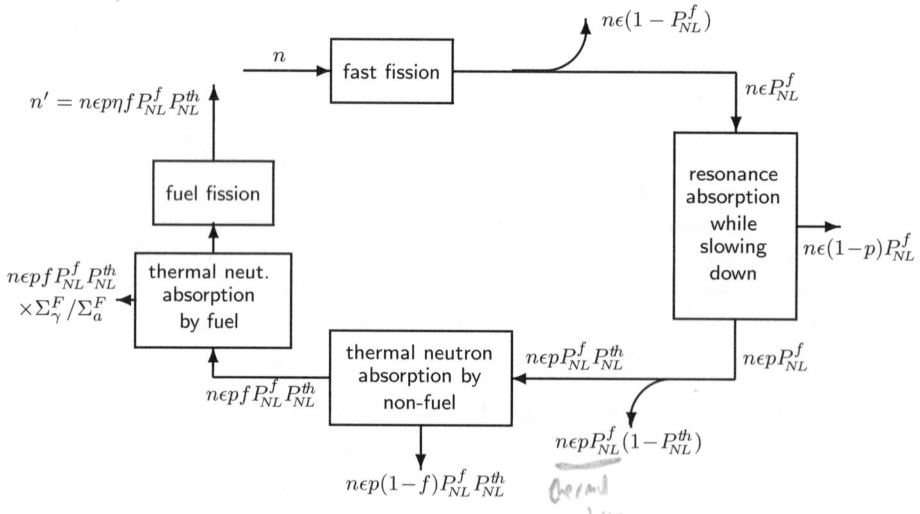

[handwritten: η = fast neutrons per thermal absorption. Attract n'. ϵ = neutrons from fast + thermal / thermal]

Figure 10.2. The neutron life cycle in a thermal reactor showing the calculation of the major contributions to the loss and gain of second-generation neutrons.

10.3.2   Effective Multiplication Factor

From Fig. 10.2 the neutron gain or *effective multiplication factor* per neutron cycle is seen to be

$$k_{eff} = \frac{\text{no. of neutrons at some point in the cycle}}{\text{no. of neutrons at the same point in the previous cycle}}$$

$$= \frac{n'}{n} = \frac{n\epsilon\eta pf P_{NL}^f P_{NL}^{th}}{n}$$

or

$$\boxed{k_{eff} = \epsilon p\eta f P_{NL}^f P_{NL}^{th}.} \qquad (10.7)$$

Clearly, if $k_{eff} < 1$, the initial neutron population decreases with each cycle until no neutrons are left. Such a system is said to be *subcritical*, requiring an independent source of neutrons to maintain a steady neutron population. However, if $k_{eff} > 1$, the number of neutrons continually increases with each cycle and the system is said to be *supercritical*. For the special case that $k_{eff} = 1$, the number of neutrons remains constant cycle after cycle and the system is said to be self-sustaining or *critical*.

For an infinite medium, there is no neutron leakage and P_{NL}^f and P_{NL}^{th} are unity. Thus the multiplication factor is the *infinite medium multiplication factor*

$$\boxed{k_\infty = \eta\epsilon pf.} \qquad (10.8)$$

This result is known as the "four-factor formula" of reactor physics. Notice that k_∞ is a property of only the core material and is independent of the size and shape of the core.

These formulas for k_∞ and k_{eff} allow us to calculate the criticality state of an assembly. More important, they provide a conceptual framework for assessing the criticality effect of some proposed change to an assembly. For example, inserting a thermal-neutron absorbing material such as a control rod into a critical core primarily affects f by making it smaller and, consequently, making the core subcritical.

The typical variation of k_{eff} and its four factors with the fuel-to-moderator ratio of the core material is shown in Fig. 10.3. Notice that there is a optimal fuel-to-moderator ratio that maximizes the value of k_{eff}. This maximum value of k_{eff} must be greater than unity to compensate for neutron leakage if a core of finite size is to become critical.

Example 10.4: What is k_∞ of a homogeneous mixture of ^{235}U and graphite with an atomic uranium to carbon ratio of 1 to 40,000? For such a dilute mixture of fully enriched uranium and carbon, $\epsilon p \simeq 1$, so that $k_\infty = \eta f$. From Table 10.1, $\eta = 2.07$ for pure ^{235}U . With Eq. (10.2) and data from Tables 10.1 and C.1

$$f = \frac{\sigma_a^{235}}{\sigma_a^{235} + \sigma_a^C(N^C/N^{235})} = \frac{687}{687 + 0.0034(40{,}000)} = 0.835.$$

Then,

$$k_\infty = \eta f = 2.07 \times 0.835 = 1.728.$$

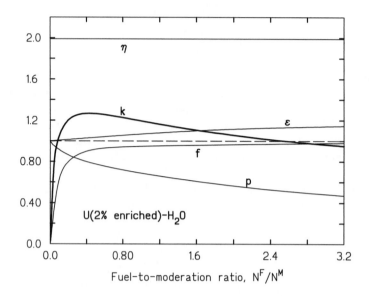

Figure 10.3. Variation of k_{eff} and its factors with the fuel-to-moderator ratio. This example is for a homogeneous mixture of water and 2%-enriched uranium. Here N^F/N^M = atom density of uranium to molecular density of water.

Example 10.5:   What is the radius R of a critical bare sphere composed of a homogeneous mixture of ^{235}U and graphite with a uranium to carbon atom ratio of 1 to 40,000? For criticality, we require

$$k_{eff} = \frac{\epsilon p \eta f}{1 + L^2 B^2} \exp(-B^2 \tau) = 1.$$

From Example 10.4, $\epsilon p = 1$ and $\eta f = 1.73$. From Example 10.3, $L^2 = 578$ cm^2 and $\tau = 368$ cm^2 so that the criticality condition is

$$k_{eff} = 1 = \frac{1.728}{1 + 578 B_c^2} \exp(-368 B_c^2).$$

Solution for $B_c^2 \equiv (\pi/R)^2$ by "trial and error" yields $B_c^2 = 6.358 \times 10^{-4}$ cm^{-2}. From this result we find the critical radius to be $R = 125$ cm.

Example 10.6: What is the mass of ^{235}U needed for the critical assembly of Example 10.5? Since $N_i = \rho_i N_a / A_i = (m_i / V_i) N_a / A_i)$ and for a homogeneous system $V^{235} = V^C \equiv V$ we have

$$\frac{N^{235}}{N^C} = \frac{m^{235}}{m^C} \frac{A^C}{A^{235}} \qquad \text{or} \qquad m^{235} = m^C \frac{N^{235}}{N^C} \frac{A^{235}}{A^C}.$$

Since we a have a very dilute mixture of ^{235}U in graphite, $m^C \simeq \rho^C V$ so that for the bare spherical core of Example 10.5 with a radius $R = 125$ cm

$$m^{235} = \left(\frac{4}{3} \pi R^3 \rho^C \right) \frac{N^{235}}{N^C} \frac{A^{235}}{A^C} = \left(\frac{4}{3} \pi (125)^3 1.60 \right) \frac{1}{40,000} \frac{235}{12} = 6.41 \text{ kg}.$$

10.4 Homogeneous and Heterogeneous Cores

The least expensive fuel to use in a reactor assembly is natural uranium (0.72 atom-% ^{235}U). However, k_∞ for a pure natural uranium core would be very small since a fast fission neutron would lose so little energy in each scatter from a uranium nucleus that over 2000 scatters would be required to slow the neutron to thermal energies (see Table 6.1). Such a neutron would thus spend considerable time in the energy regions at which ^{238}U has large absorption cross sections, and, hence, the neutron would have little chance of reaching thermal energies. Thus for a pure natural uranium core, the resonance escape probability p would be very small so that k_∞ would be much less than unity.

To use natural uranium as a fuel, it is necessary to slow the fission neutrons more quickly to thermal energies by using a light mass material as a moderator. Since fewer scatters from nuclides with small mass number are needed to slow fission neutrons, it is less likely that neutrons are absorbed while slowing down. Thus p is increased significantly, and k_∞ can approach unity. Conceptually, the simplest assembly is a homogeneous mixture of natural uranium and a moderator material. For such a mixture, there is an optimum ratio of moderator-to-uranium atomic concentrations, $(N^M/N^U)_{\text{opt}}$, that produces the maximum possible value of k_∞. If N^M/N^U is too small, there is too little moderation and p is very small. On the other hand, if N^M/N^U is too large, the thermal neutrons are not absorbed easily by the fuel and f is small. In Table 10.3 the maximum k_∞ achievable with the optimum N^M/N^U ratio is shown. Only for heavy water as the moderator is it possible, in principle, to build a critical homogeneous reactor using natural uranium as fuel.

Table 10.3. Optimum moderator-fuel ratios for a homogeneous mixture of natural uranium and moderator.

Moderator	$(N^M/N^U)_{\text{opt}}$	ϵ	η	f	p	k_∞
H_2O	1.64	1.057	1.322	0.873	0.723	0.882
D_2O	272	1.000	1.322	0.954	0.914	1.153
Be	181	1.000	1.322	0.818	0.702	0.759
C	453	1.000	1.322	0.830	0.718	0.787

However, small critical reactors using all these moderators have been built. The trick for obtaining a system with $k_{eff} > 1$ is (1) to use fuel enriched in ^{235}U and/or (2) construct a heterogeneous core in which the fuel is lumped together and embedded in a matrix of the moderator. The first technique is expensive (even though enrichments of only a few percent are sufficient), because special facilities must be constructed to separate ^{238}U from ^{235}U. Sometimes, simply lumping the fuel together is sufficient to create a core with a value of k_∞ sufficiently greater than unity.

This increase in k_∞ for a heterogeneous core arises primarily as a result of a great increase in the resonance escape probability. By lumping the fuel (and, hence, the ^{238}U responsible for resonance absorption), the fast neutrons are thermalized in the moderator away from the ^{238}U and, hence, they can slow through the energy ranges of the ^{238}U resonances with little likelihood of being captured. Only those neutrons reaching the resonance energies near a fuel lump are in danger of being absorbed. The lumping of fuel can increase p from about 0.7 for a homogeneous core to values greater than 0.9.

The fast fission factor ϵ also tends to increase slightly in a heterogeneous core. Fast neutrons born in the fuel lumps have a greater probability of causing fast fissions in ^{238}U if they are surrounded by only uranium atoms. In a homogeneous system, a fast fission neutron may first encounter a moderator atom, scatter, and lose so much energy that it is no longer capable of causing fast fission. However, the increase in ϵ for a heterogeneous core is small, typically only a few percent.

For a given ratio N^F/N^{NF}, f in a heterogeneous core is smaller that in a homogeneous core. This can be seen from Eq. (10.1) since in a heterogeneous core, the average thermal flux in the fuel is less than in the nonfuel (moderator) so that $\phi^{NF}/\phi^F > 1$. For a homogeneous core $\phi^{NF}/\phi^F = 1$ and f is consequently greater.

The one factor in k_∞ that decreases in a heterogeneous core is the thermal utilization f. From Eq. (10.1)

$$f = \frac{\Sigma_a^F}{\Sigma_a^F + \Sigma_a^{NF}(V^{NF}/V^F)(\phi^{NF}/\phi^F)}$$

$$= \frac{\sigma_a^F}{\sigma_a^F + \sigma_a^{NF}(N^{NF}/N^F)(V^{NF}/V^F)(\phi^{NF}/\phi^F)}. \tag{10.9}$$

The value of $(N^{NF}/N^F)(V^{NF}/V^F)$ is the same as in a homogeneous core with the same fuel-to-nonfuel ratio; however, in a heterogeneous core, the thermal flux in the fuel lumps is depressed compared to the moderator (i.e., $\phi^{NF}/\phi^F) < 1$, as a consequence of the relatively large absorption rate in the fuel lump compared to the moderator. In essence, the inside of the fuel lump is shielded by the outer layers of the fuel and, hence, does not have as good a chance at absorbing neutrons as would be the case in the homogeneous core.

Finally, the thermal fission factor η is a property of the fuel alone and is unaffected by the lumping of fuel. As an example of the importance of lumping the fuel, consider the heterogeneous square lattice shown in Fig. 10.4. The center of each graphite lattice cell, of size $a \times a$, contains a cylindrical fuel rod of natural uranium metal. The fuel rod is infinitely long and has a radius of 1.25 cm. For this graphite-uranium lattice, the fast fission factor does not vary significantly with cell

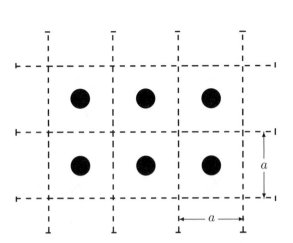

Figure 10.4. Cross-section of a heterogeneous core. Each unit cell of pitch a contains a 1.25-cm radius fuel rod (black circles) of natural uranium metal. The remainder of each lattice cell is graphite.

Table 10.4. Variation of core parameters with cell size for the natural uranium and graphite core shown in Fig. 10.4.

Pitch a (cm)	ϵ	η	f	p	k_∞
12	1.027	1.322	0.972	0.742	0.979
16	1.027	1.322	0.947	0.848	1.090
20	1.027	1.322	0.916	0.900	1.120
21	1.027	1.322	0.907	0.909	1.121
22	1.027	1.322	0.898	0.917	1.119
26	1.027	1.322	0.860	0.940	1.098
30	1.027	1.322	0.818	0.955	1.060

radius and is constant at $\epsilon = 1.027$. The thermal fission factor is that for natural uranium, $\eta = 1.32$. In Table 10.4, the variation with the cell radius of k_∞ and its factors is shown. Notice that the optimum cell size yields a value for k_∞ greater than unity.

10.5 Reflectors

Most reactor cores are surrounded by some material that has a high scattering-to-absorption cross section ratio (typical of moderators). This material, called a *reflector*, is used for two purposes. First, it reflects some of the neutrons which would escape or leak from a bare core back into the core, thereby increasing the non-leakage probabilities P_{NL}^{th} and P_{NL}^{f}. This effect is important for small experimental assemblies so as to reduce the amount of fissile fuel needed for criticality.

However, for large power reactors, the non-leakage probabilities are very close to unity, and the presence of a reflector has only a very small influence on k_{eff}. More important for power reactors, a reflector tends to raise the thermal flux density (and hence the power density) near the core edges, thereby decreasing the peak-to-average power density in the core (see Fig. 10.5). For heat-transfer purposes it is desirable to maintain as constant a thermal flux profile (and hence a power density profile) across the core as possible. To flatten the power density profile even further, most power reactors place fuel with higher concentrations of ^{235}U near the periphery of the core.

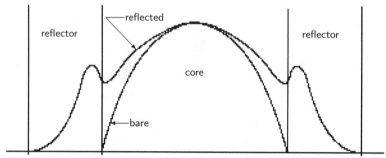

Figure 10.5. The thermal neutron flux profile in a bare and reflected reactor.

10.6 Reactor Kinetics

One of the most important aspects in the design of nuclear reactors is the dynamic response of a reactor to changes in k_{eff}. To vary the power level of a reactor or to shut the reactor down, there must be a mechanism to make k_{eff} vary about the critical value of unity. For example, the insertion (or withdrawal) of rods composed of material with a large absorption cross section for thermal neutrons (e.g., cadmium or indium) decreases (or increases) the thermal utilization factor f, thereby making k_{eff} decrease (or increase). Such *control rods* are used in most reactors to control the neutron chain reaction. Similarly, an accident or unforeseen event may cause some property of the core to change which, in turn, alters k_{eff}. For example, the formation of steam bubbles in a water coolant will decrease the amount of moderator in the core and thus alter k_{eff}.

No matter the mechanism that causes k_{eff} to change, it is crucial that the resulting power transient be kept within strict limits so as to avoid core damage from excessive heat generation. The transient response and the manner in which k_{eff} changes as a result of various core changes is thus of great importance for the safe design of a nuclear reactor.

In this chapter, kinetic equations are developed to describe how the neutron population in a core varies in time as k_{eff} changes. Then various mechanisms which cause k_{eff} to change as a result of power production are discussed.

10.6.1 A Simple Reactor Kinetics Model

Consider a core in which the neutron cycle takes ℓ' seconds to complete. The change Δn in the total number of thermal neutrons in one cycle at time t is $(k_{eff} - 1)n(t)$, where $n(t)$ is the number of neutrons at the beginning of the cycle. Thus

$$\Delta n(t) \equiv \ell' \frac{dn(t)}{dt} = (k_{\text{eff}} - 1)n(t) \qquad (10.10)$$

or equivalently

$$\frac{dn(t)}{dt} = \frac{k_{\text{eff}} - 1}{\ell'} n(t). \qquad (10.11)$$

The solution of this first-order differential equation is

$$n(t) = n(0) \exp\left[\frac{k_{\text{eff}} - 1}{\ell'} t\right], \qquad (10.12)$$

where $n(0)$ is the neutron population at $t = 0$. Notice that in this simple model, the neutron population (and hence the reactor power) varies exponentially in time if $k_{\text{eff}} \neq 1$.

However, this simple kinetics model is not very realistic. Consider, a reactor with $\ell' = 10^{-4}$ s and operating at a steady-state level of n_o neutrons. At $t = 0$, k_{eff} is increased slightly from 1 to 1.001 (a very small change). After 1 second, the neutron population, according to Eq. (10.12), would be

$$n(1 \text{ s}) = n_o \exp\left[\frac{1.001 - 1}{10^{-4}} 1\right] = n_o e^{10} \simeq 20,000 \, n_o. \qquad (10.13)$$

Clearly, such a reactor would be uncontrollable by a human operator!

10.6.2 Delayed Neutrons

If all fission neutrons were emitted at the time of fission, then the simple kinetics model of Eq. (10.11) would apply, and controllable nuclear reactors could not be built. Fortunately, a small fraction β (0.65% for ^{235}U) of fission neutrons are emitted, not during the fission event, but by the radioactive decay of daughters of certain fission products at times up to minutes after the fission event that created the

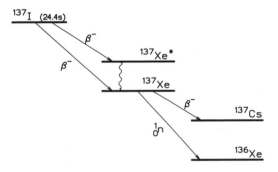

Figure 10.6. An example of a fission product whose decay leads to a delayed neutron.

fission products. The fission products, whose daughters decay by neutron emission, are called *delayed neutron precursors* and the emitted neutrons are called *delayed neutrons*. An example of a delayed neutron precursor is ^{137}I whose decay chain, shown in Fig. 10.6, includes ^{137}Xe which decays either by β^- or neutron emission. The half-life of ^{137}Xe is exceedingly small, so that the apparent emission rate of the delayed neutron is determined by the half-life of the delayed neutron precursor ^{137}I.

There are many fission-products which lead to delayed-neutron emission. About twenty such nuclides have been identified; many others remain to be identified. Because of the importance of delayed neutrons in slowing down the neutron cycle,

Table 10.5. Half-lives and yield-fractions β_i of a six delayed-neutron group representation for thermal fission of three important fissile nuclides. Here β_i is the fraction of the ν fission neutrons emitted by the ith delayed neutron precursor group. Source: [Keepin 1965].

Group	^{235}U		^{233}U		^{239}Pu	
	Half-life (s)	fraction β_i	Half-life (s)	fraction β_i	Half-life (s)	fraction β_i
1	55.7	0.00021	55.0	0.00022	54.3	0.00007
2	22.7	0.00142	20.6	0.00078	23.0	0.00063
3	6.22	0.00127	5.00	0.00066	5.60	0.00044
4	2.30	0.00257	2.13	0.00072	2.13	0.00068
5	0.610	0.00075	0.615	0.00013	0.618	0.00018
6	0.230	0.00027	0.277	0.00009	0.257	0.00009
total	-	0.0065	-	0.0026	-	0.0021

many experiments have been performed to identify the delayed-neutron yields and the rate at which they are emitted after the fission events. The results are grouped by the apparent half-lives of the observed emission rates, i.e., delayed neutron precursors with similar half-lives are placed in the same *delayed-neutron group*. The delayed-neutron fraction or yield, β_i for group i, is the fraction of the total fission neutrons that are eventually emitted by the decay of the ith type of delayed neutron precursor. Typically, six delayed-neutron groups are used in most transient calculations, although a single group approximation is often used for simplification of the kinetic equations. The yields and half-lives of a six delayed-neutron group model for thermal fission of the three fissile isotopes are shown in Table 10.5.

From the table, it is seen that the total delayed neutron fraction β (the sum of the β_i) is considerably less for ^{233}U and ^{239}Pu than for ^{235}U. In a power reactor, as ^{238}U is converted to ^{239}Pu and more and more power is generated from the fission of ^{239}Pu, the delayed neutron fraction decreases and the reactor becomes more responsive to reactivity changes.

Delayed neutrons are emitted with a range of energies, the average energy being about one-half of that for prompt fission neutrons. As a consequence, delayed neutrons have less chance of leaking from the core as they slow down since they have less energy to lose before becoming thermal than do the more energetic prompt fission neutrons. As a consequence, the *effective* delayed neutron fraction $\overline{\beta}$, i.e., the delayed neutron fraction that would be needed if delayed neutrons were emitted with the same energy as prompt fission neutrons, is typically 10–15% larger than the physical delayed neutron fraction β.

10.6.3   Reactivity and Delta-k

Although the single parameter k_{eff} clearly defines the criticality state of a reactor, the value of k_{eff} for subcritical or supercritical reactors seldom varies by more than a few percent from its critical value of unity. Similarly, changes such as the movement of a control rod often produces changes in k_{eff} only in the third or fourth significant figure. To emphasize the degree of departure from criticality, several slightly differ-

Big δk bad!

ent measures of how much k_{eff} departs from the critical value of unity have come into wide use.

One such parameter called "delta-k" is defined simply as $\delta k \equiv k_{eff} - 1$. A closely related measure is the *reactivity*

$$\rho \equiv \frac{k_{eff} - 1}{k_{eff}} = \frac{\delta k}{k_{eff}}. \tag{10.14}$$

For low-power research reactors with little change in fuel composition over time, reactivity is usually measured in multiples of the delayed neutron fraction β in units called "dollars," i.e., $k(\$) \equiv \rho/\beta$. Clearly, any of the parameters k_{eff}, δk, ρ, and k can be obtained from any other.

The great advantage of using δk, ρ, or k is that the criticality state of the reactor is immediately apparent from the sign of these parameters. They are all positive for a supercritical system, negative for a subcritical reactor, and zero at criticality. The reactivity in dollars k has the added advantage that the dangerous super prompt critical state (discussed in the next section) is readily recognized, i.e., when $k > 1\$$.

Example 10.7: A reactivity of 0.1\$ is inserted into an initially critical uranium-fueled reactor. What is k_{eff} of the reactor after the insertion? The reactivity in dollars is related to k_{eff} by

$$k(\$) = \frac{\rho}{\beta} = \frac{k_{eff} - 1}{\beta k_{eff}}.$$

$h(\$) = \dfrac{\rho}{\beta} \qquad \rho = \dfrac{h(\$)}{\beta}$

Solving for k_{eff} we find

$$k_{eff} = \frac{1}{1 - \beta k(\$)} = \frac{1}{1 - 0.0065 \times 0.1} = 1.00065.$$

$\dfrac{\delta h}{h_{eff}} = \dfrac{h(\$)}{\beta}$

$\dfrac{h_{eff} - 1}{h_{eff}} = \dfrac{h(\$)}{\beta}$

10.6.4 Revised Simplified Reactor Kinetics Models

Small Deviations from Criticality

The great importance of delayed neutrons is that they effectively slow down the neutron cycle time in reactors with k_{eff} only slightly different from the critical value of unity. This lengthening of the neutron cycle time, in turn, causes the neutron population and reactor power to vary sufficiently slowly that control of the chain reaction is possible. Consider a thermal reactor fueled with ^{235}U. The average half-life of the delayed neutron precursors is about $T_{1/2} = 8.8$ s so that their average lifetime is $\tau = T_{1/2}/\ln 2 \simeq 12.8$ s. A fraction β of the fission neutrons requires a cycle time of $\ell' + \tau$, while a fraction $(1 - \beta)$ is the prompt-neutron fraction and requires a cycle time of only ℓ'. The average or *effective generation time* required for all the neutrons produced in a single neutron cycle is thus

$$\bar{\ell} = (1 - \beta)\ell' + \beta(\ell' + \tau) = \ell' + \beta\tau. \tag{10.15}$$

Because $\ell' \leq 10^{-4}$ s and $\beta\tau = 0.0065 \times 12.8 = 0.083$ s, the effective generation time is $\bar{\ell} \simeq \beta\tau$.

The simple reactor kinetics model developed in Section 10.6.1 thus must be modified by replacing the prompt generation lifetime ℓ' with the effective generation time $\bar{\ell}$. Again, the neutron population varies exponentially in time as

$$n(t) = n_o \exp\left[\frac{k_{\mathit{eff}} - 1}{\bar{\ell}}t\right] = n_o \exp(t/T), \qquad (10.16)$$

where the *period* or e-folding time is seen to be

$$\boxed{T = \bar{\ell}/(k_{\mathit{eff}} - 1) = \bar{\ell}/\delta k = \beta\tau/\delta k.} \qquad (10.17)$$

This result is valid provided $|\delta k| << \beta$. Notice that this period for small departures from criticality is independent of the prompt generation lifetime ℓ'.

Consider the example of Section 10.6.1 in which $\delta k = 0.001$. For this case, the period is $T = \beta\tau/\delta k = 0.0065 \times 12.8/0.001 = 83$ s, so that after one second the neutron population is

$$n(1 \text{ s}) = n_o \exp(1/83) = 1.012\, n_o, \qquad (10.18)$$

a much more manageable variation in neutron population.

Super Prompt Critical Response

As k_{eff} increases past the critical value of unity, the reactor period becomes shorter and the neutron population increases more quickly. When k_{eff} becomes greater than $1 + \beta$, the chain reaction can be sustained by the prompt neutrons alone. In such a *super prompt critical* reactor, the delayed neutrons can be ignored. Then there is a prompt gain of $k_{\mathit{eff}} - 1 - \beta = \delta k - \beta > 0$ in a prompt cycle time of ℓ'. The increase in the neutron population in one cycle is thus

$$\ell'\frac{dn(t)}{dt} = (\delta k - \beta)n(t). \qquad (10.19)$$

The population thus increases exponentially as $n(t) = n_o \exp(t/T)$ where the period is

$$\boxed{T = \ell'/(\delta k - \beta), \qquad \delta k >> \beta.} \qquad (10.20)$$

For this extreme case, the reactor period is seen to be directly proportional to the prompt generation lifetime.

Subcritical Reactors

If k_{eff} were reduced from its critical value of unity, the neutron population would decrease, and would do so more rapidly for larger reductions in k_{eff}. Although the neutron population produced by prompt fission neutrons can be made to decrease rapidly by a large reduction in k_{eff}, the delayed neutron population can decrease no faster than the decay rate of the delayed neutron precursors. After a few minutes following a large decrease in k_{eff}, the neutron population decreases at the rate of decay of the longest lived delayed neutron precursor which has a half-life of about 55 s (see Table 10.5) or a mean lifetime of $\ln 2/T_{1/2} \simeq 80$ s. Thus, even if all control rods of a reactor were inserted to shut the reactor down, the neutron population and, hence, the power, would after a short time decrease exponentially with a period of -80 s. This inability to quickly reduce a reactor's power, as well as the decay heat produced by the radioactive decay of fission products in the fuel, require power reactors to maintain cooling of the core for long periods following a reactor shutdown.

Reactor Kinetic Models with Delayed Neutrons

The above cases are for extreme changes in k_{eff}. To obtain the kinetic response of a reactor to an arbitrary variation of k_{eff}, it is necessary to describe the dynamic variation of the delayed neutron precursors as well as the neutrons in a core. Such a treatment leads to a set of coupled first-order differential equations, one for each delayed neutron precursor group and one for the neutron population or power of the reactor. The derivation of these so-called *point reactor kinetic equations* is provided in Addendum 2 of this chapter. These point reactor kinetic equations have been found to describe very accurately the time variation of the neutron population in a reactor caused by changes in k_{eff}.

Example 10.8:   If the reactor of Example 10.7 were initially operating at a power of 10 W, how long after the 0.1\$ reactivity insertion is it before the reactor power reaches 10 kW? From Example 10.7, the reactivity insertion is $\delta k = 0.00065$ and from Eq. (10.17) we find the resulting reactor period is

$$T = \beta\tau/\delta k = [0.0065 \times 12.8 \text{ s}]/0.00065 = 128 \text{ s}.$$

Since the reactor power $P(t)$ is proportional to the neutron population $n(t)$, we have from Eq. (10.16), $P(t) = P(0)\exp(t/T)$. Solving for t, we find

$$t = T\ln[P(t)/P(0)] = 128\ln(10,000/10) = 884 \text{ s} = 14.7 \text{ min}.$$

10.6.5   Power Transients Following a Reactivity Insertion

The response of a reactor to a change in k_{eff} is of great importance in the design of a reactor. For example, the rate at which power can be decreased in an emergency situation dictates the design of several complex systems to maintain core cooling in an emergency. Likewise, safety features must be designed into a reactor to insure than k_{eff} cannot become so much greater than unity that rapid and uncontrollable power increases occur. In this section, the power transients following a reactivity insertion into the core are considered.

Steady-State Operation

One of the most important design and operational problems is the transient caused by the insertion of reactivity (positive or negative) into a reactor operating at some steady-state neutron level n_o. For a reactor in which there are negligible neutron sources that are independent of the neutron field, steady-state power occurs only if the reactivity is zero ($k_{eff} = 1$).

However, at low power levels (γ, n) or (α, n) reactions produce a background source of neutrons. Let the steady-rate of production of neutrons by sources that do not depend on the neutron field be denoted by S_o. In steady-state, the time derivatives in the point reactor kinetic equation, Eqs. (10.71), vanish and the following steady-state value of the neutron population is obtained:

$$n_o = -\frac{\ell}{\beta}\frac{S_o}{k_o}, \qquad (10.21)$$

where the effective neutron lifetime $\ell \equiv \ell'/k_{eff} \simeq \ell'$. Notice from this last result, that for a reactor in steady state with an non-fission source S_o, the reactivity must necessarily be negative (i.e., the reactor is subcritical). As $S_o \to 0$, k_o must also vanish (i.e., the reactor becomes critical) to maintain a steady-state neutron population.

Response to a Step Reactivity Insertion

When the reactivity of a reactor varies about its critical value of zero, the neutron population and, hence, the power, varies according to the point reactor kinetics equations, Eqs. (10.70) or Eqs. (10.71). One of the most fundamental kinetics problems is the transient produced by a constant reactivity insertion (positive or negative) into a source-free, critical reactor operating at a steady neutron level n_o. In particular, at $t = 0$ the reactivity is changed from its steady-state value of zero to k_o, i.e.,

$$k(t) = \begin{cases} 0 & t < 0 \\ k_o & t \geq 0 \end{cases}. \qquad (10.22)$$

If k_o were positive, we would expect the neutron level to rise; by contrast, for negative k_o, we would expect the level to decrease and eventually vanish. Example transients are shown in Fig. 10.7. As seen from this figure, the neutron level (or power) experiences a rapid *prompt jump* followed by an exponential variation (the straight lines in the semi-log plot). The time required for the neutron level to change by a factor of e is called the *reactor period*. Of great operational importance is the period of the asymptotic neutron variation, i.e., the slope of the transient on a semi-log plot, at long times after the reactivity insertion.

For negative insertions, the initial prompt drop arises from the decrease in prompt neutrons in each succeeding generation. However, the decay of the existing delayed neutron precursors acts as a neutron source which decays more slowly with the half-lives of the precursors. Even for very large negative reactivity insertions, the neutron level after the prompt jump cannot decrease any faster than the longest lived precursor group (with about a 55 second half-life, or about 80 seconds to decrease by a factor of e). This physical limit on how quickly a reactor can be shut down is a major safety concern and necessitates that considerable attention be given to emergency core cooling systems.

For a positive reactivity insertion, there is an initial rapid rise in the neutron population because of the increased multiplication of the prompt neutrons. For $k_o < 1\$$, however, there is less than one prompt neutron produced per generation and, hence, there must be delayed neutrons introduced to keep the system supercritical, i.e., we must wait for the decay of newly created delayed neutron precursors, thereby

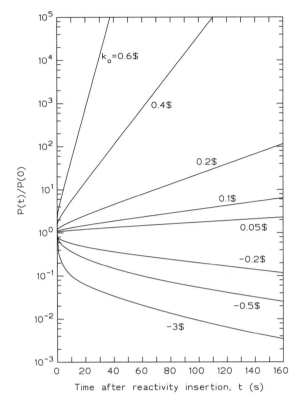

Figure 10.7. Variation of neutron level caused by various step reactivity insertions into a ^{235}U fueled reactor with $\beta = 0.007$ and $\ell = 0.0004$.

slowing down the rate of rise in the neutron level. However, if $k_o > 1\$$, the neutron population is self-sustaining on prompt neutrons alone, and the rate of increase in the neutron level is very rapid, limited by only the neutron generation lifetime.

Since the point reactor kinetics equations are first-order differential equations, we expect the mathematical solution for $n(t)$ and for the delayed neutron precursor densities $\widehat{C}_i(t)$ to be of the form $e^{\omega t}$. If we substitute this assumed form into Eqs. (10.71), we find that the only permissible values of ω must satisfy

$$\omega \left[\frac{\ell}{\beta} + \sum_{i=1}^{G} \frac{a_i}{\lambda_i + \omega} \right] = k_o. \tag{10.23}$$

This equation is known as the *inhour equation* since the units of ω are inverse time, e.g., h^{-1}. In this equation, λ_i is the decay constant for the ith delayed neutron precursor group, $a_i \equiv \beta_i/\beta$, and G is the number of delayed neutron precursor groups. There are $G + 1$ solutions of this equation, whose values are denoted by ω_j, $j = 0, \ldots G$ where we order the solutions such that $\omega_o > \omega_1 > \ldots > \omega_G$.

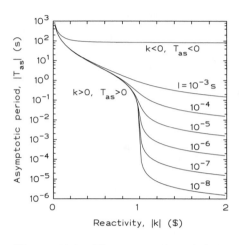 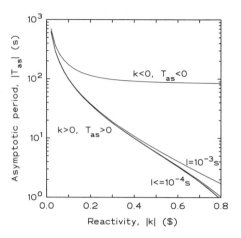

Figure 10.8. The asymptotic period for a constant reactivity insertion into a ^{235}U-fueled reactor ($\beta = 0.007$) for various neutron lifetimes.

Figure 10.9. The asymptotic period for a constant reactivity insertion into a ^{235}U-fueled reactor with $\beta = 0.007$. Enlarged view of upper left of Fig. 10.8

Thus the most general solution of Eqs. (10.71) is of the form

$$n(t) = \sum_{j=0}^{G} A_j e^{\omega_j t} \tag{10.24}$$

where A_j are constants. A rigorous derivation for a one delayed-neutron group model is given in Addendum 3 to this chapter. It can be shown that the largest root of the inhour equation, ω_o, is positive if $k_o > 0$ and negative if $k_o < 0$. The other ω_j are always negative regardless of k_o. The $e^{\omega_j t}$, $j = 1, \ldots, G$, terms decay away more rapidly than the $e^{\omega_o t}$ term. Thus at long times after the reactivity insertion

$$n_{as}(t) = A_o \exp[\omega_o t] = A_o \exp[t/T_{as}] \tag{10.25}$$

where the *asymptotic period* $T_{as} \equiv 1/\omega_o$. Hence, we see that the asymptotic (large time) neutron variation is purely exponential with a period (or e-folding time) of T_{as}.

The asymptotic period as a function of the reactivity insertion is shown in Figs. 10.8 and 10.9. Notice that the asymptotic period is almost independent of the neutron lifetime ℓ for $-\infty < k_o < 1\$$. For small reactivity insertions ($|k_o| \leq 0.1\$$), it can be shown that

$$T_{as} \simeq \ell/(\beta k_o), \tag{10.26}$$

which is in agreement with Eq. (10.17). However, as k_o becomes greater than 1$, the asymptotic period changes dramatically and becomes proportional to the neutron lifetime. It can be shown that for $k_o \gg 1\$$, $T_{as} \simeq (\ell/\beta)/(k_o - 1)$, which is in agreement with Eq. (10.20).

10.7 Reactivity Feedback

The energy and nuclear reactions that occur during the operation of a nuclear reactor change the material properties of the core and thus the multiplication factor k_{eff}. This change in the reactivity of the reactor is called *feedback reactivity* and is illustrated schematically in Fig. 10.10.

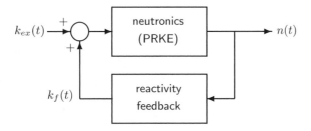

Figure 10.10. The total reactivity driving the neutron cycle is the sum of externally applied reactivity k_{ex} and the feedback reactivity k_f.

The physical causes of feedback reactivity can be grouped into two categories: reactivities resulting from isotopic changes and those from temperature changes. Some of the most common feedback mechanisms are discussed below. In general, the amount of feedback reactivity $k_f(t)$ at any time t depends on the entire past power history of the reactor, i.e., on $n(t')$, $t' < t$.

10.7.1 Feedback Caused by Isotopic Changes

As a reactor generates power, the neutron field causes changes in the isotopic composition of the core material and thus in k_{eff}. Reactivity transients produced by isotopic changes are usually very slow, occurring over a period of hours at the fastest to years. Some of the important isotopic changes are described below.

Fuel Burnup

To sustain the neutron chain reaction, the fissile fuel must be consumed. Through this fuel consumption, ηf decreases slowly in time and causes a negative reactivity feedback which must be compensated for by removing other negative reactivity from the core. This compensation for fuel burnup can be accomplished in a variety of ways, e.g., by diluting the boric acid concentration in the water moderator, by withdrawing control rods, or replacing high-burnup fuel with fresh fuel.

Fuel Breeding

In any reactor there are fertile isotopes which absorb neutrons and, through subsequent radioactive decays, produce atoms of fissile fuel. For example, in ^{235}U-fueled reactors, the ^{238}U atoms can absorb neutrons and, after two radioactive decays, yield fissile ^{239}Pu. This breeding mechanism causes a positive reactivity feedback, and is an important feedback mechanism in reactors with low enrichments, particularly in reactors fueled with natural uranium. In LWRs with typically 2–3% enrichments, at the end of a fuel-cycle (the time at which the reactor must be refueled), almost as much power is generated from the fission of the bred ^{239}Pu as from ^{235}U.

Fission Product Poisons

As the fission products and their daughters accumulate in a reactor, these new isotopes may also absorb neutrons and decrease the thermal utilization factor f. Two particular fission product poisons, ^{135}Xe and ^{149}Sm have particularly large absorption cross sections for thermal neutrons and can have severe negative reactivity effects on reactor operations. These fission-product poisons are discussed in detail in the next section.

Burnable Poisons

In many power reactors, small amounts of material with large absorption cross sections (e.g., samarium or gadolinium in fuel rods, or boric acid in a water coolant) are incorporated into the core material. Such material initially decreases f, but during reactor operation it is consumed and, thereby, increases f to offset the decrease in ηf caused by fuel consumption.

10.7.2   Feedback Caused by Temperature Changes

The temperature of the core material affects the rate at which neutrons interact with the material. These changes in interaction rates arise from changes in atomic concentrations, reaction probabilities, and geometry changes. The reactivity transients caused by temperature changes are usually very fast, as fast as the temperature itself changes. Crucial to safe reactor operation is a reactor design that has an overall negative reactivity feedback as the core temperature increases. Some of the important mechanisms are discussed below.

Changes in Atomic Concentrations

Most materials expand as their temperature increases, thereby decreasing their atomic concentration and their macroscopic cross sections. Most LWRs are slightly undermoderated, i.e., the moderator-fuel ratio is less than the ratio that produces a maximum in k_∞. Thus, as the moderator/coolant heats and expands, some is expelled from the core, making the reactor even more undermoderated and thus decreases k_{eff}. An even more severe change in atom densities is caused by the formation of steam bubbles. Steam generation expels liquid from the core and introduces a large negative reactivity. In solid graphite moderated cores, the expansion of the moderator decreases the thermalization of neutrons and thus also causes a negative reactivity feedback.

In most reactors, the fuel is in a ceramic form that expands very little with increasing fuel temperature, and, consequently, the negative fuel expansion reactivity effect is very small in most reactors. Only in small experimental reactors composed of fissile metal, does the fuel expansion effect cause a large negative reactivity.

Changes in the Neutron Energy Distribution

As the temperature of the core material increases, the thermal neutrons maintain a Maxwellian energy distribution one but shifted increasingly towards higher energies. The thermal spectrum is said to *harden*. If all the cross sections in the core had exactly a $1/v$ dependence, there would be no change in the thermal-neutron interaction rates. However, most heavy nuclides have resonances near the upper end of the thermal energy range and their cross sections, consequently, are not exactly $1/v$ in the thermal energy region. For example, ^{239}Pu has a large resonance at about

0.3 eV, and, as the thermal neutrons shift in energy towards this resonance, more neutrons are absorbed by ^{239}Pu and thus the fission rate increases, thus producing a positive reactivity feedback. By contrast, hardening the thermal neutron spectrum causes a very slight decrease in absorption by ^{235}U.

This hardening of the thermal neutron spectrum with increasing temperature is taken to an extreme in the TRIGA class of reactors which use as fuel enriched ^{235}U blended in zirconium hydride. As the fuel temperature increases, vibrating hydrogen atoms trapped in the zirconium-hydride crystal lattice can transfer some of their vibrational energy (about 0.13 eV) to thermal neutrons, thereby removing them from the thermal energy region so they are less likely to be absorbed by the fuel. This very rapid negative reactivity feedback effect is the reason these reactors can be operated in a pulse mode, in which a large positive reactivity is inserted into the core by rapidly removing a control rod to make the reactor very super prompt critical. However, the ZrH negative temperature feedback acts within a few ms to stop the runaway chain reaction, which has increased the reactor power by many thousands of times the initial power, and brings the reactor power back to safe limits.

Changes in Resonance Interactions

The most important negative temperature feedback mechanism in most reactors is that provided by the change in neutron interaction rates with materials having large cross section resonances just above the thermal region. Only epithermal neutrons (neutrons with energies above the thermal region) with energies equal to the resonance energy interact with the fuel. However, increasing the fuel temperature causes the fuel atoms to move more rapidly and the *relative* energy between a neutron and a fuel atom changes. This spreading out of the relative kinetic energy allows neutrons with energies near the resonance energy to interact with the fuel atoms, thereby, increasing the neutron interaction rate. The effective cross section is said to be *Doppler broadened*, and allows more neutrons to interact with the fuel.

This Doppler effect is the dominant feedback mechanism in many thermal reactors with low enrichment fuel. As the fuel temperature increases, more thermalizing neutrons are absorbed in the ^{238}U resonances as the neutrons slow down. The resonance escape probability p thus decreases causing a rapid negative reactivity feedback. However, an increase in fuel temperature also causes an increase in neutron absorption by the fissile atoms (thus increasing f), particularly by ^{239}Pu which has a very low energy resonance at 0.3 eV. This positive reactivity feedback places a restriction on how much ^{239}Pu can be used in a reactor, since the overall reactivity feedback must be negative.

Changes in Geometry

When thermal gradients are produced in a core, the fuel rods undergo differential expansions resulting in, for example, bowing, which changes the core geometry slightly and produces small positive or negative reactivity changes. A more severe geometry change is that caused by an accident condition in which the fuel undergoes unexpected deformations, melting, or even fuel dispersal. Reactivity changes produced by core disruptions are a major consideration in safety analyses.

10.8   Fission Product Poisons

The fission products that accumulate in a reactor core are of concern for two reasons. First, they act as long-term heat sources through their radioactive decays. Second, they act as parasitic neutron absorbers or *poisons* that, over time, decrease the thermal utilization factor f and, thus, introduce negative reactivity into a core.

The reactivity ρ_p introduced by a fission product poison is directly proportional to its average concentration N_p in the core. To see this, we start with

$$\rho_p = \frac{k'_{eff} - 1}{k'_{eff}} - \frac{k_{eff} - 1}{k_{eff}} \tag{10.27}$$

where k'_{eff} indicates the core with the poison included and k_{eff} refers to the same core without the poison. Since the poison changes only the thermal utilization factor, the two multiplication factors are related to each other by $k'_{eff} = k_{eff} f'/f$. If we assume the unpoisoned core is critical ($k_{eff} = 1$), Eq. (10.27) becomes

$$\rho_p = \frac{k'_{eff} - 1}{k'_{eff}} = 1 - \frac{1}{k'_{eff}} = 1 - \frac{f}{f'}\frac{1}{k_{eff}} = 1 - \frac{f}{f'}$$

$$= 1 - \frac{\Sigma_a^F/\Sigma_a}{\Sigma_a^F/(\Sigma_a + \Sigma_a^p)} = -\frac{\Sigma_a^p}{\Sigma_a} = -\frac{\sigma_a^p N_p}{\Sigma_a} = -\sigma_a^p \frac{\Sigma_f}{\Sigma_a}\frac{N_p}{\Sigma_f}, \tag{10.28}$$

where σ_a^p is the thermal microscopic absorption cross section for the poison.

The ratio $\Sigma_f/\Sigma_a = (\Sigma_f/\Sigma_a^F)(\Sigma_a^F/\Sigma_a) = (\eta/\nu)f$. For most light water reactors, fuel enriched in ^{235}U to about 2.5% is used so that $\eta \simeq 1.8$ and $\nu = 2.43$. For these reactors $f \simeq 0.8$, so that $\Sigma_f/\Sigma_a \simeq 0.6$. The poison reactivity is given by

$$\boxed{\rho_p \simeq -0.6\, \sigma_a^p\, \frac{N_p}{\Sigma_f}.} \tag{10.29}$$

Thus, to determine the reactivity transient caused by a particular fission product poison, we need to find $N_p(t)/\Sigma_f$, a quantity that is found from the decay and buildup equations for the poison decay chain.

There are hundreds of different fission products and their decay daughters that are created in a core over time, each with a different absorption cross section for thermal neutrons. With the exception of ^{135}Xe and ^{149}Sm, which have unusually large cross sections for absorbing a thermal neutron, it is impractical to keep track of the individual isotopes separately. Rather, the fission products and their daughters are treated collectively. For fission products produced from the fission of ^{235}U, it is often assumed that each fission produces 1 atom of stable poison with an absorption cross section of 50 barns. While this simplistic rule-of-thumb works for long-term calculations of burnup effects, the two particular poisons ^{135}Xe and ^{149}Sm have such large absorption cross sections that they must be treated separately.

10.8.1   Xenon Poisoning

^{135}Xe is a member of the fission product decay chain shown below. It is of importance because it has the largest thermal neutron absorption cross section of all isotopes, namely, $\sigma_a^x = 2.7 \times 10^6$ barns. Thus, a very small atomic concentration of this nuclide can have a considerable reactivity effect.

$$^{135}\text{Te} \xrightarrow[19 \text{ s}]{\beta^-} {}^{135}\text{I} \xrightarrow[6.7 \text{ h}]{\beta^-} {}^{135}\text{Xe} \xrightarrow[9.2 \text{ h}]{\beta^-} {}^{135}\text{Cs} \xrightarrow[2.6 \text{ My}]{\beta^-} {}^{135}\text{Ba (stable)}.$$

The isotopes ^{135}Te, ^{135}I and ^{135}Xe are all produced as fission products. However, because of the very short half-life of ^{135}Te compared to the other members of the chain, it is usually assumed that ^{135}Te immediately decays to ^{135}I. The fission product yield γ_I of ^{135}I for ^{235}U (including that of ^{135}Te) is 0.061 and, for ^{135}Xe, $\gamma_X = 0.003$. The decay constants for ^{135}I and ^{135}Xe are 2.9×10^{-5} s^{-1} and 2.1×10^{-5} s^{-1}, respectively.

The production rate (per unit volume) of ^{135}I as a fission product is $\gamma_I \Sigma_f \phi(t)$ where $\phi(t)$ is the average thermal flux density in the core at time t. Its disappearance rate through radioactive decay is $-\lambda_I I(t)$ where $I(t)$ is the average ^{135}I atomic concentration in the core. The absorption cross section for ^{135}I is negligible and burnup of ^{135}I does not contribute to its disappearance. For ^{135}Xe, the production rate is by the radioactive decay of ^{135}I and as a fission product, namely, $\lambda_I I(t) + \gamma_X \Sigma_f \phi(t)$. ^{135}Xe disappears by radioactive decay and by neutron absorption at a rate $-\lambda_X X(t) - \sigma_a^x \phi(t) X(t)$, where $X(t)$ is the average atomic concentration of ^{135}Xe in the core at time t. Thus the decay/buildup equations for ^{135}I and ^{135}Xe are

$$\frac{dI(t)}{dt} = -\lambda_I I(t) + \gamma_I \Sigma_f \phi(t) \tag{10.30}$$

$$\frac{dX(t)}{dt} = \lambda_I I(t) + \gamma_X \Sigma_f \phi(t) - \lambda_X X(t) - \sigma_a^x \phi(t) X(t). \tag{10.31}$$

The solution of these coupled differential equations subject to arbitrary initial values $I(0)$ and $X(0)$ and for a constant flux density ϕ_o is

$$\frac{I(t)}{\Sigma_f} = \frac{I(0)}{\Sigma_f} \exp[-\lambda_I t] + \frac{\gamma_I \phi_o}{\lambda_I} \{1 - \exp[-\lambda_I t]\} \tag{10.32}$$

$$\frac{X(t)}{\Sigma_f} = \frac{X(0)}{\Sigma_f} \exp[-(\lambda_X + \sigma_a^x \phi_o)t] + \frac{(\gamma_I + \gamma_X)\phi_o}{\lambda_X + \sigma_a^x \phi_o} \{1 - \exp[-(\lambda_X + \sigma_a^x \phi_o)t]\}$$
$$+ \frac{(\lambda_I I(0)/\Sigma_f) - \gamma_I \phi_o}{\lambda_X - \lambda_I + \sigma_a^x \phi_o} \{\exp[-\lambda_I t] - \exp[-(\lambda_X + \sigma_a^x \phi_o)t]\}. \tag{10.33}$$

Equilibrium Xenon Poisoning

For a reactor operating at a constant flux density ϕ_o, the equilibrium concentrations of ^{135}I and ^{135}Xe are found from Eqs. (10.30) and (10.31) by setting the time derivative to zero. The result is

$$I_o = \frac{\gamma_I \Sigma_f \phi_o}{\lambda_I} \tag{10.34}$$

$$X_o = \frac{(\gamma_I + \gamma_X)\Sigma_f \phi_o}{\lambda_X + \sigma_a^X \phi_o} \tag{10.35}$$

The equilibrium concentrations are shown in Fig. (10.11). Notice that, while the ^{135}Xe concentration is independent of ϕ_o at high flux density levels, the ^{135}I concentration continues to increase linearly with ϕ_o. This has profound consequences for reactors operating at high power.

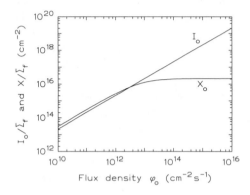

Figure 10.11. Equilibrium ^{135}I and ^{135}Xe concentrations as a function of the steady-state flux density ϕ_o.

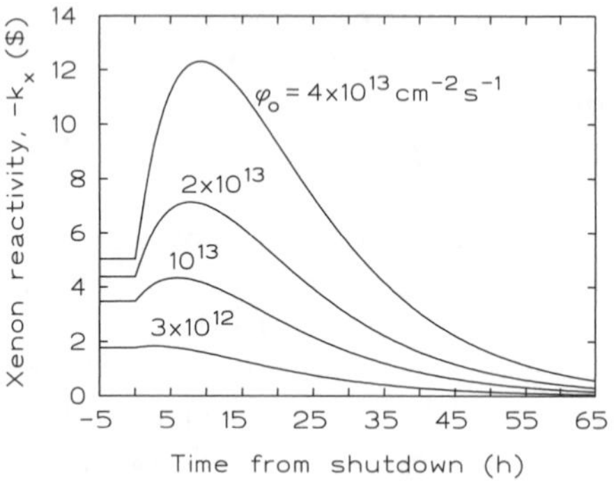

Figure 10.12. ^{135}Xe transients following shutdowns from equilibrium at various constant flux densities.

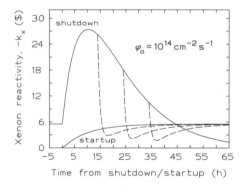

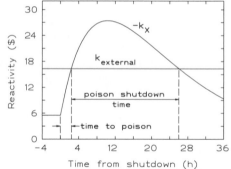

Figure 10.13. ^{135}Xe transient for the buildup to equilibrium, the transient following shutdown from equilibrium, and three transients (dashed lines) following restart during the shutdown transient.

Figure 10.14. ^{135}Xe transient following a shutdown from equilibrium showing the time-to-poison and the poison shutdown time. Equilibrium flux density before shutdown is $\phi_o = 10^{14}$ cm^{-2} s^{-1}.

Transient Following Shutdown from Equilibrium

In reactors that operate at high power, $\phi_o > 10^{13}$ cm^{-2} s^{-1}, there are higher equilibrium levels of ^{135}I (potential ^{135}Xe) than of ^{135}Xe, which has a slower decay rate. Thus, if the reactor should suddenly shut down, the ^{135}I would decay to ^{135}Xe faster than the ^{135}Xe could decay away, producing initially an *increase* in the ^{135}Xe concentration. Eventually, the ^{135}I would decay away, and the ^{135}Xe concentration would finally begin to decrease as it decays. Fig. (10.12) shows the buildup of ^{135}Xe following the shutdown from various flux levels.

If during the shutdown transient, the reactor were started up again, the large absorption cross section for ^{135}Xe would cause this nuclide to be burned up very rapidly, reducing the xenon reactivity temporarily to below its equilibrium values. Examples of these restart transients are shown in Fig. (10.13).

One consequence of the large increase in the negative ^{135}Xe reactivity following a shutdown from equilibrium, is that very quickly the xenon reactivity may become greater than offsetting positive reactivity available by removing all control rods from the core. Should this occur, it is impossible to restart the reactor, and the reactor is said to have *poisoned out*. The time from the shutdown until the reactor poisons out is called the *time-to-poison* (see Fig. (10.14)). Once the reactor has poisoned out, it is necessary to wait until the negative ^{135}Xe reactivity has peaked and descended back to a level that can be offset by all controllable positive reactivities. The time interval during which the reactor cannot be restarted is called the *poison shutdown time* and is typically of 15–25 hours duration. In many power or propulsion reactors, the time-to-poison is usually only a few tens of minutes, and the operator may experience considerable pressure to get the reactor restarted before it poisons out so as to avoid a lengthy period of lost production.

When the reactor is restarted during a shutdown transient, there is a rapid burnoff of ^{135}Xe and, since most of the ^{135}I has already decayed away, the negative xenon reactivity quickly drops below equilibrium levels (see the dashed lines in Fig. 10.13). However, over the next day of operation, the ^{135}Xe resumes its equilibrium level.

Xenon Spatial Oscillations

^{135}Xe transients generally vary so slowly that they are usually easily controlled by reactor operators. However, in very large reactors with a small negative temperature reactivity coefficient, ^{135}Xe can induce a very complex instability in which the maximum of the flux density moves around the core with periods of about ten hours. The phenomenological sequence of events is as follows:

1. An initial asymmetry in the core flux distribution (e.g., caused by a control rod misalignment) causes a change in the ^{135}I buildup and ^{135}Xe burnup.

2. In the high-flux region, increased ^{135}Xe burnup initially allows the flux to increase even further causing an increase in ^{135}I production. Meanwhile in the low-flux region, ^{135}I decays to ^{135}Xe causing the flux to be even further depressed while the ^{135}I continues to decay.

3. Eventually, in the high-flux region, the ^{135}I reaches a new higher equilibrium, and its decay to ^{135}Xe causes the ^{135}Xe concentration to begin to rise, thereby reversing the pattern. Likewise, in the low-flux region, the ^{135}Xe becomes depleted by decay thereby allowing the flux to start to rise.

4. Repetition of this pattern, causes the flux profile to oscillate spatially with a period of around 10 hours.

These xenon spatial oscillations, while producing little overall reactivity change for the core, can cause the local flux and power density to vary by factors of three.

To avoid oscillations in reactors that are susceptible to them, it is necessary to implement a very complex series of control rod adjustments which are usually determined by a computer.

10.8.2   Samarium Poisoning

The second fission product poison which must be accounted for explicitly in power reactors is ^{149}Sm. This stable nuclide is a daughter of the fission products ^{149}Nd and ^{149}Pm. The decay chain for this nuclide is shown below.

$$^{149}\text{Nd} \xrightarrow[1.7\text{ h}]{\beta^-} {}^{149}\text{Pm} \xrightarrow[53\text{ h}]{\beta^-} {}^{149}\text{Sm (stable)}.$$

Because the half-life of ^{149}Nd is so small compared to that of its daughter ^{149}Pm, it is usually assumed to decay immediately to ^{149}Pm. Thus, the effective fission product yield γ_P for ^{149}Pm is the sum of the actual yields for ^{149}Nd and ^{149}Pm. For ^{235}U, $\gamma_P = 0.0113$. Thus, the production rate of ^{149}Pm (per unit volume) is $\gamma_P \Sigma_f \phi(t)$ and the rate of decay is $\lambda_P P(t)$, where $P(t)$ is the core-averaged ^{149}Pm atomic concentration. The generation rate of ^{149}Sm is the decay rate of ^{149}Pm. There is negligible production of ^{149}Sm as a direct fission product. Since ^{149}Sm is stable, the only way it can vanish is for it to absorb a neutron, which it does at a volumetric rate of $\sigma_a^S \phi(t) S(t)$ where $S(t)$ is the average ^{149}Sm concentration. Thus, the decay/buildup equations for ^{149}Pm and ^{149}Sm are

$$\frac{dP(t)}{dt} = -\lambda_P P(t) + \gamma_P \Sigma_f \phi(t) \tag{10.36}$$

$$\frac{dS(t)}{dt} = \lambda_P P(t) - \sigma_a^S \phi(t) S(t). \tag{10.37}$$

The solution of these coupled differential equations subject to arbitrary initial values $P(0)$ and $S(0)$ and for a constant flux density ϕ_o is

$$\frac{P(t)}{\Sigma_f} = \frac{P(0)}{\Sigma_f} \exp[-\lambda_P t] + \frac{\gamma_P \phi_o}{\lambda_P} \left\{ 1 - \exp[-\lambda_P t] \right\} \tag{10.38}$$

$$\frac{S(t)}{\Sigma_f} = \frac{S(0)}{\Sigma_f} \exp[-\sigma_a^S \phi_o t] + \frac{\gamma_P \phi_o}{\sigma_a^S \phi_o} \left\{ 1 - \exp[-\sigma_a^S \phi_o t] \right\}$$
$$+ \frac{(\lambda_P P(0)/\Sigma_f) - \gamma_P \phi_o}{\sigma_a^S \phi_o - \lambda_P} \left\{ \exp[-\lambda_P t] - \exp[-\sigma_a^x \phi_o t] \right\}. \tag{10.39}$$

Because of the long half-lives of ^{149}Pm and ^{149}Sm, the buildup of ^{149}Sm to its equilibrium level takes many tens of hours, especially for reactors operating at low average flux densities ϕ_o. From Eq. (10.36), it is seen that the equilibrium ^{149}Sm is $S_o = (\gamma_P \Sigma_f)/\sigma_a^s$, a level that is independent of the flux density. Thus, at equilibrium, all reactors have the same amount of ^{149}Sm poisoning. This buildup to equilibrium is shown in Figs. (10.15) and (10.16).

Following a shutdown from equilibrium, ^{149}Sm increases as the ^{149}Pm decays. This buildup of negative reactivity can be especially severe for reactors operating

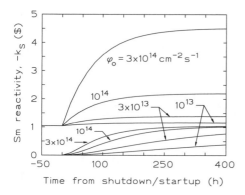

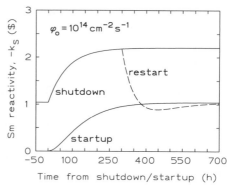

Figure 10.15. The buildup of ^{149}Sm to equilibrium (four lower curves) and the transient following shutdown from equilibrium for four different flux levels.

Figure 10.16. ^{149}Sm transient for the buildup to equilibrium during a startup, the transient following shutdown from equilibrium, and the transient (dashed line) following restart during the shutdown transient.

at high flux densities. However, unlike the ^{135}Xe transient following a reactor shutdown, the ^{149}Sm transient does not decay away with time since ^{149}Pm is stable. If a reactor does not have sufficient positive reactivity to compensate for this buildup following a scram, the reactor will permanently *poison out* and cannot be restarted until additional positive reactivity is added (e.g., by adding fresh fuel).

10.9   Addendum 1: The Diffusion Equation

To calculate the power distribution $P(\mathbf{r})$ in a reactor core, it is first necessary to know the distribution of the neutron flux density $\phi(\mathbf{r})$. If E_r is the recoverable energy per fission (roughly 200 MeV), these two distributions are related by

$$P(\mathbf{r}) = E_r \Sigma_f(\mathbf{r})\phi(\mathbf{r}) \qquad \text{MeV cm}^{-3}\text{ s}^{-1}. \tag{10.40}$$

To calculate the flux density, it is necessary to solve the neutron diffusion equation. This equation is simply a mathematical relation balancing the rate of neutron loss and production at every point in the core. In this section, we derive the diffusion equation for the special case of one-speed neutrons diffusing through a homogeneous material. Such one-speed neutrons are descriptive of the thermal neutron population in a thermal reactor.

Consider an arbitrary volume ΔV in some medium through which one-speed neutrons diffuse by scattering randomly from the medium's nuclei. Under steady-state conditions, the rate of neutron gain must equal the rate of neutron loss from the volume ΔV, i.e.,

$$\boxed{\text{leakage rate} + \text{absorption rate} = \text{production rate.}} \tag{10.41}$$

To obtain the diffusion equation, we simply have to express each term mathematically.

As neutrons scatter repeatedly, they change directions in each scatter, and thus, perform a random walk as they move through the medium. In a medium containing many neutrons, each moving and scattering randomly, more neutrons will, on average, tend to move from regions of high concentration to regions of lower concentration than in the reverse direction. To express this mathematically, consider the flux profile shown in Fig. 10.17. The neutron density (and hence ϕ) is higher to the left than to the right of the unit area perpendicular to the x-axis. It is reasonable to assume that the *net* flow across this unit area in a unit time, J_x, is proportional to the difference between the neutron densities on the low density side to that on the high density side. In other words, the net flow should be proportional to the negative slope of the neutron density or flux at the unit area, i.e., J_x is proportional to $[-d\phi/dx]$. Thus, we assume

$$J_x = -D\frac{d\phi}{dx},\qquad(10.42)$$

where D is the constant of proportionality known as the *diffusion coefficient*. This equation was first proposed by Adolf Fick in 1855 to describe the diffusion of randomly moving molecules in a fluid, and is consequently known as *Fick's law*. This empirical "law" is reminiscent of an earlier 1823 empirical observation by J.J.B. Fourier describing the net flow of heat energy (a manifestation of the random motion of a medium's atoms), namely, the heat flow $= -k\,dT/dx$, where k the is so-called thermal conductivity of the medium.

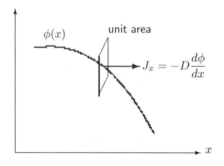

Figure 10.17. The net flow of neutrons J_x is away from the region of higher concentration.

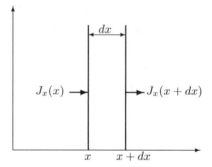

Figure 10.18. The leakage from the differential slab is the difference between the net flows at the two faces.

We now use Fick's law to express the leakage term in the balance relation of Eq. (10.41). Consider the differential slab of width dx shown in Fig. 10.18. The net leakage rate from this slab is

$$\text{net leakage rate} = [J_x(x+dx) - J_x(x)]\mathcal{A},\qquad(10.43)$$

where $\mathcal{A}$ is the area of the slab surfaces (in the y-z plane). If we assume that the flux density $\phi(x)$ does not vary in the y or z directions, there would be no net leakage in these directions.

The rate of absorption of neutrons in the differential slab is the absorption rate per unit volume ($\Sigma_a\phi$) times the volume of the slab $\Delta V = \mathcal{A}dx$, i.e.,

$$\text{absorption rate} = \Sigma_a\phi(x)\mathcal{A}\,dx. \tag{10.44}$$

Neutrons are produced by fission ($\nu\Sigma_f\phi$ per unit volume per unit time) and by any "external" sources, such as (γ,n) or (α,n) reactions, which are independent of the neutron field. We define Q as the rate of neutron production per unit volume by such external sources. Then the rate of neutron production is

$$\text{production rate} = [\nu\Sigma_f\phi(x) + Q(x)]\mathcal{A}\,dx. \tag{10.45}$$

Substitution of Eqs. (10.43), (10.44), and (10.45) into the balance relation Eq. (10.41) and simplification of the result give

$$\frac{J_x(x+dx) - J_x(x)}{dx} + \Sigma_a\phi(x) = \nu\Sigma_f\phi(x) + Q(x)$$

or

$$\frac{dJ_x(x)}{dx} + \Sigma_a\phi(x) = \nu\Sigma_f\phi(x) + Q(x). \tag{10.46}$$

Finally, with J_x expressed in terms of ϕ using Fick's law, Eq. (10.42), this equation becomes the diffusion equation

$$\boxed{-D\frac{d^2\phi(x)}{dx^2} + \Sigma_a\phi(x) = \nu\Sigma_f\phi(x) + Q(x).} \tag{10.47}$$

When the flux depends on three dimensions, Eq. (10.47) becomes

$$-D\nabla^2\phi(\mathbf{r}) + \Sigma_a\phi(\mathbf{r}) = \nu\Sigma_f\phi(\mathbf{r}) + Q(\mathbf{r}) \tag{10.48}$$

where $\nabla^2 \equiv \frac{\partial^2}{\partial x^2} + \frac{\partial^2}{\partial y^2} + \frac{\partial^2}{\partial z^2}$. This equation may be rearranged as

$$\nabla^2\phi(\mathbf{r}) + \frac{\nu\Sigma_f/\Sigma_a - 1}{D/\Sigma_a}\phi(\mathbf{r}) = -\frac{Q(\mathbf{r})}{D}. \tag{10.49}$$

For the one-speed model, $\epsilon = 1$ and $p = 1$ since there are no fast neutron effects, and $k_\infty = \eta f = \nu\Sigma_f/\Sigma_a$. Then, with the diffusion length L defined as $L^2 \equiv D/\Sigma_a$, Eq. (10.49) becomes

$$\nabla^2\phi(\mathbf{r}) + \frac{k_\infty - 1}{L^2}\phi(\mathbf{r}) = -\frac{Q(\mathbf{r})}{D}. \tag{10.50}$$

There are two main classes of problems that can be addressed with the neutron diffusion equation. First, a nuclear engineer may want to know the spatial distribution of the flux density $\phi(\mathbf{r})$ produced by a specified external neutron source $Q(\mathbf{r})$. This is known as the *fixed source problem*. For example, a PuBe (α, n) source may be stored in some medium, and the neutron field established in the medium by the source is needed (1) to estimate the rate of creation of radioactive material caused by neutron absorption and (2) to estimate the number of neutrons escaping from the storage container.

The second type of problem is the so-called *criticality problem*. Here the nuclear engineer seeks to determine the conditions under which a given mass of material is critical. In a critical configuration, a steady state neutron field can be maintained without the need for external source neutrons ($Q = 0$), i.e., the chain reaction is sustained by fission neutrons alone. Thus, the analyst seeks the conditions under which Eq. (10.48) has a non-zero steady-state solution when the external source Q is absent.

Examples of these two types of neutron diffusion problems are given below.

10.9.1   An Example Fixed-Source Problem

Consider an infinite homogeneous medium of non-fissionable material (i.e., $\Sigma_f = 0$) which contains an infinite plane source of neutrons with a strength such that each unit area of the source plane emits S_o neutrons isotropically in a unit time. Clearly, the neutron flux density $\phi(x)$ depends on only the distance x from the source plane, which we place in the y-z plane at $x = 0$. The diffusion equation for $\phi(x)$ to the right of the source is

$$-D\frac{d^2\phi(x)}{dx^2} + \Sigma_a\phi(x) = 0, \qquad x > 0 \tag{10.51}$$

or, upon dividing through by $-D$,

$$\frac{d^2\phi(x)}{dx^2} - \frac{1}{L^2}\phi(x) = 0, \qquad x > 0. \tag{10.52}$$

Because of the problem symmetry, the flux to the left of the source is the mirror image of the flux density to the right, i.e., $\phi(x) = \phi(-x)$.

The most general solution of Eq. (10.52), a homogeneous, second-order, ordinary, differential equation, is

$$\phi(x) = Ae^{-x/L} + Ce^{x/L} \tag{10.53}$$

where A and C are arbitrary constants whose values must be obtained by applying this general solution to boundary conditions for the problem. For example, as $x \to \infty$, the flux density must vanish since all neutrons will be absorbed before they can diffuse an infinite distance from the source. The only way in which the right-hand side of Eq. (10.53) can vanish at infinity is for the constant C to vanish. Thus, the form of the flux for this problem must have the form

$$\phi(x) = Ae^{-x/L}. \tag{10.54}$$

To determine A we need a condition on ϕ at the other boundary, i.e., next to the source plane. Consider, a unit area parallel to the source and just to the right of it. Through this unit area, many neutrons will pass in unit time, some traveling to the right and some to the left. However, because of symmetry, those neutrons diffusing from the right halfspace through the unit area into the left halfspace are exactly balanced by those neutrons diffusing in the other direction. Thus, the *net* flow $J_x(0^+)$ into the right halfspace is due only to neutrons emitted by the source plane into the right halfspace, namely $S_o/2$. With Fick's law, this condition is expressed mathematically as

$$J_x(o^+) = -D\left.\frac{d\phi(x)}{dx}\right|_{x=0^+} = \frac{S_o}{2}. \tag{10.55}$$

Substitution of $\phi(x) = Ae^{-x/L}$ into this condition and solving the resulting equation for A yields $A = (S_o L)/(2D)$. With this value for A, using the problem symmetry, the flux density is

$$\phi(x) = \frac{S_o L}{2D} e^{-|x|/L}. \tag{10.56}$$

From this explicit solution for the flux density, we note that the diffusion length L determines the gradient at which the neutron flux density decreases with distance from the source.

10.9.2 An Example Criticality Problem

Consider a bare homogeneous slab of multiplying material ($\Sigma_f > 0$) which is infinite in the y- and z-directions and has a thickness a in the x-direction. If there is no external source ($Q = 0$), the flux density is given by Eq. (10.50), namely, with the origin at the center of the slab,

$$\frac{d^2\phi(x)}{dx^2} + B_{\text{mat}}^2 \phi(\mathbf{r}) = 0 \qquad -\frac{a}{2} < x < \frac{a}{2}, \tag{10.57}$$

where the *material buckling* is $B_{\text{mat}}^2 \equiv (k_\infty - 1)/L^2$. For the slab to be critical, k_∞ must be greater than unity to allow for leakage and, hence, $B_{\text{mat}}^2 > 0$.

The solution of Eq. (10.57) also requires us to specify appropriate boundary conditions. Near the slab faces, the flux density can be expected to decrease very rapidly as $x \to \pm a/2$ because of the very strong probability that a neutron diffusing near the surface will leak from the slab, never to return. Hence, if the flux profile were extrapolated past the slab surface, it would appear to vanish at some *extrapolation distance* beyond the surfaces. It can be shown that this extrapolation distance is generally very small compared to the size of the reactor, and so, for this example, we assume that the flux density vanishes at the actual surfaces, i.e., we impose on Eq. (10.57) the boundary conditions $\phi(\pm a/2) \simeq 0$.

Generally Eq. (10.57), subject to these boundary conditions, has only the trivial $\phi = 0$ solution. This makes sense since, for a slab of arbitrary composition, it is very unlikely we would just happen to choose the exact thickness to make the slab critical. Mathematically, ignoring the boundary conditions for the moment, the most general solution of Eq. (10.57) is

$$\phi(x) = A\cos(B_{\text{mat}}x) + C\sin(B_{\text{mat}}x) \tag{10.58}$$

where A and C are arbitrary constants. Because the critical flux profile must be symmetric about the origin (letting $x \to -x$ gives the same slab), the constant C must be identically zero. Finally, we apply the boundary conditions to this general solution, namely,

$$\phi(\pm a/2) = A\cos(\pm B_{\text{mat}}a/2) = 0. \tag{10.59}$$

From this requirement, we see that either $A \equiv 0$ (which gives us the trivial $\phi = 0$ solution) or the argument $B_{\text{mat}}a/2$ must equal an odd multiple of $\pi/2$. Thus, to obtain a non-trivial solution, we require

$$B_{\text{mat}} = \frac{n\pi}{a} \qquad n = 1, 3, 5, \dots , \tag{10.60}$$

which yields the non-trivial solutions

$$\phi(x) = A\cos\left(\frac{n\pi}{a}x\right), \qquad n = 1, 3, 5, \ldots . \tag{10.61}$$

From this infinity of non-trivial solutions, we pick the $n = 1$ solution for two reasons. First, we want B_{mat}^2 (and hence k_∞) to be as small as possible. Second, and more important, all the non-trivial solutions for $n > 1$ become negative in some regions between $-a/2$ and $a/2$, a physical impossibility since a negative neutron density is meaningless. The only physically realistic non-trivial solution, which is *the* critical flux profile ϕ_c, is

$$\phi_c(x) = A\cos\left(\frac{\pi}{a}x\right), \tag{10.62}$$

and occurs whenever the following *criticality condition* is achieved:

$$B_{\text{mat}} = \left(\frac{n\pi}{a}\right). \tag{10.63}$$

Notice that the left-hand side of this criticality condition depends on only the material properties of the slab, while the right-hand side depends on only the geometry (a slab and its thickness). The right-hand side is often referred to as the *geometric buckling* B_g. Thus, the criticality condition, as seems reasonable, balances material and geometric properties of the core. For a bare core of any shape, the criticality condition can be shown to be

$$B_{\text{mat}} = B_g, \tag{10.64}$$

where B_g depends only on the shape and size of the assembly. The critical flux profile and geometric bucklings for some simple bare homogeneous cores are listed in Table 10.6.

10.9.3 More Detailed Neutron-Field Descriptions

For detailed analyses of reactor cores, it is necessary to account for the wide distribution of neutron energies. The one-speed diffusion equation, Eq. (10.47) or Eq. (10.48), can be easily generalized to give the energy-dependent flux density $\phi(\mathbf{r}, E)$. This energy-dependent diffusion equation is then most often approximated by a series of coupled, one-speed, diffusion equations, known as the *multigroup diffusion equations*, by averaging the energy-dependent equation over a series of contiguous energy intervals (or *groups*) that cover the entire energy range of neutrons encountered in a reactor. Numerical methods must usually be used to solve these energy-multigroup diffusion equations.

On occasion, the diffusion equation is inadequate, for example, when a detailed flux profile near a control rod is needed. The difficulty with the diffusion equation is that Fick's law is not exact, but rather, it is only an approximate description of how neutrons behave. In most instances, it is a good approximation, as verified by its wide-spread use for routine reactor analyses; however, for certain specialized calculations, it is quite inaccurate.

For such difficult problems, it is necessary to use the *neutron transport equation*, an "exact" equation which gives the neutron flux density, $\phi(\mathbf{r}, E, \mathbf{\Omega})$, as a function of position, energy, and directions of travel $\mathbf{\Omega}$ of the neutrons. The transport

Table 10.6. Critical flux profiles and bucklings for some homogeneous bare assemblies. Origin is at the assembly's center.

Geometry	Dimensions	Flux Profile	B_g^2
Slab	Thickness a	$A\cos\left(\dfrac{\pi x}{a}\right)$	$\left(\dfrac{\pi}{a}\right)^2$
Rect. Parallelepiped	$a \times b \times c$	$A\cos\left(\dfrac{\pi x}{a}\right)\cos\left(\dfrac{\pi y}{b}\right)\cos\left(\dfrac{\pi z}{c}\right)$	$\left(\dfrac{\pi}{a}\right)^2+\left(\dfrac{\pi}{b}\right)^2+\left(\dfrac{\pi}{c}\right)^2$
Sphere	Radius R	$A\dfrac{1}{r}\sin\left(\dfrac{\pi r}{R}\right)$	$\left(\dfrac{\pi}{R}\right)^2$
Infinite Cylinder	Radius R	$AJ_o\left(\dfrac{2.405r}{R}\right)$	$\left(\dfrac{2.405}{R}\right)^2$
Cylinder	Rad. R, Ht H	$AJ_o\left(\dfrac{2.405r}{R}\right)\cos\left(\dfrac{\pi z}{H}\right)$	$\left(\dfrac{2.405}{R}\right)^2+\left(\dfrac{\pi}{H}\right)^2$

equation is very difficult to solve for practical problems, and, consequently, computationally expensive methods, such as the discrete-ordinates method, must be used. An alternative approach to solving neutron transport problems is to use numerical simulation of the random interactions experienced by the neutrons. This alternative approach, called the *Monte Carlo method*, is also very computationally expensive for practical reactor analyses.

10.10   Addendum 2: Kinetic Model with Delayed Neutrons

To account properly for delayed neutrons, it is necessary to account for the time variation of the delayed neutron precursors as well as the number of neutrons $n(t)$ in the core. For simplicity, we assume a one delayed-neutron group approximation, i.e., all the delayed neutron precursors have the same half-life or decay constant λ. Let $C(t)$ be the total number of delayed neutron precursors in the core at time t.

In each neutron cycle, $k_{\text{eff}}n(t)$ new fission neutrons are eventually produced, $(1-\beta)k_{\text{eff}}n(t)$ as prompt neutrons at the end of the cycle and $\beta k_{\text{eff}}n(t)$ as delayed neutrons which appear upon the decay of the precursors that are produced during the cycle. Thus, in each cycle, $\beta k_{\text{eff}}n(t)$ precursors are created, and, since each cycle takes ℓ' seconds to complete, the number of precursors created per unit time is $\beta k_{\text{eff}}n(t)/\ell'$. The rate of disappearance of precursors by radioactive decay is $\lambda C(t)$. The net rate of increase in the number of precursors is the production rate minus the decay rate, i.e.,

$$\frac{dC(t)}{dt} = \frac{\beta k_{\text{eff}}}{\ell'}n(t) - \lambda C(t). \tag{10.65}$$

In each neutron life cycle, $(1-\beta)k_{\text{eff}}n(t)$ prompt neutrons are produced and $n(t)$ neutrons disappear. Thus, the net rate of increase by prompt neutrons is

$[k_{\mathit{eff}}(1 - \beta) - 1]n(t)/\ell'$. In addition, neutrons are produced by the decay of the delayed neutron precursors at the rate $\lambda C(t)$ and by non-fission sources at a rate $S(t)$. The rate of increase in the neutron population is thus the sum of these three neutron production mechanisms, i.e.,

$$\frac{dn(t)}{dt} = \frac{1}{\ell'}[k_{\mathit{eff}}(1 - \beta) - 1]n(t) + \lambda C(t) + S(t)$$

$$= \frac{k_{\mathit{eff}}}{\ell'}\left[\frac{k_{\mathit{eff}} - 1}{k_{\mathit{eff}}} - \beta\right]n(t) + \lambda C(t) + S(t). \tag{10.66}$$

Finally, we define the *effective neutron lifetime* $\ell \equiv \ell'/k_{\mathit{eff}}$ and the reactor *reactivity* ρ as

$$\rho \equiv \frac{k_{\mathit{eff}} - 1}{k_{\mathit{eff}}}. \tag{10.67}$$

The reactivity, like k_{eff}, is a measure of the critical state of the reactor. A positive reactivity indicates a supercritical system, and a negative value a subcritical system, while a critical system has zero reactivity. With these definitions, the kinetic equations Eqs. (10.65) and (10.66) become

$$\frac{dn(t)}{dt} = \frac{\rho - \beta}{\ell}n(t) + \lambda C(t) + S(t) \tag{10.68}$$

and

$$\frac{dC(t)}{dt} = \frac{\beta}{\ell}n(t) - \lambda C(t). \tag{10.69}$$

These coupled kinetic equations are readily generalized to the case of G groups of delayed neutron precursors. Each precursor group has its own kinetic equation, and the delayed neutron production term in the neutron equation is the sum of the decay rates of all the precursors. The result is

$$\boxed{\begin{aligned} \frac{dn(t)}{dt} &= \frac{\rho - \beta}{\ell}n(t) + \sum_{i=1}^{G}\lambda_i C_i(t) + S(t) \\[2mm] \frac{dC_i(t)}{dt} &= \frac{\beta_i}{\ell}n(t) - \lambda_i C_i(t), \qquad i = 1, \ldots, G \end{aligned}} \tag{10.70}$$

These kinetic equations are known as the *point reactor kinetic equations* and are widely used to describe the transient response of reactors.

For low-power research reactors, which maintain nearly constant composition over a long time, it is customary in the U.S. to measure reactivity in units of "dollars" defined as $k(\$) \equiv \rho/\beta$. Multiplication of Eqs. (10.70) by (ℓ/β) and defining

$\widehat{C}_i(t) \equiv (\ell/\beta)C_i(t)$ yields the equivalent form of the point reactor kinetic equations

$$
\boxed{
\begin{aligned}
\frac{\ell}{\beta}\frac{dn(t)}{dt} &= [k-1]n(t) + \sum_{i=1}^{G}\lambda_i\widehat{C}_i(t) + \frac{\ell}{\beta}S(t) \\[2mm]
\frac{d\widehat{C}_i(t)}{dt} &= a_i n(t) - \lambda_i\widehat{C}_i(t), \qquad i = 1, \ldots, G
\end{aligned}
}
\tag{10.71}
$$

where the relative precursor yield $a_i \equiv \beta_i/\beta$. Noted that $\sum_{i=1}^{G} a_i = 1$.

10.11 Addendum 3: Solution for a Step Reactivity Insertion

Consider a source-free reactor operating at a steady power level n_o. At $t = 0$, the reactivity is changed from its steady-state value of zero to k_o, i.e.,

$$
k(t) = \begin{cases} 0 & t < 0 \\ k_o & t \geq 0 \end{cases}.
\tag{10.72}
$$

For $t > 0$ and a one delayed-neutron group model ($a_1 = 1$), Eqs. (10.71) become

$$
\frac{\ell}{\beta}\frac{dn(t)}{dt} = [k_o - 1]n(t) + \lambda\widehat{C}(t)
\tag{10.73}
$$

$$
\frac{d\widehat{C}(t)}{dt} = n(t) - \lambda\widehat{C}(t).
\tag{10.74}
$$

Solutions to these two first-order, coupled, ordinary differential equations are of the form $n(t) = Ae^{\omega t}$ and $\widehat{C}(t) = Be^{\omega t}$. Substitution of these assumed forms into Eq. (10.74) yields

$$
B = \frac{A}{(\lambda + \omega)}.
\tag{10.75}
$$

Substitution of this result into Eq. (10.73) gives

$$
\frac{\ell}{\beta}\omega Ae^{\omega t} = (k_o - 1)Ae^{\omega t} + \frac{\lambda A}{(\lambda + \omega)}e^{\omega t}
\tag{10.76}
$$

or, upon simplification,

$$
A\left\{\frac{\ell}{\beta}\omega + \frac{\omega}{\lambda + \omega} - k_o\right\} = 0.
\tag{10.77}
$$

Thus for $A \neq 0$, we see that ω must be such that

$$
\omega\left[\frac{\ell}{\beta} + \frac{1}{\lambda + \omega}\right] = k_o.
\tag{10.78}
$$

This equation, which determines the permissible values of ω is known as the "in-hour" equation, since ω has units of inverse time, typically measured in "inverse hours."

For the one delayed-neutron-group model, the inhour equation is a quadratic equation in ω, i.e., there are two permissible values of ω. Simplification of Eq. (10.78) gives

$$\omega^2 + \omega \left[\lambda + \frac{\beta}{\ell}(1 - k_o) \right] - k_o \lambda \frac{\beta}{\ell} = 0. \tag{10.79}$$

The two solutions ω_1 and ω_2 of this quadratic equation are

$$\omega_{1,2} = \frac{1}{2} \left\{ -\left(\lambda + \frac{\beta}{\ell}(1 - k_o) \right) \pm \sqrt{\left(\lambda + \frac{\beta}{\ell}(1 - k_o) \right)^2 + 4k_o \lambda \frac{\beta}{\ell}} \right\}. \tag{10.80}$$

Thus, the most general solution of Eqs. (10.73) and (10.74) is the linear combination of the two possible solutions, namely,

$$n(t) = A_1 e^{\omega_1 t} + A_2 e^{\omega_2 t} \tag{10.81}$$

$$\widehat{C}(t) = \frac{A_1}{\lambda + \omega_1} e^{\omega_1 t} + \frac{A_2}{\lambda + \omega_2} e^{\omega_2 t}. \tag{10.82}$$

To determine the arbitrary constants A_1 and A_2, we use the initial conditions

$$n(0) = n_o \tag{10.83}$$

and, from Eq. (10.73) and Eq. (10.74),

$$\frac{\ell}{\beta} \frac{dn(t)}{dt} \bigg|_{0+} = (k_o - 1)n_o + n_o = k_o n_o. \tag{10.84}$$

Substitution of Eqs. (10.81) and (10.82) into these two initial conditions gives the following two algebraic equations for A_1 and A_2:

$$A_1 + A_2 = n_o \tag{10.85}$$

$$\omega_1 A_1 + \omega_2 A_2 = \frac{\beta}{\ell} k_o n_o. \tag{10.86}$$

The solution of these equations is

$$A_1 = \frac{\beta k_o / \ell - \omega_2}{\omega_1 - \omega_2} n_o \tag{10.87}$$

$$A_2 = \frac{\omega_1 - \beta k_o / \ell}{\omega_1 - \omega_2} n_o. \tag{10.88}$$

Substitution of A_1 and A_2 into Eq. (10.81) gives

$$n(t) = n_o \left\{ \frac{\beta k_o / \ell - \omega_2}{\omega_1 - \omega_2} e^{\omega_1 t} + \frac{\omega_1 - \beta k_o / \ell}{\omega_1 - \omega_2} e^{\omega_2 t} \right\}. \tag{10.89}$$

For the special case of "small" reactivities, i.e., $(1 - k_o)^2 >> 4(\ell/\beta)k_o\lambda$, the following approximations hold:

$$\omega_1 \simeq \frac{\lambda k_o}{1 - k_o} \qquad \text{and} \qquad \omega_2 \simeq -\frac{1 - k_o}{\ell/\beta} < 0 \tag{10.90}$$

$$A_1 \simeq \frac{n_o}{1 - k_o} \qquad \text{and} \qquad A_2 \simeq -\frac{n_o k_o}{1 - k_o}. \tag{10.91}$$

With this approximation, the transient variation of the neutron population is

$$
n(t) \simeq \frac{n_o}{1 - k_o} \left\{ \exp\left[\frac{\lambda k_o t}{1 - k_o} \right] - k_o \exp\left[-\frac{(1 - k_o)t}{\ell/\beta} \right] \right\}. \qquad (10.92)
$$

BIBLIOGRAPHY

ANL, *Reactor Physics Constants*, ANL-5800, 2nd ed., Argonne National Laboratory, Argonne, IL, 1963.

BELL G.I. AND S. GLASSTONE, *Nuclear Reactor Theory*, Van Nostrand Reinhold, New York, 1970.

DUDERSTADT, J.J. AND W.R. MARTIN, *Transport Theory*, Wiley, New York, 1979.

KEEPIN, G.R., *Physics of Nuclear Kinetics*, Addison-Wesley, Reading, MA, 1965.

LAMARSH, J.R., *Nuclear Reactor Theory*, Addison-Wesley, Reading, MA, 1966.

LAMARSH, J.R. AND A.J. BARATTA, *Introduction to Nuclear Engineering*, Prentice Hall, Upper Saddle River, NJ, 2001.

STACEY, W.M., *Nuclear Reactor Physics*, Wiley, New York, 2001.

PROBLEMS

1. In a liquid metal fast breeder reactor, no neutron moderation is desired and sodium is used as a coolant to minimize fission-neutron thermalization. How many scatters with sodium, on the average, would it take for 2-Mev neutrons to reach an average thermal energy of 0.025 eV? HINT: review Section 6.5.1.

2. Discuss the relative merits of water and graphite for use in a thermal reactor.

3. List five desirable properties of a moderator for a thermal reactor. Explain the importance of each property.

4. What is the thermal fission factor η for 5 atom-% enriched uranium?

5. What atom-% enrichment of uranium is needed to produce a thermal fission factor of $\eta = 1.85$?

6. Plot the thermal fission factor for uranium as a function of its atom-% enrichment in ^{235}U.

7. A soluble salt of fully enriched uranium is dissolved in water to make a solution containing 1.5×10^{-3} atoms of ^{235}U per molecule of water.

   (a) Explain why $\epsilon p \simeq 1$ for this solution.

   (b) What is k_∞ for this solution? Neglect any neutron absorption by other elements in the uranium salt.

(c) This solution is used to fill a bare spherical tank of radius R. Plot k_{eff} versus R and determine the radius of the tank needed to produce a critical reactor.

8. Consider a homogeneous, bare, spherical, source-free, critical, uranium-fueled reactor operating at a power P_o. Explain how and why the power increases, decreases, or remains unchanged as a result of each of the separate changes to the reactor.

   (a) The reactor is deformed into the shape of a football (ellipsoid).

   (b) A person stands next to the core.

   (c) The temperature of the core is raised.

   (d) A neutron source is brought close to the core.

   (e) An energetic electron beam impacts the core.

   (f) The reactor is run at high power for a long time.

   (g) The core is launched into outer space.

   (h) A sheet of cadmium is wrapped around the core.

   (i) The enrichment of the fuel is increased.

9. For a given amount of multiplying material with $k_\infty > 1$, what is the shape of a bare core with the smallest mass of this material? Explain.

10. If the uranium fuel enrichment in a reactor is increased, what is the effect on k_∞? Explain.

11. Consider a homogeneous mixture of fully enriched ^{235}U and graphite. Plot k_∞ versus N^{235}/N^C. What is the fuel-to-moderator ratio that yields the maximum value of k_∞?

12. What is the optimum fuel-to-moderator ratio of a homogeneous mixture of ^{235}U and (a) light water, (b) heavy water, (c) beryllium, and (d) graphite to produce a mixture with the maximum k_∞? Data: the thermal (0.0253 eV) absorption cross sections for water, heavy water, beryllium, and carbon are 0.664, 0.00133, 0.0092, and 0.0034 b per molecule or atom, respectively.

13. A spherical tank with a radius of 40 cm is filled with a homogeneous mixture of ^{235}U and light water. The mixture has a moderator-to-fuel ratio N^{H_2O}/N^{235} of 800. (a) What is k_∞ of the mixture? (b) What is k_{eff} for this bare core?

14. What should the radius of the tank in the previous problem be to produce a critical configuration? What is the critical mass of ^{235}U needed?

15. A control rod is dropped into a critical, source-free, uranium-fueled reactor and the asymptotic period of the resulting exponentially decreasing power is observed to be –200 s. (a) What is the value of k_{eff} of the reactor after the control rod drop? (b) What was the reactivity insertion in dollars?

16. What is the asymptotic period resulting from a reactivity insertion of (a) 0.08$ and (b) –0.08$?

17. Following a reactor scram, in which all the control rods are inserted into a power reactor, how long is it before the reactor power decreases to 0.0001 of the steady-state power prior to shutdown?

18. At time $t = 0$, a reactivity of 0.15$ is inserted into a critical reactor operating at 100 W. How long is it before the reactor power reaches 1 MW?

19. A reactivity insertion into an initially critical reactor operating at steady state causes the power to increase from 100 W to 10 kW in 6 minutes. What was the value of the reactivity insertion in $?

20. Explain how a decrease in the boiling rate inside a boiling water reactor affects the reactivity of the reactor.

21. Explain why it is reasonable that ^{135}Xe should have a very large cross section for neutron absorption.

22. Following a reactor shutdown ($\phi_o \to 0$), show from Eq. (10.33) that the time t_m for ^{135}Xe to reach a maximum is given by

$$t_m = -\frac{1}{\lambda_I - \lambda_X} \ln \left\{ r \left[1 + (1 - r) \frac{X(0)}{I(0)} \right] \right\},$$

where $X(0)$ and $I(0)$ are arbitary ^{135}Xe and ^{135}I concentrations at shutdown and $r \equiv \lambda_X / \lambda_I$. From this result show, that for an increase in ^{135}Xe following shutdown, the initial values must satisfy

$$\frac{X(0)}{I(0)} \leq \frac{\lambda_I}{\lambda_X} = 1.38.$$

23. In a fast reactor, does ^{135}Xe produce feedback? Why?

24. Many phenomena produce reactivity feedback. Within orders of magnitude, over what time interval would you expect the following reactivity feedback causes to take effect? (a) temperature increase in the fuel, (b) temperature increase in the moderator/coolant, (c) ^{135}Xe increase, (d) fuel burnup, (e) boiling in the core, and (f) increased coolant flow through the core.

Chapter 11

Nuclear Power

Nuclear reactors allow us to produce enormous amounts of thermal energy through fission chain reactions releasing no CO_2 or particulates and using small amounts of fuel compared to conventional fossil-fuel combustion reactions. Production of electricity and propulsion of ships are two major needs in our modern society for large thermal sources. Both needs, for which nuclear energy has served as an alternative to fossil fuels, are discussed in this chapter.

As these words are written, nuclear power plants of a third generation are in stages of planning, construction, and operation. The first generation of plants comprised those plants built in the 1950s and 1960s. Examples are the Magnox gas-cooled plants in England, the early CANDU heavy-water reactors in Canada, the Shippingport PWR (pressurized water reactor) and the first Dresden BWR (boiling water reactor) in the U.S. Few of these plants remain in operation. Second generation plants include the fleets of PWR and BWR plants built largely since 1970 and now in operation in the U.S., Russia, and Western Europe. The third generation of reactors are characterized by standardized designs with low probabilities and minimized consequences of accidents, and economies brought about by improved utilization of fuel, increased reliability, and longer operating lives. Reactors of a fourth generation are being planned for operation beyond year 2020. These plants operate at very high temperature, affording process heat for hydrogen production or other applications. There is some interest in modular plants with units designed for simplified operations and sized for off-grid electricity generation. A good example is the PBMR (pebble-bed modular reactor) helium-cooled design being developed in South Africa based in part on experience in Germany with the pebble-bed technology.

11.1 Nuclear Electric Power

During the last half of the twentieth century, many types of nuclear reactors have been designed and built to convert the thermal power produced from a fission chain reaction into electrical power. Such *power reactors* are today an important source of electrical energy in many countries with limited native resources of fossil fuels to use in conventional fossil-fired power plants. For example, France produces over 70% of its electrical energy from fission energy. Table 11.1 shows how various countries depend on nuclear power. Although many countries with large reserves of fossil fuels

340

available to fire conventional electrical power plants have suspended expansion of nuclear capacity, many other countries are continuing to build and plan for new nuclear power plants. The existing and future nuclear power plant capacities for various countries are also shown in Table 11.1.

11.1.1 Electricity from Thermal Energy

Most of the energy produced by fission reactions in a nuclear reactor is quickly converted to thermal energy. This heat can be converted to electrical energy in a variety of ways. By far the most common method used to produce electricity from a thermal energy source is to use the thermal energy to produce a hot pressurized gas which is then allowed to expand through a turbine causing it to turn. The rotating turbine shaft is common with the shaft of a generator which converts the energy of rotation into electrical energy. The most common gas or *working fluid* used to transfer the thermal energy to rotational energy of the turbine is steam. Such a *steam cycle* is illustrated in Fig. 11.1. In a conventional fossil-fueled power plant, the steam is generated in a boiler in which oil, natural gas, or, more usually, pulverized coal is burned. In a nuclear power plant, thermal energy produced in a reactor is used, either directly or indirectly, to boil water.

Although the steam turbines are the principal devices used to generate electricity, there is no reason why other hot gases cannot be used to turn a turbo-generator. Indeed, the hot combustion gases from burning natural gas are sometimes used directly in a gas turbine. Because combustion gases are much hotter than the steam used in conventional steam turbines, these direct fired gas turbines are much smaller (and hence less expensive) than steam turbines.

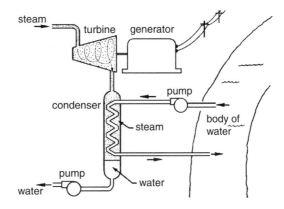

Figure 11.1. A source of steam is used to produce electricity.

Capital costs for direct-fired gas turbine units are less than those for nuclear plants, and gas turbines historically have been used as peaking units to supply electrical energy during periods of high electrical demand.

11.1.2 Conversion Efficiency

Turbo-generator systems are *heat engines* that convert thermal energy to electrical energy. From the laws of thermodynamics, the maximum conversion efficiency of any heat engine is the *Carnot efficiency*, namely,

$$\eta = \frac{T_{in} - T_{out}}{T_{in}}, \tag{11.1}$$

Table 11.1. Percent of 2004 electrical energy generated from nuclear power in different countries. Also shown is the nuclear electric power capacity existing, under construction, and planned.

Country or region	Percent electricity from nuclear in 2004	Operating at start 2006		Under construction		Planned		Total	
		GW(e)	units	GW(e)	units	GW(e)	units	GW(e)	units
Argentina	8.2	0.94	2	0.69	1			1.63	3
Armenia	39	0.38	1					0.38	1
Belgium	55	5.73	7					5.73	7
Brazil	3	1.90	2			1.25	1	3.15	3
Bulgaria	42	2.72	4			1.90	2	4.62	6
Canada	15	12.60	18			1.54	2	14.14	20
China	2.2	6.59	9	1.90	2	8.20	9	16.69	20
Czech Republic	31	3.47	6					3.47	6
Finland	27	2.68	4	1.60	1			4.28	5
France	78	63.47	59					63.47	59
Germany	32	20.30	17					20.30	17
Hungary	34	1.76	4					1.76	4
India	2.8	2.99	15	3.64	8			6.63	23
Iran				0.95	1	1.90	2	2.85	3
Japan	29	47.70	55	0.87	1	14.78	12	63.35	68
Korea DPR (N)				0.95	1	0.95	1	1.90	2
Korea RO (S)	38	16.84	20			9.20	8	26.04	28
Lithuania	72	1.19	1					1.19	1
Mexico	5.2	1.31	2					1.31	2
Netherlands	3.8	0.45	1					0.45	1
Pakistan	2.4	0.43	2	0.30	1			0.73	3
Romania	10	0.66	1	0.66	1			1.31	2
Russia	16	21.74	31	3.60	4	0.93	1	26.27	36
Slovakia	55	2.47	6					2.47	6
Slovenia	38	0.68	1					0.68	1
South Africa	6.6	1.84	2			0.17	1	2.01	3
Spain	23	7.58	9					7.58	9
Sweden	52	8.94	10					8.94	10
Switzerland	40	3.22	5					3.22	5
Taiwan	21	4.88	6	2.60	2			7.48	8
Ukraine	51	13.17	15			1.90	2	15.07	17
United Kingdom	19	11.85	23					11.85	23
United States	20	97.92	103	1.07	1			98.99	104
WORLD	16	368.39	441	18.82	24	42.71	41	0.43	506

Source: IAEA PRIS and World Nuclear Authority databases.

where T_{in} is the absolute temperature (K) of the gas entering the turbine and T_{out} is the absolute temperature of the gases leaving the turbine. Clearly, the higher the entering temperature and/or the lower the outlet temperature, more of the thermal energy is converted to electrical energy. The inlet temperature is limited by the water/steam pressure rating of the boiler or reactor vessel in a steam cycle, or by the temperature limitation of the turbine blades in a direct-fired gas turbine. The outlet temperature is usually limited by the ambient temperature of the cooling water used in the condenser of a steam cycle, while in a direct-fired gas turbine, the exit temperature is determined by the exit pressure. An important measure of a power plant's performance is the conversion efficiency, i.e., the ratio of electrical power to thermal power, MW(e)/MW(t). In modern nuclear power plants, conversion efficiencies of about 40% can be achieved, while fossil-fired units can achieve only slightly greater efficiencies. However, many older power plants have efficiencies in the range of 30–35%.

11.1.3   Some Typical Power Reactors

Many different designs for power reactors have been proposed and many different prototypes built. Most countries that have developed nuclear power started with graphite or heavy-water moderated systems, since only these moderators allow criticality with natural uranium containing 0.711 wt% ^{235}U. However, most power reactors now use slightly enriched uranium, typically 3% enriched. With such enrichments, other moderators, notably light water, can be used. The *steam supply system* for some important types of power reactors are illustrated in Figs. 11.2–11.7.

This section introduces different types of power reactors that have been used to generate electricity. Later in the chapter, the two most widely used types are discussed in detail.

Pressurized Water Reactor

Pressurized water reactors (PWR), Fig. 11.2, the most widely used type of power reactors, employ two water loops. The water in the primary loop is pumped through the reactor to remove the thermal energy produced by the core. The primary water is held at sufficiently high pressure to prevent the water from boiling. This hot pressurized water is then passed through a steam generator where the secondary-loop water is converted to high temperature and high pressure steam that turns the turbo-generator unit. The use of a two-loop system ensures that any radioactivity produced in the primary coolant does not pass through the turbine.

Boiling Water Reactors

In a boiling water reactor (BWR), cooling water is allowed to boil while passing through the core. The steam then passes directly to the turbine. The low pressure steam leaving the turbine is then condensed and pumped back to the reactor. By having a single loop, the need for steam-generators and other expensive equipment in a PWR is avoided.

Cheap + cheaful .

Pressu vessl

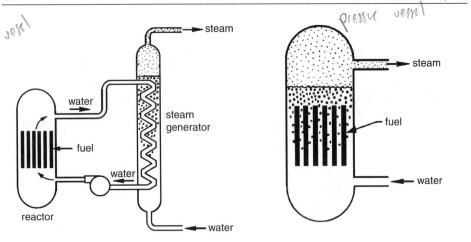

Figure 11.2. Pressurized water reactor (PWR).

Pressu vessl

Figure 11.3. Boiling water reactor (BWR).

Cold ux cw + drectly ux steam!

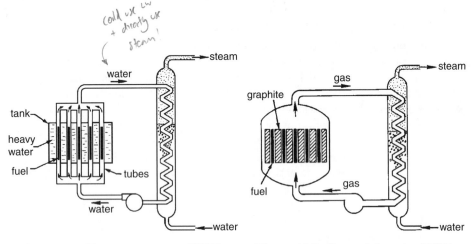

Figure 11.4. Heavy water reactor (HWR).

Figure 11.5. Gas cooled reactor (GCR).

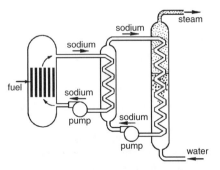

Figure 11.6. Liquid metal fast breeder reactor (LMFBR).

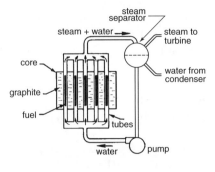

Figure 11.7. Graphite-moderated water-cooled reactor (RMBK).

Heavy Water Reactors

In one design of a heavy water reactor (HWR), Fig. 11.4, pressurized heavy water in the primary loop is used to cool the core. The fuel is contained in pressure tubes through which the heavy water coolant passes. These pressure tubes pass through the moderator vessel, which is also filled with heavy water. The heavy water in the primary loop then passes through steam generators to boil the secondary-loop light water. By using pressure tubes for the coolant, the need for an expensive high-pressure reactor vessel is avoided.

Another variation of the HWR uses light water in the primary loop, allowing it to boil as the coolant flows through the pressurized fuel tubes. In this design, no secondary water loops or steam generators are needed. The steam produced in the pressure tubes, after eliminating entrained moisture, flows directly to the turbo-generator.

Gas Cooled Reactor

In a gas cooled reactor (GCR), Fig. 11.5, carbon dioxide or helium gas is used as the core coolant by pumping it through channels in the solid graphite moderator. The fuel rods are placed in these gas cooling channels. The use of graphite, which remains solid up to very high temperatures, eliminates the need for an expensive pressure vessel around the core. The hot exit gas then passes through steam generators.

In another design known as the high-temperature gas cooled reactor (HTGR), the fuel is packed in many fuel channels in graphite prisms. Helium coolant is pumped through other channels bored through the graphite prisms. The hot exit helium then goes to a steam generator.

Liquid Metal Fast Breeder Reactors

In a fast reactor, the chain reaction is maintained by fast neutrons. Consequently, moderator materials cannot be used in the core. To avoid materials of low atomic mass, the core coolant is a liquid metal such as sodium or a mixture of potassium and sodium. Liquid metals have excellent heat transfer characteristics and do not require pressurization to avoid boiling. However, sodium becomes radioactive when it absorbs neutrons and also reacts chemically with water. To keep radioactive sodium from possibly interacting with the water/steam loop, an intermediate loop of non-radioactive sodium is used to transfer the thermal energy from the primary sodium loop to the water/steam loop (see Fig. 11.6). The great advantage of such fast liquid-metal power reactors is that it is possible to create a breeder reactor, i.e., one in which more fissile fuel is produced than is consumed by the chain reaction. In such a breeder reactor, ^{238}U is converted to fissile ^{239}Pu or ^{232}Th into fissile ^{233}U (see Section 6.5.3). Although fissile fuel breeding also occurs in water and graphite moderated reactors, the ratio of new fuel to consumed fuel is less than one (typically 0.6 to 0.8).

Pressure-Tube Graphite Reactors

A once widely used Russian designed power reactor is the (RBMK) *reactory bolshoi moshchnosti kanalnye* (translated: high-powered pressure-tube reactor). In this reactor (see Fig. 11.7), fuel is placed in fuel channels in graphite blocks that are stacked to form the core. Vertical pressure tubes are also placed through the graphite core and light water coolant is pumped through these tubes and into an overhead steam drum where the two phases are separated and the steam passes directly to the turbine.

11.1.4   Coolant Limitations

The thermal properties of a power reactor coolant greatly affect the reactor design. By far the most widely used coolant is water. It is inexpensive and engineers have a wealth of experience in using it as a working fluid in conventional fossil-fueled power plants. The great disadvantage of water as a coolant is that it must be pressurized to prevent boiling at high temperatures. If water is below the boiling point, it is called *subcooled*. Water is *saturated* when vapor and liquid coexist at the boiling point, and it is *superheated* when the vapor temperature is above the boiling temperature. Above the *critical temperature*, the liquid and gas phases are indistinguishable, and no amount of pressure produces phase transformation.

To maintain criticality in a water moderated core, the water must remain in liquid form. Moreover, steam is a much poorer coolant than liquid water. Thus, for water to be used in a reactor, it must be pressurized to prevent significant steam formation. For water, the critical temperature is 375 °C, above which liquid water cannot exist. Thus, in water moderated and cooled cores, temperatures must be below this critical temperature. Typically, coolant temperatures are limited to about 340 °C. This high temperature limit for reactor produced steam together with normal ambient environmental temperatures limit the thermal efficiency for such plants to about 34%.

Because steam produced by nuclear steam supply systems is saturated or very slightly superheated, expensive moisture separators (devices to remove liquid droplets) and special turbines that can operate with "wet steam" must be used. These turbines are larger and more expensive than those used in power plants that can produce superheated steam.

When gases or liquid metals are used as reactor coolants, liquid-gas phase transitions are no longer of concern. These coolants can reach much higher temperatures than in water systems, and can produce "dry" steam much above the critical temperature. Dry superheated steam at temperatures around 540 °C allows smaller, less expensive, turbines to be used and permits thermal efficiencies up to 40%.

11.1.5   Industrial Infrastructure

Recent years have seen significant changes in ownership of the industrial base for construction of nuclear power plants. During the decades of rapid expansion of nuclear power, the 1960s and 1970s, there were three U.S. firms marketing PWR reactor systems — Westinghouse, Combustion Engineering, and Babcock & Wilcox — and one Russian vendor producing the VVER system. Framatome, Siemens, and Mitsubishi were licensees for Westinghouse technology.

After the Three-Mile Island accident, Babcock & Wilcox no longer marketed nuclear steam supply systems and their fuel and nuclear services activities were acquired by Framatome. Combustion Engineering merged with ASEA Brown Boveri Ltd. in 1990 to form ABB Combustion Engineering. Framatome and Siemens, independent of Westinghouse, merged in 2000, forming Framatome ANP. That company is owned by the French AREVA group (66%) and the German Company Siemens (34%) and markets the EPR (European pressurized water reactor). In the United States, AREVA and Constellation Energy have formed the joint enterprise Unistar Nuclear in order to market the EPR as the U.S. Evolutionary Power Reactor (U.S. EPR).

Westinghouse left its affiliation with the Columbia Broadcasting System in 1998 when it was purchased by the government corporation British Nuclear Fuels Limited (BNFL) for $1.2 billion plus assumption of debt. Westinghouse BNFL then acquired ABB Combustion Engineering, and that acquisition brought to Westinghouse the System 80+ design which had been certified by the NRC in May, 1997, and which forms the basis for the APR 1400 design developed for the Korean nuclear power industry. Privatization of the nuclear industry in Britain led in 2006 to the purchase of Westinghouse for $5 billion by the Japanese Company Toshiba. Marketing of the Westinghouse AP1000 reactor in China was a major consideration in the transaction.

The U.S. company General Electric has been the world suppler of the boiling water reactor system, with Siemens, Hitachi, ASEA Atom, and Toshiba as licensees. The current generation of BWR designs is the ABWR, the advanced BWR. Four units are in Japan and two in Taiwan. The next generation plant is likely to be the ESBWR, the Economic Simplified Boiling Water Reactor.

The world supplier of heavy water reactor technology is the Canadian company AECL, Atomic Energy of Canada, Ltd. Plants have been supplied to Canada, of course, as well as Argentina, Romania, Korea, and China. A 1975 design continues to be built in India.

11.1.6   Evolution of Nuclear Power Reactors

Beginning in the 1950s, the first generation of nuclear power reactors were designed, built, and operated to demonstrate the feasibility of civilian nuclear electric power. These first pioneering plants were followed by the second generation, often called Gen II, nuclear power plants. Most of the world's current nuclear power is provided by these Gen II reactors which were built, for the most part, in the 1970s and 1980s. Today, many of these Gen II reactors are extending their operating licenses so they will continue to provide nuclear power for another 20 years or more.

Today, reactor vendors have designed and will soon begin construction of Gen III power reactors. These reactors are designed to be simpler in design, more economical to construct and operate, and have greatly enhanced safety systems by incorporating many passive safety features that do not rely on the proper operation of active safey systems.

At the present time, the U.S. is also embarking on inital designs of Gen IV power reactors that will explore various reactor designs with the goal of producing electricity more efficiently. These new power reactors are being developed in consort with advanced nuclear fuel cycles and with a potential for the production of hydrogen as an alternative national energy source.

These various generations of nuclear power reactors are discussed in the following sections.

11.2   Generation II Pressurized Water Reactors

Westinghouse Electric Company developed the first commercial pressurized water nuclear reactor (PWR) using technology developed for the U.S. nuclear submarine program. Many designs of various power capacities were developed by Westinghouse and built in several countries. The first, the Shippingport 60-MW(e) PWR, began operation in 1957 and operated until 1982. Following this first demonstration of commercial nuclear power, many other Gen II PWR plants, ranging in capacity from 150 MW(e) to almost 1500 MW(e), were built and operated in the U.S. and other countries. The characteristics of a typical 1000 MW(e) PWR power plant are given in Table 11.2.

11.2.1   The Steam Cycle of a PWR

The two-loop water/steam cycle of a PWR is shown in Fig. 11.8. In the primary loop, liquid water at pressures of about 2250 psi (15.5 MPa) is circulated through the core to remove the fission energy. The water leaving the core typically has a temperature of about 340 °C and the flow is regulated by the reactor coolant pumps. There are two to four separate primary coolant loops in PWRs, each with its own steam generator and recirculation pump. To avoid overpressures and pressure surges in the primary water, a *pressurizer*, which contains both liquid and saturated vapor, is included in the primary loop.

The primary water is passed through 2 to 4 *steam generators*, one for each primary loop, in which some of the thermal energy in the primary loop is transferred to the water in the secondary loop. The secondary water entering the steam generators is converted to saturated steam at about 290 °C (550 °F) and 1000 psi (7.2 MPa). This steam then expands through the turbine and, after leaving the turbine, is condensed back to the liquid phase in the *condenser*. The liquid condensate then passes through a series of 5 to 8 *feedwater heaters*, which use steam extracted from various stages of the turbine to heat the condensate before it cycles back to the steam generators. With this dual cycle, thermal energy efficiencies of about 34% can be achieved.

11.2.2   Major Components of a PWR

A nuclear power plant is a very complex system. Literally thousands of valves and pumps, miles of tubing and electrical wiring, and many tons of rebar and structural steel are needed. However, a few major components are of paramount importance. These include the pressurizer, steam generators, main recirculation pumps, reactor pressure vessel, turbo-generator, reheater, condenser, feedwater heaters, and the containment structure. Some of these items, unique to a PWR, are discussed below. The primary components of a 4-loop PWR are shown in Fig. 11.9.

The Pressure Vessel

A typical PWR pressure vessel is shown in Fig. 11.10. It is about 13 meters tall with a diameter of about 4 to 6 m. The vessel is built from low-alloy carbon steel and has a wall thickness of about 23 cm, which includes a 3-mm stainless steel clad on the inner surface. Such a thick wall is necessary to withstand the high operating pressure of about 2300 psi (158 bar). The primary coolant enters the vessel through two or more inlet nozzles, flows downward between the vessel and core barrel, flows

Table 11.2. Parameters for a typical 1000 MW(e) PWR sold in the early 1970s.

POWER		**REACTOR PRESSURE VESSEL**	
thermal output	3800 MW	inside diameter	4.4 m
electrical output	1300 MW(e)	total height	13.6 m
efficiency	0.34	wall thickness	22.0 cm
CORE		**FUEL**	
length	4.17 m	cylindrical fuel pellets	UO_2
diameter	3.37 m	pellet diameter	8.19 mm
specific power	33 kW/kg(U)	rod outer diameter	9.5 mm
power density	102 kW/L	zircaloy clad thickness	0.57 mm
av. linear heat rate	17.5 kW/m	rod lattice pitch	12.6 mm
rod surface heat flux		rods/assembly (17×17)	264
average	0.584 MW/m^2	assembly width	21.4 cm
maximum	1.46 MW/m^2	fuel assemblies in core	193
		fuel loading	115×10^3 kg
REACTOR COOLANT SYSTEM		initial enrichment %^{235}U	1.5/2.4/2.95
operating pressure	15.5 MPa	equil. enrichment % ^{235}U	3.2
	(2250 psia)	discharge fuel burnup	33 GWd/tU
inlet temperature	292 °C		
outlet temperature	329 °C	**REACTIVITY CONTROL**	
water flow to vessel	65.9×10^6 kg/h	no. control rod assemblies	68
		shape	rod cluster
STEAM GENERATOR (SG)		absorber rods per assembly	24
number	4	neutron absorber	Ag-In-Cd
outlet steam pressure	1000 psia		and/or B_4C
outlet steam temp.	284 °C	soluble poison shim	boric acid
steam flow at outlet	1.91×10^6 kg/h		H_3BO_3

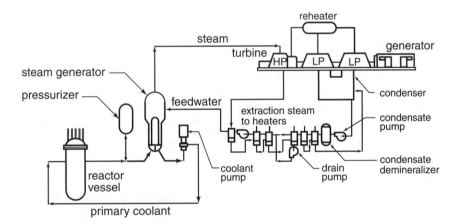

Figure 11.8. The steam cycle of a pressurized water reactor.

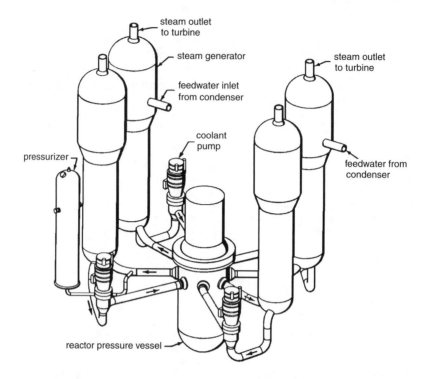

Figure 11.9. The steam cycle of a pressurized water reactor. [Westinghouse Electric Corp.]

upward through the reactor core removing the heat from the fuel pins, and then leaves the vessel through outlet nozzles. The fabrication and transportation of the roughly 500 tonne pressure vessel is a daunting task. Figure 11.11 shows components being fabricated. Notice the size of the worker standing beside the pressure vessel.

Recirculation Pumps

Flow through the reactor core is controlled by the recirculation pump in each primary loop. These large pumps are vertical single-stage centrifugal type pumps designed to operate for the 30–40 year lifetime of the plant with minimal maintenance. All parts that contact the water are made from stainless steel. The pump is driven by a large 7000 HP, vertical, squirrel cage, induction motor. A flywheel is incorporated to increase the rotational inertia, thereby prolonging pump coastdown and assuring a more gradual loss of main coolant flow should power to the pump motor be lost.

The Pressurizer

The primary system of a PWR is very nearly a constant-volume system. As the temperature of the primary water increases or decreases, the water expands or contracts. However, water is almost incompressible and a small temperature change would lead to very large pressure changes if the primary loop were totally filled with liquid water. To prevent such dangerous pressure surges, one primary loop of a PWR contains a pressure-regulating surge tank called a *pressurizer* (see Fig. 11.10).

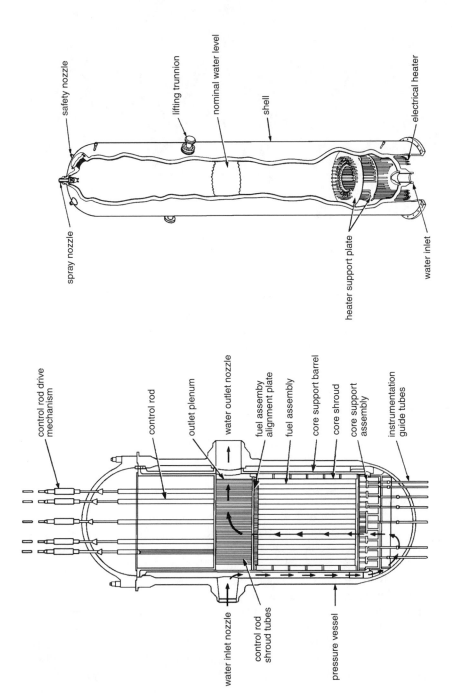

Figure 11.10. Generation II PWR Pressure vessel (left) and pressurizer (right) [Westinghouse Electric Corporation].

Figure 11.11. A PWR pressure vessel during manufacture. Source: [CE 1974].

Figure 11.12. A PWR steam generator during fabrication. Source: [CE 1974].

The pressurizer is a large cylindrical tank with steam in the upper portion and water in the lower as shown in Fig. 11.10. The steam, being compressible, can absorb any sudden pressure surges. The pressurizer also is used to maintain the proper pressure in the primary system. If the primary water temperature should decrease (say from increased steam demand by the turbine) water would flow out of the pressurizer causing the steam pressure in it to drop. This drop in pressure in turn causes some of the water to flash to steam, thereby mitigating the pressure drop. At the same time, the pressure drop actuates electrical heaters in the base of the pressurizer to restore the system pressure. Likewise, an increase in water temperature causes the water to expand and to flow into the pressurizer and to increase the steam pressure, which, in turn, actuates valves to inject spray water into the top of the pressurizer. This cooling spray condenses some of the steam thereby reducing the system pressure to normal.

Steam Generators

Steam generators are very large devices in which thermal energy of the primary water is transferred to the secondary water to produce steam for the turbo-generator. Several different designs have been used, although the most common type of steam generator is that evolved from a Westinghouse design. These steam generators, which are larger than the pressure vessel (compare Figs. 11.11 and 11.12), are nearly 21-m tall and 5.5 m in diameter at the upper end. The internal structure of such a steam generator is shown in Fig. 11.13. The hot high-pressure water of the primary loop (2250 psia, 345 °C, 262 tonnes/minute) is passed through a bank of U-shaped pressure tubes where some of the thermal energy of the primary water is transferred to the secondary water at 1000 psi on the outside of the pressure tubes and allowed to boil. The saturated steam above the tubes contains many small water droplets and the bulbous top portion of the steam generator contains cyclone separators to remove this entrained water and to allow only "dry" hot steam (0.25% moisture, 290 °C) to reach the high-pressure turbine. The primary water leaves the steam generator at a temperature of about 325 °C and reenters the reactor.

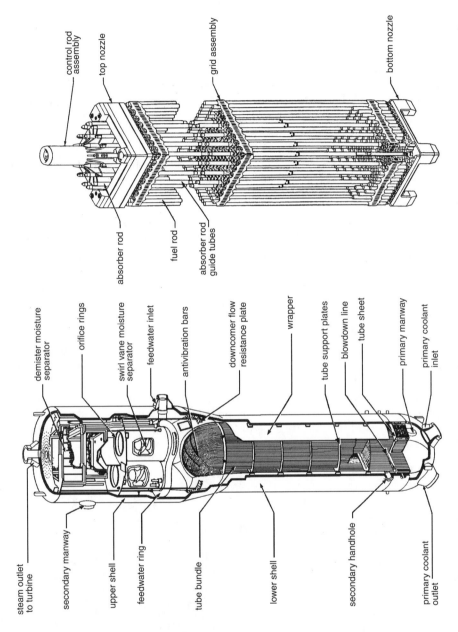

Figure 11.13. Generation II PWR steam generator (left) and fuel assembly (right) [Westinghouse Electric Corporation].

Nuclear Fuel

The uranium fuel used in a PWR is contained in many thousands of thin long fuel rods or *fuel pins*. Slightly enriched (3.3%) UO_2 pellets about 9 mm in diameter are stacked inside a Zircaloy tube 3.8-m long with a wall thickness of about 0.64 mm. The small diameter is needed to allow fission thermal energy produced in the UO_2 pellets to transfer rapidly to the water surrounding the fuel pins.

The fuel pins are assembled into *fuel assemblies* each containing an array of typically 17 × 17 pin locations. However, in many of these assemblies, 24 locations are occupied by guide tubes in which control-rod "fingers," held at the top by a "spider," move up and down in the assembly to provide coarse reactivity control. A typical fuel assembly is shown in Fig. 11.13. Fuel assemblies with 15×15 and 16×16 arrays with fewer control rod locations have also been used.

Some 200 to 300 of these fuel assemblies are loaded vertically in a cylindrical configuration to form the reactor core. The assemblies typically spend three or more years in the core before they are replaced by new assemblies.

Reactivity Control

Short term or emergency reactivity control is provided by the 24 control rod fingers in many of the assemblies. These control fingers usually contain B_4C or, more recently, a mixture of silver (80%), indium (15%), and cadmium (5%) to produce slightly weaker absorbers. Generally, 4–9 adjacent control-rod spiders (which connect all the control rod fingers in an assembly) are grouped together and moved together as a single *control-rod bank*. The various control rod banks then provide coarse reactivity control.

Intermediate to long term reactivity control is provided by varying the concentration of boric acid in the primary water. Such a soluble neutron absorber is called a *chemical shim*. By varying the boron concentration in the primary water, excessive movement of control rods can be avoided.

For long-term reactivity control, *burnable poisons* are placed in some of the lattice positions of the fuel assemblies. These *shim rods*, from 9 to 20 per assembly, are stainless steel clad boro-silicate glass or Zircaloy clad diluted boron in aluminum oxide pellets.

The Containment Building

Of paramount importance in the safe use of nuclear power is the isolation of the radioactive fission products from the biosphere. To prevent the leakage of fission products into the environment from a nuclear power plant, three principal isolation barriers are used in every nuclear power plant. First the cladding of the fuel pins prevent almost all of the fission products from leaking into the primary coolant. However, with many thousands of fuel rods in a reactor,

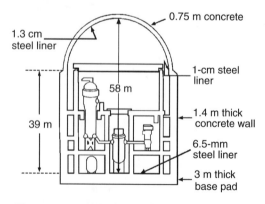

Figure 11.14. PWR containment building.

a few have small pin-holes through which some radioactive fission products escape. Elaborate clean-up loops are used to continuously purify the primary coolant and collect the fission products that leak from the fuel rods.

A second level of fission product confinement is provided by the pressure vessel and the isolated primary loop. Finally, should there be a leak in the primary system, the reactor and all the components through which the primary water flows are enclosed in a *containment building* designed to withstand tornados and other natural phenomena, as well as substantial overpressures generated from within the containment from accidental depressurization of the primary coolant. A typical confinement structure is shown in Fig. 11.14.

11.3 Generation II Boiling Water Reactors

The boiling water reactor (BWR) was developed by General Electric Company (GE) which, since its first 200 MW(e) Dresden unit in 1960, has built units as large as 1250 MW(e) in both the USA and in many other countries. Some operating parameters of a large BWR, typical of those now in operation, are given in Table 11.3.

11.3.1 The Steam Cycle of a BWR

A BWR uses a single direct-cycle steam/water loop as shown in Fig. 11.15. Its flow is regulated by the recirculation jet pumps, and feedwater entering the reactor pressure vessel is allowed to boil as it passes through the core. After passing through a complex moisture separation system in the top of the pressure vessel, saturated steam at about 290 °C and 6.9 MPa (1000 psi) leaves the reactor pressure vessel and enters the high-pressure (HP) turbine. The exit steam from the HP turbine then is reheated and moisture removed before entering low-pressure (LP) turbines from which it enters a condenser, is liquefied, and pumped through a series of feedwater heaters back into the reactor vessel. Such a conventional regenerative cycle typically has a thermal efficiency of about 34%.

11.3.2 Major Components of a BWR

The basic layout of a BWR is considerably simpler than that of a PWR. By producing steam in the reactor vessel, a single water/steam loop can be used, eliminating the steam generators and pressurizer of the PWR. However, radioactivity produced in the water as it passes through the core (notably ^{16}N with a 7-s half life) passes through the turbine, condenser, and feedwater heaters. By contrast, the same radioactivity in a PWR is confined to the primary loop, so that non-radioactive steam/water is used in the secondary loop. Thus, in a BWR plant, considerably greater attention to radiation shielding is needed. Because the pressure-vessel containment in a BWR serves also as a steam generator, the internals of a BWR vessel are somewhat more complex.

The BWR Pressure Vessel

A typical BWR pressure vessel and its internal structures are shown in Fig. 11.16. The 3.6-m high core occupies a small fraction of the 22-m high pressure vessel. The pressure vessel is about 6.4 m in diameter with a 15-cm, stainless-steel clad wall of

Table 11.3. Parameters for a typical 1000 MW(e) BWR sold in the early 1970s.

POWER		**REACTOR PRESSURE VESSEL**	
thermal output	3830 MW	inside diameter	6.4 m
electrical output	1330 MW(e)	total height	22.1 m
efficiency	0.34	wall thickness	15 cm
CORE		**FUEL**	
length	3.76 m	cylindrical fuel pellets	UO_2
diameter	4.8 m	pellet diameter	10.57 mm
specific power	25.9 kW/kg(U)	rod outer diameter	12.52 mm
power density	56 kW/L	zircaloy clad thickness	0.864 mm
av. linear heat rate	20.7 kW/m	rod lattice pitch	16.3 mm
rod surface heat flux		rods/assembly (8×8)	62
average	0.51 MW/m^2	assembly width	13.4 cm
maximum	1.12 MW/m^2	assembly height	4.48 m
		fuel assemblies in core	760
REACTOR COOLANT SYSTEM		fuel loading	168×10^3 kg
operating pressure	7.17 MPa	av. initial enrichment %^{235}U	2.6%
	(1040 psia)	equil. enrichment % ^{235}U	1.9%
feedwater temperature	216 °C	discharge fuel burnup	27.5 GWd/tU
outlet steam temperature	290 °C		
outlet steam flow rate	7.5×10^6 kg/h	**REACTIVITY CONTROL**	
core flow rate	51×10^6 kg/h	no. control elements	193
core void fraction (av.)	0.37	shape	cruciform
core void fraction (max.)	0.75	overall length	4.42 m
no. in-core jet pumps	24	length of poison section	3.66 m
no. coolant pumps/loops	2	neutron absorber	boron carbide
		burnable poison in fuel	gadolinium

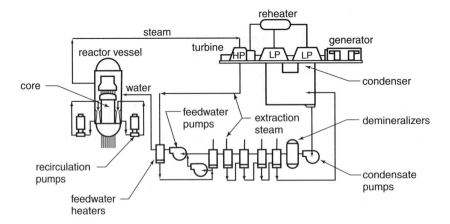

Figure 11.15. The steam cycle of a boiling water reactor.

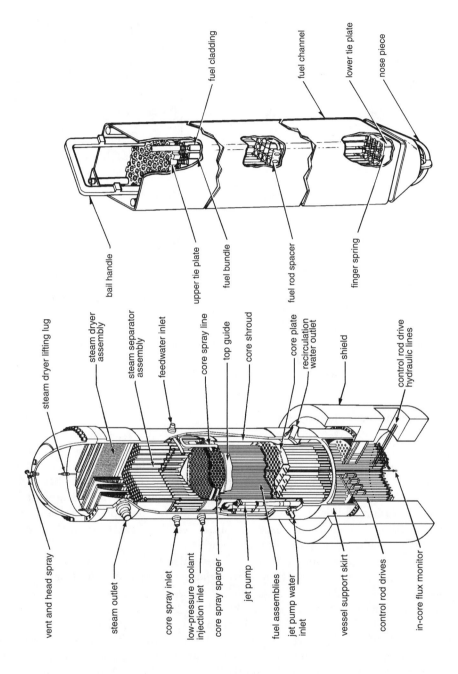

Figure 11.16. Generation II BWR. Pressure vessel (left) and fuel assembly (right) [General Electric Corporation].

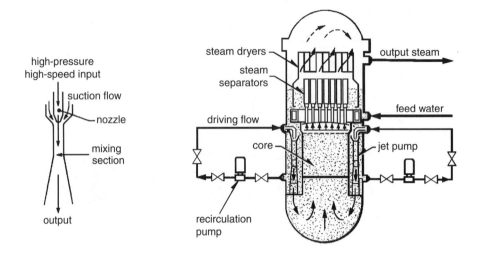

Figure 11.17. BWR jet pump.   **Figure 11.18.** Recirculation flow in a BWR.

carbon steel. Above the core are moisture separators and steam dryers designed to remove almost all of the entrained liquid from the steam before it leaves the top of the pressure vessel. Because of the steam conditioning systems in the top half of the vessel, the reactor control rods must be inserted through the bottom of the vessel. Hydraulic force rather than gravity must be relied upon to ensure that the control rods are fully inserted into the core if electrical power to the plant is lost.

Jet Pumps and Recirculation Flow

A unique feature of BWRs is the recirculation flow control provided by the jet pumps located around the periphery of the core. By varying the flow of water through the core, the amount of steam, or void fraction of the water, in the core can be controlled. This allows fine control of the reactivity of the core, since increasing steam production decreases the amount of liquid water and, thereby, the amount of neutron moderation and hence the reactivity. During normal operation, the water flow through the core is normally used to control reactivity rather than the coarse reactivity control provided by the 100–150 control rods.

The jet pump recirculation system used to control the water flow through the core is shown in Fig. 11.17. The recirculation pumps control the injection of high pressure and high-velocity water to the venturi nozzles of the jet pumps located around the periphery of the core (up to 21). This forced, high-velocity, injection water flow (see Fig. 11.17) creates a suction flow of vessel water downward between the vessel wall and the core shroud and then upward through the core. This reactivity controlling water flow through the recirculation loop is shown in Fig. 11.18. Up to about 30% of the feedwater to the BWR vessel is diverted from the vessel to the two recirculation loops used to operate the jet pumps around the periphery of the vessel.

BWR Fuel

The thousands of fuel pins, composed of enriched UO_2 pellets in Zircaloy tubes, are much like those in a PWR. However, they are arranged in square subassembly arrays of 8×8 to 10×10 pins (see Fig. 11.16). Four subassemblies, each contained in a Zircaloy shroud, make up a *fuel module* (see Fig. 11.19). Subassemblies are individually orificed to control water/steam flow and assure a uniform elevation at which boiling commences and a uniform steam quality as coolant leaves the core. In contrast, PWR fuel assemblies are open. Cross flow is minimal because of the lack of driving force, and, absent boiling, the mass flow rate throughout the core cross section is very uniform.

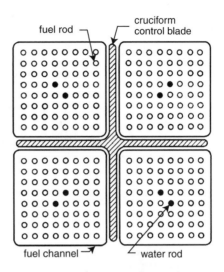

Figure 11.19. A BWR fuel module.

The four subassemblies in a fuel module are separated by water gaps in which a cruciform control blade moves up and down to control the core reactivity (see Fig. 11.19). Because the thermal flux peaks near water channels, lower enrichment fuel pins are used near the water channels. Pins of up to four different enrichments are used in each subassembly to flatten the power profile across the assembly.

Reactivity Control

In a BWR, three different mechanisms are used to control the core's reactivity. Short term reactivity changes are made by adjusting the recirculation flow through the jet pumps. As the water flow through the core is increased (decreased), the amount of boiling and voiding in the core decreases (increases) and the reactivity increases (decreases) because of increased (decreased) neutron moderation. Flow modulation can accommodate power variations of as much as 25%.

Longer term reactivity control is provided by the cruciform control blades that are raised and lowered from the bottom of the vessel, both to avoid the steam separator/dryers in the top of the vessel and to use their greater effectiveness in the liquid water in the bottom part of the core than in the vapor region in the top part.

Reactivity control to compensate for fuel burnup is also provided by burnable gadolinium oxide (GdO_2) mixed in the UO_2 pellets of the fuel pins. As the gadolinium absorbs neutrons, it is transformed into an isotope with a low neutron absorption cross section and thus allows more neutrons to be absorbed in the remaining fuel.

Although PWRs use soluble boron, a strong thermal neutron absorber, in the primary cooling water to regulate the core's reactivity, this method of reactivity control is not available to BWRs. Boiling would cause boron to be precipitated as a solid on the fuel pin surfaces, thus making reactivity control impossible.

11.4   Generation III Nuclear Reactor Designs

During the 1990s, a major effort was made in the United States to design, license, and install a new generation of nuclear reactors for central-station power generation. Efforts proceeded along three avenues. Electric utility requirements and specifications were developed under the auspices of the Electric Power Research Institute as the Advanced Light Water Reactor (ALWR) Program. Design work was sponsored by the U.S. Department of Energy in partnership with vendors and utilities. Design certification was done by the Nuclear Regulatory Commission under streamlined licensing procedures. A path similar to that of the USA was followed in the European Pressurized Water Reactor (EPR) development program.

Two design scopes were followed. One, *evolutionary design*, built directly upon an installed capacity of about 100,000 MW(e) from 103 nuclear plants, stressing safety, efficiency and standardization. The other, stressing ultra-safe features, called for passive safety features involving gravity, natural circulation, and pressurized gas as driving forces for cooling and residual heat removal. Two evolutionary plant designs (ABWR and System 80+) and two passive designs (AP600 and AP1000) have received design certification by the U.S. Nuclear Regulatory Commission. A third passive design (ESBWR) has received design approval and the design certification process is under way.

11.4.1   The ABWR and ESBWR Designs

GE Advanced Boiling Water Reactor (ABWR)
The design effort for this plant was accomplished by GE Nuclear Energy, in cooperation with Hitachi and Toshiba. It is a 1350-MW(e) single-cycle, forced circulation plant operable with a mixed-oxide (PuO_2/UO_2) fuel cycle. The ABWR uses digital logic and control, and improved electronics, turbine and fuel technology all of which lead to improved plant availability, operating capacity, safety, and reliability. The Tokyo Electric Power Company has two ABWRs in operation, the second of which was constructed in a record 48 months. Two similar plants are in construction in Taiwan for the Taiwan Power Company. The ABWR is adapted to USA utility needs and conforms with the Electric Power Research Institute (EPRI) evolutionary design requirements. The ABWR was the first such design to receive the NRC final design approval.

GE Economic and Simplified Boiling Water Reactor (ESBWR)
Thes ESBWR plant utilizes features of its predecessor, the SBWR, as well as the ABWR. The Simplified Boiling Water Reactor, SBWR, is no longer under active consideration in the U.S., GE's attention being shifted to higher power systems. Key features of the SBWR design, however, were coolant flow produced by natural circulations, gravity-driven emergency core cooling, simplifying piping, and the elimination of the need for coolant pumps and standby diesel generators, all of which were major improvements on Gen II reactors.

The Economic and Simplified Boiling Water Reactor, shown in Fig. 11.20, is a 1,390 MW(e), natural circulation boiling water reactor that incorporates passive safety features. As in the SBWR, natural-convection coolant flow is used, enhanced by use of a taller vessel, a shorter core, and reduced flow restrictions. With no large piping connections to the reactor vessel near or below the core elevation, the core is

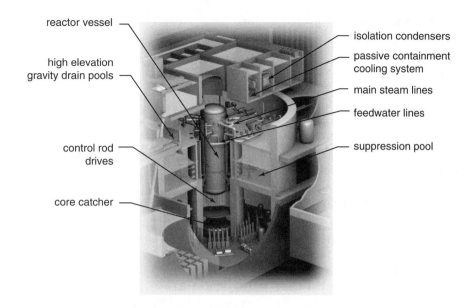

reactor vessel

high elevation
gravity drain pools

control rod
drives

core catcher

isolation condensers

passive containment
cooling system

main steam lines

feedwater lines

suppression pool

Figure 11.20. Cutaway view of the ESBWR reactor building.

assured of being covered by water for all design-basis accidents. This safety feature
is a key design intention. In emergency conditons resulting in reactor isolation,
an isolation condenser is designed as a safety-related system to remove decay heat
from the reactor core by natural circulation and with minimal loss of reactor coolant
inventory. In addition, a core catcher and passive lower drywell flooder is in place
to limit impact of molten fuel in the event of a very low probability severe accident.

Characteristics of the Gen II BWR/6, the Gen III ABWR, and ESBWR boiling
water reactor systems are compared in Table 11.4.

Table 11.4. Comparison of the key properties of generation II and III boiling water reactors.

Property	BWR/6	ABWR	ESBWR
Power MW(t)/MW(e)	3900/1360	3925/1350	4500/1550
Vessel height/dia (m)	21.8/6.4	21.1/7.1	27.7/7.1
No. fuel bundles	800	872	1132
Active fuel height (m)	3.7	3.7	3.0
Power density (kW/L)	54.2	51	54
No. control rod drives	193	205	269
No. recirculation pumps	2 (external)	10 (internal)	0
No. safety system pumps	9	18	0
No. safety diesel generators	3	3	0

Source: Nuclear News, January 2006.

11.4.2   The System 80+ Design

The System 80+ plant is a 1350-MW(e) pressurized water reactor. It was designed by Combustion Engineering and ASEA Brown Boveri (ABB), was acquired by Westinghouse-BNFL (British Nuclear Fuels), and then by Framatome ANP. Features of the System 80+ design are incorporated in eight units completed or under construction in the Republic of Korea. Important design features include a dual spherical steel confinement for accident mitigation and a cavity-flood system with an in-containment refueling-water storage supply. While not expected to be promoted for domestic sale in the U.S., the System 80+ design evolved to the APR1400 design developed for installation in Korea.

11.4.3   AP600 and AP1000 Designs

The AP600 Reactor

The AP600 is a 600-MW(e) modular pressurized water reactor that relies on passive systems for both emergency core cooling and residual heat removal. As part of the cooperative U.S. Department of Energy (DOE) Advanced Light Water Reactor (ALWR) Program and the Electric Power Research Institute (EPRI), the AP600 team developed a simplified, safe, and economic plant designed to satisfy the standards set by DOE and defined in the ALWR Utility Requirements Document. Implementation of the passive safety features greatly reduces the operation, maintenance, and testing requirements of the AP600. Through the use of modular construction techniques, similar to those applied in ship construction, the design objective is a 36-month construction schedule from the first concrete pour to the fuel load.

The AP1000 Reactor

The Westinghouse AP1000 is a two-loop configuration (see Fig. 11.21), capable of producing over 1000 MW(e), and is based on the earlier AP600 design. The passive safety systems (see Fig. 11.22) are significantly simpler than those of traditional Gen II PWR safety systems. They do not require the large network of safety support systems needed in present day nuclear plants, such as emergency electric power, HVAC, cooling water systems, and seismically hardened buildings to house these emergency components. Sim-

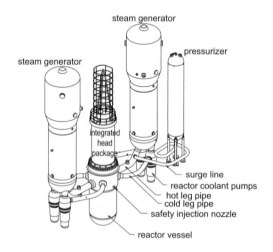

Figure 11.21. AP1000 nuclear steam supply system.

plification of plant systems, combined with increased plant operating margins, reduces the actions required by the operator. The AP1000 has 50 percent fewer valves, less piping, less control cable, fewer pumps, and less seismic building volume than

a similarly sized conventional Gen II plant, thus resulting in major savings in plant
costs and construction schedules.

As shown in Fig. 11.21, the AP1000 NSSS plant configuration consists of two
steam generators, each connected to the reactor pressure vessel by a single hot leg
and two cold legs. There are four reactor coolant pumps that provide circulation
of the reactor coolant for heat removal. Major component changes incorporated
into the AP1000 design include a taller reactor vessel, larger steam generators, a
larger pressurizer, and slightly taller, canned reactor coolant pumps with higher
reactor coolant flows. The AP1000 fuel assembly design is based on the 17×17
pin array (14 foot long) design used successfully at most Gen II plants in the U.S.
and Europe. Studies have shown that the AP1000 can operate with a full core
loading of mixed-oxide (PuO_2/UO_2) fuel. The AP1000 has received final design
approval and standard design certification by the U.S. Nuclear Regulatory Commission. Characteristics of the AP1000 and EPR reactor systems are compared in
Table 11.5.

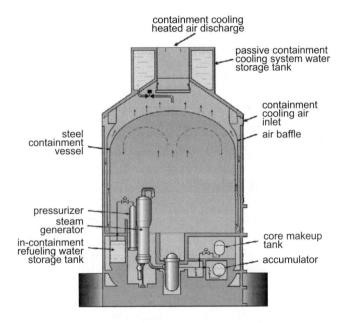

Figure 11.22. AP1000 passive safety injection system.

11.4.4 Other Evolutionary LWR Designs

The IRIS system (International Reactor Innovative and Secure) is a modular system,
100 to 350-MW(e) per module, being developed by an international design team
headed by Westinghouse. The design emphasizes proliferation resistance and safety
enhancements. IRIS uses PWR technology, but is newly engineered to include a
five-to-eight-year core reloading schedule. The reactor vessel houses not only the
reactor core but also pressurizer and steam generators. Thus, the reactor vessel is
larger than that of a conventional PWR, but the containment volume is much less
than that of the conventional PWR.

Table 11.5. Comparison of the key properties of EPR and AP1000 pressurized water reactors.

Property	AP1000	EPR
Power (MWt/MWe)	3400/1117	4324/1600
Number of loops	2	4
Hot leg temp. (^oC)	321	327
No. fuel bundles	157	241
Type of fuel assembly	17x17	17x17
Active fuel height (m)	4.3	4.2
Linear heat rate (W/cm)	187	156
Control rod cluster assemblies	53 (16 gray)	89
RV inside dia. (cm)	399	489
Vessel flow (m^3/h)	68,100	113,320

Source: Framatome ANP, March 2005; Nuclear News, Apr. & Nov. 2004.

11.4.5   Heavy Water Reactors

The ACR700 is the evolutionary design derived from the conventional Canadian CANDU design. The plant is a 700-MW(e) reactor employing slightly enriched fuel, heavy water moderation, and light water cooling. Control is digital and, unlike most reactors, refueling can be performed while the reactor is at power. Such on-line refueling capability allows for very uniform burnup of the fuel allowing an optimal economical fuel-cycle.

11.4.6   Gas-Cooled Reactors

Two designs of gas-cooled reactors are currently being pursued actively. One is the Pebble Bed Modular Reactor (PBMR) under development in South Africa by the utility Eskom, by the Industrial Development Corporation of South Africa, by British Nuclear Fuels, and the U.S. firm, Exelon. The other design is the General Atomics Gas Turbine-Modular Helium Reactor (GT-MHR).

The PBMR is an evolutionary design based on HTR (helium cooled high temperature reactor) technology developed jointly by Siemens and ABB. It is a modular system, 120-MW(e) per module, with an estimated 18 to 24 month construction time. The core of the reactor is a cylindrical annulus, reflected on the outside by a layer of graphite bricks and on the inside by approximately 110,000 graphite spheres. The core contains approximately 330,000 60-mm diameter fuel spheres. Each has a graphite-clad, 50-mm diameter, graphite matrix containing 0.5-mm diameter UO_2 fuel particles surrounded by refractory layers of graphite and silicon carbide. The fuel is stable at temperatures as high as 2000 °C, well above core temperatures even in a worst-case loss-of-coolant accident. Helium flows through the pebble bed and drives a gas turbine for power generation. Because high temperature helium serves both as the coolant and the working fluid, the gas turbines are considerably smaller, and, hence, less expensive than turbines used in LWR plants.

General Atomics (GA) is the industrial pioneer of the Gas Turbine–Modular Helium Reactor (GT-MHR), an ultra-safe, meltdown-proof, helium-cooled reactor, with refractory-coated particle fuel. In early 1995, GA and Russia's Ministry of Atomic Energy (MINATOM), in cooperation with Framatome and Fuji Electric, began a joint program to design and develop the GT-MHR for use in Russia in the destruction of weapons-grade plutonium and replacement of plutonium production reactors in the Russian Federation. A typical GT-MHR module, rated at 600 MW(t), yields a net output of about 285-MW(e). The reactor can be fueled with uranium or plutonium.

11.5 Generation IV Nuclear Reactor Designs

As called for in the U.S. *National Energy Policy* [NEPD 2001], the United States and nine other countries agreed to develop six Generation IV nuclear energy concepts [DOE 2002] with the goals of expansion of nuclear energy and development of advanced nuclear fuel cycles, next generation reactor technologies, and advanced reprocessing and fuel treatment technologies. To achieve this, the Department of Energy Office of Advanced Nuclear Research adopted an integrated strategy formulated in three mutually complementary programs: the Generation IV Nuclear Energy Systems Initiative (Generation IV), the Nuclear Hydrogen Initiative (NHI), and the Advanced Fuel Cycle Initiative (AFCI).

Several of the Generation IV systems are particularly well suited to meet U.S. national energy needs. The U.S. strategy includes development of reactor systems [DOE 2003] as well as leveraging international cooperation through the Generation IV International Forum and bi- and multi-lateral collaborations. The implementation strategy focuses the program on two principal goals: The first is to develop a Next Generation Nuclear Plant (NGNP) to achieve economically competitive hydrogen and electricity production in the mid-term. The second is to develop a fast reactor to achieve significant advances in sustainability for the long term. A target schedule for the development and construction of the NGNP includes starting preconceptual designs in 2006, completing preliminary design in 2009, completing major R&D activities by 2012, and starting construction in 2012. NGNP operations are scheduled to begin in 2017 [DOE 2005].

As described in the program plan [DOE 2005], two systems employ a thermal neutron spectrum with coolants and temperatures that enable hydrogen or electricity production with high efficiency (the Supercritical Watercooled Reactor [SCWR] and the Very High Temperature Reactor [VHTR]). Three employ a fast neutron spectrum to enable more effective management of actinides through recycling of most components in the discharged fuel (the Gas-cooled Fast Reactor [GFR], the Lead-cooled Fast Reactor [LFR], and the Sodium-cooled Fast Reactor [SFR]). The Molten Salt Reactor (MSR) employs a circulating liquid fuel mixture that offers considerable flexibility for recycling actinides and may provide an alternative to accelerator-driven systems for waste transmutation. The six Gen IV reactor systems are shown in Figs. 11.23 to Fig. 11.28.

The descriptions of the six Generation IV reactors presented below are taken verbatim from the U.S. Technology Roadmap [DOE 2002].

11.5.1 Supercritical Water-Cooled Reactors

SCWRs, Fig. 11.23, are high-temperature, high-pressure watercooled reactors that operate above the thermodynamic critical point of water (374 °C, 22.1 MPa). These systems may have a thermal

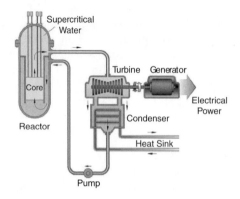

Figure 11.23. Generation IV supercritical-water cooled reactor (SCWR).

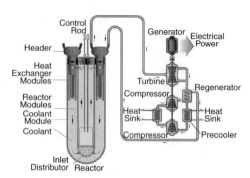

Figure 11.24. Generation IV lead-cooled fast reactor (LFR).

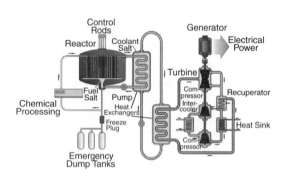

Figure 11.25. Generation IV molten-salt reactor (MSR).

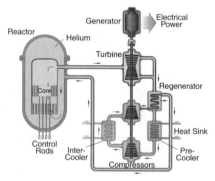

Figure 11.26. Generation IV gas cooled fast reactor (GFR).

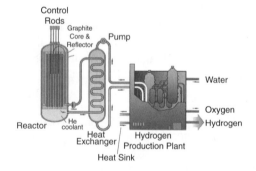

Figure 11.27. Generation IV very hight temperature reactor (VHTR).

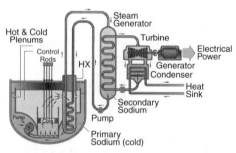

Figure 11.28. Generation IV sodium cooled fast reactor (SFR).

or fast-neutron spectrum, depending on the core design. SCWRs have unique features that may offer advantages compared to state-of-the-art LWRs in the following:

- SCWRs offer increases in thermal efficiency relative to current-generation LWRs. The efficiency of a SCWR can approach 44%, compared to 33–35% for LWRs.

- A lower-coolant mass flow rate per unit core thermal power results from the higher enthalpy content of the coolant. This offers a reduction in the size of the reactor coolant pumps, piping, and associated equipment, and a reduction in the pumping power.

- A lower-coolant mass inventory results from the once-through coolant path in the reactor vessel and the lower-coolant density. This opens the possibility of smaller containment buildings.

- No boiling crisis (i.e., departure from nucleate boiling or dry out) exists due to the lack of a second phase in the reactor, thereby avoiding discontinuous heat transfer regimes within the core during normal operation.

- Steam dryers, steam separators, recirculation pumps, and steam generators are eliminated. Therefore, the SCWR can be a simpler plant with fewer major components.

The Japanese supercritical light water reactor (SCLWR) with a thermal spectrum has been the subject of the most development work in the last 10 to 15 years and is the basis for much of the reference design. The SCLWR reactor vessel is similar in design to a PWR vessel (although the primary coolant system is a direct-cycle, BWR-type system). High-pressure (25.0 MPa) coolant enters the vessel at 280 °C. The inlet flow splits, partly to a downcomer and partly to a plenum at the top of the core to flow down through the core in special water rods. This strategy provides moderation in the core. The coolant is heated to about 510 °C and delivered to a power conversion cycle, which blends LWR and supercritical fossil plant technology; high-, intermediate-, and low-pressure turbines are employed with two reheat cycles. The overnight capital cost for a 1700-MWe SCLWR plant may be as low as $900/kWe (about half that of current ALWR capital costs), considering the effects of simplification, compactness, and economy of scale. The operating costs may be 35% less than current LWRs.

The SCWR can also be designed to operate as a fast reactor. The difference between thermal and fast versions is primarily the amount of moderator material in the SCWR core. The fast spectrum reactors use no additional moderator material, while the thermal spectrum reactors need additional moderator material in the core.

11.5.2 Lead-Cooled Fast Reactors

LFR systems are Pb or Pb-Bi alloy-cooled reactors with a fast-neutron spectrum and closed fuel cycle. One LFR system is shown in Fig. 11.24. Options include a range of plant ratings, including a long refueling interval battery ranging from 50–150 MWe, a modular system from 300–400 MWe, and a large monolithic plant at 1200 MW(e). These options also provide a range of energy products.

The LFR battery option is a small factory-built turnkey plant operating on a closed fuel cycle with very long refueling interval (15 to 20 years) cassette core or replaceable reactor module. Its features are designed to meet market opportunities for electricity production on small grids, and for developing countries that may not wish to deploy an indigenous fuel cycle infrastructure to support their nuclear energy systems. Its small size, reduced cost, and full support fuel cycle services can be attractive for these markets. It had the highest evaluations to the Generation IV goals among the LFR options, but also the largest R&D needs and longest development time.

The options in the LFR class may provide a time-phased development path: The nearer-term options focus on electricity production and rely on more easily developed fuel, clad, and coolant combinations and their associated fuel recycle and refabrication technologies. The longer term option seeks to further exploit the inherently safe properties of Pb and raise the coolant outlet temperature sufficiently high to enter markets for hydrogen and process heat, possibly as merchant plants. LFR holds the potential for advances compared to state-of-the-art liquid metal fast reactors in the following:

- Innovations in heat transport and energy conversion are a central feature of the LFR options. Innovations in heat transport are afforded by natural circulation, lift pumps, in-vessel steam generators, and other features. Innovations in energy conversion are afforded by rising to higher temperatures than liquid sodium allows, and by reaching beyond the traditional superheated Rankine steam cycle to supercritical Brayton or Rankine cycles or process heat applications such as hydrogen production and desalination.

- The favorable neutronics of Pb and Pb-Bi coolants in the battery option enable low power density, natural circulation-cooled reactors with fissile selfsufficient core designs that hold their reactivity over their very long 15- to 20-year refueling interval. For modular and large units more conventional higher power density, forced circulation, and shorter refueling intervals are used, but these units benefit from the improved heat transport and energy conversion technology.

- Plants with increased inherent safety and a closed fuel cycle can be achieved in the near- to mid-term. The longer-term option is intended for hydrogen production while still retaining the inherent safety features and controllability advantages of a heat transport circuit with large thermal inertia and a coolant that remains at ambient pressure. The favorable sustainability features of fast spectrum reactors with closed fuel cycles are also retained in all options.

- The favorable properties of Pb coolant and nitride fuel, combined with high temperature structural materials, can extend the reactor coolant outlet temperature into the 750–800 °C range in the long term, which is potentially suitable for hydrogen manufacture and other process heat applications. In this option, the Bi alloying agent is eliminated, and the less corrosive properties of Pb help to enable the use of new high-temperature materials. The required R&D is more extensive than that required for the 550 °C options because the higher reactor outlet temperature requires new structural materials and nitride fuel development.

11.5.3 Molten-Salt Reactors

The MSR, Fig. 11.25, produces fission power in a circulating molten salt fuel mixture. MSRs are fueled with uranium or plutonium fluorides dissolved in a mixture of molten fluorides, with Na and Zr fluorides as the primary option. MSRs have the following unique characteristics, which may afford advances:

- MSRs have good neutron economy, opening alternatives for actinide burning and/or high conversion.

- High-temperature operation holds the potential for thermochemical hydrogen production.

- Molten fluoride salts have a very low vapor pressure, reducing stresses on the vessel and piping.

- Inherent safety is afforded by fail-safe drainage, passive cooling, and a low inventory of volatile fission products in the fuel.

- Refueling, processing, and fission product removal can be performed online, potentially yielding high availability.

- MSRs allow the addition of actinide feeds of widely varying composition to the homogenous salt solution without the blending and fabrication needed by solid fuel reactors.

There are four fuel cycle options: (1) Maximum conversion ratio (up to 1.07) using a Th-^{233}U fuel cycle, (2) denatured Th-^{233}U converter with minimum inventory of nuclear material suitable for weapons use, (3) denatured once-through actinide burning (Pu and minor actinides) fuel cycle with minimum chemical processing, and (4) actinide burning with continuous recycling. The fourth option with electricity production is favored for the Generation IV MSR. Fluoride salts with higher solubility for actinides such as NaF/ZrF$_4$ are preferred for this option. Salts with lower potential for tritium production would be preferred if hydrogen production were the objective. Lithium and beryllium fluorides would be preferred if high conversion were the objective. Online processing of the liquid fuel is only required for high conversion to avoid parasitic neutron loses of ^{232}Pa that decays to ^{233}U fuel. Offline fuel salt processing is acceptable for actinide management and hydrogen or electricity generation missions. To achieve conversion ratios similar to LWRs, the fuel salt needs only to be replaced every few years.

The reactor can use ^{238}U or ^{232}Th as a fertile fuel dissolved as fluorides in the molten salt. Due to the thermal or epithermal spectrum of the fluoride MSR, ^{232}Th achieves the highest conversion factors. All of the MSRs may be started using low-enriched uranium or other fissile materials. The range of operating temperatures of MSRs ranges from the melting point of eutectic fluorine salts (about 450 °C) to below the chemical compatibility temperature of nickel-based alloys ($\sim$800°C).

11.5.4 Gas-Cooled Fast Reactors

The GFR system, Fig. 11.26, features a fast-spectrum helium-cooled reactor and closed fuel cycle. Like thermal-spectrum helium-cooled reactors such as the GT-MHR and the PBMR, the high outlet temperature of the helium coolant makes it possible to deliver electricity, hydrogen, or process heat with high conversion efficiency. The GFR uses a direct-cycle helium turbine for electricity and can use process heat for thermochemical production of hydrogen. Through the combination of a fast-neutron spectrum and full recycle of actinides, GFRs minimize the production of long-lived radioactive waste isotopes. The GFRs fast spectrum also makes it possible to utilize available fissile and fertile materials (including depleted uranium from enrichment plants) two orders of magnitude more efficiently than thermal spectrum gas reactors with once-through fuel cycles. The GFR reference assumes an integrated, on-site spent fuel treatment and refabrication plant.

11.5.5 Very High-Temperature Fast Reactors

The VHTR is a next step in the evolutionary development of high-temperature gas-cooled reactors. The VHTR can produce hydrogen from only heat and water by using thermochemical iodine-sulfur (I-S) process or from heat, water, and natural gas by applying the steam reformer technology to core outlet temperatures greater than about 1000 °C. A reference VHTR system that produces hydrogen is shown in Fig. 11.27. A 600 MWth VHTR dedicated to hydrogen production can yield over 2 million normal cubic meters per day. The VHTR can also generate electricity with high efficiency, over 50% at 1000 °C, compared with 47% at 850 °C in the GT-MHR or PBMR. Co-generation of heat and power makes the VHTR an attractive heat source for large industrial complexes. The VHTR can be deployed in refineries and petrochemical industries to substitute large amounts of process heat at different temperatures, including hydrogen generation for up-grading heavy and sour crude oil. Core outlet temperatures higher than 1000 °C would enable nuclear heat application to such processes as steel, aluminum oxide, and aluminum production. The VHTR is a graphite-moderated, helium-cooled reactor with thermal neutron spectrum. It can supply nuclear heat with core-outlet temperatures of 1000 °C. The reactor core type of the VHTR can be a prismatic block core such as the operating Japanese HTTR, or a pebble-bed core such as the Chinese HTR-10. For electricity generation, the helium gas turbine system can be directly set in the primary coolant loop, which is called a direct cycle. For nuclear heat applications such as process heat for refineries, petrochemistry, metallurgy, and hydrogen production, the heat application process is generally coupled with the reactor through an intermediate heat exchanger (IHX), which is called an indirect cycle.

11.5.6 Sodium-Cooled Fast Reactors

The Sodium-Cooled Fast Reactor (SFR) system features a fast-spectrum reactor, Fig. 11.28, and closed fuel recycle system. The primary mission for the SFR is management of high-level wastes and, in particular, management of plutonium and other actinides. With innovations to reduce capital cost, the mission can extend to electricity production, given the proven capability of sodium reactors to utilize almost all of the energy in the natural uranium versus the 1% utilized in thermal spectrum systems.

A range of plant size options are available for the SFR, ranging from modular systems of a few hundred MWe to large monolithic reactors of 1500–1700 MWe. Sodium core outlet temperatures are typically 530–550 °C. The primary coolant system can either be arranged in a pool layout (a common approach, where all primary system components are housed in a single vessel), or in a compact loop layout, favored in Japan. For both options, there is a relatively large thermal inertia of the primary coolant. A large margin to coolant boiling is achieved by design, and is an important safety feature of these systems. Another major safety feature is that the primary system operates at essentially atmospheric pressure, pressurized only to the extent needed to move fluid. Sodium reacts chemically with air, and with water, and thus the design must limit the potential for such reactions and their consequences. To improve safety, a secondary sodium system acts as a buffer between the radioactive sodium in the primary system and the steam or water that is contained in the conventional Rankine-cycle power plant. If a sodium-water reaction occurs, it does not involve a radioactive release. Two fuel options exist for the SFR: (1) MOX and (2) mixed uranium-plutonium-zirconium metal alloy (metal). The experience with MOX fuel is considerably more extensive than with metal.

SFRs require a closed fuel cycle to enable their advantageous actinide management and fuel utilization features. There are two primary fuel cycle technology options: (1) an advanced aqueous process, and (2) the pyroprocess, which derives from the term, pyrometallurgical process. Both processes have similar objectives: (1) recovery and recycle of 99.9% of the actinides, (2) inherently low decontamination factor of the product, making it highly radioactive, and (3) never separating plutonium at any stage. These fuel cycle technologies must be adaptable to thermal spectrum fuels in addition to serving the needs of the SFR. This is needed for two reasons: First, the startup fuel for the fast reactors must come ultimately from spent thermal reactor fuel. Second, for the waste management advantages of the advanced fuel cycles to be realized (namely, a reduction in the number of future repositories required and a reduction in their technical performance requirements), fuel from thermal spectrum plants will need to be processed with the same recovery factors. Thus, the reactor technology and the fuel cycle technology are strongly linked. Consequently, much of the research recommended for the SFR is relevant to crosscutting fuel cycle issues.

11.6   The Nuclear Fuel Cycle

The several stages involved in the processing of nuclear fuel from its extraction from uranium ore to the ultimate disposal of the waste from a reactor is called the *nuclear fuel cycle*. A schematic of this cycle for LWRs is shown in Fig. 11.29. In this figure, the cycle shown by the solid boxes and arrows (i.e., ignoring the dashed line components) is the *once-through* cycle, currently used by all power plants in the United States.

The once-through cycle begins with the mining of uranium ore by shaft mining or, more commonly, by open-pit mining. The uranium is extracted from the ore in a milling process to produce "yellow cake" which is about 80% U_3O_8. Before the uranium can be used in modern LWRs, the ^{235}U content must be enriched from its natural isotopic abundance of 0.720 a% to about 3 a%. The first step in the enrichment process is the conversion of U_3O_8 to UF_6, a substance that

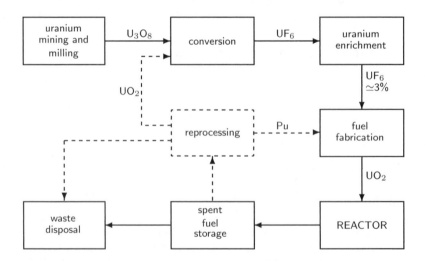

Figure 11.29. The nuclear fuel cycle for LWRs without fuel recycle (solid boxes) and with uranium and plutonium recycle (solid plus dashed elements).

becomes gaseous at relatively low temperatures and pressures. The gaseous UF_6 is then processed to separate $^{235}UF_6$ molecules from the far more abundant $^{238}UF_6$ molecules.

The enrichment techniques are discussed later in Section 11.6.2. After UF_6 has been isotopically enriched to about 3 a%, it is converted to ceramic UO_2 pellets which are then used in the fuel rods of LWR reactors.

Every one to two years, a LWR is shut down for several weeks, during which time about one-third of a PWR's fuel and about one-quarter of a BWR's fuel is replaced and the older remaining fuel is shuffled inward towards the center of the core. The fuel assemblies removed from the core are initially submerged for a few years in a spent-fuel storage pool where the water safely removes the decay heat produced by the radioactive decay of the fission products. After several years, the spent fuel assemblies may be transferred to a long-term spent-fuel storage pool or to dry spent-fuel storage casks at a facility outside the plant. By now, the greatly reduced decay heat can be convectively removed by only the gas in the casks. Eventually, it is planned to place the waste in these spent-fuel assemblies into a permanent waste repository (see Section 11.6.3). The annual uranium needs and production of other elements in a typical 1000 MW(e) PWR are listed in Table 11.6. The data in this table are based on an assumed plant capacity factor of 0.75, i.e., the production of 750 GWy of electrical power, and on an assumed 0.2% enrichment in the tailings discarded by the enrichment process.

Table 11.6. The annual material requirements and production (in kg) for a typical 1000 MW(e) PWR. Source: Lamarsh and Baratta [2001].

Mining/Milling output:	
U in U_3O_8	150,047
Conversion output:	
U in UF_6	149,297
Enrichment output:	
^{235}U	821
^{238}U	27,249
U tails (0.2%)	121,227
Reactor Output:	
^{235}U	220
total U	25,858
Pu (fissile)	178
total Pu	246
total U+Pu	26,104
fission products	873

The fuel removed from a reactor, besides containing radioactive fission products, also contains significant amounts of residual ^{235}U and fissile ^{239}Pu and ^{241}Pu (see Table 11.6). In principle, these fissile isotopes can be extracted from the spent fuel and recycled back into new *mixed oxide* fuel for a LWR. The fuel cycle with such recycling is shown in Fig. 11.29 with the addition of the dashed components. The use of recycling of fissile isotopes in spent fuel can reduce the lifetime U_3O_8 requirements by about 45%. However, the economics of such recycling are uncertain and political concern over the possible diversion of recycled plutonium for terrorist bombs so far has prevented recycling in the United States. Nevertheless, the annual discarding from a LWR of 220 kg of ^{235}U and the 180 kg of fissile plutonium isotopes represents an energy equivalent of about 1.3 million tons of coal. Thus there are strong energy incentives to adopt recycling.

11.6.1 Uranium Requirements and Availability

To fuel existing nuclear power plants, uranium must first be extracted from natural deposits and converted into a form suitable for use in a reactor. In this section the uranium needs of LWRs and the availability of uranium are discussed.

Uranium Needs for LWRs

Currently, uranium is extracted from ores containing uranium-bearing minerals of complex composition. High grade ores contain about 2% uranium with medium-grade ores ranging from 0.1 to 0.5% uranium. To extract uranium from its ore, mills near the mining areas use either chemical leaching or solvent extraction techniques to produce yellow cake, which is about 80% U_3O_8. The yellow cake is then shipped to other facilities where it is purified and converted into uranium dioxide (UO_2) or uranium hexafluoride (UF_6). The tailings left at the milling site still contain considerable uranium and, consequently, emit relatively large amounts of radon. To mitigate radon releases, the tailing can be either placed underground or capped with a thick earthen barrier.

From the data in Table 11.6, it is seen that a 1000-MW(e) LWR with a 75% capacity factor requires about 150 tonnes of new natural uranium each year if the once-through fuel cycle is used. This amounts to 4500 tonnes of uranium over the 30-y lifetime of such a plant. With plutonium and uranium recycling, about 80 tonnes of new uranium are needed per year or 2400 tonnes over the lifetime of the plant.[1] By contrast, a liquid-metal fast breeder reactor of the same capacity has a lifetime requirement for natural or depleted uranium of about only 40 kg (assuming mixed oxide fuel recycling), since such a reactor produces more fissile fuel than it consumes.

Uranium Availability

In 2000, there was a worldwide installed nuclear power capacity of about 340 GWe. Since a 1 GWe LWR has a lifetime need of about 4500 tonnes of uranium, and since most reactors are LWRs, this installed capacity represents a uranium need of about $340 \times 4500 = 1.5 \times 10^6$ tonnes The global nuclear capacity is expected to increase to between 415 and 490 GWe by 2010, and even more uranium will be needed. An obvious question is how much uranium will be available for the future needs of nuclear power.

Most of the uranium needed for reactor fuel during the next 20 to 30 years will come from reasonably assured deposits from which uranium can be produced for no more than $130 per kg. Such resources as estimated to total about 4×10^6 tonnes, 80% of which are in North America, Africa, and Australia [Lamarsh and Baratta 2001]. In addition, there are resources that are more speculative whose estimate is based on geological similarities of similar ore deposits and other indirect evidence, that suggest additional uranium resources can be realized for production costs of less than $130 per kg. Such speculative resources have been estimated at about 11×10^6 tons worldwide with about 1.4×10^6 tons in the United States [Lamarsh and Baratta 2001]. However, few of these speculative reserves will be developed within the next thirty decades.

At higher production costs, much more uranium can be obtained since ore with more dilute uranium concentrations can be exploited. There are many low grade uranium deposits containing enormous quantities of uranium. For example, a shallow geological formation known as Chattanooga shale underlies six Midwestern

---

[1]These uranium requirements are based on the depleted uranium produced in the enrichment process having a ^{235}U enrichment of 0.2 wt%. Less new uranium would be required if a lower, but more expensive, tailing enrichment were used.

states. This shale contains an estimated 5×10^6 tonnes of uranium with concentrations up to 66 ppm. Indeed, an area of just 7 miles square of this shale contains the energy equivalent of all the world's oil [Lamarsh and Baratta 2001]. Other similar formations containing significant amounts of low-concentration uranium exist in many other countries. However, none of these resources are presently economically useful.

From the elemental abundances listed in Table A.3, we see that uranium is almost 700 times more abundant than gold and 6 times more abundant than iodine in the earth's crust. Within the first 10 km of the earth's crust, there are about 5×10^{13} tonnes of uranium. Of course, only a small fraction of this can be recovered economically. Uranium, because of its high solubility in an oxidizing environment, is also relatively abundant in sea water (0.0032 mg/L). All the oceans of the world contain about 4×10^9 tonnes of uranium. However, no economically feasible method has yet been devised to extract uranium from the sea.

11.6.2 Enrichment Techniques

There are many methods by which elements can be enriched in a particular isotope. Many techniques have been proposed for enrichment or separation of an isotope from its element. Several have been developed and demonstrated to be economically feasible. Generally, the lighter the element, the easier (and less costly) it is to separate the isotopes. In this section several of the more important methods for enriching uranium in ^{235}U are summarized.

Gaseous Diffusion

The gaseous diffusion method was the first method used to enrich uranium in ^{235}U to an exceedingly high level ($> 90\%$) so that atomic bombs could be constructed. Today it is the primary method used to enrich uranium to about 3% as needed for LWRs.

Uranium, whose isotopes are to be separated, must first be incorporated into a molecule such as UF_6, which is gaseous at only slightly elevated temperatures. Fluorine has only the single stable isotope ^{19}F so that the mass difference of $^{235}UF_6$ and $^{238}UF_6$ is due only to the mass difference in the uranium isotopes. The basis of gaseous diffusion enrichment is to find a porous membrane with pores sufficiently large to allow passage of the molecules but prevent bulk gas flow. In a container with two regions separated by such a membrane, the UF_6 molecules are pumped into one region. The $^{235}UF_6$ molecules travel faster than the $^{238}UF_6$ molecules, strike the membrane more often, and preferentially are transmitted through the membrane into the second region. The gas extracted from the second region is thus enriched in ^{235}U.

In gaseous diffusion, enrichment of UF_6 membranes of nickel or of austenitic stainless steel are used. However, because the mass difference between $^{235}UF_6$ and $^{238}UF_6$ is so small, each gaseous diffusion cell or *stage* has but a small enrichment capability. Thus, hundreds of such diffusion-separation stages must be interconnected such that the output UF_6 from one cell becomes the input of another cell. This cascade technology, which requires enormous amounts of electricity to pump and cool the gas, has been used by France, China, and Argentina as well as the United States to obtain enriched uranium.

Gas Centrifuge

Gas molecules of different masses can be separated by placing the gas in a rotor container and spinning it at high speed. The centrifugal force causes the molecules to move toward the outer wall of the rotating container where the lighter molecules are buoyed up and moved slightly away from the rotor wall by the heavier molecules adjacent to the wall. This technique works more effectively when there is a large percentage difference between the molecular masses. Indeed, it was first used in the 1930s to separate isotopes of chlorine.

Because of the small mass difference between $^{235}UF_6$ and $^{238}UF_6$, the capability of a single centrifuge to separate $^{235}UF_6$ from $^{238}UF_6$ is not large. Like gaseous diffusion cells, many gas centrifuges must be connected together in cascades to achieve the necessary enrichment for nuclear fuel. The great advantage of uranium enrichment by gas centrifuges is that it requires only a few percent of the electrical energy needed by gaseous diffusion plants of the same enrichment capacity. At least nine countries have developed uranium enrichment facilities using gas centrifuges.

Aerodynamic Separation

In this enrichment process, a mixture of hydrogen and uranium hexafluoride is subjected to strong aerodynamic forces to separate $^{235}UF_6$ from $^{238}UF_6$. In one variation, the gas flows at high speed through a curved nozzle. During passage through this curved nozzle, the heavier $^{238}UF_6$ preferentially moves towards the outer wall (surface with the larger radius) of the nozzle. An appropriately placed sharp divider at the nozzle exit then separates the two uranium isotopes. Although this curved nozzle technique has a better separation ability than a gas centrifuge cell, it has not yet proven to be economically superior.

An alternative aerodynamic technique introduces a mixture of hydrogen and uranium hexafluoride at high speed through holes in the side wall of a tube. The tube narrows towards the exit, and as the gas flows down the tube, it spirals with increasing angular speed so that $^{238}UF_6$ is preferentially forced to the tube surface where it is extracted. The lighter $^{235}UF_6$ is left to exit the tube. A plant using this technique has been successfully operated in South Africa, but was found to be economically impractical.

Electromagnetic Separation

In this process, ionized uranium gas is accelerated by an electric potential through a perpendicular magnetic field. The magnetic field deflects the circular trajectories of $^{235}UF_6$ and $^{238}UF_6$ to different extents. Two appropriately placed graphite catchers receive the two uranium isotopes. The great advantage of this method is that complete separation can be achieved by a single machine.

This technique was developed at Oak Ridge, TN during World War II for obtaining highly enriched uranium for the atom bomb project. However, it has proven to be more costly than gaseous diffusion enrichment and is no longer used for enriching nuclear reactor fuel. However, such magnetic separation devices, known as *cyclotrons* are routinely used today in medical facilities to extract radioactive isotopes of lighter elements for use in nuclear medicine.

Laser Isotope Separation

A recent enrichment technique uses the small difference in the electron energy levels of ^{235}U and ^{238}U caused by the different masses of their nuclei. For excitation of a particular electron energy level in ^{235}U and ^{238}U, this *isotope shift* is about 0.1×10^{-8} cm (10 nm) for light of wavelength 5027.3×10^{-8} cm, corresponding to a photon energy difference of 49.1 μeV about an energy of 0.0246 eV. Lasers with a bandwidth of 10 pm at this frequency are available, so that one of the isotopes can be selectively excited.

To separate ^{235}U from ^{238}U, uranium in a vacuum chamber is first vaporized with an energetic electron beam. Then a laser beam of precisely the correct frequency to excite ^{235}U, but not ^{238}U, is passed back and forth through the chamber. A second laser beam is then used to ionize the excited ^{235}U whose ion is subsequently removed by electric fields in the chamber without disturbing the vaporized, but still neutral, ^{238}U atoms. The same technique can also be applied to the selective ionization of UF$_6$ gas. Laser enrichment technologies are currently being actively developed.

11.6.3 Radioactive Waste

Radioactive waste is generated in all portions of the nuclear fuel cycle, ranging from slightly contaminated clothing to highly active spent fuel. The goal of radioactive waste management is to prevent any significant waste activity from entering the biosphere before the radionuclides have decayed to stable products. Clearly, the sophistication of the technology required for the safe sequestration of the radwaste depends on both the number, or activity, and the half-lives of the radionuclides in the waste. For example, low-activity radwaste with half-lives of a few days can be retained for several weeks to allow the activity to decay to negligible levels and then disposed of as ordinary (non-radioactive) waste. By contrast, some waste from spent fuel must be contained safely for hundreds of thousands of years.

Classification of Radioactive Wastes

There is no universally accepted classification of the many different types of radioactive wastes generated in our modern technological world. Several different national standards exist. However, for our discussion of the nuclear fuel cycle, the following classification scheme is useful to distinguish among the different radioactive wastes encountered.

High Level Waste (HWL): These are the fission products produced by power reactors. They are separated from spent fuel in the first stage of fuel reprocessing and are appropriately named because of their very large activity. In the once-through fuel cycle, spent fuel itself is discarded as waste and hence also classified as HLW, even though it also contains fissile fuel and transuranic isotopes.

Transuranic Waste (TRU): These wastes are composed of plutonium and higher Z-number actinides and have an activity concentration greater than 100 nCi/g. Such wastes are generated primarily by fuel reprocessing plants where transuranic fissile isotopes are separated from the fission products in spent fuel.

Mine and Mill Tailings: These are wastes from mining and milling operations and consist of low levels of naturally occurring radioactivity. The primary concern is the radioactive radon gas emitted from these wastes.

Low Level Waste (LLW): This is waste that has low actinide content ($<$ 100 nCi/g) and sufficiently low activity of other radionuclides that shielding is not required for its normal handling and transportation. This waste can have up to 1 Ci activity per waste package, but is generally of lesser activity distributed over a large volume of inert material. Such waste is usually placed in metal drums and stored in near-surface disposal sites.

Intermediate Level Waste (ILW): This category is rather loosely defined as wastes not belonging to any other category. This waste may contain $>$ 100 nCi/g of transuranic actinides and, generally, requires shielding when handled or transported. Such wastes typically are activated reactor materials or fuel cladding from reprocessing.

11.6.4   Spent Fuel

By far the most problematic of all radioactive wastes is that of spent fuel. During the three to four years a uranium fuel rod spends in a power reactor, much of the ^{235}U and a small amount of the ^{238}U are converted into fission products and transuranic isotopes. The typical conversion of uranium to these products in a LWR is summarized in Table 11.7.

Table 11.7. Composition (in atom-%) of the uranium in LWR fuel before use and the residual uranium and nuclides created by fission and transmutation after use. After Murray [2001].

New Fuel		Spent Fuel	
Nuclide	Percent	Nuclide	Percent
^{238}U	96.7	^{238}U	94.3
^{235}U	3.3	^{235}U	0.81
		^{236}U	0.51
		^{239}Pu	0.52
		^{240}Pu	0.21
		^{241}Pu	0.10
		^{242}Pu	0.05
		fiss. products	3.5

Of the hundreds of different radionuclides produced as fission products, only seven have half-lives greater than 25 years: ^{90}Sr (29.1 y), ^{137}Cs (30.2 y), ^{99}Te (0.21 My), ^{79}Se (1.1 My), ^{93}Zr (1.5 My), ^{135}Cs (2.3 My), and ^{129}I (16 My). The latter five, with such long half-lives, are effectively stable, and thus the long-term activity of fission-product waste is determined solely by ^{137}Cs and ^{90}Sr (in secular equilibrium with their daughters ^{137m}Ba and ^{90}Y). After 1000 years, the fission-product activity will have decreased by a factor of $\exp[-(1000 \text{ y} \ln 2)/30 \text{ y}] \simeq 10^{-10}$, an activity less than the ore from which the uranium was extracted.

However, some transuranic isotopes in the spent fuel have much longer half-lives than ^{90}Sr and ^{137}Cs, notably ^{239}Pu with a 24,000 y half-life. It is these transuranic actinides that pose the greatest challenge for permanent disposal of spent fuel, requiring isolation of this HWL from the biosphere for several hundred thousand years.

Spent fuel reprocessing, as currently practiced by some countries such as France and England, allows fissile isotopes to be recovered and used in new fuel. In addition, this reprocessing of spent fuel allows the fission products to be separated from the transuranic radionuclides. As discussed above, storage of the fission products requires isolation for only about a thousand years; the transuranic nuclides can be recycled into new reactor fuel and transmuted or fissioned into radionuclides with much shorter half-lives. Although spent-fuel reprocessing can greatly reduce the length of time the radioactive waste must be safely stored, it poses nuclear weapon proliferation problems since one of the products of reprocessing is plutonium from which nuclear weapons can be fabricated. By keeping the plutonium in the highly radioactive spent fuel, it is less likely that it will be diverted and used for weapons. For this reason the U.S. currently elects not to reprocess its spent fuel.

In addition to fission products and transuranic nuclides, spent fuel accumulates radioactive daughters of the almost stable ^{238}U and ^{235}U, which were also present in the original uranium ore (see Figs. 5.20 and 5.21).

HLW Disposal

If the spent fuel is reprocessed, the moist chemical slurries containing the separated fission products are first solidified by mixing the waste with pulverized glass, heating and melting the mixture, and pouring it into canisters where it solidifies into a glass-like substance from which the radionuclides resist leaching by water. If the spent fuel rods are not to be reprocessed, the fuel assemblies can be placed in a container for ultimate disposal or, alternatively, the rods can be bundled together in a container and consolidated into a single mass by pouring in a liquid metal such as lead to fill the void regions.

The resulting solidified HLW is then placed in a permanent waste repository where it will be safely contained until virtually all the radionuclides have decayed into stable products. How such isolation may be achieved for the many tens of thousand of years has been the subject of much study and public debate. Some proposed HLW disposal techniques are summarized in Table 11.8.

Although no permanent HLW repository has yet been placed in service, most countries, including the USA, are planning on using geological isolation in mines. In the USA, several sites around the country were investigated for geological suitability as a national HLW repository. Congress mandated in 1987 that the national USA HLW repository is to be established at Yucca Mountain about 160 km north of Las Vegas, Nevada, near the Nevada test site for nuclear weapons. Favorable characteristics of this site include a desert environment with less than seven inches of rain a year, a very stable geological formation, with the repository 2000 ft above the water table, and a very low population density around the site. Although no HLW waste has been stored at the Yucca Mountain repository, extensive site characterization and numerous experiments have been performed to validate the suitability of this site.

Table 11.8. Possible permanent HLW disposal strategies.

HLW Disposal Concept	Comment
Geological Disposal in Mines: Put waste in underground mined chambers, backfill chambers, and eventually backfill and close the mine.	This is the current U.S. planned disposal method. It requires long-term seismically stable geological formation and the exclusion of ground water from the waste.
Seabed Disposal: Let waste canisters fall into the thick sediment beneath the seabed floor in deep ocean waters. Modifications include placing waste in deep holes drilled in the seabed or in subduction zones at edges of tectonic plates so that the waste is eventually drawn deep into the earth's mantle.	Tests in the 1980s showed such seabed disposal is feasible with very low diffusion of radionuclides in the sediment. There are obvious environmental concerns. Also using international waters presents legal/political difficulties, and inaccessibility makes monitoring or recovery difficult.
Deep Hole Disposal: Place waste in deep holes, e.g., 10 km deep, so the great depth will isolate the waste from the bioshpere.	Drilling such deep holes is likely to be very expensive and is currently beyond current drilling technology.
Space Disposal: Launch waste into interstellar space, into a solar orbit, or into the sun.	Weight of the encapsulation to prevent vaporization in the atmosphere should launch fail makes this option very expensive.
Ice Cap Disposal: Place waste canisters on the Antartic ice cap. The decay heat will cause the canister to melt deep into the ice cap coming to rest on the bed rock. Refreezing behind the canister isolates the waste from the biosphere.	There are important economic uncertainties and obvious environmental concerns. Moreover, the use of Antartica poses difficult political problems.
Rock Melting Disposal: Place waste in a deep hole where the decay heat will melt the surrounding rock and waste. Upon eventual refreezing, the waste will be in a stable solid form.	This technology is not well developed. Geological and environmental concerns are not yet addressed.
Injection into Wells: Inject wastes as liquids or slurries into deep wells using technology similar to that used in the oil and gas industry.	This scheme is used by some countries for LLW; but liquid waste can migrate in underground formations and their long-term safe isolation from the biosphere is not certain.
Waste Processing and Transmutation: Chemically separate the fission products from the actinides (TRU). Then use a reactor or accelerator to transmute the TRU into higher actinides that decay more rapidly by spontaneous fission into relatively short-lived fission products.	This process converts the long-lived waste into fission products that need be stored safely for only several hundred years. This technology requires fuel reprocessing and the economic costs may be prohibitive.

As presently envisioned, spent fuel assemblies would arrival by rail at Yucca Mountain, be placed in storage containers consisting of a 2-cm inner shell of a nickel alloy that is very resistant to corrosion and a 10-cm thick outer shell of carbon steel. The waste containers would then be placed in concrete-lined horizontal tunnels in the repository atop support piers that allow uniform heat flow from the containers. Initially, the waste packages could be retrieved; but, at some future date, the storage chambers would be backfilled, after possibly coating the packages with a ceramic shield. With these multiple barriers around the spent fuel, it is expected that no water could reach the waste for at least 10,000 years.

Disposal of LLW and ILW

Besides the HWL of spent fuel rods, a nuclear power plant also generates a much greater volume of solid LLW, about 1 m^3 per 10 MW(e)y of electrical energy production. This waste consists of slightly contaminated clothing, tools, glassware, and such, as well as higher activity waste from resins, demineralizers, air filters, and so on. This solid waste is usually placed in drums and transported to a LLW repository where the drums are placed in near-surface trenches designed to prevent surface water from reaching the waste. A 1000 MW(e) nuclear power plant typically generates several hundred LLW drums a year.

Nuclear power plants also produce liquid LLW containing primarily tritium, which is readily incorporated into water as HTO molecules. It is not economically feasible to concentrate or separate the very small amount of HTO involved and, consequently, such tritiated waste water is usually diluted to reduce the activity concentration and then dispersed safely into the environment since the amounts involved are dwarfed by the natural production of tritium.

Finally, it should be mentioned that a modern hospital using nuclear medical procedures also generates large quantities of LLW, often more than a nuclear power plant. Such hospital wastes consist of ^{3}H, ^{14}C, and other radioisotopes that decay rapidly. Thus, after storage for a few weeks, only ^{3}H and ^{14}C remain. The amounts of these radioisotopes are negligibly small compared to their natural occurrence in the environment and could safely be disposed of by incineration and dilution of the exhaust gases. However, they are usually disposed of by shipping them to a LLW repository.

11.7 Nuclear Propulsion

Small power sources that can operate for extended periods without refueling are ideal for propulsion of vehicles that must travel large distances. Nuclear reactors are such power sources, and much development has gone into their use for propulsion. In the 1950s, the U.S. developed many designs and even some prototype reactors for nuclear powered aircraft. However, the obvious hazards posed by airborne nuclear reactors precluded the construction of such aircraft and this effort is now remembered as an interesting footnote in the history of nuclear power.

Today, nuclear power reactors are being planned as power sources for space missions, both as an electrical power source (see the next chapter) as well as a source of propulsion for deep space missions. However, by far the most successful use of nuclear power for propulsion has been in ships, particularly in modern navies.

11.7.1   Naval Applications

The potential of nuclear power for ships was immediately recognized. The ability of a nuclear powered ship to travel long distances at high speeds without refueling was highly attractive to the military. Moreover, a nuclear submarine could remain submerged almost indefinitely since the reactor needed no air, unlike the diesel engines used to charge batteries in a conventional submarine.

In 1946, the legendary Admiral Rickover assembled a team to design and build the first nuclear-powered ship, the submarine *Nautilus*. This prototype submarine used a small water-moderated pressurized reactor with highly enriched uranium fuel. In this design, the primary pressurized liquid water passes through a steam generator where secondary water boils and the steam is used to turn a turbine which, in turn, drives the propeller shafts of the submarine. The *Nautilus* was launched in 1954 and soon broke many submarine endurance records. It was the first submarine to reach the north pole by traveling under the Arctic ice cap, it travelled extended distances at speeds in excess of 20 knots, and travelled almost 100,000 miles on its second fuel loading.

Nuclear submarines can travel faster underwater than on the surface and can travel submerged at speeds between 20–25 knots for weeks on end. By comparison, World War II submarines could travel only a maximum speed of eight knots submerged for only an hour before surfacing to recharge their batteries. In 1960, the nuclear submarine *Triton* followed the route taken by Magellan in the sixteenth century to circumnavigate the world. The 36,000 mile voyage took Magellan nearly three years; the *Triton* completed the trip, entirely submerged, in 83 days! Nuclear power has revolutionized the strategic importance of submarines with their capability to launch missiles while submerged and to hunt submerged enemy submarines. At the end of the cold war in 1989, there were 400 nuclear-powered submarines, either operating or being built. Russia and the U.S. had in operation over 100 each, with the UK and France less than 20 each and China six. About 250 of these submarines have been scrapped or their construction cancelled as a result of weapons reduction programs. Today there are about 160 nuclear submarines in operation.

Nuclear powered surface vessels have also been added to the navy. Both the Soviet Union and the U.S. have deployed nuclear-powered cruisers, and several countries, including the U.S., have nuclear-powered aircraft carriers. The U.S. has the most nuclear-powered aircraft carriers, the first being the 1960 USS *Enterprise* powered by eight reactors, followed by nine other carriers with two reactors each.

All naval reactors are PWRs with compact cores fueled by rods composed of a uranium-zirconium alloy using highly enriched uranium (originally about 93% but today about 20–25% in the U.S. cores and about 50% in Russian cores). The cores can operate for 10 years without refueling, and newer designs produce cores with lifetimes of 30–40 years in submarines and 50 years in aircraft carriers. Maximum thermal power of these cores ranges up to 190 MW in large submarines and surface ships. The Russian, U.S., and British vessels use secondary-loop steam to drive a turbine which is connected, through a gearbox, to the propeller shafts. By contrast, the French and Chinese use the turbine to generate electricity for motor driven propeller shafts. All surface vessels since the *Enterprise* and all Russian submarines use two reactors; all other submarines are powered by a single reactor.

11.7.2 Other Marine Applications

The same benefits that nuclear power gives naval ships also apply to civilian ships. For cargo ships, nuclear power eliminates the need for oil fuel tanks or coal bins thereby making more space available for cargo. Also the higher cruising speeds and the greatly reduced time needed for refueling allow better ship utilization. Three nuclear merchant ships have been built and commissioned.

In 1959 the U.S. launched the *NS Savannah* a demonstration freighter which could also carry passengers (60 cabins). The *Savannah* was almost 600 feet long with a displacement, when fully loaded, of 20,000 tons. Her cruising speed was 21 knots and she was powered by a pressurized reactor using 4.4% enriched fuel and with a maximum thermal power of 80 MW. This first nuclear-powered commercial vessel was intended as a demonstration of the peaceful uses of nuclear energy and made many goodwill voyages to ports around the world for several years in the 1960s. It was decommissioned in 1970.

Germany built and operated Western Europe's first nuclear merchant ship, the *Otto Hahn*. This demonstration vessel used a pressurized water reactor, which was very similar to that used in the *Savannah*. The *Otto Hahn* sailed some 650,000 miles on 126 voyages over ten years without experiencing any technical problems. However, because of its high operation expense, it was converted to diesel power.

Japan launched the nuclear-power merchant ship *Mutsu* in 1962. This merchant ship also used a pressurized water reactor and was operated for several years. However, it suffered both technical and political problems, and was prematurely decommissioned and now resides at a naval museum.

From these three demonstration merchant vessels, several other advanced and larger nuclear-powered merchant marine vessels were designed. However, no nuclear-powered merchant ship operates today, primarily because of the large capital costs associated with nuclear ships compared to diesel-powered ships.

There is one country that has found nonmilitary ships to be technically and economically feasible. To keep its northern shipping lanes open in the winter, Russia operates several icebreakers. Ice breaking requires powerful ships which consume large amounts of energy. Diesel-powered icebreakers need frequent refueling and thus cannot navigate the entire Arctic basin. In 1959 the world's first nuclear-powered icebreaker, the *Lenin*, joined the Arctic fleet and remained in service for 30 years, although new reactors were fitted in 1970. Russia has since built several other nuclear-powered icebreakers. The large, two-reactor, Arktika-class icebreakers are used in deep Arctic waters. Such an icebreaker was the first surface ship to reach the North Pole. For shallower waters, Russia is now building the one-reactor, Taymyr-class icebreakers.

Although today non-military marine nuclear propulsion is not economically feasible, the U.S. nuclear navy has benefited the nuclear-electric industry in two important ways. Much of the technology developed for naval reactors has been widely used in the design of civilian pressurized-water power reactors. Also many highly skilled personnel in the civilian nuclear power industry have obtained their nuclear background from earlier service in the nuclear navy.

11.7.3   Nuclear Propulsion in Space

Because the energy content of nuclear fuel in a reactor is about 10^8 times that in an equal mass of chemical reactants, nuclear fission power offers far greater propulsive capability than conventional chemical rockets and, consequently, is ideally suited for deep space missions. Although no nuclear space propulsion systems have yet been launched, there have been extensive design studies and even ground tests of various ways of converting nuclear fission energy into propulsive thrust.

Two basic approaches are being pursued. The first called *nuclear thermal propulsion* mimics conventional chemical rockets in which reactants are combined to produce high-temperature gases which are allowed to escape at high speed from the rear of the rocket. An alternative, called *electric propulsion*, is first to convert nuclear thermal energy into electrical energy and then to use this energy in electromagnet devices to eject atoms at very high speeds from the rear of the spacecraft.

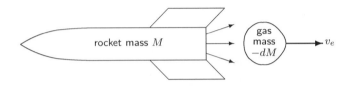

Figure 11.30. Rocket of mass M ejecting a mass of $-dM$ of hot gas with speed v_e in a time interval dt.

Thermal Propulsion

In thermal propulsion, a hot gas, whose atoms or molecules are moving with high speeds, is allowed to expand through the rocket nozzle and escape into space behind the rocket, thereby, providing a forward thrust. To understand the basic physics, consider a rocket of mass M that emits, in time dt, a mass $-dM$ from its rear with a speed v_e (see Fig. 11.30). Here dM is the mass *loss* of the rocket, a negative quantity. This gas release increases the rocket's speed by dv. From the principle of conservation of linear momentum, the gain in momentum of the space craft, $M\,dv$, must equal the momentum of the emitted gas, $(-dM)v_e$, namely, $M\,dv = (-dM)v_e$. From this relation we obtain

$$-\frac{dM}{M} = \frac{1}{v_e}dv. \tag{11.2}$$

Integration of this differential equation from the rocket's initial state, when it had mass M_o and speed v_o, to the time it has mass $M(v)$ and speed v, yields

$$-\int_{M_o}^{M(v)} \frac{dM}{M} = \frac{1}{v_e}\int_{v_o}^{v} dv, \tag{11.3}$$

from which we obtain

$$\frac{M(v)}{M_o} = \exp[-(v - v_o)/v_e]. \tag{11.4}$$

The fuel mass consumed to give the rocket a speed v is

$$M_f(v) = M_o - M(v) = M_o\left\{1 - \exp[-(v - v_o)/v_e]\right\}. \tag{11.5}$$

From this result we see that the fuel mass needed for a given increase in speed of the rocket, $v - v_o$, decreases as the speed of the ejected exhaust atoms or molecules increases. The average molecular speed in a gas in thermal equilibrium at an absolute temperature T is $v_e = \sqrt{(8kT)/(\pi M_e)}$, where M_e is the mass of the exhaust gas molecules and k is Boltzmann's constant $(1.3806503 \times 10^{-23} \text{ J K}^{-1})$. Because v_e is proportional to $\sqrt{T/M_e}$, to reduce the amount of fuel needed to achieve a rocket speed v, the temperature of the ejected gas must be increased and/or the molecular weight of the exhaust molecules must be decreased.

Present chemical rockets used for space launches combine liquid hydrogen with the oxidizer O_2 to produce water (H_2O) as the exhaust gas with a molecular weight of 18. By using nuclear reactors to heat hydrogen molecules (molecular weight 2) to the same temperature, the mass of the escaping hydrogen molecules would be nine times smaller, requiring almost three times less fuel for the rocket to achieve the same speed.

A very useful figure of merit for a rocket engine is the *specific impulse* I_{sp} defined as the thrust or force F exerted by the engine to accelerate the rocket divided by the fuel mass consumption rate (equal to the rate at which mass is lost by the rocket $-dM/dt$). The thrust is found from Eq. (11.2) by dividing by dt, namely,

$$F \equiv M\frac{dv}{dt} = \left(-\frac{dM}{dt}\right)v_e. \tag{11.6}$$

Hence the specific impulse is

$$I_{sp} \equiv \frac{F}{-dM/dt} = v_e. \tag{11.7}$$

To create a thermal nuclear rocket engine, a reactor core composed of heat resistant material such as graphite is cooled by pumping liquid hydrogen through components around the core to keep them cool, and then the resulting hydrogen gas enters channels in the core where it is heated to a temperature as high as the core can withstand. The hot hydrogen gas then expands through a nozzle to produce the rocket thrust.

Thermal nuclear engines were considered as early as 1946 for the initial designs of Intercontinental Ballistic Missiles (ICBMs); however, conventional chemical rockets were selected. In the early 1950s, nuclear engines were again considered by the U.S. Air Force for nuclear powered aircraft as well as for rocket use. In 1956, project Rover was begun at Los Alamos National Laboratory culminating in several tests of the Kiwi reactor engine. In the late 1950s, the Air Force lost interest in nuclear rockets since the chemically propelled ICBMs had proven themselves. With the launching by the Soviet Union in 1957 of Sputnik I, the National Aeronautics and Space Administration (NASA) developed an interest in nuclear rockets and the NERVA (Nuclear Engine for Rocket Vehicle Application) Program was started in 1960. A series of rocket engines, named Kiwi, NRX, Phoebus, Pewee, and XE', were built and tested throughout the 1960s at the Nuclear Rocket Development Station in Nevada. The NERVA program convincingly demonstrated the technical feasibility of thermal nuclear rocket engines, but it was terminated in 1973 when NASA withdrew its support.

During the NERVA program, several fuels were developed for graphite-moderated, once-through, hydrogen-cooled reactor cores.

- Beaded loaded graphite consisted of a graphite matrix containing a multitude of very small spheres of fuel coated with pyrocarbon. Reactor tests with this fuel achieved a temperature of 2500 K for 1 hour.

- A composite fuel of 30–35 volume% of UC.ZrC dispersed in graphite. This fuel could sustain temperatures of 2700 K for at least an hour.

- A pure carbide fuel such as UC.ZrC could maximize the reactor's time–temperature performance. However, this fuel is difficult to fabricate and insufficient testing was done. Temperatures around 3000 K are thought possible with this fuel, yielding a specific impulse I_{sp} of nearly 9.5 km/s, compared to chemical rocket engines which have specific impulses ranging from 1.5 to 4.5 km/s.

Although nuclear rockets are not used today, partially because of the present emphasis on near-earth manned space missions, NASA has a manned mission to Mars scheduled for 2014. To move the spacecraft between Earth and Mars, a NERVA-type nuclear engine would greatly reduce the transit time to Mars and back compared to using chemical engines.

Electric Propulsion

An alternative to thermal nuclear rockets is first to convert thermal energy from a reactor to electrical energy (see the next chapter), and then use this electrical energy to accelerate ions to very high speeds, passing them through a neutralizer in the rocket nozzle, to produce a beam of very fast moving neutral atoms leaving the rocket engine and producing a forward thrust. Several technologies have been proposed for creating the fast ions: (1) thermoelectric heating in which electricity is passed through the propellant to ionize and heat the gas, (2) electromagnetic devices to confine, heat, and accelerate a plasma of the propellant ions, and (3) electrostatic devices that accelerate the ions between two electric grids as in an ion accelerator.

Such electric rocket engines produce very low thrusts but also have very small fuel flows and high exhaust speeds v_e and, hence, high specific impulses I_{sp} (see Table 11.9). Design and construction of electric propulsion systems began in the 1940s and by 1990 more than 80 such systems were tested in orbital missions, the majority by the Soviet Union.

Table 11.9. Specific impulses of different rocket engines. After [Niehoff and Hoffman].

Engine Type	I_{sp} (km/s)
chemical	1.5–4.5
thermonuclear	8.3–9.2
electrothermal	8–12
electromagnetic	20–50
electrostatic	35–100

BIBLIOGRAPHY

ANGELO, J.A. AND D. Buden, *Space Nuclear Power*, Orbit Book Co., Malabar, FL, 1985.

BERLIN, R.E. AND C.C. STANTON, *Radioactive Waste Management*, Wiley, New York, 1989.

CE, *Nuclear Power Systems*, Combustion Engineering, Windsor, CN, 1974.

COCHRAN, R.G. AND N. TOULFANIDIS, *The Nuclear Fuel Cycle: Analysis and Management*, American Nuclear Soc., La Grange Park, IL, 1999.

CORLISS, W.R., *Nuclear Propulsion for Space*, USAEC, Div. of Tech. Information, 1969.

DOE, *A Technology Roadmap for Generation IV Nuclear Energy Systems, Generation IV International Forum*, GIF-002-00, U.S. Department of Energy, 2002, available at `http://www.inel.gov/initiatives/generation.shtml`.

DOE, *The U.S. Generation IV Implementation Strategy*, U.S. Department of Energy, 2003.

DOE, *Generation IV Nuclear Energy Systems, Ten-Year Program Plan, Fiscal Year 2005*, U.S. Department of Energy, Idaho National Laboratory, 2005.

EL-WAKIL, M.M., *Nuclear Energy Conversion*, International Textbook Co., Scranton, PA, 1971.

FOSTER, A.R. AND R.L. WRIGHT, *Basic Nuclear Engineering*, 2nd ed., Allyn and Bacon, Boston, MA, 1973.

GE, *BWR-6: General Description*, Nuclear Energy Div., General Electric, San Jose, CA, 1973.

GLASSTONE, A. AND A. SESONSKE, *Nuclear Reactor Engineering*, 4th ed., Vols. 1 & 2, Chapman & Hall, New York, 1994.

KNIEF, R.A., *Nuclear Engineering: Theory and Technology of Commercial Nuclear Power*, Hemisphere, Washington, 1992.

LAHEY, R.T. AND F.J. MOODY, *The Thermal Hydraulics of a Boiling Water Reactor*, 2nd. ed., American Nuclear Society, La Grange Park, IL, 1997.

LAMARSH, J.R. AND A.J. BARATTA, *Introduction to Nuclear Engineering*, Prentice Hall, Upper Saddle River, NJ, 2001.

LYERLY, R.L. AND W. MITCHELL, *Nuclear Power Plants*, Div. of Tech. Info., USAEC, 1969.

NEPD *National Energy Policy*, National Energy Policy Development Group, available at `http://www.whitehouse.gov/energy/National-Energy-Policy.pdf`, May 2001.

MURRAY, R.L., *Nuclear Energy*, Butterworth-Heineman, Boston, MA, 2001.

NIEHOFF, J.C. AND S.J. HOFFMAN, *Strategies for Mars*, Ch. 8; as reported by `http//www.marsacademy.com/propul`.

SIMPSON, J.W., *Nuclear Power from Undersea to Outer Space*, American Nuclear Society, La Grange Park, IL, 1994.

TONG, L.S. AND J. WEISMAN, 3rd. ed., *Thermal Analysis of Pressurized Water Reactors*, American Nuclear Society, La Grange Park, IL, 1996.

PROBLEMS

1. In a BWR or PWR, steam is generated with a temperature of about 290 °C. If river water used to receive waste heat has a temperature of 20 °C, what is the maximum possible (ideal) conversion efficiency of the reactor's thermal energy into electrical energy? Nuclear power plants typically have conversion efficiencies of 34%. Why is this efficiency less than the ideal efficiency?

2. A 1000 MW(e) nuclear power plant has a thermal conversion efficiency of 33%. (a) How much thermal power is rejected through the condenser to cooling water? (b) What is the flow rate (kg/s) of the condenser cooling water if the temperature rise of this water is 12 °C? Note: specific heat of water is about 4180 J kg^{-1} C^{-1}.

3. What are the advantages and disadvantages of using (a) light water, (b) heavy water, and (c) graphite as a moderator in a power reactor.

4. Why are the blades of a low-pressure turbine larger than those of a high-pressure turbine?

5. Why can a heavy-water moderated reactor use a lower enrichment uranium fuel than a light-water moderated reactor?

6. Explain whether the turbine room of a BWR is habitable during normal operation.

7. If the demand on the generator increases (i.e., a greater load is placed on the turbine), explain what happens to the reactor power of (a) a PWR and (b) a BWR if no operator-caused reactivity changes are made. Which reactor *follows* the load?

8. Although the steam cycle is simpler in a BWR, explain why the capital costs of BWR and PWR plants are very competitive.

9. Explain the advantages and disadvantages of using helium instead of water as a coolant for a power reactor.

10. During the April 1986 accident in the 1000-MW(e) RMBK Chernobyl reactor, the water in the cooling tubes of the graphite-moderated reactor was allowed, through operator error, to boil into steam and cause a supercritical, run-away chain reaction. The resulting energy excursion resulted in the destruction of the reactor containment and a large amount of the fission products in the fuel elements were released to the environment as the reactor containment ruptured. Explain why the boiling in the cooling tubes led to supercriticality.

11. How many years are required for the initial activity to decrease by a factor of 10^{10} for (a) ^{137}Cs, (b) ^{90}Sr, and (c) ^{239}Pu?

12. Over a period of one year, what mass (in kg) of fission products is generated by a 1000 MW(e) power reactor?

13. Discuss possible environmental, technical, and politcal problems associated with each of the disposal options listed in Table 11.8.

14. A nuclear drive in a submarine delivers 25,000 shaft horse power at a cruising speed of 20 knots (1 knot = 1.15 miles/h). If the power plant has an efficiency of 25%, how much (in kg) of the ^{235}U fuel is consumed on a 40,000 mile trip around the world?

15. Reactors for naval vessels are designed to have very long lifetimes without the need to refuel. Discuss possible techniques that can be used to maintain criticality over the core lifetime as ^{235}U is consumed.

16. A thermal nuclear rocket using hydrogen as the propulsive gas operates for one hour at a thermal power of 4000 MW and a temperature of 2700 K. Estimate (a) the amount (in grams) of fissile material consumed, (b) the specific impulse of the engine.

17. Estimate the specific impulse I_{sp} for (a) a chemical rocket burning hydrogen and oxygen at a temperature of 4000 K, and (b) a thermal nuclear rocket emitting hydrogen at 3000 K.

18. A nuclear rocket propulsion system uses an ion drive to accelerate mercury atoms to energies of 50 keV. Estimate the specific impulse of this drive.

Chapter 12

Fusion Reactors and Other Conversion Devices

At the present time, nuclear fission reactors are the only practical devices for producing large amounts of electrical power from nuclear energy. However, the potentially immense source of energy available through nuclear fusion has motivated much research to develop a practical fusion reactor. Today the first fusion device is being constructed that is expected to produce more fusion energy than is required to operate the reactor. Although it will be many decades from now before fusion reactors generate commercial electricity, the promise of almost limitless energy justifies the present large research effort devoted to this energy source. In this chapter, we review some of the basic physics, history, and devices being contemplated for converting fusion energy to electricity.

Methods are also discussed in this chapter for directly converting nuclear radiation or the thermal energy produced by the radiation into electrical energy. Such direct conversion has been an attractive engineering challenge because converters based on this technology have the advantage of simplicity, reliability, quietness of operation, and a long lifetime since no moving parts are generally needed. Over the past several decades, a few such converters have been developed. Unlike nuclear power plants which produce thousands of megawatts of electrical power, these direct conversion devices typically produce only hundreds of watts to microwatts of electrical power. Nevertheless, these converters have found specialized applications for which low power levels are adequate, such as space satellites, remote meteorological weather stations, and heart pacemakers.

Finally in this chapter, specialized nuclear fission reactor concepts are presented for converting thermal fission energy into electrical power either through conventional thermodynamic power cycles or by direct conversion. The development of these specialized reactors has been motivated by the need for electrical power for space applications at levels in excess of the capabilities of direct conversion devices.

12.1 Fusion Reactors

In Section 6.7 we saw that the fusing of two light nuclei releases energy. Since the late 1940s, there has been sustained efforts at trying to harness this fusion energy as a practical power source. As discussed earlier, it is necessary to create a plasma of ionized nuclei and electrons heated to high temperatures so that the Coulomb repulsion between the ions can be overcome and allow nuclear forces to fuse two colliding nuclei. Although stars use gravity to contain the plasma in their cores, a practical fusion device must rely on either magnetic fields or inertial confinement techniques to hold the plasma together long enough for the ions to fuse. Because of the huge potential of fusion energy, major efforts have been made in the past sixty years to devise fusion reactors that produce more energy than is required to create and heat the plasma. Progress has be agonizingly slow, however, and only in 2006 has an international consortium agreed to build the world's first fusion reactor that is expected to be a net energy multiplier. In this section, we present some basic properties of plasma physics, the history of fusion devices, and the future directions of fusion reactors.

12.1.1 Energy Production in Plasmas

To create a plasma, a gas such as deuterium or tritium must be heated to high temperatures so that atomic collisions can ionize one of the colliding neutral atoms. Consider a hydrogen gas at temperature T containing n_n and n_i neutral atoms and ions per cubic meter. The free electron density n_e is assumed to equal the ion density n_i. The total density of protons, either in neutral atoms or as ions, is $n = n_n + n_i = n_n + n_e$. The ionization fraction $f \equiv n_i/n$ of such a gas in thermal equilibrium is given by the Saha equation

$$\frac{f^2}{1-f} = \frac{1}{n} \left(\frac{2\pi m_e kT}{h^2} \right)^{3/2} \exp(-I/kT), \tag{12.1}$$

where T is the gas temperature in K, k is Boltzmann's constant, and I is the ionization energy required to remove the outermost electron from a neutral atom. For example, for hydrogen isotopes $I = 13.06$ eV and with $n = 2 \times 10^{21}$, 95% of the atoms are ionized at $T = 13,150$ K. By contrast, at room temperature, the ionization fraction predicted by Eq. (12.1) is a ridiculously small 1.5×10^{-106}.

Although many fusion reactions are possible (see Section 6.7), the reactions currently of most interest for thermonuclear power generation are the D-T fusion reaction

$$^2_1\text{H} + {}^3_1\text{H} \longrightarrow {}^4_2\text{He} \ (3.5 \text{ MeV}) + {}^1_0\text{n} \ (14.1 \text{ MeV}), \qquad Q_{fus} = 17.6 \text{ MeV} \tag{12.2}$$

and the D-D fusion reaction

$$^2_1\text{H} + {}^2_1\text{H} \longrightarrow \begin{cases} {}^3_2\text{He} \ (0.82 \text{ MeV}) + {}^1_0\text{n} \ (2.45 \text{ MeV}), & Q_{fus} = 3.27 \text{ MeV} \\ {}^1_1\text{H} \ (3.02 \text{ MeV}) + {}^3_1\text{H} \ (1.01 \text{ MeV}), & Q_{fus} = 4.03 \text{ MeV} \end{cases} . \tag{12.3}$$

In the plasma, the ions have a distribution of energies well-described by a Maxwellian distribution. If n_1 and n_2 are the density of the two interacting ion species, the reaction rate is $R = n_1 n_2 \langle v\sigma \rangle$, where $\langle v\sigma \rangle$ is the value of $v(E)\sigma(E)$ averaged over the Maxwellian energy distribution of the ions, namely [Angulo 1999],

$$\langle v\sigma \rangle = \left[\frac{8}{\pi\mu k^3 T^3} \right]^{3/2} \int_0^\infty \sigma(E) E \exp(-E/kT) \, dE, \tag{12.4}$$

where μ is the reduced mass of the system. In terms of masses of the reactants m_1 and m_2, $\mu = m_1 m_2/(m_1 + m_2)$. The variation of $\langle v\sigma \rangle$ with the plasma temperature is shown in Fig. 12.1 for the above three fusion reactions. For each fusion reaction, the energy released is the Q_{fus}-value. The fusion power density generated by the plasma is thus

$$P_{fus} = n_1 n_2 \langle v\sigma \rangle Q_{fus}. \tag{12.5}$$

From the results of Fig. 12.1 we see that, for the same ion densities, the deuterium-tritium (D-T) reaction is "easier" to achieve than the two deuterium-deuterium (D-D) reactions, which from the figure are seen to occur with almost equal probabilities.[1] For this reason it is the D-T reaction that is the basis of most experimental fusion research.

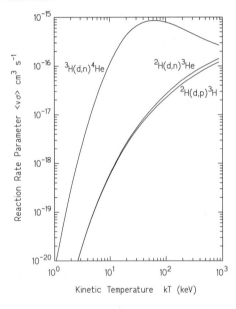

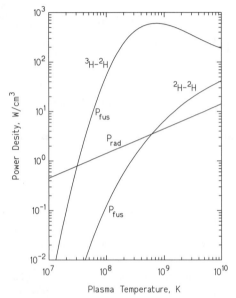

Figure 12.1. The Maxwellian averaged reaction rate parameter $\langle v\sigma \rangle$ for the D-D and D-T fusion reactions as a function of the plasma kinetic temperature kT. Data source: Angulo et al.[1996].

Figure 12.2. Fusion power production and bremsstrahlung power loss in a plasma with an ion density of $n_i = 10^{15}$ cm^{-3} as a function of the plasma temperature.

A plasma continuously loses energy. Some energy is lost in heating and ionizing the incoming gas which refuels the plasma. A plasma also radiates away some of its energy, primarily by bremsstrahlung production from interacting electrons and ions. At very high plasma temperatures, energy is also lost by synchrotron radiation produced by the energetic charged particles, primarily the electrons, moving in the magnetic fields used to confine the plasma. The rate at which bremsstrahlung energy loss occurs, per unit volume, is given by [El-Wakil 1977]

$$P_{rad} = 1.42 \times 10^{-34} Z^2 n_i n_e \sqrt{T}, \quad \text{W cm}^{-3} \text{ s}^{-1}, \tag{12.6}$$

---

[1] For the D-T reaction, $n_1 = n_2 = n_i/2$; however, for the D-D reaction $n_1 = n_2 = n_i$ and the right-hand side of Eq. (12.5) must be multiplied by $(1/2)$ to avoid counting the same reactions twice.

where n_i and n_e are the density or ions and electrons (cm^{-3}), respectively, and normally $n_i = n_e$.

12.2 Magnetically Confined Fusion (MCF)

For a fusion reactor to produce useful amounts of energy, the fusion power P_{fus} must exceed the energy needed to heat the plasma to compensate for radiative energy losses. In Fig. 12.2 the fusion power and radiative power loss are shown for a D-D and D-T plasma with an ion density of 10^{15} cm^{-3}. When the power loss equals the power generated, the plasma is said to be at the *critical ignition temperature* T_c. From Fig. 12.2 it is seen that once again the D-T reaction is superior to the D-D reaction, having a $T_c \simeq 3 \times 10^7$ K compared to about 6×10^8 K for a D-D plasma.

12.2.1 Fusion Energy Gain Factor

A useful measure of a fusion reactor is the *fusion energy gain factor*,[2] usually given the symbol Q, is the ratio of the fusion power density P_{fus} to the externally supplied power required to heat a unit volume of the plasma and maintain the plasma in a steady state. Some of the plasma heating power comes from the charged fusion products that transfer their initial kinetic energies to the thermal energy of the plasma. The remainder, P_{heat}, must be supplied by external power sources. The Q factor is thus defined as $Q = P_{fus}/P_{heat}$.

The density of the reaction-product heating power is $f_c P_{fus}$, where f_c is the fraction of fusion energy carried away by the charged products. For the D-T reaction, $f_c = 1/5$. The externally supplied power density P_{heat} is eventually lost by various processes that occur mostly at relatively low temperature, precluding any recovery as electrical power. In a D-T fusion power reactor, electrical power comes from recovering the energy of the fusion neutrons which stream through the plasma containment and deposit their kinetic energy in a *blanket* surrounding the plasma chamber. In the blanket, the neutrons heat a working fluid, such as liquid lithium, which is then used to produce electrical power at an efficiency[3] η_{elect}. The electric power produced is $P_{elect} = \eta_{elect}(1 - f_c)P_{fus}$. A fraction f_{recirc} of this power is then used to run the fusion reactor. Some is used for pumps, magnetic fields, lighting, etc., but the bulk is used to heat the plasma and the incoming new fuel needed to maintain the plasma. The efficiency with which P_{elect} is converted to the power needed to maintain the plasma is denoted by η_{heat}. Thus, $P_{heat} = \eta_{heat} f_{recirc} \eta_{elect}(1 - f_c)P_{fus}$, giving a gain factor of

$$Q \equiv \frac{P_{fus}}{P_{heat}} = \frac{1}{\eta_{heat} f_{recirc} \eta_{elect}(1 - f_c)}. \tag{12.7}$$

Because a fusion power plant is designed to produce electricity for external consumption, f_{recirc} must be substantially less than unity, say 0.25. If we assume $\eta_{heat} \simeq 0.7$, $\eta_{elec} \simeq 0.35$, then for a D-T plasma, Q must be around 20. The goal

---

[2]This must not be confused with the a nuclear reaction Q-value.

[3]The neutrons can be absorbed by lithium nuclei to release additional nuclear energy and also to produce tritium. Thus, the blanket becomes an "energy multiplier" and a source of tritium to fuel the fusion reactor.

that a plasma heat itself without *any* externally supplied energy requires an infinite Q factor. The case of $Q = 1$ is called the *break-even* condition and requires a D-T plasma to generate 20% of the plasma heating power from the alpha particles produced, since $E_\alpha/Q_{fus} = 3.5$ MeV/17.6 MeV $\simeq 0.20$.

12.2.2   Confinement Times

One method to confine a plasma is to use magnetic fields generated by currents flowing in wire coils placed around the plasma chamber. The electrons and ions spiral around the magnetic lines of force. By applying differently shaped and time varying magnetic fields, the plasma can be constrained spatially and forced to move in a closed path that keeps the plasma from coming into contact with the walls of the vacuum chamber containing the plasma. Should an ion or elctron hit the chamber wall, it would be quickly cooled and be lost from the plasma.

Plasmas confined by magnetic fields are susceptible to a variety of instabilities and eventually the high energy particles will leak from the confining field and be lost. However, the confinement time must be long enough to generate a significant amount of fusion energy and to help heat, with the charged fusion reaction products, new fuel needed to replenish the plasma. By contrast, the confinement time should not be so long that the plasma particles, particularly the electrons, gain so much energy that radiative energy losses become very large. Thus, there should be an optimum confinement time for a plasma of a given composition, density, and temperature.

A simple expression for the optimal confinement time was proposed by Lawson [1957]. He argued that, in a plasma that has ignited, the fusion power density that goes into heating the plasma, P_{heat}, must exceed the power density lost to the environment, P_{loss}. The *energy confinement time* τ_E is defined as

$$\tau_E = \frac{\text{energy content of the plasma}}{P_{loss}} = \frac{1}{P_{loss}}(n + n_e)\frac{3}{2}kT = \frac{3nkT}{P_{loss}}, \qquad (12.8)$$

where we have used the fact that the average kinetic energy of the electrons and ions with a Maxwellian distribution is $3kT/2$ and that $n_e = n$. The heating power P_{heat} is obtained from Eq. (12.5) with Q_{fus} replaced by the kinetic energy E_c of all charged fusion products, e.g., $E_c = 3.5$ MeV for the D-T reaction. Thus, Lawson's criterion becomes

$$n_1 n_2 \langle \sigma v \rangle E_c \geq 3nkT/\tau_E. \qquad (12.9)$$

For the D-T reaction, $n_1 = n_2 = n/2$, and rearrangement of this equation gives[4]

$$n\tau_E \geq F \equiv \frac{12}{E_c}\frac{kT}{\langle \sigma v \rangle}. \qquad (12.10)$$

Figure 12.3 shows the variation of F with kT for both the D-D and D-T fusion reactions. From this figure it is seen that for a plasma to reach break-even, $n\tau_E > 10^{14}$ s cm^{-3} for D-T fusion and $n\tau_E > 10^{16}$ s cm^{-3} for T-T fusion. Once again, it is easier to make a break-even device with a D-T plasma than with a D-D plasma.

---

[4]For the D-D reaction, $F \equiv 6kT/(E_c\langle\sigma v\rangle)$ and $E_c = (4.03 + 0.82)/2 = 2.43$ MeV (assuming the two branching ratios have equal probability).

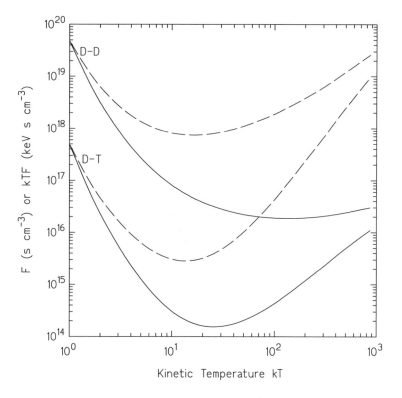

Figure 12.3. The variation of F (solid lines) and kTF (broken lines) with the plasma temperature for the D-D and D-T fusion reactions.

12.2.3 Triple Product Figure-of-Merit

As fusion research matured, a more useful figure-of-merit was found to be $n\tau_E T$, which from Eq. (12.10) is given by

$$n\tau_E T \geq F \equiv \frac{12}{E_c} \frac{kT^2}{\langle \sigma v \rangle}. \tag{12.11}$$

To see why, we observe that the plasma temperature and density can vary widely. But the maximum plasma outward pressure p_{max}, which must be balanced by the applied magnetic fields, is limited. From the ideal gas law $p = nkT$, and Eq. (12.5) can be written as

$$P_{fus} = \frac{1}{4} \frac{p_{max}^2}{k^2 T^2} \langle \sigma v \rangle Q_{fus}. \tag{12.12}$$

Hence, for a given reactor, the maximum power is obtained at a plasma temperature that minimizes $T^2/\langle \sigma v \rangle$.

In particular, for tokamak-like devices,[5] this triple product is found to be rather insensitive to ion density and temperature and, thus, is a good measure of the

---

[5]Tokamak is a transliteration of Russian words meaning "toroidal chamber in magnetic coils."

plasma confinement arrangement. This insensitivity follows from the empirical observation that the confinement time is proportional to $(n/P_{heat}^2)^{1/3}$, where P_{heat} is the plasma heating power. For a plasma operating at its optimum temperature and that has reached break-even, the heating power equals the fusion power P_{fus} which is proportional to n^2/T^2. Thus,

$$n\tau_E T \propto nT \left(\frac{n}{P_{heat}^2} \right)^{1/3} \propto nT \left(\frac{n}{n^4/T^4} \right)^{1/3} \propto T^{-1/3}, \qquad (12.13)$$

showing that $n\tau_E T$ is independent of ion density and only weakly dependent on the plasma temperature. The triple product $n\tau_E T$ has a minimum value at slightly lower temperature than does $n\tau_E$ as seen in Fig. 12.3. For a D-T plasma,

$$n\tau_E kT \gtrsim 2 \times 10^{15} \text{ keV s cm}^{-3}. \qquad (12.14)$$

This inequality is now often referred to as "Lawson's criterion."

12.2.4   Plasma Heating

In an operating fusion reactor, some of the charged particles produced by fusion give up their energy to the plasma and thus heat it. To start up a fusion reactor and keep the plasma at its operating temperature, it is necessary to supply external energy to the plasma. Several techniques can be used.

Ohmic Heating

Because a plasma can conduct electricity, a current can be passed through the plasma to heat it. The heating power is simply I^2R, where I is the current and R is the resistance. The most common way to force a current to flow in a closed plasma is through inductive coupling. A plasma current is induced by slowly increasing the current in a primary transformer coil linked with the plasma torus which forms the secondary winding of the transformer. However, with this method, the primary current cannot continue to increase forever, and the reactor can be operated only for short periods. Also as the plasma becomes hotter, ohmic heating becomes less effective because the plasma resistance decreases. Above about 10-20 MK, other heating methods must be used.

Neutral-Beam Injection

The gaseous fuel needed by the plasma can be ionized externally to the machine, accelerated to high energies (hundreds of keV) by a particle accelerator, neutralized by adding back the missing electrons, and then directed into the plasma where they quickly give up their kinetic energy to the plasma and also become reionized and trapped by the confining magnetic fields. The reason the injected atoms must be neutralized is that, otherwise, the magnetic fields confining the plasma would prevent the atoms from reaching the plasma.

Magnetic Compressions

Just as a gas is heated when compressed so is a plasma. By quickly increasing the confining magnetic field, a plasma is heated. In a tokamak, for instance, this is accomplished by moving the plasma radially inward to a region of higher magnetic fields. This also increases the ion density and thereby increases the fusion reaction density.

Radio-Frequency Heating

Because ions and electrons spiral or rotate about the confining magnetic field lines, electromagnetic waves matched to the ion or electron frequencies can resonate with the plasma particles who then absorb the electromagnetic energy. These absorbing particles then collide with other plasma particles thereby increasing the plasma temperature. Radio-frequency radiation (tens of MHz) is generated outside the plasma with gyrotrons or klystrons and radiated into the plasma by placing antennas inside the vacuum chamber. This heating scheme has the advantage of occuring at precise locations in the plasma.

12.2.5   History of Magnetically Confined Fusion Reactors

Although the first fusion experiments were done in England in the 1930s, it was then thought that obtaining energy from such reactions was very unlikely. However, the highly successful World War II Manhattan Project, which developed the atomic bomb, motivated renewed interest in nuclear physics and fusion in particular. In Britain, a small toroidal pinch device was constructed in 1952 by Cousins and Ware, followed later by a large stabilized toroidal pinch device called the Zero Energy Toroidal Assembly (ZETA). With ZETA, much useful data was gathered during 1954 to 1958 to enable the construction of even larger devices.

In the U.S., Lyman Spitzer at Princeton proposed in 1951 the *stellerator* concept. This type of machine uses strong magnetic fields to confine the plasma in a torus shaped reaction vessel as modern tokamaks do but without the tokamak's current drive system, which can lead to plasma disruptions and instabilities. Stellerators were used in the 1950s and 1960s in the U.S. and Europe to develop better understanding of plasma physics. About the same time, Richard Tuck began to explore magnetic pinch devices at Los Alamos National Laboratory.

Secret work on fusion devices and plasma physics began in the Soviet Union soon after World War II. With a thaw in the cold war in 1956, Soviet fusion researchers in Russia and Britain began the world's first fusion collaborations. Fusion research also began in other countries such as France and Germany. Prompted by the Geneva *Atoms for Peace Conference* in 1958, secret fusion research was declassified and, since then, fusion research has operated in the open with the normal exchange of scientific information. This openness led immediately to the development of a facility at Culham Laboratory which later became the home of the *Joint European Torus* (JET).

Ten years later in 1968, two Russians Sakharov and Tamm startled the fusion community by showing breakthrough results from their *tokamak* machine that produced remarkably stable plasmas with temperatures ten times higher than previously produced.[6] Many such devices were then used in many fusion experiments and now it has become the dominant concept for fusion reactors. The tokamak JET was begun in 1978 and began operation in 1983. In Japan, the tokamak JT-60 began operation in 1985.

The promise of the tokamak design was made stronger when, in 1991, the JET device produced a power of 1.7 MW from fusion reactions, and later in 1993 the

---

[6]The name tokamak is an acronym of a Russian phrase that translates to "toroidal chamber in magnetic coils."

Princeton Tokamak Fusion Test Reactor (TTFT) produced 10 MW of fusion power. Today there are 16 tokamak-like experimental devices, constructed between 1975 and 2006, that are still in operation.

12.2.6   The ITER Fusion Reactor

In the early 1990s, an international effort was undertaken to design a new generation of tokamak reactors that were intended to produce, for the first time, more fusion energy than the energy needed to create the plasma. The first such research fusion reactor is called ITER, with the name associated with various acronyms such as "International Thermonuclear Energy Reactor" but now simply Latin for "the way."

The international membership has changed over the years, but in June 2005 the U.S., Russia, European Union, India, China, Japan, and South Korea agreed to finance and construct ITER at Cadarache in the south of France. Construction has already begun and first operation is expect in 2016. The facility is then expected to operate for the next 20 years. ITER will cost about $6 billion to construct and an equal amount to operate it over its lifetime. The ITER project, after the International Space Station, becomes the second most expensive scientific experiment in history. Although ITER will not produce any electricity, it will provide a wealth of technical information about many critical components of a practical fusion reactor that can produce electricity. An artist's cutaway view of the ITER is shown in Fig. 12.4. Its size dwarfs the human in this picture.

The ITER is designed to produce 500 MW of fusion power from D-T reactions for extended periods of time (300 to 500 s) and to produce 10 times more power than is needed to maintain the plasma, i.e., to have a Q-factor of 10. There are roughly 85 different modules or subsystems that have to be developed and integrated and that are necessary for future fusion power stations. For example, technology for building and operating large superconducting magnets, eventually using liquid lithium for cooling the chamber wall and to breed tritium fuel, demonstrating a tritium closed cycle, and remote handling to replace major components, to name just a few.

The ITER Magnets

Superconducting magnets are used exclusively to produce the magnetic fields that confine the plasma inside the toroidal vacuum chamber. The different fields and magnets are shown in Fig. 12.5. ITER uses 18 toroidal magnets surrounding the vacuum chamber producing closed magnetic lines of force parallel to the chamber walls and around which the ions and electrons spiral. This toroidal field is the primary mechanism for confining the plasma. Six poloidal magnetic coils surround the perimeter of the torus and a central solenoid coil provides an inner poloidal field. The poloidal field is used to keep the plasma away from the chamber walls and to maintain the shape and stability of the plasma. In addition, The inner poloidal field coils are wrapped around an iron transformer core and act as the primary winding of a transformer circuit so that current flowing in the central solenoid induces a large secondary transformer current of 15 MA to flow in the plasma around the torus chamber. This inductive current is a primary mechanism for heating the plasma. Though not shown in Fig. 12.5, there are correction coils around the machine outside the toroidal field coils. These correction magnets are used to adjust errors in the magnetic fields cause by manufacturing or alignment inaccuracies as well as to increase the plasma stability.

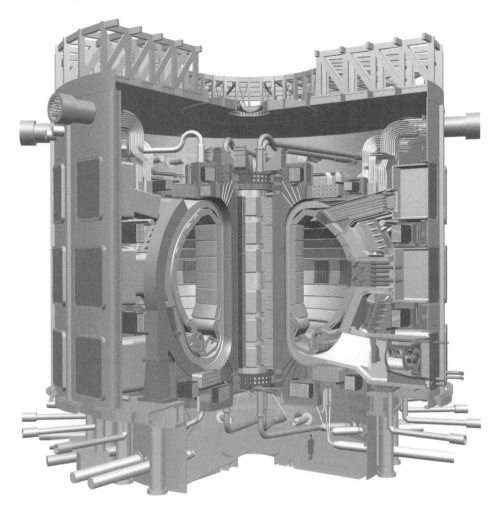

Figure 12.4. The proposed ITER fusion reactor. The D-shaped toroidal plasma chamber will contain a circular D-T plasma with a radius of 6.2 m and maximum half width of 2 m. [Published with permission of ITER.]

Blanket Modules

The fusion energy is absorbed by the components lining the inside surface of the vacuum chamber, namely, the blanket, divertor, and the port plugs. There are 440 blanket modules which form the bulk of the inner chamber surface. Facing the plasma, a module has beryllium armor attached to a copper substrate mounted on a stainless steel support cooled by pressurized water. The water and stainless steel also act as a neutron shield. Eventually, the water cooled component may be replaced by lithium cooled components to demonstrate a tritium breeding capability. The heat load on the inner chamber walls is about 0.4 MW m^{-2} and the incident fast neutron (14.1 MeV) flux density is about 4×10^{13} neutrons cm^2 s^{-1}. Over time, this neutron bombardment of the blanket modules makes them very radioactive

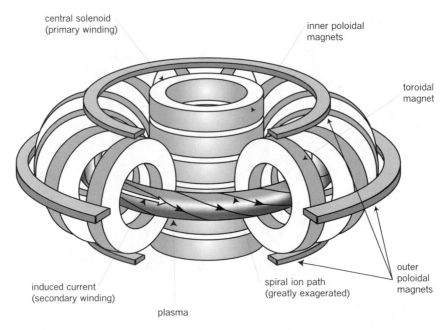

Figure 12.5. The major magnetic coils used in the ITER. The toroidal field confines the plasma to move around the torus while the poloidal field keeps the plasma away from the chamber walls and also provides the current that heats the plasma.

and compromises their structural integrity. Consequently, remote replacement of the blanket modules is necessary.

Divertors

At the bottom of the plasma chamber are 54 divertor modules that remove helium formed in the fusion process and other impurities produced by fusion products reacting with the chamber walls. These modules also remove some of the fusion power and, hence, like the blanket modules, must also be water cooled. Because of uncertainties in the heat flux, these modules have an uncertain lifetime and, consequently, they must also be able to be replaced remotely.

Port Plugs

Because of the experiment purpose of the ITER project, it is necessary to be able to have access to the plasma to make diagnostic measurement as well as to replacement chamber components and to replenish and heat the fuel. Consequently, the ITER plasma chamber will have many ports that allow access to the plasma. There will be 17 equatorial ports used for radio-frequency plasma heating, neutral-beam injection, testing different blanket modules, plasma diagnostics, and allowing remote handling of chamber components. There will be 9 divertor ports on the bottom of the chamber used for cryopumps, diagnostics, a glow-discharge cleaning system, pellet and gas injection, and in-vessel viewing. Finally, there are 18 upper ports used for diagnostics and a cyclotron current drive for plasma stability control.

Bioshield

A concrete bioshield surrounds the magnets and plasma vessel. This shield lowers the neutron doses outside the shield so that equipment there is not appreciably activated and allows personnel access to this equipment soon after the reactor is shut down. Inside the shield, however, the radiation dose rates can be quite extreme even with the reactor shut down, and, generally, remotely controlled machines are needed to work in these high radiation areas.

12.3   Inertial Confinement Fusion (ICF)

In this approach for extracting fusion energy, a small pellet about 1 to 3 mm in diameter and containing several milligrams of tritium-deuterium fuel is compressed and heated rapidly by short pulses (1 to 10 ns long) of energetic laser beams, ion beams, or x rays that simultaneously hit the pellet from many directions. The hot outer layer quickly blows off the pellet, producing a reaction force that accelerates the rest of the pellet inward and sends a shock wave into the center. This shock wave compresses the fuel in the center to many times the density of lead and heats it to temperatures sufficiently high (e.g., 10^8 K) that fusion reactions occur. The charged particles produced by the fusion reactions then further heat the surrounding fuel and causes a chain reaction throughout the compressed fuel that produces many more fusions before the pellet explodes outward. The purpose is to have the pellet "ignite" so that an exploding pellet produces many times more fusion energy than is needed to produce the compressive input energy. In practice, only a small portion of the pellet fuel actually fuses.

The goal of ICF is to have a large *target gain* G, defined as the ratio of the energy yield from fusion to the energy supplied by the compressive beams. To obtain a large G, there must be a large fraction of the fuel fused during the inertial confinement time τ_E. For a compressed spherical target of radius R, the confinement time can be approximated by the time it takes an ion to travel a distance R, its thermal speed taken as the speed of sound $v = \sqrt{kT/\overline{m}}$, namely,

$$\tau_E = R/v = R\sqrt{\frac{\overline{m}}{kT}}. \tag{12.15}$$

Here $\overline{m}$ is the average mass of the fusing ions. The burn criterion, known as the ρR condition or the *high-gain condition*, requires that the number of fusion reactions during the confinement time τ_E must nearly equal the number of deuterons or tritons, i.e., that nearly all the fuel is consumed. From Eq. (12.5) with $n_1 = n_2 = n/2$ for a D-T plasma, the maxmium reaction rate density is $n^2\langle v\sigma \rangle/4$. At the end of the burn, the reaction rate density is almost zero, so the *average* reaction rate density is $\overline{RR} = n^2\langle v\sigma \rangle/8$. The requirement that all the ions fuse is $\overline{RR}\tau_E V \simeq (n/2)V$, where the pellet volume V cancels out. Thus,

$$\frac{n^2}{8}\langle v\sigma \rangle \tau_E \simeq \frac{n}{2}. \tag{12.16}$$

Substitution of Eq. (12.15) into this result, recognition that the plasma mass density is $\rho = \overline{m}n$, and rearrangement yields the ρR criterion [Rosen 1999]

$$\rho R \gtrsim 4\frac{\sqrt{\overline{m}kT}}{\langle v\sigma \rangle}. \tag{12.17}$$

For a D-T plasma $\rho R \gtrsim 3$ g/cm^2 at $kT = 50$ keV. From Eq. (12.16), Lawson's criterion for ICF is seen to be $n\tau_E \gtrsim 4/\langle v\sigma \rangle$, and the triple product criterion is

$$nT\tau_E \gtrsim \frac{4T}{\langle v\sigma \rangle}. \tag{12.18}$$

The triple product $nT\tau_E$ for ICF is 10–20 times greater than that for mangetically confined fusion (MCF). This occurs because assembling the fuel is less efficient in ICF. Finally, it should be noted that the term "ignition" in ICF has a different meaning than in MCF. In ICF, ignition refers to the condition that all alpha particles are stopped and thus heat the plasma. This requires a ρR value of at least 0.2 g/cm^2 [Lindl 1995].

The system that produces the compressive input energy is known as the *driver*. Drivers such as lasers and ion accelerators that deliver the compressive energy pulse to the fuel pellet directly are known as *direct-drive* devises. There is also an *indirect-drive* ICF technique in which the fuel pellet is placed inside a small metal cylinder or sphere, known as a *holraum*,[7] or a "burning chamber." Lasers are then focused on the outside of the holraum heating it into a very hot plasma which then radiates x rays. These x rays are then absorbed by the pellet compressing it in the same ways as if hit directly by the laser. This indirect-drive approach is considerably less efficient at delivering compressive energy to the fuel pellet. However, because thermonuclear bombs use a similar technique to compress the thermonuclear fuel, the indirect-drive technique is used mostly to simulate nuclear weapon tests.

To produce electricity from ICF, fuel pellets would be dropped into a vacuum reaction chamber where they would be ignited by a brief pulse of compressive energy. The radiation from the hot plasma and the fusion neutrons would travel to the reaction chamber wall where they and their energy would be absorbed by liquid lithium. Besides being a heat conductor and thermalizer for neutrons, lithium also absorbs the neutrons producing tritium which can be recycled as new fuel. The hot liquid lithium can then be pumped through a steam generator whose steam then runs a conventional turbogenerator system, as is done in liquid metal fast breeder reactors. Pellets could be dropped and exploded at a rate governed by how rapidly the driver system can be recharged, typically every few seconds.

12.3.1　History of ICF

The vast majority of the ICF research has been performed in the U.S. as a result of its potential applications in thermonuclear weapon testing. Edward Teller soon after World War II recognized the potential for such research and soon started initial classified efforts at Los Alamos and Lawrence Livermore National Laboratories. After the discovery of the ruby laser in the early 1960s, initial experiments were conducted by shining high-intensity laser light into a hydrogen plasma. But the power available with these early lasers was far from what was needed to produce abundant fusion reactions, although they did provide a basis for developing the fundamental theory describing laser-plasma interactions. It was recognized that the major challenge for successful ICF was the development of a powerful laser with sufficiently uniform beams to implode the target symmetrically.

---

[7]German for "empty room."

However, laser development progressed rapidly and, by the early 1970s, several U.S. laboratories started experiments using a variety of new lasers such as the krypton fluoride excimer laser at the Naval Research Laboratory (NRL) and the neodymium glass laser at Lawrence Livermore National Laboratory (LLNL). In the early 1970s, there was great concern about future energy resources and high funding for both magnetically confined and inertially confined fusion was provided. LLNL was especially funded to develop an ICF program. Starting with the neodymium-glass Janus laser in 1974, followed by the Long Path, Cyclops, and Argus lasers, many technical problems such as focusing the multiple beams were explored and confidence was gained that a laser much larger than the Cyclops could ignite fuel pellets. ICF experiments using lasers producing multiple hundred joule pulses allowed rapid gains in the fusion energy produced by ICF and, by the end of the 1970s, ICF devices were producing about the same below break-even results as the best magnetically confined devices.

In the late 1970s and early 1980s, however, laser instabilities and coupled energy loss mechanisms were discovered and gradually understood. The estimate of the energy needed to ignite a fuel pellet soared and it was realized that the "exploding-pusher" fuel pellet (gold spheres filled with D-T gas) and kilojoule lasers would not lead to ignition. As a consequence, new designs for 100 kJ lasers and the ablative cryogenic D-T ice pellets were formulated.

The first major effort at a large ICF driver was the Shiva system, a 20-beam, neodymium-dope, glass laser, which went into operation LLNL in 1978. Shiva succeeded in compressing a target to 100 times liquid hydrogen density, but fusion yields were low as a result of the laser's strong interaction with the target electrons that produce premature plasma heating. The solution was to use optical frequency multipliers to triple the frequency of the infrared laser light into the ultraviolet. In the 1980s such frequency tripling was used with the 24-beam OMEGA laser, later with the NOVETTE laser, and then with the Nova laser, a device ten times more powerful than the Shiva laser and one that was thought could cause target ignition.

But once again the ICF community was disappointed. Large intensity imbalances among the beams and within each beam resulted in assymetric implosion of the target and hydrodynamic instabilities in the plasma. This failure led to a better understanding of the implosion process and the need for better beam uniformity and intensity balance among the beams. Although funding for fusion research decreased in the 1980s, enough information from the Nova system was obtained that designs for an even better ICR device were started.

This latest ICF machine is the National Ignition Facility (NIF), currently under construction at LLNL, and is now scheduled to be completed in 2009. When fully complete, the NIF laser will have 192 beams delivering 1.8 MJ of ultraviolet laser energy and 500 TW of power to a mm-sized pellet at the center of a 10-m diameter target chamber. To eliminate wavefront aberrations, adaptive optics are used, and to smooth and filter the beams so they retain their specific spatial and temporal characteristics, opto-mechanical components are employed. With the NIF it is hoped the way is set to demonstrate an ICF device that will produce significantly more fusion energy than is used to create the implosion of the target.

Finally, another concept has recently been proposed. This method called *fast ignition* first compresses the target using a conventional laser drive system. Then when the target has reach its maximum density, a second ultra-short ultra-high energy PW laser delivers a single pulse focused on one side of the target. In the simple *plasma bore-through* method, this second pulse will burn its way through the outer corona plasma and impinge on the dense core raising it to extremely high temperature and leading, it is hoped, to pellet ignition. In another approach, called the *cone in shell* method, the target is mounted on the vertex of a high-Z cone such that the tip projects into the core of the target. Then when the pellet is imploded, the second laser pulse hits the dense pellet core from within the cone and does not need to waste energy boring through the coronal plasma.

12.3.2   ICF Technical Problems

To achieve successful ICF, a number of technical problems must be solved. First the fuel pellet must have a very precise spherical shape if the shock wave is to be accurately focused to the pellet's center. Spherical abberations of the inner and outer surfaces must be less than a few microns. Also to achieve a properly focused shock wave, the laser/ion beams must be aimed extremely precisely to hit the pellet symmetrically all within a picosecond. With delay lines in the optic paths of the multiple laser beams, the beams can be made to reach the target within the required picosecond accuracy. Less easy to solve are the so-called "beam-beam" imbalance and the beam anisotropy problems. A beam-beam imbalance occurs when beams deliver slightly different amounts of energy to the target surface, thereby producing asymmetric implosion of the fuel pellet and reducing the central temperature/density buildup. The beam anisotropy problem refers to nonuniform energy distribution across a beam diameter. Such nonuniformity produced hot spots on the illuminated particle surface which, in turn, lead to uneven compression that forms Raleigh-Taylor plasma instabilities in the fuel. These hydrodynamic instabilities in the fuel plasma cause premature mixing of the hot and cool plasma and, thus, reduce the plasma temperature at maximum compression.

Over the past twenty-five years, ways have been found to mitigate, to some extent, most of these difficulties. Target design has improved significantly. Early targets were hollow plastic or gold spheres filled with high-pressure tritium-deuterium gas. Today, cryogenic tritium-deuterium ice is used to fill hollow plastic spheres. Infrared lasers or the natural beta decay of tritium are used to smooth the inner surface and microscope cameras are used to ensure the smoothness of the pellet. Likewise various beam smoothing techniques have been developed to minimize nonuniform energy distribution in the beam, and diagnostic procedures have been developed to balance the energy among the beams. Not so easy to solve, however, are the plasma hydrodynamic instabilities.

The other major development over the past few decades of research into ICF has been in the development of ever more energetic drivers. Early ICF laser drivers had energies of a few joules and several kilowatts. Today lasers that can produce picosecond pulses with energies of megajoules and power of terrawatts are planned. These improvements arise from using frequency doubling or tripling techniques using neodymium glass amplifiers.

Heavy ion-beam drivers are also attractive because they are easy to use, control, and focus the beam. However, it is difficult to achieve a high energy density as a result of the mutual repulsion of the ions in the beam. Most ion-beam drivers use a holraum to smooth out the radiation incident on the target, thereby reducing the energy transfered to the target from the beam.

12.3.3 Prospects for Commercial Fusion Power

The path to producing commercial electric power from fusion machines is daunting. Many steps, each requiring decades of effort will be required. First, break-even machines that produce as much fusion energy as is needed to operate the machine must be demonstrated, i.e., machines with a Q-factor ≥ 1. The ITER reactor is expected do this around 2016. ICF machines will take much longer. Because the conversion of electrical energy into laser energy is only 1–1.5% efficient and the conversion of thermal energy into electricity is at best 35% efficient, the fusion energy must be hundreds of times greater than the energy of the laser beams compressing the fuel pellet. Although new ways to excite laser, such as replacing the present xenon flash tubes by frequency-tuned laser diodes whose light is more readily absorbed by the lasing medium, may increase the laser efficiencies to 10% or even higher. Such lasers when combined with the *fast ignition* method, which has a energy gain of about 100, may be able to produce a break-even ICF machine. Moreover, these break-even machines must be able to produce power either continuously or in short bursts many hundreds of times a day. In ICF machnes these would mean tens of implosions per second; but many existing laser drivers can operate only on a daily basis.

After a $Q = 1$ machine has been demonstrated, machines with a high gain $G > 15 - 20$ must be built and operated. Then an industrial demonstration machine must be developed that validates all the techical options and components needed for an electric power station. Finally, a commerical demonstration plant will have to be built that shows a fusion power plant can operate for many years in a safe and reliable manner. Although, fusion power is often said to be superior to fission power because no long lived fission products are produced, the intense fast neutron fields produced by D-T fusion will quickly compromise the integrity of materials exposed to such radiation, and frequent remote replacement of highly radioative components will have to be done routinely and safely in such a commercial demostration plant.

Clearly, the realization of commercial electrical power from fusion machines is a great many decades away. But the immense power potentially available from fusion power motivates mankind to continue the jouney along this lengthy path.

12.4 Thermoelectric Generators

If two wires or rods, made of different metals, are joined and their junction placed at a different temperature than their opposite ends, a voltage is produced across the unjoined ends. This effect was discovered by Seebeck in 1822; however, because only milliamperes of current at a fraction of a volt are produced by metal wires, the Seebeck effect was used only in thermocouples to measure and control temperatures. Only with the discovery of semiconductors in the late 1950s were materials discovered that could produce useful amounts of electrical power.

The basic operation of a thermoelectric cell is shown in Fig. 12.6. Both p-type and n-type semiconductors are used. These semiconductors are made by introducing impurity atoms into the crystal matrix. In a p-type semiconductor, the

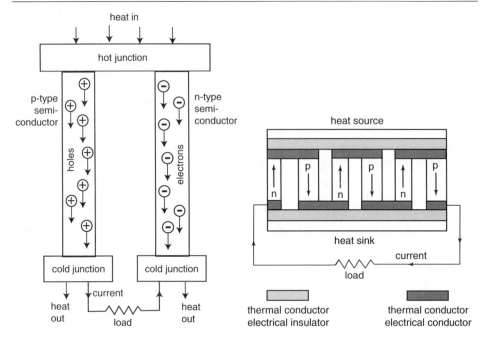

Figure 12.6. A thermoelectric converter cell. After El-Wakil [1982].

Figure 12.7. A three-cell thermopile. After El-Wakil [1982].

impurity atoms have fewer valence electrons than the matrix-lattice atoms so that the resulting crystal lattice has positive *holes* which move easily through the lattice as positive charges. By contrast, in an n-type semiconductor, the impurity atoms have more valence electrons than the lattice atoms. As a result, the lattice has extra free negative electrons. When n-type and p-type materials are joined and the junction heated, the holes and free electrons tend to move away from the hot junction towards the cold junction, much like a heated gas expands and diffuses away from hot regions. This flow of charge, in turn, produces a current through an external load attached to the two cold junctions of a thermoelectric cell.

A thermoelectric converter cell is a low-voltage (a few tenths of a volt), high-current (tens of amperes) device. To obtain useful amounts of power and reasonable voltages, several cells are connected together in series to form a *thermopile*. A simple thermopile is shown in Fig. 12.7.

With the discovery of tellurides and selenides, thermoelectric devices with conversion efficiencies up to 10 percent have been constructed. Very high efficiency is achieved with n-type PbTe and p-type BiTe-SbTe semiconductors, which can operate at temperatures up to 680 °C. Higher operating temperatures can be achieved with silicon and germanium semiconductors, and research continues in finding better materials for higher efficiency thermoelectric generators.

Because there are no moving parts in thermoelectric cells, they tend to be very reliable. However, they have several limitations. It is generally difficult to fasten the semiconductors to the hot junction and encapsulation is needed to prevent chemical contamination of the semiconductor elements. The cell also tends to be fragile and needs appropriate containment.

The use of thermoelectric cells to convert thermal energy into electrical energy has been very successful for specialized applications in which small compact power units with a long lifetime are needed and for which high cost compared to conventional electric power is acceptable. They have found application in remote navigational buoys, weather stations, and space satellites and probes.

12.4.1 Radionuclide Thermoelectric Generators

Any high-temperature thermal energy source can be used for a thermopile. One source is the decay heat from radionuclides. The layout of such a *radioisotope thermoelectric generator* (RTG) is shown in Fig. 12.8. Many RTGs of varying designs and power capacities have been made and tested in the last 40 years. For example, in Fig. 12.9, the SNAP-7B RTG is shown. This 60-W(e) RTG uses 14 tubular fuel capsules containing pellets of ^{90}Sr-titanate whose radioac-

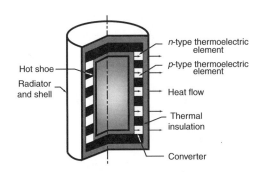

Figure 12.8. A thermoelectric isotopic power generator. After Mead and Corliss [1966].

tive decay (half-life 29.1 y) provides the input thermal energy. Around these central fuel tubes are 120 pairs of lead telluride thermoelectric cells that convert the thermal energy into electrical energy. The acronym SNAP stands for *S*ystem for *N*uclear *A*uxiliary *P*ower. Many other RTGs (the odd numbered SNAP series) have been built and used for a variety of terrestrial and space applications (see Table 12.1).

Since 1961, the U.S. has launched 26 space missions that have carried over 40 RTGs to provide part or all of the electrical power needs. These RTGs had electrical power capacities ranging from a few W(e) up to 285 W(e). Many of the early RTGs used in these missions are listed in Table 12.1. Although designed for lifetimes generally of 5 years or less, many of these early deep space probes still continue to operate. For example, Pioneer 10, launched in 1972, carried four SNAP-19 RTGs producing an initial total power of about 155 W(e). Having survived transits of the asteroid belt and the intense radiation field of Jupiter, by (2002) it was about 80 AU[8] and was heading away from the sun at 2.6 AU/y. Although routine tracking and collection of data were stopped in 1997 for budgetary reasons, the ship was still tracked occasionally for training purposes. However, the last signal from Pioneer 10 was received on 23 January 2003.

For the two Voyager missions to the outer planets (1977), a multihundred-watt (MHW) RTG was developed. Using decay heat from ^{238}Pu dioxide, 312 SiGe thermoelectric couples (replacing the lead-telluride couples used in the earlier SNAP RTGs) produced about 157 W(e) at the beginning of the mission. Three such MHW-RTGs were used on each Voyager mission. For the Galileo mission to Jupiter (1989), the Ulysses Solar polar orbit mission (1990), and the Cassini mission to Sat-

---

[8]One astronomical unit (AU) equals the average distance of the earth from the sun or about 1.50×10^8 km.

Table 12.1. Early (1960–1975) U.S. SNAP radioisotope power generators.

SNAP No.	Function	Fuel	Power (We)	Dia. × Ht. (cm)	Mass (kg)	Design Life
3	demonstration	^{210}Po	2.5	12×14	1.82	90 d
3A	satellite power	^{238}Pu	2.7	12×14	2.10	5 y
–	weather station	^{90}Sr	5	46×51	764	2 y min
7A	navigation buoy	^{90}Sr	10	51×53	850	2 y min
7B	navigation light	^{90}Sr	60	56×88	2100	2 y min
7C	weather station	^{90}Sr	10	51×53	850	2 y min
7D	floating weather station	^{90}Sr	60	56×88	2100	2 y min
7E	ocean bottom beacon	^{90}Sr	7.5	51×53	273	2 y min
7F	offshore oil rig	^{90}Sr	60	56×88	2100	2 y min
9A	satellite power	^{238}Pu	25	51×24	12	5 y
11	moon probe	^{242}Cm	23	51×30	14	90 d
13	demonstration	^{242}Cm	12	6.4×10	1.8	90 d
15	military	^{238}Pu	0.001	7.6×7.6	0.5	5 y
17	communication satellite	^{90}Sr	25	61×36	14	5 y
19	Nimbus weather satellites	^{238}Pu	30	56×25	14	5 y
19	Viking/Pioneer missions	^{238}Pu	45			5 y
21	deep sea application	^{90}Sr	10	41×61	230	5 y
23	terrestrial uses	^{90}Sr	60	64×64	410	5 y
27	Apollo lunar modules	^{238}Pu	60	46×46	14	5 y
29	various missions	^{210}Po	500		230	90 d

Data Sources: Mead and Corliss [1966], Furlong and Wahlquist [1999].

urn (1997) even larger power modules were needed. The heat source for this refined RTG was a so-called general purpose heat source (GPHS) which contains 4 ^{238}Pu dioxide radioisotope pellets, each weighing 150 g and encapsulated in iridium and graphite and placed in a module 5.4×9.4×9.4 cm in size. The GPHS-RTG, shown in Fig. 12.10, uses 18 of these GPHS modules to initially produce about 4300 W of thermal energy. This 56-kg GPHS-RTG unit has a length of 1.13 m and a diameter of 0.43 m.

Each GPHS-RTG produces about 290 W(e) at the beginning of service from the 4.3 kW(t) produced by the decay heat of ^{238}Pu. The temperatures of the hot/cold junctions of the SiGe thermoelectric semiconductors are about 1000/300 K. The radioisotope modules are clad in iridium and then encased in graphite to provide integrity of the plutonium should accidental reentry into the atmosphere occur. The graphite also provides impact protection and the iridium post-impact containment. The 1989 Galileo (Jupiter) mission contained two of these GPHS-RTGs, the 1990 Ulysses probe in a large-radius solar polar orbit used one, and the 1997 Cassini Saturn mission used three to produce about 890 W(e) of power at launch.

Besides the RTGs used on a space mission, many low-power radioisotope heat sources are also used to provide heat to sensitive components of the space craft. For example, in the 1989 Galileo Jupiter mission, 120 light weight radioisotope

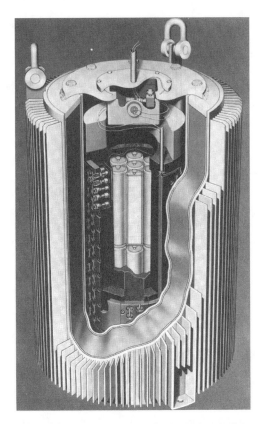

Figure 12.9. The SNAP-7B 60-W thermoelectric isotopic generator.

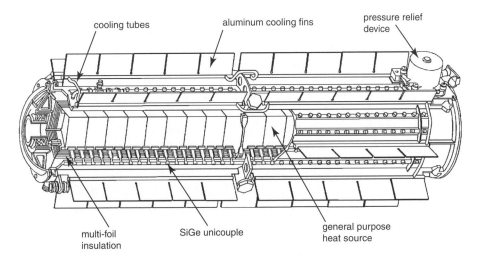

Figure 12.10. A general purpose heat source radioisotope thermoelectric generator (GPHS-RTG) currently used in the U.S. space program. After Furlong and Wahlquist [1999].

heater units (LWRHU) were used to warm critical components and instruments. Each LWRHU contained 2.68 g of ^{238}Pu dioxide and produced 1 W(t) of heat. The fuel pellet has a platinum-rhodium clad and is encased in a multilayer graphite containment to ensure the integrity of the pellet should a launch fail.

Further details of the use of radioisotopes in U.S. space missions are provided in a recent article by Furlong and Wahlquist [1999].

Reactor Thermoelectric Generators

The even-numbered SNAP series of thermoelectric devices use a small nuclear reactor to heat a liquid coolant that, in turn, heats the hot junctions of the thermoelectric cells. Such a thermal energy source can provide far more energy than a radionuclide source. Several preliminary designs were made, but never used. The first nuclear reactor to provide thermal energy for thermoelectric energy conversion was the SNAP-10A system, which was used to provide power to a space satellite launched in April of 1965. The SNAP-10A system is shown in Fig. 12.11. Its liquid metal NaK coolant is used as the thermal energy source for the thermoelectric conversion cells (see Fig. 12.12). The cold side of the thermoelectric cells is connected to a conical thermal radiator (shown in Fig. 12.13 below the SNAP-10A reactor).

12.5 Thermionic Electrical Generators

A thermionic generator converts thermal energy directly into electrical energy. In its simplest form (see Fig. 12.14), it consists of two closely spaced metal plates. One plate (the *emitter* or *cathode*) is heated to a high temperature to boil electrons from its surface into the gap between the plates. The second plate, at a much lower temperature, collects the electrons and is called the *anode* or *collector*. In this manner, a potential difference is developed between the two plates, which, in turn, can be used to produce a current through an external electrical load.

The minimum thermal energy required to boil an electron from the emitter surface is called the *work function* and equals the work that must be done against the electric field produced by the atoms near the surface of the emitter. For example, in tungsten the work function is about 4.5 eV. For a thermionic generator to function, the work function of the emitter must be greater than that of the collector.

As electrons boil into the gap between the emitter and collector, a negative space charge is created which inhibits the flow of electrons forcing some back towards the emitter. To mitigate this *space-charge* effect, the gap between electrodes is made very narrow (typically, 0.02 cm), and, more effectively, a gas such as cesium vapor, which readily ionizes to form a plasma, is placed between the electrodes. The positive ions of the interelectrode gas counteract the negative electric field of the electrons, thereby allowing more of the electrons boiled from the emitter surface to reach the collector. One possible design of a thermionic cell is shown in Fig. 12.15.

Any source of heat can be used in a thermionic generator. For low power applications, the decay heat from radioisotopes can be used, while, for higher power applications, the heated coolant from a compact nuclear reactor can be used.

12.5.1 Conversion Efficiency

A thermionic generator is a heat engine, in which thermal energy Q_e is added to the emitter and thermal energy Q_c is rejected at the collector (see Fig. 12.14). If there

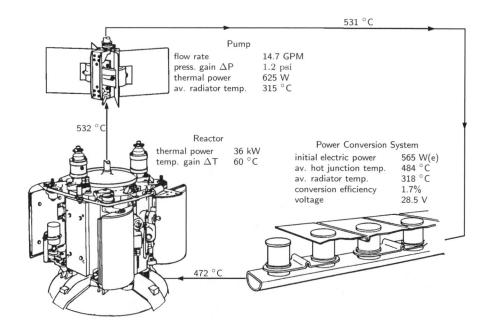

Figure 12.11. The SNAP-10A power system. Power for the NaK-coolant Faraday electromagnetic pump is provided by integrated thermoelectric elements. The compact 25×30 cm core containing 37 ZrH fuel/moderator elements of 10% enrichment is surrounded by 4 movable beryllium reflectors used for control. After El-Wakil [1982].

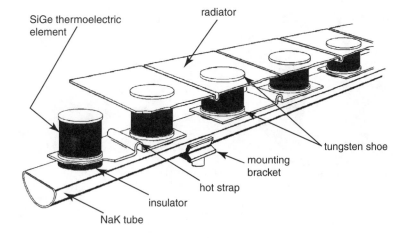

Figure 12.12. The SNAP-10A thermoelectric conversion pile. Forty such piles comprise the conical radiator shown below the reactor in Fig. 12.13. After Corliss [1966].

Figure 12.13. The SNAP-10 reactor atop the conical radiator of the thermoelectric conversion pile. Atop the reactor is the NaK coolant pump. From Corliss [1966].

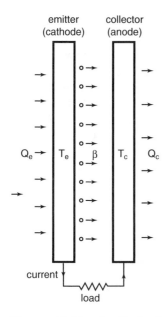

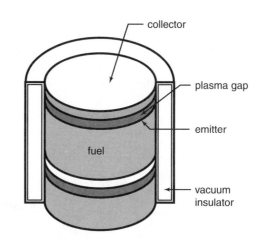

Figure 12.14. An idealized planar thermionic energy cell. From El-Wakil [1982].

Figure 12.15. A simple thermionic radioisotope power generator. After Foster and Wright [1973].

were no energy losses besides Q_c, then the amount of energy converted to electrical energy would be $Q_e - Q_c$ and a thermal conversion efficiency of $\eta = (Q_e - Q_c)/Q_e$ would be achieved.

From the laws of thermodynamics, such a heat engine can have an efficiency no greater than the ideal Carnot efficiency $\eta_C = (T_e - T_c)/T_e$, where T_e is the absolute temperature of the surface at which the heat energy is added (the emitter) and T_c is the absolute temperature of the surface at which heat is rejected into the environment (the collector).

In the absence of energy losses (other than Q_c), the conversion efficiency of a thermionic cell would approach that of the ideal Carnot efficiency. Thus, to achieve high conversion efficiencies, very high emitter temperatures need to be used, typically in excess of 1400 K. However, the ideal Carnot conversion efficiency cannot be realized in practice because of energy losses in the cell. These include thermal radiative and conductive losses between the two electrodes, and electrical losses in the electrode contacts. Because the emitter and collector are necessarily very close, a major challenge in designing thermionic generators is to insulate the high-temperature emitter from the collector. As much as half the heat produced by the radioisotopes may escape the emitter. In addition, the buildup of impurities on the surface of the collector can seriously degrade its electron absorption ability.

Typical conversion efficiencies range from less than 1% to as much as 10% and produce potential differences ranging from 0.3 to 1.2 V per cell. To obtain higher voltages and greater power, several cells may be connected in series.

Radionuclide Thermionic Generators

Many thermionic generators using the decay heat of radioisotopes have been designed. One of the earliest type of thermionic cells is the so-called *isomite battery* produced by the McDonnell Douglas Co. (see Fig. 12.16). Several prototype isomite cells have been made and tested. These small cells, which are only 1 to 3 cm in height and diameter, have relatively low emitter temperatures (700 K to 1400 K) and, consequently, low conversion efficiencies of less than 1%. Although cesium vapor is present, it is at such low pressures ($< 10^{-2}$ torr) that the cells operate as if there were a vacuum between the electrodes. The cesium is used only to improve the work functions of the emitter and collector surfaces.

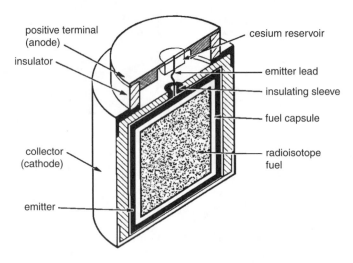

Figure 12.16. An isomite thermionic power cell. From El-Wakil [1982].

Eleven to 87 grams of ^{147}Pm or ^{238}Pu are used to provide a thermal heat source of between 0.3 and 3.5 W. An interelectrode gap of between 0.025 and 0.25 mm is used. The current densities in these isomite batteries are low (0.1 to 400 mA/cm^2), with an output voltage of between 0.1 and 0.15 V to yield power outputs of between 1 and 20 mW(e).

12.5.2   In-Pile Thermionic Generator

To produce higher temperatures for thermionic cells and, thus, greater conversion efficiencies, several efforts have been made to incorporate thermionic cells into a nuclear reactor core. The type of reactor most suitable for in-pile thermionics is a small compact fast reactor using liquid metal as the coolant. Thermal reactors are not suitable because the high temperatures needed for thermionics are not compatible with moderators such as water and beryllium. Moreover, reactors for space applications must be small and have low mass to reduce the launch cost.

In a reactor thermionic system, the cladding around the fuel rod can serve as the cathode. The outer anode, cooled by the reactor's liquid metal coolant, is separated by a small cesium-vapor filled gap from the cathode. Such an integrated thermionic cell is shown in Fig. 12.17. Alternatively, the thermionic cells can be placed externally to the core, heated by the reactor coolant, and cooled by a radiator which emits waste heat into the coldness of space.

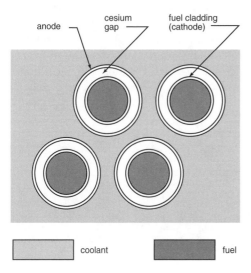

anode — cesium gap — fuel cladding (cathode) —

coolant   fuel

Figure 12.17. An in-pile thermionic cell with internal fuel. After El-Wakil [1982].

Several U.S. and English in-pile experiments have been performed. However, only Russia has produced an in-pile thermionic system that has been deployed in its MIR space programs. Between 1970 and 1984, prototypes were ground tested, and two TOPAZ units were sent into space with the COSMOS satellite. Advanced TOPAZ-II reactors using liquid lithium coolant have been constructed for possible future missions to Mars. The Russian TOPAZ technology has been purchased and tested by the U.S. for evaluation of potential use in its space program. However, the U.S. has yet to launch any reactor-based thermionic system.

The great advantage of a reactor thermionic generator is the realization of power as great as several kW(e). By contrast, thermionic systems, fueled by decay heat from radioisotopes, usually achieve power outputs of, at most, several tens of watts. Moreover, with UO_2 ceramic fuel, very high operation temperature of around 1700 K can be achieved so that conversion efficiencies near 10% can be realized.

Nevertheless, construction of in-pile thermionic fuel assemblies is a daunting challenge since such an assembly must include an emitter, a collector, insulators, spacers, sheathing, cladding, and fission-product purge lines. All of these components must maintain their close tolerances over extended operating times. So far, experience has shown these thermionic assemblies can indeed operate successfully for several years. However, longer lifetimes will be needed for some space applications. Finally, the intense radiation environment produced by the reactor requires careful consideration be given to material selection and configuration design. Typically, a reactor thermionic system is placed at the end of a long boom with a shadow shield between the reactor and the rest of the payload.

12.6 AMTEC Conversion

A relatively new method for converting thermal energy into electrical energy uses an *alkali metal thermal to electric converter* (AMTEC). This technology is based on the unique properties of the ceramic β-alumina. This material is a solid electrolyte (BASE) that, while being an electrical insulator, readily conducts sodium ions. When sodium vapor is present at two different pressures separated by this electrolyte, an electrochemical potential is established. Typically, the β-alumina is formed into a cylindrical tube with metal electrodes attached to the inner and outer surfaces. Eight such BASE tubes are shown in Fig. 12.18. Hot pressurized sodium vapor is fed into the central cavity of the BASE tube where it gives up an electron at the inner electrode (cathode) and the resulting sodium ion Na^+ is thermally driven through the electrolyte as a result of cooler lower-pressure sodium vapor on the outside of the tube. As the Na^+ ion reaches the outer electrode (anode), it acquires an electron and becomes a neutral sodium atom. The potential difference thus established between the two electrodes can be used to supply electrical power to an external load. To increase the output voltage, several BASE cells may be connected in series as in Fig. 12.18.

Figure 12.18. The BASE cells and sodium artery in an AMTEC converter. Each cylindrical cell is 1 cm in diameter with a 2.5 cm active electrode length. Courtesy of Advanced Modular Power Systems, Inc.

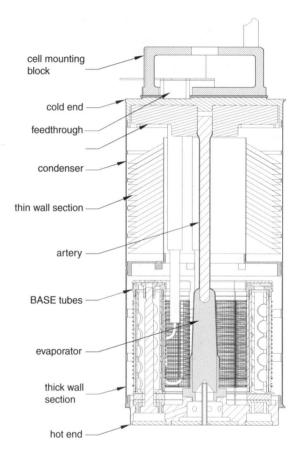

Figure 12.19. An AMTEC converter. This design is 10 cm by 5.1 cm, weighs about 400 g, and produces about 8.5 W(e) at 3 A. Courtesy of Advanced Modular Power Systems, Inc.

In an AMTEC converter, the sodium is the working fluid whose cycle provides the thermal energy that is converted to electricity. The cycle begins with liquid sodium being heated and vaporized by some external heat source. The hot pressurized sodium vapor flows into the cavity of a BASE cell and then migrates through the electrolyte wall creating a potential difference between the inner and outer electrodes. The sodium vapor leaving the outer BASE cell surface then diffuses to a cold surface where it is cooled and condensed to a liquid in order to reduce the sodium vapor pressure at the outer surface of the cells. The liquid sodium is then transported back to the heat source where it is vaporized and the cycle is repeated.

A recent design of an AMTEC converter is shown in Fig. 12.19. In this converter, designed for use in deep space missions, liquid sodium is heated and vaporized in an evaporator attached to the converter's "hot end" which is heated by a radioisotope source. The hot pressurized sodium vapor flows from the evaporator into the inner cavity of the BASE cells. As the sodium ions migrate through the electrolyte wall, some of the thermal energy is given up to create the potential difference between the BASE electrodes. The sodium vapor, upon reaching the outer surface of a BASE cell, diffuses towards the cold end of the converter where it condenses. The liquid sodium is then returned to the evaporator through an "artery." The artery is simply a cylindrical heat pipe containing a porous tungsten-coated molybdenum substrate which allows capillary forces or wick action to draw the liquid sodium from the cold end to the evaporator. In this way the sodium cycle is maintained without any mechanical pumps or moving parts.

The AMTEC converter of Fig. 12.19 contains about 3 to 5 grams of sodium which flows through its cycle at about 20 g/h. With this design, a thermal to electric efficiency of about 16% has been achieved. The efficiency is limited primarily by heat loss. To minimize radiative heat loss to the cold end, this AMTEC converter has many heat shields placed between the BASE cells and the cold end as shown in Fig. 12.19. The power output of an AMTEC converter depends primarily on the temperature of the hot end and on the current drawn from the converter. For the design shown in Fig. 12.19, a peak power of about 8.5 W(e) (3 A at 2.8 V) has been realized.

Multiple converters can be connected to the same heat source to produce power systems with higher electrical output. For example, it has been proposed, for space applications, to incorporate twelve to sixteen of these small AMTEC converters to produce a small light-weight power module that can produce over 100 W(e) of power for up to 15 years. The heat source for such an advanced radioisotope power source (ARPS) would be ^{238}Pu. AMTEC generators can also be used for terrestrial applications using a small combustion heat source as an alternative to much heavier batteries. The lack of moving parts make AMTEC converters reliable, noise free, and vibrationless, and, because of their small size and light weight, they are an attractive technology for space applications, auxiliary and remote power, portable power, and self-power appliances. The current state of development of these converters is discussed by Giglio et al. [2001].

12.7 Stirling Converters

Stirling engines have been used for over sixty years to convert thermal power into mechanical power. In a Stirling engine, gas in a closed chamber is alternately heated and cooled thereby causing the gas pressure to rise and fall. This cyclic variation of the gas pressure can in turn be used to move a piston back and forth, thereby extracting mechanical power from the thermal energy used to heat the gas. The piston can in turn be connected by a direct mechanical linkage to drive a linear alternator to produce electrical power.

A somewhat refined version of such a Stirling converter is shown in Fig. 12.20. The gas volume above the piston is called the working space, and that below the piston, the bounce space. Because the gas in the working space generally cannot be heated and cooled sufficiently quickly, a "displacer" is incorporated into a Stirling engine to shuttle or displace the working gas alternately between the heated and cooled ends of the working space. In the engine of Fig. 12.20, the displacer moves the gas back and forth through a heat exchanger/regenerator loop. The regenerator stores heat as the working gas flows from the hot (expansion) end of the working space to the cool (compression) end, and then reheats the gas as the gas flow is reversed. The gas in the bounce space also alternately increases and decreases in pressure and acts as a spring on the displacer/piston assemblies. With selection of proper masses, areas, loads, damping pressures and heat flow to and from the regenerator, the Stirling engine can be made to resonate at a well-defined frequency as a "free piston" Stirling converter.

Free-piston Stirling converters, similar to that of Fig. 12.20, have been constructed for terrestrial applications and operated for extended periods of time with power output of 10 to 350 W(e). These prototypes demonstrated reliable operation with thermal efficiencies between 20 and 25%. Currently, efforts continue to develop a 55-W(e) Stirling converter for space applications. In Fig. 12.21 a sketch of a mars landing craft powered by a Stirling converter is shown. Because a Stirling converter has moving parts, the changing center of mass causes the platform to which it is attached to vibrate. Such vibrations are unacceptable for most space crafts such as reconnaissance satellites which must remain still as photographs are taken. To minimize such vibrations, two 55-W(e) converters operating in a synchronous, dynamically-balanced, opposed arrangement can be used. Such a dual converter system powered by a radioisotope source is shown in Fig. 12.22. This dual configuration augmented by adaptive vibrational control makes Stirling converters very attractive for satellite and deep space missions.

12.8 Direct Conversion of Nuclear Radiation

In this conversion technique, electricity is produced directly from the radiation emitted in nuclear or atomic reactions such as radioactive decay and fission. Devices using this approach are not heat engines and, thus, are not limited by the Carnot thermodynamic efficiency. However, in practice their conversion efficiencies are still low, and typically, are low power (milliwatts) and small current (micro- to milliamperes) devices with voltages ranging from a fraction of a volt to thousands of volts. These small devices are generally trouble free and have specialized applications in medicine, space research, and other fields.

12.8.1 Types of Nuclear Radiation Conversion Devices

There are many concepts for converting nuclear radiation directly into electricity. They can be classified into the following broad categories.

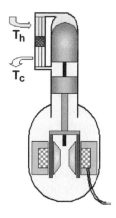

Figure 12.20. Stirling engine powering a linear alternator. Courtesy of NASA Glenn Research Center.

Figure 12.21. A 3 kW(e)-Stirling engine power system proposed for a mars surface experiment package. Courtesy of USDOE.

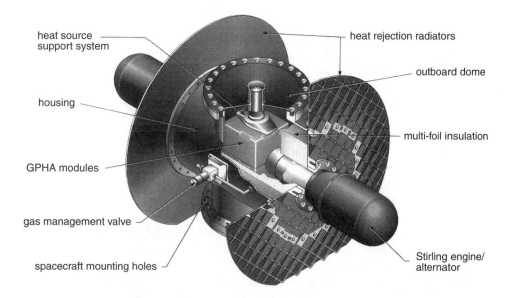

Figure 12.22. Two synchronous Stirling power generators powered by a radioisotope thermal energy source. Source: Cockfield [2000]. Courtesy of Lockheed Martin Astronautics.

Radiation-Induced Ionization: As radiation passes through matter, it produces ion-electron pairs in the material. If, for example, the ionization is produced in a gas between two electrodes, each with a different *work function*, the electrons preferentially migrate to one electrode and the positive ions to the other, thereby establishing a potential difference.[9]

Radiation Excitation of Semiconductors: When radiation (photons, neutrons, beta particles, protons, etc.) interacts with a semiconductor material, charge carriers are created and electrical energy is produced. Such devices can be termed *radiation-voltaic* energy-conversion devices. In particular, if beta particles are used the devices are called *betavoltaic* devices, while if photons are used they are called *photovoltaic* devices. For example, solar cells use photons of visible light to produce electricity. Another device is the betavoltaic nuclear battery, which uses beta particles emitted in radioactive decay. This device is discussed at length in the next section.

Direct Collection of Charged-Particle Radiation: If the charged particles emitted from nuclear reactions, such as radioactive decay or fission, are collected, a potential can be established between the collecting anode and the source producing the reactions. Such devices include the *alpha* or *beta* cell which collect the alpha or beta particles emitted by a radioisotope source, the *fission electric cell* which collects the positively charged fission fragments emitted from a fissioning surface, and the *gamma electric cell* which collects the recoil electrons produced in Compton scattering of photons.

Conversion through Intermediate Energy Forms: Energy can be transformed into different forms until the form we seek (here electrical energy) is reached. One such device for transforming nuclear energy into electricity is the so-called *nuclear battery* or *double-conversion* device. In this device, radiation emitted from a radioactive source impinges on a phosphor that then emits visible light. This light is then absorbed by carefully placed solar cells that, in turn, cause a current to flow through a connected load. Although efficiency is lost at each step in this double-conversion process, overall efficiencies of several percent can be achieved, and powers of milliwatts to a few watts can be produced.

12.8.2   Betavoltaic Batteries

In a betavoltaic cell (see Fig. 12.23) beta particles emitted by radionuclides deposited on a support plate impinge on a pair of joined n-p semiconductor plates with the n-type plate facing the radioactive source. Such a pair of n-p semiconductors is called a *n-p diode*. As the beta particles move through the semiconductor material, losing their energy by ionization and excitation of the semiconductor atoms, electron-hole pairs are produced in the diode. Such electron-hole production then causes a current to flow through an attached external load as shown in Fig. 12.23.

The thickness of the beta emitting layer is important. Too thin and too few beta particles are emitted. Too thick and many of the emitted beta particles are

---

[9]This ionization can also be used for non-electrical purposes, such as serving as a catalyst for chemical reactions, pumping a laser, and neutralizing space charge in thermionic devices.

absorbed or degraded in energy before they can reach the n-p junction. With an optimal thickness of the radionuclide layer, the percent of the emitted beta energy converted into electrical power can be between 1 and 2%.

Semiconductors that have been used for betavoltaic cells include Si, Ge, Se, and GaAs. The choice of radioisotope is also very important. Ideally, few, if any, x rays or gamma rays should accompany the radioactive decay to minimize the shielding needed around the battery. Two choices have been tried, namely ^{147}Pm (2.62 year half-life) and ^{90}Sr-^{90}Y (28.6 y half-life).[10] The ^{90}Sr-^{90}Y radionuclides are pure beta emitters (i.e., no associated gamma rays are produced.) ^{147}Pm, however, also produces a 121 keV gamma ray and, thus, requires a few mm of tantalum shielding to reduce the photon dose rate outside the betavoltaic device. Other possible pure beta emitters include ^{3}H (half-life 12.3 y) and ^{63}Ni (half-life 100 y). The shorter the half-life the more beta particles are emitted per second in the n-p junction and hence the greater power produced. For this reason ^{147}Pm is the more attractive radioisotope despite the associated gamma ray emission.

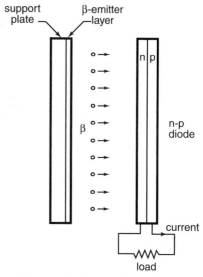

Figure 12.23. A planar betavoltaic cell. After El-Wakil [1982].

Betavoltaic batteries are constructed by combining several betavoltaic cells in a network of series and parallel connections. Such batteries typically produce micro- to milliwatts of electric power and a potential of a few volts. Short-circuit currents are tens of microamperes. The energy conversion efficiency is about 1%. In Fig. 12.24 the internal structure of one type of betavoltaic battery is shown. Although these batteries have small power outputs, they may find application in specialized applications, such as biomedical uses to provide power for pacemakers, telemetry, and monitoring devices.

12.9 Radioisotopes for Thermal Power Sources

Although all radioisotopes emit energy when they decay, only a handful are of practical importance as thermal energy sources for radioisotope power generators. There are several important criteria for the selection of a particular radionuclide as a thermal power source. First, the radionuclide must be available in a stable chemical form that can be encapsulated to prevent any radionuclide leakage into the environment. The radiation emitted by the radioactive decay must be absorbed in the source material, or by a relatively thin layer of surrounding shielding material,

---

[10]The nuclide ^{90}Y, with a 64 h half-life, is a daughter of ^{90}Sr and is in secular equilibrium with its parent.

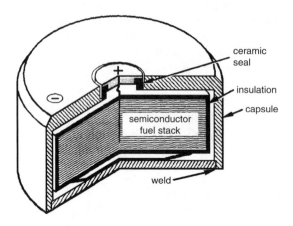

Figure 12.24. General structure of a betavoltaic nuclear battery. From El-Wakil [1982].

to prevent excessive exposures outside the source. The lifetime of the radionuclide must be comparable to or greater than the expected lifetime of the application using the power source. To reduce weight for space applications, the source material must have a high thermal power density. Finally, the cost and availability of the radioisotope must be reasonable.

Thus, practical radionuclide thermal power sources should have a half-life of at least several years, low gamma-ray emission, a power density of at least 0.1 W/g, a relatively low cost, and desirable chemical and physical properties (such as stability, chemical inertness, high melting point, etc.). These constraints limit the selection from the hundreds of different radionuclides to the nine practical radionuclides listed in Table 12.2. Four of these are beta-particle emitters and can be easily recovered by chemically extracting them from the fission products in the spent fuel produced by nuclear reactors. Four are alpha-particle emitters that must be produced in nuclear reactors by non-fission reactions and, generally, are of low concentration in spent fuel and, consequently, more expensive. The ninth useful radioisotope is ^{60}Co which is plentiful (produced easily by neutron activation of stable ^{59}Co in a reactor). However, ^{60}Co emits energetic gamma photons (1.17 and 1.33 MeV) and thus requires considerable shielding to reduce external exposure.

The two radionuclides ^{90}Sr and ^{238}Pu have been used the most extensively. ^{90}Sr is inexpensive and plentiful, has a long half-life (28.8 y), and, as a pure beta emitter, requires very little shielding. Because ^{90}Sr has an affinity for bone, it is important that it be well encapsulated to prevent its leakage into the biosphere. The chemical form chosen for ^{90}Sr is usually strontium titanate ($SrTiO_3$) because it is insoluble in water, has a high melting point, and is resistant to shock. Most terrestrial radionuclide sources used to date employ ^{90}Sr.

For space applications, solar cells have been a primary electrical energy conversion device for long, near-earth missions. However, radioisotope thermoelectric generators (RTG) have distinct advantages for certain types of space missions. RTGs do not deteriorate like solar cells when passing through the radiation belts that

Table 12.2. Properties of some radionuclide thermal power sources. Recoverable energy excludes the energetic gamma photons produced by ^{137}Cs and ^{60}Co. Energy from daughter radionuclides in secular equilibrium are included for ^{144}Ce and ^{90}Sr.

Property	^{144}Ce	^{90}Sr	^{137}Cs	^{147}Pm	^{60}Co	^{242}Cm	^{244}Cm	^{210}Po	^{238}Pu
Half-life	284.9 d	28.84 y	30.07 y	2.623 y	5.271 y	162.8 d	18.101 y	138.4 d	87.7 y
Recov. en. (MeV/dec.)	1.30^a	1.132^a	0.187	0.062	0.0962	6.11	5.803	5.411	5.495
Sp. activity (Ci/g)	3190	136	87.0	927	1131	3307	80.9	4494	17.1
Sp. power (W/g)	24.6	0.916	0.0966	0.341	0.644	120	2.78	144	0.558
1-W activity (Ci/W)	130	149	901	2722	1755	27.6	29.1	31.2	30.7
1-W activity (TBq/W)	4.81	5.52	33.3	101	64.9	1.02	1.08	1.15	1.14
U-Shielding^b	3.6	negl.	3.8	negl.	9.9	neut.	neut.	negl.	neut.
Pb-Shielding^b	6.5	negl.	7.5	negl.	18.0	neut.	neut.	negl.	neut.
Compound	Ce$_2$O$_3$	SrTiO$_3$		Pm$_2$O$_2$	metal	Cm$_2$O$_3$	Cm$_2$O$_3$	GdPo	PuO$_2$
Melting pt., °C	1692	1910		2350	1495	1950	1950	590	

^{a} In secular equilibrium with its short lived daughter.
^{b} Thickness (cm) to attenuate γ-dose rate to 10 rads/h at 100 cm from a 100-W source.

surround the earth. They are also well suited for deep space missions where solar energy is weak, for moon applications for which solar cells would require large heavy batteries during the long periods of darkness, and for planetary atmospheric probes.

Most space applications of RTGs have used ^{238}Pu in the form of plutonium dioxide (PuO$_2$). This radionuclide has a relatively high power density and its long half-life (88 y) makes it quite suitable for long duration space probes. In the past several years, the U.S. Department of Energy (DOE) has fabricated several general purpose RTGs and radioisotope heater units for deep space missions. These units all used the decay heat from ^{238}Pu. However, ^{238}Pu is much more expensive than ^{90}Sr. To provide the necessary amounts of ^{238}Pu, the DOE in 1995 has obtained about 10 kg of this radionuclide from Russia to supplement its inventory.

12.10 Space Reactors

For long lunar, planetary and deep-space missions, space vehicles often need more electrical power than can be provided by radioisotope electric generators or solar cells. Nuclear reactors, however, can provide both the thermal source and the duration needed for such missions. The United States and the former USSR embarked in the 1960s on the development of small, light-weight nuclear reactors whose heat could be converted into electrical energy by a variety of conversion devices. These devices included thermoelectric generators, thermionic generators, and several turbo-generator systems.

The design of nuclear reactors for space applications has several unique requirements, which are not usually important for earth-based electricity generation from nuclear energy. Foremost is the requirement to reduce the mass of the power system to be as small as possible in order to minimize the launch thrust. Second, there must be minimal risk to earth inhabitants should the launch be unsuccessful and the satellite fall back to earth. Other considerations include minimizing the cost and maximizing the reliability of the conversion device.

12.10.1   The U.S. Space Reactor Program

The United States began in the late 1950s the development of a series of space reactors, designated SNAP-n, where n is an even number to distinguish these thermal power sources from the odd-n SNAP series used for radioisotope generators. Several early prototype SNAP reactors were built and tested, leading to the last of the series, the SNAP-10A, which was launched into space in 1965. The SNAP-10A is shown in Fig. 12.13 sitting atop its conical thermal radiator used to emit waste heat into space. The SNAP-10 reactors used thermoelectric conversion piles composed of SiGe semiconductors heated by liquid metal NaK used to cool the reactor. The cold junction of the thermoelectric pile is cooled by the thermal radiator (see Figs. 12.11 and 12.12).

The 650-W(e) SNAP-10A reactor, which was launched April 3, 1965 into a polar orbit, functioned for 43 days providing power for telemetry and a small experimental ion-propulsion unit. Then a series of failures in electrical components caused the reactor to automatically shutdown. An earth-bound prototype operated for well over a year.

The need for a general-purpose space reactor to provide electrical power levels of a few to several hundred kilowatts of electrical power for many different civilian and military space applications led the U.S. Dept. of Energy (DOE), NASA, and the Dept. of Defense to develop the SP-100 Space Nuclear Reactor. This compact fast reactor (the size of a small waste basket) uses niobium-clad, highly enriched, uranium nitride ceramic fuel rods so that it can operate at temperatures up to 1400 K. Besides high temperature performance, these fuel elements provide shock resistance and long operational lifetimes (tens of years). The cooling system of the SP-100 reactor used liquid lithium pumped by electromagnetic pumps with the waste heat rejected through radiation fins into space. The SP-100 reactor-electrical system is shown in Fig. 12.25. The SP-100 heat exchange module was designed to accommodate a variety of heat-to-electricity conversion devices. The first SP-100 units used the same thermoelectric technology of the earlier SNAP-10 space reactors.

Safety considerations for the launch of a space satellite or probe containing a nuclear reactor is a major concern. The SP-100 can be launched in a cold shut-down mode with the lithium coolant as a solid. Once in orbit, the reactor can be started remotely by adjusting the movable reflector panels around the core to increase the reflection of neutrons back into the core. A special reentry shield is used to keep the reactor intact should it fall from orbit, and a special radiation shield is used to protect other onboard modules from the radiation emitted from the reactor.

The SP-100 space reactor developed in the early 1990s, has yet to be launched. Nevertheless, many new technologies were developed that can be used for other commercial applications. The SP-100 system, affixed to a space probe, as originally envisioned, is shown in Fig. 12.26. Some of the characteristics of the U.S. space nuclear reactors are listed in Table 12.3.

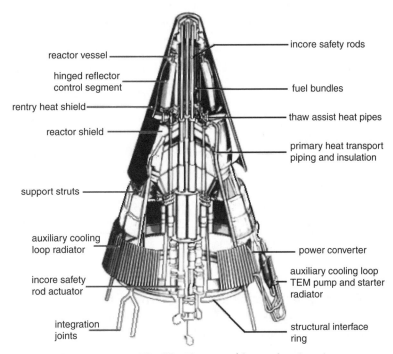

Figure 12.25. The SP-100 reactor/thermoelectric unit.

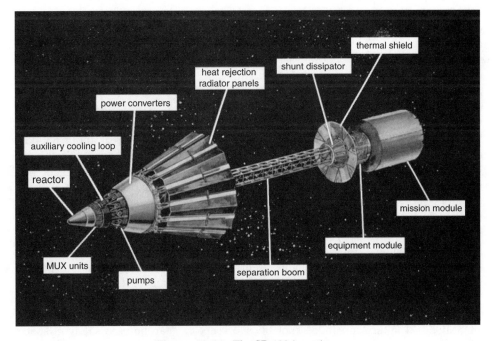

Figure 12.26. The SP-100 in action.

Table 12.3. Some characteristics of recent space power reactors.

Characteristic	United States		Former Soviet Union		
	SNAP-10A	SP-100	BUK	TOPAZ-I	TOPAZ-II
Date	1965	1990–	1974–88	1987-88	1991–
Power (kWt)	45	2,000	100	150	135
Power (kWe)	0.65	100	3	5	5.5
Convertor[a]	TE	TE	TE	TI	TI
Fuel	UZr_x	UN	UMo	UO_2	UO_2
Mass ^{235}U (kg)	4.3	140	30	11.5	25
System mass (kg)	435	5422	930	980	1061
Coolant	NaK	Li	NaK	NaK	NaK
No. space flights	1	0	31	2	0

[a] TE for thermoelectric; TI for thermionic.

12.10.2   The Russian Space Reactor Program

The former Soviet Union (USSR) also developed a number of compact nuclear reactors for space applications. Some characteristics of the Soviet space nuclear reactors are listed in Table 12.3.

Between 1970 and 1988, 31 so-called BUK nuclear power systems (NPS) were used to supply electrical power for a series of COSMOS spacecrafts in low earth orbits (about 300 km altitude) used for marine radar surveillance. The BUK NPS used thermoelectric conversion based on SiGe semiconductors. The core contained 30 kg of 90% enriched uranium in the form of 37 rods composed of a uranium-molybdenum alloy. A NaK liquid metal transferred heat from the core to the thermoelectric generator whose waste heat was rejected to space by a large conical radiator, much as in the U.S. SNAP-10 NPS.

Of considerable concern for using nuclear reactors in near-earth orbits is the eventually reentry to the earth's atmosphere of the reactor (and its accumulated fission products). The BUK NPS was designed to separate from the spacecraft at the end of the mission and be injected into a high circular orbit of about 850-km altitude. A reactor placed in such an orbit takes several hundreds years to fall back into the atmosphere, by which time the radioactivity of the fission products will have decayed to acceptable levels. However, this reactor disposal system failed for the COSMOS-954 satellite, which fell out of orbit over Canada in 1978 and scattered its radioactive debris over a large portion of Canada's Northwest Territory. Subsequently, a secondary backup safety system was incorporated into the BUK NPS. This secondary system can eject the fuel assembly from the NPS, either while still in low-earth orbit or just before reentry, so that the fuel is widely dispersed into minuscule particles in the upper atmosphere as a result of its thermal destruction upon reentry. Such dispersion prevents resulting doses at any place on the earth's surface exceeding ICRP recommended levels. A UN report issued after the COSMOS-954 failure concluded that nuclear reactors "can be used safely in space provided that all necessary safety requirements are met" [UN 1981].

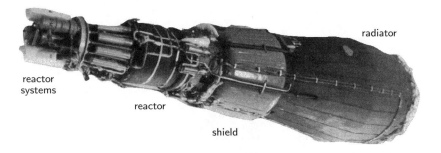

Figure 12.27. The TOPAZ Nuclear Power System. The reactor has a thermal output of 150 kW producing 5 kW of electrical power. The mass of the system is 980 kg. Source: Ponomarev-Stepnoi et al. [2000].

An advanced NPS, known as TOPAZ, was begun in the early 1970s. In this NPS, thermionic energy conversion was used. The thermionic cells were made an integral part of each fuel rod. Seventy-nine thermionic fuel elements of 90% enriched uranium oxide formed the TOPAZ reactor core, which was cooled by NaK and waste heated rejected into space by a radiator module. To increase the efficiency of the thermionic conversion, a once-through cesium vapor supply system was incorporated. With a reactor thermal power of 150 kW, 5–6 kW of electrical power was produced. Several TOPAZ systems were ground tested and refinements made, leading to the flight testing of two TOPAZ units in 1987-88. The COMOS-1818 test lasted 142 days and the COSMOS-1867 test lasted 342 days. In both cases, the tests were ended when the cesium vapor generator ran out of cesium. Nevertheless, these tests confirmed the reliable operation of thermionic conversion in a space NPS. A TOPAZ NPS is shown in Fig. 12.27

A more advanced NPS known as ENISEY or TOPAZ-2 has also been developed by Russia (see Fig. 12.28). This NPS, originally designed for powering television relay satellites, also uses thermionic conversion technology and produces 5.5 kW(e) from a reactor power of about 135 kW(t). Although this unit has yet to be flown in space, it represents the most advanced space reactor power system yet developed. In 1991 the U.S. and Russia began collaborate work on testing the advance TOPAZ-2 system for use as a alternative to solar power systems for civil, commercial, and scientific applications. Six TOPAZ-2 units have been delivered to the U.S. and tested under this cooperative program. Although this program has since been canceled by the U.S. government, the use of NPS for space applications is still considered essential, especially for deep space missions that cannot use solar power.

Figure 12.28. The TOPAZ-II space nuclear reactor. Source: Ponomarev-Stepnoi et al. [2000].

BIBLIOGRAPHY

ANGULO, C., et al., "A Compilation of Charged-Particle Induced Thermonuclear Reaction Rates," *Nuclear Physics A*, **656**, 3, 1999.

CHEN, F.F., *Introduction to Plasma Physcis and Controlled Fusion*, Plenum Press, New York, 1984.

COCKFIELD, R.D., "Radioisotope Stirling Generator Concepts for Planetary Missions," Paper 2000-2843, 35th Intersociety Energy Conversion Engineering Conference, AIAA, July, 2000.

CORLISS, W.R. *SNAP Nuclear Space Reactors*, USAEC, Washington, DC, 1966.

EL-GENK AND HOOVER, *Space Nuclear Power Systems*, Orbit Book Co., Malabar, FL, 1985.

EL-WAKIL, M.M., *Nuclear Energy Conversion*, American Nuclear Society, La Grange Park, IL, 1982.

FOSTER, A.R. AND R.L. WRIGHT, *Basic Nuclear Engineering*, 2nd ed., Allyn and Bacon, Boston, MA, 1973.

FURLONG, R.R. AND E.J. WAHLQUIST, "U.S. Space Missions Using Radioisotope Power Systems," *Nuclear News*, Am. Nuclear Soc., La Grange Park, IL, April 1999.

GIGLIO, J.C., R.K. SIEVERS AND E.F. MUSSI, "Update of the Design of the AMTEC Converter for Use in AMTEC Radioisotopes Power Systems," *Proc. 18th Symposium on Space Power and Propulsion*, Albuquerque, NM, 2001.

LAWSON, J.D., "Some Criteria for a Power Producing Thermonuclear Reactor, *Proc. Phys. Soc. B*, **70**, 6-10, 1957.

LINDL, J., *Physics of Plasmas*, **2**, 1690, 1995.

MEAD, R.L. AND W.R. CORLISS, *Power from Radioisotopes*, USAEC, Washington, DC, 1966.

PONOMAREV-STEPNOI, N.N., V.M. TALYZIN AND V.A. USOV, "Russian Space Nuclear Power and Nuclear Thermal Propulsion Systems," *Nuclear News*, Dec. 2000, American Nuclear Society, La Grange Park, IL.

ROSEN, M., *Physics of Plasmas*, **6**, 1690, 1999.

SIEVERS, R.K. AND J.E. PANTOLIN, "Advanced AMTEC Cell Development," *Proc. 18th Symposium on Space Power and Propulsion*, Albuquerque, NM, 2001.

THIEME, L.G., S. QIU AND M.A. WHITE, "Technology Development for a Stirling Radioisotope Power System," *Proc. Technology and Applications International Forum*, 2000.

UN, United Nations Committee on the Peaceful Uses of Outer Space, *Report of the Working Group of the Uses of Nuclear Power Sources in Outer Space*, Annex II to the report *Report of the Scientific and Technical Subcommittee on the Work of the Eighteenth Session*, UN Document A/AC.105/287, Feb., 1981.

PROBLEMS

1. In Section 6.7 some fusion reactions that do not produce neutrons as a reaction product are shown. Discuss the advantages and disadvantages of using such reactions to convert nuclear energy to electricity.

2. Starting with Eq. (12.17) verify that, for an ICF D-T plasma at $kT = 50$ keV, the ρR criterion is $\rho R \gtrsim 3$ g/cm^2.

3. Derive an expression for the ratio of the triple product $nT\tau_E$ for ICF to that for magnetically confined fusion. For a plasma at $kT = 50$ keV, estimate the value of this ratio.

4. If all the neutrons produced in a D-T reactor were slowed and absorbed in a natural lithium medium, what is the thermal energy that would be produced per incident 14.1 neutron? How many tritium atoms would be produced per incident thermal neutron?

5. For an ICF device, estimate the energy gain, i.e., fusion energy to incident laser energy, that would be needed with present day lasers (1.5% electric-to-light conversion efficiency) to achieve break-even. Assume the conversion of captured fusion energy to electrical energy has an efficiency of 32%.

6. A pellet for ICF contains 4 mg of a 50:50 mixture of deuterium and tritium. If all of the atoms fused, how much energy would be released? How many gallons of gasoline is this equivalent to? The heat content of gasoline is 1.42×10^8 J/gallon.

7. Discuss the advantages and disadvantages of a fusion power plant to a conventional coal-fired power plant.

8. The energy of decay alpha particles from ^{226}Ra is shown in Fig. 5.2. What initial mass of ^{226}Ra is needed to provide a thermal power of 100 W(t) after 10 years?

9. Show that ^{90}Sr in secular equilibrium with its daughter ^{90}Y has (a) a specific activity of 136 Ci/g, and (b) a specific thermal power of 0.916 W/g.

10. What initial mass of ^{210}Po is needed in a space RTG with a 19% thermal to electric conversion efficiency that is to have a power output of 65 W(e) one year after the start of the mission?

11. Many thermal energy conversion devices, such as thermoelectric, AMTEC and Stirling converters developed for space applications, also have terrestrial uses. Rather than use radioisotopes as a heat, a combustion flame can be used. Consider a converter producing 100 W(e) with a 15% thermal to electric conversion efficiency designed to produce power continuously for one year. (a) What mass of ^{238}Pu would be needed? (b) What mass of a petroleum-based fuel, with density 0.9 g/cm^3 and a heat of combustion of 40 MJ/L, would be needed?

12. Discuss the advantages and disadvantages of using RTGs versus solar cells for the following space missions: (a) communication satellites, (b) lunar surface experiments, (c) a mars lander, and (d) a planetary mission to Neptune. Consider issues such as launch constraints, safety, reliability, space environment, mass, size, auxiliary systems.

13. A particular heart pacemaker requires 200 μW of power from its betavoltaic battery. (a) How many curies of ^{147}Pm are needed for a betavoltaic battery that has a conversion efficiency of 0.95%? (b) If the battery is to have a lifetime of 5 years, what must be the initial loading of ^{147}Pm?

14. In a betavoltaic battery, the energy of the beta particles should be below about 0.2 MeV. Explain why betas with significantly greater energy are not as attractive for the battery's operation.

15. The ^{147}Pm used in betavoltaic batteries usually contains ^{146}Pm as an impurity. This impurity is undesirable because it is radioactive (5 y half-life) and emits energetic gamma rays (0.75 MeV). To avoid bulky shielding around the battery, it is preferable to eliminate much of this impurity. Suggest how this could be done. HINT: ^{146}Pm has a large thermal-neutron absorption cross section.

16. Discuss the relative advantages and disadvantages of using fission reactors versus radioisotopes as thermal energy sources for use in thermal electric converters.

17. Show that ^{238}Pu has (a) a specific activity of 17.1 Ci/g, and (b) a specific thermal power of 0.558 W/g.

Chapter 13

Nuclear Technology in Industry and Research

In the past 50 years, nuclear technology has found ever increasing use in medicine, industry, and research. Such applications depend on either the unique characteristics of radioisotopes or the radiation produced by various nuclear and atomic devices. The ingenuity of nuclear technology applications developed over the past half century is extraordinary. The applications are myriad, and any comprehensive discussion would itself require at least a large book. In this and the next chapter, only a brief survey is attempted with some of the more important applications presented.

By far the most complex and highly developed applications of nuclear technology have occurred in the medical field. Almost all of us, sometimes in our lives, will benefit from such medical applications. These medical applications of nuclear technology are the subject of the chapter that follows this one. However, we encounter the benefits of nuclear technology far more frequently in our everyday lives through industrial and research applications. Smoke detectors in our homes, paper of uniform thickness, properly filled soda cans, and food from new strains of crops are just a few examples of how we benefit from nuclear technology. We present these and other industrial and research applications of nuclear technology in this chapter.

Finally, a rather lengthy section is included in this chapter discussing the different types of particle accelerators and how they work. Particle accelerators have many important applications in industry, research, and medicine. Accelerators are routinely used to create radionuclides, change material properties by ion implantation, and for medical therapy. Enormous accelerators have been used for many years in research to reveal the fundamental physics of our universe.

13.1 Production of Radioisotopes

Nuclear technology almost always involves the use of nuclear radiation.[1] Such radiations can be produced by special generators such as nuclear reactors, x-ray machines, and particle accelerators. Alternatively, radiation produced by the decay

---

[1]An exception is the separation of stable isotopes such as ^{2}H and ^{18}O and their use as labels in biological reagents or other material. Stable isotopic labels avoid both radiation effects and time constraints inherent in the use of more traditional radionuclide labels. However, such stable isotopic labels are far more difficult to detect and more costly to produce than are radioactive labels.

of radionuclides is frequently used. Because of the very small size of radionuclide radiation sources, the great variety of available radionuclides and their radiations, and their relatively low cost, radionuclides have been used in a vast number of applications.

The production of radioisotopes is today a multi-billion dollar industry. Three methods are used for the production of radioisotopes.

Nuclear Reactor Irradiation: Stable nuclides placed in or near the core of a nuclear reactor can absorb neutrons, transforming them into radioisotopes. These radioisotopes, being neutron rich, generally decay by β^- emission often accompanied by gamma-ray emission. Examples of important radioisotopes produced by different neutron induced reactions include ^{59}Co(n,γ)^{60}Co, ^{14}N(n,p)^{14}C, and ^{6}Li(n,α)^{3}H.

Recovery of Fission Products: Many useful radionuclides are produced copiously as fission products and can be obtained by chemically processing spent fuel from reactors. These include ^{137}Cs and ^{90}Sr. Since fission products are neutron rich, they almost always decay by β^- emission. Spent nuclear fuel also contains important transuranic isotopes produced by (multiple) neutron absorption(s) and radioactive decay reactions. Important transuranic radionuclides include ^{238}Pu, ^{244}Cm, and ^{252}Cf. These heavy radionuclides usually decay by α emission or by spontaneous fission.

Accelerator Production: To produce radioisotopes that are proton rich and that generally decay by positron emission, particle accelerators are used. Proton beams with energies up to 10 MeV can be produced by linear accelerators or cyclotrons. Bombarding a target with such energetic protons can produce radioisotopes by a variety of nuclear reactions. Some examples are ^{65}Cu(p,n)^{65}Zn, ^{68}Zn(p,2n)^{67}Ga, ^{55}Mn(p,pn)^{54}Mn, ^{25}Mg(p,α)^{22}Na, and ^{58}Ni(p,2p)^{57}Co. Since the radionuclide produced is a different element, chemical extraction techniques can be used to separate the newly formed radionuclides from the unreacted target material.

In some cases the radionuclide of interest is the daughter of another much longer lived radionuclide. In this case the long-lived parent radionuclide is placed in a *radionuclide generator* from which the daughter can be extracted, as needed, by solvent extraction techniques. With such a generator, the need for rapid shipment of a short-lived radioisotope is eliminated. The most widely used radionuclide generator is a ^{99}Mo ($T_{1/2} = 65.9$ d) "cow" which can be "milked" to extract the short-lived ^{99m}Tc daughter ($T_{1/2} = 6.01$ h). The radionuclide ^{99m}Tc is the most widely used radioisotope in medical diagnoses. Another example is a ^{137}Cs ($T_{1/2} = 30.1$ y) radionuclide generator from which the daughter ^{137m}Ba ($T_1/2 = 2.55$ m) can be extracted. These generators are used to produce inexpensive, short-lived radionuclide samples for laboratory or medical use.

13.2   Industrial and Research Uses of Radioisotopes and Radiation

Radioisotopes and radiation from x-ray machines and other devices have found widespread use in many industrial and research activities. These applications can be categorized into four broad types of applications.

Tracer Applications: Material can be tagged with minute amounts of a radioisotope to allow it to be easily followed as it moves through some process.

Materials Affect Radiation: Radiation is generally altered, either in intensity or energy, as it passes through a material. By measuring these changes, properties of the material can be determined.

Radiation Affects Materials: As radiation passes through a material, it produces ionization or changes in the electron bonds of the material, which, in turn, can alter the physical or chemical properties of the material.

Energy from Radioisotopes: The energy released as a radionuclide decays is quickly transformed into thermal energy when the emitted radiation is absorbed in the surrounding medium. This thermal energy, as we discussed in the last chapter, can be used as a specialized thermal energy source or converted directly into electrical energy.

A brief summary of some uses in these four categories is given in Table 13.1. The use of the energy from radionuclides has been discussed at length in Chapter 12. In this chapter, we briefly review some representative applications in the other three categories.

Table 13.1. Summary of the industrial and research uses of radioisotopes and radiation. After Baker [1967].

Tracer Applications:	
flow measurements	wear and friction studies
isotope dilution	labeled reagents
tracking of material	preparing tagged materials
radiometric analysis	chemical reaction mechanisms
metabolic studies	material separation studies
Materials Affecting the Radiation:	
density gauges	liquid level gauges
thickness gauges	neutron moisture gauges
radiation absorptiometry	x-ray/neutron radiography
x-ray and neutron scattering	bremsstrahlung production
Radiation Affecting Materials:	
radioactive catalysis	modification of fibers
food preservation	increasing biological growth
biological growth inhibition	sterile-male insect control
insect disinfestation	luminescence
Mössbauer effect	polymer modification
radiolysis	biological mutations
static elimination	bacterial sterilization
synthesis	x-ray fluorescence
Use of Radiation's Energy:	
thermal power sources	electric power sources

13.3 Tracer Applications

The use of some easily detected material to tag or label some bulk material allows the bulk material to be followed as it moves through some complex process. Fluorescent dyes, stable isotopes, and radionuclides have all been used as tags. The amount of tagging material needed to trace some bulk material depends on both the sensitivity of instrumentation used to identify the tracer and the dilution experienced by the tagged material as it moves through the process.

Radionuclides are ideal tracers because few atoms are generally needed to identify the tracer in a sample. Suppose the radiation emitted by the decay of a particular radionuclide with a half-life of $T_{1/2}$ can be detected in a sample with an efficiency of ϵ. If a sample contains N atoms of the radionuclide, the observed count rate is

$$CR = \epsilon\lambda N = \epsilon\frac{\ln 2}{T_{1/2}}N. \tag{13.1}$$

To detect the presence of the radionuclide tag, this count rate must be greater than some minimum count rate CR_{min} that is above the background count rate. Then the minimum number of radioactive atoms in the sample needed to detect the presence of the radionuclide is

$$N_{min} = \frac{CR_{min}\, T_{1/2}}{\epsilon\,\ln 2} \quad \text{atoms.}$$

If the atomic weight of the radionuclide is A, the minimum mass of radionuclides in the sample is $M_{min} = N_{min}/(N_a/A)$ or

$$M_{min} = \frac{CR_{min}\, T_{1/2}\, A}{N_a\,\epsilon\,\ln 2} \quad \text{grams.} \tag{13.2}$$

A typical gamma-ray detector efficiency is $\epsilon \simeq 0.1$ and a minimum count rate is $CR_{min} \simeq 30$ min^{-1} = 0.5 s^{-1}. Thus, for ^{14}C ($T_{1/2} = 5730$ y $= 1.18 \times 10^{11}$ s), the minimum detectable mass of ^{14}C in a sample is, from Eq. (13.2),

$$M_{min}(^{14}\text{C}) = \frac{0.5 \text{ (s}^{-1}) \times 1.18 \times 10^{11} \text{ (s)} \times 14 \text{ (g/mol)}}{6.024 \times 10^{23} \text{ (atoms/mol)} \times 0.1 \times \ln 2} \simeq 2 \times 10^{-11} \text{ g.}$$

Even smaller masses of radionuclides are needed if a radionuclide is selected that has a half-life comparable to the process being followed. For example, ^{32}P ($T_{1/2} = 14.26$ d) is often used in plant studies to follow the uptake of phosphorus by plants. In this instance, the minimum mass of ^{32}P is about 5×10^{-16} g.

13.3.1 Leak Detection

A common problem associated with complex flow systems is detection of a leak that cannot be observed directly. To find the location of a leak in a shallowly buried pipe without excavation, for example, a radionuclide tracer can be injected into the pipe flow. The tracer will subsequently leak from the pipe and the position of the leak can readily be identified, from above the soil, by finding the position over the pipe at which gamma radiation released by the tracer radionuclide is emitted over an extended area. This same technique can be used to determine the position of buried pipes by following the radiation signature of the tracer as it flows through the pipes.

This use of radionuclide tracers to find leaks or flow paths has wide application. These include (1) finding the location of leaks in oil-well casings, (2) determining the tightness of abandoned slate quarries for the temporary storage of oil, (3) locating the positions of freon leaks in refrigeration coils, (4) finding leaks in heat exchanger piping, and (5) locating leaks in engine seals.

13.3.2 Pipeline Interfaces

Oil from different producers is often carried in the same pipeline. To distinguish the oil from one producer from that of another, an oil-soluble radioisotope can be injected into the head of each batch of oil. Then, a simple radiation monitor can be used at the pipeline terminus to identify the arrival of a new batch of oil.

13.3.3 Flow Patterns

Radionuclide tracer techniques are very effective for determining complex flow patterns. The diffusion patterns can be observed by tagging the material that is to diffuse or move through some medium and by subsequently measuring the spatial distribution of the activity concentration,

The determination of flow patterns by radionuclide tracers has found wide use. A few examples are (1) ocean current movements, (2) atmospheric dispersion of airborne pollutants, (3) flow of glass lubricants in the hot extrusion of stainless steel, (4) dispersion of sand along beaches, (5) mixing of pollutant discharges into receiving bodies or water, and (6) gas flow through a complex filtration system.

13.3.4 Flow Rate Measurements

Tracer techniques have been used to determine the flow rates in complex fluid systems. Based on measurements of the activity concentrations of a radioactive tag in the fluid medium, secondary techniques for flow measurements can be calibrated as, for example, a flow meter in a pipeline. This method is also used where no other flow rate method is available, for example, measuring flow rates in a river estuary. Many variations are possible. Two basic techniques to determine flow rates as follows.

Peak-to-Peak Method: If the flow occurs at a relatively constant cross sectional area (e.g., in a pipeline), an injection of a radionuclide will travel downstream with the flow. The time required for the radionuclide (and the flowing material) to travel to a downstream location is given by the time for the activity to reach a maximum at the downstream location.

Tracer Rate Balance Scheme: If a tracer is injected at a constant rate Q_o (Bq s^{-1}) into a complex flow system (e.g., a river with a variable cross section), then (assuming complete mixing, no losses or additional injections, and equilibrium conditions) the tracer must appear at the same rate at any downstream location, i.e., $Cq = Q_o$ where C is the average activity concentration (Bq m^{-3}) and q is the river flow (m^3 s^{-1}) at some downstream location.

13.3.5 Labeled Reagents

By labeling various chemicals with radionuclides, i.e., replacing one or more stable atoms of an element in a molecule by radioactive isotopes of the element, it is possible to analyze complex chemical reactions. For example, to determine which of the many hydrocarbons in gasoline are responsible for engine deposits, the various hydrocarbons can be successively tagged with radioactive ^{14}C and the engine deposits subsequently examined for this radioisotope.

This use of radionuclides as material labels or tracers is ideally suited for measuring physical and chemical constants, particularly if the process is slow or if very small amounts of a substance are involved. For example, diffusion constants, solubility constants, and equilibrium constants are easily determined by radiotracer techniques.

13.3.6 Tracer Dilution

By measuring how a radioisotope is diluted after being injected into a fluid of unknown volume V, the volume can easily be determined. A radionuclide is first dissolved in a small sample of the fluid to produce an activity concentration C_o (Bq/cm^3). A small volume V_o (cm^3) of this sample is then injected into the fluid system whose volume is to be determined. The total activity injected is V_oC_o (Bq) and must equal the total activity circulating in the system, which, with the assumptions of complete mixing and no losses of the radionuclides, must equal VC where C (Bq/m^3) is the concentration of radionuclide activity in the system after mixing. Thus, the fluid volume is calculated as $V = V_o(C_o/C)$. Notice only the ratio C_o/C or the *dilution* is needed. This is much more easily measured than the absolute activity concentrations C and C_o.

This tracer dilution technique has been applied widely. For example, it can be used to determine the amount of a catalyst in a chemical process, the amount of molten iron in an ore-refining process, the volume of an irregularly shaped container, and the amount of blood in a person.

13.3.7 Wear Analyses

Radioisotope tracers are exceptionally useful in studies of wear, friction, corrosion, and similar phenomena in which very small amounts of material are involved. For example, to determine the wear of piston rings in an engine, the rings can be tagged with a radioisotope. Then, by measuring the rate at which the tag appears in the engine oil, the rate of wear of the piston rings is easily measured. Similarly, by tagging different moving parts of an engine with different radionuclide tags, the wear on the tagged components can be determined simultaneously by measuring the concentration of the different tags in the oil. Such analysis results in better scheduling of preventive engine maintenance.

13.3.8 Mixing Times

A common problem in batch mixing operations is to determine the time required for the components to become completely mixed. By tagging one of the components and measuring the tag concentration in test samples taken at various times after the start of mixing, the time for complete mixing is easily found.

13.3.9   Residence Times

In continuous manufacturing processes, such as oil refining, the times some material spends in the various stages of the system are important for optimal design and operation of the system. By injecting quickly a tagged sample of the material in some stage of the system and following the decline of the activity in that stage, the mean residence time (or transfer rate to the next stage of the process) can easily be determined.

13.3.10   Frequency Response

For proper operation of chemical reactors, adsorption towers, distillation systems, purification processes, and other complex systems, the frequency response or *transfer function* of the system is needed. By injecting a "pulse" of radioactively-tagged input material and measuring the transients of the resulting activity concentrations in the various stages of the process, the frequency response of the system can be determined from a Fourier analysis of the concentration transients.

13.3.11   Surface Temperature Measurements

In many machines, the peak temperature on the metallic surface of some critical component, such as a turbine blade in a jet engine, is often a vital operational parameter. Direct measurement is often not possible. One indirect method is to label the surface of the metal with ^{85}Kr ($T_{1/2} = 10.7$ y) by implanting the Kr atoms into the metal lattice by either Kr ion bombardment or by diffusion of Kr atoms into the metal by applying high temperature and pressure. Such kryptonated surfaces have the property that the krypton atoms are released in a reproducible and controlled manner at high temperatures. By measuring the surface distribution of the remaining ^{85}Kr, after a thermal stress test, the maximum temperature distribution over the surface can be determined.

13.3.12   Radiodating

By observing how much of a long-lived naturally occurring radionuclide in a sample has decayed, it is possible to infer the age of the sample. Many radiodating schemes based on different radionuclides have been developed and applied by researchers to estimate ages of many ancient items, ranging from bones to moon rocks. The theory behind radiodating is discussed at length in Section 5.8.

13.4   Materials Affect Radiation

Radiation produced by the decay of radioisotopes or from devices such as x-ray machines interacts with the matter through which it passes. From measurements of the uncollided radiation passing through a material or of the secondary radiation produced by interactions in the material, properties of the material can be determined.

13.4.1   Radiography

The first practical application of nuclear technology was radiography. Soon after Roentgen discovered x rays, he demonstrated in 1895 that x rays passing downward through his wife's hand produced an image on a photographic plate beneath her hand that showed the hand bones in great detail. The passage of radiation through an object and its measurement as a 2- or 3-dimensional image is the basis of the science of radiography. Many types of radiography have been developed, some of which are discussed below, and others in the next chapter discussing medical uses of nuclear technology.

X-Ray and Gamma-Ray Radiography: Almost all of us have had x-ray pictures made of our teeth or some part of our bodies. This common radiographic procedure, usually uses an x-ray machine to generate a beam of high-energy photons. In such a machine, a beam of electrons is accelerated by a high voltage in a vacuum tube. The electron beam impinges on a target of high-Z material, such as tungsten, and, as they slow down, they generate bremsstrahlung radiation, and, through ionization and excitation of the target atoms, fluorescence. The bremsstrahlung photons have a continuous distribution of energies up to the maximum voltage used in the machine. By contrast, the fluorescence photons are monoenergetic, characteristic of the electron energy levels of the target material.

As x rays pass through a heterogeneous specimen, dense high-Z objects preferentially absorb or scatter x rays from the beam so that fewer photons reach the detection surface (e.g., a sheet of film or a fluorescent screen) behind these objects. Each image point has an intensity inversely proportional roughly to the integrated electron density along a straight-line path from the x-ray source to the image point. In effect, an x-ray image is a "shadowgram" cast by all the electrons along a photon's path.

Besides obvious medical applications to detect broken bones, x-ray radiography is widely used in industry and research. Detection of cracks in welds or voids is one such application of industrial radiography.

In many applications of x-ray radiography, it is impractical to use an x-ray machine, either because of space or weight limitations or a lack of electrical power. As an alternative to an x-ray machine, radioisotopes can be used to generate the photons needed to create x-ray images. Radioisotopic gamma-ray sources can be used to create small portable x-ray devices that have no tubes to burn out and, consequently, are very reliable. Moreover, because gamma-ray sources can be far smaller than x-ray tube sources, higher resolutions in the image can be obtained.

X-ray film is most sensitive to low energy gamma rays and, consequently, radionuclides that emit low-energy gamma rays are best suited for such portable x-ray devices. Unfortunately, such radionuclides generally have short half-lives. When radiation doses are of little concern, such as in radiographing welds for a pipeline, longer-lived and higher-energy gamma-rays can be used. Table 13.2 lists some properties of radioisotopes used for gamma-ray radiography.

Neutron Radiography: Other radiations besides photons can be used in radiography. Parallel beams of thermal neutrons streaming from beam-ports of nuclear reactors are used to image objects with different thermal-neutron cross sections. Behind the sample, a "converter" screen absorbs transmitted neutrons and emits fluorescence and/or beta particles as a result. These secondary radiations, in turn, expose a photographic film or, in modern applications, a charged-coupled transistor array behind the converter screen to create an image of regions in the specimen through which neutrons are preferentially transmitted.

Table 13.2. Radioisotopes that have been used in portable gamma-ray devices.

Isotope	Half-Life	$E_{\gamma,x}$ (MeV)
^{125}I	60.1 d	0.027, 0.031, 0.035
^{192}Ir	74 d	0.02 + others
^{170}Th	120 d	0.084
^{153}Ga	236 d	0.097
^{145}Sm	360 d	0.06
^{55}Fe	2.73 y	0.006
^{60}Co	5.27 y	1.33, 1.17
^{137}Cs/^{137m}Ba	30.1 y	0.667

Materials such as water and hydrocarbons readily scatter thermal neutrons from the incident beam; by contrast, gases and many metals are relatively transparent to thermal neutrons. Voids, of course are totally transparent to the neutrons. Thus, neutron radiographs are ideally suited for distinguishing voids from water and hydrocarbons and for measuring density variations in neutron absorbing and scattering materials, even when inside metal enclosures that usually have a negligible effect on the neutron beam. X-ray radiography is poorly suited for such studies. With neutron radiography, water droplets inside metal tubes, air passages inside insects, burning gun powder in a cartridge being fired by a gun, and oil filled or void channels inside an engine block are readily observed. Figure 13.1 shows a neutron radiograph of a valve in which the rubber gaskets, o-rings, and even the teflon tape in the threads are readily observed.

Beta Radiography: To detect material variations in thin specimens, beta particles with ranges greater than the specimen thickness can be used. The number of beta particles transmitted through the specimen and exposing film behind the specimen is inversely proportional to the mass thickness of material in the specimen. Beta radiography, for example, can be used to obtain images of watermarks in paper and density variations in biological specimens.

Autoradiography: In autoradiography, the radiation used to form an image is generated within the specimen. In some biological applications, molecules tagged with radionuclides are concentrated in various portions of a specimen. By placing the specimen in contact with film and waiting for some period of time, the radiation emitted by the decay of tagged radionuclides exposes the film adjacent to the radionuclides. In this way, an image is created showing regions of high concentration of the tagged biological molecules. This technique is used extensively in biological studies ranging from plant uptake studies to DNA characterization.

A variation is to expose an ordinary specimen to thermal neutrons in order to convert some stable nuclides into radioactive isotopes. Materials in the sample with high neutron absorption cross sections preferentially generate radionuclides. After the neutron irradiation, placement of the sample against a photographic film will create an image of regions with high concentrations of radionuclides characteristic of the elements in the sample.

Figure 13.1. A neutron radiograph of a valve. The iron casing is relatively transparent to neutrons, but the plastic gaskets and teflon tape around the threads are readily apparent. Courtesy of Mike Whaley, Kansas State University.

13.4.2 Thickness Gauging

In many manufacturing processes involving continuous sheets of materials, the accurate real-time measurement and control of the thickness of the material, or coatings on a substrate, is a critical aspect of quality control. One widely used technique for making such thickness measurements, without requiring contact with the product, is to irradiate the material with radiation from a radioisotope or x-ray machine and to measure the intensity of transmitted or reflected radiation. Radiation of all types (α, γ, β, and x rays) have been used, depending on the material and thicknesses involved, and several variations of radiogauging have been developed.

In a transmission thickness gauge, see Fig. 13.2, radiation from a radioactive source passes through the test material and, as the thickness of the material increases, the intensity of radiation reaching a detector on the other side of the material decreases. A variation in the detector signal indicates a variation of the material thickness, and such a signal can be used to activate a servomechanism controlling the pressure exerted by rollers to adjust the thickness of the material being processed. Such gauging devices are used in the production of many thin products such as paper (with various added coatings, e.g., sandpaper), metal extrusions, rubber, plastics, and many other basic products.

In some production applications, it is not possible to have access to both sides of the product (e.g., extrusion of steel pipes, or measuring the thickness of egg shells). In one-sided thickness gauging (see Fig. 13.3), radiation is emitted toward the surface of the material. As the material's thickness increases, the amount of radiation scattered also increases. The measured variation in the intensity of backscattered radiation can be used to control the product thickness.

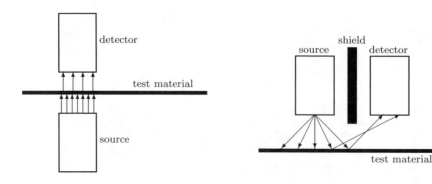

Figure 13.2. Thickness gauging by radiation transmission.

Figure 13.3. Thickness gauging by backscatter transmission.

An alternative to thickness gauging by measuring backscattered radiation is a fluorescence radiation measurement (see Fig. 13.4). As radiation from the source interacts in the material, it causes ionization of some of the material's atoms. Subsequently, as electrons cascade to fill vacancies in the inner electron orbits, characteristic x rays (or *fluorescence*) are emitted. The intensity of fluorescence increases as the material irradiated by the source radiation increases in thickness. Moreover, the energy of fluorescent photons is unique for each element.

Fluorescence thickness gauging is ideally suited for measuring and controlling the thickness of a coating on a substrate of a different material. Radiation from the source produces fluorescence in both the coating and the substrate. As the coating layer increases in thickness, more x rays with energies characteristic of the coating material are produced. The substrate x rays, with their unique energies, must pass through the coating and are increasingly absorbed as the coating thickness increases. Thus, by measuring both the coating and substrate x rays, very accurate control of the coating thickness can be obtained. For example, the thickness of a thin tin layer applied to a steel sheet to produce galvanized steel or of an adhesive coating on a cloth fabric to produce adhesive tape is readily measured and controlled by fluorescence thickness gauging.

Thickness gauging with radiation is used to manufacture a wide array of products: adhesive tape, aluminum strips, asphalt shingles, brass plate, floor coverings, galvanized zinc, many types of paper, plastic films, roofing felt, rubber products, soda-cracker dough, titanium sheets, and vinyl wall coverings, to name just a few items. These radiogauges are small, inexpensive, require little power and maintenance, can be placed deep inside machinery, and are both accurate and reliable.

13.4.3   Density Gauges

In some applications, the density of a material is of interest. By placing a detector a fixed distance from a source of radiation, the intensity of the detected radiation decreases as the density of the material between the source and detector increases. Such density gauges are used to manufacture many products including cement slur-

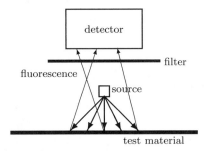

Figure 13.4. Thickness gauging using stimulated fluorescence.

ries, condensed milk, foam rubber, petroleum products, acids, ore slurries, resins, pulverized coal, sugar solutions, synthetic rubber, and tobacco in cigarettes.

13.4.4   Level Gauges

Determining the level of some liquid in a closed container is a common problem in many industries. The use of radiation offers an effective method for determining such liquid levels. The simplest such level gauge floats a radioactive source on top of the liquid. Then a detector is moved up and down the outside of the container until a maximum in the detection signal indicates the top of the liquid. Many variations are available. For example, if the source cannot be placed inside the container, the source and detector can be placed on opposite sides of the container and moved up and down together. A sharp change in the detector signal indicates the liquid surface. Another variation is to mount two pairs of source and detector on opposite sides of the tank, one pair mounted near the bottom and the second near the top. Such fixed pairs can then be used to alarm low and high liquid levels in the tank.

The attenuation of radiation by the liquid in closed containers is used, for example, in the beverage industry to reject improperly filled cans in a high-speed filling line. A can with a low fill will permit more radiation from an x-ray source to pass through the can. An abnormally high signal from a detector on the other side of the can is then used to remove automatically the under filled can.

Radiation level gauges are used to determine the levels of many materials including bauxite ores, chemicals of many types, petroleum liquids, grain syrups, molten glass and metals, and pulp slurries, to name a few.

13.4.5   Radiation Absorptiometry

Low energy photons passing through a material are absorbed strongly by photoelectric interactions. The photoelectric cross section varies with the material, and by measuring the degree of absorption of x and gamma rays of various low energies, it is possible to determine the composition of a material. For example, the value of crude oil depends strongly on its sulfur content. Since sulfur has a photon absorption coefficient almost ten times that of oil, low energy photons will be strongly absorbed by oil with a high sulfur content. By using a radionuclide source that emits low energy photons, such as ^{55}Fe which emits a 6 keV x ray following decay by electron capture, and by measuring the degree of photon absorption in the oil it is possible to monitor accurately in real time the sulfur content of oil flowing in a pipeline.

13.4.6   Oil-Well Logging

Radioisotopes are used extensively in oil exploration. By lowering a probe containing a gamma-ray or neutron source down a test well and measuring the characteristics of the radiation scattered back to a detector on the probe, it is possible to infer the type of rock formations adjacent to the probe. By making a vertical profile or *log* of the type of rocks and soils encountered, a geologist is able to identify the location and extent of possible oil-bearing formations.

13.4.7   Neutron Activation Analysis (NAA)

In many research and quality control applications, the determination of very small amounts of impurities or trace elements in a sample is needed. One technique that can be used for some elements with exquisite sensitivity is that of neutron activation analysis. In this technique, a small amount of the sample is placed in a nuclear reactor where neutrons transform some of the stable atoms into radioactive isotopes of the same elements. After irradiation, the gamma radiation emitted by the now radioactive sample is analyzed by a gamma spectrometer that sorts the emitted gamma photons by their energy and creates an energy spectrum of the detected photons. An example gamma spectrum is shown in Chapter 8. Peaks in such a spectrum indicate a photon of a particular energy, and since photons emitted by radioactive decay have energies unique to the decay of each radionuclide species, the radionuclide and its element can be readily identified. Moreover, the intensity or amplitude of a gamma-ray peak in the spectrum is proportional to the number of associated radioisotopes emitting this gamma photon, which, in turn, is proportional to the number of atoms of the element that were originally present in the sample and from which the radioisotope was produced.

Unlike other sensitive techniques for identifying trace elements in a sample, neutron activation analysis requires little sample preparation and is relatively inexpensive. Its major limitation is that it is useful for identifying only elements with large neutron absorption cross sections and whose isotopes produced by neutron absorption reactions are radioactive and emit gamma rays with energies distinguishable from those emitted by other activation products. About 75 elements can be detected in amounts as low as 10^{-6} g, about 50 at 10^{-9} g, and 10 at 10^{-12} g. Table 13.3 shows the sensitivity of this technique for identifying various elements.

Neutron activation analysis is routinely used in many research and industrial areas. Geologists use it to identify trace elements in rock samples. It is widely used in forensic investigations to obtain a unique trace-element signature in a sample to identify its source, for example, hair samples, paint chips, and spilled oil. Similarly, several elemental toxins can be identified in biological samples by neutron activation analysis. Pesticide residue of crops can be identified by their bromine and chlorine signatures. Petroleum refiners use neutron activation analysis to verify the natural contaminant vanadium has been removed from oil in the initial distillation process

Table 13.3. Approximate minimum detectable masses of various elements using the neutron activation technique. Limits are based on a one hour irradiation in a thermal neutron flux of 10^{13} cm^{-2}s^{-1} with no interfering elements. Source: Weaver [1978].

El.	Min. Mass (μg)	El.	Min. Mass (μg)	El.	Min. Mass (μg)	El.	Min. Mass (μg)
Ag	0.0001	Eu	0.0000009	Mo	0.01	Sm	0.00009
Al	0.001	Fe	10.0	Na	0.0004	Sn	0.02
Ar	0.0001	F	0.2	Nb	0.1	Sr	0.0009
As	0.0002	Ga	0.004	Nd	0.02	Ta	0.009
Au	0.0002	Gd	0.002	Ne	0.4	Tb	0.005
Ba	0.005	Ge	0.00005	Ni	0.004	Te	0.01
Bi	0.05	Hf	0.007	Os	0.004	Th	0.007
Br	0.009	Hg	0.0002	Pd	0.0002	Ti	0.001
Ca	0.5	Ho	0.00004	Pr	0.0001	Tl	0.5
Cd	0.009	I	0.001	Pt	0.004	Tm	0.002
Ce	0.02	In	0.000005	Rb	0.04	U	0.0004
Cl	0.002	Ir	0.00002	Re	0.00004	V	0.0001
Co	0.0009	K	0.002	Rh	0.00002	W	0.0002
Cr	0.2	Kr	0.001	Ru	0.002	Xe	0.01
Cs	0.0004	La	0.0004	Sb	0.0009	Y	0.0002
Cu	0.002	Lu	0.000009	Sc	0.002	Yb	0.0004
Dy	0.00000004	Mg	0.05	Se	0.02	Zn	0.02
Er	0.004	Mn	0.000005	Si	0.0009	Zr	0.2

before being "cracked" in further refining processes in which expensive catalysts would be "poisoned" by vanadium.

13.4.8 Neutron Capture-Gamma Ray Analysis

Just as radionuclides emit uniquely identifying radiation, many nuclear reactions also produce secondary radiations with unique characteristics. In particular, when a neutron is absorbed, *capture gamma rays* are usually emitted. These gamma rays have energies that are unique to the nuclide produced in the (n,γ) reaction. Thus, by observing the energies and intensities of capture gamma rays produced by neutron irradiation of a sample, the amounts of various elements in the sample can be determined. This method, for example, is being investigated as a means of detecting plastic explosives in airport luggage or for finding buried land mines or unexploded ordinance.

13.4.9 X-Ray Fluoresence Analysis

In many research areas, ranging from material science to archaeology, the elemental composition of some sample or specimen is desired. Many techniques can be used for such analyses such as mass spectrometry, chemical analysis, and neutron activation analysis. However, these techniques either require the destruction of the sample or a small piece be removed for analysis. For many artifacts, such as valuable art objects, this is not acceptable. X-ray fluoresence analysis provides a non-destructive technique for elemental analysis of important samples or artifacts.

The basic idea is to knock out inner electrons in some of the sample atoms and observe the *characteristic x rays* that are emitted as outer-shell electrons fall into the inner shell vacancies. Each element has characteristic x rays of unique energies (see Section 7.3.1), so by measuring the x-ray energies and their relative intensities, the elemental composition of the sample can, in principle, be determined. However, the energies of the characteristic x rays are relatively low. For instance, the $K_{\alpha1}$ energies are 0.52 keV for oxygen ($Z = 8$), 6.4 keV for iron ($Z = 26$), 22.2 keV for silver ($Z = 47$), and 98.2 keV for uranium ($Z = 92$). These x rays, particularly those from low Z elements, are readily absorbed as they travel to the sample surface where their energies and intensities are measured by a photon spectrometer (see Chapter 8). Consequently, x-ray fluorescence analysis can be used to give the elemental composition only in the surface layer of a sample. Several methods can be used to create inner shell vacancies in the sample atoms; two are discussed below.

Traditional X-Ray Fluorescence (XRF)

In this method an x-ray machine or a radioactive source that emits low energy gamma rays is used to irradiate the sample surface. Ideally, the energies of these source photons should be slightly above the K-edges of the elements of interest so as to maximize the creation of electron vacancies in the sample via the photoelectric effect. This method is widely used in industry and research. It has the great advantage that the equipment can be portable and brought to the object to be analyzed and can probe a sample to a depth of about 150 μm. However, because of the relatively large background produced by the source photons, the detection limit is restricted to about 10 μg/g.

Proton Induced X-Ray Emission (PIXE)

In this more recent method, a proton beam produced by a particle accelerator (see Section 13.6) is used to bombard the sample. These proton quickly slow, ionizing some of the sample atoms which results in x-ray emissions. Protons with energies less than about 4 MeV, typical of those produced by small accelerators, can probe a sample to depths of about 70 μm. Because of the relatively low photon background, PIXE can determine concentration down to about 1 μg/g and, by varying the energy of the protons, elemental compostions can be determined as a function of depth. For accelerators that can produce pro-

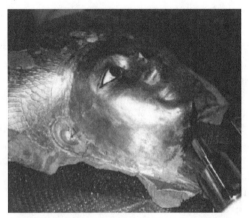

Figure 13.5. A gilded wood mask of Queen Satdjehuti (1600 BCE) excavated about 100 years ago. Courtesy of Dr. Andrea Denker, Hahn-Meitner Institute, Berlin.

tons with energies up to several tens of MeV, even greater depths in the sample can be probed. The major difficulty with PIXE is that the sample must be brought to the accelerator and beam time must be available without interfering with the accelerator's primary function, usually medical therapy or radioisotope production. To avoid such conflicts, the Louvre museum in Paris has a dedicated 2-MeV ion accelerator used for establishing the age, authenticity, provenance, manufacturing methods, and restoration questions about art objects.

As an example, Fig. 13.5 shows an ancient Egyptian gilded mask that was excavated about 100 years ago and kept in a private collection. Recently the mask was acquired by a German museum. The mask is in excellent condition, and the question arose whether the gilding had been restored. The gilding was analyzed with PIXE at 21 different points. At nearly all points the gold composition was that of typical river gold ($> 92\%$ Au, about 6% Ag, and 1% Cu) that would have been used by the Egytians; moreover, the thickness of the gilding was about 1.5 μm, typical of the gilding used on other ancient Egyptian artifact. From the thickness and composition it was concluded that the mask has its original gilding.

13.4.10 Proton Induced Gamma-Ray Emission (PIGE)

With high energy protons, nuclear reactions can be produced in the sample leading to a variation of the PIXE method called *proton induced gamma-ray emission* (PIGE). For example, the most difficult challenge for XFR or PIXE is to determine the amount of a light element in a matrix of a heavy element, such as copper in a silver coin. Because the 8.98 keV K-shell copper x rays are strongly absorbed by the silver, the probe depth is very limited. With PIGE, the ^{63}Cu$(n, \alpha)^{63}$Zn reaction produces a 670-keV gamma ray, thereby allowing a nearly one hundred times greater probe depth than could be obtained with PIXE.

13.4.11 Molecular Structure Determination

Arrangements of atoms in crystals or in complex molecules are often determined by using intense beams of low energy photons or neutrons. These photons or neutrons, with wavelengths comparable to the distances between atoms in a sample, scatter simultaneously from several atoms and, because of their wave-like properties, create a complex interference pattern in the scattered radiation. By analyzing the scattered interference pattern, the three-dimensional structure of the sample can be determined.

13.4.12 Smoke Detectors

Inexpensive and reliable smoke detectors use a small amount of an alpha emitter, typically ^{241}Am. Two metal plates are separated by a few centimeters. A small amount of the alpha emitter is deposited in the center of one plate so that alpha particles will be emitted into the air between the two plates. By placing a voltage difference between the two plates, the ion-electron pairs created in the air as the alpha particles are slowed are collected by the plates causing a small but steady current to flow between the plates. Smoke particles diffusing into the air between the plates stop some of the alpha particles and reduce the amount of air ionization, thereby causing the current flowing between the plates to decrease. This decrease in current in turn triggers an alarm.

13.5   Radiation Affects Materials

Radiation is also used to process materials in order to alter their properties. Because of the typically large volumes of material involved and the high radiation doses needed for radiation processing, the amount of radionuclides used in such activities far exceeds that for all other radionuclide uses. In some instances requiring large radiation doses, large x-ray machines are used as an alternative to large radionuclide sources.

13.5.1   Food Preservation

Perhaps the greatest potential for radiation processing is in the preservation of food. Three basic approaches are of importance. First, irradiation of packaged food products, such as milk and poultry, can greatly extend their shelf life by destroying many of the harmful bacteria naturally contained in the food. This mechanism for pasteurization of food does not require heating the product, as in conventional thermal pasteurization processes which often changes the flavor of the food product. Typical gamma or x-ray doses required for such pasteurization are a few kGy. With higher doses, generally between 20 and 50 kGy, complete sterilization of food products can be achieved and produce an indefinite shelf life for sealed products without refrigeration. However, complete radiation sterilization of food products often is accompanied by unacceptable changes in flavor, smell, color and texture, although these changes can be reduced by irradiating the product at low temperatures.

A second application of radiation processing of food is to extend the storage life of root crops. For example, potatoes or onions can be prevented from sprouting by irradiating them with photon doses of between 60 and 150 Gy. Finally, irradiation of grain and cereals with photon doses between 200 and 500 Gy before storage can greatly reduce subsequent insect damage.

Although food preservation by irradiation is not yet routinely used, the need to increase the safety of our food supply has led to increasing use of this technology. Today only a few food products processed by radiation are available; but there is currently much research interest in expanding the use of radiation preservation of food. Astronauts regularly dine on a wide variety of irradiated foods when on space missions, and have made many favorable comments on these food products.

13.5.2   Sterilization

Many medical supplies require complete sterilization, and many of these products cannot withstand conventional autoclaving (heat) sterilization. Gamma radiation can be used as an alternative to sterilize sealed packages containing delicate medical items such as plastic hypodermic syringes and pharmaceuticals which cannot withstand high temperatures.

Other consumer products, which contain pathogens, can be sterilized by large doses of radiation. For example, goat hair may contain the deadly anthrax bacterium. This bacterium can be destroyed by gamma irradiation before turning the hair into cloth or rugs. Anthrax as well as other pathogens may be used in biological warfare or terrorism, and gamma irradiation has been used to treat potentially contaminated letters in the postal system.

13.5.3   Insect Control

By irradiating insect larvae with enough radiation, typically many tens of grays, the resulting insect is rendered sterile. By releasing such sterilized insects into the environment who subsequently mate with normal "wild" partners and produce sterile eggs, the population of the insect species can be reduced or even eliminated.

This approach was first applied to control the screwworm fly which lays eggs in an animal host, such as cattle. The resulting maggots burrow into the host inevitably resulting in the host's death. This insect problem was particularly severe in the southeast United States. Between 1958 and 1960, a few billion radiation-sterilized screwworm flies were released by planes over Florida and neighboring states. The result was the complete eradication of this pest.

13.5.4   Polymer Modification

As radiation passes through a polymer, electronic bonds are broken and new bonds formed between adjacent polymer molecules. For example, polyethylene film is often used as a packaging material that can be shrink wrapped around the product. To increase the strength of such polyethylene wrap, gamma or x-ray irradiation of the plastic can greatly increase its strength. Many have seen plastic spoons deform when immersed in a hot liquid. By irradiating these plastics, they can be made stronger so that such deformations do not occur. Wood flooring products saturated in simple chemicals and subsequently irradiated to produce new bonds between the wood fibers and the plastic resins produce a very durable surface without any loss of the beauty of the wood grain.

13.5.5   Biological Mutation Studies

The development of new plant strains with desirable traits, such as disease resistance or enhanced grain yields, has been a major emphasis in agricultural research. New strains result from a modification of the genome of a plant. Such mutations can result from cross breeding, the laboratory microscopic insertion of new genes into the DNA of a plant cell, or by stochastic mutations in a cell's DNA induced by radiation. Radiation induced mutations have been a mainstay in producing new crop strains. By irradiating seeds of a particular crop, mutations are generated. From subsequent field trials of the irradiated seeds, those few plants exhibiting new desirable characteristics can be selected and propagated. Such radiation induced mutations have resulted in dozens of new crop strains which are now grown throughout the world.

13.5.6   Chemonuclear Processing

Many chemical reactions are greatly enhanced by the presence of a catalyst, which promotes the interaction of the reacting species. Traditionally catalysts are expensive noble metals, such as the platinum used in the catalytic converters in our cars to reduce the hydrocarbon emissions.

Radiation can also act as a catalyst. For example, the production of ethyl bromide by reacting ethylene with hydrogen bromide is greatly enhanced by the ionization produced by a gamma-ray source such as ^{60}Co.

13.6   Particle Accelerators

Beginning in the 1930s, machines were developed to accelerate charged subatomic particles such as protons and alpha particles to speeds that could induce nuclear reactions. The first such *particle accelerator* produced protons with 700 keV of energy. Soon an accelerator will be operational that produces 7-TeV protons or heavy ions such as lead with 1200 TeV of kinetic energy.

Today the development of ever more energetic particle accelerators is driven by the high-energy physics community. With these enormous and costly machines physicists will perform experiments that will reveal information about the fundamental physics governing the subatomic world. Accelerators with lower energies are also central to other areas of research such as the study of atomic and nuclear physics. A 1 GeV proton accelerator is now used at the Spallation Neutron Source at Oak Ridge National Laboratory to bombard a liquid mercury target. The resulting spallation reactions release copious neutrons which are ideal probes to determine molecular structures. Accelerators can also be used to produce intense x-ray beams that in turn can be used in fundamental research on materials. Finally, as in most areas of fundamental research, accelerator technology has spun off many practical applications such as cancer therapy, production of important radionuclides, ion implanting, and food preservation to name a few.

13.6.1   Cockcroft-Walton Accelerator

One obvious way to accelerate a charged particle is to release the particle in a strong electric field. The kinetic energy gained by the particle as it moves through the potential of V volts between its point of release (source) and an electrode (target) is qV, where q is the charge on the particle. To produce a source of electrons, a hot filament can be used while a gas discharge tube containing hydrogen, deuterium, or helium can be used as an ion source of protons, deuterons, and alpha particles. Thus, a simple accelerator can be made by placing an ion/electron source at one end of an evacuated tube and a target or exit hole at the other end. The challenge is to produce a very large voltage between the two ends of the tube.

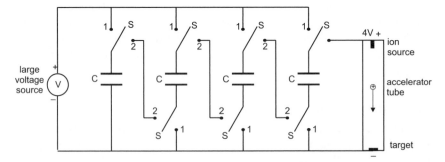

Figure 13.6. A circuit to charge capacitors C in parallel and discharge them in series. When all switches S are in position 1, all the capacitors are charged in parallel. When all the switches are in position 2, the capacitors are connected across the accelerator tube in series, thereby quadrupling the voltage of the original source. After Kernan [1969].

John Cockcroft and Ernest Walton [1932] built a machine that accelerated protons to 700 keV. To obtain the 700 kV voltage source, they used a circuit proposed in 1919 by Heinrich Greit. The circuit, a simplified form of which is shown in Fig. 13.6, uses large banks of capacitors that are charged with the capacitors connected in parallel, but are connected in series between the ion source and target of the accelerator tube. A high multiple of the source voltage can be obtained by using a large number of capacitors. In their accelerator, Cockcroft and Walton used an alternating voltage from a transformer and vacuum-tube rectifiers as the switches. With this machine, they achieved for the first time the alchemists' dream of transmuting one element into another by converting lithium into helium through the $^{7}_{3}$Li$(p, \alpha)^{4}_{2}$He reaction. For this accomplishment they were awarded the Nobel prize in 1951.

Figure 13.7. A Cockcroft-Walton accelerator in which hydrogen gas is negatively ionized and accelerated by a positive voltage to an energy of 750 keV. Courtesy of the Intense Pulsed Neutron Source at Argonne National Laboratory.

Today Cockcroft-Walton voltage multipliers are found in a diverse array of technologies that use high voltage, such as laser systems, x-ray systems, ion pumps, electrostatic systems, to name a few. They are also routinely used as the first accelerator stage for particle accelerators. They can produce fairly large ion currents and have no moving parts, but can produce energies of at most a few MeV. Fig. 13.7 shows a modern Cockcroft-Walton accelerator that is used in the first stage of a much larger accelerator.

13.6.2 Van de Graaff Accelerator

Robert Van de Graaff [1931] proposed and subsequently built a new type of electrostatic accelerator. This type of charged-particle accelerator is much like a Cockcroft-Walton accelerator except that, instead of a electronic circuit to produce the large voltage, charge is mechanically moved from one electrode to the other. The charge transfer depends on the fact that, if a charged conductor makes contact with the interior of a second hollow conductor, all of the charge is transferred to the hollow conductor no matter how much charge is already on the hollow conductor. Limited only by electrical insulation, the hollow conductor can be raised to any potential by adding more and more charges to it through the internal contact.

A schematic of the Van de Graaff accelerator is shown in Fig. 13.8. One terminal with many fine points is maintained at a positive potential of many thousands of volts. At these *spray points*, coronal discharge fills the nearby air with thousands of positive and negative ions. The positive ions are repelled by the spray points and attach to a belt made of a flexible nonconductor such as rubber, silk, or rayon. The belt is driven over pulleys by a motor so that the positive charges travel upward to a large highly polished metal shell where a set of conducting points removes the charge from the belt and transfers it to the shell, which is the other electrode of the electrostatic accelerator. The Faraday cage effect assures that positive charge is transferred to the shell regardless of the shell's existing potential. The buildup of charge on the shell continues until the charge leakage through the support insulators and surrounding air equals the rate of charge transfer to the shell. A gas discharge tube with a small opening creates ions inside the shell where they are then accelerated down a vacuum tube to a target at the low-voltage end of the tube. The accelerator tube is composed of annuli of insulating material, such as ceramics or glass, joined together with vacuum tight seals. The acceleration tube must be sufficiently long to prevent sparks or other discharges to pass from one end to the other.

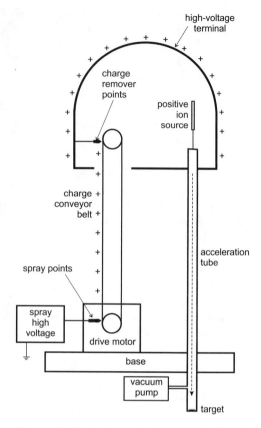

Figure 13.8. A schematic diagram of a Van de Graaff generator.

The maximum attainable voltage can be increased several times if the whole assembly is enclosed in a chamber filled with a pressurized nonconductive gas such as air, nitrogen, or methane. In this way, voltages of several MV can be achieved. These machines can produce a relatively large current and are able to maintain the voltage, and hence the energy of the ions, to well within a 1% variation.

A novel way of doubling the energy of a Van de Graaff accelerator is to use two acceleration stages. In such a *tandem* Van de Graaff machine, *negative* ions are produced at the low voltage end and accelerated toward the positive high-voltage terminal located in the middle of the pressure shell. Just before reaching this positive electrode, the beam passes through a carbon foil where the ions are stripped of their electrons. The now positively charged ions pass through a hole in the electrode, are now repelled by the high-voltage electrode and accelerated toward the other low-voltage end where they leave the machine with twice the energy of a one-stage Van de Graaff accelerator. Such tandem machines, which typically produce protons with up to 20 MeV of energy, are in wide use today for low-energy

Figure 13.9. A tandem, van de Graaff, ion accelerator that is used for atomic physics studies. In the foreground is the ion source that injects negative ions of various elements into tandem machine. Half way down the high-pressure enclosing chamber, just before reaching the van de Graaff's high voltage positive electrode, the ions pass through thin film that removes electrons to create positive ions that are further accelerated by repulsion from the positive electrode. Magnetic fields at the far end select beams of positive ions with the desired degree of ionization. Courtesy of the James MacDonald Laboratory at Kansas State University.

atomic and nuclear physics research and for injecting ions into much higher energy accelerators. Such an accelerator is shown in Fig. 13.9.

13.6.3 Linear Accelerators

The idea of an accelerator that accelerates ions or electrons in a straight line by an oscillating electric field connected to a series of electrodes dates to the 1920s, but such major linear accelerators were constructed only in the 1950s because of the technical problems encountered with coordinating the power supply oscillations to the ion motion. Its simplest form is illustrated in Fig. 13.10. A linear accelerator, or *linac* for short, consists of a vacuum chamber containing a colinear series of conductive cylindrical electrodes with alternate electrodes connected electrically to each other. Small holes are drilled into the faces of electrodes so the ion or electron beam can pass. Applied to the electrodes is a radiofrequency oscillating voltage which accelerates the ions or electrons as they pass from one electrode into the next one. Axial holes in the electrodes allow ions to pass through the electrodes in which they travel at constant speed because there is no electric field inside a hollow metal conductor. By using electrodes of increasing length, the time required for the charged particles of increasing speed to pass through each cylinder is kept constant so that, when the particle emerges from an electrode, the oscillating voltage has changed across the gap and the particle is accelerated through the gap before entering the next electrode. With such a design, the particles can be kept in phase with the reversals of potential as they pass from one electrode to the next.

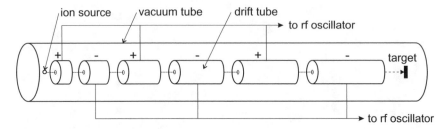

Figure 13.10. A schematic diagram a linear accelerator. The drift tubes between the acceleration gaps increase in length as the charged particles increase in speed. The time to traverse a drift tube equals one half of the cycle of the applied oscillating electric potential. At all times adjacent drift tubes have opposite charges.

Early linear accelerators were used only for very heavy, slowly moving ions because of the limited frequency range of available voltage oscillators. But radar development during the second world war created power supplies with much higher radiofrequencies that allowed protons and electrons to be accelerated to hundreds of MeV.

As charged particles become relativistic, their increasing kinetic energy does not produce an appreciable increase of speed. Thus, the design of a linear accelerator for relativistic particles is considerably simplified since the constant length electrodes can be used with a fixed-frequency voltage source. The charged particles can be introduced to such an accelerator with relativistic energies. For electrons, this type of accelerator is particularly attractive since electrons become relativistic at relatively low energies (1% mass increase at 5 keV). In Fig. 13.11 the inside of a linac used to accelerate protons is shown.

Linear accelerators have many advantages, the primary being the ease with which charged particles can be introduced at one end and extracted at the other. Also, unlike circular accelerators (see the following sections), the energetic particle beam does not radiate away any of the input power as bremsstrahlung radiation.

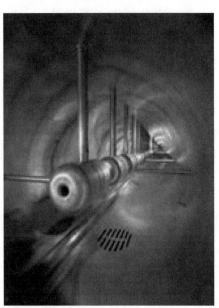

Figure 13.11. Drift tubes inside the vacuum chamber of a linear accelerator used to accelerate protons. Courtesy of the Fermi National Laboratory.

But, ultra high energy linear accelerators present many technical challenges. The synchronization of the many oscillating power sources over a long baseline is one. Alignment of the electrodes to maintain the beam path over the length of the accelerator, in the presence of earth compression and settling, is another challenge. Maintaining a vacuum in the accelerator tube is yet another challenge.

Today, many linear accelerators are designed to accelerate electrons. In medical therapy, electron linear accelerators are widely used to produce bremsstrahlung photons with several MeV of energy by impinging the electron beam onto a heavy metal target. See Chapter 14.

At the other extreme, the world's highest energy electron accelerator is at Stanford Linear Accelerator Center (SLAC). Here over 80,000 electrodes, each about an inch long, in a beam tube 3-km long are used to accelerate pulses of electrons to almost 30 BeV with about 1.5×10^{10} electrons per pulse. With this machine, many fundamental discoveries have been made in particle physics.

13.6.4 The Cyclotron

To produce even higher ion energies, Lawrence and Livingston [1932] proposed an accelerator that accelerated ions in many small stages instead of using a single stage as in the Van de Graaff accelerator.[2] With many stages, the voltage between adjacent stages need be only a few thousand volts. The novel feature about Lawrence's scheme is the use of a magnetic field oriented perpendicular to the ion trajectories so that the ions are forced to move in circles between stages. In this manner, the machine could be kept relatively compact.

In a cyclotron two flat semi-circular metal boxes, each called a "dee" because of its resemblance to the letter D, are positioned with their diameter edges parallel and slightly apart (see Fig. 13.12). Each dee is connected to a terminal of a high-frequency oscillating high-voltage source. In this way the direction of the electrical field across the gap between the dees changes direction millions of times per second. A small source of ions is placed in the center midway between the two dees. The space inside each dee, however, has no electric field to affect the motion of the ions because of the electric shielding by the metal dees. The dees are enclosed in, but insulated from, an evacuated box. This box is then placed between the poles of an electromagnet which produces a strong magnetic field perpendicular to the plane of the dees.

The centripetal force that makes the ion trajectory curve inside a dee is provided by the magnetic field B and equals Bqv, where q and v are the ion's charge and speed, respectively. Thus

$$\frac{mv^2}{r} = Bqv. \tag{13.3}$$

From this result, the radius of the semicircle made by an ion in a dee is

$$r = \frac{mv}{Bq}. \tag{13.4}$$

---

[2]Less well know is that Sándor Gaál of Hungary earlier described the operation of this type of particle accelerator.

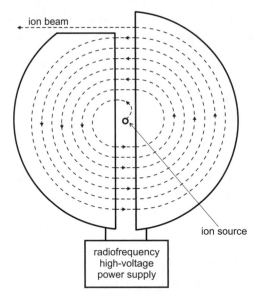

Figure 13.12. A schematic diagram of how ions
are accelerated in a cyclotron. The magnetic field
is directed into the plane of the paper.

If during the time the ion travels through a semicircle the voltage applied to the
dees has reversed, the ion will be accelerated as it crosses the gap between the dees
so that it enters the other dee with a greater speed and, from Eq. (13.4), travels in
a semicircle with a larger radius. In this manner the ions spiral outwards until they
come to the periphery where the positive ion beam is deflected out of the cyclotron
by a highly negatively charged deflection plate and directed to an appropriately
placed target.

From Eq. (13.3) the angular speed of the ions and of the oscillating voltage
applied to the dees is

$$\omega \equiv \frac{v}{r} = B\frac{q}{m}, \tag{13.5}$$

and the frequency of the oscillating voltage applied to the dees must be

$$f = \frac{\omega}{2\pi} = \frac{B}{2\pi}\frac{q}{m}. \tag{13.6}$$

Note that this frequency is independent of the ion speed and radius of the ion
semicircles. The amplitude of the oscillating voltage determines the energy boost
to the ion as it crosses the gap between the dees. The smaller the voltage amplitude,
the more spiral orbits the ion makes before reaching the exit port. This property
allows cyclotrons to accelerate ions to several MeV with relatively small applied
voltages.

An ion reaching the exit port at radius R with speed v has an energy given by

$$E = \frac{1}{2}mv^2 = qV, \tag{13.7}$$

where V is the equivalent potential difference through which the ion has been accelerated. Use of Eq. (13.3) to eliminate v (and with $r = R$) gives

$$V = \frac{1}{2}B^2 R^2 \frac{q}{m}. \tag{13.8}$$

Thus, for a given ion (q/m value), doubling the radius of the cyclotron quadruples the energy of the ions.

The fixed frequency cyclotron described above can produce a continuous stream of ions with a high average power. The device is relatively compact thereby reducing costs for shielding, foundations, and the enclosing building. Such machines can produce protons up to 10 MeV and deuterons up to 20 MeV. Figure 13.13 shows a cyclotron in operation.

Figure 13.13. The white horizontal misty line is a beam of deuterons emerging from the 60-inch cyclotron at Argonne National Laboratory. The beam is visible because of excitation of air molecules by the energetic deuterons. From Kernan [1969].

13.6.5 The Synchrocyclotron and the Isochronous Cyclotron

The energies to which ions can be accelerated in a fixed-frequency cyclotron are limited by relativistic effects. Recall from Chapter 2, the mass m increases as the speed v increases according to

$$m = \frac{m_o}{\sqrt{1 - v^2/c^2}}. \tag{13.9}$$

For less than 1% increase in mass the energies of a proton, deuteron and triton must be less than 10, 20, and 40 MeV, respectively. Such are the upper limits for fixed-frequency cyclotrons. Above these energies, relativistic effects become important.

From Eq. (13.5) angular speed of an ion in a cyclotron is

$$\omega = B \frac{q}{m_o} \sqrt{1 - \frac{v^2}{c^2}}, \tag{13.10}$$

which decreases as the velocity increases. Thus, as an ion starts to become relativistic it takes longer to complete its semicircular journey through a dee and becomes out of phase with the fixed frequency of the oscillating electric field. Thus, to keep the ions in phase with the oscillating voltage, the oscillation frequency must decrease as the ion energy and orbital radius increases. Such frequency-modulated machines are termed *synchrocyclotrons* and are capable of producing protons with energies of hundreds of MeV. However, unlike constant-frequency cyclotron which produce a continuous beam of ions, a synchrocyclotron produces bursts of ions since the ions must be together to stay in phase with the changing oscillating electric field. The repetition rate typically is between 60 and 300 pulses per second. Although the average beam intensity is considerably less than that for a constant-frequency cyclotron, the goal of obtaining much high energy ions is realized.

As can be seen from Eq. (13.10), an alternative way to keep the relativistic ions in phase with the oscillating voltage is to increase the strength of the magnetic field with increasing radius rather than to vary it with time. Such fixed-frequency cyclotrons with radially varying magnetic fields are called *isochronous cyclotrons*. The radial gradient of the magnetic field, however, tends to defocus the ion beam. To compensate for this defocusing effect, ridges on the magnet faces, which vary the magnetic field in the azimuthal direction (see Fig. 13.14), are used. These isochronous cyclotrons can produce a continuous ion beam current and today are widely used for radioisotope production in many hospitals.

Because of the centripetal acceleration applied to the spiraling ions in a cyclotron, some of the ion energy is radiated away as bremsstrahlung thereby slowing the emitting ion. In high energy cyclotrons, most of the input electric power is dissipated by bremsstrahlung with a consequence that very few high energy ions are produced. Also, extensive radiation shielding is required for such research class cyclotrons. However, sometimes the emission of bremsstrahlung is desired. Some modern cyclotrons are purposely constructed to maximize bremsstrahlung production so as to generate very bright and spectrally pure far-ultraviolet light and soft x rays which are hard to produce any other way.

13.6.6   Proton Synchrotrons

Although there is no theoretical limit to the energy protons can acquire in a synchrocyclotron, the construction of such a device to produce protons with energies of BeV and beyond is prohibitively expensive because of the enormous magnets that would be required. To avoid this cost, a new accelerator design, the *synchrotron*, was used to first produce protons above 1 BeV. In a synchrotron, the strength of the magnetic field is increased as the proton's relativistic mass increases so that the protons move in a circle of almost constant radius rather than in the widening spirals as occurs in a synchrocyclotron. In this manner the entire center portion of the synchrocyclotron's magnet can be eliminated. Now only many small magnets around the doughnut-shaped beam tube are needed to keep the protons moving in a circle.

Figure 13.14. A 16.5 MeV GE PETrace Cyclotron, used for radiopharmaceutical production, opened and showing the elaborate magnet face used to produce the radially varying magnetic field. Courtesy of Paul E. Christian, Huntsman Cancer Institute, University of Utah.

Typically, bursts of protons from a preaccelerator, such as a linear accelerator, that already have started to become relativistic are introduced into the circular beam tube. In this way the time variation of the confining magnetic field does not have to change nearly as much as it would if the synchrotron had to also bring the protons up to relativistic speeds. Placed at one or more locations around the beam tube are pairs of electrodes to which radiofrequency oscillating voltage is applied across the acceleration gap, between the electrodes. The frequency of this oscillating electric field must be synchronized with the speed of the proton bunch so that every time the protons pass through the acceleration gap they receive a small boost in energy. The radiofrequency electric field also keeps the protons bunched together, since a proton that is going slightly faster than the others will reach the gap too soon and receive less of an energy kick than the protons behind it. After the proton bunch has made hundreds of thousand of orbits, they can acquire enormous kinetic energies. Finally, switching magnets are used to extract the proton bunch and send it down a beam line to a target.

The rapidity with which cyclotrons can accelerate another burst of protons is known as the cycle time. In small synchrotrons this cycle time can be tens of pulses per second; however, the largest synchrotrons require many seconds between successive pulses.

The Tevatron

The most energetic proton accelerator facility in the world today is the Tevatron at the Fermi Laboratory located near Chicago. In this huge complex, both protons and antiprotons are created with 0.98 TeV of kinetic energy. The protons and antiprotons can be forced to collided with each other with a reaction energy of 1.96 TeV.

A series of five accelerators are used. First, protons are produced by ionizing hydrogen and accelerating the resulting protons to 750 keV in a Cockcroft-Walton accelerator. The 750-keV protons are injected into a 130-m long linear accelerator and accelerated to 400 MeV. The pulses of 400-MeV protons are then injected into a "booster" synchrotron 475 m in diameter in which they revolve about 16,000 times in 0.033 s and reach an energy of 8 GeV.

Next the protons are switched into the "main injector," another synchrotron but with a circumference of nearly 2 miles. In the main injector, protons are accelerated to energies of 150 GeV. Protons from the main injector are also used to produce anti-protons by smashing the energetic protons into a nickel target. The antiprotons are collected, focused, pre-accelerated to 8 GeV in two other synchrotrons, and then accelerated to 150 GeV in the main injector.

Finally, protons and antiprotons are injected into the *Tevatron*, the world's first synchrotron with all superconducting magnets (about 1000 of them). Because protons and antiprotons have opposite charge, they orbit in opposite directions. Every revolution adds 650 keV to the their kinetic energy, and, after 20 seconds and after traveling around the ring more that a million times, the particles reach almost 1 TeV. During this 20 second period, the magnetic field increases from about 0.66 to 3.64 tesla to keep the particles in the same circular orbit as their relativistic mass increases. The counter rotating proton and antiproton beams are allowed to intersect producing a few head-on collisions with a reaction energy of almost 2 TeV. Immense particle detectors around the collision sites reveal the numerous particles created and allow physicists to test fundamental theories of matter.

Although the Fermi Lab is currently (2006) home to the most powerful particle accelerator, the Large Hadron Collider (LHC) at CERN[3] will soon come into operation and produce 7 TeV protons.

13.6.7 Betatron

The use of magnetic fields to accelerate electrons was first proposed in 1922 by the mathematician Joseph Slepian who was working for Westinghouse at the time [Kernan 1969]. The first betatron, a machine based on Slepian's ideas, was built in 1940 by D.W. Kerst at the University of Illinois. His machine produced electrons with energies of 2.5 MeV.

In a betatron a changing magnetic field is used to make electrons move in a circle and also to accelerate them as they move in circular orbits. The betatron is based on the principle of magnetic induction for a transformer, in which an alternating current applied to one coil produces an oscillating magnetic field that, in turn, induces an oscillating current in the secondary coil. In a betatron, a cloud of electrons in a doughnut-shaped vacuum tube acts as the secondary coil with a single turn (see Fig. 13.15). The chamber is placed between the poles of an electromagnet which

---

[3]Originally "Conseil Européenne pour Recherches Nucléaires" it is now called the "European Laboratory for Particle Physics." It is located near Geneva, Switzerland.

is excited by external coils in which an oscillating current flows, thereby producing a strong time-varying magnetic field in the center of the toroidal beam tube. The poles also extend over the beam tube to produce the magnetic field necessary to make the electrons revolve in circles.

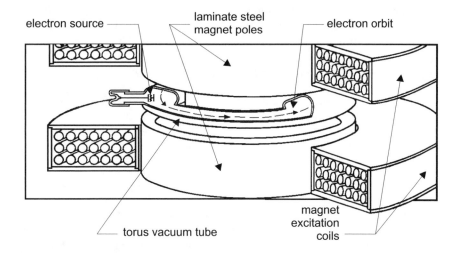

Figure 13.15. In a betatron, electrons are accelerated in a circular vacuum chamber by a rapidly increasing magnetic field produced by a pulsed current passing through the magnetic excitation coils. When the electrons reach their maximum speed, they are magnetically switched either to a target in the chamber or switched to an exit port. From Kernan [1969].

An electron gun, consisting of a hot filament and a several thousand volt potential, is used to inject electrons with several keV of energy into the beam tube at the instant the magnetic field is just starting to increase from zero in the first quarter of it sinusoidal cycle. The increasing magnetic field induces an increasing electric field in the torus which accelerates the electrons. As the electrons increase in speed, their masses increase and an ever stronger magnetic field is necessary to keep them rotating in the same circular orbits. When the magnetic field strength reaches its maximum (a quarter period into the oscillation cycle), the electrons are deflected from their circular paths into a target or exit port by a switching magnet which is a secondary coil into which a pulse of current is applied at the appropriate time.

Key to a betatron's operation is a proper balance between the strength of the magnetic field used to keep the electrons moving in the same circular orbit, B_{orb}, and the average magnetic field $\overline{B}$ between the poles of the magnetic inside the torus that induces the accelerating electric field. To derive this relationship, first consider the condition necessary to keep an electron in a circular orbit of radius R. The centripetal force must equal the Lorentz magnetic force, i.e.,

$$\frac{mv^2}{R} = evB_{orb} \qquad \text{or} \qquad p(t) \equiv mv = eRB_{orb}(t), \qquad (13.11)$$

where $p(t)$ is the electron's momentum at time t. Hence the force required to keep the electron in a circular orbit as its mass increases is

$$F_c(t) \equiv \frac{d(mv)}{dt} = eR\frac{dB_{orb}(t)}{dt}. \tag{13.12}$$

Now consider the accelerating force. From Faraday's law of electromagnetic induction, the electromagnetic induction in a secondary coil is $V = NA(d\overline{B}/dt)$. In the betatron, the number of secondary turns is $N = 1$ and the area over which the magnetic induction flux flows in $A = \pi R^2$. The electric field applied to an orbiting electron is $E = V/(2\pi R)$. Combining these two results yields

$$E(t) = \frac{V(t)}{2\pi R} = \frac{1}{2}R\frac{d\overline{B}(t)}{dt}. \tag{13.13}$$

The force on the electron produced by this electric field and which accelerates the electrons is

$$F_a(t) \equiv \frac{d(mv)}{dt} = eE(t) = \frac{1}{2}eR\frac{d\overline{B}(t)}{dt}. \tag{13.14}$$

For there to be acceleration by magnetic induction at constant radius, the time rate of momentum change by the magnetic field must be the same as that produced by magnetic induction, i.e., F_c must equal F_a. Equating Eq. (13.12) and Eq. (13.14) and simplifying yields the *betatron condition*

$$B_{orb}(t) = \frac{1}{2}\overline{B}(t). \tag{13.15}$$

This condition says that, for an electron to be accelerated at constant radius, the magnetic field keeping the electrons moving in circles must be one-half of the central magnetic induction field used to accelerate the electrons. This difference in the two magnetic fields is produced by having the poles inside the toroidal beam tube much closer together than at the radius of the electron orbits. Note that the betatron condition applies to relativistic electrons since the forces of Eq. (13.12) and Eq. (13.14) are in terms of the rate of momentum change, as required by special relativity.

Betatrons have been constructed to produce electrons with several hundred MeV of energy. As with cyclotrons, such high energy electron accelerators must be heavily shielded because of the bremsstrahlung radiated away as the electrons are accelerated. Lower energy betatrons can be constructed that are very compact and such betatrons are still used in industry and medicine.

As an alternative to betatrons, electron cyclotrons have also been built to produce electrons with several hundred MeV of energy. Such cyclotrons have the advantage that only smaller orbital magnets are needed since the acceleration is performed in a radiofrequency electric field inside the torus beam tube. If electrons are injected at a few MeV of energy by, say, a small betatron, their speeds change little as they are accelerated in the cyclotron, so that nearly constant if power supplies can be used. Of course the magnetic field producing circular orbits must increase in strength as the mass of the electrons increases.

BIBLIOGRAPHY

BAKER, P.B., D.A. FUCCILLO, M.A. GERRARD, AND R.H. LAFFERTY, *Radioisotopes in Industry*, USAEC, Washington, DC (1967).

COCKCROFT J.D. AND E.T.S. WALTON, "Experiments with High Velocity Positive Ions: I. Further Developments in the Method of Obtaining High Velocity Positive Ions," *Proc. Roy. Soc.*, (London) **A137**. 619 (1932).

FOLDIAK, G. (Ed.), *Industrial Applications of Radioisotopes*, Elsevier, Amsterdam, 1986.

HALMSHAW, R., *Industrial Radiology: Theory and Practice*, Chapman and Hall, London, 1995.

IAEA, *Guidebook on Radiotracers in Industry*, International Atomic Energy Agency, Vienna, 1990.

KERNAN, W.J., *Accelerators*, Understanding the Atom series, US Atomic Energy Comm., 1969.

KERST, D.W., "Acceleration of Electrons by Magnetic Induction," *Phys. Rev.*, **60**, 47 (1941).

LAWRENCE E.O. AND M.S. LIVINGSTON, "The Production of High Speed Light Ions without the Use of High Voltages," *Phys. Rev.*, **40**, 19, (1932).

MURRAY, R.L., *Nuclear Energy*, Butterworth-Heineman, Boston (2000).

URBAIN, W.M., *Food Irradiation*, Academic Press, New York (1986).

URROWS, G.M., *Food Preservation by Irradiation*, USAEC, Washington, DC (1968).

VAN DE GRAAFF, R.J., "A 1.5 MeV Electrostatic Generator," *Phys. Rev.*, **38**, 1919 (1931).

WEAVER, J.N., *Analytic Methods for Coal and Coal Products*, Ch. 12, Academic Press, New York (1978).

WILKINSON, V.M. AND G.W. Gould, *Food Irradiation: A Reference Guide*, Butterworth-Heinemann, Oxford, 1996.

WOODS, R.J. AND A.K. PIKAEV, *Applied Radiation Chemistry: Radiation Processing*, Wiley, New York, 1994.

PROBLEMS

1. You are to select a radionuclide to indicate the arrival of a new batch of petroleum in a 700-km long oil pipeline in which the oil flows with a speed of 1.4 m/s. Which of the following radioisotopes would you pick and explain why: ^{24}Na, ^{35}S, ^{60}Co, or ^{59}Fe.

2. A biological tissue contains ^{131}I ($T_{1/2} = 8.04$ d) with an activity concentration of 3 pCi/g. What is the smallest mass of this tissue for which the ^{131}I can be detected? State and explain all assumptions made.

3. What is the atomic concentration (atoms/g) of ^{60}Co in a tagged 5-gram sample of iron that produces a photon count rate (1.17 and 1.33 MeV photons) of 335 counts/minute over background. The detector has an efficiency of 0.15 for ^{60}Co photons.

4. Iron pistons rings in a motor have ^{60}Co imbedded as a label with a concentration of 5.0 nCi/kg. The engine is lubricated by recirculation of 8 kg of oil. After 1,000 hours of operation, the oil is found to have a ^{60}Co concentration of 2.7 fCi/g. How much iron has been worn from the piston rings?

5. A radioisotope with a half-life of 30 days is injected into the center of a river at a steady rate of 10 μCi/min. At 10 km downstream, the water activity is found to be 7.3 nCi/g. What is the river flow rate in m^3/h? What assumptions did you make?

6. Five grams of water containing a radionuclide with a concentration of 10^7 Bq/L and a half life of 1.3 d are injected into a small pond without an outlet. After 10 days, during which the radioisotope is uniformly mixed with the pond water, the concentration of the water is observed to be 1.75 μBq/cm^3. What is the volume of water in the pond?

7. A radioactive tracer, with a half life of 10 days, is injected into an underground aquifer at $t = 0$. Eighty-five days later the radioisotope is first observed in a monitoring well 500 m from the injection point. What is the speed of water flowing in the aquifer between the two wells?

8. The sizes of objects imbedded in a homogeneous material are to be determined by radiography. Explain what physical properties the objects and the homogeneous material should have for (a) x-ray radiography, (b) neutron radiography, and (c) beta-radiography to be the preferred choice.

9. What type of a radioactive source should be used, and how should it be used in thickness gauging processes if it is to be used for (a) gauging paper thickness, (b) the control of sheet metal thickness between 0.1 and 1 cm, and (c) controlling the coating thickness of adhesive on a cloth substrate?

10. It is proposed to a use a ^{90}Sr source (half-life 29.1 y) in a thickness gauge for aluminum sheets. This radionuclide source is in secular equilibrium with its ^{90}Y daughter (half-life 64 h). Both radioisotopes emit beta particles with no accompanying gamma rays. The average energies of the ^{90}Sr and ^{90}Y beta particles are 195 keV and 602 keV, respectively. What is the maximum thickness of aluminum for which gauging by ^{90}Sr is practical? Ranges of electrons in aluminum are shown in Fig. 7.16.

11. It has been suggested that Napoleon died as a result of arsenic poisoning since the growing end of his hair has been found, through NAA analysis, to have abnormally high levels of arsenic. What hair sample mass (in grams) is needed to detect arsenic with a concentration of 0.001 ppm?

12. Mercury pollution in water is of concern since fish often concentrate this element in their tissues. To measure such mercury contamination by neutron activation analysis, a fish sample is irradiated in a reactor in a thermal neutron flux $\phi = 1.5 \times 10^{12}$ cm^{-2}s^{-1}. The stable mercury isotope ^{196}Hg has a neutron absorption cross section for thermal neutrons of 3.2 kb. The resulting ^{197}Hg has a half life of 2.67 d and can be detected in a fish sample at a minimum activity of 15 Bq. What irradiation time is needed to detect mercury contamination at a level of 30 pg/g (30 ppb) in a 10-g fish sample?

13. Neutron transmutation doping is a process in which high purity silicon is placed in a nuclear reactor where some of the silicon atoms are transmuted to stable phosphorus atoms through the reactions

$$^{30}_{14}\text{Si} + ^{1}_{0}\text{n} \longrightarrow ^{31}_{14}\text{Si} \xrightarrow[2.62 \text{ h}]{\beta^-} ^{31}_{15}\text{P}.$$

After annealing to remove crystalline defects caused by the fast neutrons, the resulting semiconductor can be used in devices such as power thyristors which require a large area of a semiconductor with uniform resistivity. The thermal-neutron capture cross section in ^{30}Si is 0.107 b. What thermal flux is needed to produce a phosphorus impurity concentration of 8 parts per billion following a 20-hour irradiation?

14. If potatoes receive gamma-ray doses between 60 and 150 Gy, premature sprouting is inhibited. Such irradiation can be done in an irradiator with a large ^{60}Co source. The maximum dose the potatoes receive from a given source in a given time can be estimated by assuming all the gamma rays are absorbed in the potatoes. What is the minimum activity of ^{60}Co needed in an irradiator to deliver such a dose of 250 kGy to 100,000 kg of potatoes in 8 hours? How realistic is it to suppose all the gamma-ray energy is deposited in the potatoes?

15. Bacteria such as salmonella can be killed by radiation doses of gamma rays of several kGy. Estimate the irradiation time needed to sterilize a small delicate medical instrument sealed in plastic if the instrument is placed 30 cm from a small ^{60}Co source with an activity of 10 kCi.

Note: In the remaining problems you may assume that the proton, deuteron, and helium nucleus have masses of, respectively, 1.6726×10^{-27}, 3.3436×10^{-27}, and 6.6447×10^{-27} kg. The unit of charge is 1.6022×10^{-19} C.

16. How long does it take for a helium nucleus to make one revolution in a magnetic field of strength 1.5 Wb/m^2? If the speed of the particle is 2×10^7 m/s, what is the radius of the circular trajectory?

17. Consider a fixed-frequency cyclotron with a radius of 30 cm, a radiofrequency voltage supply of 10^7 cycles/s, and a magnetic field of 1.3 Wb/m^2 (telsa). What is the energy of deuterons than can be produced in such a machine?

18. If the frequency of the applied voltage to the dees of a cyclotron is 1.5×10^7 s^{-1}, what are the magnetic field strengths needed to accelerator protons, deuterons, and helium nuclei? If the radius at ejection is 40 cm, what is the energy of each of the particles?

19. What is the mass (compared to its rest mass) of a 1 GeV electron? At what fraction of the speed of light are such electrons traveling?

20. How many orbits per second do the 1 TeV protons make in the Tevatron?

21. Consider a betatron with the following properties: maximum field strength at orbit of 0.4 Wb/m^2, operational frequency of 60 cycles/s, and a stable orbit radius of 84 cm. Show that the energy gain per orbit is about 400 eV and the final energy is about 100 MeV.

Chapter 14

Medical Applications of Nuclear Technology

The exploitation of nuclear technology in medical applications began almost from the moment of Roentgen's discovery of x rays in 1895 and Becquerel's discovery of radioactivity in 1896.[1] The importance of x rays in medical diagnosis was immediately apparent and, within months of their discovery, the bactericidal action of x rays and their ability to destroy tumors were revealed. Likewise, the effectiveness of the newly discovered radioactive elements of radium and radon in treatment of certain tumors was discovered early and put to use in medical practice. Today, both diagnostic and therapeutic medicine as well as medical research depend critically on many clever and increasingly sophisticated applications of nuclear radiation and radioisotopes.

In diagnostic medicine, the radiologist's ability to produce images of various organs and tissues of the human body is extremely useful. Beginning with the use of x rays at the start of the twentieth century to produce shadowgraphs of the bones on film, medical imaging technology has seen continuous refinement. By the 1920s, barium was in use to provide contrast in x-ray imaging of the gastrointestinal system. Intravenous contrast media such as iodine compounds were introduced in the 1930s and in angiography applications by the 1940s. Fluoroscopic image intensifiers came into use in the 1950s and rare-earth intensifier screens in the 1960s. Computed tomography (CT), positron emission tomography (PET), and single-photon emission tomography (SPECT) saw their beginnings for clinical use in the 1960s and 1970s. Picture archive and communication systems (PACS) began to see common use in the 1990s. The 1990s also saw interventional radiology becoming well established in medical practice, especially in coronary angioplasty procedures.

Along with advances in diagnostic medicine, corresponding advances have been made in using nuclear technology for therapy. There are three general classes of radiation therapy. In *brachytherapy*, direct implants of a radioisotope are made into a tumor to deliver a concentrated dose to that region. In *teletherapy*, a beam delivers radiation to a particular region of the body or even to the whole body. In *radionuclide therapy*, unsealed radiopharmaceuticals are directly administered to patients for curative or palliative purposes.

---

[1]The history of medical uses of radiation from 1895 through 1995 is addressed thoroughly in a series of articles published in the November 1995 issue of *Health Physics*, the Radiation Protection Journal. Extensive reviews of medical health physics and shielding of x-ray facilities appear in the June 2005 issue of *Health Physics*.

Table 14.1. Global use of medical radiology (1991-1996). From UN [2000].

Quantity		No. per 10^6 population	
		Level I[a]	Global
Physicians			
All physicians		2800	1100
Radiological physicians		110	80
X-ray Imaging			
Equipment:	medical	290	110
	dental	440	150
	mammography	24	7
	CT	17	6
Exams per year:	medical	920,000	330,000
	dental	310,000	90,000
Radionuclide Imaging			
Equipment:	gamma cameras	7.2	2.1
	rectilinear scanners	0.9	0.4
	PET scanners	0.2	0.05
Exams per year		19,000	5600
Radionuclide Therapy			
Patients per year:		170	65
Teletherapy			
Equipment:	x ray	2.8	0.9
	radionuclide	1.6	0.7
	LINAC	3.0	0.9
Patients per year		1500	820
Brachytherapy			
Afterloading units		1.7	0.7
Patient per year		200	70

[a]Level I represents countries with one or more physicians per 1000 population.

Indeed, nuclear applications have become such a routine part of modern medical practice that almost all of us at one time or another have encountered some of them. Table 14.1 lists personnel and frequencies for medical radiology and radiation therapy procedures, both globally and in developed countries.

Today we see many changes and new applications of nuclear technology in medicine. The use of rectilinear scanners is declining rapidly while the use of gamma-ray cameras, PET, and CT scanners is growing. Diagnostic radiology has long been used for imaging and study of human anatomy. In recent years, using CT, PET, and gamma-ray scanners, as well as MRI (magnetic resonance imaging), medical science has advanced to imaging the physiology and metabolism of the human body.

14.1 Diagnostic Imaging

Diagnostic radiology using x rays, both dental and medical, including mammography, dominates radiology. This is true for both numbers of patients and numbers of procedures. Each year in the United States, for example, more than 130 million persons annually receive diagnostic x rays [NAS 1980] and more than 250 million examinations are performed [UN 2000].

In the past thirty years, alternatives to traditional x-ray imaging have become available and are being increasingly used to image organs and tissues not easily seen by conventional x-ray diagnostics. In the 1970s, digital methods of processing and displaying x-ray images led to the clinical use of digital radiology and computed tomography (CT). Positron emission tomography (PET), and single photon emission computed tomography (SPECT), both radiological procedures, were realized in the 1980s.

Magnetic resonance imaging (MRI) matured and became a widely used medical imaging technique in the 1990s. MRI does not use x rays or radionuclides; however, it does depend on the spin (angular momentum) properties of the atomic nuclei in the body tissues and also employs the sophisticated image processing techniques used in nuclear imaging methods. Indeed, PET and MRI or PET and CT are often used together, with images superimposed to reveal physiological processes, particularly in the brain.

Readers interested in a comprehensive review of imaging technology will be well served by examining Oppelt's [2005] review of imaging systems for medical diagnostics. Those interested in an in-depth study of the mathematical theory and procedures of imaging will find invaluable the treatise of Barrett and Myers [2003].

14.1.1 X-Ray Projection Imaging

Projection x-ray imaging is by far the most common diagnostic imaging technique used. In this method, a beam of x rays illuminates some part of the body and a film (or digital imaging detector) behind the body records the transmitted x rays. Areas of the film behind dense high-Z materials such as bone, which preferentially absorb x rays, receive little exposure on the film compared to areas behind soft tissue, which more readily transmit the photons. In essence, an x-ray image records the "shadows" cast or *projected* by the irradiated specimen onto the film.

During the fifty years following the discovery of x rays, advances were made in the design and standardization of x-ray sources, and in the use of contrast agents. Notable advances in design include the 1913 invention of the hot-cathode x-ray tube by William Coolidge, the invention of the anti-scatter grid by Gustav Bucky and H.E. Potter in 1917, and the invention of the x-ray image intensifier by John Coltman in 1948 [Webster 1995]. Contrast agents are fluids containing high-Z atoms that strongly absorb x rays compared to normal tissues. Barium compounds flooding the gastrointestinal system were found to give definition in an x-ray image of the volume of the system. After clearance and distention of the system by gas, residual barium defined the walls of the system. Similarly, iodine compounds in the blood were found to define the circulatory system in detail. Advances continue to be made in the availability of contrast agents and in their applications. Film subtraction angiography, which began in the 1930s, relies on contrast agents. Subtraction angiography requires two images, one positive, one negative, recorded before and after injection of a contrast agent. Subtraction of the images by superimposing the two film images reveals vascular structure, absent interference caused by superposition of extraneous images of bones and other structures. In recent decades, digital-imaging methods have greatly enhanced the effectiveness of this application by performing the subtraction digitally.

The X-Ray Source

The production of x-ray photons as bremsstrahlung and fluorescence occurs in any device that produces high-energy electrons. Devices that can produce significant amounts of x rays are those in which a high voltage is used to accelerate electrons, which then strike an appropriate target material. Such is the basic principle of all x-ray tubes used in medical diagnosis and therapy, industrial applications, and research laboratories.

Although there are many different designs of x-ray sources for different applications, most early designs for low to medium voltage sources ($\lesssim$ 180 kV) placed the electron source (cathode) and electron target (anode) in a sealed glass tube. The glass tube acts as both an insulator between the anode and cathode and a container for the necessary vacuum through which the electrons are accelerated by the high voltage between the anode and cathode. The anodes of x-ray tubes incorporate a suitable metal upon which the electrons impinge and generate bremsstrahlung and characteristic x rays. Most of the electron energy is deposited in the anode as heat rather than being radiated away as x rays and, thus, heat removal is an important aspect in the design of x-ray tubes. For example, the x-ray tube shown in Fig. 14.1 has a rotating anode that spreads the heat over a large area and thereby allows higher beam currents and greater x-ray output. Relatively few of the emitted x rays are of high energy and useful in imaging. The more abundant low-energy x rays have to be shielded or filtered out of the beam to protect personnel. Modern tubes have metal envelopes with x-ray windows made of beryllium or titanium. Figures 14.2 and 14.3 illustrate the Straton tube used with modern Siemens x-

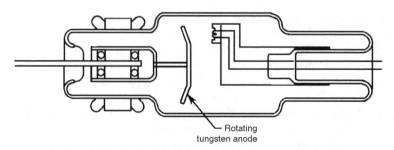

Figure 14.1. Schematic diagram of a typical x-ray tube used for projection radiography. From Kaelble (1967).

Table 14.2. Characteristic x-ray properties of anode and filter materials used in x-ray tubes. The x-ray line notation refers to the specific electron transition to the K, L, or M shell that produces the characteristic x ray. The wavelength and energy of the resulting characteristic x ray is listed. The excitation voltage is the energy required to create (ionize) a shell vacancy whose repopulation generates the x ray.

Element	X-ray line	Wavelength $(10^{-10}$ m)	Energy (keV)	Excitation voltage (kV)
Tungsten	$K_{\alpha 1}$	0.2090	59.3182	69.525
	$K_{\beta 1}$	0.1844	67.2443	69.525
	$L_{\alpha 1}$	1.4764	8.3976	10.207
	$L_{\beta 1}$	1.2818	9.6724	11.514
Molybdenum	$K_{\alpha 1}$	0.7093	17.4793	20.000
	$K_{\beta 1}$	0.6323	19.6083	20.000
	$L_{\alpha 1}$	5.4066	2.2932	2.520
Rhodium	$K_{\alpha 1}$	0.6134	20.2158	23.230
	$K_{\beta 1}$	0.5456	22.7236	23.230
	$L_{\alpha 1}$	4.5971	2.6973	3.014

ray tomography systems. The metal-walled tube is surrounded by a housing that provides a window for x rays, oil cooling, and a power train for anode rotation. Windows and collimation shape the beam and remove low-energy x rays, thereby limiting patient exposure.

Tungsten is the most commonly used target material because of its high atomic number and because of its high melting point, high thermal conductivity, and low vapor pressure. Occasionally, other target materials are used when different characteristic x-ray energies are desired (see Table 14.2). Generally, the operating conditions of a particular tube (current, voltage, and operating time) are limited by the rate at which heat can be removed from the anode. For most medical and dental diagnostic units, voltages between 40 and 150 kV are used, while medical therapy units may use 6 to 150 kV for superficial treatment or 180 kV to 50 MV for treatment requiring very penetrating radiation.

The energy spectrum of x-ray photons emitted from an x-ray tube has a con-

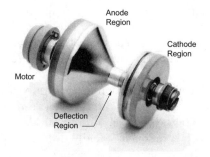

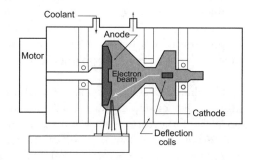

Figure 14.2. View of the Straton tube used with Siemens CT systems. Image courtesy of Siemens Medical Systems (annotation added).

Figure 14.3. Schematic diagram of the housing for the Straton tube.

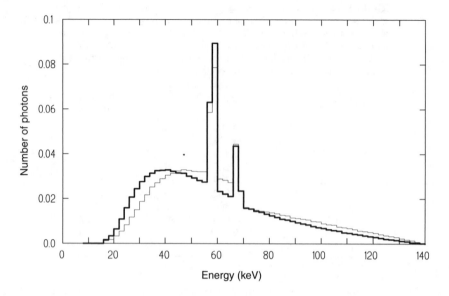

Figure 14.4. Measured photon spectra from a Machlett Aeromax x-ray tube (tungsten anode) operated at a constant 140 kV potential. This tube has an inherent filter thickness of 2.50-mm aluminum equivalent and yields a spectrum (thick line) with a HVL quality of 5.56-mm Al equivalent. The addition of an external 6-mm aluminum filter hardens the spectrum (thin line) to a HVL quality of 8.35-mm Al equivalent. Both spectra are normalized to unit area. Data are from Fewell, Shuping, and Hawkins (1981).

tinuous bremsstrahlung component up to the maximum electron energy (i.e., the maximum voltage applied to the tube). If the applied voltage is sufficiently high as to cause ionization in the target material, there will also be characteristic x-ray lines superimposed on the continuous bremsstrahlung spectrum. In Fig. 14.4 two calculated exposure spectra of x rays are shown for the same operating voltage but for different amounts of beam filtration (i.e., different amounts of material attenuation in the x-ray beam). As the beam filtration increases, the low-energy x rays are preferentially attenuated and the x-ray spectrum hardens and becomes more penetrating. Also readily apparent in these spectra are the tungsten $K_{\alpha 1}$ and $K_{\beta 1}$ characteristic x rays.

The characteristic x rays may contribute a substantial fraction of the total x-ray emission. For example, the L-shell radiation from a tungsten target is between 20 and 35% of the total energy emission when voltages between 15 and 50 kV are used. Above and below this voltage range, the L component rapidly decreases in importance. However, even a small degree of filtering of the x-ray beam effectively eliminates the low-energy portion of the spectrum containing the L-shell x rays. The higher-energy K-series x rays from a tungsten target contribute a maximum of 12% of the total x-ray exposure for operating voltages between 100 and 200 kV [ICRU 1970].

The X-Ray Receiver

The receiver in x-ray projection imaging is normally a film cassette, although photostimulable phosphor plates, with digital output, have seen growing use since the 1980s. Within the cassette is a film sheet with a polyester base and silver halide emulsion on one or both sides. Likewise, on one or both sides is found a fluorescent screen which absorbs the x rays and emits visible light matched to the sensitivity of the emulsion. The screen most frequently used from 1896 through the 1970s is $CaWO_4$, which typically stops 20 to 40% of the x rays. In recent years screens made of lanthanum, gadolinium, and yttrium compounds have come into use. These screens, which absorb 40 to 60% of the x rays, greatly improves sensitivity and typically reduces patient doses by a factor of 50.

In the late 1990s, flat detectors became the prime standard in digital imaging, replacing film-screen combinations for projection radiography, image-intensifier systems in fluoroscopy and general angiography. They are capable of generating real time images at 30 to 60 frames per second. X rays interact in a scintillation layer (typically Tl activated CsI). A photodiode converts the scintillation photons to electrical charge pairs. An amorphous Si layer provides an active readout matrix. Absorption of one x ray photon at 60 keV results in collection of about 1000 electrons. A typical configuration for flat detectors is a square active area 43 cm on a side, with pixels 0.143 mm on a side yielding 9 million pixels. Analog-to-digital conversion at 14 bits yields linear response over a factor of 10,000 variation in x-ray intensity. This broad range of linearity fosters minimal patient exposures and makes exposure errors rare.

Contrast in the X-Ray Image

A basic, ideal measure of contrast is the *subject contrast* associated with the visibility of a feature in the x-ray subject. This is illustrated in Fig. 14.5. The subject is irradiated with a parallel beam of x rays with intensity I_o, which are stopped, scattered, or reach the receiver without interaction. From Eq. (7.4), uncollided rays that reach the receiver without passing through the feature have an intensity $I_b = I_o \exp[-\mu_b L]$, where μ_b is the effective linear attenuation coefficient for the background material. Uncollided rays that pass through the feature of thickness x

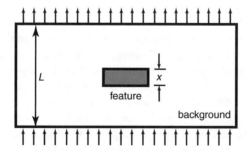

Figure 14.5. Background space of thickness L containing a feature of thickness x exposed to a parallel beam of x rays.

and a thickness $L - x$ of background material have an intensity $I_f = I_o \exp[-(L - x)\mu_b - \mu_f x]$, where μ_f is the effective linear attenuation coefficient for the feature material. The subject contrast is defined as

$$C_s \equiv \frac{I_b - I_f}{I_b} = 1 - \frac{I_f}{I_b} = 1 - \exp[-(\mu_f - \mu_b)x]. \tag{14.1}$$

To a first approximation, the attenuation coefficient is proportional to the density of the material and only very secondarily dependent on the energy spectra of the x rays. Thus,

$$C_s = 1 - \exp\left(\left[\left(\frac{\mu}{\rho}\right)_b \rho_b - \left(\frac{\mu}{\rho}\right)_f \rho_f \right] x \right). \tag{14.2}$$

Values of μ/ρ are relatively insensitive to the type of biological material, so the subject contrast depends primarily on relative densities. For example, for 100-keV x rays, μ/ρ (cm^2/g) varies from 0.149 for air to 0.164 for tissue or water, to 0.175 for bone, while densities vary from 0.0012 g/cm^3 for air to 1.0 for tissue, to about 1.85 for bone. For a 1-cm thickness of bone in tissue, the contrast would be about 15%. Very dense substances such as barium or iodine are excellent choices for contrast agents, not only because of their relatively high densities but also because of the greater photoelectric absorption of x rays in these high-Z materials.

Image contrast is less than subject contrast for several reasons. One is the contribution of scattered x rays to the "background." For this reason, and also to minimize patient exposure, the x-ray field of view should be kept to a minimum. At low exposures, statistical variations in the x-ray fluence at the image result in "quantum mottle" in the image. Loss of contrast due to scattered x rays can be reduced greatly by the use of an *antiscatter grid* adjacent to and in front of the receiver. Such a grid consists of alternating slats of lead and a low-Z filler material such as carbon fiber or aluminum, oriented so as to collimate the uncollided x ray beam through the filler and to absorb in the lead the scattered x rays, which mostly travel in directions other than the direction of the primary beam. Because the slats can yield structure in the image, the grid is sometimes placed in motion during the x-ray exposure. Grids are characterized by the number of slats (lines) per inch or per cm, and by the grid ratio, the ratio of the height (attenuation thickness) of the slats to the distance separating them.

14.1.2 Fluoroscopy

There are many circumstances calling for viewing the x-ray image in real time or for recording a time sequence of images. Examples include the placement of catheters in coronary artery angioplasty and the observation of peristalsis in the gastrointestinal system. Fluoroscopy procedures involve lower dose rates than those used in film imaging, but greater total patient doses.

In early years, radiologists directly viewed fluorescent-screen receivers in darkened rooms, a process requiring dark-adapted eyes and long exposures. Indeed, conventional photographic images were sometimes taken of the screen images. In the 1950s, invention of the image intensifier revolutionized fluoroscopy. In the intensifier, the transmitted x-ray beam strikes a thin input phosphor plate, typically CsI, 33 cm or greater in diameter. Visible light is emitted from this phosphor, with many photons released per x ray absorbed. The visible light, in turn, falls on a photocathode, typically SbS_3, which releases photoelectrons. These electrons are accelerated and focused onto an output phosphor, typically 1-in. square $ZnCdS(Ag)$. In early applications, the phosphor was viewed by a video camera and, using a beam splitter, photographed on conventional 35-mm movie film. Modern systems use a high-resolution charge-coupled device(CCD) satisfying nearly all diagnostic imaging requirements. Indeed, modern fluoroscopy systems may also be used for radiological purposes, supplementing projection imaging.

14.1.3   Mammography

The use of x rays to screen for malignancies of the breast is technically very demanding. Two features are of interest: microcalcifications that are sometimes indicative of cancer, and actual tumors a fraction of a cm in size, usually in lymphatic tissue, and with a composition similar to the surrounding breast tissue. Microcalcification imaging requires high resolution and tumor imaging requires high contrast. Breast compression improves resolution and, ideally, low-energy and nearly monoenergetic x rays would be used. This is approached by using an x-ray anode such as Mo, and by restricting the electron beam to a very small focal spot on the anode. The first practical work in mammography dates from the introduction of the Mo target by Gros in 1995. This and other advances are described by Hendee [1995] and Oppelt [2005]. Rhodium filtration is sometimes used for thicker or denser breast tissue and x-ray units are provided with various options for anode and filter materials. From Table 14.2 we see the K-shell x rays for Mo have energies of 17.5 and 19.6 keV, and the K-shell binding energy is about 20 keV. The Rh binding energy is 23.2 keV and x rays have energies of 20.2 and 22.7 keV. A Mo or Rh filter is placed in the x-ray beam. This filter passes the K-shell x rays but attenuates those of higher or lower energy. Thus, the beam reaching the breast consists mostly of characteristic x rays.

To increase resolution, single-emulsion films are used with single intensifying screens. Antiscatter grids with 30–50 lines per cm and 5:1 ratios are also used to substantially improve contrast by reducing the scattered radiation reaching the film. Digital radiography, however, offers significant advantages over film. Linear response is offered over a much wider exposure range, allowing for minimal patient exposure. Furthermore, detection and display may be optimized independently and computer-aided detection algorithms may be applied. Image storage and retrieval are simplified as well. This is important because each examination generally requires two images of each breast, with those four images compared to four from a previous examination.

14.1.4   Bone Densitometry

Monitoring of bone density is essential in the diagnosis and treatment of osteoporosis and related disorders. Several methods are available to measure bone density, but the most widely used technique is dual energy x-ray absorptiometry (DEXA). Other techniques include ultrasound and quantitative computed tomography (see subsequent sections). DEXA is thought to be more reproducible and more able to predict bone strength. DEXA, which involves an x-ray beam containing photons of two distinct energies, has largely replaced dual photon absorptiometry (DPA), which involves a low-energy gamma ray source with two distinct energies. An example is ^{153}Gd, which emits 44 and 100-keV photons.

In measurement of bone mass with a DEXA machine, the patient rests on a flat padded table and remains motionless while the "arm" of the instrument passes over the whole body or over selected areas. A beam of x rays, of two energies, passes from below the table through the area being measured. These x-rays are detected by a device in the instrument's arm. The machine converts the information received by the detector into an image of the skeleton and analyzes the quantity of bone the skeleton contains. The results are usually reported as bone mineral density (BMD), the amount of bone per unit of skeletal area.

Consider a subject region containing bone, with mass thickness $\rho_b x_b$ and soft tissue, with mass thickness $\rho_s x_s$, perpendicular to the dual beam, with incident intensities identified respectively as I_{1o} and I_{2o}. The two transmitted beams would have the following uncollided intensities after attenuation:

$$I_1 = I_{1o} \exp\left[-\left(\frac{\mu_1}{\rho}\right)_b \rho_b x_b - \left(\frac{\mu_1}{\rho}\right)_s \rho_s x_s\right], \tag{14.3}$$

$$I_2 = I_{2o} \exp\left[-\left(\frac{\mu_2}{\rho}\right)_b \rho_b x_b - \left(\frac{\mu_2}{\rho}\right)_s \rho_s x_s\right]. \tag{14.4}$$

From these equations, the unknown mass thickness of the bone may be computed as

$$\rho_b x_b = \frac{\mathcal{R}\ln[I_1/I_{1o}] - \ln[I_2/I_{2o}]}{(\mu_2/\rho)_b - \mathcal{R}(\mu_1/\rho)_s}, \tag{14.5}$$

in which $\mathcal{R} \equiv (\mu_2/\rho)_s/(\mu_1/\rho)_s$.

14.1.5 X-Ray Computed Tomography (CT)

X-ray tomography produces a two-dimensional cross sectional image of an object from x-ray transmission data consisting of projections of the cross section collected from many directions. Consider the cross section shown in Fig. 14.6. This figure is associated with the first generation of CT scanners. In generating a single image, a pencil beam of x rays and a detector are translated along the trajectory t. This is followed by rotation of, say, one degree in θ, and then the translation-rotation process repeated. Generating a single image required several minutes.[2] Suppose, then, that the object is traversed by x rays in a uniform, parallel beam in the plane of the cross section. X-ray computed tomography (CT) generates an image of the *object function* $f(x, y) \equiv \mu(x, y)$, the effective linear attenuation coefficient of the material at position (x, y). Three dimensional images may be generated in principle from successive plane slices.

Suppose the x-ray beam travels along direction s, normal to the projection-traverse direction t, and at angle θ to the x axis. The relation between the (s, t) rotated coordinated system and the original (x, y) coordinate system is

$$t = x\cos\theta + y\sin\theta \quad \text{and} \quad s = -x\sin\theta + y\cos\theta. \tag{14.6}$$

---

[2]A second generation of CT scanners used a small-angle fan beam and a plane array of detectors. The first-generation translation-rotation process was also used, but the scan time for a single image was reduced to about 20 seconds [Cho et al. 1993].

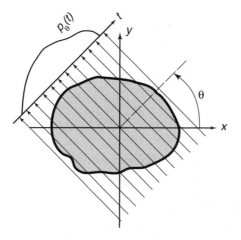

Figure 14.6. Plane cross section with the object function $f(x,y)$, and the projection generated by a parallel beam of x rays at angle θ.

If I_o is the intensity of the incident beam, then the intensity of uncollided photons passing through the object plane at transverse distance t and angle θ is

$$I_\theta(t) = I_o \exp\left[-\oint ds f(x(s), y(s))\right], \tag{14.7}$$

where $\oint ds\, f$ is the line-integral of μ along the ray s that intersects the projection axis at t. Sometimes this integral is called the optical thickness, and represents the number of mean-free-path lengths an uncollided photon must traverse to reach t. The projection data used to construct an image of $f(x,y)$ is

$$p_\theta(t) \equiv -\ln\left[\frac{I_\theta(t)}{I_o}\right] = \oint ds f(x(s), y(s))$$

$$= \int_{-\infty}^{\infty} dx \int_{-\infty}^{\infty} dy\, f(x,y)\delta(x\cos\theta + y\sin\theta - t). \tag{14.8}$$

The challenge of CT is to reconstruct $f(x,y)$ from a set of projections $p_\theta(t)$, each one taken at a different value of θ. Implementation of a practical CT scanner dates from the efforts of Hounsfield and Cormack, working independently, who shared the Nobel prize in 1972 for their work.

Of critical importance to image reconstruction is the Fourier transform which converts spatial domain data into spatial frequency data. The one-dimensional transform and its inverse are defined as

$$F(\omega) = \mathcal{F}_1\left[f(x)\right] = \int_{-\infty}^{\infty} dx\, f(x)e^{+j2\pi\omega x}, \tag{14.9}$$

$$f(x) = \mathcal{F}_1^{-1}\left[F(\omega)\right] = \int_{-\infty}^{\infty} d\omega\, F(\omega)e^{-j2\pi\omega x}. \tag{14.10}$$

Similarly, the two-dimensional Fourier transform is used to convert (x, y) data into spatial frequency (u, v) data, namely,

$$F(u, v) = \mathcal{F}_2\left[f(x, y)\right] = \int_{-\infty}^{\infty} dx \int_{-\infty}^{\infty} dy \, f(x, y) e^{+j2\pi(ux+vy)}. \tag{14.11}$$

$$f(x, y) = \mathcal{F}_2^{-1}\left[F(u, v)\right] = \int_{-\infty}^{\infty} du \int_{-\infty}^{\infty} dv \, F(u, v) e^{-j2\pi(ux+vy)}. \tag{14.12}$$

Image Reconstruction by Filtered Backprojection

The filtered backprojection algorithm (called the convolution backprojection algorithm) is the most popular and most frequently used technique for constructing an image from projection data. This method is based on the slice or projection theorem, which relates a projection to a "slice" of the two-dimensional Fourier transform $F(u, v)$ of the object function. To obtain this key relationship, take the 1-D Fourier transform of $p_\theta(t)$, as given by Eq. (14.8), namely,

$$P_\theta(\omega) = \mathcal{F}_1[p_\theta(t)] = \int_{-\infty}^{\infty} dt \, p_\theta(t) e^{+j2\pi\omega t}$$

$$= \int_{-\infty}^{\infty} dt \, e^{+j2\pi\omega t} \int_{-\infty}^{\infty} dx \int_{-\infty}^{\infty} dy \, f(x, y) \delta(x \cos\theta + y \sin\theta - t)$$

$$= \int_{-\infty}^{\infty} dx \int_{-\infty}^{\infty} dy \, f(x, y) e^{2\pi j(x\omega \cos\theta + y\omega \sin\theta)}$$

$$= F(u, v)\bigg|_{\theta} = F(\omega, \theta). \tag{14.13}$$

Here $F(u, v)$ is the 2-D Fourier transform of $f(x, y)$, and $u = \omega \cos\theta$ and $v = \omega \sin\theta$ are the spatial frequencies. The pairs (u, v) and (ω, θ) thus represent the Cartesian and polar coordinates of (x, y) in the spatial frequency domain.

This projection theorem, which is central to 2-D image reconstruction, shows that a 1-D Fourier transform of the projection data for a given value of θ yields a slice of the 2-D Fourier transform of $f(x, y)$ in the frequency domain along a frequency radius $u = \omega \cos\theta$ and $v = \omega \sin\theta$. With values of $F(u, v)$ obtained from this theorem, the object functions can be recovered by taking a 2-D inverse Fourier transform of $F(u, v)$, i.e.,

$$f(x, y) = \int_{-\infty}^{\infty} du \int_{-\infty}^{\infty} dv \, F(u, v) e^{-j2\pi(ux+vy)}, \tag{14.14}$$

which, in polar frequency coordinates, can be written as

$$f(x, y) = \int_0^{2\pi} d\theta \int_{-\infty}^{\infty} d\omega \, |\omega| F(\omega, \theta) e^{-j2\pi\omega(x \cos\theta + y \sin\theta)}. \tag{14.15}$$

Filtered backprojection for the geometry of Fig. 14.6 may be described as was done by Kak and Slaney [1988]: Make K measurements equally spaced over $0 \leq \theta \leq 2\pi$.

For each, obtain the Fourier transform $F(\omega, \theta)$, as in Eq. (14.13). Multiply each by the filter or weighting function $2\pi|\omega|/K$. Take the inverse transform of each product and sum over all measurements, i.e.,

$$f(x, y) = \frac{2\pi}{K} \sum_{i=1}^{K} \mathcal{F}^{-1}(|\omega| F(\omega, \theta_i)). \tag{14.16}$$

This result is simply a discretization approximation of Eq. (14.15). Since $t = x \cos \theta + y \sin \theta$ and with Eq. (14.13) used to express $F(u, v)$ in terms of the projection data p_θ, this result can be written as

$$f(x, y) = \int_0^{2\pi} d\theta \int_{-\infty}^{\infty} d\omega \, |\omega| \left\{ \int_{-\infty}^{\infty} dt' \, p_\theta(t') e^{+j2\pi\omega t'} \right\} e^{-j2\pi\omega(x \cos \theta + y \sin \theta)}$$

$$= \int_0^{2\pi} d\theta \int_{-\infty}^{\infty} dt' \, p_\theta(t') g(t - t'), \tag{14.17}$$

where $g(\tau) \equiv \int_{-\infty}^{\infty} d\omega \, |\omega| e^{-j2\pi\omega\tau} = \mathcal{F}_1^{-1}[|\omega|]$. Finally, the projection data has the symmetry $p_\theta(t) = p_{\theta+\pi}(t)$, so that Eq. (14.17) can be written as

$$f(x, y) = \int_0^{\pi} d\theta \int_{-\infty}^{\infty} dt' \, \widehat{p}_\theta(t') g(t - t'), \tag{14.18}$$

with $\widehat{p}_\theta(t') = p_\theta(t') + p_{\theta+\pi}(t')$ and $t = x \cos \theta + y \sin \theta$.

The convolution kernel $g(\tau)$ in Eq. (14.18) is an inverse Fourier transform of $|\omega|$, whose exact form cannot be achieved in practice. Consequently, several different filter functions are used to obtain numerical results, each with its own resolution and contrast characteristics. Finally, the above equations have been written as though data were continuous in the dependent variables, and without noise or error. In fact, data are digital and not without uncertainty. Many methods are available for generating $f(x, y)$ from projections $P_\theta(t)$. As advances are made in computing power and parallel processing methods, even more algorithms will become available for reconstruction of images from projection data.

Fan-Beam CT

The parallel-beam projection geometry just described is very cumbersome and not very practical. For each angle θ, both the source and the detector must move. Large scanning times are required, and there is undue patient exposure to x rays. In fan-beam tomography, introduced in 1973, the x-ray beam is collimated to a thin fan-shaped beam of rays, which pass through the object and are received in an array of detectors in the plane of the beam. In the third generation of CT scanners, Fig. 14.7, the fan beam is broadened and a array of detectors sufficient to encompass the scanning object are used. Both source and detectors are rotated simultaneously, generally for 360 degrees. The time required for a single scan plane is reduced to

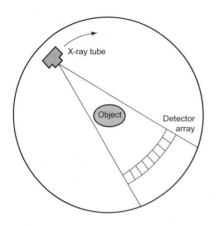

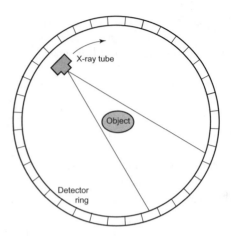

Figure 14.7. Third generation fan-beam CT scanner.

Figure 14.8. Fourth generation fan-beam CT scanner.

a few seconds. Figure 14.9 shows the fan-beam x-ray source in the GE HiSpeed Advance CT and Fig. 14.10 shows the array of detectors used in this device.

Fourth generation scanning is illustrated in Fig. 14.8. A stationary ring array of 600 or more detectors is used, and only the x-ray source is rotated. This is a faster and more stable configuration than that for the third generation and sets the stage for spiral and multi-slice spiral scanning.

Image reconstruction for fan-beam systems is similar to that for first generation systems, namely, filtered back-projection. However, a more complicated weighted back-projection procedure is required [Kak and Slaney 1988].

Figure 14.9. GE HiSpeed Advance CT x-ray source.

Figure 14.10. GE HiSpeed Advance CT x-ray receiver.

Spiral CT

In fan-beam CT, a series of slices are imaged, with the object translated axially a few mm between successive slices. Individual slice thicknesses are 2 to 10 mm. Each slice may be imaged in about a second, but tens of seconds are required for translation and preparation. In thorax imaging the breath must be held for each slice, but uneven breathing yields differences between images and artifacts in 3-D volume reconstruction. Thorax imaging is thus difficult and angiography procedures are precluded [Hiriyannaiah 1997]. In spiral CT, which was introduced in 1989 there is continuous rotation of the source and continuous translation of the object normal to the (x, y) image plane. It is possible to scan the entire thorax or abdomen in one breath hold, in a fraction of a minute. With this technique, it is possible to generate 3-D images very nearly isotropic in voxel (3-D pixel) size. Figure 14.11 illustrates a modern spiral CT scanner used in diagnostic radiology and in treatment planning for radiation therapy.

Multi-Slice CT

Systems were introduced in the late 1990s permitting acquisition of data for not one but multiple "slices," or image planes. These systems, with accompanying image-processing software permit, by interpolation, exquisitely detailed three-dimensional images that may be viewed in many modes, even with surface rendering. The Siemens Somatom Sensation unit, for example, collects data for 64 slices, with rotation time as low as 0.33 s and table speed as fast as 87 mm/s. Images have three-dimensional isotropic resolution of 0.33 mm at any speed, and 0.24 mm for high-resolution imaging. Representative examples of two dimensional renderings are shown in Fig. 14.13.

Cone-Beam CT

Originally developed for industrial radiography, this type of imaging uses a cone-shaped x-ray beam and a plane detector. Since the late 1990s, cone-beam CT has found wide use in rotating C-arm gantry systems for angiography and image-guided radiation therapy. A typical system involves a kV X-ray source, an amorphous-Si based kV image detector and 2 robotic arms that independently position the kV source and imager. In therapy, a similar robotic arm positions a MV portal imager allowing both to be used in concert. Applications include kV and MV planar radiographic imaging for patient positioning, kV planar fluoroscopic imaging for gating and tracking, and kV volumetric cone beam CT imaging for patient positioning. One such radiation therapy unit is shown in Fig. 14.12. Similar cone-beam imaging is available for the treatment unit shown in Fig. 14.24. Oppelt [2005] makes the following distinctions between the C-arm x-ray system and a regular CT scanner:

- Standard multislice scanning uses up to 64 rows of detectors and 700 to 800 columns of detectors, whereas the two dimensional plane detector used in cone-beam CT involves up to 1920 rows by 2480 columns of detectors.
- Cone-beam image generation requires approximate reconstruction techniques.
- A C-arm system field of view is smaller than that of a regular CT system.
- Planar detectors are optimized for projection radiography rather than CT.
- The C-arm geometry is not as well defined as that of a conventional scanner, and thus calibration presents a technical challenge.

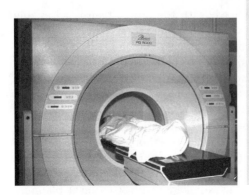

Figure 14.11. Marconi Medical System PQ5000 continuous spiral CT scanner.

Figure 14.12. Elekta Axesse image guided stereotactic radiosurgery accelerator. Image courtesy of Elekta AB; all rights reserved.

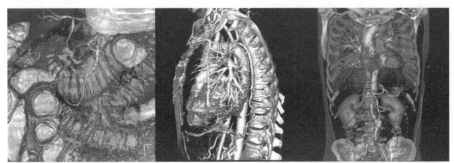

Figure 14.13. Abdominal, thoracic, and vascular CT images obtained using the Somatom Sensation multi-slice scanner. Images courtesy of Siemens Medical Systems.

14.1.6 CT Detector Technology

Severe demands are placed on detectors for x-ray computed tomography. As pointed out by Seeram [2001], detectors must be stable in detection efficiency, highly efficient in absorption of x rays, very short in response time, and very broad in dynamic range. High-pressure xenon-filled ion chambers with ceramic housing were introduced as detectors for fan-beam tomography units. They are fast and stable but less efficient than the scintillation detectors provided in detector modules for the multi-slice helical scanners. Such detectors are typically cadmium tungstate, gadolinia, or yttria scintillators optically bonded to photodiodes.

14.1.7 Single Photon Emission Computed Tomography (SPECT)

The goal of SPECT is the determination of the 3-D spatial distribution within the body of a radiopharmaceutical administered to a patient. Because a radiopharmaceutical concentrates in regions in which it undergoes biological use, SPECT is capable of measuring quantitatively biological and metabolic functions in the body. This is in contrast to CT which primarily produces images of anatomical structures in the human body.

The radionuclide used as a tag in the administered pharmaceutical emits a gamma ray upon its radioactive decay. Frequently used radioisotopes include ^{99m}Tc, ^{125}I, and ^{131}I, all widely used in nuclear medicine. The basic idea behind SPECT is very similar to CT. At some detector plane, the intensity of uncollided gamma rays leaving the body in a well defined direction is recorded. By moving the detector plane around the subject, gamma-ray emission projections are obtained at many different angles. From these projections, the spatial distribution of the radionuclide activity in the body can be obtained in much the same way as the x-ray projections in CT are used to reconstruct an image of the different materials in the body.

SPECT was first developed in 1963 by Kuhl and Edwards, well before x-ray CT and the advent of modern tomographic image reconstruction methods. Since then, the method has been considerably extended by the use of the gamma camera, invented by Anger. Today SPECT machines with up to four gamma cameras rotating around the patient are available. Work continues on improving both the sensitivity and resolution of this imaging technology.

The Gamma Camera

Central to SPECT is a so-called *gamma camera* which detects the gamma rays emitted from within the patient. The gamma camera consists of a NaI(Tl) scintillator crystal, circular or rectangular in shape, with a collimator on the side facing the patient and an array of photomultiplier tubes, optically coupled to the crystal, on the other side. Circular crystals are typically 25 to 50 cm in diameter; rectangular crystals have sides of 15 to 50 cm. Thickness is typically 1/4 to 5/8 in., with 3/8 in. being most common. Thin crystals provide the best resolution; but for imaging with high energy gamma rays or 0.511-MeV annihilation photons, the greater efficiency of a thicker crystal is often needed.

As many as 100 photomultiplier tubes are used in the array. Each has a different response (pulse height output) resulting from a scintillation event at position (x, y) in the crystal plane. A weighted average of the responses (Anger position logic) yields the (x, y) coordinates of the event in the crystal. Pulse height discriminators are used, with a 15 to 20% window set to record events corresponding to photoelectric absorptions in the crystal from uncollided source photons. In this way, preponderantly uncollided photons coming from the patient are recorded. This energy discrimination, along with use of the front-end collimator, minimizes "out of focus" registration of events caused by photons that have scattered one or more times in the body of the patient. Figure 14.14 shows a SPECT scanner with two cameras and Fig. 14.15 shows a three-camera scanner.

Collimators

The position logic used to register the location of a gamma-ray event in the scintillator crystal establishes an intrinsic spatial resolution R_I of 2.5 to 4.5 mm. However, to identify the site in the patient where the gamma ray originated requires the use of a collimator. The simplest and earliest used is a pinhole collimator (see Fig. 14.16). With a pinhole diameter d, distances f and b from the pinhole to the image and object plane, and image diameter I, the image magnification is $M = f/b$, and the point-spread, i.e., the image width of a point emitting photons, is $R_{ph} = (d/f)(f + b)$. The system resolution, accounting for both intrinsic resolution

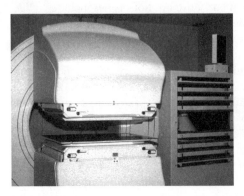

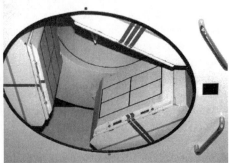

Figure 14.14. Two-headed SPECT scanner.   **Figure 14.15.** Three-headed SPECT scanner.

and the point spread, is

$$R_{sys} = \sqrt{R_{ph}^2 + (R_I/M)^2}. \tag{14.19}$$

Because a pinhole collimator has very poor sensitivity as a result of the relatively few photons admitted to the NaI scintillator, a parallel-hole collimator is more frequently used. This collimator has densely packed tubes, usually hexagonal in shape and perpendicular to the detector surface, with absorbing walls that ideally absorb all gamma rays except those that travel in the direction of the tubes. With a the thickness of the collimator (hole length), b the source to collimator distance, c

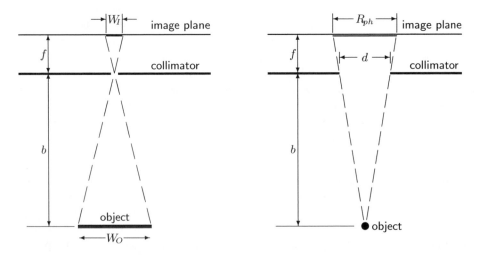

Figure 14.16. Left: The magnification of an object produced by a pin-hole collimator is $\equiv W_I/W_O = f/b$. Right: The resolution R_{ph} of a pin-hole collimator is given by $R_{ph}/(f + b) = d/b$ or $R_{ph} = (d/b)/(f + b)$.

the distance from the collimator to the point of interaction in the crystal, and d the effective diameter of the collimator hole, the collimation point-spread resolution is $R_{ph} = (d/a)(a+b+c)$, and the system resolution is as above. The system resolution may be improved by reducing d, reducing c, and/or reducing b. However, improving the resolution decreases the sensitivity, and collimators are designed to be either high resolution or high sensitivity. Focusing collimators, converging or diverging holes, are also used to better match the image size to the size of the organ being imaged. Focal length as well as sensitivity versus resolution are important choices in the selection of focusing collimators.

Image Reconstruction

Suppose $f(x,y)$ in Fig. 14.6 represents the activity distribution in the (x,y) object plane. Multiple projections $p_\theta(t)$ are gathered singly or simultaneously for successive object planes, i.e., for different values of θ. Reconstruction of the spatial activity distribution from the projections generally makes use of the filtered backprojection algorithm. However, this reconstruction process is complicated because of the need to compensate for three-dimensional effects of attenuation and scatter of photons in the patient's body and the spatial response of the collimator and detector. Advances in computer power have prompted the use of iterative reconstruction algorithms. As described by the Institute of Medicine [NRC 1996], a typical algorithm starts with an initial image estimate of $f(x,y)$, from which projections $p_\theta(t)$ are generated mathematically, based on the imaging process. Calculated and measured projections are compared and various statistical processes are used to update the image estimate and the process is repeated until differences between measured and generated projections meet acceptance criteria. A favored statistical process is the expectation maximum–maximum likelihood (EM-ML) algorithm.

14.1.8 Positron Emission Tomography (PET)

As early as the 1950s there was the realization that radionuclides emitting positrons offered enhanced medical imaging possibilities over those of SPECT. The emitted positron, within at most a few mm in tissue, annihilates with an ambient electron producing simultaneously two annihilation photons, equal in energy (0.511 MeV) and, most importantly, moving in almost opposite directions. It was recognized that detection of these photons, using the property that they are emitted simultaneously in opposite directions, would permit description, in three dimensions, of the distribution of the radionuclides in the body. Decades of development have given us positron emission tomography (PET), an indispensable tool in medical diagnosis and physiologic imaging. PET is very similar to SPECT in how images are reconstructed; however, the use of two detectors on opposite sides of the patient with coincidence photon detection logic allows finer spatial resolution of the emission location in the patient than does SPECT and, hence, better tomographic image reconstruction of the activity distribution in the patient.

PET imaging was first undertaken in the late 1960s, with imaging of a single plane per acquisition and requiring patient repositioning between acquisitions. By the late 1990s, PET scanners were available with more than 18,000 independent scintillation crystals, capable of imaging three-dimensional object regions 6-in. axially and 23-in. transversely, with resolution better than 5 mm in both directions.

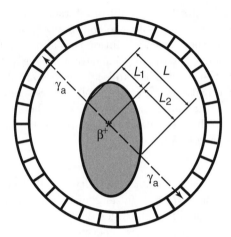

Figure 14.17. PET imaging of an object surrounded by
a ring of detectors. Annihilation photons γ_a are recorded
by detectors on opposite sides of the ring, and the relative
intensities allow determination of the mass thickness of the
distances L_1 and L_2 in the patient through which these
photons travel.

Table 14.3. Characteristics of radionuclides used for PET.

Nuclide	E_{max} (MeV)	E_{av} (MeV)	Frequency	Half-life	Production Reaction
^{11}C	0.960	0.386	0.998	20.5 m	^{14}N(p,α)
^{13}N	1.199	0.492	0.998	9.97 m	^{16}O(p,α),^{13}C(p,n)
^{15}O	1.732	0.735	0.999	122 s	^{15}N(p,n)
^{18}F	0.634	0.250	1.000	110 m	^{18}O(p,n)

Principles

Positrons have but a fleeting existence as anti-matter, usually coming to rest before
annihilating with an electron. The annihilation produces two 0.511-MeV photons
traveling exactly in opposite direction. Rarely, annihilation may occur in flight,
resulting in slight deviations in the colinearity of photon emission and subsequent
loss in resolution on the order of 1 mm. Positrons have a very short path in tissue
as they come to rest. For example, a 1-MeV positron has a path length of about
4 mm. The radionuclide ^{18}F, the most frequently used isotope for PET, emits
positrons with an average energy of 0.25 MeV, which have a range of less than 1
mm. Characteristics of ^{18}F and other useful positron emitters are listed in Table
14.3. All these emitters are short lived, only ^{18}F having a sufficiently long life to
permit transport of radiopharmaceuticals to sites a few hours from the point of
preparation. These radionuclides are normally produced by cyclotrons. A typical
cyclotron (Siemens/CTI RDS 112) uses a beam of 11-MeV negative hydrogen ions
to induce the nuclear reactions listed in the table.

Annihilation photons leaving the body are detected by an array of detectors that
surround the patient, as shown in Fig. 14.17, which illustrates imaging in the plane of

the paper. Events are recorded only when two detectors each detect an annihilation photon simultaneously, i.e., within 10 to 25 ns of each other. Events separated further in time are not recorded. The line joining the two recording detectors is a *line of response* (LOR) along which the annihilation photons have traveled and on which the positron decay occurred. This coincident detection technique allows a determination of the direction of the annihilation photons without the physical collimation needed in SPECT. For this reason coincident detection is often called electronic collimation.

Along a given LOR (see Fig. 14.17), the attenuation factor, i.e., the reduction in probability of detecting positron emission due to scattering or absorption of one or both the annihilation photons, is given by

$$A = \exp\left(-\oint_{L_1} ds\, \mu(s)\right) \exp\left(-\oint_{L_2} ds\, \mu(s)\right) = \exp\left(-\oint_{L} ds\, \mu(s)\right), \quad (14.20)$$

in which μ is the total linear interaction coefficient for 0.511-MeV photons. Thus we see that the attenuation factor is the same no matter where on the LOR the positron decay occurs. Although, with knowledge of the anatomy of the patient being imaged, it would be possible to compute the attenuation factor for each LOR, measured attenuation factors are preferred. One method, for example, uses, within the detector ring, a rotating source of annihilation photons, such as an equilibrium mixture of ^{68}Ge and ^{68}Ga, to collect blank scans with and without the patient present. Attenuation factors for each LOR may be derived from these scans and applied to scans of the patient containing positron-emitting radiopharmaceuticals.

The total number of coincidence events recorded by a given pair of detectors, corrected by the appropriate attenuation factor between the detectors, is a measure of the activity in the patient integrated along the LOR between the detectors. From a complete set of such line integrals of the activity between all pairs of detectors surrounding a patient, the activity distribution within the plane of the detector ring can be reconstructed using methods similar to those used in CT image reconstruction.

PET Detectors

Detectors used for annihilation coincidence counting must be small because their size defines the resolution of the scanner. To improve sensitivity, they must present a high photoelectric cross section for 0.511-MeV photons and a high scintillation output (fluorescence efficiency). For these reasons, bismuth germanate (BGO) or lutetium oxyorthosilicate (LSO) have largely replaced sodium iodide as scintillator crystals for PET. For annihilation photons, bismuth has a photoelectric cross section about seven times that of iodine and BGO has a density twice that of NaI although a lower fluorescence efficiency. However, while lutetium has a cross section only half that of bismuth, LSO has a fluorescent efficiency five times that of BGO.

Figure 14.18 shows the organization of the ring of scintillation detectors in a modern PET scanner (GE PET Advance). In the ring are 56 modules circumferentially, each supplied with its own electronics package. In each module there are six detector blocks, each consisting of a 6 by 6 array of BGO crystals, for a total of 12,096 crystals. As illustrated in Fig. 14.19, each block of 36 crystals is observed by two dual photomultiplier tubes (PMT). By a weighted average of signals from

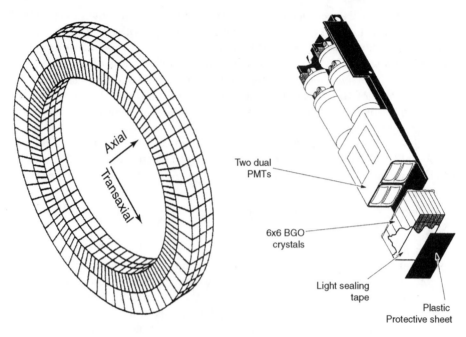

Figure 14.18. Ring structure for the GE Advance PET scanner showing the 336 detector blocks.

Figure 14.19. Detector block structure for the GE Advance PET scanner. Each block contains 36 BGO detector crystals.

Table 14.4. Comparison of PET scanner features.

Characteristic	Siemens ECAT Exact	GE Advance
Detector material	BGO	BGO
Crystal dimensions (mm)	6.75 x 6.75 x 20	4 x 8.5 x 30
Crystals per detector	64	36
PM tubes per detector	4	4
Detector ring dia. (mm)	824	939
Number of rings	24	18
Number of crystrals	9,216	12,096
Axial field of view (mm)	162	153
Portal diameter (mm)	562	590

the PMTs, the position of an absorption event in the block may be determined. There are thus 18 direct image planes axially, each 8.5 mm wide, giving an axial field of view of 15 cm. Other characteristics of the GE Advance scanner, along with characteristics of the Siemens ECAT scanner are listed in Table 14.4.

In multi-ring systems such as the GE and Siemens scanners, sensitivity may be improved by accepting coincidences not only for detectors within the same ring but also for detectors in adjacent rings, even across several rings. In the Advance scanner, for example, there are 18 direct planes and 17 crossed planes, for a total of 35 planes over the 153 mm axial field of view. Axial and transaxial resolution are thus about the same, 4.8 mm.

Applications

PET is ideally suited to generating images that reveal physiology. As we have seen, low-Z radionuclides are rich in positron emitters, which are easily substituted into molecules such as water, glucose, ammonia, etc. that move through metabolism and circulation within the human body. There are two broad avenues of application. One is studies of changes in brain stimulation and cognitive activation associated with Alzheimer's disease, Parkinson's disease, epilepsy, and coronary artery disease. The other is location of tumors and metastases in the brain, breast, lung, and gastrointestinal system. PET is distinguished by its very great sensitivity, allowing display of nanomolar trace concentrations.

Examples of PET Images

Figure 14.20 shows whole-body PET scans of a cancer patient suffering from lymphoma. The left panel is pre-chemotherapy, the right two months post chemotherapy, the two images showing that the disease has been almost totally cured. Dark regions, displaying excess radiopharmaceutical activity, identify uptake by cancer cells (left panel) or experiencing normal excretion (right panel). In each case, patients were administered mCi quantities of [^{18}F] 2-fluoro-2-deoxy-d-glucose (FDG).

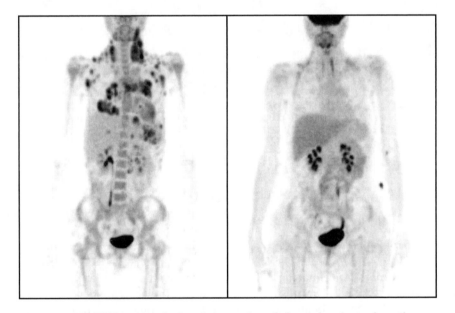

Figure 14.20. PET images of a lymphoma patient. Left panel: prior to chemotherapy. Right panel: two months post chemotherapy. Illustration courtesy of Dr. Frederic Fahey, Children's Hospital, Boston, MA.

Figure 14.21 illustrates use of combined imaging techniques in cancer diagnosis and treatment planning, in this case using a hybrid scanner (CT and PET combined into a single scanner). The patient had a known right lung mass. The left image is a coronal CT slice through the thorax. The central image is an FDG PET image in geometric registration with the CT image. The right image displays merger of the CT and PET scans. With the PET scan alone the tumor is evident, but anatomical information is weak. The combined image on the right, drawing on the strengths of both CT and PET, reveals the tumor in relation to nearby anatomical features. Other examples, and details of the procedure for superimposing PET and CT images are discussed in a paper by Yu, Fahey, and Harkness [1995].

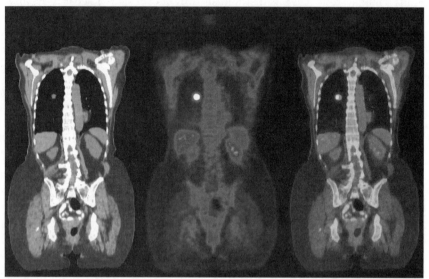

Figure 14.21. Coronal slice images of a patient with a lung tumor obtained with a hybrid PET/CT scanner. Left: CT, center: PET, right: PET and CT images superimposed. Illustration courtesy of Matthew Palmer, Ph.D., Beth Israel Deaconess Medical Center, Boston, MA.

14.1.9   Magnetic Resonance Imaging (MRI)

Magnetic resonance imaging (MRI), based on the phenomenon of nuclear magnetic resonance, produces high quality images of the soft tissues of the human body.[3] Initially a tomographic technique, MRI is now capable of slice images in any orientation as well as three dimensional and cinematographic imaging. MRI is used not only for anatomical imaging but also for functional studies of blood flow and neurological activity. Although it does not involve radioactive nuclides or nuclear radiation, it depends critically on the spin (angular momentum) properties of the atomic nucleus. Thus, its inclusion in this chapter of applications of nuclear technology.

---

[3]The MRI technique has undergone a significant acronym and name change since its discovery. Initially, it was called nuclear magnetic resonance imaging (NMRI) or nuclear magnetic resonance computerized tomography (NMR-CT); but the word "nuclear" has since been thought to be unsettling to many patients, and this word has now been eliminated, even though the technique fundamentally relies on nuclear magnetic properties.

Historical Development

Working independently, Felix Bloch and Edward Purcell discovered nuclear magnetic resonance phenomena in 1946 for which they were awarded the Nobel Prize in 1952. Although NMR was first used for chemical and physical molecular analysis, it was shown in 1971 by Raymond Damadian that different tissues, normal and abnormal, showed differing nuclear magnetic responses. This, and the 1973 discovery of computed tomography opened the way for use of MRI in the diagnosis of disease. Magnetic resonance imaging was first demonstrated in 1973 by Paul Lauterbur who used a back projection technique similar to that used in x-ray CT. In 1975 Richard Ernst proposed magnetic resonance imaging using phase and frequency encoding together with the Fourier image reconstruction technique, the basis of current MRI techniques. In 1977, Raymond Damadian demonstrated MRI of the whole body and, in the same year, Peter Mansfield developed the echo-planar imaging (EPI) technique. Edelstein and coworkers demonstrated imaging of the body using Ernst's technique in 1980, and showed that a single image could be acquired in approximately five minutes by this technique. By 1986, the imaging time was reduced to about five seconds. In 1987 echo-planar imaging was used to perform real-time movie imaging of a single cardiac cycle. In this same year, Charles Dumoulin was perfecting magnetic resonance angiography (MRA), which allowed imaging of flowing blood without the use of contrast agents. In 1991, Richard Ernst was awarded the Nobel Prize in Chemistry for his achievements in pulsed Fourier transform NMR and MRI. In 1993 functional MRI (fMRI) was developed, allowing the functional mapping of the various regions of the human brain. Advances in the capabilities of MRI are a continuing area of research.

Principles

All nuclei with an odd number of neutrons and/or protons possess an intrinsic angular momentum (spin) and, consequently, a magnetic moment. In the presence of a strong magnetic field, such nuclei experience a torque and tend to align either in a parallel or antiparallel direction to the magnetic field. The spinning nuclei respond to a strong external magnetic field by precessing around the direction of the field much like the spin axis of a gyroscope precesses around the direction of the gravitation field. The precession frequency is directly proportional to the magnetic field strength H_o and is unique to each nuclear species. A precessing nucleus has two possible orientations each with a slight difference in energy: in the lowest energy state the nuclear spin precesses around the direction of the external magnetic field (parallel), and in a slightly higher energy state, the nuclear spin precesses around the direction opposite the magnetic field (antiparallel). The difference between these two energy states is $2\mu H_o$ where μ is the nuclear magnetic moment. For protons in a magnetic field of 0.5 to 20 kG, this energy difference is relatively small, typically that of photons in the radio frequency (RF) range (2 to 85 MHz).

In MRI, the patient is placed in a large static magnetic field, often produced by a superconducting magnet. This field causes the nuclei to align and precess about the magnetic field in their lowest (parallel) energy state. A pulse of electromagnetic energy provided by an RF magnetic field causes many of the aligned nuclei to flip to the higher energy state in which they precess antiparallel to the static magnetic field. The excited precessing nuclei then spontaneously return or relax to their lower energy state by emitting RF photons which can be detected by the same RF coils used to produce the RF pulse that originally excited the spinning nuclei. It is empirically observed that the relaxation times are sensitive to the molecular structures and environments surrounding the nuclei. For example, average proton relaxation times in normal tissue are significantly less than in many malignant tissues. The strength of RF signal emitted by relaxing nuclei also depends on the density of precessing nuclei or the spin density. Thus, in MRI, magnetic field gradients are also applied, in addition to the large uniform static field H_o, to resolve the spatial distribution of spin densities. The different spin relaxation times of materials and the ability to measure the spatial distribution of spin densities make MRI a unique and robust technique in diagnostic imaging. A scanner opened for repair, shown in Fig. 14.22, reveals gradient and radiofrequency coils that surround a subject during scanning. An example MRI result is shown in Fig. 14.23.

Figure 14.22. MRI scanner opened for maintenance, revealing gradient and radiofrequency coils. (GE Echo Speed Plus scanner with Excite technology.)

MRI is unusual compared to other nuclear medicine imaging techniques in that there is a great diversity in the way data are collected and images are reconstructed. For example, MRI images can be formed by direct mappings, projection reconstruction, and Fourier imaging. Also there are many combinations of different magnetic gradient coils and magnetic field and pulse strengths that can be used. Work continues on improving and extending MRI capabilities by applying different variations in the measurement and image reconstruction technology.

14.2   Radioimmunoassay

Radioimmunoassay (RIA) is a technique utilizing radionuclides for measuring the concentration of almost any biological molecule with exquisite sensitivity. It was originally developed for use in studies of the human immune system in which a foreign protein (antigen) provokes an immune response causing special molecules (antibodies) to be created that chemically attach themselves to the antigen at some

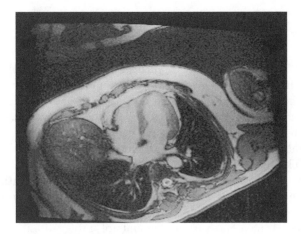

Figure 14.23. MRI image of the thorax, displaying a four-chamber view of the heart, generated using gated MRI imaging. Courtesy of Dr. Craig A. Hamilton, Wake Forest University Baptist Medical Center.

specific site on the antigen's surface, thereby deactivating the antigen. Today it can be used to measure concentrations of enzymes, hormones, or almost any other molecule to which antigens can be chemically attached. Measurement of concentrations as small as a few ng/ml to pg/ml can often be obtained.

RIA is extensively used today in many areas of biomedical research and diagnostic medicine. It is used in clinical chemistry, endocrinology, and toxicology. In some laboratories, it finds use in tests for steroid and peptide hormones and in markers for hepatitis-B infection [Koneman et al. 1997]. It also finds use in diagnosis of histoplasmosis and streptococcus pneumonia [Rose et al. 1997]. The first use of RIA, in 1959, was for detection of insulin in human plasma [Joklik et al. 1992].

Principles

Suppose we want to measure the amount or concentration of some particular antigen in a sample. There are two basic approaches used in RIA. In *capture* RIA, we first obtain two different antibodies that attach themselves to different sites on the antigen in question. The number of both kinds of antibodies to be used in the assay is assumed to be greater than the unknown number of antigens in the sample. One of the antibody species is tagged or labeled with a radionuclide such as ^{3}H or ^{125}I. The non-radioactive antibodies are attached to a solid surface, e.g., plastic beads, polystyrene microtubes, or a substrate in a sample vial. To begin the assay, the antigens, of unknown concentration, and the solid-phase bound antibodies are mixed and incubated to allow the antigens to bind to the antibodies. The tagged antibodies are then added to the mixture and also allowed to bind to the antigens. Extraneous material is then washed away, leaving all the antigens bound to the solid material and now tagged with radioactive antibodies. The activity of the resulting complex is proportional to the number of tagged antibodies bound to the solid surface. This number equals the unknown number of antigens in the sample. By performing the same assay procedure with different *known* amounts of the antigen, a "standard" or calibration plot can be obtained that relates activity to antigen concentration. From this plot, the unknown antigen concentration is then readily determined from the measured sample activity.

In the RIA approach, known as *competitive binding* RIA, antibodies are again attached to a solid phase. The sample with the unknown amount of antigen, the antibodies attached to a solid phase, and a solution of known concentration of the same antigen that has been tagged with a radionuclide are all mixed together. The antigens, of unknown amount, in the sample being assayed and the tagged antigens, of known amount, compete in attaching themselves to the antibodies fixed to the solid surfaces. The ratio of tagged to untagged antigens eventually binding to the antibodies equals their ratio in the initial mixture. After incubating, the sample is washed to remove unbound antigens and the activity of the antigens bound to the solid phase is measured. By performing this same procedure using the same amount of tagged antigen and varying amounts of untagged antigen, a calibration curve can be obtained relating activity to untagged antigen concentration.

Instead of using a radionuclide as a tag, enzymes can also be used to tag antigens or antibodies. This technique is referred to as enzyme-linked immunosorbent assay (ELISA) and the detection of the enzyme may be colorimetric. Whereas RIA presents difficulties in dealing with radioactive materials, it is more sensitive than ELISA and is widely used in research.

14.3 Diagnostic Radiotracers

Radioactive tracer techniques see wide use in the physical, biological, and medical sciences. Because of the ease of quantitative determination of very small quantities of a radionuclide, it is quite feasible to incorporate trace quantities of the radionuclide in reactants of a process and to measure yields, time constants, etc., by measuring activities in process intermediates and products. The first use of diagnostic radionuclides was in 1924, when George de Hevesy and, independently, Hermann Blumgart used ^{214}Bi to study circulation. The first artificially produced tracer, ^{24}Na was used in 1935 [Ice 1995]. The same year, Hevesy began using the physiologic radionuclide ^{32}P produced by bombarding ^{31}P with neutrons. For this work he received the Nobel prize in 1943. First used in 1939, with production in a cyclotron, was the nuclide ^{131}I. Later, this and a host of other radionuclides for pharmaceutical application were produced in nuclear reactors. Key among these was ^{99m}Tc, which became commercially available in 1957 through generation from a ^{99}Mo "cow." The generator was developed at Brookhaven National Laboratory and the product is ideal in half-life and gamma-ray energy. The ^{99}Mo, reactor produced, is loaded on an aluminum oxide column and the ^{99m}Tc is eluted (or "milked" from the Mo cow) using a saline solution, and is sterile and free of pyrogens. First used on humans in 1957, ^{99m}Tc is the mainstay of some 90 percent of diagnostic nuclear medicine. Other radionuclide generators include ^{68}Ge/^{68}Ga, which delivers a positron source, ^{81}Rb/^{81m}Kr, ^{82}Sr/^{82}Rb, ^{87}Y/^{87m}Sr, ^{113}Sn/^{113m}In, and ^{191}Os/^{191m}Ir [Ice 1995].

The availability of radioisotopes and the newly invented rectilinear scanner of Benedict Cassen in 1949 and the gamma-ray (scintillation) camera of Hal O. Anger in 1957 opened the way to dramatic advances in nuclear medicine during the past 50 years. A comprehensive history of the use of diagnostic radionuclides is found in the

Table 14.5. Selected examples of tracers used in nuclear medicine. From Sorenson and Phelps [1987].

Process	Tracer
Blood Flow	
Diffusable	^{133}Xe, ^{15}O, [^{11}C]alcohols, CH$_3$^{18}F
Non-Diffusable	[^{99m}Tc]macroaggregated albumin microspheres
Blood Volume	
Red Cells (RBC)	[^{99m}Tc]-, [^{51}Cr]-,[^{15}CO]-RBC
Plasma	[^{125}I]-, [^{111}C]-albumin
Transport/Metabolism	
Oxygen	[^{15}O]O$_2$
Glucose	2-deoxy-2-[^{18}F]fluoro-D-glucose, etc.
Protein Synthesis	L-[1-^{11}C]leucine,-methionine, etc.

review article by Early [1995]. A comprehensive listing of radionuclides produced in North America, and their suppliers, was published by Silberstein et al. [2000].

Uptake studies and multi-compartment kinetic modeling are among the most common and most important medical applications of radioactive tracers. A good example of uptake studies is the diagnosis of hyperthyroidism, most commonly Graves' disease, which involves abnormal release of thyroid hormones. Abnormal iodine uptake in the thyroid may be determined by measurement of ^{123}I or ^{131}I using a radiation detector such as a NaI scintillation counter external to the body in the thyroid region of the neck.

The ICRP Report 53 [1988] on the use of radiopharmaceuticals lists 75 radionuclides used in tracer studies. The most commonly used radionuclide is ^{99m}Tc. Next most common are the ^{123}I, ^{125}I, and ^{131}I isotopes of iodine. Some two dozen radiopharmaceuticals are listed for ^{99m}Tc, ranging from sulfur colloids used for liver studies and polyphosphates used for bone scans. Some example radiopharmaceutical tracers and their applications are given in Table 14.5.

14.4 Radioimmunoscintigraphy

The radioimmunoscintigraphy technique, abbreviated RIS, employs radiolabeled antibodies to image and characterize disease *in vivo*. The technique is used for malignant pathology as well as benign, e.g., myocardial infarction. In RIS, a radiolabeled monoclonal antibody is targeted to a particular line of cells such as tumor cells. Gamma-ray camera, PET, or SPECT techniques are then used to image the spatial distribution of the radionuclide, and hence the pathology of interest.

Several factors are required [Britton & Granowska 1998]: The antibody must be capable of detecting the cancer; the radionuclide must be capable of being imaged; and a radiolabeling method suitable for human use must be available. The production of a pure antibody against a selected antigen has been made possible through the development of monoclonal antibody (Moab) technology. Mettler and Guiberteau [1998] describe the production of monoclonal antibodies as follows: An animal such as a mouse is immunized with the antigen. In the animal, B lymphocytes begin producing antibodies. These are harvested and mixed with mouse myeloma cells. Fusion of the myeloma cells with the B lymphocytes produces a hybridoma which can continue producing antigen-specific antibodies and can perpetuate itself. The hybridomas are cloned, from which one is selected that produces the antibody of interest. The antibodies are then purified and labeled.

14.5   Radiation Therapy

Nuclear therapeutic medicine had its beginnings with Marie and Pierre Curie's discovery in 1898 of radioactive polonium and radium and with Roentgen's discovery in 1896 of x rays. These events, and a thorough review of the history of therapeutic nuclear medicine are found in the review articles by Early and Landa [1995] and Coursey and Nath [2000]. Here we highlight very early uses of radioactivity and nuclear radiation, then address modern techniques of radiation therapy. These include teletherapy, employing accelerator sources of electrons and x rays, brachytherapy, employing radionuclide sources of beta particles and gamma rays, and boron neutron capture therapy, employing nuclear-reactor sources of thermal neutrons.

Radiation-oncology needs drive the development of hardware for radiation therapy as well as methods and software for treatment planning. Each year in the United States, some 1.37 million new cases of cancer are reported and 564 thousand deaths are reported, as recorded in Table 14.6. Nearly half of the population of the United States may expect to experience cancer of one form or another in the course of their lives. Only about half of those persons encountering cancer survive the experience, as shown in Table 14.7. Advances in radiation oncology, surgical oncology, and chemotherapy offer hope that cancer survival probabilities will improve.

Because of time factors in cellular response to ionizing radiation, therapy is most often delivered in a series of dose fractions. A typical fractionation scheme dictates delivery of a tumor dose of 2 Gy each day for five days, repeated for six weeks, for a total dose of 60 Gy. Relevant dose time factors include (1) repair of sublethal and potentially lethal damage, (2) repopulation of cells between fractions, (3) redistribution of cells throughout the cell cycle, including radiation-induced syncrony, and (4) reoxygenation. The advantages of dose fractionation include (1) enhanced re-oxygenation of tumor cells relative to normal cells, (2) reduction in absolute numbers of clonogenic tumor cells, (3) decompression of blood vessels originally compressed by tumor, (4) radiation induced syncrony sensitizing highly proliferating cells, and (5) decrease in normal-tissue toxicity [Halperin et al. 2004].

14.5.1   Early Applications

X-Ray Therapy

The first therapeutic uses of x rays took place in 1896, shortly after Roentgen's discovery. As described by Orton [1995], Grubbe, in Chicago, treated breast cancer, Voight in Germany, treated nasopharyngeal cancer, and Despeignes, in France, treated stomach cancer. The first successful therapeutic uses were in Sweden in 1899, in treatments of both squamous-cell and basal-cell skin cancer. Early x ray tubes were not standardized and were largely unfiltered, resulting in excessive skin doses. Dose fractionation was used, but only for the convenience of the patient or physician, not for scientific reasons. The work of Bergonié and Tribondeau in 1906

Table 14.6. Estimated U.S. 2004 cancer cases and deaths by sex for all sites. Source: ACS [2004].

Cancer site	Estimated New Cases		Estimated Deaths	
	Males	Females	Males	Females
Digestive system	135,410	120,230	73,240	61,600
Respiratory system	102,730	83,820	95,460	69,670
Breast	1,450	215,990	470	40,110
Genital system	240,660	82,550	30,530	28,720
Urinary system	68,290	30,110	17,060	8,820
Lymphoma	33,180	29,070	11,090	9,640
Leukemia	19,020	14,420	12,990	10,310
Other	98,820	92,280	50,050	43,940
Total	699,560	668,470	290,890	272,810

Table 14.7. Lifetime risks of cancer incidence and mortality, cases or deaths per 100,000. Source: NAS [2005].

Cancer site	Incidence		Mortality	
	Males	Females	Males	Females
Solid cancer	45,500	36,900	22,100	17,500
Thyroid	230	550	40	60
Leukemia	830	590	710	530

showed that cancer cells were most susceptible to destruction by radiation if in the process of mitosis. This led to the expectation that a fractionated dose would be much more effective than a single dose, but many years would pass before dose fractionation was applied scientifically.

Radionuclide Therapy

Until about 1950, only ^{226}Ra and ^{222}Rn saw use for therapy. First used for treatment of skin lesions in 1900 and cancer therapy in 1903, radium was being used by 1920 for treatment of lung cancer, with a sealed source positioned in the thoracic cavity. Sealed 10-100 mg sources of ^{226}Ra were used near the skin or within body cavities. Gaseous ^{222}Rn, daughter product of ^{226}Ra, was first used in 1914, sealed in glass capillary tubes. In later years, the long-lived radium was used in fixed installations as a generator for the short-lived radon.

By the late 1920s, ^{226}Ra was used in treatment of deep lying tumors. This practice continued until the 1950s, when ^{60}Co, ^{198}Au, and ^{192}Ir became available.

Internal Therapy

The exploitation of radium and radon in the first few decades after discovery was a disgraceful display of quackery and excess. These radionuclides were misused in treatment of a host of ailments and conditions ranging from barber's itch to hemorrhoids. Only in the 1930s, after some well publicized deaths from misuse of radiation sources, and after the health effects of radium on the World War I radium dial painters were made known, were efforts made to regulate and control radium use. The Department of Agriculture, the Federal Trade Commission, and the American Medical Association led the way. Nevertheless, well into the 1980s, appliances containing uranium were being promoted as agents for arthritis cures.

14.5.2   Early Teletherapy

The use of beams of radiation to destroy malignant growths and other lesions has a long history. Many types of radiation, all generated externally to the patient, can be used.

In 1913, William D. Coolidge invented the tungsten-anode, vacuum x-ray tube. This Coolidge tube was the prototype of modern x-ray tubes and allowed standardization and reliable dosimetry and treatment planning. Higher energy x-ray machines were introduced in the 1920s, with 80 to 140-kVp units common. Depth-dose tables and isodose curves greatly improved treatment planning. Ionization dosimetry was introduced and the Roentgen unit of radiation exposure was established in 1928. The 1930s saw 500-kVp units introduced in 1934. In 1937, 1.25-MeV Van de Graaf accelerator sources came into use. The greater depth of penetration of the higher energy x rays, and the application of fractionation, improved treatment and avoided severe damage to superficial tissues.

In the 1960s intense beams of gamma rays produced by high activity radionuclide sources were used also for teletherapy. ^{60}Co was the radioisotope most frequently used. It emits 1.17 and 1.33-MeV gamma rays and can be produced in quantity in a nuclear reactor. Teletherapy devices with many kCi of ^{60}Co in a rotating head which, through careful collimation and filtering, could produce an intense beam of gamma rays, came into widespread use.

14.5.3   Accelerator Based Teletherapy

Linear accelerators were pioneered in the 1930s and, by the 1980s, became the standard sources of x rays for radiation therapy in developed countries, largely replacing ^{60}Co radioisotope units. Common are machines with lower energies ranging from 4 to 6 MeV and upper energies ranging from 10 to 20 MeV. Multi-leaf collimators came into use in the 1990s, as did lower-energy treatment simulators used in the design and testing of patient-specific collimators. Spiral CT scanners, with three-dimensional imaging, are now used for treatment-specific collimation and dosimetry planning. Figure 14.11 shows such a scanner. Figures 14.24 and 14.25 show the Clinac medical linear accelerator, representative of accelerators that rotate 360 degrees around the patient to deliver radiation beams from many different angles, as well as the beam delivery system. The theory and operational principles of linear accelerators is presented in Section 13.6.3 and their application to radiation therapy is well described in the monograph by Karzmark and Morton [1998].

14.5.4   Three Dimensional Conformal Radiation Therapy (CRT)

This type of therapy, 3-D CRT, was pioneered in the 1960s and 1970s and developed in the 1980s and 1990s. It makes use of three dimensional anatomical information to devise and apply dose distributions conforming to tumor volumes. Tumor control probability, TCP, is maximized under the constraint of minimizing normal-tissue complication probability, NTCP. Treatment planning relies on closely spaced transverse CT images that are reconstructed digitally in planes relevant to planning,

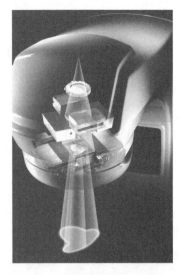

Figure 14.24. The Clinac medical linear accelerator, with 360 degree rotation around the patient to deliver beams of radiation from many angles.

Figure 14.25. Clinac beam delivery, showing the x-y and multileaf collimators and the radiation beam.

namely, beam's eye views. Point-kernel or ray-theory methodology may be used to compute radiation dose distributions at selected angles, with selected beam modifiers such as wedges, blocks, and compensators. Increasingly the dose calculation methodology is shifting to Monte Carlo radiation transport calculations because, thereby, dose distributions may be computed in the absence of charged particle equilibrium. Using computer graphics techniques, digitally reconstructed radiographs are superimposed on dose distributions at selected beam directions and dose-volume histograms are generated for target volumes, adjacent volumes, and critical structures such as the spinal column. The planning process is repeated in a labor intensive optimization process ultimately leading to maximized TCP and minimized NTCP. A tumor is treated with an array of x-ray beams, each of which conforms in shape to a two dimensional projection of target volume. Tumors with convex surfaces are well suited for treatment. Prostate cancer and lung cancer are

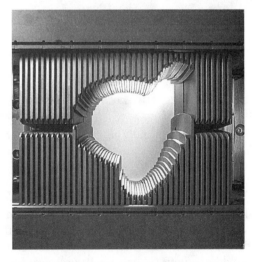

Figure 14.26. The Millenium MLC multileaf collimator.

good candidates for treatment. The reader will find more information on 3-D CRT in the textbook by Khan [2003] and the review article by Purdy [2004].

14.5.5   Intensity Modulated Radiation Therapy

This is a form of 3-D CRT, but with non-uniform treatment beams varying with angle of application. Computer controlled dynamic multileaf collimators, MLC, permit custom dose patterns at each angle (see Fig. 14.26). In each treatment direction, the radiation field is subdivided into segments, each of which may have a custom intensity. It is thus possible to treat irregularly shaped, even concave surfaced, tumors with a high degree of protection for critical structures such as the spine (for treatment of lung cancer) or the rectum (for treatment of prostate cancer). However, because of greater beam-on times in any one treatment, IMRT does result in higher whole-body doses from leakage radiation. Clinical implementation of multileaf collimators is addressed by the AAPM [2001a]. Treatment plans for treatment of lung and prostate cancer are shown in Figs. 14.27 and 14.28.

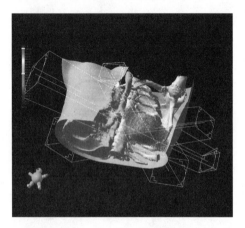

Figure 14.27. The Eclipse IMRT treatment plan for the lung. Image courtesy of Varian Medical Systems. All rights reserved.

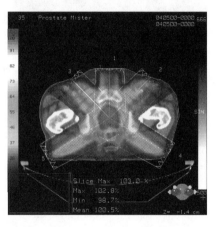

Figure 14.28. The Eclipse IMRT treatment plan for the prostate. Image courtesy of Varian Medical Systems. All rights reserved.

14.5.6   Electron Beam Therapy

Accelerators used for x-ray teletherapy may be used as well for electron beam therapy, which is especially useful for treatment of skin, head, neck, and certain breast cancers. Electron beam therapy is also useful in providing booster doses to lymph nodes as an adjunct to x-ray therapy. The useful range of beam energies is about 5 to 20 MeV. Figure 14.29 illustrates the variation with depth in a water phantom of the radiation dose delivered along the axes of broad beams of electrons. The dependent variable is the ratio of the local dose to the maximum dose along the beam path. It is quite evident that, as beam energy increases, the spatial distribution of the dose becomes more uniform axially. For 20-MeV electrons, for example, the dose is uniform, within 10 percent of the maximum, for about the first 7.5 cm of beam penetration and then decreases to 10 percent of the maximum within the next 2.5 cm. In comparison to x-ray beam therapy, electron beams suffer some loss

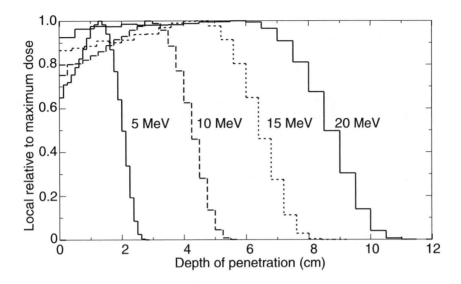

Figure 14.29. Attenuation of a broad beam of electrons in a water phantom, as computed using the MCNP5 code.

of resolution at beam lateral boundaries due to electron scattering. Treatment is ordinarily accomplished using a single beam axis.

14.5.7   Proton Beam Therapy

Beams of protons or, indeed, other heavy charged particles may have characteristics that make them superior to x-ray or electron beams for radiation therapy. As seen in Fig. 14.29, high-energy electrons lose energy quite uniformly along the beam path until stopping rather abruptly. Beams of high energy x rays used in therapy are also rather uniform in energy deposition along the beam path within a subject. In contrast, within a beam of high energy heavy charged particles, almost all the energy of the particles is deposited in the last tiny fraction of the particle paths. For example, a 200-MeV proton beam has a range of about 26 cm in tissue. Initially, about 4.5 Mev/cm is lost by a proton in the beam. When the proton reaches 10 MeV, only about 2 cm of beam path remains, and about 45 MeV/cm is lost. The peak energy loss rate is about 830 MeV/cm, when the proton has but a fraction of a millimeter path remaining. This exquisite targeting by high-energy protons is illustrated in the "Bragg curves" of Fig. 14.30. For comparison, see the curves for low-energy protons demonstrated in Chapter 7, Fig. 7.15, along with a discussion of charged particle range and stopping power. Consequently, by selection of the energy of incident particles, one may dictate the very narrow region of depth in which the radiation dose is delivered in a subject's body. Collimators dictate the lateral beam extent, and thus, the treatment volume is precisely defined.

The first effort at proton-beam therapy took place in 1954 [Denker 2006]. From then until 1990, nuclear-physics accelerators were used in advancing charged-particle beam therapy. The first hospital based treatment was established at the Loma Linda University Medical Center in California. At this writing there are 27 proton and 3 ion therapy facilities worldwide [Denker 2006]. Other U.S. treatment accelerators are located at Massachusetts General Hospital, the University of Indiana, the University of Florida, and the M.D. Anderson Cancer Center of the University of Texas. In planning or construction are centers at Hampton University in Virginia and Washington University in St. Louis. More than 40,000 patients have been treated.

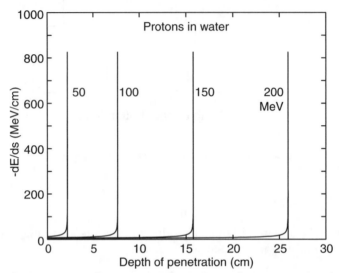

Figure 14.30. "Bragg curves" for high energy protons in liquid water: stopping power vs. depth of penetration. Data obtained using the PSTAR program [Berger 1992].

Proton beams have especial promise for treatment of prostate cancer and other lesions next to vital structures. Among these are pediatric cancers, spine and central-nervous-system tumors, head and neck cancers, post-operative cancers and liver metastases. Treatment planning requires careful consideration of proton linear energy transfer and associated relative biological effectiveness.

Denker [2006] describes the very successful use of proton accelerators for treatment of optical melanomas and hemangiomas. Regions treated, in the choroid for example, may be mm thick, mm in lateral extent, and a few cm deep. For example, a 60-Mev proton beam delivers its energy at a depth of about 3.1 cm, a 50-MeV beam at about 2.2 cm. With the use of range (energy) shifters and intensity modulators in the proton accelerator, it is possible in the course of treatment to cycle energies between 50 and 60 MeV, simultaneously adjusting the beam intensity. Thereby it is possible in this example to deliver a uniform dose between 2.2 and 3.1 cm depth, with very little dose at greater or lesser depths. Treatment of choroidal melanoma, for example, is done in four daily doses of 15 Gy, for a total of 60 Gy.

In therapy, synchrotrons deliver beams of up to 250-MeV protons allowing treatment depths up to about 30 cm and beam diameters up to 20 cm. Beams may be diverted to several treatment rooms. Hitachi, Ltd. is a principal vendor of the

Figure 14.31. Leksell Gamma Knife for stereotactic radiosurgery.

Figure 14.32. Close-up view of the collimator helmet for the Leksell Gamma Knife.

treatment delivery systems and Varian Medical Systems, Inc. is a principal vendor of treatment planning systems. High energy proton beams require substantial radiation shielding, a subject addressed in NCRP Report 144 [2003].

14.5.8 Stereotactic Radiation Therapy

Lesions in the brain may be treated with the use of multiple, non-coplanar beams of x rays or gamma rays focused to a very small region. When radiation is delivered in a single fraction, the treatment is called stereotactic radiosurgery (SRS). When multiple dose fractions are applied, the treatment is called stereotactic radiation therapy (SRT). The lesion and surrounding structures must be carefully mapped in three dimensions by using CT scans. The patient's head must be fixed rigidly in a frame during treatment. The focus of the radiation beams is fixed in space. Thus, during treatment, the head and body are moved so that the beam focus traverses the precise volume of the lesion. Because the positioning must be so rigid and precise, the procedure is painful to the patient, and because changes in lesion volume may occur between fractionated treatments, single-fraction radiosurgery is the usual practice. When delivered by accelerator, the term *x-ray knife* is used to describe the equipment set up. Similarly, the term *gamma-ray knife* describes equipment for radiosurgery using gamma rays from fixed radionuclide sources.

The neurosurgeon Lars Leksell developed stereotactic radiosurgery in the 1940s and later pioneered the use of a gamma-ray knife using radiation from multiple ^{60}Co sources. Figures 14.31 and 14.32 show the Leksell Gamma Knife, which delivers highly accurate external radiation to intracranial structures from an array of collimated beams of ionizing radiation. The Knife utilizes a single dose of radiation directed through the 201 ports of a collimator helmet to the target within the brain. Some collimators may be blocked to protect sensitive structures such as the eyes. Beam focal regions are spheres as small as 3 mm, with positioning precision of ± 0.1 mm. During irradiation there are no moving parts within the Knife and therefore safety, stability, and accuracy are inherent features. The cost of the Gamma Knife is several million dollars. Some 18 tonnes of shielding are required and sources must be replaced about every five years.

14.5.9   Clinical Brachytherapy

In brachytherapy, or short-distance therapy, sealed radiation sources are placed in or near tumors. They may be placed in surface molds outside the skin, implanted in tissues (interstitial treatment), or placed in body cavities (intra-cavity treatment). In surface molds, sources are typically 0.5 to 1 cm from the skin. Interstitial sources are in the form of needles, wires, or seeds. Placement may be permanent or may be temporary, with permanent placement generally favored for treatment of prostate carcinoma. Sources may be placed in tubes previously implanted in tissues. This *afterloading* process facilitates precise source placement and dosimetry. Intracavity treatment, especially suitable for gynecological therapy, makes use of applicators for fixed source placement.

There are several advantages of brachytherapy over beam therapy. Lower energy and lower strength sources can be used, leading to precise control of the spatial distribution of radiation dose. Furthermore, using longer irradiation times allows treatment over several days during which cancer cells pass through each phase of the cell cycle. This assures that each cell is treated during the more radiosensitive phases of the cycle.

In brachytherapy planning, source configurations are determined using dummy sources, with actual sources emplaced later. This procedure, which limits personnel doses, is known as *afterloading*. Remotely controlled afterloaders are available for source placement in applicators or catheters. Remote afterloading is of great benefit in high dose rate brachytherapy by which several hundred Gy/h dose rates at 1 cm are delivered. Such treatment allows for fractionated delivery in outpatient treatment. Implants for low dose rate brachytherapy typically deliver 2 Gy/h or less at 1 cm, and treatments require several days of inpatient attention.

Common Therapies

Sealed radium or radon sources were among the first sources used in interstitial treatment of solid tumors of many types. Intra-cavity treatments for uterine, rectal, and gynecological cancer make use of sources such as ^{137}Cs and ^{192}Ir. Each year in the United States, some 10,000 males are treated for prostate cancer interstitially using sources such as ^{125}I and ^{103}Pd. Some 60 to 100 "seeds" containing these radionuclides are surgically implanted in the prostate [Coursey and Nath 2000].

Intravascular Therapy

Balloon angioplasty, used to open occluded arteries, is one form of brachytherapy. In the procedure, artery walls may be damaged and in the healing process the artery walls may reclose, a phenomenon called restenosis. This may be minimized by applying a radiation dose to arterial lesions using catheter-borne radiation sources. Similarly, arterial stents may incorporate radiation sources in the stent material to minimize restenosis.

14.5.10   Radionuclide Therapy

In the United States alone, hundreds of thousands of patients annually receive therapy using radionuclides. In many cases, sealed sources are used, often in the close proximity of cancers for treatment purposes, and sometimes for short-range treatment of other conditions. In other cases, unsealed sources are ingested or injected in a form as to be concentrated in cancer tissue, and thereby attacking cancer cells from within the affected tissue. Maisey et al. [1998] and Coursey and Nath [2000] are good sources of information on radionuclide therapy and The AAPM [2001] has published a primer on the subject.

Direct Radionuclide Therapy

A good example of the use of a radionuclide in therapy involves the treatment of thyroid disease. Iodine entering the bloodstream either directly or after ingestion is strongly concentrated in the thyroid and radioactive iodine, commonly ^{131}I, is used to ablate all or a part of the thyroid in treatment of thyrotoxicosis, goiter, or thyroid cancer. Other examples include the use of beta-particle emitters such as ^{32}P or ^{90}Y in the ablation of the synovial lining in the treatment of synovial disease. A common palliative use of beta-particle emitters such as ^{32}P and ^{89}Sr is found in the treatment of cancer metastases in bone. The radiation dose causes both tumor shrinkage and hormonal mechanisms of pain reduction.

Radioimmunotherapy

In this technique, called RIT, a radiolabeled monoclonal antibody is targeted to a particular line of tumor cells. The antibody is produced as described in Section 14.4. The radionuclide is attached to the antibody or a smaller protein fragment. In the attachment, the radionuclide is first chemically bound to a small precursor molecule called a ligand. The ligand is then attached to the antibody, which is then injected into the bloodstream. Coursey and Nath [2000] characterize candidate radionuclides. The mainstays are beta-particle emitters ^{131}I and ^{90}Y. In many instances, nuclides emitting short-range radiations are required. Among these are alpha-particle emitters such as ^{211}At, ^{225}Ac, ^{213}Bi, and ^{212}Bi, and Auger-electron emitters such as ^{125}I and ^{111}In.

14.5.11   Boron Neutron Capture Therapy

This therapy, called BNCT, is a hybrid or binary therapy, employing a beam of radiation to generate an interstitial source of highly ionizing radiation within cancer cells. The therapy has reached the stage of clinical trials for treatment of glioblastoma multiforme (GBM), the most highly malignant and persistent of brain tumors.

In 1936, G.L. Locher introduced the concept of neutron capture therapy (NCT), i.e., introduction into a tumor of an element that reacts with particles in an incident radiation beam to produce short-range secondary charged particles capable of cell destruction. Boron, and the reaction ^{10}B(n,α)^{7}Li, or lithium, and the reaction ^{7}Li(p,n)^{7}Be, are possible candidates. In 1952, Sweet suggested that the boron reaction might be useful in the treatment of GBM.

In the treatment of GBM, a drug such as borocaptate sodium (BCS) is used to transfer boron through the blood-brain barrier to tumor cells. The barrier protects normal brain cells from blood-borne substances but is ineffective in tumor cells. The patients head is irradiated by a beam of epithermal neutrons, ideally with energies sufficient to assure that the neutrons are just thermalized as they reach the glioblastoma. The nuclear reaction produces an alpha particle of energy about 1.5 MeV and a ^{7}Li atom of energy about 0.84 MeV. These charged particles give up their energy in tracks only about one cell diameter in length. They are thus very effective in destruction of tumor cells. Unfortunately, blood-vessel walls are also damaged by products of the (n,α) reactions, and this damage places a limit on the dose that can be delivered to tumor cells.

Figure 14.33 is a view of the 5-MW MIT research reactor used for neutron therapy. Located beneath the reactor is a thermal-neutron irradiation facility used for research and for clinical irradiation of shallow malignancies. Transverse from the reactor core is an epithermal-neutron irradiation facility used for neutron capture therapy. The maximum thermal and epithermal-neutron flux densities are, respectively, 5.9×10^9 and 4.6×10^9 cm^{-2}s^{-1}.

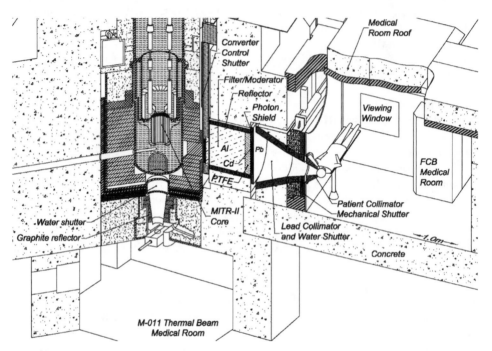

Figure 14.33. Isometric view of the MITR with the thermal neutron irradiation facility, M-011, directly beneath the core of the 5 MW research reactor and the epithermal beam facility, FCB, shown on the right-hand side of the figure. Source: Harling et al. [2005].

ACKNOWLEDGMENT

The authors are grateful to Dr. Frederic H. Fahey Children's Hospital, Boston, MA, and Dr. Craig A. Hamilton, Wake Forest University Medical School and Baptist Medical Center, Winston-Salem, NC, for assistance in preparation of this chapter.

BIBLIOGRAPHY

AAPM, *A Primer for Radioimmunotherapy and Radionuclide Therapy*, Report 71, American Association of Physicists in Medicine, Medical Physics Publishing, Madison, WI, 2001.

AAPM, *Basic Applications of Multileaf Collimators*, Report 72, American Association of Physicists in Medicine, Medical Physics Publishing, Madison, Wisconsin, 2001a.

ACS, *Cancer Facts and Figures 2004*, American Cancer Society, Atlanta, GA, 2004.

BARRETT, H.D. AND K. MYERS, *Foundations of Image Science*, John Wiley & Sons, New Jersey, 2003.

BERGER, M.J., *ESTAR, PSTAR, and ASTAR: Computer Programs for Calculating Stopping-Power and Range Tables for Electrons, Protons and Helium Ions*, Report NISTIR 4999, National Institute of Standards and Technology, 1992.

BRITTON, K.E., AND M. GRANOWSKA, "Radioimmunoscintigraphy," in *Clinical Nuclear Medicine*, 3d. ed., M.N. MAISY, K.E. BRITTON, AND B.D. COLLIER (Eds), Chapman & Hall Medical Publishers, London, 1998.

CHO, Z.H., J.P. JONES AND M. SINGH, *Foundations of Medical Imaging*, John Wiley & Sons, New York, 1993.

COURSEY, B.M. AND R. NATH, "Radionuclide Therapy," *Physics Today*, April, 25–30 (2000).

DENKER, A, *Ion Accelerator Applications in Medicine and Cultural Heritage*, 10th International Symposium on Radiation Physics, Coimbra, Portugal, 2006.

EARLY, P.J., "Use of Diagnostic Radionuclides in Medicine," *Health Physics*, **69**, 649–661 (1995).

EARLY, P.J. AND E.R. LANDA, "Use of Therapeutic Radionuclides in Medicine," *Health Physics*, **69**, 677–694 (1995).

FEWELL, T.R., R.E. SHUPING AND K.R. HAWKINS, *Handbook of Computed Tomography X-Ray Spectra*, HHS (FDA) 81-8162, U.S. Department of Health and Human Services, Food and Drug Administration, Rockville, MD, 1981.

HALPERIN, E.C. et al., "The Discipline of Radiation Oncology," in *Principles and Practice of Radiation Oncology*, 4th ed, PEREZ, C.A., et al. (Eds), Lippincot, Williams and Wilkins, Philadelphia, 2004.

HARLING, O.K., K.J. RILEY, P.J. BINNS, H. PATEL AND J.A. CODERRE, "The MIT User Center for Neutron Capture Therapy Research," *Radiat. Res.*, **164**, 221-229 (2005).

HENDEE, W.R., "History and Status of X-Ray Mammography," *Health Physics*, **69**, 636-648 (1995).

HIRIYANNAIAH, H.P., "X-ray Computed Tomography for Medical Imaging," *IEEE Signal Processing Magazine*, March (1997).

ICE, R.D., "History of Medical Radionuclide Production," *Health Physics*, **69**, 721–727 (1995).

ICRP, *Radiation Dose to Patients from Radiopharmaceuticals*, Report 53, International Commission on Radiological Protection, Pergamon Press, Oxford, 1988.

ICRU, *Radiation Dosimetry: X Rays Generated at Potentials of 5 to 150 kV*, Report 17, International Commission on Radiation Units and Measurements, Washington, DC, 1970.

JOKLIK, W.K. et al. (Eds), *Zinsser Microbiology*, 20th ed, Appleton & Lange, Norwalk, CT, 1992.

KAELBLE, E.F. (Ed), *Handbook of X Rays*, McGraw-Hill, New York, 1967.

KAK, A.C. AND M. SLANEY, *Principles of Computerized Tomographic Imaging*, IEEE Press, New York, 1988.

KARSMARK, C.J. AND R.J. MORTON, *A Primer on Theory and Operation of Linear Accelerators in Radiation Therapy*, 2d ed, Medical Physics Publishing, Madison, WI, 1998.

KHAN, F.M., *The Physics of Radiation Therapy*, 3d ed, Lippincott Williams & Wilkins, Philadelphia, 2003.

KONEMAN, E.W. et al., *Color Atlas and Textbook of Diagnostic Microbiology*, 5th ed, Lippincott, Philadelphia, 1997.

LEONDES, C.T. (Ed), *Medical Imaging Systems Techniques and Applications: Modalities,*, Gordon & Breach, Amsterdam, 1997.

METTLER, F.A., AND M.J. GUIBERTEAU, *Essentials of Nuclear Medicine Imaging*, 5th ed, W.B. Saunders, Philadelphia, 2006.

MAISEY, M.N., K.E. BRITTON AND B.D. COLLIER (Eds), *Clinical Nuclear Medicine*, Chapman & Hall, London 1998.

NAS, National Research Council, Advisory Committee on the Biological Effects of Ionizing Radiations. *The Effects on Populations of Exposure to Low Levels of Ionizing Radiation*, [The BEIR-III Report], Report of the BEIR Committee, National Academy of Sciences, Washington, D.C., 1980.

NAS, National Research Council, Advisory Committee on the Biological Effects of Ionizing Radiations. *Health Risks from Exposure to Low Levels of Ionizing Radiation* [The BEIR-VII Report, Phase 2], Report of the Committee to Assess Health Risks from Exposure to Low Levels of Ionizing Radiation, National Research Council, National Academy of Sciences, Washington, D.C., 2005.

NCRP, *Radiation Protection for Particle Accelerator Facilities*, Report 144, National Council on Radiation Protection and Measurements, Washington, D.C., 2003.

NRC, National Research Council, Institute of Medicine, *Mathematics and Physics of Emerging Biomedical Imaging*, National Academy Press, Washington, D.C., 1996.

ORTON, C.G., "Uses of Therapeutic X Rays in Medicine," *Health Physics*, **69**, 662–676 (1995).

OPPELT, A., *Imaging Systems for Medical Diagnostics: Fundamentals, Technical Solutions and Applications for Systems Applying Ionizing Radiation, Nuclear Magnetic Resonance and Ultrasound*, John Wiley-Interscience, New Jersey, 2005.

PURDY, J.A., "Three Dimensional Conformal Radiation Therapy: Physics, Treatment Planning, and Clinical Aspects," in *Principles and Practice of Radiation Oncology*, 4th ed, PEREZ, C.A., et al. (Eds), Lippincot, Williams, and Wilkins, Philadelphia, 2004.

ROSE, N.R. et al. (Eds), *Manual of Clinical Immunology*, 5th ed, American Society for Microbiology, Washington, D.C., 1997.

SEERAM, E., *Computed Tomography*, 2nd ed, W.B. Saunders Co., Philadeplphia, 2001.

SILBERSTEIN, E.B. et al., "Availability of Radioisotopes Produced in North America", *J. Nucl. Med.*, **41**, 10N-13N (2000).

SORENSON, J.A. AND M.E. PHELPS, *Physics in Nuclear Medicine*, 2nd ed, W.B. Saunders Company, Philadelphia, 1987.

UN, *Sources and Effects of Ionizing Radiation, Vol. I: Sources, Vol. II: Effects*, United Nations Scientific Committee on the Effects of Atomic Radiation, UNSCEAR 2000 Report to the General Assembly, with Scientific Annexes, United Nations, New York, NY, 2000.

YU, J.,N., F.H. FAHEY, B.A. HARKNESS et al., "Intermodality, Retrospective Image Registration in the Thorax," *J. Nucl. Med.*, **36**, 2333–2338 (1995).

WEBSTER, E.W., "X Rays in Diagnostic Radiology," *Health Physics,* **69**, 610–635 (1995).

PROBLEMS

1. Relate a personal experience with a diagnostic-radiology or nuclear-medicine procedure. Were risks and benefits of the procedure explained to you? Were you given information about radiation doses associated with the procedure? In reading this chapter, have you gained a better understanding of the procedure? What insights have been gained and what new questions have come to mind?

2. Consider the two x-ray spectra depicted in Fig. 14.2 produced by the same x-ray machine. Explain why the spectra are shifted in energy.

3. Consider an x-ray examination of bone, approximated as a 2-cm diameter cylinder. In the image, what is the bone's range of *subject contrast* for 100-keV x rays? Explain why the *image contrast* of the bone is less than the subject contrast. What can be done to improve the image contrast?

4. Many source and detector configurations can be used in x-ray tomography. Explain how the machine design is influenced by (a) minimizing costs, (b) increasing image resolution, (c) minimizing patient exposure, (d) minimizing the exposure time, and (e) increasing image contrast.

5. Why is ^{99m}Tc useful in diagnostic nuclear medicine but not therapeutic? Describe how decay characteristics such as half-life, atomic number, type of radiation, and energy of radiation affect the choice of a radionuclide for a particular nuclear-medicine procedure.

6. Explain why positron-emitting isotopes are not usually produced in nuclear reactors.

7. To treat thyroid cancer ^{131}I is injected in the patient where it rapidly accumulates in the thyroid. ^{131}I with a half-life of 8.0 d emits beta particles with an average energy of 182 keV/decay and gamma rays with an average energy of 382 keV/decay. In addition, ^{131}I has a biological half-life in the thyroid of 4.1 d. (a) What is the effective half-life of ^{131}I in the thyroid. (b) How many millicuries of this radioisotope should be injected to deliver a 250 Gy dose to the 20-gram thyroid?

8. Explain why SPECT must use a physical collimator, while PET has no need of such a collimator. Also explain what determines the image resolution in both procedures.

9. Explain why nuclear wastes *per unit activity* is of less societal *and* physical concern for medical applications than those produced in nuclear power plants.

Appendix A

Fundamental Atomic Data

Fundamental Physical Constants

Although there are many physical constants, which determine the nature of our universe, the following values are of particular importance when dealing with atomic and nuclear phenomena.

Table A.1. Some important physical constants as internationally recommended in 2002. These and other constants can be obtained through the web from http://physics.nist.gov/cuu/Constants/index.html

Constant	Symbol	Value
Speed of light (in vacuum)	c	2.99792458×10^8 m s^{-1}
Electron charge	e	$1.60217653 \times 10^{-19}$ C
Atomic mass unit	u	$1.6605389 \times 10^{-27}$ kg (931.494043 MeV/c^2)
Electron rest mass	m_e	$9.1093826 \times 10^{-31}$ kg (0.51099892 MeV/c^2) ($5.48579909 \times 10^{-4}$ u)
Proton rest mass	m_p	$1.67262171 \times 10^{-27}$ kg (938.27203 MeV/c^2) (1.0072764669 u)
Neutron rest mass	m_n	$1.6749273 \times 10^{-27}$ kg (939.56536 MeV/c^2) (1.0086649156 u)
Planck's constant	h	$6.6260693 \times 10^{-34}$ J s $4.1356674 \times 10^{-15}$ eV s
Avogadro's constant	N_a	6.0221415×10^{23} mol^{-1}
Boltzmann constant	k	$1.3806505 \times 10^{-23}$ J K^{-1} (8.617343×10^{-5} eV K^{-1})
Ideal gas constant (STP)	R	8.314472 J mol^{-1} K^{-1}
Electric constant	ϵ_o	$8.854187817 \times 10^{-12}$ F m^{-1}

Source: P.J. Mohr and B.N. Taylor, "CODATA Recommended Values of the Fundamental Physical Constants," *Rev. Modern Physics*, **77**, 1, 2005.

The Periodic Table

The periodic table presented in Table A.2 shows the new IUPAC group format numbers from 1 to 18, while the numbering system used by the Chemical Abstract Service is given in parentheses at the top of each column. In the left of each elemental square, the number of electrons in each of the various electron shells are given. In the upper right of each square, the melting point (MP), boiling point (BP), and critical point (CP) temperatures are given in degrees celsius. Sublimation and critical temperatures are indicated by s and t. In the center of each elemental square the oxidation states, atomic weight, and natural abundance is given for each element. For elements that do not exist naturally, (e.g., the transuranics), the mass number of the longest-lived isotope is given in square brackets. The abundances are based on meteorite and solar wind data.

The table is from Firestone et al. [1999], and may be found on the web at http://isotopes.lbl.gov/isotopes/toi.html. Data for the table are from Lide [1997], Leigh [1990], Anders and Grevesse [1989], and CE News [1985].

Physical Properties and Abundances of Elements

Some of the important physical properties of the naturally occurring elements are given in Table A.3. In this table the atomic weights, densities, melting and boiling points, and abundances of the elements are given.

The atomic weights are for the elements as they exist naturally on earth, or, in the cases of thorium and protactinium, to the isotopes which have the longest half-lives. For elements whose isotopes are all radioactive, the mass number of the longest lived isotope is given in square brackets.

Mass densities for solids and liquids are given at 25 °C, unless otherwise indicated by a superscript temperature (in °C). Densities for normally gaseous elements are for the liquids at their boiling points. The melting and boiling points at normal pressures are in degrees celsius. Superscripts "t" and "s" are the critical and sublimation temperature (in degrees celsius).

The solar system elemental abundances (atomic %) are based on meteorite and solar wind data. The elemental abundances in the earth's crust and in the oceans represent the median values of reported measurements. The concentrations of the less abundant elements may vary with location by several orders of magnitude.

Data are from the 78th edition of the *Handbook of Chemistry and Physics* [Lide 1997], and have been extracted from tables on the web at http://isotopes.lbl.gov/isotopes/toi.html.

Properties of Stable and Long-Lived Nuclides

In Table A.4, the percent abundance of each naturally occurring isotope of each element is given. Also tabulated are the half-life and decay modes of all the radionuclides with half-lives greater than one hour. In this table, percent abundances of stable nuclides and naturally occurring radionuclides are displayed in bold face.

For a half-life of a radionuclide expressed in years (or multiples of years), the year is the tropical year (1900) equal to 365.242 198 78 d or 31 556 925.9747 s. The percent frequencies of the principle decay modes for radioactive nuclides are given in parentheses.

The data in Table A.4 are taken from the *NuBase* data evaluation published by
G. Audi, O. Bersillon, J. Blachot, and A.H. Wapstra, "The NuBase Evaluation of
Nuclear Decay Properties," *Nuclear Physics A*, **729**, 3–128, (2003). The data for
this table (and for short-lived radionuclides) are available from the Atomic Mass
Data Center on the web at `http://amdc.in2p3.fr/web/nubase_en.html`.

Internet Data Sources

The data presented in this appendix are sufficient to allow you to do most assigned
problems. All of these data have been taken from the web, and you are encouraged
to become familiar with these important resources. Some sites, which have many
links to various sets of nuclear and atomic data are

```
http://www.nndc.bnl.gov/
http://physics.nist.gov/PhysRefData/contents.html
http://isotopes.lbl.gov/
http://wwwndc.tokai.jaeri.go.jp/index.html
http://wwwndc.tokai.jaeri.go.jp/nucldata/index.html
http://nucleardata.nuclear.lu.se/nucleardata/toi/
http://atom.kaeri.re.kr/
```

BIBLIOGRAPHY

ANDERS, E. AND N. GREVESSE, "Abundances of the Elements: Meteoritic and Solar,"
  Geochimica et Cosmochimica Acta, **53**, 197 (1989).

Chemical and Engineering News, **63**(5), 27 (1985).

DeBIEVRE, P. AND P.D.P. TAYLOR, *Int. J. Mass Spectrom. Ion Phys.*, **123**, 149, 1993.

FIRESTONE, R.B., C.M. BAGLIN, and S.Y.F. CHU, *Table of Isotopes (1999)*, Wiley, New
  York, 1999.

LEIGH, G.J., *Nomenclature of Inorganic Chemistry*, Blackwells Scientific Publ., Oxford
  (1990).

LIDE, D.R., Ed., *Handbook of Chemistry and Physics*, CRC Press (1997).

1 (IA)

Hydrogen
H₁ −259.34° / −252.87° / −240.18°
+1-1
1.00794
91.0%

Key to Table

Group

	Element
K	M.P.°
L	E_z B.P.°
M	C.P.°
N	Ox.States
O	At.Weight
P	
Q	Abundance%

2 (IIA)

Lithium
Li₃ 180.5° / 1342°
+1
6.941
1.86×10⁻⁷%

Beryllium
Be₄ 1287° / 2471°
+2
9.012182
2.38×10⁻⁹%

Sodium
Na₁₁ 97.80° / 883°
+1
22.989770
0.000187%

Magnesium
Mg₁₂ 650° / 1090°
+2
24.3050
0.00350%

3 (IIIB) 4 (IVB) 5 (VB) 6 (VIB) 7 (VIIB) 8 (VIII) 9 (VIII)

Potassium K₁₉ 63.38° / 759° +1 39.0983 0.0000123%

Calcium Ca₂₀ 842° / 1484° +2 40.078 0.000199%

Scandium Sc₂₁ 1541° / 2836° +3 44.955910 1.12×10⁻⁷%

Titanium Ti₂₂ 1668° / 3287° +2+3+4 47.867 7.8×10⁻⁶%

Vanadium V₂₃ 1910° / 3407° +2+3+4+5 50.9415 9.6×10⁻⁷%

Chromium Cr₂₄ 1907° / 2671° +2+3+6 51.9961 0.000044%

Manganese Mn₂₅ 1246° / 2061° +2+3+4+7 54.938049 0.000031%

Iron Fe₂₆ 1538° / 2861° +2+3 55.845 0.00294%

Cobalt Co₂₇ 1495° / 2927° +2+3 58.933200 7.3×10⁻⁶%

Rubidium Rb₃₇ 39.31° / 688° +1 85.4678 2.31×10⁻⁸%

Strontium Sr₃₈ 777° / 1382° +2 87.62 7.7×10⁻⁸%

Yttrium Y₃₉ 1522° / 3345° +3 88.90585 1.51×10⁻⁸%

Zirconium Zr₄₀ 1855° / 4409° +4 91.224 3.72×10⁻⁸%

Niobium Nb₄₁ 2477° / 4744° +3+5 92.90638 2.28×10⁻⁹%

Molybdenum Mo₄₂ 2623° / 4639° +6 95.94 8.3×10⁻⁹%

Technetium Tc₄₃ 2157° / 4265° [98] 6.1×10⁻⁹%

Ruthenium Ru₄₄ 2334° / 4150° 101.07 6.1×10⁻⁹%

Rhodium Rh₄₅ 1964° / 3695° +3 102.90550 1.12×10⁻⁹%

Cesium Cs₅₅ 28.44° / 671° +1 132.90545 1.21×10⁻⁹%

Barium Ba₅₆ 727° / 1897° +2 137.327 1.46×10⁻⁸%

Lanthanum La₅₇ 918° / 3464° +3 138.9055 1.45×10⁻⁸%

Hafnium Hf₇₂ 2233° / 4603° +4 178.49 5.02×10⁻¹⁰%

Tantalum Ta₇₃ 3017° / 5458° +5 180.9479 6.75×10⁻¹¹%

Tungsten W₇₄ 3422° / 5555° +6 183.84 4.34×10⁻¹⁰%

Rhenium Re₇₅ 3186° / 5596° +4+6+7 186.207 1.69×10⁻¹⁰%

Osmium Os₇₆ 3033° / 5012° +3+4 190.23 2.20×10⁻⁹%

Iridium Ir₇₇ 2446° / 4428° +3+4 192.217 1.69×10⁻⁹%

Francium Fr₈₇ 27° +1 [223]

Radium Ra₈₈ 700° +2 [226]

Actinium Ac₈₉ 1051° / 3198° +3 [227]

Rutherfordium Rf₁₀₄ +4 [261]

Hahnium Ha₁₀₅ [262]

Seaborgium Sg₁₀₆ [266]

Nielsbohrium Ns₁₀₇ [264]

Hassium Hs₁₀₈ [269]

Meitnerium Mt₁₀₉ [268]

² Lanthanides

Cerium Ce₅₈ 798° / 3443° +3+4 140.116 3.70×10⁻⁹%

Praseodymium Pr₅₉ 931° / 3520° +3 140.90765 5.44×10⁻¹⁰%

Neodymium Nd₆₀ 1021° / 3074° +3 144.24 2.70×10⁻⁹%

Promethium Pm₆₁ 1042° / 3000° +3 [145]

Samarium Sm₆₂ 1074° / 1794° +2+3 150.36 8.42×10⁻¹⁰%

Europium Eu₆₃ 822° / 1596° +2+3 151.964 3.17×10⁻¹⁰%

Gadolinium Gd₆₄ 1313° / 3273° +3 157.25 1.076×10⁻⁹%

³ Actinides

Thorium Th₉₀ 1750° / 4788° +4 232.0381 1.09×10⁻¹⁰%

Protactinium Pa₉₁ 1572° +5+4 231.03588

Uranium U₉₂ 1135° / 4131° +3+4+5+6 238.0289 2.94×10⁻¹¹%

Neptunium Np₉₃ 644° +3+4+5+6 [237]

Plutonium Pu₉₄ 640° +3+4+5+6 [244]

Americium Am₉₅ 1176° / 2011° +3+4+5+6 [243]

Curium Cm₉₆ 1345° +3 [247]

Table A.2. The periodic table of the elements. The new IUPAC group format numbers are from 1 to 18, while the numbering system used by the Chemical Abstract Service is given in parentheses. For elements that are not naturally abundant (e.g., the transuranics), the mass number of the longest-lived isotope is given in square brackets. The abundances are based on meteorite and solar wind data. The melting point (MP), boiling point (BP), and critical point (CP) temperatures are given in degrees celsius. *Source*: Firestone, Baglin, and Chu [1999].

18 (VIIIA)

Helium
^{2}He$_2$ −272.2° / −268.93° / −267.96°
0
4.002602
8.9%

13 (IIIA) 14 (IVA) 15 (VA) 16 (VIA) 17 (VIIA)

Element	Config	MP°	BP°/sub	CP°	Ox. states	Mass	Abundance
Boron $^2_3$B$_5$	2,3	2075	4000		+3	10.811	6.9×10^{-8}%
Carbon $^2_4$C$_6$	2,4	4492t	3642s		+2+4−4	12.0107	0.033%
Nitrogen $^2_5$N$_7$	2,5	−210.00	−195.79	−146.94	±1±2±3+4+5	14.00674	0.0102%
Oxygen $^2_6$O$_8$	2,6	−218.79	−182.95	−118.56	−2	15.9994	0.078%
Fluorine $^2_7$F$_9$	2,7	−219.62	−188.12	−129.02	−1	18.9984032	2.7×10^{-6}%
Neon $^2_8$Ne$_{10}$	2,8	−248.59	−246.08	−228.7	0	20.1797	0.0112%

10 (VIII)   11 (IB)   12 (IIB)

Element	Config	MP°	BP°/sub	CP°	Ox. states	Mass	Abundance
Aluminum $^2_8$Al$_{13}$ (…3)	2,8,3	660.32	2519		+3	26.981538	0.000277%
Silicon $^2_8$Si$_{14}$ (…4)	2,8,4	1414	3265		+2+4−4	28.0855	0.00326%
Phosphorus $^2_8$P$_{15}$ (…5)	2,8,5	44.15	280.5	721	+3+5−3	30.973761	0.000034%
Sulfur $^2_8$S$_{16}$ (…6)	2,8,6	115.21	444.60	1041	+4+6−2	32.066	0.00168%
Chlorine $^2_8$Cl$_{17}$ (…7)	2,8,7	−101.5	−34.04	143.8	+1+5+7−1	35.4527	0.000017%
Argon $^2_8$Ar$_{18}$ (…8)	2,8,8	−189.35	−185.85	−122.28	0	39.948	0.000329%
Nickel $^{2\,8}_{16\,2}$Ni$_{28}$	2,8,16,2	1455	2913		+2+3	58.6934	0.000161%
Copper $^{2\,8}_{18\,1}$Cu$_{29}$	2,8,18,1	1084.62	2562		+1+2	63.546	1.70×10^{-6}%
Zinc $^{2\,8}_{18\,2}$Zn$_{30}$	2,8,18,2	419.53	907		+2	65.39	4.11×10^{-6}%
Gallium $^{2\,8}_{18\,3}$Ga$_{31}$	2,8,18,3	29.76	2204		+3	69.723	1.23×10^{-7}%
Germanium $^{2\,8}_{18\,4}$Ge$_{32}$	2,8,18,4	938.25	2833		+2+4	72.61	3.9×10^{-7}%
Arsenic $^{2\,8}_{18\,5}$As$_{33}$	2,8,18,5	817t	614s	1400	+3+5−3	74.92160	2.1×10^{-8}%
Selenium $^{2\,8}_{18\,6}$Se$_{34}$	2,8,18,6	221	685	1493	+4+6−2	78.96	2.03×10^{-7}%
Bromine $^{2\,8}_{18\,7}$Br$_{35}$	2,8,18,7	−7.2	58.8	315	+1+5−1	79.904	3.8×10^{-8}%
Krypton $^{2\,8}_{18\,8}$Kr$_{36}$	2,8,18,8	−157.36	−153.22	−63.74	0	83.80	1.5×10^{-7}%
Palladium $^{2\,8\,18}_{18\,0}$Pd$_{46}$	2,8,18,18,0	1554.9	2963		+2+4	106.42	4.5×10^{-9}%
Silver $^{2\,8\,18}_{18\,1}$Ag$_{47}$	2,8,18,18,1	961.78	2162		+1	107.8682	1.58×10^{-9}%
Cadmium $^{2\,8\,18}_{18\,2}$Cd$_{48}$	2,8,18,18,2	321.07	767		+2	112.411	5.3×10^{-9}%
Indium $^{2\,8\,18}_{18\,3}$In$_{49}$	2,8,18,18,3	156.60	2072		+3	114.818	6.0×10^{-10}%
Tin $^{2\,8\,18}_{18\,4}$Sn$_{50}$	2,8,18,18,4	231.93	2602		+2+4	118.710	1.25×10^{-8}%
Antimony $^{2\,8\,18}_{18\,5}$Sb$_{51}$	2,8,18,18,5	630.63	1587		+3+5−3	121.760	1.01×10^{-9}%
Tellurium $^{2\,8\,18}_{18\,6}$Te$_{52}$	2,8,18,18,6	449.51	988		+4+6−2	127.60	1.57×10^{-8}%
Iodine $^{2\,8\,18}_{18\,7}$I$_{53}$	2,8,18,18,7	113.7	184.4	546	+1+5+7−1	126.90447	2.9×10^{-9}%
Xenon $^{2\,8\,18}_{18\,8}$Xe$_{54}$	2,8,18,18,8	−111.75	−108.04	16.58	0	131.29	1.5×10^{-8}%
Platinum $^{2\,8\,18\,32}_{16\,2}$Pt$_{78}$	2,8,18,32,16,2	1768.4	3825		+2+4	195.078	4.4×10^{-9}%
Gold $^{2\,8\,18\,32}_{18\,1}$Au$_{79}$	2,8,18,32,18,1	1064.18	2856		+1+3	196.96655	6.1×10^{-10}%
Mercury $^{2\,8\,18\,32}_{18\,2}$Hg$_{80}$	2,8,18,32,18,2	−38.83	356.73	1477	+1+2	200.59	1.11×10^{-9}%
Thallium $^{2\,8\,18\,32}_{18\,3}$Tl$_{81}$	2,8,18,32,18,3	304	1473		+1+3	204.3833	6.0×10^{-10}%
Lead $^{2\,8\,18\,32}_{18\,4}$Pb$_{82}$	2,8,18,32,18,4	327.46	1749		+2+4	207.2	1.03×10^{-8}%
Bismuth $^{2\,8\,18\,32}_{18\,5}$Bi$_{83}$	2,8,18,32,18,5	271.40	1564		+3+5	208.98038	4.7×10^{-10}%
Polonium $^{2\,8\,18\,32}_{18\,6}$Po$_{84}$	2,8,18,32,18,6	254	962		+2+4	[209]	
Astatine $^{2\,8\,18\,32}_{18\,7}$At$_{85}$	2,8,18,32,18,7	302				[210]	
Radon $^{2\,8\,18\,32}_{18\,8}$Rn$_{86}$	2,8,18,32,18,8	−71	−61.7	104	0	[222]	
Element-110 110_{110}	2,8,18,32,32,16,2					[271]	
Element-111 111_{111}	2,8,18,32,32,17,2					[272]	
Element-112 112_{112}	2,8,18,32,32,18,2					[277]	

Element	Config	MP°	BP°	Ox.	Mass	Abundance
Terbium $^{2\,8\,18\,27}_{8\,2}$Tb$_{65}$	2,8,18,27,8,2	1356	3230	+3	158.92534	1.97×10^{-10}%
Dysprosium $^{2\,8\,18\,28}_{8\,2}$Dy$_{66}$	2,8,18,28,8,2	1412	2567	+3	162.50	1.286×10^{-9}%
Holmium $^{2\,8\,18\,29}_{8\,2}$Ho$_{67}$	2,8,18,29,8,2	1474	2700	+3	164.93032	2.90×10^{-10}%
Erbium $^{2\,8\,18\,30}_{8\,2}$Er$_{68}$	2,8,18,30,8,2	1529	2868	+3	167.26	8.18×10^{-10}%
Thulium $^{2\,8\,18\,31}_{8\,2}$Tm$_{69}$	2,8,18,31,8,2	1545	1950	+3	168.93421	1.23×10^{-10}%
Ytterbium $^{2\,8\,18\,32}_{8\,2}$Yb$_{70}$	2,8,18,32,8,2	819	1196	+2+3	173.04	8.08×10^{-10}%
Lutetium $^{2\,8\,18\,32}_{9\,2}$Lu$_{71}$	2,8,18,32,9,2	1663	3402	+3	174.967	1.197×10^{-10}%
Berkelium $^{2\,8\,18\,32\,27}_{8\,2}Bk_{97}$	2,8,18,32,27,8,2	1050		+3+4	[247]	
Californium $^{2\,8\,18\,32\,28}_{8\,2}Cf_{98}$	2,8,18,32,28,8,2	900		+3	[251]	
Einsteinium $^{2\,8\,18\,32\,29}_{8\,2}Es_{99}$	2,8,18,32,29,8,2	860		+3	[252]	
Fermium $^{2\,8\,18\,32\,30}_{8\,2}Fm_{100}$	2,8,18,32,30,8,2	1527		+3	[257]	
Mendelevium $^{2\,8\,18\,32\,31}_{8\,2}Md_{101}$	2,8,18,32,31,8,2	827		+2+3	[258]	
Nobelium $^{2\,8\,18\,32\,32}_{8\,2}No_{102}$	2,8,18,32,32,8,2	827		+2+3	[259]	
Lawrencium $^{2\,8\,18\,32\,32}_{9\,2}Lr_{103}$	2,8,18,32,32,9,2	1627		+3	[262]	

Table A.2. *(cont.)* The periodic table of the elements. The new IUPAC group format numbers are from 1 to 18, while the numbering system used by the Chemical Abstract Service is given in parentheses. For elements that are not naturally abundant (e.g., the transuranics), the mass number of the longest-lived isotope is given in square brackets. The abundances are based on meteorite and solar wind data. The melting point (MP), boiling point (BP), and critical point (CP) temperatures are given in degrees celsius. *Source*: Firestone, Baglin, and Chu [1999].

Table A.3. Some physical properties and abundances of the elements.

						Elemental Abundances		
Z	El	Atomic Weight	Mass density (g/cm^3)	Melting Point (°C)	Boiling Point (°C)	Solar System (%)	Crustal Average (mg/kg)	Earth's Oceans (mg/L)
1	H	1.00794	0.0708	-259.34	-252.87	91.0	1400	1.08×10^5
2	He	4.002602	0.124901	-272.2	-268.93	8.9	0.008	7×10^{-6}
3	Li	6.941	0.534	180.5	1342	1.86×10^{-7}	20	0.18
4	Be	9.012182	1.85	1287	2471	2.38×10^{-9}	2.8	5.6×10^{-6}
5	B	10.811	2.37	2075	4000	6.9×10^{-8}	10	4.44
6	C^a	12.0107	$2.2670^{15°}$	4492^t	3842s^s	0.033	200	28
7	N	14.00674	0.807	-210.00	-195.79	0.0102	19	0.5
8	O	15.9994	1.141	-218.79	-182.95	0.078	4.61×10^5	8.57×10^5
9	F	18.9984032	1.50	-219.62	-188.12	2.7×10^{-6}	585	1.3
10	Ne	20.1797	1.204	-248.59	-246.08	0.0112	0.005	1.2×10^{-4}
11	Na	22.989770	0.97	97.80	883	0.000187	2.36×10^4	1.08×10^4
12	Mg	24.3050	0.74	650	1090	0.00350	2.33×10^4	1290
13	Al	26.981538	2.70	660.32	2519	0.000277	8.23×10^4	0.002
14	Si	28.0855	2.3296	1414	3265	0.00326	2.82×10^5	2.2
15	P	30.973761	1.82	44.15	280.5	3.4×10^{-5}	1050	0.06
16	S	32.066	2.067	115.21	444.60	0.00168	350	905
17	Cl	35.4527	1.56	-101.5	-34.04	1.7×10^{-5}	145	1.94×10^4
18	Ar	39.948	1.396	-189.35	-185.85	0.000329	3.5	0.45
19	K	39.0983	0.89	63.38	759	1.23×10^{-5}	2.09×10^4	399
20	Ca	40.078	1.54	842	1484	0.000199	4.15×10^4	412
21	Sc	44.955910	2.99	1541	2836	1.12×10^{-7}	22	6×10^{-7}
22	Ti	47.867	4.5	1668	3287	7.8×10^{-6}	5650	0.001
23	V	50.9415	6.0	1910	3407	9.6×10^{-7}	120	0.0025
24	Cr	51.9961	7.15	1907	2671	4.4×10^{-5}	102	3×10^{-4}
25	Mn	54.938049	7.3	1246	2061	3.1×10^{-5}	950	2×10^{-4}
26	Fe	55.845	7.875	1538	2861	0.00294	5.63×10^4	0.002
27	Co	58.933200	8.86	1495	2927	7.3×10^{-6}	25	2×10^{-5}
28	Ni	58.6934	8.912	1455	2913	0.000161	84	5.6×10^{-4}
29	Cu	63.546	8.933	1084.62	2562	1.70×10^{-6}	60	2.5×10^{-4}
30	Zn	65.39	7.134	419.53	907	4.11×10^{-6}	70	0.0049
31	Ga	69.723	5.91	29.76	2204	1.23×10^{-7}	19	3×10^{-5}
32	Ge	72.61	5.323	938.25	2833	3.9×10^{-7}	1.5	5×10^{-5}
33	As	74.92160	$5.776^{26°}$	817^t	614^s	2.1×10^{-8}	1.8	0.0037
34	Se	78.96	$4.809^{26°}$	221	685	2.03×10^{-7}	0.05	2×10^{-4}
35	Br	79.904	3.11	-7.2	58.8	3.8×10^{-8}	2.4	67.3
36	Kr	83.80	2.418	-157.36	-153.22	1.5×10^{-7}	1×10^{-4}	2.1×10^{-4}
37	Rb	85.4678	1.53	39.31	688	2.31×10^{-8}	90	0.12
38	Sr	87.62	2.64	777	1382	7.7×10^{-8}	370	7.9
39	Y	88.90585	4.47	1522	3345	1.51×10^{-8}	33	1.3×10^{-5}
40	Zr	91.224	6.52	1855	4409	3.72×10^{-8}	165	3×10^{-5}
41	Nb	92.90638	8.57	2477	4744	2.28×10^{-9}	20	1×10^{-5}
42	Mo	95.94	10.2	2623	4639	8.3×10^{-9}	1.2	0.01
43	Tc	[98]	11	2157	4265			
44	Ru	101.07	12.1	2334	4150	6.1×10^{-9}	0.001	7×10^{-7}
45	Rh	102.90550	12.4	1964	3695	1.12×10^{-9}	0.001	
46	Pd	106.42	12.0	1554.9	2963	4.5×10^{-9}	0.015	

^{a}graphite ^{t}critical temperature ^{s}sublimation temperature

Z	El	Atomic Weight	Mass density (g/cm³)	Melting Point (°C)	Boiling Point (°C)	Elemental Abundances Solar System (%)	Crustal Average (mg/kg)	Earth's Oceans (mg/L)
47	Ag	107.8682	10.501	961.78	2162	1.58×10^{-9}	0.075	4×10^{-5}
48	Cd	112.411	8.69	321.07	767	5.3×10^{-9}	0.15	1.1×10^{-4}
49	In	114.818	7.31	156.60	2072	6.0×10^{-10}	0.25	0.02
50	Sn	118.710	$7.287^{26°}$	231.93	2602	1.25×10^{-8}	2.3	4×10^{-6}
51	Sb	121.760	$6.685^{26°}$	630.63	1587	1.01×10^{-9}	0.2	2.4×10^{-4}
52	Te	127.60	6.232	449.51	988	1.57×10^{-8}	0.001	
53	I	126.90447	$4.93^{20°}$	113.7	184.4	2.9×10^{-9}	0.45	0.06
54	Xe	131.29	2.953	-111.75	-108.04	1.5×10^{-8}	3×10^{-5}	5×10^{-5}
55	Cs	132.90545	1.93	28.44	671	1.21×10^{-9}	3	3×10^{-4}
56	Ba	137.327	3.62	727	1897	1.46×10^{-8}	425	0.013
57	La	138.9055	6.15	918	3464	1.45×10^{-9}	39	3.4×10^{-6}
58	Ce	140.116	8.16	798	3443	3.70×10^{-9}	66.5	1.2×10^{-6}
59	Pr	140.90765	6.77	931	3520	5.44×10^{-10}	9.2	6.4×10^{-7}
60	Nd	144.24	7.01	1021	3074	2.70×10^{-9}	41.5	2.8×10^{-6}
61	Pm	[145]	7.26	1042	3000			
62	Sm	150.36	7.52	1074	1794	8.42×10^{-10}	7.05	4.5×10^{-7}
63	Eu	151.964	5.24	822	1596	3.17×10^{-10}	2.0	1.3×10^{-7}
64	Gd	157.25	7.90	1313	3273	1.076×10^{-9}	6.2	7×10^{-7}
65	Tb	158.92534	8.23	1356	3230	1.97×10^{-10}	1.2	1.4×10^{-7}
66	Dy	162.50	8.55	1412	2567	1.286×10^{-9}	5.2	9.1×10^{-7}
67	Ho	164.93032	8.80	1474	2700	2.90×10^{-10}	1.3	2.2×10^{-7}
68	Er	167.26	9.07	1529	2868	8.18×10^{-10}	3.5	8.7×10^{-7}
69	Tm	168.93421	9.32	1545	1950	1.23×10^{-10}	0.52	1.7×10^{-7}
70	Yb	173.04	6.90	819	1196	8.08×10^{-10}	3.2	8.2×10^{-7}
71	Lu	174.967	9.84	1663	3402	1.197×10^{-10}	0.8	1.5×10^{-7}
72	Hf	178.49	13.3	2233	4603	5.02×10^{-10}	3.0	7×10^{-6}
73	Ta	180.9479	16.4	3017	5458	6.75×10^{-11}	2.0	2×10^{-6}
74	W	183.84	19.3	3422	5555	4.34×10^{-10}	1.25	1×10^{-4}
75	Re	186.207	20.8	3186	5596	1.69×10^{-10}	7×10^{-4}	4×10^{-6}
76	Os	190.23	22.5	3033	5012	2.20×10^{-9}	0.0015	
77	Ir	192.217	22.5	2446	4428	2.16×10^{-9}	0.001	
78	Pt	195.078	21.46	1768.4	3825	4.4×10^{-9}	0.005	
79	Au	196.96655	19.282	1064.18	2856	6.1×10^{-10}	0.004	4×10^{-6}
80	Hg	200.59	13.5336	-38.83	356.73	1.11×10^{-9}	0.085	3×10^{-5}
81	Tl	204.3833	11.8	304	1473	6.0×10^{-10}	0.85	1.9×10^{-5}
82	Pb	207.2	11.342	327.46	1749	1.03×10^{-8}	14	3×10^{-5}
83	Bi	208.98038	9.807	271.40	1564	4.7×10^{-10}	0.0085	2×10^{-5}
84	Po	[209]	9.32	254	962		2×10^{-10}	1.5×10^{-14}
85	At	[210]		302				
86	Rn	[222]	4.4	-71	-61.7		4×10^{-13}	6×10^{-16}
87	Fr	[223]		27				
88	Ra	[226]	5	700			9×10^{-7}	8.9×10^{-11}
89	Ac	[227]	10.07	1051	3198		5.5×10^{-10}	
90	Th	232.0381	11.72	1750	4788	1.09×10^{-10}	69.6	1×10^{-6}
91	Pa	231.03588	15.37	1572			1.4×10^{-6}	5×10^{-11}
92	U	238.0289	18.95	1135	4131	2.94×10^{-11}	2.7	0.0032

Table A.4. Percent isotopic abundances (bold) of naturally occurring nuclides and decay characteristics of radionuclides with $T_{1/2} > 1$ h. Branching ratios are in percent.

Nucl.	Half-life	Abund. / Decay Mode	Nucl.	Half-life	Abund. / Decay Mode
^{1}H	stable	**99.9885**	^{42}Ar	32.9 y	$\beta^-(100)$
^{2}H	stable	**0.0115**	^{42}K	12.360 h	$\beta^-(100)$
^{3}H	12.32 y	$\beta^-(100)$	^{42}Ca	stable	**0.647**
^{3}He	stable	**0.000137**	^{43}K	22.3 h	$\beta^-(100)$
^{4}He	stable	**99.999863**	^{43}Ca	stable	**0.135**
^{6}Li	stable	**7.59**	^{43}Sc	3.891 h	$\beta^+(100)$
^{7}Li	stable	**92.41**	^{44}Ca	stable	**2.086**
^{7}Be	53.22 d	EC(100)	^{44}Sc	3.97 h	$\beta^+(100)$
^{9}Be	stable	**100.**	^{44m}Sc	58.61 h	IT(98.80) $\beta^+(1.20)$
^{10}Be	1.51 My	$\beta^-(100)$	^{44}Ti	60.0 y	EC(100)
^{10}B	stable	**19.9**	^{45}Ca	162.67 d	$\beta^-(100)$
^{11}B	stable	**80.1**	^{45}Sc	stable	**100.**
^{12}C	stable	**98.93**	^{46}Ca	stable	**0.004**
^{13}C	stable	**1.07**	^{46}Sc	83.79 d	$\beta^-(100)$
^{14}C	5.70 ky	$\beta^-(100)$	^{46}Ti	stable	**8.25**
^{14}N	stable	**99.632**	^{47}Ca	4.536 d	$\beta^-(100)$
^{15}N	stable	**0.368**	^{47}Sc	3.3492 d	$\beta^-(100)$
^{16}O	stable	**99.757**	^{47}Ti	stable	**7.44**
^{17}O	stable	**0.038**	^{48}Ca	53 Ey	**0.187**
^{18}O	stable	**0.205**	^{48}Sc	43.67 h	$\beta^-(100)$
^{19}F	stable	**100.**	^{48}Ti	stable	**73.72**
^{20}Ne	stable	**90.48**	^{48}V	15.9735 d	$\beta^+(100)$
^{21}Ne	stable	**0.27**	^{48}Cr	21.56 h	$\beta^+(100)$
^{22}Ne	stable	**9.25**	^{49}Ti	stable	**5.41**
^{22}Na	2.6019 y	$\beta^+(100)$	^{49}V	330 d	EC(100)
^{23}Na	stable	**100.**	^{50}Ti	stable	**5.18**
^{24}Na	14.9590 h	$\beta^-(100)$	^{50}V	150 Py	**0.250** $\beta^+(83)$ $\beta^-(17)$
^{24}Mg	stable	**78.99**	^{50}Cr	stable	**4.345**
^{25}Mg	stable	**10.00**	^{51}V	stable	**99.750**
^{26}Mg	stable	**11.01**	^{51}Cr	27.703 d	EC(100)
^{26}Al	717 ky	$\beta^+(100)$	^{52}Cr	stable	**83.789**
^{27}Al	stable	**100.**	^{52}Mn	5.591 d	$\beta^+(100)$
^{28}Mg	20.915 h	$\beta^-(100)$	^{52}Fe	8.275 h	$\beta^+(100)$
^{28}Si	stable	**92.23**	^{53}Cr	stable	**9.501**
^{29}Si	stable	**4.683**	^{53}Mn	3.7 My	EC(100)
^{30}Si	stable	**3.087**	^{54}Cr	stable	**2.365**
^{31}P	stable	**100.**	^{54}Mn	312.03 d	EC(100)
^{32}Si	132 y	$\beta^-(100)$	^{54}Fe	stable	**5.845**
^{32}P	14.263 d	$\beta^-(100)$	^{55}Mn	stable	**100.**
^{32}S	stable	**95.02**	^{55}Fe	2.737 y	EC(100)
^{33}P	25.34 d	$\beta^-(100)$	^{55}Co	17.53 h	$\beta^+(100)$
^{33}S	stable	**0.76**	^{56}Mn	2.5789 h	$\beta^-(100)$
^{34}S	stable	**4.29**	^{56}Fe	stable	**91.754**
^{35}S	87.51 d	$\beta^-(100)$	^{56}Co	77.23 d	$\beta^+(100)$
^{35}Cl	stable	**75.78**	^{56}Ni	6.075 d	$\beta^+(100)$
^{36}S	stable	**0.02**	^{57}Fe	stable	**2.119**
^{36}Cl	301 ky	$\beta^-(98.1)$ $\beta^+(1.9)$	^{57}Co	271.74 d	EC(100)
^{36}Ar	stable	**0.3365**	^{57}Ni	35.60 h	$\beta^+(100)$
^{37}Cl	stable	**24.22**	^{58}Fe	stable	**0.282**
^{37}Ar	35.04 d	EC(100)	^{58}Co	70.86 d	$\beta^+(100)$
^{38}Ar	stable	**0.0632**	^{58m}Co	9.04 h	IT(100)
^{39}Ar	269 y	$\beta^-(100)$	^{58}Ni	stable	**68.077**
^{39}K	stable	**93.2581**	^{59}Fe	44.495 d	$\beta^-(100)$
^{40}Ar	stable	**99.6003**	^{59}Co	stable	**100.**
^{40}K	1.251 Gy	**0.0117** $\beta^-(89.28)$ $\beta^+(10.72)$	^{59}Ni	101 ky	$\beta^+(100)$
^{40}Ca	stable	**96.941**	^{60}Fe	1.5 My	$\beta^-(100)$
^{41}K	stable	**6.7302**	^{60}Co	5.2714 y	$\beta^-(100)$
^{41}Ca	102 ky	EC(100)	^{60}Ni	stable	**26.223**

Table A.4. *(cont.)* Percent isotopic abundances (bold) of naturally occurring nuclides and decay characteristics of radionuclides with $T_{1/2} > 1$ h. Branching ratios are in percent.

Nucl.	Half-life	Abund. / Decay Mode	Nucl.	Half-life	Abund. / Decay Mode
^{61}Co	1.650 h	$\beta^-(100)$	^{79}Kr	35.04 h	$\beta^+(100)$
^{61}Ni	stable	**1.140**	^{80}Se	stable	**49.61**
^{61}Cu	3.333 h	$\beta^+(100)$	^{80m}Br	4.4205 h	IT(100)
^{62}Ni	stable	**3.6345**	^{80}Kr	stable	**2.28**
^{62}Zn	9.186 h	$\beta^+(100)$	^{81}Br	stable	**49.31**
^{63}Ni	100.1 y	$\beta^-(100)$	^{81}Kr	229 ky	EC(100)
^{63}Cu	stable	**69.17**	^{81}Rb	4.576 h	$\beta^+(100)$
^{64}Ni	stable	**0.9256**	^{82}Se	97 Ey	**8.73**
^{64}Cu	12.700 h	$\beta^+(61.0)$ $\beta^-(39.0)$	^{82}Br	35.282 h	$\beta^-(100)$
^{64}Zn	stable	**48.63**	^{82}Kr	stable	**11.58**
^{65}Ni	2.5172 h	$\beta^-(100)$	^{82m}Rb	6.472 h	$\beta^+(\simeq100)$ IT(<0.33)
^{65}Cu	stable	**30.83**	^{82}Sr	25.36 d	EC(100)
^{65}Zn	244.06 d	$\beta^+(100)$	^{83}Br	2.40 h	$\beta^-(100)$
^{66}Ni	54.6 h	$\beta^-(100)$	^{83}Kr	stable	**11.49**
^{66}Zn	stable	**27.9**	^{83m}Kr	1.83 h	IT(100)
^{66}Ga	9.49 h	$\beta^+(100)$	^{83}Rb	86.2 d	EC(100)
^{66}Ge	2.26 h	$\beta^+(100)$	^{83}Sr	32.41 h	$\beta^+(100)$
^{67}Cu	61.83 h	$\beta^-(100)$	^{84}Kr	stable	**57.0**
^{67}Zn	stable	**4.10**	^{84}Rb	32.77 d	$\beta^+(96.2)$ $\beta^-(3.8)$
^{67}Ga	3.2612 d	EC(100)	^{84}Sr	stable	**0.56**
^{68}Zn	stable	**18.75**	^{85}Kr	10.776 y	$\beta^-(100)$
^{68}Ge	270.95 d	EC(100)	^{85m}Kr	4.480 h	$\beta^-(78.6)$ IT(21.4)
^{69m}Zn	13.86 h	IT($\simeq100$) $\beta^-(0.033)$	^{85}Rb	stable	**72.17**
^{69}Ga	stable	**60.108**	^{85}Sr	64.853 d	EC(100)
^{69}Ge	39.05 h	$\beta^+(100)$	^{85}Y	2.68 h	$\beta^+(100)$
^{70}Zn	stable	**0.62**	^{85m}Y	4.86 h	$\beta^+(\simeq100)$ IT(<0.002)
^{70}Ge	stable	**20.84**	^{86}Kr	stable	**17.3**
^{71m}Zn	3.96 h	$\beta^-(\simeq100)$	^{86}Rb	18.642 d	$\beta^-(\simeq100)$ EC(0.0052)
^{71}Ga	stable	**39.892**	^{86}Sr	stable	**9.86**
^{71}Ge	11.43 d	EC(100)	^{86}Y	14.74 h	$\beta^+(100)$
^{71}As	65.28 h	$\beta^+(100)$	^{86}Zr	16.5 h	$\beta^+(100)$
^{72}Zn	46.5 h	$\beta^-(100)$	^{87}Rb	49.23 Gy	**27.83** $\beta^-(100)$
^{72}Ga	14.10 h	$\beta^-(100)$	^{87}Sr	stable	**7.00**
^{72}Ge	stable	**27.54**	^{87m}Sr	2.815 h	IT($\simeq100$) EC(0.30)
^{72}As	26.0 h	$\beta^+(100)$	^{87}Y	79.8 h	$\beta^+(100)$
^{72}Se	8.40 d	EC(100)	^{87m}Y	13.37 h	IT(98.43) $\beta^+(1.57)$
^{73}Ga	4.86 h	$\beta^-(100)$	^{87}Zr	1.68 h	$\beta^+(100)$
^{73}Ge	stable	**7.73**	^{88}Kr	2.84 h	$\beta^-(100)$
^{73}As	80.30 d	EC(100)	^{88}Sr	stable	**82.58**
^{73}Se	7.15 h	$\beta^+(100)$	^{88}Y	106.65 d	$\beta^+(100)$
^{74}Ge	stable	**36.28**	^{88}Zr	83.4 d	EC(100)
^{74}As	17.77 d	$\beta^+(66)$ $\beta^-(34)$	^{89}Sr	50.53 d	$\beta^-(100)$
^{74}Se	stable	**0.89**	^{89}Y	stable	**100.**
^{75}As	stable	**100.**	^{89}Zr	78.41 h	$\beta^+(100)$
^{75}Se	119.79 d	EC(100)	^{89}Nb	2.03 h	$\beta^+(100)$
^{76}Ge	1.58 Zy	**7.61**	^{89m}Nb	1.10 h	$\beta^+(100)$
^{76}As	1.0778 d	$\beta^-(\simeq100)$ EC(<0.02)	^{90}Sr	28.79 y	$\beta^-(100)$
^{76}Se	stable	**9.37**	^{90}Y	64.00 h	$\beta^-(100)$
^{76}Br	16.2 h	$\beta^+(100)$	^{90m}Y	3.19 h	IT($\simeq100$) $\beta^-(0.0018)$
^{76}Kr	14.8 h	$\beta^+(100)$	^{90}Zr	stable	**51.45**
^{77}Ge	11.30 h	$\beta^-(100)$	^{90}Nb	14.60 h	$\beta^+(100)$
^{77}As	38.83 h	$\beta^-(100)$	^{90}Mo	5.56 h	$\beta^+(100)$
^{77}Se	stable	**7.63**	^{91}Sr	9.63 h	$\beta^-(100)$
^{77}Br	57.036 h	$\beta^+(100)$	^{91}Y	58.51 d	$\beta^-(100)$
^{78}Se	stable	**23.77**	^{91}Zr	stable	**11.22**
^{78}Kr	stable	**0.35**	^{91}Nb	680 y	EC($\simeq100$) $e^+(0.0138)$
^{79}Se	295 ky	$\beta^-(100)$	^{91m}Nb	60.86 d	IT(96.6) EC(3.4) $e^+(0.0028)$
^{79}Br	stable	**50.69**	^{92}Sr	2.66 h	$\beta^-(100)$

Table A.4. *(cont.)* Percent isotopic abundances (bold) of naturally occurring nuclides and decay characteristics of radionuclides with $T_{1/2} > 1$ h. Branching ratios are in percent.

Nucl.	Half-life	Abund. / Decay Mode	Nucl.	Half-life	Abund. / Decay Mode
^{92}Y	3.54 h	β^-(100)	^{105}Ru	4.44 h	β^-(100)
^{92}Zr	stable	**17.15**	^{105}Rh	35.36 h	β^-(100)
^{92}Nb	34.7 My	β^+($\simeq$100) β^-($<$0.05)	^{105}Pd	stable	**22.33**
^{92m}Nb	10.15 d	β^+(100)	^{105}Ag	41.29 d	β^+(100)
^{92}Mo	stable	**14.84**	^{106}Ru	373.59 d	β^-(100)
^{93}Y	10.18 h	β^-(100)	^{106}Pd	stable	**27.33**
^{93}Zr	1.53 My	β^-(100)	^{106m}Ag	8.28 d	β^+(100)
^{93}Nb	stable	**100.**	^{106}Cd	stable	**1.25**
^{93m}Nb	16.13 y	IT(100)	^{107}Pd	6.5 My	β^-(100)
^{93}Mo	4.0 ky	EC(100)	^{107}Ag	stable	**51.839**
^{93m}Mo	6.85 h	IT($\simeq$100) β^+(0.12)	^{107}Cd	6.50 h	β^+(100)
^{93}Tc	2.75 h	β^+(100)	^{108}Pd	stable	**26.46**
^{94}Zr	stable	**17.38**	^{108m}Ag	418 y	β^+(91.3) IT(8.7)
^{94}Nb	20.3 ky	β^-(100)	^{108}Cd	stable	**0.89**
^{94}Mo	stable	**9.25**	^{109}Pd	13.7012 h	β^-(100)
^{95}Zr	64.032 d	β^-(100)	^{109}Ag	stable	**48.161**
^{95}Nb	34.991 d	β^-(100)	^{109}Cd	461.4 d	EC(100)
^{95m}Nb	86.6 h	IT(94.4) β^-(5.6)	^{109}In	4.2 h	β^+(100)
^{95}Mo	stable	**15.92**	^{110}Pd	stable	**11.72**
^{95}Tc	20.0 h	β^+(100)	^{110m}Ag	249.95 d	β^-(98.64) IT(1.36)
^{95m}Tc	61 d	β^+(96.12) IT(3.88)	^{110}Cd	stable	**12.49**
^{95}Ru	1.643 h	β^+(100)	^{110}In	4.9 h	β^+(100)
^{96}Zr	24 Ey	**2.80**	^{110}Sn	4.11 h	EC(100)
^{96}Nb	23.35 h	β^-(100)	^{111m}Pd	5.5 h	IT(73) β^-(27)
^{96}Mo	stable	**16.68**	^{111}Ag	7.45 d	β^-(100)
^{96}Tc	4.28 d	β^+(100)	^{111}Cd	stable	**12.80**
^{96}Ru	stable	**5.54**	^{111}In	2.8047 d	EC(100)
^{97}Zr	16.90 h	β^-(100)	^{112}Pd	21.03 h	β^-(100)
^{97}Mo	stable	**9.55**	^{112}Ag	3.130 h	β^-(100)
^{97}Tc	2.6 My	EC(100)	^{112}Cd	stable	**24.13**
^{97m}Tc	90.1 d	IT($\simeq$100) EC($<$0.34)	^{112}Sn	stable	**0.97**
^{97}Ru	2.9 d	β^+(100)	^{113}Ag	5.37 h	β^-(100)
^{98}Mo	stable	**24.13**	^{113}Cd	7.7 Py	**12.22** β^-(100)
^{98}Tc	4.2 My	β^-(100)	^{113m}Cd	14.1 y	β^-($\simeq$100) IT(0.14)
^{98}Ru	stable	**1.87**	^{113}In	stable	**4.29**
^{99}Mo	65.94 h	β^-(100)	^{113m}In	1.6579 h	IT(100)
^{99}Tc	211.1 ky	β^-(100)	^{113}Sn	115.09 d	β^+(100)
^{99m}Tc	6.015 h	IT($\simeq$100) β^-(0.0037)	^{114}Cd	stable	**28.73**
^{99}Ru	stable	**12.76**	^{114m}In	49.51 d	IT(96.75) β^+(3.25)
^{99}Rh	16.1 d	β^+(100)	^{114}Sn	stable	**0.65**
^{99m}Rh	4.7 h	β^+($\simeq$100) IT($<$0.16)	^{115}Cd	53.46 h	β^-(100)
^{100}Mo	8.5 Ey	**9.63**	^{115m}Cd	44.56 d	β^-(100)
^{100}Ru	stable	**12.6**	^{115}In	441 Ty	**95.71** β^-(100)
^{100}Rh	20.8 h	β^+(100)	^{115m}In	4.486 h	IT(95.0) β^-(5.0)
^{100}Pd	3.63 d	EC(100)	^{115}Sn	stable	**0.34**
^{101}Ru	stable	**17.06**	^{116}Cd	30 Ey	**7.49**
^{101}Rh	3.3 y	EC(100)	^{116}Sn	stable	**14.54**
^{101m}Rh	4.34 d	EC(93.6) IT(6.4)	^{116}Te	2.49 h	β^+(100)
^{101}Pd	8.47 h	β^+(100)	^{117}Cd	2.49 h	β^-(100)
^{102}Ru	stable	**31.55**	^{117m}Cd	3.36 h	β^-(100)
^{102}Rh	207.0 d	β^+(78) β^-(22)	^{117}Sn	stable	**7.68**
^{102m}Rh	3.742 y	β^+($\simeq$100) IT(0.233)	^{117m}Sn	13.76 d	IT(100)
^{102}Pd	stable	**1.02**	^{117}Sb	2.80 h	β^+(100)
^{103}Ru	39.26 d	β^-(100)	^{118}Sn	stable	**24.22**
^{103}Rh	stable	**100.**	^{118m}Sb	5.00 h	β^+(100)
^{103}Pd	16.991 d	EC(100)	^{118}Te	6.00 d	EC(100)
^{104}Ru	stable	**18.62**	^{119}Sn	stable	**8.59**
^{104}Pd	stable	**11.14**	^{119m}Sn	293.1 d	IT(100)

Table A.4. *(cont.)* Percent isotopic abundances (bold) of naturally occurring nuclides and decay characteristics of radionuclides with $T_{1/2} > 1$ h. Branching ratios are in percent.

Nucl.	Half-life	Abund. / Decay Mode	Nucl.	Half-life	Abund. / Decay Mode
^{119}Sb	38.19 h	EC(100)	^{130}I	12.36 h	β^-(100)
^{119}Te	16.05 h	β^+(100)	^{130}Xe	stable	**4.08**
^{119m}Te	4.70 d	$\beta^+(\simeq100)$ IT($<$0.008)	^{130}Ba	stable	**0.106**
^{120}Sn	stable	**32.58**	^{131m}Te	30 h	β^-(77.8) IT(22.2)
^{120m}Sb	5.76 d	β^+(100)	^{131}I	8.02070 d	β^-(100)
^{120}Te	stable	**0.09**	^{131}Xe	stable	**21.18**
^{121}Sn	27.03 h	β^-(100)	^{131m}Xe	11.84 d	IT(100)
^{121m}Sn	43.9 y	IT(77.6) β^-(22.4)	^{131}Cs	9.689 d	EC(100)
^{121}Sb	stable	**57.21**	^{131}Ba	11.50 d	β^+(100)
^{121}Te	19.16 d	β^+(100)	^{132}Te	3.204 d	β^-(100)
^{121m}Te	154 d	IT(88.6) β^+(11.4)	^{132}I	2.295 h	β^-(100)
^{121}I	2.12 h	β^+(100)	^{132m}Ix	1.387 h	IT(86) β^-(14)
^{122}Sn	stable	**4.63**	^{132}Xe	stable	**26.89**
^{122}Sb	2.7238 d	β^-(97.59) β^+(2.41)	^{132}Cs	6.479 d	β^+(98.13) β^-(1.87)
^{122}Te	stable	**2.55**	^{132}Ba	stable	**0.101**
^{122}Xe	20.1 h	EC(100)	^{132}La	4.8 h	β^+(100)
^{123}Sn	129.2 d	β^-(100)	^{132}Ce	3.51 h	β^+(100)
^{123}Sb	stable	**42.79**	^{133}I	20.8 h	β^-(100)
^{123}Te	$>$600 Ty	**0.89** EC(100)	^{133}Xe	5.2475 d	β^-(100)
^{123m}Te	119.25 d	IT(100)	^{133m}Xe	2.19 d	IT(100)
^{123}I	13.2235 h	β^+(100)	^{133}Cs	stable	**100.**
^{123}Xe	2.08 h	β^+(100)	^{133}Ba	10.51 y	EC(100)
^{124}Sn	stable	**5.79**	^{133m}Ba	38.9 h	IT($\simeq$100) EC(0.0096)
^{124}Sb	60.20 d	β^-(100)	^{133}La	3.912 h	β^+(100)
^{124}Te	stable	**4.74**	^{133m}Ce	4.9 h	β^+(100)
^{124}I	4.1760 d	β^+(100)	^{134}Xe	stable	**10.4**
^{124}Xe	stable	**0.10**	^{134}Cs	2.0648 y	β^-(100) EC(0.0003)
^{125}Sn	9.64 d	β^-(100)	^{134m}Cs	2.903 h	IT(100)
^{125}Sb	2.7586 y	β^-(100)	^{134}Ba	stable	**2.417**
^{125}Te	stable	**7.07**	^{134}Ce	3.16 d	EC(100)
^{125m}Te	57.40 d	IT(100)	^{135}I	6.57 h	β^-(100)
^{125}I	59.400 d	EC(100)	^{135}Xe	9.14 h	β^-(100)
^{125}Xe	16.9 h	β^+(100)	^{135}Cs	2.3 My	β^-(100)
^{126}Sn	230 ky	β^-(100)	^{135}Ba	stable	**6.592**
^{126}Sb	12.35 d	β^-(100)	^{135m}Ba	28.7 h	IT(100)
^{126}Te	stable	**18.84**	^{135}La	19.5 h	β^+(100)
^{126}I	12.93 d	β^+(52.7) β^-(47.3)	^{135}Ce	17.7 h	β^+(100)
^{126}Xe	stable	**0.09**	^{136}Xe	stable	**8.87**
^{127}Sn	2.10 h	β^-(100)	^{136}Cs	13.16 d	β^-(100)
^{127}Sb	3.85 d	β^-(100)	^{136}Ba	stable	**7.854**
^{127}Te	9.35 h	β^-(100)	^{136}Ce	stable	**0.185**
^{127m}Te	109 d	IT(97.6) β^-(2.4)	^{137}Cs	30.1671 y	β^-(100)
^{127}I	stable	**100.**	^{137}Ba	stable	**11.232**
^{127}Xe	36.345 d	EC(100)	^{137}La	60 ky	EC(100)
^{127}Cs	6.25 h	β^+(100)	^{137}Ce	9.0 h	β^+(100)
^{128}Sb	9.01 h	β^-(100)	^{137m}Ce	34.4 h	IT(99.22) β^+(0.78)
^{128}Te	2.2 Yy	**31.74**	^{137}Pr	1.28 h	β^+(100)
^{128}Xe	stable	**1.92**	^{138}Ba	stable	**71.698**
^{128}Ba	2.43 d	EC(100)	^{138}La	102 Gy	**0.090** β^+(65.6) β^-(34.4)
^{129}Sb	4.40 h	β^-(100)	^{138}Ce	stable	**0.251**
^{129m}Te	33.6 d	IT(63) β^-(37)	^{138m}Pr	2.12 h	β^+(100)
^{129}I	15.7 My	β^-(100)	^{138}Nd	5.04 h	β^+(100)
^{129}Xe	stable	**26.44**	^{139}La	stable	**99.910**
^{129m}Xe	8.88 d	IT(100)	^{139}Ce	137.641 d	EC(100)
^{129}Cs	32.06 h	β^+(100)	^{139}Pr	4.41 h	β^+(100)
^{129}Ba	2.23 h	β^+(100)	^{139m}Nd	5.50 h	β^+(88.2) IT(11.8)
^{129m}Ba	2.16 h	$\beta^+(\simeq100)$	^{140}Ba	12.752 d	β^-(100)
^{130}Te	790 Ey	**34.08** $2\beta^-$(100)	^{140}La	1.6781 d	β^-(100)

Table A.4. *(cont.)* Percent isotopic abundances (bold) of naturally occurring nuclides and decay characteristics of radionuclides with $T_{1/2} > 1$ h. Branching ratios are in percent.

Nucl.	Half-life	Abund. / Decay Mode	Nucl.	Half-life	Abund. / Decay Mode
^{140}Ce	stable	**88.450**	^{152}Eu	13.537 y	$\beta^+(72.1)$ $\beta^-(27.9)$
^{140}Nd	3.37 d	EC(100)	^{152m}Eu	9.3116 h	$\beta^-(72)$ $\beta^+(28)$
^{141}La	3.92 h	$\beta^-(100)$	^{152}Gd	108 Ty	**0.20** $\alpha(100)$
^{141}Ce	32.508 d	$\beta^-(100)$	^{152}Tb	17.5 h	$\beta^+(100)$
^{141}Pr	stable	**100.**	^{152}Dy	2.38 h	EC($\simeq$100) $\alpha(0.100)$
^{141}Nd	2.49 h	$\beta^+(100)$	^{153}Sm	46.284 h	$\beta^-(100)$
^{142}Ce	stable	**11.114**	^{153}Eu	stable	**52.19**
^{142}Pr	19.12 h	$\beta^-(\simeq100)$ EC(0.0164)	^{153}Gd	240.4 d	EC(100)
^{142}Nd	stable	**27.2**	^{153}Tb	2.34 d	$\beta^+(100)$
^{143}Ce	33.039 h	$\beta^-(100)$	^{153}Dy	6.4 h	$\beta^+(\simeq100)$ $\alpha(0.0094)$
^{143}Pr	13.57 d	$\beta^-(100)$	^{154}Sm	stable	**22.75**
^{143}Nd	stable	**12.2**	^{154}Eu	8.593 y	$\beta^-(\simeq100)$ EC(0.02)
^{143}Pm	265 d	EC(100) $e^+(<5.7e\text{-}6)$	^{154}Gd	stable	**2.18**
^{144}Ce	284.91 d	$\beta^-(100)$	^{154}Tb	21.5 h	$\beta^+(\simeq100)$ $\beta^-(<0.1)$
^{144}Nd	2.29 Py	**23.80** $\alpha(100)$	^{154m}Tb	9.4 h	$\beta^+(78.2)$ IT(21.8) $\beta^-(<0.1)$
^{144}Pm	363 d	EC(100) $e^+(<8e\text{-}5)$	^{154n}Tb	22.7 h	$\beta^+(98.2)$ IT(1.8)
^{144}Sm	stable	**3.07**	^{154}Dy	3.0 My	$\alpha(100)$
^{145}Pr	5.984 h	$\beta^-(100)$	^{155}Eu	4.7611 y	$\beta^-(100)$
^{145}Nd	stable	**8.3**	^{155}Gd	stable	**14.80**
^{145}Pm	17.7 y	EC(100) $\alpha(2.8e\text{-}7)$	^{155}Tb	5.32 d	EC(100)
^{145}Sm	340 d	EC(100)	^{155}Dy	9.9 h	$\beta^+(100)$
^{145}Eu	5.93 d	$\beta^+(100)$	^{156}Sm	9.4 h	$\beta^-(100)$
^{146}Nd	stable	**17.2**	^{156}Eu	15.19 d	$\beta^-(100)$
^{146}Pm	5.53 y	EC(66.0) $\beta^-(34.0)$	^{156}Gd	stable	**20.47**
^{146}Sm	103 My	$\alpha(100)$	^{156}Tb	5.35 d	$\beta^+(100)$
^{146}Eu	4.61 d	$\beta^+(100)$	^{156m}Tb	24.4 h	IT(100)
^{146}Gd	48.27 d	EC(100)	^{156}Dy	stable	**0.06**
^{147}Nd	10.98 d	$\beta^-(100)$	^{157}Eu	15.18 h	$\beta^-(100)$
^{147}Pm	2.6234 y	$\beta^-(100)$	^{157}Gd	stable	**15.65**
^{147}Sm	106 Gy	**15.0** $\alpha(100)$	^{157}Tb	71 y	EC(100)
^{147}Eu	24.1 d	$\beta^+(\simeq100)$ $\alpha(0.0022)$	^{157}Dy	8.14 h	$\beta^+(100)$
^{147}Gd	38.06 h	$\beta^+(100)$	^{158}Gd	stable	**24.84**
^{147}Tb	1.64 h	$\beta^+(100)$	^{158}Tb	180 y	$\beta^+(83.4)$ $\beta^-(16.6)$
^{148}Nd	stable	**5.7**	^{158}Dy	stable	**0.10**
^{148}Pm	5.368 d	$\beta^-(100)$	^{158}Er	2.29 h	EC(100)
^{148m}Pm	41.29 d	$\beta^-(95.8)$ IT(4.2)	^{159}Gd	18.479 h	$\beta^-(100)$
^{148}Sm	7 Py	**11.24** $\alpha(100)$	^{159}Tb	stable	**100.**
^{148}Eu	54.5 d	$\beta^+(100)$ $\alpha(9.4e\text{-}7)$	^{159}Dy	144.4 d	EC(100)
^{148}Gd	74.6 y	$\alpha(100)$	^{160}Gd	stable	**21.86**
^{149}Nd	1.728 h	$\beta^-(100)$	^{160}Tb	72.3 d	$\beta^-(100)$
^{149}Pm	53.08 h	$\beta^-(100)$	^{160}Dy	stable	**2.34**
^{149}Sm	stable	**13.82**	^{160m}Ho	5.02 h	IT(65) $\beta^+(35)$
^{149}Eu	93.1 d	EC(100)	^{160}Er	28.58 h	EC(100)
^{149}Gd	9.28 d	$\beta^+(100)$ $\alpha(4.3e\text{-}4)$	^{161}Tb	6.906 d	$\beta^-(100)$
^{149}Tb	4.118 h	$\beta^+(83.3)$ $\alpha(16.7)$	^{161}Dy	stable	**18.91**
^{150}Nd	6.7 Ey	**5.6** $2\beta^-(100)$	^{161}Ho	2.48 h	EC(100)
^{150}Pm	2.68 h	$\beta^-(100)$	^{161}Er	3.21 h	$\beta^+(100)$
^{150}Sm	stable	**7.38**	^{162}Dy	stable	**25.51**
^{150}Eu	36.9 y	$\beta^+(100)$	^{162}Er	stable	**0.14**
^{150m}Eu	12.8 h	$\beta^-(89)$ $\beta^+(11)$	^{163}Dy	stable	**24.90**
^{150}Gd	1.79 My	$\alpha(100)$	^{163}Ho	4.570 ky	EC(100)
^{150}Tb	3.48 h	$\beta^+(\simeq100)$ $\alpha(<0.05)$	^{163}Tm	1.810 h	$\beta^+(100)$
^{151}Pm	28.40 h	$\beta^-(100)$	^{164}Dy	stable	**28.18**
^{151}Sm	90 y	$\beta^-(100)$	^{164}Er	stable	**1.61**
^{151}Eu	stable	**47.81**	^{165}Dy	2.334 h	$\beta^-(100)$
^{151}Gd	124 d	EC(100)	^{165}Ho	stable	**100.**
^{151}Tb	17.609 h	$\beta^+(\simeq100)$ $\alpha(0.0095)$	^{165}Er	10.36 h	EC(100)
^{152}Sm	stable	**26.75**	^{165}Tm	30.06 h	$\beta^+(100)$

Table A.4. *(cont.)* Percent isotopic abundances (bold) of naturally occurring nuclides and decay characteristics of radionuclides with $T_{1/2} > 1$ h. Branching ratios are in percent.

Nucl.	Half-life	Abund. / Decay Mode	Nucl.	Half-life	Abund. / Decay Mode
^{166}Dy	81.6 h	$\beta^-(100)$	^{178m}Ta	2.36 h	$\beta^+(100)$
^{166}Ho	26.83 h	$\beta^-(100)$	^{178}W	21.6 d	EC(100)
^{166m}Ho	1.20 ky	$\beta^-(100)$	^{179}Lu	4.59 h	$\beta^-(100)$
^{166}Er	stable	**33.61**	^{179}Hf	stable	**13.62**
^{166}Tm	7.70 h	$\beta^+(100)$	^{179n}Hf	25.05 d	IT(100)
^{166}Yb	56.7 h	EC(100)	^{179}Ta	1.82 y	EC(100)
^{167}Ho	3.1 h	$\beta^-(100)$	^{180}Hf	stable	**35.08**
^{167}Er	stable	**22.93**	^{180m}Hf	5.5 h	IT($\simeq$100) $\beta^-(0.3)$
^{167}Tm	9.25 d	EC(100)	^{180}Ta	8.152 h	EC(86) $\beta^-(14)$
^{168}Er	stable	**26.78**	^{180m}Ta	stable	**0.012**
^{168}Tm	93.1 d	$\beta^+(\simeq100)$ $\beta^-(0.010)$	^{180}W	stable	**0.12**
^{168}Yb	stable	**0.13**	^{181}Hf	42.39 d	$\beta^-(100)$
^{169}Er	9.40 d	$\beta^-(100)$	^{181}Ta	stable	**99.988**
^{169}Tm	stable	**100.**	^{181}W	121.2 d	EC(100)
^{169}Yb	32.026 d	EC(100)	^{181}Re	19.9 h	$\beta^+(100)$
^{169}Lu	34.06 h	$\beta^+(100)$	^{182}Hf	9 My	$\beta^-(100)$
^{170}Er	stable	**14.93**	^{182}Ta	114.43 d	$\beta^-(100)$
^{170}Tm	128.6 d	$\beta^-(\simeq100)$ EC(0.131)	^{182}W	stable	**26.50**
^{170}Yb	stable	**3.04**	^{182}Re	64.0 h	$\beta^+(100)$
^{170}Lu	2.012 d	$\beta^+(100)$	^{182m}Re	12.7 h	$\beta^+(100)$
^{170}Hf	16.01 h	EC(100)	^{182}Os	22.10 h	EC(100)
^{171}Er	7.516 h	$\beta^-(100)$	^{183}Hf	1.067 h	$\beta^-(100)$
^{171}Tm	1.92 y	$\beta^-(100)$	^{183}Ta	5.1 d	$\beta^-(100)$
^{171}Yb	stable	**14.28**	^{183}W	stable	**14.31**
^{171}Lu	8.24 d	$\beta^+(100)$	^{183}Re	70.0 d	EC(100)
^{171}Hf	12.1 h	$\beta^+(100)$	^{183}Os	13.0 h	$\beta^+(100)$
^{172}Er	49.3 h	$\beta^-(100)$	^{183m}Os	9.9 h	$\beta^+(85)$ IT(15)
^{172}Tm	63.6 h	$\beta^-(100)$	^{184}Hf	4.12 h	$\beta^-(100)$
^{172}Yb	stable	**21.83**	^{184}Ta	8.7 h	$\beta^-(100)$
^{172}Lu	6.70 d	$\beta^+(100)$	^{184}W	stable	**30.64**
^{172}Hf	1.87 y	EC(100)	^{184}Re	38.0 d	$\beta^+(100)$
^{173}Tm	8.24 h	$\beta^-(100)$	^{184m}Re	169 d	IT(75.4) EC(24.6)
^{173}Yb	stable	**16.13**	^{184}Os	stable	**0.020**
^{173}Lu	1.37 y	EC(100)	^{184}Ir	3.09 h	$\beta^+(100)$
^{173}Hf	23.6 h	$\beta^+(100)$	^{185}W	75.1 d	$\beta^-(100)$
^{173}Ta	3.14 h	$\beta^+(100)$	^{185}Re	stable	**37.40**
^{174}Yb	stable	**31.83**	^{185}Os	93.6 d	EC(100)
^{174}Lu	3.31 y	$\beta^+(100)$	^{185}Ir	14.4 h	$\beta^+(100)$
^{174m}Lu	142 d	IT(99.38) EC(0.62)	^{186}W	stable	**28.43**
^{174}Hf	2.0 Py	**0.162** $\alpha(100)$	^{186}Re	3.7183 d	$\beta^-(92.53)$ EC(7.47)
^{174}Ta	1.14 h	$\beta^+(100)$	^{186}Os	2.0 Py	**1.59** $\alpha(100)$
^{175}Yb	4.185 d	$\beta^-(100)$	^{186}Ir	16.64 h	$\beta^+(100)$
^{175}Lu	stable	**97.41**	^{186m}Ir	1.92 h	$\beta^+(\simeq75)$IT($\simeq25$)
^{175}Hf	70 d	EC(100)	^{186}Pt	2.08 h	$\beta^+(100)$
^{175}Ta	10.5 h	$\beta^+(100)$	^{187}W	23.72 h	$\beta^-(100)$
^{176}Yb	stable	**12.7**	^{187}Re	41.2 Gy	**62.60** $\beta^-(100)$
^{176}Lu	38.5 Gy	**2.59** $\beta^-(100)$	^{187}Os	stable	**1.96**
^{176m}Lu	3.664 h	$\beta^-(\simeq100)$ EC(0.095)	^{187}Ir	10.5 h	$\beta^+(100)$
^{176}Hf	stable	**5.26**	^{187}Pt	2.35 h	$\beta^+(100)$
^{176}Ta	8.09 h	$\beta^+(100)$	^{188}W	69.78 d	$\beta^-(100)$
^{176}W	2.5 h	EC(100)	^{188}Re	17.0040 h	$\beta^-(100)$
^{177}Yb	1.911 h	$\beta^-(100)$	^{188}Os	stable	**13.24**
^{177}Lu	6.647 d	$\beta^-(100)$	^{188}Ir	41.5 h	$\beta^+(100)$
^{177m}Lu	160.44 d	$\beta^-(78.6)$ IT(21.4)	^{188}Pt	10.2 d	EC(100)
^{177}Hf	stable	**18.60**	^{189}Re	24.3 h	$\beta^-(100)$
^{177}Ta	56.56 h	$\beta^+(100)$	^{189}Os	stable	**16.15**
^{178}Hf	stable	**27.28**	^{189m}Os	5.8 h	IT(100)
^{178n}Hf	31 y	IT(100)	^{189}Ir	13.2 d	EC(100)

Table A.4. *(cont.)* Percent isotopic abundances (bold) of naturally occurring nuclides and decay characteristics of radionuclides with $T_{1/2} > 1$ h. Branching ratios are in percent.

Nucl.	Half-life	Abund. / Decay Mode	Nucl.	Half-life	Abund. / Decay Mode
^{189}Pt	10.87 h	$\beta^+(100)$	^{198}Pb	2.40 h	$\beta^+(100)$
^{190m}Re	3.2 h	$\beta^-(54.4)$ IT(?)	^{199}Au	3.139 d	$\beta^-(100)$
^{190}Os	stable	**26.4**	^{199}Hg	stable	**16.87**
^{190}Ir	11.78 d	$\beta^+(100)$	^{199}Tl	7.42 h	$\beta^+(100)$
^{190m}Ir	1.120 h	IT(100)	^{200}Pt	12.5 h	$\beta^-(100)$
^{190n}Ir	3.087 h	$\beta^+(91.4)$ IT(8.6)	^{200m}Au	18.7 h	$\beta^-(82)$ IT(18)
^{190}Pt	650 Gy	**0.014** $\alpha(100)$	^{200}Hg	stable	**23.10**
^{191}Os	15.4 d	$\beta^-(100)$	^{200}Tl	26.1 h	$\beta^+(100)$
^{191m}Os	13.10 h	IT(100)	^{200}Pb	21.5 h	EC(100)
^{191}Ir	stable	**37.3**	^{201}Hg	stable	**13.18**
^{191}Pt	2.802 d	EC(100)	^{201}Tl	72.912 h	EC(100)
^{191}Au	3.18 h	$\beta^+(100)$	^{201}Pb	9.33 h	$\beta^+(100)$
^{192}Os	stable	**40.78**	^{202}Pt	44 h	$\beta^-(100)$
^{192}Ir	73.831 d	$\beta^-(95.13)$ EC(4.87)	^{202}Hg	stable	**29.86**
^{192n}Ir	241 y	IT(100)	^{202}Tl	12.23 d	$\beta^+(100)$
^{192}Pt	stable	**0.782**	^{202}Pb	52.5 ky	EC($\simeq$100)
^{192}Au	4.94 h	$\beta^+(100)$	^{202m}Pb	3.53 h	IT(90.5) EC(9.5)
^{192}Hg	4.85 h	EC(100)	^{202}Bi	1.72 h	$\beta^+(100)$
^{193}Os	30.11 h	$\beta^-(100)$	^{203}Hg	46.612 d	$\beta^-(100)$
^{193}Ir	stable	**62.7**	^{203}Tl	stable	**29.524**
^{193m}Ir	10.53 d	IT(100)	^{203}Pb	51.873 h	EC(100)
^{193}Pt	50 y	EC(100)	^{203}Bi	11.76 h	$\beta^+(100)$
^{193m}Pt	4.33 d	IT(100)	^{204}Hg	stable	**6.87**
^{193}Au	17.65 h	$\beta^+(100)$	^{204}Tl	3.78 y	$\beta^-(97.10)$ EC(2.90)
^{193}Hg	3.80 h	$\beta^+(100)$	^{204}Pb	stable	**1.4**
^{193m}Hg	11.8 h	$\beta^+(92.8)$ IT(7.2)	^{204}Bi	11.22 h	$\beta^+(100)$
^{194}Os	6.0 y	$\beta^-(100)$	^{204}Po	3.53 h	$\beta^+(99.34)$ $\alpha(0.66)$
^{194}Ir	19.28 h	$\beta^-(100)$	^{205}Tl	stable	**70.476**
^{194n}Ir	171 d	$\beta^-(100)$	^{205}Pb	15.3 My	EC(100)
^{194}Pt	stable	**32.67**	^{205}Bi	15.31 d	$\beta^+(100)$
^{194}Au	38.02 h	$\beta^+(100)$	^{205}Po	1.66 h	$\beta^+(\simeq$100) $\alpha(0.04)$
^{194}Hg	440 y	EC(100)	^{206}Pb	stable	**24.1**
^{195}Ir	2.5 h	$\beta^-(100)$	^{206}Bi	6.243 d	$\beta^+(100)$
^{195m}Ir	3.8 h	$\beta^-(95)$ IT(5)	^{206}Po	8.8 d	$\beta^+(94.55)$ $\alpha(5.45)$
^{195}Pt	stable	**33.832**	^{207}Pb	stable	**22.1**
^{195m}Pt	4.02 d	IT(100)	^{207}Bi	32.9 y	$\beta^+(100)$
^{195}Au	186.10 d	EC(100)	^{207}Po	5.80 h	$\beta^+(\simeq$100) $\alpha(0.021)$
^{195}Hg	10.53 h	$\beta^+(100)$	^{207}At	1.80 h	$\beta^+(91.4)$ $\alpha(8.6)$
^{195m}Hg	41.6 h	IT(54.2) $\beta^+(45.8)$	^{208}Pb	stable	**52.4**
^{195}Tl	1.16 h	$\beta^+(100)$	^{208}Bi	368 ky	$\beta^+(100)$
^{196m}Ir	1.40 h	$\beta^-(\simeq$100) IT(<0.3)	^{208}Po	2.898 y	$\alpha(\simeq$100) $\beta^+(0.00223)$
^{196}Pt	stable	**25.242**	^{208}At	1.63 h	$\beta^+(99.45)$ $\alpha(0.55)$
^{196}Au	6.1669 d	$\beta^+(92.8)$ $\beta^-(7.2)$	^{209}Pb	3.253 h	$\beta^-(100)$
^{196n}Au	9.6 h	IT(100)	^{209}Bi	19 Ey	**100** $\alpha(100)$
^{196}Hg	stable	**0.15**	^{209}Po	102 y	$\alpha(\simeq$100) $\beta^+(0.48)$
^{196}Tl	1.84 h	$\beta^+(100)$	^{209}At	5.41 h	$\beta^+(95.9\ 5)$ $\alpha(4.1)$
^{196m}Tl	1.41 h	$\beta^+(95.5)$ IT(4.5)	^{210}Pb	22.2 y	$\beta^-(100)$
^{197}Pt	19.8915 h	$\beta^-(100)$	^{210}Bi	5.012 d	$\beta^-(100)$
^{197}Au	stable	**100.**	^{210m}Bi	3.04 My	$\alpha(100)$
^{197}Hg	64.94 h	EC(100)	^{210}Po	138.376 d	$\alpha(100)$
^{197m}Hg	23.8 h	IT(91.4) EC(8.6)	^{210}At	8.1 h	$\beta^+(\simeq$100) $\alpha(0.175)$
^{197}Tl	2.84 h	$\beta^+(100)$	^{210}Rn	2.4 h	$\alpha(96)$ $\beta^+(?)$
^{198}Pt	stable	**7.163**	^{211}At	7.214 h	EC(58.20) $\alpha(41.80)$
^{198}Au	2.69517 d	$\beta^-(100)$	^{211}Rn	14.6 h	$\beta^+(72.6)$ $\alpha(27.4)$
^{198m}Au	2.27 d	IT(100)	^{212}Pb	10.64 h	$\beta^-(100)$
^{198}Hg	stable	**9.97**	^{222}Rn	3.8235 d	$\alpha(100)$
^{198}Tl	5.3 h	$\beta^+(100)$	^{223}Ra	11.43 d	$\alpha(100)$
^{198m}Tl	1.87 h	$\beta^+(54)$ IT(46)	^{224}Ra	3.66 d	$\alpha(100)$

Table A.4. *(cont.)* Percent isotopic abundances (bold) of naturally occurring nuclides and decay characteristics of radionuclides with $T_{1/2} > 1$ h. Branching ratios are in percent.

Nucl.	Half-life	Abund. / Decay Mode	Nucl.	Half-life	Abund. / Decay Mode
^{224}Ac	2.78 h	$\beta^+(90.6)$ $\alpha(9.4)$ $\beta^-(<8.4)$	^{242m}Am	141 y	IT($\simeq$100) $\alpha(0.45)$
^{225}Ra	14.9 d	$\beta^-(100)$			SF($<$4.7e-9)
^{225}Ac	10.0 d	$\alpha(100)$	^{242}Cm	162.8 d	$\alpha(100)$ SF(6.2e-6)
^{226}Ra	1.600 ky	$\alpha(100)$	^{243}Pu	4.956 h	$\beta^-(100)$
^{226}Ac	29.37 h	$\beta^-(83)$ EC(17) $\alpha(0.006)$	^{243}Am	7.37 ky	$\alpha(100)$ SF(3.7e-9)
^{227}Ac	21.772 y	$\beta^-(98.62)$ $\alpha(1.38)$	^{243}Cm	29.1 y	$\alpha(\simeq$100) EC(0.29)
^{227}Th	18.68 d	$\alpha(100)$			SF(5.3e-9)
^{228}Ra	5.75 y	$\beta^-(100)$	^{243}Bk	4.5 h	$\beta^+(\simeq$100) $\alpha(\simeq$0.15)
^{228}Ac	6.15 h	$\beta^-(100)$	^{244}Pu	80.0 My	$\alpha(\simeq$100) SF(0.121)
^{228}Th	1.9116 y	$\alpha(100)$	^{244}Am	10.1 h	$\beta^-(100)$
^{228}Pa	22 h	$\beta^+(98.0)$ $\alpha(2.0)$	^{244}Cm	18.10 y	$\alpha(100)$ SF(1.347e-4)
^{229}Th	7.34 ky	$\alpha(100)$	^{244}Bk	4.35 h	$\beta^+(?)$ $\alpha(0.006)$
^{229}Pa	1.50 d	EC($\simeq$100) $\alpha(0.48)$	^{245}Pu	10.5 h	$\beta^-(100)$
^{230}Th	75.38 ky	$\alpha(100)$	^{245}Am	2.05 h	$\beta^-(100)$
^{230}Pa	17.4 d	$\beta^+(91.6)$ $\beta^-(8.4)$ $\alpha(0.0032)$	^{245}Cm	8.5 ky	$\alpha(100)$ SF(6.1e-7)
^{230}U	20.8 d	$\alpha(100)$ SF($<$1.4e-10)	^{245}Bk	4.94 d	EC($\simeq$100) $\alpha(0.12)$
^{231}Th	25.52 h	$\beta^-(100)$	^{246}Pu	10.84 d	$\beta^-(100)$
^{231}Pa	32.76 ky	SF($<$3e-10) $\alpha(100)$	^{246}Cm	4.73 ky	$\alpha(\simeq$100) SF(0.02615)
^{231}U	4.2 d	EC($\simeq$100) $\alpha(0.004)$	^{246}Bk	1.80 d	$\beta^+(\simeq$100) $\alpha(<$0.1)
^{232}Th	14.05 Gy	**100.** $\alpha(100)$ SF(11e-10)	^{246}Cf	35.7 h	$\alpha(100)$ SF(2.5e-4)
^{232}Pa	1.31 d	$\beta^-(\simeq$100) EC(0.003)	^{247}Pu	2.27 d	$\beta^-(100)$
^{232}U	68.9 y	$\alpha(100)$	^{247}Cm	15.6 My	$\alpha(100)$
^{233}Pa	26.967 d	$\beta^-(100)$	^{247}Bk	1.38 ky	$\alpha(\simeq$100) SF(?)
^{233}U	159.2 ky	$\alpha(100)$ SF($<$6e-9)	^{247}Cf	3.11 h	EC($\simeq$100) $\alpha(0.035)$
^{234}Th	24.10 d	$\beta^-(100)$	^{248}Cm	348 ky	$\alpha(91.61)$ SF(8.39)
^{234}Pa	6.70 h	$\beta^-(100)$ SF($<$3e-10)	^{248m}Bk	23.7 h	$\beta^-(70)$ EC(30) $\alpha(0.001)$
^{234}U	245.5 ky	**0.0055** $\alpha(100)$ SF(1.73e-9)	^{248}Cf	334 d	$\alpha(100)$ SF(0.0029)
^{234}Np	4.4 d	$\beta^+(100)$	^{249}Bk	330 d	$\beta^-(\simeq$100) $\alpha(0.00145)$
^{234}Pu	8.8 h	EC($\simeq$94) $\alpha(\simeq$6)			SF(47e-9)
^{235}U	704 My	**0.720** $\alpha(100)$ SF(7e-9)	^{249}Cf	351 y	$\alpha(100)$ SF(5.0e-7)
^{235}Np	396.1 d	EC($\simeq$100) $\alpha(0.00260)$	^{250}Cm	8300 y	SF($\simeq$74) $\alpha(\simeq$18) $\beta^-(\simeq$8)
^{236}U	23.42 My	$\alpha(100)$ SF(9.6e-8)	^{250}Bk	3.212 h	$\beta^-(100)$
^{236}Np	154 ky	EC(87.3) $\beta^-(12.5)$ $\alpha(0.16)$	^{250}Cf	13.08 y	$\alpha(\simeq$100) SF(0.077)
^{236m}Np	22.5 h	EC(52) $\beta^-(48)$	^{250}Es	8.6 h	$\beta^+(>$97) $\alpha(?)$
^{236}Pu	2.858 y	$\alpha(100)$ SF(1.36e-7)	^{250m}Es	2.22 h	EC($>$99) $\alpha(?)$
^{237}U	6.75 d	$\beta^-(100)$	^{251}Cf	900 y	$\alpha(100)$ SF(?)
^{237}Np	2.144 My	SF($<$3e-10) $\alpha(100)$	^{251}Es	33 h	EC(?) $\alpha(0.5)$
^{237}Pu	45.2 d	EC($\simeq$100) $\alpha(0.0042)$	^{251}Fm	5.30 h	$\beta^+(98.20)$ $\alpha(1.80)$
^{238}U	4.468 Gy	**99.2745** $\alpha(100)$ SF(5.45e-5)	^{252}Cm	$<$1 d	$\beta^-(?)$
^{238}Np	2.117 d	$\beta^-(100)$	^{252}Cf	2.645 y	$\alpha(96.908)$ SF(3.092)
^{238}Pu	87.7 y	$\alpha(100)$ SF(1.9e-7)	^{252}Es	471.7 d	$\alpha(78)$ EC(22) $\beta^-(\simeq$0.01)
^{238}Cm	2.4 h	EC(?) $\alpha(<$10)	^{252}Fm	25.39 h	$\alpha(\simeq$100) SF(0.0023)
^{239}Np	2.3565 d	$\beta^-(100)$ $\alpha(5e-10)$	^{253}Cf	17.81 d	$\beta^-(\simeq$100) $\alpha(0.31)$
^{239}Pu	24.11 ky	$\alpha(100)$ SF(3.1e-10)	^{253}Es	20.47 d	$\alpha(100)$ SF(8.7e-6)
^{239}U	11.9 h	EC($\simeq$100) $\alpha(0.010)$	^{253}Fm	3.00 d	EC(88) $\alpha(12)$
^{239}Cm	2.9 h	$\beta^+(\simeq$100) $\alpha(<$0.1)	^{254}Cf	60.5 d	SF($\simeq$100) $\alpha(0.31)$
^{240}U	14.1 h	$\beta^-(100)$ $\alpha(<$1e-10)	^{254}Es	275.7 d	$\alpha(100)$ EC(0.03) SF(3e-6)
^{240}Pu	6.564 ky	$\alpha(100)$ SF(5.7e-6)			$\beta^-(1.74$e-4)
^{240}Am	50.8 h	$\beta^+(100)$ $\alpha(\simeq$1.9e-4)	^{254m}Es	39.3 h	$\beta^-(98)$ IT($<$3) $\alpha(0.32)$
^{240}Cm	27 d	$\alpha(\simeq$100) EC($<$0.5)			EC(0.076) SF($<$0.045)
		SF(3.9e-6)	^{254}Fm	3.240 h	$\alpha(\simeq$100) SF(0.0592)
^{241}Pu	14.35 y	$\beta^-(\simeq$100) $\alpha(0.00245)$	^{255}Es	39.8 d	$\beta^-(92)$ $\alpha(8.0)$ SF(0.0041)
		SF($<$2.4e-14)	^{255}Fm	20.07 h	$\alpha(100)$ SF(2.4e-5)
^{241}Am	432.2 y	$\alpha(100)$ SF(4.3e-10)	^{256m}Es	7.6 h	$\beta^-(\simeq$100) SF(0.002)
^{241}Cm	32.8 d	EC(99.0) $\alpha(1.0)$	^{257}Fm	100.5 d	$\alpha(\simeq$100) SF(0.210)
^{242}Pu	375 ky	$\alpha(100)$ SF(5.50e-4)	^{257}Md	5.52 h	EC(85) $\alpha(15)$ SF($<$4)
^{242}Am	16.02 h	$\beta^-(82.7)$ EC(17.3)	^{258}Md	51.50 d	$\alpha(\simeq$100) $\beta\pm(<$0.0015)

Appendix B

Atomic Mass Table

Atomic Mass Tables

In this appendix a series of tables of atomic masses is given. These data are from the "The 1995 update to the atomic mass evaluation" by G. Audi and A.H. Wapstra, *Nuclear Physics*, A595, Vol. 4 p. 409–480, December 25, 1995. Revisions are issued periodically, e.g., G. Audi and A.H. Wapstra, *Nuclear Physics*, A729, 337-376 (2003). The latest data as well as searching programs are available at http://ie.lbl.gov/toimass.html.

Table B.1. Atomic mass tables

N	Z	A	El	Atomic Mass (μu)
1	0	1	n	1 008664.9233
0	1		H	1 007825.0321
1	1	2	H	2 014101.7780
2	1	3	H	3 016049.2675
1	2		He	3 016029.3097
3	1	4	H	4 027830
2	2		He	4 002603.2497
1	3		Li	4 027180
4	1	5	H	5 039540
3	2		He	5 012220
2	3		Li	5 012540
1	4		Be	5 040790
5	1	6	H	6 044940
4	2		He	6 018888.1
3	3		Li	6 015122.3
2	4		Be	6 019726
5	2	7	He	7 028030
4	3		Li	7 016004.0
3	4		Be	7 016929.2
2	5		B	7 029920
6	2	8	He	8 033922
5	3		Li	8 022486.7
4	4		Be	8 005305.09
3	5		B	8 024606.7
2	6		C	8 037675
7	2	9	He	9 043820
6	3		Li	9 026789.1
5	4		Be	9 012182.1
4	5		B	9 013328.8
3	6		C	9 031040.1
8	2	10	He	10 052400
7	3		Li	10 035481
6	4		Be	10 013533.7
5	5		B	10 012937.0
4	6		C	10 016853.1
3	7		N	10 042620
8	3	11	Li	11 043796
7	4		Be	11 021658
6	5		B	11 009305.5
5	6		C	11 011433.8
4	7		N	11 026800
9	3	12	Li	12 053780
8	4		Be	12 026921
7	5		B	12 014352.1
6	6		C	12 000000.0
5	7		N	12 018613.2
4	8		O	12 034405
9	4	13	Be	13 036130
8	5		B	13 017780.3
7	6		C	13 003354.8378
6	7		N	13 005738.58
5	8		O	13 024810
10	4	14	Be	14 042820
9	5		B	14 025404
8	6		C	14 003241.988
7	7		N	14 003074.0052
6	8		O	14 008595.29
5	9		F	14 036080
10	5	15	B	15 031097
9	6		C	15 010599.3
8	7		N	15 000108.8984
7	8		O	15 003065.4
6	9		F	15 018010
11	5	16	B	16 039810
10	6		C	16 014701
9	7		N	16 006101.4
8	8		O	15 994914.6221
7	9		F	16 011466
6	10		Ne	16 025757
12	5	17	B	17 046930
11	6		C	17 022584
10	7		N	17 008450
9	8		O	16 999131.50
8	9		F	17 002095.24
7	10		Ne	17 017700
13	5	18	B	18 056170
12	6		C	18 026760
11	7		N	18 014082
10	8		O	17 999160.4
9	9		F	18 000937.7
8	10		Ne	18 005697.1
7	11		Na	18 027180
14	5	19	B	19 063730
13	6		C	19 035250
12	7		N	19 017027
11	8		O	19 003579
10	9		F	18 998403.20
9	10		Ne	19 001879.8
8	11		Na	19 013879
14	6	20	C	20 040320
13	7		N	20 023370
12	8		O	20 004076.2
11	9		F	19 999981.32
10	10		Ne	19 992440.1759
9	11		Na	20 007348
8	12		Mg	20 018863
15	6	21	C	21 049340
14	7		N	21 027090
13	8		O	21 008655
12	9		F	20 999948.9
11	10		Ne	20 993846.74
10	11		Na	20 997655.1
9	12		Mg	21 011714
8	13		Al	21 028040
16	6	22	C	22 056450
15	7		N	22 034440
14	8		O	22 009970
13	9		F	22 002999
12	10		Ne	21 991385.51
11	11		Na	21 994436.8
10	12		Mg	21 999574.1
9	13		Al	22 019520
8	14		Si	22 034530
16	7	23	N	23 040510
15	8		O	23 015690
14	9		F	23 003570
13	10		Ne	22 994467.34
12	11		Na	22 989769.67
11	12		Mg	22 994124.9
10	13		Al	23 007265
9	14		Si	23 025520
17	7	24	N	24 050500
16	8		O	24 020370
15	9		F	24 008100
14	10		Ne	23 993615
13	11		Na	23 990963.33
12	12		Mg	23 985041.90
11	13		Al	23 999941
10	14		Si	24 011546
9	15		P	24 034350
17	8	25	O	25 029140
16	9		F	25 012090
15	10		Ne	24 997790
14	11		Na	24 989954.4
13	12		Mg	24 985837.02
12	13		Al	24 990428.6
11	14		Si	25 004107
10	15		P	25 020260
18	8	26	O	26 037750
17	9		F	26 019630
16	10		Ne	26 000460
15	11		Na	25 992590
14	12		Mg	25 982593.04
13	13		Al	25 986891.66
12	14		Si	25 992330
11	15		P	26 011780
10	16		S	26 027880
18	9	27	F	27 026890
17	10		Ne	27 007620
16	11		Na	26 994010
15	12		Mg	26 984340.74
14	13		Al	26 981538.44
13	14		Si	26 986704.76
12	15		P	26 999190
11	16		S	27 018800
19	9	28	F	28 035670
18	10		Ne	28 012110
17	11		Na	27 998890
16	12		Mg	27 983876.7
15	13		Al	27 981910.18
14	14		Si	27 976926.5327
13	15		P	27 992312
12	16		S	28 004370
11	17		Cl	28 028510
20	9	29	F	29 043260
19	10		Ne	29 019350
18	11		Na	29 002810
17	12		Mg	28 988550
16	13		Al	28 980444.8
15	14		Si	28 976494.72
14	15		P	28 981801.4
13	16		S	28 996610
12	17		Cl	29 014110
20	10	30	Ne	30 023870

Table B.1. Atomic mass tables *(cont.)*

N	Z	A	El	Atomic Mass (μu)	N	Z	A	El	Atomic Mass (μu)	N	Z	A	El	Atomic Mass (μu)
19	11		Na	30 009230	22	14		Si	35 986690	28	14	42	Si	42 016100
18	12		Mg	29 990460	21	15		P	35 978260	27	15		P	42 000090
17	13		Al	29 982960	20	16		S	35 967080.88	26	16		S	41 981490
16	14		Si	29 973770.22	19	17		Cl	35 968306.95	25	17		Cl	41 973170
15	15		P	29 978313.8	18	18		Ar	35 967546.28	24	18		Ar	41 963050
14	16		S	29 984903	17	19		K	35 981293	23	19		K	41 962403.1
13	17		Cl	30 004770	16	20		Ca	35 993090	22	20		Ca	41 958618.3
12	18		Ar	30 021560	15	21		Sc	36 014920	21	21		Sc	41 965516.8
21	10	31	Ne	31 033110	25	12	37	Mg	37 031240	20	22		Ti	41 973032
20	11		Na	31 013600	24	13		Al	37 010310	19	23		V	41 991230
19	12		Mg	30 996550	23	14		Si	36 993000	18	24		Cr	42 006430
18	13		Al	30 983946	22	15		P	36 979610	28	15	43	P	43 003310
17	14		Si	30 975363.27	21	16		S	36 971125.72	27	16		S	42 986600
16	15		P	30 973761.51	20	17		Cl	36 965902.60	26	17		Cl	42 974200
15	16		S	30 979554.4	19	18		Ar	36 966775.9	25	18		Ar	42 965670
14	17		Cl	30 992420	18	19		K	36 973376.91	24	19		K	42 960716
13	18		Ar	31 012130	17	20		Ca	36 985872	23	20		Ca	42 958766.8
22	10	32	Ne	32 039910	16	21		Sc	37 003050	22	21		Sc	42 961151.0
21	11		Na	32 019650	25	13	38	Al	38 016900	21	22		Ti	42 968523
20	12		Mg	31 999150	24	14		Si	37 995980	20	23		V	42 980650
19	13		Al	31 988120	23	15		P	37 984470	19	24		Cr	42 997710
18	14		Si	31 974148.1	22	16		S	37 971163	29	15	44	P	44 009880
17	15		P	31 973907.16	21	17		Cl	37 968010.55	28	16		S	43 988320
16	16		S	31 972070.69	20	18		Ar	37 962732.2	27	17		Cl	43 978540
15	17		Cl	31 985689	19	19		K	37 969080.1	26	18		Ar	43 965365
14	18		Ar	31 997660	18	20		Ca	37 976319	25	19		K	43 961560
13	19		K	32 021920	17	21		Sc	37 994700	24	20		Ca	43 955481.1
22	11	33	Na	33 027390	16	22		Ti	38 009770	23	21		Sc	43 959403.0
21	12		Mg	33 005590	26	13	39	Al	39 021900	22	22		Ti	43 959690.2
20	13		Al	32 990870	25	14		Si	39 002300	21	23		V	43 974400
19	14		Si	32 978001	24	15		P	38 986420	20	24		Cr	43 985470
18	15		P	32 971725.3	23	16		S	38 975140	19	25		Mn	44 006870
17	16		S	32 971458.50	22	17		Cl	38 968007.7	30	15	45	P	45 015140
16	17		Cl	32 977451.8	21	18		Ar	38 964313	29	16		S	44 994820
15	18		Ar	32 989930	20	19		K	38 963706.9	28	17		Cl	44 979700
14	19		K	33 007260	19	20		Ca	38 970717.7	27	18		Ar	44 968090
23	11	34	Na	34 034900	18	21		Sc	38 984790	26	19		K	44 960700
22	12		Mg	34 009070	17	22		Ti	39 001320	25	20		Ca	44 956185.9
21	13		Al	33 996930	26	14	40	Si	40 005800	24	21		Sc	44 955910.2
20	14		Si	33 978576	25	15		P	39 991050	23	22		Ti	44 958124.3
19	15		P	33 973636	24	16		S	39 975470	22	23		V	44 965782
18	16		S	33 967866.83	23	17		Cl	39 970420	21	24		Cr	44 979160
17	17		Cl	33 973761.97	22	18		Ar	39 962383.123	20	25		Mn	44 994510
16	18		Ar	33 980270	21	19		K	39 963998.67	19	26		Fe	45 014560
15	19		K	33 998410	20	20		Ca	39 962591.2	31	15	46	P	46 023830
14	20		Ca	34 014120	19	21		Sc	39 977964	30	16		S	45 999570
24	11	35	Na	35 044180	18	22		Ti	39 990500	29	17		Cl	45 984120
23	12		Mg	35 017490	17	23		V	40 011090	28	18		Ar	45 968090
22	13		Al	34 999940	27	14	41	Si	41 012700	27	19		K	45 961976
21	14		Si	34 984580	26	15		P	40 994800	26	20		Ca	45 953692.8
20	15		P	34 973314.2	25	16		S	40 980030	25	21		Sc	45 955170.3
19	16		S	34 969032.14	24	17		Cl	40 970650	24	22		Ti	45 952629.5
18	17		Cl	34 968852.71	23	18		Ar	40 964500.8	23	23		V	45 960199.5
17	18		Ar	34 975256.7	22	19		K	40 961825.97	22	24		Cr	45 968362
16	19		K	34 988012	21	20		Ca	40 962278.3	21	25		Mn	45 986720
15	20		Ca	35 004770	20	21		Sc	40 969251.3	20	26		Fe	46 000810
24	12	36	Mg	36 022450	19	22		Ti	40 983130	31	16	47	S	47 007620
23	13		Al	36 006350	18	23		V	40 999740	30	17		Cl	46 987950

Table B.1. Atomic mass tables *(cont.)*

N	Z	A	El	Atomic Mass (μu)	N	Z	A	El	Atomic Mass (μu)	N	Z	A	El	Atomic Mass (μu)
29	18		Ar	46 972190	33	19		K	51 982610	25	31		Ga	55 994910
28	19		K	46 961678	32	20		Ca	51 965100	37	20	57	Ca	56 992360
27	20		Ca	46 954546.5	31	21		Sc	51 956650	36	21		Sc	56 977040
26	21		Sc	46 952408.0	30	22		Ti	51 946898	35	22		Ti	56 962900
25	22		Ti	46 951763.8	29	23		V	51 944779.7	34	23		V	56 952360
24	23		V	46 954906.9	28	24		Cr	51 940511.9	33	24		Cr	56 943750
23	24		Cr	46 962907	27	25		Mn	51 945570.1	32	25		Mn	56 938287
22	25		Mn	46 976100	26	26		Fe	51 948117	31	26		Fe	56 935398.7
21	26		Fe	46 992890	25	27		Co	51 963590	30	27		Co	56 936296.2
32	16	48	S	48 012990	24	28		Ni	51 975680	29	28		Ni	56 939800
31	17		Cl	47 994850	23	29		Cu	51 997180	28	29		Cu	56 949216
30	18		Ar	47 975070	35	18	53	Ar	53 006230	27	30		Zn	56 964910
29	19		K	47 965513	34	19		K	52 987120	26	31		Ga	56 982930
28	20		Ca	47 952534	33	20		Ca	52 970050	37	21	58	Sc	57 983070
27	21		Sc	47 952235	32	21		Sc	52 959240	36	22		Ti	57 966110
26	22		Ti	47 947947.1	31	22		Ti	52 949730	35	23		V	57 956650
25	23		V	47 952254.5	30	23		V	52 944343	34	24		Cr	57 944250
24	24		Cr	47 954036	29	24		Cr	52 940653.8	33	25		Mn	57 939990
23	25		Mn	47 968870	28	25		Mn	52 941294.7	32	26		Fe	57 933280.5
22	26		Fe	47 980560	27	26		Fe	52 945312.3	31	27		Co	57 935757.6
21	27		Co	48 001760	26	27		Co	52 954225	30	28		Ni	57 935347.9
33	16	49	S	49 022010	25	28		Ni	52 968460	29	29		Cu	57 944540.7
32	17		Cl	48 999890	24	29		Cu	52 985550	28	30		Zn	57 954600
31	18		Ar	48 982180	35	19	54	K	53 993990	27	31		Ga	57 974250
30	19		K	48 967450	34	20		Ca	53 974680	26	32		Ge	57 991010
29	20		Ca	48 955673	33	21		Sc	53 963000	38	21	59	Sc	58 988040
28	21		Sc	48 950024	32	22		Ti	53 950870	37	22		Ti	58 971960
27	22		Ti	48 947870.8	31	23		V	53 946444	36	23		V	58 959300
26	23		V	48 948516.9	30	24		Cr	53 938884.9	35	24		Cr	58 948630
25	24		Cr	48 951341.1	29	25		Mn	53 940363.2	34	25		Mn	58 940450
24	25		Mn	48 959623	28	26		Fe	53 939614.8	33	26		Fe	58 934880.5
23	26		Fe	48 973610	27	27		Co	53 948464.1	32	27		Co	58 933200.2
22	27		Co	48 989720	26	28		Ni	53 957910	31	28		Ni	58 934351.6
33	17	50	Cl	50 007730	25	29		Cu	53 976710	30	29		Cu	58 939504.1
32	18		Ar	49 985940	24	30		Zn	53 992950	29	30		Zn	58 949270
31	19		K	49 972780	36	19	55	K	54 999390	28	31		Ga	58 963370
30	20		Ca	49 957518	35	20		Ca	54 980550	27	32		Ge	58 981750
29	21		Sc	49 952187	34	21		Sc	54 967430	38	22	60	Ti	59 975640
28	22		Ti	49 944792.1	33	22		Ti	54 955120	37	23		V	59 964500
27	23		V	49 947162.8	32	23		V	54 947240	36	24		Cr	59 949730
26	24		Cr	49 946049.6	31	24		Cr	54 940844.2	35	25		Mn	59 943190
25	25		Mn	49 954244.0	30	25		Mn	54 938049.6	34	26		Fe	59 934077
24	26		Fe	49 962990	29	26		Fe	54 938298.0	33	27		Co	59 933822.2
23	27		Co	49 981540	28	27		Co	54 942003.1	32	28		Ni	59 930790.6
22	28		Ni	49 995930	27	28		Ni	54 951336	31	29		Cu	59 937368.1
34	17	51	Cl	51 013530	26	29		Cu	54 966050	30	30		Zn	59 941832
33	18		Ar	50 993240	25	30		Zn	54 983980	29	31		Ga	59 957060
32	19		K	50 976380	36	20	56	Ca	55 985790	28	32		Ge	59 970190
31	20		Ca	50 961470	35	21		Sc	55 972660	27	33		As	59 993130
30	21		Sc	50 953603	34	22		Ti	55 957990	39	22	61	Ti	60 982020
29	22		Ti	50 946616.0	33	23		V	55 950360	38	23		V	60 967410
28	23		V	50 943963.7	32	24		Cr	55 940645	37	24		Cr	60 954090
27	24		Cr	50 944771.8	31	25		Mn	55 938909.4	36	25		Mn	60 944460
26	25		Mn	50 948215.5	30	26		Fe	55 934942.1	35	26		Fe	60 936749
25	26		Fe	50 956825	29	27		Co	55 939843.9	34	27		Co	60 932479.4
24	27		Co	50 970720	28	28		Ni	55 942136	33	28		Ni	60 931060.4
23	28		Ni	50 987720	27	29		Cu	55 958560	32	29		Cu	60 933462.2
34	18	52	Ar	51 998170	26	30		Zn	55 972380	31	30		Zn	60 939514

Table B.1. Atomic mass tables *(cont.)*

N	Z	A	El	Atomic Mass (μu)	N	Z	A	El	Atomic Mass (μu)	N	Z	A	El	Atomic Mass (μu)
30	31		Ga	60 949170	40	27		Co	66 940610	38	34		Se	71 927112
29	32		Ge	60 963790	39	28		Ni	66 931570	37	35		Br	71 936500
28	33		As	60 980620	38	29		Cu	66 927750	36	36		Kr	71 941910
39	23	62	V	61 973140	37	30		Zn	66 927130.9	35	37		Rb	71 959080
38	24		Cr	61 955800	36	31		Ga	66 928204.9	45	28	73	Ni	72 946080
37	25		Mn	61 947970	35	32		Ge	66 932738	44	29		Cu	72 936490
36	26		Fe	61 936770	34	33		As	66 939190	43	30		Zn	72 929780
35	27		Co	61 934054	33	34		Se	66 950090	42	31		Ga	72 925170
34	28		Ni	61 928348.8	32	35		Br	66 964790	41	32		Ge	72 923459.4
33	29		Cu	61 932587	42	26	68	Fe	67 952510	40	33		As	72 923825
32	30		Zn	61 934334	41	27		Co	67 944360	39	34		Se	72 926767
31	31		Ga	61 944180	40	28		Ni	67 931845	38	35		Br	72 931790
30	32		Ge	61 954650	39	29		Cu	67 929640	37	36		Kr	72 938930
29	33		As	61 973200	38	30		Zn	67 924847.6	36	37		Rb	72 950370
40	23	63	V	62 976750	37	31		Ga	67 927983.5	35	38		Sr	72 965970
39	24		Cr	62 961860	36	32		Ge	67 928097	46	28	74	Ni	73 947910
38	25		Mn	62 949810	35	33		As	67 936790	45	29		Cu	73 940200
37	26		Fe	62 940120	34	34		Se	67 941870	44	30		Zn	73 929460
36	27		Co	62 933615	33	35		Br	67 958250	43	31		Ga	73 926940
35	28		Ni	62 929672.9	43	26	69	Fe	68 957700	42	32		Ge	73 921178.2
34	29		Cu	62 929601.1	42	27		Co	68 945200	41	33		As	73 923929.1
33	30		Zn	62 933215.6	41	28		Ni	68 935180	40	34		Se	73 922476.6
32	31		Ga	62 939140	40	29		Cu	68 929425	39	35		Br	73 929891
31	32		Ge	62 949640	39	30		Zn	68 926553.5	38	36		Kr	73 933260
30	33		As	62 963690	38	31		Ga	68 925581	37	37		Rb	73 944470
40	24	64	Cr	63 964200	37	32		Ge	68 927972	36	38		Sr	73 956310
39	25		Mn	63 953730	36	33		As	68 932280	47	28	75	Ni	74 952970
38	26		Fe	63 940870	35	34		Se	68 939560	46	29		Cu	74 941700
37	27		Co	63 935814	34	35		Br	68 950180	45	30		Zn	74 932940
36	28		Ni	63 927969.6	33	36		Kr	68 965320	44	31		Ga	74 926501
35	29		Cu	63 929767.9	43	27	70	Co	69 949810	43	32		Ge	74 922859.5
34	30		Zn	63 929146.6	42	28		Ni	69 936140	42	33		As	74 921596.4
33	31		Ga	63 936838	41	29		Cu	69 932409	41	34		Se	74 922523.6
32	32		Ge	63 941570	40	30		Zn	69 925325	40	35		Br	74 925776
31	33		As	63 957570	39	31		Ga	69 926028	39	36		Kr	74 931034
41	24	65	Cr	64 970370	38	32		Ge	69 924250.4	38	37		Rb	74 938569
40	25		Mn	64 956100	37	33		As	69 930930	37	38		Sr	74 949920
39	26		Fe	64 944940	36	34		Se	69 933500	48	28	76	Ni	75 955330
38	27		Co	64 936485	35	35		Br	69 944620	47	29		Cu	75 945990
37	28		Ni	64 930088.0	34	36		Kr	69 956010	46	30		Zn	75 933390
36	29		Cu	64 927793.7	44	27	71	Co	70 951730	45	31		Ga	75 928930
35	30		Zn	64 929245.1	43	28		Ni	70 940000	44	32		Ge	75 921402.7
34	31		Ga	64 932739.3	42	29		Cu	70 932620	43	33		As	75 922393.9
33	32		Ge	64 939440	41	30		Zn	70 927727	42	34		Se	75 919214.1
32	33		As	64 949480	40	31		Ga	70 924705.0	41	35		Br	75 924542
31	34		Se	64 964660	39	32		Ge	70 924954.0	40	36		Kr	75 925948
41	25	66	Mn	65 960820	38	33		As	70 927115	39	37		Rb	75 935071
40	26		Fe	65 945980	37	34		Se	70 932270	38	38		Sr	75 941610
39	27		Co	65 939830	36	35		Br	70 939250	49	28	77	Ni	76 960830
38	28		Ni	65 929115	35	36		Kr	70 950510	48	29		Cu	76 947950
37	29		Cu	65 928873.0	34	37		Rb	70 965320	47	30		Zn	76 937090
36	30		Zn	65 926036.8	45	27	72	Co	71 956410	46	31		Ga	76 929280
35	31		Ga	65 931592	44	28		Ni	71 941300	45	32		Ge	76 923548.5
34	32		Ge	65 933850	43	29		Cu	71 935520	44	33		As	76 920647.7
33	33		As	65 944370	42	30		Zn	71 926861	43	34		Se	76 919914.6
32	34		Se	65 955210	41	31		Ga	71 926369.4	42	35		Br	76 921380
42	25	67	Mn	66 963820	40	32		Ge	71 922076.2	41	36		Kr	76 924668
41	26		Fe	66 950000	39	33		As	71 926753	40	37		Rb	76 930407

Table B.1. Atomic mass tables *(cont.)*

N	Z	A	El	Atomic Mass (μu)	N	Z	A	El	Atomic Mass (μu)	N	Z	A	El	Atomic Mass (μu)
39	38		Sr	76 937760	44	38		Sr	81 918401	48	39		Y	86 910877.8
38	39		Y	76 949620	43	39		Y	81 926790	47	40		Zr	86 914817
50	28	78	Ni	77 963800	42	40		Zr	81 931090	46	41		Nb	86 920360
49	29		Cu	77 952810	41	41		Nb	81 943130	45	42		Mo	86 927330
48	30		Zn	77 938570	52	31	83	Ga	82 946870	44	43		Tc	86 936530
47	31		Ga	77 931660	51	32		Ge	82 934510	43	44		Ru	86 949180
46	32		Ge	77 922853	50	33		As	82 924980	55	33	88	As	87 944560
45	33		As	77 921829	49	34		Se	82 919119	54	34		Se	87 931420
44	34		Se	77 917309.5	48	35		Br	82 915180	53	35		Br	87 924070
43	35		Br	77 921146	47	36		Kr	82 914136	52	36		Kr	87 914447
42	36		Kr	77 920386	46	37		Rb	82 915112	51	37		Rb	87 911319
41	37		Rb	77 928141	45	38		Sr	82 917555	50	38		Sr	87 905614.3
40	38		Sr	77 932179	44	39		Y	82 922350	49	39		Y	87 909503.4
39	39		Y	77 943500	43	40		Zr	82 928650	48	40		Zr	87 910226
50	29	79	Cu	78 955280	42	41		Nb	82 936700	47	41		Nb	87 917960
49	30		Zn	78 942680	41	42		Mo	82 948740	46	42		Mo	87 921953
48	31		Ga	78 932920	53	31	84	Ga	83 952340	45	43		Tc	87 932830
47	32		Ge	78 925400	52	32		Ge	83 937310	44	44		Ru	87 940420
46	33		As	78 920948	51	33		As	83 929060	56	33	89	As	88 949230
45	34		Se	78 918499.8	50	34		Se	83 918465	55	34		Se	88 936020
44	35		Br	78 918337.6	49	35		Br	83 916504	54	35		Br	88 926390
43	36		Kr	78 920083	48	36		Kr	83 911507	53	36		Kr	88 917630
42	37		Rb	78 923997	47	37		Rb	83 914385	52	37		Rb	88 912280
41	38		Sr	78 929707	46	38		Sr	83 913425	51	38		Sr	88 907452.9
40	39		Y	78 937350	45	39		Y	83 920390	50	39		Y	88 905847.9
39	40		Zr	78 949160	44	40		Zr	83 923250	49	40		Zr	88 908889
51	29	80	Cu	79 961890	43	41		Nb	83 933570	48	41		Nb	88 913500
50	30		Zn	79 944410	42	42		Mo	83 940090	47	42		Mo	88 919481
49	31		Ga	79 936590	53	32	85	Ge	84 942690	46	43		Tc	88 927540
48	32		Ge	79 925445	52	33		As	84 931810	45	44		Ru	88 936110
47	33		As	79 922578	51	34		Se	84 922240	44	45		Rh	88 949380
46	34		Se	79 916521.8	50	35		Br	84 915608	56	34	90	Se	89 939420
45	35		Br	79 918530.0	49	36		Kr	84 912527	55	35		Br	89 930630
44	36		Kr	79 916378	48	37		Rb	84 911789.3	54	36		Kr	89 919524
43	37		Rb	79 922519	47	38		Sr	84 912933	53	37		Rb	89 914809
42	38		Sr	79 924525	46	39		Y	84 916427	52	38		Sr	89 907737.6
41	39		Y	79 934340	45	40		Zr	84 921470	51	39		Y	89 907151.4
40	40		Zr	79 940550	44	41		Nb	84 927910	50	40		Zr	89 904703.7
51	30	81	Zn	80 950480	43	42		Mo	84 936590	49	41		Nb	89 911264
50	31		Ga	80 937750	42	43		Tc	84 948940	48	42		Mo	89 913936
49	32		Ge	80 928820	54	32	86	Ge	85 946270	47	43		Tc	89 923560
48	33		As	80 922133	53	33		As	85 936230	46	44		Ru	89 929780
47	34		Se	80 917992.9	52	34		Se	85 924271	45	45		Rh	89 942870
46	35		Br	80 916291	51	35		Br	85 918797	57	34	91	Se	90 945370
45	36		Kr	80 916592	50	36		Kr	85 910610.3	56	35		Br	90 933970
44	37		Rb	80 918994	49	37		Rb	85 911167.1	55	36		Kr	90 923440
43	38		Sr	80 923213	48	38		Sr	85 909262.4	54	37		Rb	90 916534
42	39		Y	80 929130	47	39		Y	85 914888	53	38		Sr	90 910210
41	40		Zr	80 936820	46	40		Zr	85 916470	52	39		Y	90 907303
40	41		Nb	80 949050	45	41		Nb	85 925040	51	40		Zr	90 905645.0
52	30	82	Zn	81 954840	44	42		Mo	85 930700	50	41		Nb	90 906991
51	31		Ga	81 943160	43	43		Tc	85 942880	49	42		Mo	90 911751
50	32		Ge	81 929550	54	33	87	As	86 939580	48	43		Tc	90 918430
49	33		As	81 924500	53	34		Se	86 928520	47	44		Ru	90 926380
48	34		Se	81 916700.0	52	35		Br	86 920711	46	45		Rh	90 936550
47	35		Br	81 916805	51	36		Kr	86 913354.3	45	46		Pd	90 949480
46	36		Kr	81 913484.6	50	37		Rb	86 909183.5	58	34	92	Se	91 949330
45	37		Rb	81 918208	49	38		Sr	86 908879.3	57	35		Br	91 939260

Table B.1. Atomic mass tables *(cont.)*

N	Z	A	El	Atomic Mass (μu)	N	Z	A	El	Atomic Mass (μu)	N	Z	A	El	Atomic Mass (μu)
56	36		Kr	91 926153	50	46		Pd	95 918220	62	39		Y	100 930310
55	37		Rb	91 919725	49	47		Ag	95 930680	61	40		Zr	100 921140
54	38		Sr	91 911030	48	48		Cd	95 939770	60	41		Nb	100 915252
53	39		Y	91 908947	61	36	97	Kr	96 948560	59	42		Mo	100 910347
52	40		Zr	91 905040.1	60	37		Rb	96 937340	58	43		Tc	100 907314
51	41		Nb	91 907193.2	59	38		Sr	96 926149	57	44		Ru	100 905582.2
50	42		Mo	91 906810	58	39		Y	96 918131	56	45		Rh	100 906164
49	43		Tc	91 915260	57	40		Zr	96 910951	55	46		Pd	100 908289
48	44		Ru	91 920120	56	41		Nb	96 908097.1	54	47		Ag	100 912800
47	45		Rh	91 931980	55	42		Mo	96 906021.0	53	48		Cd	100 918680
46	46		Pd	91 940420	54	43		Tc	96 906365	52	49		In	100 926560
58	35	93	Br	92 943100	53	44		Ru	96 907555	51	50		Sn	100 936060
57	36		Kr	92 931270	52	45		Rh	96 911340	65	37	102	Rb	101 959210
56	37		Rb	92 922033	51	46		Pd	96 916480	64	38		Sr	101 943020
55	38		Sr	92 914022	50	47		Ag	96 924000	63	39		Y	101 933560
54	39		Y	92 909582	49	48		Cd	96 934940	62	40		Zr	101 922980
53	40		Zr	92 906475.6	61	37	98	Rb	97 941700	61	41		Nb	101 918040
52	41		Nb	92 906377.5	60	38		Sr	97 928471	60	42		Mo	101 910297
51	42		Mo	92 906812	59	39		Y	97 922220	59	43		Tc	101 909213
50	43		Tc	92 910248	58	40		Zr	97 912746	58	44		Ru	101 904349.5
49	44		Ru	92 917050	57	41		Nb	97 910331	57	45		Rh	101 906843
48	45		Rh	92 925740	56	42		Mo	97 905407.8	56	46		Pd	101 905608
47	46		Pd	92 935910	55	43		Tc	97 907216	55	47		Ag	101 912000
59	35	94	Br	93 948680	54	44		Ru	97 905287	54	48		Cd	101 914780
58	36		Kr	93 934360	53	45		Rh	97 910716	53	49		In	101 924710
57	37		Rb	93 926407	52	46		Pd	97 912721	52	50		Sn	101 930490
56	38		Sr	93 915360	51	47		Ag	97 921760	65	38	103	Sr	102 948950
55	39		Y	93 911594	50	48		Cd	97 927580	64	39		Y	102 936940
54	40		Zr	93 906315.8	49	49		In	97 942240	63	40		Zr	102 926600
53	41		Nb	93 907283.5	62	37	99	Rb	98 945420	62	41		Nb	102 919140
52	42		Mo	93 905087.6	61	38		Sr	98 933320	61	42		Mo	102 913200
51	43		Tc	93 909656	60	39		Y	98 924635	60	43		Tc	102 909179
50	44		Ru	93 911360	59	40		Zr	98 916511	59	44		Ru	102 906323.7
49	45		Rh	93 921700	58	41		Nb	98 911618	58	45		Rh	102 905504
48	46		Pd	93 928770	57	42		Mo	98 907711.6	57	46		Pd	102 906087
47	47		Ag	93 942780	56	43		Tc	98 906254.6	56	47		Ag	102 908972
59	36	95	Kr	94 939840	55	44		Ru	98 905939.3	55	48		Cd	102 913419
58	37		Rb	94 929319	54	45		Rh	98 908132	54	49		In	102 919914
57	38		Sr	94 919358	53	46		Pd	98 911768	53	50		Sn	102 928130
56	39		Y	94 912824	52	47		Ag	98 917600	52	51		Sb	102 940120
55	40		Zr	94 908042.7	51	48		Cd	98 925010	66	38	104	Sr	103 952330
54	41		Nb	94 906835.2	50	49		In	98 934610	65	39		Y	103 941450
53	42		Mo	94 905841.5	63	37	100	Rb	99 949870	64	40		Zr	103 928780
52	43		Tc	94 907656	62	38		Sr	99 935350	63	41		Nb	103 922460
51	44		Ru	94 910413	61	39		Y	99 927760	62	42		Mo	103 913760
50	45		Rh	94 915900	60	40		Zr	99 917760	61	43		Tc	103 911440
49	46		Pd	94 924690	59	41		Nb	99 914181	60	44		Ru	103 905430
48	47		Ag	94 935480	58	42		Mo	99 907477	59	45		Rh	103 906655
60	36	96	Kr	95 943070	57	43		Tc	99 907657.6	58	46		Pd	103 904035
59	37		Rb	95 934284	56	44		Ru	99 904219.7	57	47		Ag	103 908628
58	38		Sr	95 921680	55	45		Rh	99 908117	56	48		Cd	103 909848
57	39		Y	95 915898	54	46		Pd	99 908505	55	49		In	103 918340
56	40		Zr	95 908276	53	47		Ag	99 916070	54	50		Sn	103 923190
55	41		Nb	95 908100	52	48		Cd	99 920230	53	51		Sb	103 936290
54	42		Mo	95 904678.9	51	49		In	99 931150	66	39	105	Y	104 945090
53	43		Tc	95 907871	50	50		Sn	99 938950	65	40		Zr	104 933050
52	44		Ru	95 907598	64	37	101	Rb	100 953200	64	41		Nb	104 923930
51	45		Rh	95 914518	63	38		Sr	100 940520	63	42		Mo	104 916970

Table B.1. Atomic mass tables *(cont.)*

N	Z	A	El	Atomic Mass (μu)
62	43		Tc	104 911660
61	44		Ru	104 907750
60	45		Rh	104 905692
59	46		Pd	104 905084
58	47		Ag	104 906528
57	48		Cd	104 909468
56	49		In	104 914673
55	50		Sn	104 921390
54	51		Sb	104 931530
67	39	106	Y	105 950220
66	40		Zr	105 935910
65	41		Nb	105 928190
64	42		Mo	105 918134
63	43		Tc	105 914355
62	44		Ru	105 907327
61	45		Rh	105 907285
60	46		Pd	105 903483
59	47		Ag	105 906666
58	48		Cd	105 906458
57	49		In	105 913461
56	50		Sn	105 916880
55	51		Sb	105 928760
54	52		Te	105 937700
67	40	107	Zr	106 940860
66	41		Nb	106 930310
65	42		Mo	106 921690
64	43		Tc	106 915080
63	44		Ru	106 909910
62	45		Rh	106 906751
61	46		Pd	106 905128
60	47		Ag	106 905093
59	48		Cd	106 906614
58	49		In	106 910292
57	50		Sn	106 915670
56	51		Sb	106 924150
55	52		Te	106 935040
68	40	108	Zr	107 944280
67	41		Nb	107 935010
66	42		Mo	107 923580
65	43		Tc	107 918480
64	44		Ru	107 910190
63	45		Rh	107 908730
62	46		Pd	107 903894
61	47		Ag	107 905954
60	48		Cd	107 904183
59	49		In	107 909720
58	50		Sn	107 911970
57	51		Sb	107 922160
56	52		Te	107 929490
55	53		I	107 943290
68	41	109	Nb	108 937630
67	42		Mo	108 927810
66	43		Tc	108 919630
65	44		Ru	108 913200
64	45		Rh	108 908736
63	46		Pd	108 905954
62	47		Ag	108 904756
61	48		Cd	108 904986
60	49		In	108 907154
59	50		Sn	108 911287
58	51		Sb	108 918136
57	52		Te	108 927460
56	53		I	108 938190
69	41	110	Nb	109 942680
68	42		Mo	109 929730
67	43		Tc	109 923390
66	44		Ru	109 913970
65	45		Rh	109 910950
64	46		Pd	109 905152
63	47		Ag	109 906110
62	48		Cd	109 903006
61	49		In	109 907169
60	50		Sn	109 907853
59	51		Sb	109 916760
58	52		Te	109 922410
57	53		I	109 935210
56	54		Xe	109 944480
69	42	111	Mo	110 934510
68	43		Tc	110 925050
67	44		Ru	110 917560
66	45		Rh	110 911660
65	46		Pd	110 907640
64	47		Ag	110 905295
63	48		Cd	110 904182
62	49		In	110 905111
61	50		Sn	110 907735
60	51		Sb	110 913210
59	52		Te	110 921120
58	53		I	110 930280
57	54		Xe	110 941630
70	42	112	Mo	111 936840
69	43		Tc	111 929240
68	44		Ru	111 918550
67	45		Rh	111 914610
66	46		Pd	111 907313
65	47		Ag	111 907004
64	48		Cd	111 902757.2
63	49		In	111 905533
62	50		Sn	111 904821
61	51		Sb	111 912395
60	52		Te	111 917060
59	53		I	111 927970
58	54		Xe	111 935670
57	55		Cs	111 950330
71	42	113	Mo	112 942030
70	43		Tc	112 931330
69	44		Ru	112 922540
68	45		Rh	112 915420
67	46		Pd	112 910150
66	47		Ag	112 906566
65	48		Cd	112 904400.9
64	49		In	112 904061
63	50		Sn	112 905173
62	51		Sb	112 909378
61	52		Te	112 915930
60	53		I	112 923640
59	54		Xe	112 933380
58	55		Cs	112 944540
71	43	114	Tc	113 935880
70	44		Ru	113 924000
69	45		Rh	113 918850
68	46		Pd	113 910365
67	47		Ag	113 908808
66	48		Cd	113 903358.1
65	49		In	113 904917
64	50		Sn	113 902782
63	51		Sb	113 909100
62	52		Te	113 912060
61	53		I	113 921850
60	54		Xe	113 928150
59	55		Cs	113 941420
58	56		Ba	113 950940
72	43	115	Tc	114 938280
71	44		Ru	114 928310
70	45		Rh	114 920120
69	46		Pd	114 913680
68	47		Ag	114 908760
67	48		Cd	114 905431
66	49		In	114 903878
65	50		Sn	114 903346
64	51		Sb	114 906599
63	52		Te	114 911580
62	53		I	114 917920
61	54		Xe	114 926540
60	55		Cs	114 935940
59	56		Ba	114 947710
72	44	116	Ru	115 930160
71	45		Rh	115 923710
70	46		Pd	115 914160
69	47		Ag	115 911360
68	48		Cd	115 904755
67	49		In	115 905260
66	50		Sn	115 901744
65	51		Sb	115 906797
64	52		Te	115 908420
63	53		I	115 916740
62	54		Xe	115 921740
61	55		Cs	115 932910
60	56		Ba	115 941680
73	44	117	Ru	116 934790
72	45		Rh	116 925350
71	46		Pd	116 917840
70	47		Ag	116 911680
69	48		Cd	116 907218
68	49		In	116 904516
67	50		Sn	116 902954
66	51		Sb	116 904840
65	52		Te	116 908634
64	53		I	116 913650
63	54		Xe	116 920560
62	55		Cs	116 928640
61	56		Ba	116 938860
60	57		La	116 950010
74	44	118	Ru	117 937030

Table B.1. Atomic mass tables *(cont.)*

N	Z	A	El	Atomic Mass (μu)	N	Z	A	El	Atomic Mass (μu)	N	Z	A	El	Atomic Mass (μu)
73	45		Rh	117 929430	74	48		Cd	121 913500	73	53		I	125 905619
72	46		Pd	117 918980	73	49		In	121 910280	72	54		Xe	125 904269
71	47		Ag	117 914580	72	50		Sn	121 903440.1	71	55		Cs	125 909448
70	48		Cd	117 906914	71	51		Sb	121 905175.4	70	56		Ba	125 911244
69	49		In	117 906355	70	52		Te	121 903047.1	69	57		La	125 919370
68	50		Sn	117 901606	69	53		I	121 907592	68	58		Ce	125 924100
67	51		Sb	117 905532	68	54		Xe	121 908550	67	59		Pr	125 935310
66	52		Te	117 905825	67	55		Cs	121 916122	66	60		Nd	125 943070
65	53		I	117 913380	66	56		Ba	121 920260	80	47	127	Ag	126 936880
64	54		Xe	117 916570	65	57		La	121 930710	79	48		Cd	126 926430
63	55		Cs	117 926555	64	58		Ce	121 938010	78	49		In	126 917340
62	56		Ba	117 933440	63	59		Pr	121 951650	77	50		Sn	126 910351
61	57		La	117 946570	77	46	123	Pd	122 934260	76	51		Sb	126 906915
74	45	119	Rh	118 931360	76	47		Ag	122 924900	75	52		Te	126 905217
73	46		Pd	118 922680	75	48		Cd	122 917000	74	53		I	126 904468
72	47		Ag	118 915670	74	49		In	122 910439	73	54		Xe	126 905180
71	48		Cd	118 909920	73	50		Sn	122 905721.9	72	55		Cs	126 907418
70	49		In	118 905846	72	51		Sb	122 904215.7	71	56		Ba	126 911120
69	50		Sn	118 903309	71	52		Te	122 904273.0	70	57		La	126 916160
68	51		Sb	118 903946	70	53		I	122 905598	69	58		Ce	126 922750
67	52		Te	118 906408	69	54		Xe	122 908471	68	59		Pr	126 930830
66	53		I	118 910180	68	55		Cs	122 912990	67	60		Nd	126 940500
65	54		Xe	118 915550	67	56		Ba	122 918850	80	48	128	Cd	127 927760
64	55		Cs	118 922371	66	57		La	122 926240	79	49		In	127 920170
63	56		Ba	118 931050	65	58		Ce	122 935510	78	50		Sn	127 910535
62	57		La	118 940990	64	59		Pr	122 945960	77	51		Sb	127 909167
61	58		Ce	118 952760	77	47	124	Ag	123 928530	76	52		Te	127 904461.4
75	45	120	Rh	119 935780	76	48		Cd	123 917650	75	53		I	127 905805
74	46		Pd	119 924030	75	49		In	123 913180	74	54		Xe	127 903530.4
73	47		Ag	119 918790	74	50		Sn	123 905274.6	73	55		Cs	127 907748
72	48		Cd	119 909851	73	51		Sb	123 905937.5	72	56		Ba	127 908309
71	49		In	119 907960	72	52		Te	123 902819.5	71	57		La	127 915450
70	50		Sn	119 902196.6	71	53		I	123 906211.4	70	58		Ce	127 918870
69	51		Sb	119 905074	70	54		Xe	123 905895.8	69	59		Pr	127 928800
68	52		Te	119 904020	69	55		Cs	123 912246	68	60		Nd	127 935390
67	53		I	119 910048	68	56		Ba	123 915088	67	61		Pm	127 948260
66	54		Xe	119 912150	67	57		La	123 924530	81	48	129	Cd	128 932260
65	55		Cs	119 920678	66	58		Ce	123 930520	80	49		In	128 921660
64	56		Ba	119 926050	65	59		Pr	123 942960	79	50		Sn	128 913440
63	57		La	119 938070	78	47	125	Ag	124 930540	78	51		Sb	128 909150
62	58		Ce	119 946640	77	48		Cd	124 921250	77	52		Te	128 906596
76	45	121	Rh	120 938080	76	49		In	124 913600	76	53		I	128 904987
75	46		Pd	120 928180	75	50		Sn	124 907784.9	75	54		Xe	128 904779.5
74	47		Ag	120 919850	74	51		Sb	124 905248	74	55		Cs	128 906063
73	48		Cd	120 912980	73	52		Te	124 904424.7	73	56		Ba	128 908674
72	49		In	120 907849	72	53		I	124 904624.1	72	57		La	128 912670
71	50		Sn	120 904236.9	71	54		Xe	124 906398.2	71	58		Ce	128 918090
70	51		Sb	120 903818.0	70	55		Cs	124 909725	70	59		Pr	128 924860
69	52		Te	120 904930	69	56		Ba	124 914620	69	60		Nd	128 933250
68	53		I	120 907366	68	57		La	124 920670	68	61		Pm	128 943160
67	54		Xe	120 911386	67	58		Ce	124 928540	82	48	130	Cd	129 933980
66	55		Cs	120 917184	66	59		Pr	124 937830	81	49		In	129 924850
65	56		Ba	120 924490	79	47	126	Ag	125 934500	80	50		Sn	129 913850
64	57		La	120 933010	78	48		Cd	125 922350	79	51		Sb	129 911546
63	58		Ce	120 943670	77	49		In	125 916460	78	52		Te	129 906222.8
62	59		Pr	120 955360	76	50		Sn	125 907654	77	53		I	129 906674
76	46	122	Pd	121 929800	75	51		Sb	125 907250	76	54		Xe	129 903507.9
75	47		Ag	121 923320	74	52		Te	125 903305.5	75	55		Cs	129 906706

Table B.1. Atomic mass tables *(cont.)*

N	Z	A	El	Atomic Mass (μu)	N	Z	A	El	Atomic Mass (μu)	N	Z	A	El	Atomic Mass (μu)
74	56		Ba	129 906310	78	56		Ba	133 904503	81	57		La	137 907107
73	57		La	129 912320	77	57		La	133 908490	80	58		Ce	137 905986
72	58		Ce	129 914690	76	58		Ce	133 909030	79	59		Pr	137 910749
71	59		Pr	129 923380	75	59		Pr	133 915670	78	60		Nd	137 911930
70	60		Nd	129 928780	74	60		Nd	133 918650	77	61		Pm	137 919450
69	61		Pm	129 940450	73	61		Pm	133 928490	76	62		Sm	137 923540
68	62		Sm	129 948630	72	62		Sm	133 934020	75	63		Eu	137 933450
82	49	131	In	130 926770	71	63		Eu	133 946320	74	64		Gd	137 939970
81	50		Sn	130 916920	85	50	135	Sn	134 934730	73	65		Tb	137 952870
80	51		Sb	130 911950	84	51		Sb	134 925170	88	51	139	Sb	138 945710
79	52		Te	130 908521.9	83	52		Te	134 916450	87	52		Te	138 934730
78	53		I	130 906124.2	82	53		I	134 910050	86	53		I	138 926090
77	54		Xe	130 905081.9	81	54		Xe	134 907207	85	54		Xe	138 918787
76	55		Cs	130 905460	80	55		Cs	134 905972	84	55		Cs	138 913358
75	56		Ba	130 906931	79	56		Ba	134 905683	83	56		Ba	138 908835
74	57		La	130 910110	78	57		La	134 906971	82	57		La	138 906348
73	58		Ce	130 914420	77	58		Ce	134 909146	81	58		Ce	138 906647
72	59		Pr	130 920060	76	59		Pr	134 913140	80	59		Pr	138 908932
71	60		Nd	130 927100	75	60		Nd	134 918240	79	60		Nd	138 911920
70	61		Pm	130 935800	74	61		Pm	134 924620	78	61		Pm	138 916760
69	62		Sm	130 945890	73	62		Sm	134 932350	77	62		Sm	138 922302
83	49	132	In	131 932920	72	63		Eu	134 941720	76	63		Eu	138 929840
82	50		Sn	131 917744	86	50	136	Sn	135 939340	75	64		Gd	138 938080
81	51		Sb	131 914413	85	51		Sb	135 930660	74	65		Tb	138 948030
80	52		Te	131 908524	84	52		Te	135 920100	88	52	140	Te	139 938700
79	53		I	131 907995	83	53		I	135 914660	87	53		I	139 931210
78	54		Xe	131 904154.5	82	54		Xe	135 907220	86	54		Xe	139 921640
77	55		Cs	131 906430	81	55		Cs	135 907306	85	55		Cs	139 917277
76	56		Ba	131 905056	80	56		Ba	135 904570	84	56		Ba	139 910599
75	57		La	131 910110	79	57		La	135 907650	83	57		La	139 909473
74	58		Ce	131 911490	78	58		Ce	135 907140	82	58		Ce	139 905434
73	59		Pr	131 919120	77	59		Pr	135 912650	81	59		Pr	139 909071
72	60		Nd	131 923120	76	60		Nd	135 915020	80	60		Nd	139 909310
71	61		Pm	131 933750	75	61		Pm	135 923450	79	61		Pm	139 915800
70	62		Sm	131 940820	74	62		Sm	135 928300	78	62		Sm	139 918991
69	63		Eu	131 954160	73	63		Eu	135 939500	77	63		Eu	139 928080
84	49	133	In	132 938340	72	64		Gd	135 947070	76	64		Gd	139 933950
83	50		Sn	132 923810	87	50	137	Sn	136 945790	75	65		Tb	139 945540
82	51		Sb	132 915240	86	51		Sb	136 935310	74	66		Dy	139 953790
81	52		Te	132 910940	85	52		Te	136 925320	89	52	141	Te	140 944390
80	53		I	132 907806	84	53		I	136 917873	88	53		I	140 934830
79	54		Xe	132 905906	83	54		Xe	136 911563	87	54		Xe	140 926650
78	55		Cs	132 905447	82	55		Cs	136 907084	86	55		Cs	140 920044
77	56		Ba	132 906002	81	56		Ba	136 905821	85	56		Ba	140 914406
76	57		La	132 908400	80	57		La	136 906470	84	57		La	140 910957
75	58		Ce	132 911550	79	58		Ce	136 907780	83	58		Ce	140 908271
74	59		Pr	132 916200	78	59		Pr	136 910680	82	59		Pr	140 907648
73	60		Nd	132 922210	77	60		Nd	136 914640	81	60		Nd	140 909605
72	61		Pm	132 929720	76	61		Pm	136 920710	80	61		Pm	140 913607
71	62		Sm	132 938730	75	62		Sm	136 927050	79	62		Sm	140 918469
70	63		Eu	132 948900	74	63		Eu	136 935210	78	63		Eu	140 924890
85	49	134	In	133 944660	73	64		Gd	136 944650	77	64		Gd	140 932210
84	50		Sn	133 928460	87	51	138	Sb	137 940960	76	65		Tb	140 941160
83	51		Sb	133 920550	86	52		Te	137 929220	75	66		Dy	140 951190
82	52		Te	133 911540	85	53		I	137 922380	90	52	142	Te	141 948500
81	53		I	133 909877	84	54		Xe	137 913990	89	53		I	141 940180
80	54		Xe	133 905394.5	83	55		Cs	137 911011	88	54		Xe	141 929700
79	55		Cs	133 906713	82	56		Ba	137 905241	87	55		Cs	141 924292

Table B.1. Atomic mass tables *(cont.)*

N	Z	A	El	Atomic Mass (μu)	N	Z	A	El	Atomic Mass (μu)	N	Z	A	El	Atomic Mass (μu)
86	56		Ba	141 916448	92	54	146	Xe	145 947300	84	65		Tb	148 923242
85	57		La	141 914074	91	55		Cs	145 940160	83	66		Dy	148 927334
84	58		Ce	141 909240	90	56		Ba	145 930110	82	67		Ho	148 933790
83	59		Pr	141 910040	89	57		La	145 925700	81	68		Er	148 942170
82	60		Nd	141 907719	88	58		Ce	145 918690	80	69		Tm	148 952650
81	61		Pm	141 912950	87	59		Pr	145 917590	79	70		Yb	148 963480
80	62		Sm	141 915193	86	60		Nd	145 913112	95	55	150	Cs	149 957970
79	63		Eu	141 923400	85	61		Pm	145 914692	94	56		Ba	149 945620
78	64		Gd	141 928230	84	62		Sm	145 913037	93	57		La	149 938570
77	65		Tb	141 938860	83	63		Eu	145 917200	92	58		Ce	149 930230
76	66		Dy	141 946270	82	64		Gd	145 918305	91	59		Pr	149 927000
75	67		Ho	141 959860	81	65		Tb	145 927180	90	60		Nd	149 920887
90	53	143	I	142 944070	80	66		Dy	145 932720	89	61		Pm	149 920979
89	54		Xe	142 934890	79	67		Ho	145 944100	88	62		Sm	149 917271
88	55		Cs	142 927330	78	68		Er	145 952120	87	63		Eu	149 919698
87	56		Ba	142 920617	77	69		Tm	145 966500	86	64		Gd	149 918655
86	57		La	142 916050	93	54	147	Xe	146 953010	85	65		Tb	149 923654
85	58		Ce	142 912381	92	55		Cs	146 943860	84	66		Dy	149 925580
84	59		Pr	142 910812	91	56		Ba	146 933990	83	67		Ho	149 933350
83	60		Nd	142 909810	90	57		La	146 927820	82	68		Er	149 937760
82	61		Pm	142 910928	89	58		Ce	146 922510	81	69		Tm	149 949670
81	62		Sm	142 914624	88	59		Pr	146 918980	80	70		Yb	149 957990
80	63		Eu	142 920287	87	60		Nd	146 916096	79	71		Lu	149 972670
79	64		Gd	142 926740	86	61		Pm	146 915134	96	55	151	Cs	150 962000
78	65		Tb	142 934750	85	62		Sm	146 914893	95	56		Ba	150 950700
77	66		Dy	142 943830	84	63		Eu	146 916741	94	57		La	150 941560
76	67		Ho	142 954690	83	64		Gd	146 919089	93	58		Ce	150 934040
91	53	144	I	143 949610	82	65		Tb	146 924037	92	59		Pr	150 928230
90	54		Xe	143 938230	81	66		Dy	146 930880	91	60		Nd	150 923825
89	55		Cs	143 932030	80	67		Ho	146 939840	90	61		Pm	150 921203
88	56		Ba	143 922940	79	68		Er	146 949310	89	62		Sm	150 919928
87	57		La	143 919590	78	69		Tm	146 961080	88	63		Eu	150 919846
86	58		Ce	143 913643	93	55	148	Cs	147 948900	87	64		Gd	150 920344
85	59		Pr	143 913301	92	56		Ba	147 937680	86	65		Tb	150 923098
84	60		Nd	143 910083	91	57		La	147 932190	85	66		Dy	150 926180
83	61		Pm	143 912586	90	58		Ce	147 924390	84	67		Ho	150 931681
82	62		Sm	143 911995	89	59		Pr	147 922180	83	68		Er	150 937460
81	63		Eu	143 918774	88	60		Nd	147 916889	82	69		Tm	150 945430
80	64		Gd	143 922790	87	61		Pm	147 917468	81	70		Yb	150 955250
79	65		Tb	143 932530	86	62		Sm	147 914818	80	71		Lu	150 967150
78	66		Dy	143 939070	85	63		Eu	147 918154	96	56	152	Ba	151 946180
77	67		Ho	143 951640	84	64		Gd	147 918110	95	57		La	151 946110
76	68		Er	143 960590	83	65		Tb	147 924300	94	58		Ce	151 936380
91	54	145	Xe	144 943670	82	66		Dy	147 927180	93	59		Pr	151 931600
90	55		Cs	144 935390	81	67		Ho	147 937270	92	60		Nd	151 924680
89	56		Ba	144 926920	80	68		Er	147 944440	91	61		Pm	151 923490
88	57		La	144 921640	79	69		Tm	147 957550	90	62		Sm	151 919728
87	58		Ce	144 917230	78	70		Yb	147 966760	89	63		Eu	151 921740
86	59		Pr	144 914507	94	55	149	Cs	148 952720	88	64		Gd	151 919788
85	60		Nd	144 912569	93	56		Ba	148 942460	87	65		Tb	151 924070
84	61		Pm	144 912744	92	57		La	148 934370	86	66		Dy	151 924714
83	62		Sm	144 913406	91	58		Ce	148 928290	85	67		Ho	151 931740
82	63		Eu	144 916261	90	59		Pr	148 923791	84	68		Er	151 935080
81	64		Gd	144 921690	89	60		Nd	148 920144	83	69		Tm	151 944300
80	65		Tb	144 928880	88	61		Pm	148 918329	82	70		Yb	151 950170
79	66		Dy	144 936950	87	62		Sm	148 917180	81	71		Lu	151 963610
78	67		Ho	144 946880	86	63		Eu	148 917926	97	56	153	Ba	152 959610
77	68		Er	144 957460	85	64		Gd	148 919336	96	57		La	152 949450

Table B.1. Atomic mass tables *(cont.)*

N	Z	A	El	Atomic Mass (μu)	N	Z	A	El	Atomic Mass (μu)	N	Z	A	El	Atomic Mass (μu)
95	58		Ce	152 940580	86	70		Yb	155 942850	94	66		Dy	159 925194
94	59		Pr	152 933650	85	71		Lu	155 952910	93	67		Ho	159 928726
93	60		Nd	152 927695	84	72		Hf	155 959250	92	68		Er	159 929080
92	61		Pm	152 924113	83	73		Ta	155 971690	91	69		Tm	159 935090
91	62		Sm	152 922094	99	58	157	Ce	156 956340	90	70		Yb	159 937560
90	63		Eu	152 921226	98	59		Pr	156 947170	89	71		Lu	159 946020
89	64		Gd	152 921746	97	60		Nd	156 939270	88	72		Hf	159 950710
88	65		Tb	152 923431	96	61		Pm	156 933200	87	73		Ta	159 961360
87	66		Dy	152 925761	95	62		Sm	156 928350	86	74		W	159 968370
86	67		Ho	152 930195	94	63		Eu	156 925419	85	75		Re	159 981490
85	68		Er	152 935093	93	64		Gd	156 923957	101	60	161	Nd	160 954330
84	69		Tm	152 942028	92	65		Tb	156 924021	100	61		Pm	160 945860
83	70		Yb	152 949210	91	66		Dy	156 925461	99	62		Sm	160 938830
82	71		Lu	152 958690	90	67		Ho	156 928190	98	63		Eu	160 933680
97	57	154	La	153 954400	89	68		Er	156 931950	97	64		Gd	160 929666
96	58		Ce	153 943320	88	69		Tm	156 936760	96	65		Tb	160 927566
95	59		Pr	153 937390	87	70		Yb	156 942660	95	66		Dy	160 926930
94	60		Nd	153 929480	86	71		Lu	156 950102	94	67		Ho	160 927852
93	61		Pm	153 926550	85	72		Hf	156 958130	93	68		Er	160 930001
92	62		Sm	153 922205	84	73		Ta	156 968150	92	69		Tm	160 933400
91	63		Eu	153 922975	99	59	158	Pr	157 951780	91	70		Yb	160 937850
90	64		Gd	153 920862	98	60		Nd	157 941870	90	71		Lu	160 943540
89	65		Tb	153 924690	97	61		Pm	157 936690	89	72		Hf	160 950330
88	66		Dy	153 924422	96	62		Sm	157 929990	88	73		Ta	160 958370
87	67		Ho	153 930596	95	63		Eu	157 927840	87	74		W	160 967090
86	68		Er	153 932777	94	64		Gd	157 924101	86	75		Re	160 977660
85	69		Tm	153 941420	93	65		Tb	157 925410	101	61	162	Pm	161 950290
84	70		Yb	153 946240	92	66		Dy	157 924405	100	62		Sm	161 941220
83	71		Lu	153 957100	91	67		Ho	157 928950	99	63		Eu	161 937040
82	72		Hf	153 964250	90	68		Er	157 929910	98	64		Gd	161 930981
98	57	155	La	154 958130	89	69		Tm	157 937000	97	65		Tb	161 929480
97	58		Ce	154 948040	88	70		Yb	157 939858	96	66		Dy	161 926795
96	59		Pr	154 939990	87	71		Lu	157 949170	95	67		Ho	161 929092
95	60		Nd	154 932630	86	72		Hf	157 954650	94	68		Er	161 928775
94	61		Pm	154 928100	85	73		Ta	157 966370	93	69		Tm	161 933970
93	62		Sm	154 924636	84	74		W	157 973940	92	70		Yb	161 935750
92	63		Eu	154 922889	100	59	159	Pr	158 955230	91	71		Lu	161 943220
91	64		Gd	154 922619	99	60		Nd	158 946390	90	72		Hf	161 947203
90	65		Tb	154 923500	98	61		Pm	158 939130	89	73		Ta	161 957150
89	66		Dy	154 925749	97	62		Sm	158 933200	88	74		W	161 963340
88	67		Ho	154 929079	96	63		Eu	158 929084	87	75		Re	161 975710
87	68		Er	154 933200	95	64		Gd	158 926385	86	76		Os	161 983820
86	69		Tm	154 939192	94	65		Tb	158 925343	102	61	163	Pm	162 953520
85	70		Yb	154 945790	93	66		Dy	158 925736	101	62		Sm	162 945360
84	71		Lu	154 954230	92	67		Ho	158 927709	100	63		Eu	162 939210
83	72		Hf	154 962760	91	68		Er	158 930681	99	64		Gd	162 933990
98	58	156	Ce	155 951260	90	69		Tm	158 934810	98	65		Tb	162 930644
97	59		Pr	155 944120	89	70		Yb	158 940150	97	66		Dy	162 928728
96	60		Nd	155 935200	88	71		Lu	158 946620	96	67		Ho	162 928730
95	61		Pm	155 931060	87	72		Hf	158 954000	95	68		Er	162 930029
94	62		Sm	155 925526	86	73		Ta	158 962910	94	69		Tm	162 932648
93	63		Eu	155 924751	85	74		W	158 972280	93	70		Yb	162 936270
92	64		Gd	155 922120	100	60	160	Nd	159 949390	92	71		Lu	162 941200
91	65		Tb	155 924744	99	61		Pm	159 942990	91	72		Hf	162 947060
90	66		Dy	155 924278	98	62		Sm	159 935140	90	73		Ta	162 954320
89	67		Ho	155 929710	97	63		Eu	159 931970	89	74		W	162 962530
88	68		Er	155 931020	96	64		Gd	159 927051	88	75		Re	162 971970
87	69		Tm	155 939010	95	65		Tb	159 927164	87	76		Os	162 982050

Table B.1. Atomic mass tables *(cont.)*

N	Z	A	El	Atomic Mass (μu)	N	Z	A	El	Atomic Mass (μu)	N	Z	A	El	Atomic Mass (μu)
102	62	164	Sm	163 948280	92	75		Re	166 962560	95	76		Os	170 963040
101	63		Eu	163 942990	91	76		Os	166 971550	94	77		Ir	170 971780
100	64		Gd	163 935860	90	77		Ir	166 981540	93	78		Pt	170 981250
99	65		Tb	163 933350	104	64	168	Gd	167 948360	92	79		Au	170 991770
98	66		Dy	163 929171	103	65		Tb	167 943640	106	66	172	Dy	171 949110
97	67		Ho	163 930231	102	66		Dy	167 937230	105	67		Ho	171 944820
96	68		Er	163 929197	101	67		Ho	167 935500	104	68		Er	171 939352
95	69		Tm	163 933451	100	68		Er	167 932368	103	69		Tm	171 938396
94	70		Yb	163 934520	99	69		Tm	167 934170	102	70		Yb	171 936377.7
93	71		Lu	163 941220	98	70		Yb	167 933894	101	71		Lu	171 939082
92	72		Hf	163 944420	97	71		Lu	167 938700	100	72		Hf	171 939460
91	73		Ta	163 953570	96	72		Hf	167 940630	99	73		Ta	171 944740
90	74		W	163 958980	95	73		Ta	167 947790	98	74		W	171 947420
89	75		Re	163 970320	94	74		W	167 951860	97	75		Re	171 955290
88	76		Os	163 977930	93	75		Re	167 961610	96	76		Os	171 960080
103	62	165	Sm	164 952980	92	76		Os	167 967830	95	77		Ir	171 970640
102	63		Eu	164 945720	91	77		Ir	167 979970	94	78		Pt	171 977380
101	64		Gd	164 939380	90	78		Pt	167 988040	93	79		Au	171 990110
100	65		Tb	164 934880	105	64	169	Gd	168 952870	107	66	173	Dy	172 953440
99	66		Dy	164 931700	104	65		Tb	168 946220	106	67		Ho	172 947290
98	67		Ho	164 930319	103	66		Dy	168 940300	105	68		Er	172 942400
97	68		Er	164 930723	102	67		Ho	168 936868	104	69		Tm	172 939600
96	69		Tm	164 932432	101	68		Er	168 934588	103	70		Yb	172 938206.8
95	70		Yb	164 935398	100	69		Tm	168 934211	102	71		Lu	172 938927
94	71		Lu	164 939610	99	70		Yb	168 935187	101	72		Hf	172 940650
93	72		Hf	164 944540	98	71		Lu	168 937649	100	73		Ta	172 943540
92	73		Ta	164 950820	97	72		Hf	168 941160	99	74		W	172 947830
91	74		W	164 958340	96	73		Ta	168 945920	98	75		Re	172 953060
90	75		Re	164 967050	95	74		W	168 951760	97	76		Os	172 959790
89	76		Os	164 976480	94	75		Re	168 958830	96	77		Ir	172 967710
88	77		Ir	164 987580	93	76		Os	168 967080	95	78		Pt	172 976500
103	63	166	Eu	165 949970	92	77		Ir	168 976390	94	79		Au	172 986400
102	64		Gd	165 941600	91	78		Pt	168 986420	107	67	174	Ho	173 951150
101	65		Tb	165 938050	105	65	170	Tb	169 950250	106	68		Er	173 944340
100	66		Dy	165 932803	104	66		Dy	169 942670	105	69		Tm	173 942160
99	67		Ho	165 932281	103	67		Ho	169 939610	104	70		Yb	173 938858.1
98	68		Er	165 930290	102	68		Er	169 935460	103	71		Lu	173 940333.5
97	69		Tm	165 933553	101	69		Tm	169 935798	102	72		Hf	173 940040
96	70		Yb	165 933880	100	70		Yb	169 934759	101	73		Ta	173 944170
95	71		Lu	165 939760	99	71		Lu	169 938472	100	74		W	173 946160
94	72		Hf	165 942250	98	72		Hf	169 939650	99	75		Re	173 953110
93	73		Ta	165 950470	97	73		Ta	169 946090	98	76		Os	173 957120
92	74		W	165 955020	96	74		W	169 949290	97	77		Ir	173 966800
91	75		Re	165 965800	95	75		Re	169 958160	96	78		Pt	173 972811
90	76		Os	165 972530	94	76		Os	169 963570	95	79		Au	173 984920
89	77		Ir	165 985510	93	77		Ir	169 975030	108	67	175	Ho	174 954050
104	63	167	Eu	166 953050	92	78		Pt	169 982330	107	68		Er	174 947930
103	64		Gd	166 945570	106	65	171	Tb	170 953300	106	69		Tm	174 943830
102	65		Tb	166 940050	105	66		Dy	170 946480	105	70		Yb	174 941272.5
101	66		Dy	166 935650	104	67		Ho	170 941460	104	71		Lu	174 940767.9
100	67		Ho	166 933126	103	68		Er	170 938026	103	72		Hf	174 941503
99	68		Er	166 932045	102	69		Tm	170 936426	102	73		Ta	174 943650
98	69		Tm	166 932849	101	70		Yb	170 936322	101	74		W	174 946770
97	70		Yb	166 934947	100	71		Lu	170 937910	100	75		Re	174 951390
96	71		Lu	166 938310	99	72		Hf	170 940490	99	76		Os	174 957080
95	72		Hf	166 942600	98	73		Ta	170 944460	98	77		Ir	174 964280
94	73		Ta	166 947970	97	74		W	170 949460	97	78		Pt	174 972280
93	74		W	166 954670	96	75		Re	170 955550	96	79		Au	174 981550

Table B.1. Atomic mass tables *(cont.)*

N	Z	A	El	Atomic Mass (μu)	N	Z	A	El	Atomic Mass (μu)	N	Z	A	El	Atomic Mass (μu)
95	80		Hg	174 991410	106	74		W	179 946706	112	73		Ta	184 955559
108	68	176	Er	175 950290	105	75		Re	179 950790	111	74		W	184 953420.6
107	69		Tm	175 946990	104	76		Os	179 952350	110	75		Re	184 952955.7
106	70		Yb	175 942568	103	77		Ir	179 959250	109	76		Os	184 954043
105	71		Lu	175 942682.4	102	78		Pt	179 963220	108	77		Ir	184 956590
104	72		Hf	175 941401.8	101	79		Au	179 972400	107	78		Pt	184 960750
103	73		Ta	175 944740	100	80		Hg	179 978320	106	79		Au	184 965810
102	74		W	175 945590	99	81		Tl	179 990190	105	80		Hg	184 971980
101	75		Re	175 951570	111	70	181	Yb	180 956150	104	81		Tl	184 979100
100	76		Os	175 954950	110	71		Lu	180 951970	103	82		Pb	184 987580
99	77		Ir	175 963510	109	72		Hf	180 949099.1	102	83		Bi	184 997710
98	78		Pt	175 969000	108	73		Ta	180 947996	114	72	186	Hf	185 960920
97	79		Au	175 980270	107	74		W	180 948198	113	73		Ta	185 958550
96	80		Hg	175 987410	106	75		Re	180 950065	112	74		W	185 954362
109	68	177	Er	176 954370	105	76		Os	180 953270	111	75		Re	185 954987
108	69		Tm	176 949040	104	77		Ir	180 957640	110	76		Os	185 953838
107	70		Yb	176 945257	103	78		Pt	180 963180	109	77		Ir	185 957951
106	71		Lu	176 943755.0	102	79		Au	180 969950	108	78		Pt	185 959430
105	72		Hf	176 943220.0	101	80		Hg	180 977810	107	79		Au	185 966000
104	73		Ta	176 944472	100	81		Tl	180 986900	106	80		Hg	185 969460
103	74		W	176 946620	99	82		Pb	180 996710	105	81		Tl	185 978550
102	75		Re	176 950270	111	71	182	Lu	181 955210	104	82		Pb	185 984300
101	76		Os	176 955050	110	72		Hf	181 950553	103	83		Bi	185 996480
100	77		Ir	176 961170	109	73		Ta	181 950152	114	73	187	Ta	186 960410
99	78		Pt	176 968450	108	74		W	181 948206	113	74		W	186 957158
98	79		Au	176 977220	107	75		Re	181 951210	112	75		Re	186 955750.8
97	80		Hg	176 986340	106	76		Os	181 952186	111	76		Os	186 955747.9
96	81		Tl	176 996880	105	77		Ir	181 958130	110	77		Ir	186 957361
109	69	178	Tm	177 952640	104	78		Pt	181 961270	109	78		Pt	186 960560
108	70		Yb	177 946643	103	79		Au	181 969620	108	79		Au	186 964560
107	71		Lu	177 945951	102	80		Hg	181 974750	107	80		Hg	186 969790
106	72		Hf	177 943697.7	101	81		Tl	181 985610	106	81		Tl	186 976170
105	73		Ta	177 945750	100	82		Pb	181 992676	105	82		Pb	186 984030
104	74		W	177 945850	112	71	183	Lu	182 957570	104	83		Bi	186 993460
103	75		Re	177 950850	111	72		Hf	182 953530	115	73	188	Ta	187 963710
102	76		Os	177 953350	110	73		Ta	182 951373	114	74		W	187 958487
101	77		Ir	177 961080	109	74		W	182 950224.5	113	75		Re	187 958112.3
100	78		Pt	177 965710	108	75		Re	182 950821	112	76		Os	187 955836.0
99	79		Au	177 975980	107	76		Os	182 953110	111	77		Ir	187 958852
98	80		Hg	177 982476	106	77		Ir	182 956810	110	78		Pt	187 959396
97	81		Tl	177 995230	105	78		Pt	182 961730	109	79		Au	187 965090
110	69	179	Tm	178 955340	104	79		Au	182 967620	108	80		Hg	187 967560
109	70		Yb	178 950170	103	80		Hg	182 974560	107	81		Tl	187 975920
108	71		Lu	178 947324	102	81		Tl	182 982700	106	82		Pb	187 981060
107	72		Hf	178 945815.1	101	82		Pb	182 991930	105	83		Bi	187 992170
106	73		Ta	178 945934	113	71	184	Lu	183 961170	115	74	189	W	188 961910
105	74		W	178 947072	112	72		Hf	183 955450	114	75		Re	188 959228
104	75		Re	178 949980	111	73		Ta	183 954009	113	76		Os	188 958144.9
103	76		Os	178 953950	110	74		W	183 950932.6	112	77		Ir	188 958716
102	77		Ir	178 959150	109	75		Re	183 952524	111	78		Pt	188 960832
101	78		Pt	178 965480	108	76		Os	183 952491	110	79		Au	188 963890
100	79		Au	178 973410	107	77		Ir	183 957390	109	80		Hg	188 968130
99	80		Hg	178 981780	106	78		Pt	183 959900	108	81		Tl	188 973690
98	81		Tl	178 991470	105	79		Au	183 967470	107	82		Pb	188 980880
110	70	180	Yb	179 952330	104	80		Hg	183 971900	106	83		Bi	188 989510
109	71		Lu	179 949880	103	81		Tl	183 981760	116	74	190	W	189 963180
108	72		Hf	179 946548.8	102	82		Pb	183 988200	115	75		Re	189 961820
107	73		Ta	179 947466	113	72	185	Hf	184 958780	114	76		Os	189 958445

Table B.1. Atomic mass tables *(cont.)*

N	Z	A	El	Atomic Mass (μu)	N	Z	A	El	Atomic Mass (μu)	N	Z	A	El	Atomic Mass (μu)
113	77		Ir	189 960590	120	76	196	Os	195 969620	116	85		At	200 988490
112	78		Pt	189 959930	119	77		Ir	195 968380	115	86		Rn	200 995540
111	79		Au	189 964699	118	78		Pt	195 964935	114	87		Fr	201 003990
110	80		Hg	189 966280	117	79		Au	195 966551	124	78	202	Pt	201 975740
109	81		Tl	189 973790	116	80		Hg	195 965815	123	79		Au	201 973790
108	82		Pb	189 978180	115	81		Tl	195 970520	122	80		Hg	201 970626
107	83		Bi	189 988520	114	82		Pb	195 972710	121	81		Tl	201 972091
106	84		Po	189 995110	113	83		Bi	195 980610	120	82		Pb	201 972144
116	75	191	Re	190 963124	112	84		Po	195 985510	119	83		Bi	201 977670
115	76		Os	190 960928	111	85		At	195 995700	118	84		Po	201 980700
114	77		Ir	190 960591	110	86		Rn	196 002310	117	85		At	201 988450
113	78		Pt	190 961685	120	77	197	Ir	196 969636	116	86		Rn	201 993220
112	79		Au	190 963650	119	78		Pt	196 967323	115	87		Fr	202 003290
111	80		Hg	190 967060	118	79		Au	196 966552	124	79	203	Au	202 975137
110	81		Tl	190 971890	117	80		Hg	196 967195	123	80		Hg	202 972857
109	82		Pb	190 978200	116	81		Tl	196 969540	122	81		Tl	202 972329
108	83		Bi	190 986050	115	82		Pb	196 973380	121	82		Pb	202 973375
107	84		Po	190 994650	114	83		Bi	196 978930	120	83		Bi	202 976868
117	75	192	Re	191 965960	113	84		Po	196 985570	119	84		Po	202 981410
116	76		Os	191 961479	112	85		At	196 993290	118	85		At	202 986850
115	77		Ir	191 962602	111	86		Rn	197 001660	117	86		Rn	202 993320
114	78		Pt	191 961035	121	77	198	Ir	197 972280	116	87		Fr	203 001050
113	79		Au	191 964810	120	78		Pt	197 967876	115	88		Ra	203 009210
112	80		Hg	191 965570	119	79		Au	197 968225	125	79	204	Au	203 977710
111	81		Tl	191 972140	118	80		Hg	197 966752	124	80		Hg	203 973476
110	82		Pb	191 975760	117	81		Tl	197 970470	123	81		Tl	203 973849
109	83		Bi	191 985370	116	82		Pb	197 971980	122	82		Pb	203 973029
108	84		Po	191 991520	115	83		Bi	197 979020	121	83		Bi	203 977805
117	76	193	Os	192 964148	114	84		Po	197 983340	120	84		Po	203 980307
116	77		Ir	192 962924	113	85		At	197 992750	119	85		At	203 987260
115	78		Pt	192 962985	112	86		Rn	197 998780	118	86		Rn	203 991370
114	79		Au	192 964132	122	77	199	Ir	198 973790	117	87		Fr	204 000590
113	80		Hg	192 966644	121	78		Pt	198 970576	116	88		Ra	204 006480
112	81		Tl	192 970550	120	79		Au	198 968748	126	79	205	Au	204 979610
111	82		Pb	192 976080	119	80		Hg	198 968262	125	80		Hg	204 976056
110	83		Bi	192 983060	118	81		Tl	198 969810	124	81		Tl	204 974412
109	84		Po	192 991100	117	82		Pb	198 972910	123	82		Pb	204 974467
108	85		At	193 000190	116	83		Bi	198 977580	122	83		Bi	204 977375
118	76	194	Os	193 965179	115	84		Po	198 983600	121	84		Po	204 981170
117	77		Ir	193 965076	114	85		At	198 990630	120	85		At	204 986040
116	78		Pt	193 962664	113	86		Rn	198 998310	119	86		Rn	204 991670
115	79		Au	193 965339	122	78	200	Pt	199 971424	118	87		Fr	204 998660
114	80		Hg	193 965382	121	79		Au	199 970720	117	88		Ra	205 006190
113	81		Tl	193 971050	120	80		Hg	199 968309	126	80	206	Hg	205 977499
112	82		Pb	193 973970	119	81		Tl	199 970945	125	81		Tl	205 976095
111	83		Bi	193 982750	118	82		Pb	199 971816	124	82		Pb	205 974449
110	84		Po	193 988280	117	83		Bi	199 978140	123	83		Bi	205 978483
109	85		At	193 998970	116	84		Po	199 981740	122	84		Po	205 980465
119	76	195	Os	194 968120	115	85		At	199 990290	121	85		At	205 986600
118	77		Ir	194 965977	114	86		Rn	199 995680	120	86		Rn	205 990160
117	78		Pt	194 964774	113	87		Fr	200 006500	119	87		Fr	205 998490
116	79		Au	194 965018	123	78	201	Pt	200 974500	118	88		Ra	206 003780
115	80		Hg	194 966640	122	79		Au	200 971641	127	80	207	Hg	206 982580
114	81		Tl	194 969650	121	80		Hg	200 970285	126	81		Tl	206 977408
113	82		Pb	194 974470	120	81		Tl	200 970804	125	82		Pb	206 975881
112	83		Bi	194 980750	119	82		Pb	200 972850	124	83		Bi	206 978455
111	84		Po	194 988050	118	83		Bi	200 976970	123	84		Po	206 981578
110	85		At	194 996550	117	84		Po	200 982210	122	85		At	206 985776

Table B.1. Atomic mass tables *(cont.)*

N	Z	A	El	Atomic Mass (μu)	N	Z	A	El	Atomic Mass (μu)	N	Z	A	El	Atomic Mass (μu)
121	86		Rn	206 990730	124	89		Ac	213 006570	133	87		Fr	220 012313
120	87		Fr	206 996860	123	90		Th	213 012960	132	88		Ra	220 011015
119	88		Ra	207 003730	122	91		Pa	213 021180	131	89		Ac	220 014750
118	89		Ac	207 012090	132	82	214	Pb	213 999798.1	130	90		Th	220 015733
128	80	208	Hg	207 985940	131	83		Bi	213 998699	129	91		Pa	220 021880
127	81		Tl	207 982005	130	84		Po	213 995186	128	92		U	220 024710
126	82		Pb	207 976636	129	85		At	213 996356	136	85	221	At	221 018140
125	83		Bi	207 979727	128	86		Rn	213 995346	135	86		Rn	221 015460
124	84		Po	207 981231	127	87		Fr	213 998955	134	87		Fr	221 014246
123	85		At	207 986583	126	88		Ra	214 000091	133	88		Ra	221 013908
122	86		Rn	207 989631	125	89		Ac	214 006890	132	89		Ac	221 015580
121	87		Fr	207 997130	124	90		Th	214 011450	131	90		Th	221 018171
120	88		Ra	208 001780	123	91		Pa	214 020740	130	91		Pa	221 021860
119	89		Ac	208 011490	132	83	215	Bi	215 001830	129	92		U	221 026350
128	81	209	Tl	208 985349	131	84		Po	214 999415	137	85	222	At	222 022330
127	82		Pb	208 981075	130	85		At	214 998641	136	86		Rn	222 017570.5
126	83		Bi	208 980383	129	86		Rn	214 998729	135	87		Fr	222 017544
125	84		Po	208 982416	128	87		Fr	215 000326	134	88		Ra	222 015362
124	85		At	208 986159	127	88		Ra	215 002704	133	89		Ac	222 017829
123	86		Rn	208 990380	126	89		Ac	215 006450	132	90		Th	222 018454
122	87		Fr	208 995920	125	90		Th	215 011730	131	91		Pa	222 023730
121	88		Ra	209 001940	124	91		Pa	215 019100	130	92		U	222 026070
120	89		Ac	209 009570	133	83	216	Bi	216 006200	138	85	223	At	223 025340
129	81	210	Tl	209 990066	132	84		Po	216 001905.2	137	86		Rn	223 021790
128	82		Pb	209 984173	131	85		At	216 002409	136	87		Fr	223 019730.7
127	83		Bi	209 984105	130	86		Rn	216 000258	135	88		Ra	223 018497
126	84		Po	209 982857	129	87		Fr	216 003188	134	89		Ac	223 019126
125	85		At	209 987131	128	88		Ra	216 003518	133	90		Th	223 020795
124	86		Rn	209 989680	127	89		Ac	216 008721	132	91		Pa	223 023960
123	87		Fr	209 996398	126	90		Th	216 011051	131	92		U	223 027720
122	88		Ra	210 000450	125	91		Pa	216 019110	138	86	224	Rn	224 024090
121	89		Ac	210 009260	133	84	217	Po	217 006250	137	87		Fr	224 023240
120	90		Th	210 015030	132	85		At	217 004710	136	88		Ra	224 020202.0
129	82	211	Pb	210 988731	131	86		Rn	217 003915	135	89		Ac	224 021708
128	83		Bi	210 987258	130	87		Fr	217 004616	134	90		Th	224 021459
127	84		Po	210 986637	129	88		Ra	217 006306	133	91		Pa	224 025610
126	85		At	210 987481	128	89		Ac	217 009333	132	92		U	224 027590
125	86		Rn	210 990585	127	90		Th	217 013070	139	86	225	Rn	225 028440
124	87		Fr	210 995529	126	91		Pa	217 018290	138	87		Fr	225 025607
123	88		Ra	211 000890	134	84	218	Po	218 008965.8	137	88		Ra	225 023604
122	89		Ac	211 007650	133	85		At	218 008681	136	89		Ac	225 023221
121	90		Th	211 014860	132	86		Rn	218 005586	135	90		Th	225 023941
130	82	212	Pb	211 991887.5	131	87		Fr	218 007563	134	91		Pa	225 026120
129	83		Bi	211 991272	130	88		Ra	218 007124	133	92		U	225 029380
128	84		Po	211 988852	129	89		Ac	218 011630	132	93		Np	225 033900
127	85		At	211 990735	128	90		Th	218 013268	140	86	226	Rn	226 030890
126	86		Rn	211 990689	127	91		Pa	218 020010	139	87		Fr	226 029340
125	87		Fr	211 996195	126	92		U	218 023490	138	88		Ra	226 025402.6
124	88		Ra	211 999783	134	85	219	At	219 011300	137	89		Ac	226 026090
123	89		Ac	212 007810	133	86		Rn	219 009475	136	90		Th	226 024891
122	90		Th	212 012920	132	87		Fr	219 009241	135	91		Pa	226 027933
131	82	213	Pb	212 996500	131	88		Ra	219 010069	134	92		U	226 029340
130	83		Bi	212 994375	130	89		Ac	219 012400	133	93		Np	226 035130
129	84		Po	212 992843	129	90		Th	219 015520	141	86	227	Rn	227 035410
128	85		At	212 992921	128	91		Pa	219 019880	140	87		Fr	227 031830
127	86		Rn	212 993868	127	92		U	219 024920	139	88		Ra	227 029170.7
126	87		Fr	212 996175	135	85	220	At	220 015300	138	89		Ac	227 027747.0
125	88		Ra	213 000350	134	86		Rn	220 011384.1	137	90		Th	227 027699

Table B.1. Atomic mass tables (cont.)

N	Z	A	El	Atomic Mass (μu)	N	Z	A	El	Atomic Mass (μu)	N	Z	A	El	Atomic Mass (μu)
136	91		Pa	227 028793	143	91		Pa	234 043302	141	99		Es	240 068920
135	92		U	227 031140	142	92		U	234 040945.6	149	92	241	U	241 060330
134	93		Np	227 034960	141	93		Np	234 042889	148	93		Np	241 058250
142	86	228	Rn	228 038080	140	94		Pu	234 043305	147	94		Pu	241 056845.3
141	87		Fr	228 035720	139	95		Am	234 047790	146	95		Am	241 056822.9
140	88		Ra	228 031064.1	138	96		Cm	234 050240	145	96		Cm	241 057646.7
139	89		Ac	228 031014.8	146	89	235	Ac	235 051100	144	97		Bk	241 060220
138	90		Th	228 028731.3	145	90		Th	235 047500	143	98		Cf	241 063720
137	91		Pa	228 031037	144	91		Pa	235 045440	142	99		Es	241 068660
136	92		U	228 031366	143	92		U	235 043923.1	150	92	242	U	242 062930
135	93		Np	228 036180	142	93		Np	235 044055.9	149	93		Np	242 061640
134	94		Pu	228 038730	141	94		Pu	235 045282	148	94		Pu	242 058736.8
142	87	229	Fr	229 038430	140	95		Am	235 048030	147	95		Am	242 059543.0
141	88		Ra	229 034820	139	96		Cm	235 051590	146	96		Cm	242 058829.3
140	89		Ac	229 032930	138	97		Bk	235 056580	145	97		Bk	242 062050
139	90		Th	229 031755	147	89	236	Ac	236 055180	144	98		Cf	242 063690
138	91		Pa	229 032089	146	90		Th	236 049710	143	99		Es	242 069700
137	92		U	229 033496	145	91		Pa	236 048680	142	100		Fm	242 073430
136	93		Np	229 036250	144	92		U	236 045561.9	150	93	243	Np	243 064270
135	94		Pu	229 040140	143	93		Np	236 046560	149	94		Pu	243 061997
143	87	230	Fr	230 042510	142	94		Pu	236 046048.1	148	95		Am	243 061372.7
142	88		Ra	230 037080	141	95		Am	236 049570	147	96		Cm	243 061382.2
141	89		Ac	230 036030	140	96		Cm	236 051410	146	97		Bk	243 063002
140	90		Th	230 033126.6	139	97		Bk	236 057330	145	98		Cf	243 065420
139	91		Pa	230 034533	147	90	237	Th	237 053890	144	99		Es	243 069630
138	92		U	230 033927	146	91		Pa	237 051140	143	100		Fm	243 074510
137	93		Np	230 037810	145	92		U	237 048724.0	151	93	244	Np	244 067850
136	94		Pu	230 039646	144	93		Np	237 048167.3	150	94		Pu	244 064198
144	87	231	Fr	231 045410	143	94		Pu	237 048403.8	149	95		Am	244 064279.4
143	88		Ra	231 041220	142	95		Am	237 049970	148	96		Cm	244 062746.3
142	89		Ac	231 038550	141	96		Cm	237 052890	147	97		Bk	244 065168
141	90		Th	231 036297.1	140	97		Bk	237 057130	146	98		Cf	244 065990
140	91		Pa	231 035878.9	139	98		Cf	237 062070	145	99		Es	244 070970
139	92		U	231 036289	148	90	238	Th	238 056240	144	100		Fm	244 074080
138	93		Np	231 038230	147	91		Pa	238 054500	151	94	245	Pu	245 067739
137	94		Pu	231 041260	146	92		U	238 050782.6	150	95		Am	245 066445
136	95		Am	231 045560	145	93		Np	238 050940.5	149	96		Cm	245 065485.6
145	87	232	Fr	232 049650	144	94		Pu	238 049553.4	148	97		Bk	245 066355.4
144	88		Ra	232 043690	143	95		Am	238 051980	147	98		Cf	245 068040
143	89		Ac	232 042020	142	96		Cm	238 053020	146	99		Es	245 071320
142	90		Th	232 038050.4	141	97		Bk	238 058270	145	100		Fm	245 075380
141	91		Pa	232 038582	140	98		Cf	238 061410	144	101		Md	245 081020
140	92		U	232 037146.3	148	91	239	Pa	239 057130	152	94	246	Pu	246 070198
139	93		Np	232 040100	147	92		U	239 054287.8	151	95		Am	246 069768
138	94		Pu	232 041179	146	93		Np	239 052931.4	150	96		Cm	246 067217.6
137	95		Am	232 046590	145	94		Pu	239 052156.5	149	97		Bk	246 068670
145	88	233	Ra	233 048000	144	95		Am	239 053018	148	98		Cf	246 068798.8
144	89		Ac	233 044550	143	96		Cm	239 054950	147	99		Es	246 072970
143	90		Th	233 041576.9	142	97		Bk	239 058360	146	100		Fm	246 075280
142	91		Pa	233 040240.2	141	98		Cf	239 062580	145	101		Md	246 081930
141	92		U	233 039628	149	91	240	Pa	240 060980	153	94	247	Pu	247 074070
140	93		Np	233 040730	148	92		U	240 056586	152	95		Am	247 072090
139	94		Pu	233 042990	147	93		Np	240 056169	151	96		Cm	247 070347
138	95		Am	233 046470	146	94		Pu	240 053807.5	150	97		Bk	247 070299
137	96		Cm	233 050800	145	95		Am	240 055288	149	98		Cf	247 070992
146	88	234	Ra	234 050550	144	96		Cm	240 055519.0	148	99		Es	247 073650
145	89		Ac	234 048420	143	97		Bk	240 059750	147	100		Fm	247 076820
144	90		Th	234 043595	142	98		Cf	240 062300	146	101		Md	247 081800

Table B.1. Atomic mass tables *(cont.)*

N	Z	A	El	Atomic Mass (μu)	N	Z	A	El	Atomic Mass (μu)	N	Z	A	El	Atomic Mass (μu)
153	95	248	Am	248 075750	154	100		Fm	254 086848	159	102	261	No	261 105740
152	96		Cm	248 072342	153	101		Md	254 089730	158	103		Lr	261 106940
151	97		Bk	248 073080	152	102		No	254 090949	157	104		Db	261 108750
150	98		Cf	248 072178	151	103		Lr	254 096590	156	105		Jl	261 112110
149	99		Es	248 075460	150	104		Db	254 100170	155	106		Rf	261 116200
148	100		Fm	248 077184	157	98	255	Cf	255 091040	154	107		Bh	261 121800
147	101		Md	248 082910	156	99		Es	255 090266	160	102	262	No	262 107520
154	95	249	Am	249 078480	155	100		Fm	255 089955	159	103		Lr	262 109690
153	96		Cm	249 075947	154	101		Md	255 091075	158	104		Db	262 109920
152	97		Bk	249 074980	153	102		No	255 093232	157	105		Jl	262 114150
151	98		Cf	249 074847	152	103		Lr	255 096770	156	106		Rf	262 116480
150	99		Es	249 076410	151	104		Db	255 101490	155	107		Bh	262 123010
149	100		Fm	249 079020	150	105		Jl	255 107400	160	103	263	Lr	263 111390
148	101		Md	249 083000	158	98	256	Cf	256 093440	159	104		Db	263 112540
147	102		No	249 087820	157	99		Es	256 093590	158	105		Jl	263 115080
154	96	250	Cm	250 078351	156	100		Fm	256 091767	157	106		Rf	263 118310
153	97		Bk	250 078311	155	101		Md	256 094050	156	107		Bh	263 123150
152	98		Cf	250 076400.0	154	102		No	256 094276	155	108		Hn	263 128710
151	99		Es	250 078650	153	103		Lr	256 098760	160	104	264	Db	264 113980
150	100		Fm	250 079515	152	104		Db	256 101180	159	105		Jl	264 117470
149	101		Md	250 084490	151	105		Jl	256 108110	158	106		Rf	264 118920
148	102		No	250 087490	158	99	257	Es	257 095980	157	107		Bh	264 124730
155	96	251	Cm	251 082278	157	100		Fm	257 095099	156	108		Hn	264 128410
154	97		Bk	251 080753	156	101		Md	257 095535	160	105	265	Jl	265 118660
153	98		Cf	251 079580	155	102		No	257 096850	159	106		Rf	265 121070
152	99		Es	251 079984	154	103		Lr	257 099610	158	107		Bh	265 125200
151	100		Fm	251 081566	153	104		Db	257 103070	157	108		Hn	265 130000
150	101		Md	251 084920	152	105		Jl	257 107860	156	109		Mt	265 136570
149	102		No	251 088960	158	100	258	Fm	258 097070	160	106	266	Rf	266 121930
148	103		Lr	251 094360	157	101		Md	258 098425	159	107		Bh	266 127010
156	96	252	Cm	252 084870	156	102		No	258 098200	158	108		Hn	266 130040
155	97		Bk	252 084300	155	103		Lr	258 101880	157	109		Mt	266 137940
154	98		Cf	252 081620	154	104		Db	258 103570	160	107	267	Bh	267 127740
153	99		Es	252 082970	153	105		Jl	258 109440	159	108		Hn	267 131770
152	100		Fm	252 082460	152	106		Rf	258 113150	158	109		Mt	267 137530
151	101		Md	252 086630	159	100	259	Fm	259 100590	157	110		Xa	267 143960
150	102		No	252 088966	158	101		Md	259 100500	160	108	268	Hn	268 132160
149	103		Lr	252 095330	157	102		No	259 101020	159	109		Mt	268 138820
156	97	253	Bk	253 086880	156	103		Lr	259 102990	158	110		Xa	268 143530
155	98		Cf	253 085127	155	104		Db	259 105630	161	108	269	Hn	269 134110
154	99		Es	253 084818	154	105		Jl	259 109720	160	109		Mt	269 139110
153	100		Fm	253 085176	153	106		Rf	259 114650	159	110		Xa	269 145140
152	101		Md	253 087280	159	101	260	Md	260 103650	161	109	270	Mt	270 140720
151	102		No	253 090650	158	102		No	260 102640	160	110		Xa	270 144630
150	103		Lr	253 095260	157	103		Lr	260 105570	162	109	271	Mt	271 141230
149	104		Db	253 100680	156	104		Db	260 106430	161	110		Xa	271 146080
157	97	254	Bk	254 090600	155	105		Jl	260 111430	162	110	272	Xa	272 146310
156	98		Cf	254 087316	154	106		Rf	260 114440	161	111		Xb	272 153480
155	99		Es	254 088016	153	107		Bh	260 121800	163	110	273	Xa	273 149250

Appendix C

Cross Sections and Related Data

The tables in this appendix present basic interaction data for thermal neutrons and for photons. These tables give but a very small fraction of the data contained in comprehensive data libraries and other voluminous compilations. The data selected for these tables are for important interactions encountered in many shield analyses. However, for many calculations, it is necessary to obtain interaction data directly from data libraries maintained at several national nuclear data centers around the world. In the United States, the National Nuclear Data Center at Brookhaven National Laboratory (nndc@bnl.gov) and the Radiation Shielding and Information Center at Oak Ridge National Laboratory (pdc@ornl.gov) provide access to extensive nuclear data libraries. The Nuclear Data Service of the International Atomic Energy Agency (online@iaeand.iaea.or.at) and the Japan Atomic Energy Research Institute (wwwndc.tokai.jaeri.go.jp/index.html) also provide nuclear data to the international shielding community.

Data Tables

The following data are provided in this appendix.

Table C.1. Thermal neutron cross sections for special isotopes.

Table C.2. Radioisotopes produced by thermal neutron absorption.

Table C.3. Photon coefficients for air, water, concrete, iron, and lead.

Sources for the neutron data tables are provided in the tables. All photon interaction data were provided courtesy of S. Seltzer.[1]

In Table C.3, listing photon interaction coefficients, the following notation is used: incoherent scattering coefficient with electron binding, μ_c; photoelectric effect coefficient, μ_{ph}; pair-production coefficient, μ_{pp}; effective total interaction coefficient $\mu \equiv \mu_c + \mu_{ph} + \mu_{pp}$; energy transfer coefficient, μ_{tr}; and energy absorption coefficient, μ_{en}.

---

[1] S.M. Seltzer, "Calculation of Photon Mass Energy-Transfer and Mass Energy-Absorption Coefficients," *Radiat. Res.*, **136**, 147–179 (1993). J.H. Hubbell and S.M. Seltzer, *Tables of X-Ray Attenuation Coefficients 1 keV to 20 MeV for Elements Z = 1 to 92 and 48 Additional Substances of Dosimetric Interest*, Report NISTIR 5632, National Institute of Standards and Technology, Gaithersburg, MD, 1995.

Table C.1. Thermal neutron (2200 m s^{-1} or 0.0253 eV) cross sections for some special isotopes at 300 K. Cross-section notation: σ_γ for (n, γ); σ_s for elastic scattering; σ_t for total; σ_f for fission; σ_α for (n, α); and σ_p for (n, p).

Isotope	Abundance (atom %)	Half-life	Reaction cross sections (b)		
^{1}H	99.985		$\sigma_\gamma = 333$ mb	$\sigma_s = 30.5$	$\sigma_t = 30.9$
^{2}H	0.015		$\sigma_\gamma = 506$ μb	$\sigma_s = 4.26$	$\sigma_t = 4.30$
^{3}H		12.33 y	$\sigma_\gamma = 6$ μb	$\sigma_s = 1.53$	$\sigma_t = 1.53$
^{6}Li	7.5		$\sigma_\alpha = 941$	$\sigma_\gamma = 38.6$ mb	$\sigma_t = 943$
^{7}Li	92.5		$\sigma_\gamma = 45.7$ mb	$\sigma_s = 1.04$	$\sigma_t = 1.09$
^{10}B	19.6		$\sigma_\alpha = 3840$	$\sigma_\gamma = 0.50$	$\sigma_t = 3847$
^{11}B	80.4		$\sigma_\gamma = 5.53$ mb	$\sigma_s = 5.08$	$\sigma_t = 5.08$
^{12}C	98.89		$\sigma_s = 4.74$	$\sigma_\gamma = 3.4$ mb	$\sigma_t = 4.74$
^{13}C	1.11		$\sigma_\gamma = 1.37$ mb	$\sigma_t = 4.19$	
^{14}C		5736 y	$\sigma_\gamma = 1.0$ μb		
^{14}N	99.64		$\sigma_p = 1.83$	$\sigma_\gamma = 75$ mb	$\sigma_t = 12.2$
^{15}N	0.36		$\sigma_\gamma = 24$ μb	$\sigma_s = 4.58$	$\sigma_t = 4.58$
^{16}O	99.756		$\sigma_\gamma = 190$ μb	$\sigma_s = 4.03$	$\sigma_t = 4.03$
^{17}O	0.039		$\sigma_\alpha = 235$ mb	$\sigma_\gamma = 3.84$ mb	$\sigma_t = 4.17$
^{18}O	0.205		$\sigma_\gamma = 160$ μb		
^{232}Th	100	1.405×10^{10} y	$\sigma_f = 2.5$ μb	$\sigma_\gamma = 5.13$	$\sigma_t = 20.4$
^{233}Th		22.3 m	$\sigma_f = 15$	$\sigma_\gamma = 1450$	$\sigma_t = 1478$
^{233}U		1.592×10^5 y	$\sigma_f = 529$	$\sigma_\gamma = 46.0$	$\sigma_t = 588$
^{234}U	0.0055	2.455×10^5 y	$\sigma_f = 0.465$	$\sigma_\gamma = 103$	$\sigma_t = 116$
^{235}U	0.7200	7.038×10^8 y	$\sigma_f = 587$	$\sigma_\gamma = 99.3$	$\sigma_t = 700$
^{236}U		2.342×10^7 y	$\sigma_f = 47$ mb	$\sigma_\gamma = 5.14$	$\sigma_t = 14.1$
^{238}U	99.2745	4.468×10^9 y	$\sigma_f = 11.8$ μb	$\sigma_\gamma = 2.73$	$\sigma_t = 12.2$
^{239}U		23.45 m	$\sigma_f = 14$	$\sigma_\gamma = 22$	
^{239}Pu		24110 y	$\sigma_f = 749$	$\sigma_\gamma = 271$	$\sigma_t = 1028$
^{240}Pu		6564 y	$\sigma_f = 64$ mb	$\sigma_\gamma = 289$	$\sigma_t = 290$
^{241}Pu		14.35 y	$\sigma_f = 1015$	$\sigma_\gamma = 363$	$\sigma_t = 1389$
^{242}Pu		3.733×10^5 y	$\sigma_f = 1.0$ mb	$\sigma_\gamma = 19.3$	$\sigma_t = 27.0$

Source: ENDF/B-VI and other data extracted from the National Nuclear Data Center Online Service, Brookhaven National Laboratory, Upton, NY, January 1995.

Table C.2. Activation radionuclides resulting from thermal neutron (2200 m s^{-1} or 0.0253 eV) capture. Activation cross sections include production of short-lived metastable daughter nuclides except for ^{59}Co. The gamma photon energies and frequencies for the decay of the activation nuclide are given in Ap. D. More extensive tabulations are provided by B. Shleien (Ed.), *The Health Physics and Radiological Health Handbook*, Scinta, Silver Spring, MD, 1992.

Activated nuclide		Parent nuclide		
Nuclide	Half-life	Nuclide	Abundance (%) or half-life	Activation cross section (b)
^{12}B	0.0202 s	^{11}B	80.4	5.5 mb
^{16}N	7.13 s	^{15}N	0.38	24 μb
^{17}N^a	4.173 s	^{17}O	0.039	5.2 μb
^{19}O	26.91 s	^{18}O	0.205	160 μb
^{24}Na	14.96 h	^{23}Na	100	0.530
^{28}Al	2.241 m	^{27}Al	100	0.231
^{31}Si	2.62 h	^{30}Si	3.12	0.107
^{38}Cl	37.24 m	^{37}Cl	24.23	0.433
^{41}Ar	1.822 h	^{40}Ar	99.59	0.660
^{42}K	12.36 h	^{41}K	6.7	1.46
^{46}Sc	83.81 d	^{45}Sc	100	27.2
^{47}Ca	4.536 d	^{46}Ca	0.0033	0.74
^{49}Ca	8.715 m	^{48}Ca	0.185	1.09
^{51}Cr	27.70 d	^{50}Cr	4.35	15.9
^{56}Mn	2.579 h	^{55}Mn	100	13.3
^{59}Fe	44.50 d	^{58}Fe	0.31	1.28
^{60m}Co	10.47 m	^{59}Co	100	20.4
^{60}Co	5.271 y	^{59}Co	100	16.8
		^{60m}Co	10.47 m	
^{75}Se	119.8 d	^{74}Se	0.87	51.8
^{76}As	1.078 d	^{75}As	100	4.5
^{95}Zr	64.02 d	^{94}Zr	17.5	49.9 mb
^{97}Zr	16.91 h	^{96}Zr	2.8	22.9 mb
^{99}Mo	65.94 h	^{98}Mo	23.4	0.130
^{99m}Tc	6.01 h	^{99}Mo	65.94 h	
^{108}Ag	2.37 m	^{107}Ag	51.83	37.6
^{116m}In	54.4 m	^{115}In	95.7	202
^{198}Au	2.6952 d	^{197}Au	100	98.65
^{203}Hg	46.61 d	^{202}Hg	29.8	4.89
^{233}Th	22.3 m	^{232}Th	100	7.37
^{233}Pa	26.97 d	^{233}Th	22.3 m	1500
^{239}U	23.45 m	^{238}U	99.2745	2.68
^{239}Np	2.357 d	^{239}U	23.45 m	22.0

^{a}Decays by neutron emission.

Source: Data extracted from the National Nuclear Data Center Online Service, Brookhaven National Laboratory, Upton, NY, January 1995.

Table C.3. Mass coefficients (cm^2/g) for dry air near sea level. Composition by weight fraction: N 0.755268, O 0.231781, Ar 0.012827, C 0.000124. Nominal density is 1.205×10^{-3} g cm^{-3}. Read entry $1.234-5$ as 1.235×10^{-5}.

E (MeV)	μ_c/ρ	μ_{ph}/ρ	μ_{pp}/ρ	μ/ρ	μ_{tr}/ρ	μ_{en}/ρ
			Air			
0.0010	1.038−2	3.605+3		3.605+3	3.599+3	3.599+3
0.0015	2.116−2	1.190+3		1.190+3	1.188+3	1.188+3
0.0020	3.340−2	5.267+2		5.267+2	5.263+2	5.262+2
0.0030	5.748−2	1.616+2		1.616+2	1.615+2	1.614+2
0.00320	6.196−2	1.331+2		1.332+2	1.330+2	1.330+2
0.00320K	6.196−2	1.476+2		1.477+2	1.460+2	1.460+2
0.0040	7.770−2	7.713+1		7.721+1	7.637+1	7.636+1
0.0050	9.331−2	3.966+1		3.975+1	3.932+1	3.931+1
0.0060	1.051−1	2.288+1		2.299+1	2.271+1	2.270+1
0.0080	1.213−1	9.505+0		9.626+0	9.448+0	9.446+0
0.010	1.316−1	4.766+0		4.897+0	4.743+0	4.742+0
0.015	1.471−1	1.335+0		1.482+0	1.334+0	1.334+0
0.020	1.556−1	5.347−1		6.904−1	5.391−1	5.389−1
0.030	1.625−1	1.451−1		3.076−1	1.538−1	1.537−1
0.040	1.631−1	5.704−2		2.202−1	6.836−2	6.833−2
0.050	1.613−1	2.755−2		1.889−1	4.100−2	4.098−2
0.060	1.586−1	1.517−2		1.738−1	3.042−2	3.041−2
0.080	1.523−1	5.912−3		1.582−1	2.408−2	2.407−2
0.10	1.460−1	2.847−3		1.489−1	2.326−2	2.325−2
0.15	1.324−1	7.602−4		1.332−1	2.497−2	2.496−2
0.20	1.217−1	3.026−4		1.220−1	2.674−2	2.672−2
0.30	1.061−1	8.604−5		1.061−1	2.875−2	2.872−2
0.40	9.511−2	3.698−5		9.514−2	2.953−2	2.949−2
0.50	8.687−2	1.998−5		8.689−2	2.971−2	2.966−2
0.60	8.039−2	1.246−5		8.040−2	2.958−2	2.953−2
0.80	7.064−2	6.296−6		7.065−2	2.889−2	2.882−2
1.00	6.352−2	3.914−6		6.353−2	2.797−2	2.789−2
1.25	5.682−2	2.545−6	1.781−5	5.684−2	2.675−2	2.666−2
1.50	5.162−2	1.798−6	9.848−5	5.172−2	2.557−2	2.547−2
2.00	4.407−2	1.128−6	3.918−4	4.446−2	2.359−2	2.345−2
3.00	3.467−2	6.276−7	1.132−3	3.580−2	2.076−2	2.057−2
4.00	2.892−2	4.297−7	1.866−3	3.079−2	1.894−2	1.870−2
5.00	2.497−2	3.252−7	2.536−3	2.751−2	1.770−2	1.740−2
6.00	2.207−2	2.611−7	3.147−3	2.522−2	1.683−2	1.647−2
8.00	1.806−2	1.869−7	4.196−3	2.225−2	1.571−2	1.525−2
10.0	1.538−2	1.453−7	5.067−3	2.045−2	1.506−2	1.450−2
15.0	1.138−2	9.323−8	6.717−3	1.810−2	1.434−2	1.353−2
20.0	9.134−3	6.859−8	7.920−3	1.705−2	1.415−2	1.311−2
30.0	6.652−3	4.483−8	9.629−3	1.628−2	1.427−2	1.277−2
40.0	5.286−3	3.329−8	1.082−2	1.610−2	1.456−2	1.262−2
50.0	4.411−3	2.647−8	1.173−2	1.614−2	1.488−2	1.252−2
60.0	3.801−3	2.197−8	1.245−2	1.625−2	1.519−2	1.242−2
80.0	2.998−3	1.640−8	1.354−2	1.654−2	1.572−2	1.220−2
100.0	2.488−3	1.308−8	1.435−2	1.683−2	1.617−2	1.195−2

(cont.)

Table C.3. *(cont.)* Mass coefficients (cm^2/g) for water with density of $1\ \text{g/cm}^3$. Read entry $1.234{-}5$ as 1.235×10^{-5}.

			Water			
E (MeV)	μ_c/ρ	μ_{ph}/ρ	μ_{pp}/ρ	μ/ρ	μ_{tr}/ρ	μ_{en}/ρ
0.0010	1.319−2	4.076+3		4.076+3	4.065+3	4.065+3
0.0015	2.673−2	1.375+3		1.375+3	1.372+3	1.372+3
0.0020	4.184−2	6.161+2		6.161+2	6.152+2	6.152+2
0.0030	7.075−2	1.919+2		1.919+2	1.917+2	1.917+2
0.0040	9.430−2	8.198+1		8.207+1	8.192+1	8.191+1
0.0050	1.123−1	4.191+1		4.203+1	4.189+1	4.188+1
0.0060	1.259−1	2.407+1		2.419+1	2.406+1	2.405+1
0.0080	1.440−1	9.919+0		1.006+1	9.918	9.915
0.0100	1.550−1	4.943+0		5.098	4.945	4.944
0.015	1.699−1	1.369+0		1.539	1.374	1.374
0.020	1.774−1	5.437−1		7.211−1	5.505−1	5.503−1
0.030	1.829−1	1.457−1		3.286−1	1.557−1	1.557−1
0.040	1.827−1	5.680−2		2.395−1	6.950−2	6.947−2
0.050	1.803−1	2.725−2		2.076−1	4.225−2	4.223−2
0.060	1.770−1	1.493−2		1.920−1	3.191−2	3.190−2
0.080	1.697−1	5.770−3		1.755−1	2.598−2	2.597−2
0.10	1.626−1	2.762−3		1.654−1	2.547−2	2.546−2
0.15	1.473−1	7.307−4		1.481−1	2.765−2	2.764−2
0.20	1.353−1	2.887−4		1.356−1	2.969−2	2.967−2
0.30	1.179−1	8.162−5		1.180−1	3.195−2	3.192−2
0.40	1.058−1	3.495−5		1.058−1	3.282−2	3.279−2
0.50	9.663−2	1.884−5		9.664−2	3.303−2	3.299−2
0.60	8.939−2	1.173−5		8.940−2	3.289−2	3.284−2
0.80	7.856−2	5.920−6		7.857−2	3.212−2	3.206−2
1.00	7.066−2	3.680−6		7.066−2	3.111−2	3.103−2
1.25	6.318−2	2.394−6	1.777−5	6.320−2	2.974−2	2.965−2
1.50	5.741−2	1.689−6	9.820−5	5.751−2	2.844−2	2.833−2
2.00	4.901−2	1.063−6	3.908−4	4.940−2	2.621−2	2.608−2
3.00	3.855−2	5.937−7	1.131−3	3.968−2	2.300−2	2.281−2
4.00	3.216−2	4.075−7	1.867−3	3.402−2	2.091−2	2.066−2
5.00	2.777−2	3.089−7	2.540−3	3.031−2	1.946−2	1.915−2
6.00	2.454−2	2.484−7	3.155−3	2.770−2	1.843−2	1.806−2
8.00	2.008−2	1.780−7	4.211−3	2.429−2	1.707−2	1.658−2
10.0	1.710−2	1.386−7	5.090−3	2.219−2	1.626−2	1.566−2
15.0	1.266−2	8.905−8	6.754−3	1.941−2	1.528−2	1.441−2
20.0	1.016−2	6.555−8	7.974−3	1.813−2	1.495−2	1.382−2
30.0	7.395−3	4.289−8	9.709−3	1.710−2	1.490−2	1.327−2
40.0	5.875−3	3.186−8	1.091−2	1.679−2	1.510−2	1.298−2
50.0	4.906−3	2.534−8	1.183−2	1.674−2	1.537−2	1.279−2
60.0	4.225−3	2.104−8	1.256−2	1.679−2	1.563−2	1.261−2
80.0	3.333−3	1.570−8	1.368−2	1.701−2	1.613−2	1.228−2
100.0	2.767−3	1.253−8	1.450−2	1.727−2	1.655−2	1.194−2
100.0	2.767−3	1.253−8	1.450−2	1.727−2	1.655−2	1.194−2

Table C.3. *(cont.)* Mass coefficients $(\mathrm{cm^2/g})$ for ANSI/ANS-6.4.3 standard concrete of density of 2.3 g/cm^3. Composition by weight fraction: H 0.005599, O 0.498250, Na 0.017098, Mg 0.002400, Al 0.045595, Si 0.315768, S 0.001200, K 0.019198, Ca 0.082592, Fe 0.012299. Read entry 1.234−5 as 1.235×10^{-5}.

	Concrete					
E (MeV)	μ_c/ρ	μ_{ph}/ρ	μ_{pp}/ρ	μ/ρ	μ_{tr}/ρ	μ_{en}/ρ
0.00100	1.117−2	3.443+3		3.443+3	3.436+3	3.436+3
0.00104	2.098−2	3.144+3		3.144+3	3.138+3	3.138+3
0.00107	1.249−2	2.871+3		2.871+3	2.865+3	2.865+3
0.00107K	1.249−2	2.972+3		2.972+3	2.964+3	2.964+3
0.00118	3.115−2	2.300+3		2.300+3	2.295+3	2.295+3
0.00131	1.701−2	1.775+3		1.775+3	1.771+3	1.771+3
0.00131K	1.701−2	1.787+3		1.787+3	1.783+3	1.783+3
0.00150	2.098−2	1.233+3		1.233+3	1.231+3	1.230+3
0.00156	2.220−2	1.110+3		1.110+3	1.108+3	1.108+3
0.00156K	2.220−2	1.274+3		1.274+3	1.266+3	1.266+3
0.00169	5.019−2	1.025+3		1.025+3	1.019+3	1.019+3
0.00184	2.790−2	8.223+2		8.223+2	8.177+2	8.176+2
0.00184K	2.790−2	1.733+3		1.733+3	1.687+3	1.687+3
0.0020	3.115−2	1.454+3		1.454+3	1.418+3	1.417+3
0.00247	4.044−2	8.479+2		8.479+2	8.299+2	8.299+2
0.00247K	4.044−2	8.501+2		8.501+2	8.320+2	8.319+2
0.0030	5.019−2	4.966+2		4.966+2	4.878+2	4.878+2
0.00361	6.036−2	2.998+2		2.999+2	2.954+2	2.953+2
0.00361K	6.036−2	3.203+2		3.204+2	3.133+2	3.133+2
0.0040	6.635−2	2.410+2		2.410+2	2.361+2	2.361+2
0.00404	6.691−2	2.347+2		2.348+2	2.301+2	2.300+2
0.00404K	6.691−2	3.094+2		3.095+2	2.938+2	2.938+2
0.0050	7.972−2	1.733+2		1.734+2	1.659+2	1.659+2
0.0060	9.069−2	1.043+2		1.044+2	1.005+2	1.005+2
0.00711	1.007−1	6.448+1		6.458+1	6.246+1	6.244+1
0.00711K	1.007−1	6.884+1		6.894+1	6.549+1	6.548+1
0.0080	1.073−1	4.937+1		4.948+1	4.718+1	4.716+1
0.010	1.190−1	2.599+1		2.611+1	2.503+1	2.502+1
0.015	1.372−1	7.891		8.028	7.689	7.685
0.020	1.470−1	3.326		3.473	3.265	3.262
0.030	1.557−1	9.629−1		1.119	9.580−1	9.571−1
0.040	1.579−1	3.942−1		5.521−1	4.011−1	4.007−1
0.050	1.573−1	1.959−1		3.533−1	2.075−1	2.072−1
0.060	1.555−1	1.103−1		2.658−1	1.246−1	1.244−1
0.080	1.503−1	4.432−2		1.947−1	6.215−2	6.206−2
0.10	1.447−1	2.181−2		1.665−1	4.208−2	4.203−2
0.15	1.319−1	6.028−3		1.379−1	3.023−2	3.019−2
0.20	1.214−1	2.447−3		1.239−1	2.892−2	2.887−2
0.30	1.060−1	7.120−4		1.068−1	2.942−2	2.937−2
0.40	9.517−2	3.100−4		9.548−2	2.985−2	2.978−2
0.50	8.699−2	1.687−4		8.716−2	2.992−2	2.983−2
0.60	8.051−2	1.057−4		8.062−2	2.974−2	2.964−2
0.80	7.077−2	5.359−5		7.083−2	2.900−2	2.888−2
1.00	6.366−2	3.332−5		6.369−2	2.806−2	2.792−2
1.25	5.693−2	2.163−5	2.904−5	5.698−2	2.683−2	2.666−2
1.50	5.174−2	1.542−5	1.575−4	5.191−2	2.567−2	2.548−2
2.00	4.416−2	9.586−6	6.212−4	4.480−2	2.376−2	2.353−2
3.00	3.474−2	5.267−6	1.775−3	3.652−2	2.124−2	2.091−2
4.00	2.898−2	3.576−6	2.899−3	3.189−2	1.975−2	1.934−2
5.00	2.503−2	2.692−6	3.915−3	2.895−2	1.884−2	1.832−2
6.00	2.212−2	2.153−6	4.833−3	2.696−2	1.826−2	1.764−2
8.00	1.810−2	1.533−6	6.400−3	2.450−2	1.766−2	1.684−2
10.0	1.541−2	1.187−6	7.696−3	2.311−2	1.745−2	1.643−2
15.0	1.141−2	7.581−7	1.013−2	2.153−2	1.754−2	1.602−2
20.0	9.156−3	5.561−7	1.189−2	2.105−2	1.794−2	1.592−2
30.0	6.665−3	3.625−7	1.438−2	2.105−2	1.887−2	1.588−2
40.0	5.296−3	2.689−7	1.611−2	2.141−2	1.973−2	1.581−2
50.0	4.422−3	2.136−7	1.742−2	2.184−2	2.046−2	1.566−2
60.0	3.808−3	1.772−7	1.845−2	2.226−2	2.109−2	1.546−2
80.0	3.004−3	1.321−7	2.000−2	2.300−2	2.211−2	1.497−2
100.0	2.494−3	1.053−7	2.112−2	2.361−2	2.288−2	1.443−2

(cont.)

Table C.3. *(cont.)* Mass coefficients (cm^2/g) for natural iron with a density of 7.874 g/cm^3. Read entry 1.234−5 as 1.235×10^{-5}.

	Iron ($Z = 26$)					
E (MeV)	μ_c/ρ	μ_{ph}/ρ	μ_{pp}/ρ	μ/ρ	μ_{tr}/ρ	μ_{en}/ρ
0.0010	8.776−3	9.080+3		9.080+3	9.052+3	9.052+3
0.0015	1.530−2	3.395+3		3.395+3	3.388+3	3.388+3
0.0020	2.124−2	1.622+3		1.622+3	1.620+3	1.620+3
0.0030	3.206−2	5.542+2		5.543+2	5.536+2	5.535+2
0.0040	4.212−2	2.538+2		2.539+2	2.536+2	2.536+2
0.0050	5.133−2	1.373+2		1.374+2	1.372+2	1.372+2
0.0060	5.966−2	8.272+1		8.278+1	8.268+1	8.265+1
0.00711	6.798−2	5.138+1		5.144+1	5.135+1	5.133+1
0.00711K	6.798−2	4.058+2		4.059+2	2.978+2	2.978+2
0.0080	7.395−2	3.040+2		3.040+2	2.316+2	2.316+2
0.010	8.541−2	1.693+2		1.694+2	1.369+2	1.369+2
0.015	1.047−1	5.623+1		5.633+1	4.899+1	4.896+1
0.020	1.162−1	2.505+1		2.516+1	2.262+1	2.260+1
0.030	1.286−1	7.762		7.891	7.266	7.251
0.040	1.338−1	3.316		3.450	3.164	3.155
0.050	1.355−1	1.698		1.833	1.643	1.638
0.060	1.355−1	9.776−1		1.113	9.591−1	9.555−1
0.080	1.332−1	4.060−1		5.391−1	4.122−1	4.104−1
0.10	1.296−1	2.044−1		3.340−1	2.189−1	2.177−1
0.15	1.200−1	5.861−2		1.786−1	8.010−2	7.961−2
0.20	1.114−1	2.432−2		1.357−1	4.856−2	4.825−2
0.30	9.788−2	7.265−3		1.051−1	3.386−2	3.361−2
0.40	8.810−2	3.209−3		9.131−2	3.064−2	3.039−2
0.50	8.065−2	1.762−3		8.241−2	2.941−2	2.914−2
0.60	7.472−2	1.109−3		7.583−2	2.866−2	2.836−2
0.80	6.574−2	5.650−4		6.631−2	2.749−2	2.714−2
1.00	5.916−2	3.515−4		5.951−2	2.642−2	2.603−2
1.25	5.292−2	2.277−4	7.031−5	5.322−2	2.517−2	2.472−2
1.50	4.811−2	1.627−4	3.580−4	4.863−2	2.410−2	2.360−2
2.00	4.107−2	1.003−4	1.364−3	4.254−2	2.258−2	2.199−2
3.00	3.232−2	5.448−5	3.792−3	3.616−2	2.122−2	2.042−2
4.00	2.697−2	3.669−5	6.085−3	3.309−2	2.094−2	1.990−2
5.00	2.329−2	2.746−5	8.124−3	3.144−2	2.112−2	1.983−2
6.00	2.058−2	2.188−5	9.951−3	3.056−2	2.154−2	1.997−2
8.00	1.684−2	1.548−5	1.305−2	2.991−2	2.264−2	2.050−2
10.00	1.434−2	1.196−5	1.559−2	2.994−2	2.381−2	2.108−2
15.0	1.061−2	7.592−6	2.030−2	3.092−2	2.646−2	2.221−2
20.0	8.520−3	5.554−6	2.371−2	3.223−2	2.869−2	2.292−2
30.0	6.202−3	3.610−6	2.848−2	3.469−2	3.215−2	2.346−2
40.0	4.929−3	2.673−6	3.173−2	3.666−2	3.466−2	2.333−2
50.0	4.114−3	2.122−6	3.416−2	3.828−2	3.663−2	2.292−2
60.0	3.544−3	1.759−6	3.607−2	3.961−2	3.820−2	2.236−2
80.0	2.796−3	1.311−6	3.892−2	4.172−2	4.063−2	2.115−2
100.0	2.321−3	1.045−6	4.097−2	4.329−2	4.239−2	1.997−2

Table C.3. *(cont.)* Mass coefficients (cm^2/g) for natural lead with a density of 11.35 g/cm^3. Read entry 1.234−5 as 1.235×10^{-5}.

E (MeV)	μ_c/ρ	μ_{ph}/ρ	μ_{pp}/ρ	μ/ρ	μ_{tr}/ρ	μ_{en}/ρ
			Lead ($Z = 82$)			
0.0010	3.586−3	5.198+3		5.198+3	5.197+3	5.197+3
0.0015	6.600−3	2.344+3		2.344+3	2.344+3	2.344+3
0.0020	9.620−3	1.274+3		1.274+3	1.274+3	1.274+3
0.00248	1.241−2	7.898+2		7.898+2	7.897+2	7.895+2
0.00248M5	1.241−2	1.386+3		1.386+3	1.367+3	1.366+3
0.00253	6.600−3	1.714+3		1.714+3	1.683+3	1.682+3
0.00259	1.298−2	1.933+3		1.933+3	1.895+3	1.895+3
0.00259M4	1.298−2	2.447+3		2.447+3	2.390+3	2.390+3
0.0030	1.525−2	1.954+3		1.955+3	1.913+3	1.913+3
0.00307	1.560−2	1.847+3		1.847+3	1.809+3	1.808+3
0.00307M3	1.560−2	2.136+3		2.136+3	2.091+3	2.090+3
0.00330	1.684−2	1.784+3		1.784+3	1.749+3	1.748+3
0.00355	1.814−2	1.486+3		1.486+3	1.459+3	1.459+3
0.00355M2	1.814−2	1.575+3		1.575+3	1.546+3	1.546+3
0.00370	1.887−2	1.431+3		1.431+3	1.405+3	1.405+3
0.00385	1.963−2	1.302+3		1.302+3	1.279+3	1.279+3
0.00385M1	1.963−2	1.359+3		1.359+3	1.335+3	1.335+3
0.0040	2.037−2	1.242+3		1.242+3	1.221+3	1.221+3
0.0050	2.515−2	7.222+2		7.222+2	7.126+2	7.124+2
0.0060	2.970−2	4.599+2		4.599+2	4.548+2	4.546+2
0.0080	3.807−2	2.226+2		2.227+2	2.208+2	2.207+2
0.0100	4.540−2	1.256+2		1.257+2	1.248+2	1.247+2
0.01304	5.441−2	6.311+1		6.316+1	6.279+1	6.270+1
0.01304L3	5.441−2	1.582+2		1.582+2	1.292+2	1.291+2
0.0150	5.920−2	1.082+2		1.083+2	9.108+1	9.100+1
0.01520	5.965−2	1.045+2		1.045+2	8.815+1	8.807+1
0.01520L2	5.965−2	1.451+2		1.452+2	1.132+2	1.131+2
0.01553	6.038−2	1.384+2		1.384+2	1.084+2	1.083+2
0.01586	6.110−2	1.312+2		1.313+2	1.033+2	1.032+2
0.01586L1	6.110−2	1.517+2		1.517+2	1.181+2	1.180+2
0.0200	6.897−2	8.395+1		8.402+1	6.906+1	6.899+1
0.0300	8.228−2	2.886+1		2.894+1	2.542+1	2.536+1
0.0400	9.019−2	1.335+1		1.344+1	1.216+1	1.211+1
0.0500	9.478−2	7.291		7.386	6.776	6.740
0.0600	9.734−2	4.433		4.531	4.178	4.149
0.0800	9.922−2	2.012		2.112	1.934	1.916
0.08800	9.928−2	1.547		1.647	1.497	1.482
0.08800K	9.928−2	7.320		7.420	2.175	2.160
0.100	9.893−2	5.238		5.337	1.990	1.976
0.150	9.484−2	1.815		1.910	1.069	1.056
0.200	8.966−2	8.463−1		9.359−1	5.975−1	5.870−1
0.300	8.036−2	2.928−1		3.732−1	2.525−1	2.455−1
0.400	7.310−2	1.417−1		2.148−1	1.419−1	1.370−1
0.500	6.734−2	8.258−2		1.499−1	9.514−2	9.128−2
0.600	6.263−2	5.407−2		1.167−1	7.143−2	6.819−2
0.800	5.537−2	2.871−2		8.408−2	4.911−2	4.644−2
1.00	4.993−2	1.809−2		6.803−2	3.896−2	3.654−2
1.25	4.476−2	1.169−2	3.781−4	5.683−2	3.226−2	2.988−2
1.50	4.075−2	8.321−3	1.806−3	5.087−2	2.873−2	2.640−2
2.00	3.482−2	5.034−3	5.449−3	4.530−2	2.604−2	2.360−2
3.00	2.744−2	2.631−3	1.193−2	4.200−2	2.629−2	2.322−2
4.00	2.290−2	1.723−3	1.716−2	4.178−2	2.837−2	2.449−2
5.00	1.978−2	1.263−3	2.155−2	4.260−2	3.082−2	2.600−2
6.00	1.749−2	9.893−4	2.535−2	4.382−2	3.327−2	2.744−2
8.00	1.431−2	6.845−4	3.171−2	4.670−2	3.788−2	2.989−2
10.0	1.219−2	5.202−4	3.698−2	4.969−2	4.205−2	3.181−2
15.0	9.021−3	3.229−4	4.722−2	5.656−2	5.073−2	3.478−2
20.0	7.243−3	2.334−4	5.457−2	6.205−2	5.728−2	3.595−2
30.0	5.272−3	1.497−4	6.479−2	7.022−2	6.668−2	3.594−2
40.0	4.188−3	1.101−4	7.180−2	7.610−2	7.326−2	3.481−2
50.0	3.496−3	8.699−5	7.697−2	8.056−2	7.818−2	3.338−2
60.0	3.014−3	7.190−5	8.099−2	8.408−2	8.203−2	3.193−2
80.0	2.376−3	5.339−5	8.691−2	8.934−2	8.772−2	2.925−2
100.0	1.972−3	4.243−5	9.108−2	9.310−2	9.176−2	2.697−2

Appendix D

Decay Characteristics of Selected Radionuclides

This tabulation is based on calculations performed using the EDISTR code.[1] Data were kindly provided by J.C. Ryman and K.F. Eckerman of the Metabolism and Dosimetry Research Group, Health and Safety Research Division, Oak Ridge National Laboratory, Oak Ridge, Tennessee.

For the most part, the tabulation is based on the data used in preparation of the MIRD compendium of radioactive decay data.[2] For those nuclides not included in the MIRD document, tabulations are based on data used in preparation of the ICRP compendium of radioactive decay data.[3] Exceptions are the nuclides ^{16}N and ^{89}Kr, for which the tabulations are based on data of Kocher.[4] Frequencies of annihilation quanta are included.

Individual radiations are listed only if two criteria are met. First, the (maximum) energy must exceed 10-keV. Second, for any one group, e.g., combined gamma and x rays, the product of the energy and the frequency of an individual radiation must exceed 1% of the sum of the products for that group. Group totals include contributions from all radiations in the group, not just those listed. When the sum of the products of those listed is less than 95% of the group total, an asterisk after the group total is used to alert the reader.

The selected radionuclides in this compilation are primarily those of most concern for reactor operations and environmental radiological assessments. Decay data and energies and frequencies of emitted radiation for all radionuclides can be found at several sites on the web. One particularly useful site is http://www.nndc.bnl.gov/nndc/formmird.html.

---

[1]Dillman, L.T., *EDISTR—A Computer Program to Obtain a Nuclear Decay Data Base for Radiation Dosimetry*, Report ORNL/TM-6689, Oak Ridge National Laboratory, Oak Ridge, Tennessee, 1980.

[2]Weber, D.A., K.F. Eckerman, L.T. Dillman, and J.C. Ryman, *MIRD: Radionuclide Data and Decay Schemes*, Society of Nuclear Medicine, New York, 1989.

[3]*Radionuclide Transformations*, Publication 38, International Commission on Radiological Protection, Annals of the ICRP 11-13, 1983.

[4]Kocher, D.C., *Radioactive Decay Tables*, Report DOE/TIC 11026, Technical Information enter, U.S. Department of Energy, Washington, D.C., 1981.

$^{3}_{1}$H (12.33 y)

Principal Beta Particles

Freq. (%)	E_{av} (keV)	E_{max} (keV)
100.0	5.67	18.6
keV/decay	5.67	

$^{11}_{6}$C (20.39 min)

Positrons

Freq. (%)	E_{av} (keV)	E_{max} (keV)
99.76	385.6	960.1
keV/decay	384.6	

Principal Gamma and X Rays

Freq. (%)	E (keV)
199.52	511.0
keV/decay	1019.5

$^{14}_{6}$C (5730 y)

Principal Beta Particles

Freq. (%)	E_{av} (keV)	E_{max} (keV)
100.00	49.5	156.5
keV/decay	49.5	

$^{13}_{7}$N (9.965 min)

Positrons

Freq. (%)	E_{av} (keV)	E_{max} (keV)
99.82	491.8	1198.5
keV/decay	490.9	

Conversion/Auger Electrons

keV/decay 4.61×10^{-4} *

Principal Gamma and X Rays

Freq. (%)	E (keV)
199.64	511.0
keV/decay	1020.2

$^{16}_{8}$N (7.13 s)

Principal Beta Particles

Freq. (%)	E_{av} (keV)	E_{max} (keV)
4.9	1461.5	3301.9
68.0	1941.2	4288.3
26.0	4979.2	10419.0
keV/decay	2692.7	

Principal Gamma and X Rays

Freq. (%)	E (keV)
69.0	6129.2
5.0	7115.1
keV/decay	4618.2

$^{15}_{8}$O (122.24 s)

Positrons

Freq. (%)	E_{av} (keV)	E_{max} (keV)
99.89	735.3	1731.8
keV/decay	734.5	

Conversion/Auger Electrons

keV/decay 4.15×10^{-6}*

Principal Gamma and X Rays

Freq. (%)	E (keV)
199.77	511.0
keV/decay	1020.8

$^{19}_{8}$O (26.91 s)

Principal Beta Particles

Freq. (%)	E_{av} (keV)	E_{max} (keV)
56.06	1442.7	3264.8
39.77	2103.5	4621.6
3.83	2216.1	4818.8
keV/decay	1732.7	

Conversion/Auger Electrons

keV/decay 0.431 *

Principal Gamma and X Rays

Freq. (%)	E (keV)
3.45	110.0
90.34	197.0
50.34	1356.0
3.35	1444.0
2.20	1550.0
keV/decay	957.0

$^{18}_{9}$F (109.77 min)

Positrons

Freq. (%)	E_{av} (keV)	E_{max} (keV)
100.00	249.8	633.5
keV/decay	249.8	

Principal Gamma and X Rays

Freq. (%)	E (keV)
200.00	511.0
keV/decay	1022.0

$^{22}_{11}$Na (2.602 y)

Positrons

Freq. (%)	E_{av} (keV)	E_{max} (keV)
89.84	215.4	545.5
keV/decay	194.0	

Conversion/Auger Electrons

keV/decay 0.088 *

more ...

$^{22}_{11}\text{Na}$ (continued)

Principal Gamma and X Rays

Freq. (%)	E (keV)
179.80	511.0
99.94	1274.5
keV/decay	2192.6

$^{24}_{11}\text{Na}$ (15.020 h)

Principal Beta Particles

Freq. (%)	E_{av} (keV)	E_{max} (keV)
99.94	553.8	1390.2
keV/decay	553.5	

Conversion/Auger Electrons

keV/decay	0.021 *

Principal Gamma and X Rays

Freq. (%)	E (keV)
100.00	1368.5
99.94	2754.1
keV/decay	4123.1

$^{32}_{15}\text{P}$ (14.26 d)

Principal Beta Particles

Freq. (%)	E_{av} (keV)	E_{max} (keV)
100.00	694.7	1710.4
keV/decay	694.7	

$^{35}_{16}\text{S}$ (87.44 d)

Principal Beta Particles

Freq. (%)	E_{av} (keV)	E_{max} (keV)
100.00	48.8	167.5
keV/decay	48.8	

$^{36}_{17}\text{Cl}$ (3.01×10^5 y)

Positrons

Freq. (%)	E_{av} (keV)	E_{max} (keV)
keV/decay	0.009 *	

Principal Beta Particles

Freq. (%)	E_{av} (keV)	E_{max} (keV)
98.10	278.8	709.5
keV/decay	273.5	

Conversion/Auger Electrons

keV/decay	0.033 *

Principal Gamma and X Rays

keV/decay	0.155 *

$^{38}_{17}\text{Cl}$ (37.21 min)

Principal Beta Particles

Freq. (%)	E_{av} (keV)	E_{max} (keV)
32.53	420.1	1107.2
11.41	1181.4	2749.6
56.06	2244.0	4917.2
keV/decay	1529.4	

Conversion/Auger Electrons

Freq. (%)	E (keV)
keV/decay	0.020 *

Principal Gamma and X Rays

Freq. (%)	E (keV)
32.50	1642.4
44.00	2167.5
keV/decay	1488.4

$^{41}_{18}\text{Ar}$ (1.827 h)

Principal Beta Particles

Freq. (%)	E_{av} (keV)	E_{max} (keV)
99.17	459.0	1198.1
keV/decay	463.7	

Conversion/Auger Electrons

keV/decay	0.089 *

Principal Gamma and X Rays

Freq. (%)	E (keV)
99.16	1293.6
keV/decay	1283.6

$^{38}_{19}\text{K}$ (7.636 min)

Positrons

Freq. (%)	E_{av} (keV)	E_{max} (keV)
99.29	1216.5	2733.4
keV/decay	1208.5	

Conversion/Auger Electrons

keV/decay	0.041 *

Principal Gamma and X Rays

Freq. (%)	E (keV)
198.93	511.0
99.80	2167.0
keV/decay	3186.9

$^{40}_{19}\text{K}$ (1.28×10^9 y)

Principal Beta Particles

Freq. (%)	E_{av} (keV)	E_{max} (keV)
89.30	585.0	1311.6
keV/decay	522.4	

Conversion/Auger Electrons

keV/decay	0.202 *

more ...

$^{40}_{19}\text{K}$ (continued)

Principal Gamma and X Rays

Freq. (%)	E (keV)
10.70	1460.7
keV/decay	156.3

$^{42}_{19}\text{K}$ (12.36 h)

Principal Beta Particles

Freq. (%)	E_{av} (keV)	E_{max} (keV)
17.49	822.0	1996.4
82.07	1563.7	3521.1
keV/decay	1429.5	

Conversion/Auger Electrons

keV/decay	0.013 *

Principal Gamma and X Rays

Freq. (%)	E (keV)
17.90	1524.7
keV/decay	275.9

$^{43}_{19}\text{K}$ (22.6 h)

Principal Beta Particles

Freq. (%)	E_{av} (keV)	E_{max} (keV)
2.24	136.9	422.4
91.97	297.9	826.7
3.69	469.1	1223.6
1.30	761.5	1817.0
keV/decay	308.8	

Conversion/Auger Electrons

keV/decay	0.292 *

Principal Gamma and X Rays

Freq. (%)	E (keV)
4.11	220.6
87.27	372.8
11.43	396.9
11.03	593.4
80.51	617.5
1.88	1021.8
keV/decay	970.0

$^{45}_{20}\text{Ca}$ (163.8 d)

Principal Beta Particles

Freq. (%)	E_{av} (keV)	E_{max} (keV)
100.00	77.2	256.9
keV/decay	77.2	

Conversion/Auger Electrons

keV/decay	2.22×10^{-4} *

Principal Gamma and X Rays

keV/decay	1.34×10^{-5} *

$^{47}_{20}\text{Ca}$ (4.536 d)

Radioactive Daughter   Half-Life
^{47}Sc (100.0%)   3.345 y

Principal Beta Particles

Freq. (%)	E_{av} (keV)	E_{max} (keV)
80.83	241.0	690.9
18.96	816.7	1988.0
keV/decay	350.5	

Conversion/Auger Electrons

keV/decay	0.069 *

Principal Gamma and X Rays

Freq. (%)	E (keV)
6.51	489.2
6.51	807.9
74.00	1297.1
keV/decay	1047.1

$^{46}_{21}\text{Sc}$ (83.83 d)

Principal Beta Particles

Freq. (%)	E_{av} (keV)	E_{max} (keV)
100.00	112.0	357.3
keV/decay	112.0	

Conversion/Auger Electrons

keV/decay	0.247 *

Principal Gamma and X Rays

Freq. (%)	E (keV)
99.98	889.3
99.99	1120.5
keV/decay	2009.5

$^{47}_{21}\text{Sc}$ (3.345 d)

Principal Beta Particles

Freq. (%)	E_{av} (keV)	E_{max} (keV)
68.30	142.7	441.2
31.70	204.1	600.6
keV/decay	162.2	

Conversion/Auger Electrons

keV/decay	0.663 *

Principal Gamma and X Rays

Freq. (%)	E (keV)
67.90	159.4
keV/decay	108.2

$^{51}_{24}\text{Cr}$ (27.704 d)

Conversion/Auger Electrons
keV/decay   3.87 *

Principal Gamma and X Rays

Freq. (%)	E (keV)
10.08	320.1
keV/decay	33.4

$^{54}_{25}\text{Mn}$ (312.5 d)

Conversion/Auger Electrons
keV/decay   4.22 *

Principal Gamma and X Rays

Freq. (%)	E (keV)
99.98	834.8
keV/decay	836.0

$^{52}_{26}\text{Fe}$ (8.275 h)

Radioactive Daughter   Half-Life
^{52m}Mn   (100.0%)   21.1 min

Positrons

Freq. (%)	E_{av} (keV)	E_{max} (keV)
56.00	339.9	803.6
keV/decay	190.4	

Conversion/Auger Electrons
keV/decay   3.15 *

Principal Gamma and X Rays

Freq. (%)	E (keV)
99.23	168.7
112.00	511.0
keV/decay	740.5

$^{55}_{26}\text{Fe}$ (2.73 y)

Conversion/Auger Electrons
keV/decay   4.20 *

Principal Gamma and X Rays
keV/decay   1.69 *

$^{59}_{26}\text{Fe}$ (44.496 d)

Principal Beta Particles

Freq. (%)	E_{av} (keV)	E_{max} (keV)
1.29	35.6	130.5
45.26	80.9	273.2
53.17	149.1	465.5
keV/decay	117.5	

Conversion/Auger Electrons
keV/decay   0.252 *

more ...

$^{59}_{26}\text{Fe}$ (continued)

Principal Gamma and X Rays

Freq. (%)	E (keV)
1.02	142.7
3.08	192.3
56.50	1099.3
43.20	1291.6
keV/decay	1188.3

$^{57}_{27}\text{Co}$ (271.80 d)

Conversion/Auger Electrons

Freq. (%)	E (keV)
6.83	13.6
1.84	114.9
1.39	129.4
keV/decay	18.6 *

Principal Gamma and X Rays

Freq. (%)	E (keV)
9.23	14.4
85.95	122.1
10.33	136.5
keV/decay	125.3

$^{58}_{27}\text{Co}$ (70.916 d)

Positrons

Freq. (%)	E_{av} (keV)	E_{max} (keV)
14.99	201.2	475.0
keV/decay	30.2	

Conversion/Auger Electrons
keV/decay   4.01 *

Principal Gamma and X Rays

Freq. (%)	E (keV)
29.98	511.0
99.45	810.8
keV/decay	975.8

$^{60}_{27}\text{Co}$ (5.2704 y)

Principal Beta Particles

Freq. (%)	E_{av} (keV)	E_{max} (keV)
99.94	95.8	317.9
keV/decay	96.1	

Conversion/Auger Electrons
keV/decay   0.359 *

Principal Gamma and X Rays

Freq. (%)	E (keV)
99.90	1173.2
99.99	1332.5
keV/decay	2504.5

$^{63}_{28}$Ni   (100.1 y)

Principal Beta Particles

Freq. (%)	E_{av} (keV)	E_{max} (keV)
100.00	17.1	65.9
keV/decay	17.1	

$^{62}_{30}$Zn   (9.26 h)

Radioactive Daughter   Half-Life
^{62}Cu   (100.0%)   9.74 min

Positrons

Freq. (%)	E_{av} (keV)	E_{max} (keV)
8.39	258.6	605.0
keV/decay	21.7	

Conversion/Auger Electrons

Freq. (%)	E (keV)
14.65	31.9
1.46	39.8
keV/decay	10.9 *

Principal Gamma and X Rays

Freq. (%)	E (keV)
25.19	40.9
2.49	243.4
1.88	247.0
1.34	260.4
2.21	394.0
14.65	507.6
16.78	511.0
15.16	548.3
25.70	596.6
keV/decay	439.0

$^{65}_{30}$Zn   (243.9 d)

Positrons

Freq. (%)	E_{av} (keV)	E_{max} (keV)
1.42	142.6	328.8
keV/decay	2.02	

Conversion/Auger Electrons

Freq. (%)	E (keV)
keV/decay	4.79 *

Principal Gamma and X Rays

Freq. (%)	E (keV)
2.83	511.0
50.70	1115.5
keV/decay	583.2

$^{69m}_{30}$Zn   (13.76 h)

Radioactive Daughter   Half-Life
^{69}Zn   (100.0%)   55.6 min

Principal Beta Particles

keV/decay	0.096 *

more ...

$^{69m}_{30}$Zn   (continued)

Conversion/Auger Electrons

Freq. (%)	E (keV)
4.53	429.0
keV/decay	22.2 *

Principal Gamma and X Rays

Freq. (%)	E (keV)
94.87	438.6
keV/decay	416.5

$^{66}_{31}$Ga   (9.49 h)

Positrons

Freq. (%)	E_{av} (keV)	E_{max} (keV)
3.80	397.0	923.7
50.24	1903.4	4153.0
keV/decay	981.0	

Conversion/Auger Electrons

keV/decay	2.31 *

Principal Gamma and X Rays

Freq. (%)	E (keV)
113.07	511.0
6.03	833.5
37.90	1039.2
1.23	1333.1
2.14	1918.6
5.71	2189.9
1.96	2422.8
23.19	2752.0
1.48	3229.2
1.40	3381.3
1.02	3791.6
1.14	4086.3
3.49	4295.9
1.48	4806.6
keV/decay	2454.6 *

$^{67}_{31}$Ga   (3.261 d)

Conversion/Auger Electrons

Freq. (%)	E (keV)
27.77	83.7
2.46	92.1
keV/decay	34.4 *

Principal Gamma and X Rays

Freq. (%)	E (keV)
2.96	91.3
37.00	93.3
20.40	184.6
2.33	209.0
16.60	300.2
4.64	393.5
keV/decay	154.8

$^{68}_{31}$Ga (68.06 min)

Positrons

Freq. (%)	E_{av} (keV)	E_{max} (keV)
1.12	352.6	821.7
87.92	835.8	1899.1
keV/decay	738.8	

Conversion/Auger Electrons

keV/decay	0.544 *

Principal Gamma and X Rays

Freq. (%)	E (keV)
178.08	511.0
3.00	1077.4
keV/decay	947.4

$^{72}_{31}$Ga (14.10 h)

Principal Beta Particles

Freq. (%)	E_{av} (keV)	E_{max} (keV)
14.97	217.1	650.2
21.65	223.7	667.0
27.64	341.9	956.4
1.87	381.0	1048.5
8.88	569.1	1477.2
2.98	773.9	1927.0
8.48	1054.8	2528.0
10.28	1354.2	3158.0
keV/decay	497.8	

Conversion/Auger Electrons

keV/decay	4.23 *

Principal Gamma and X Rays

Freq. (%)	E (keV)
5.54	601.0
24.77	630.0
3.20	786.4
2.01	810.2
95.63	834.0
9.88	894.2
1.10	970.5
6.91	1050.7
1.45	1230.9
1.45	1230.9
1.13	1260.1
1.56	1276.8
3.55	1464.0
4.24	1596.7
5.25	1861.1
1.04	2109.5
25.92	2201.7
7.68	2491.0
12.78	2507.8
keV/decay	2724.2

$^{72}_{33}$As (26.0 h)

Positrons

Freq. (%)	E_{av} (keV)	E_{max} (keV)
5.80	821.8	1865.0
64.04	1114.4	2495.0
16.29	1525.7	3329.0
keV/decay	1021.3	

Conversion/Auger Electrons

Freq. (%)	E (keV)
1.43	679.9
keV/decay	12.0 *

Principal Gamma and X Rays

Freq. (%)	E (keV)
175.27	511.0
7.92	629.9
79.50	834.0
1.11	1464.0
keV/decay	1776.4 *

$^{74}_{33}$As (17.76 d)

Positrons

Freq. (%)	E_{av} (keV)	E_{max} (keV)
25.70	407.9	944.3
3.40	700.9	1540.1
keV/decay	128.6	

Principal Beta Particles

Freq. (%)	E_{av} (keV)	E_{max} (keV)
15.38	242.8	718.3
18.78	530.9	1353.1
keV/decay	137.1	

Conversion/Auger Electrons

keV/decay	2.49 *

Principal Gamma and X Rays

Freq. (%)	E (keV)
1.45	11.0
58.20	511.0
59.22	595.8
15.39	634.8
keV/decay	758.5

$^{76}_{33}$As (26.32 h)

Principal Beta Particles

Freq. (%)	E_{av} (keV)	E_{max} (keV)
1.19	92.8	313.6
1.89	173.8	540.1
2.02	436.3	1181.2
7.60	691.6	1752.8
34.59	996.2	2409.8
50.99	1266.9	2968.9
keV/decay	1063.8	

more ...

$^{76}_{33}$As   (continued)

Conversion/Auger Electrons
keV/decay   0.578 *

Principal Gamma and X Rays

Freq. (%)	E (keV)
44.70	559.1
1.17	563.2
6.08	657.0
1.63	1212.7
3.84	1216.0
1.39	1228.5
keV/decay	430.0 *

$^{74}_{34}$Se   (119.770 d)

Conversion/Auger Electrons

Freq. (%)	E (keV)
3.02	10.2
2.48	10.4
4.48	10.4
1.05	11.6
5.32	12.7
2.70	84.9
1.55	124.1
keV/decay	14.7 *

Principal Gamma and X Rays

Freq. (%)	E (keV)
16.50	10.5
32.11	10.5
2.41	11.7
4.72	11.7
1.14	66.1
3.48	96.7
17.32	121.1
58.98	136.0
1.47	198.6
59.10	264.7
25.18	279.5
1.34	303.9
11.56	400.7
keV/decay	391.7

$^{76}_{35}$Br   (16.2 h)

Positrons

Freq. (%)	E_{av} (keV)	E_{max} (keV)
1.42	333.8	770.0
6.20	373.0	860.3
5.12	425.1	979.4
1.22	549.3	1259.8
2.76	1218.5	2713.9
2.07	1262.7	2807.7
25.41	1529.8	3370.9
5.91	1797.3	3930.0
keV/decay	642.1	

Conversion/Auger Electrons

Freq. (%)	E (keV)
1.16	10.8
1.71	11.0
keV/decay	3.57 *

Principal Gamma and X Rays

Freq. (%)	E (keV)
6.98	11.2
13.55	11.2
1.04	12.5
2.03	12.5
1.86	472.9
109.53	511.0
74.00	559.1
3.55	563.2
15.91	657.0
4.59	1129.8
1.70	1213.1
8.81	1216.1
2.09	1228.7
2.52	1380.5
2.31	1471.1
14.65	1853.7
1.36	2096.7
2.49	2111.2
4.74	2391.2
1.95	2510.8
5.62	2792.7
7.40	2950.5
1.55	3604.0
keV/decay	2790.4 *

$^{77}_{35}$Br   (57.036 h)

Positrons
keV/decay   1.14 *

Conversion/Auger Electrons

Freq. (%)	E (keV)
2.58	10.8
2.13	11.0
3.79	11.0
keV/decay	8.36 *

more ...

$^{77}_{35}$Br (continued)

Principal Gamma and X Rays

Freq. (%)	E (keV)
15.48	11.2
30.06	11.2
2.31	12.5
4.51	12.5
1.45	87.6
1.14	161.8
1.25	200.4
23.89	239.0
3.08	249.8
2.37	281.7
4.30	297.2
1.22	303.8
1.62	439.5
1.03	484.6
1.48	511.0
23.17	520.7
1.23	574.6
3.06	578.9
1.62	585.5
1.72	755.3
2.15	817.8
keV/decay	331.4 *

$^{82}_{35}$Br (35.30 h)

Principal Beta Particles

Freq. (%)	E_{av} (keV)	E_{max} (keV)
1.36	76.2	264.6
98.64	137.8	444.3
keV/decay	137.0	

Conversion/Auger Electrons
keV/decay 2.06 *

Principal Gamma and X Rays

Freq. (%)	E (keV)
2.26	221.4
70.78	554.3
1.17	606.3
43.45	619.1
28.49	698.3
83.56	776.5
24.04	827.8
1.27	1007.6
27.20	1044.0
26.52	1317.5
16.32	1474.8
keV/decay	2642.0

$^{81m}_{36}$Kr (13 s)

Conversion/Auger Electrons

Freq. (%)	E (keV)
1.10	10.6
3.42	10.8
1.05	12.4
26.86	176.1
2.51	188.5
1.03	188.7
1.04	188.8
keV/decay	59.0

Principal Gamma and X Rays

Freq. (%)	E (keV)
5.07	12.6
9.81	12.7
1.50	14.1
67.10	190.4
keV/decay	130.0

$^{85m}_{36}$Kr (4.480 h)

Radioactive Daughter	Half-Life
^{85}Kr (21.1%)	10.72 y

Principal Beta Particles

Freq. (%)	E_{av} (keV)	E_{max} (keV)
79.00	290.3	840.7
keV/decay	229.3	

Conversion/Auger Electrons

Freq. (%)	E (keV)
3.15	136.0
6.00	290.5
keV/decay	26.1 *

Principal Gamma and X Rays

Freq. (%)	E (keV)
1.13	12.6
2.19	12.7
1.18	13.4
75.37	151.2
13.76	304.9
keV/decay	156.7

$^{85}_{36}$Kr (10.72 y)

Principal Beta Particles

Freq. (%)	E_{av} (keV)	E_{max} (keV)
99.56	251.4	687.0
keV/decay	250.5	

Conversion/Auger Electrons
keV/decay 0.016 *

Principal Gamma and X Rays
keV/decay 2.24 *

$^{87}_{36}$Kr   (76.3 min)

Radioactive Daughter   Half-Life
^{87}Rb   (100.0%) 4.7×10^{10} y

Principal Beta Particles

Freq. (%)	E_{av} (keV)	E_{max} (keV)
4.39	326.0	927.4
9.48	499.9	1333.2
5.59	562.1	1473.6
6.88	1293.6	3042.6
40.69	1501.3	3485.4
30.42	1694.0	3888.0
keV/decay	1322.9	

Conversion/Auger Electrons
keV/decay   0.899 *

Principal Gamma and X Rays

Freq. (%)	E (keV)
49.50	402.6
1.91	673.9
7.28	845.4
1.12	1175.4
2.05	1740.5
2.90	2011.9
9.31	2554.8
3.91	2558.1
keV/decay	793.1 *

$^{88}_{36}$Kr   (2.84 h)

Radioactive Daughter   Half-Life
^{88}Rb   (100.0%)   17.8 min

Principal Beta Particles

Freq. (%)	E_{av} (keV)	E_{max} (keV)
2.64	109.5	364.6
66.80	165.4	520.8
9.07	226.5	681.2
1.91	441.0	1198.3
1.31	824.4	2050.7
1.79	1135.9	2716.7
13.96	1233.0	2913.0
keV/decay	358.6	

Conversion/Auger Electrons

Freq. (%)	E (keV)
1.43	11.4
10.75	12.3
1.13	25.4
1.17	181.1
keV/decay	5.70 *

more ...

$^{88}_{36}$Kr   (continued)

Principal Gamma and X Rays

Freq. (%)	E (keV)
2.40	13.3
4.63	13.4
2.07	27.5
3.10	166.0
25.98	196.3
2.25	362.2
12.97	834.8
1.31	985.8
1.28	1141.3
1.12	1250.7
1.48	1369.5
2.15	1518.4
10.93	1529.8
4.53	2029.8
3.74	2035.4
13.18	2195.8
3.39	2231.8
34.60	2392.1
keV/decay	1954.6 *

$^{89}_{36}$Kr   (3.16 m)

Radioactive Daughter   Half-Life
^{89}Rb   (100.0%)   15.44 m

Principal Beta Particles

Freq. (%)	E_{av} (keV)	E_{max} (keV)
4.00	850.0	2103.9
14.40	970.0	2371.9
5.70	1070.0	2569.1
3.09	1180.0	2810.0
2.53	1260.0	2971.5
10.20	1400.0	3276.2
2.90	1480.0	3439.8
3.60	1580.0	3645.7
1.30	1730.0	3972.5
2.30	1930.0	4384.1
4.40	1940.0	4393.0
1.20	1980.0	4472.5
23.00	2210.0	4970.0
keV/decay	1360.7 *	

Conversion/Auger Electrons
keV/decay   1.3

more ...

$^{89}_{36}$Kr   (continued)

Principal Gamma and X Rays

Freq. (%)	E (keV)
20.00	220.9
6.64	497.5
5.64	577.0
16.60	585.8
4.20	738.4
5.92	867.1
7.18	904.3
2.92	1107.8
1.66	1116.6
3.06	1324.3
6.88	1472.8
1.32	1501.0
3.32	1530.0
5.12	1533.7
4.38	1693.7
1.04	1903.4
1.56	2012.2
1.74	2866.2
1.04	3140.3
1.04	3361.7
1.34	3532.9
7.12	2181.1
keV/decay	1834.5 *

$^{81}_{37}$Rb   (4.576 h)

Radioactive Daughter

	Half-Life
^{81}Kr (100.0%)	2.13×10^5 y

Positrons

Freq. (%)	E_{av} (keV)	E_{max} (keV)
1.79	252.7	577.4
26.31	446.3	1023.5
keV/decay	123.3	

Conversion/Auger Electrons

Freq. (%)	E (keV)
1.67	10.4
3.64	10.6
2.12	10.7
11.27	10.8
3.23	10.9
2.39	12.2
1.93	12.4
3.45	12.4
25.69	176.0
2.40	188.4
1.00	188.6
keV/decay	60.9 *

more ...

$^{81}_{37}$Rb   (continued)

Principal Gamma and X Rays

Freq. (%)	E (keV)
16.73	12.6
32.35	12.7
2.54	14.1
4.95	14.1
64.03	190.3
23.20	446.1
3.02	456.7
5.34	510.5
57.05	511.0
2.23	537.6
keV/decay	644.9 *

$^{84}_{37}$Rb   (32.87 d)

Positrons

Freq. (%)	E_{av} (keV)	E_{max} (keV)
12.41	338.5	776.7
13.51	756.3	1658.1
keV/decay	144.1	

Principal Beta Particles

Freq. (%)	E_{av} (keV)	E_{max} (keV)
4.00	331.2	890.0
keV/decay	13.2	

Conversion/Auger Electrons

Freq. (%)	E (keV)
1.16	10.4
2.52	10.6
1.47	10.7
7.82	10.8
2.24	10.9
1.65	12.2
1.34	12.4
2.40	12.4
keV/decay	4.18 *

Principal Gamma and X Rays

Freq. (%)	E (keV)
11.60	12.6
22.44	12.7
1.76	14.1
3.43	14.1
51.82	511.0
67.87	881.5
keV/decay	886.6

$^{86}_{37}$Rb (18.66 d)

Principal Beta Particles

Freq. (%)	E_{av} (keV)	E_{max} (keV)
8.78	232.6	698.0
91.22	709.4	1774.6
keV/decay	667.5	

Conversion/Auger Electrons

keV/decay   0.045 *

Principal Gamma and X Rays

Freq. (%)	E (keV)
8.78	1076.6
keV/decay	94.5

$^{85}_{38}$Sr (64.84 d)

Conversion/Auger Electrons

Freq. (%)	E (keV)
1.57	11.0
3.29	11.2
2.02	11.3
10.21	11.4
3.09	11.5
2.26	12.9
1.82	13.1
3.24	13.1
keV/decay	8.99 *

Principal Gamma and X Rays

Freq. (%)	E (keV)
17.10	13.3
32.98	13.4
2.64	15.0
5.15	15.0
98.30	514.0
keV/decay	513.4

$^{87m}_{38}$Sr (2.81 h)

Radioactive Daughter	Half-Life
^{87}Rb (0.3%)	4.73×10^{10} y

Conversion/Auger Electrons

Freq. (%)	E (keV)
1.60	12.1
15.02	372.3
1.78	386.2
keV/decay	67.1 *

Principal Gamma and X Rays

Freq. (%)	E (keV)
3.01	14.1
5.79	14.2
82.26	388.4
keV/decay	321.0

$^{89}_{38}$Sr (50.5 d)

Principal Beta Particles

Freq. (%)	E_{av} (keV)	E_{max} (keV)
99.99	583.3	1492.1
keV/decay	583.2	

Conversion/Auger Electrons

keV/decay 7.10×10^{-4} *

Principal Gamma and X Rays

keV/decay   0.085 *

$^{90}_{38}$Sr (29.12 y)

Radioactive Daughter	Half-Life
^{90}Y (100.0%)	64.0 h

Principal Beta Particles

Freq. (%)	E_{av} (keV)	E_{max} (keV)
100.00	195.7	546.0
keV/decay	195.7	

$^{90}_{39}$Y (64.0 h)

Principal Beta Particles

Freq. (%)	E_{av} (keV)	E_{max} (keV)
99.98	934.8	2284.0
keV/decay	934.6	

Conversion/Auger Electrons

keV/decay   0.280 *

Principal Gamma and X Rays

keV/decay   0.002 *

$^{91}_{39}$Y (58.51 d)

Principal Beta Particles

Freq. (%)	E_{av} (keV)	E_{max} (keV)
99.70	603.7	1543.0
keV/decay	602.2	

Conversion/Auger Electrons

keV/decay   0.002 *

Principal Gamma and X Rays

keV/decay   3.61 *

$_{40}^{95}$Zr   (64.02 d)

Radioactive Daughter   Half-Life
^{95}Nb   (99.2%)   35.02 d

Principal Beta Particles

Freq. (%)	E_{av} (keV)	E_{max} (keV)
54.59	109.2	366.4
44.20	120.4	398.9
1.11	327.0	887.4
keV/decay	116.9	

Conversion/Auger Electrons
keV/decay   1.02 *

Principal Gamma and X Rays

Freq. (%)	E (keV)
44.14	724.2
54.50	756.7
keV/decay	732.1

$_{41}^{95m}$Nb   (86.6 h)

Radioactive Daughter   Half-Life
^{95}Nb   (94.5%)   35.02 d

Principal Beta Particles

Freq. (%)	E_{av} (keV)	E_{max} (keV)
2.31	334.9	957.2
3.22	437.7	1161.3
keV/decay	21.9	

Conversion/Auger Electrons

Freq. (%)	E (keV)
1.45	13.7
1.04	13.8
4.42	14.0
1.58	14.1
1.14	15.9
1.61	16.2
54.39	216.7
7.41	233.0
1.23	233.2
2.23	233.3
2.00	235.3
keV/decay	152.6

Principal Gamma and X Rays

Freq. (%)	E (keV)
11.71	16.5
22.35	16.6
1.99	18.6
3.73	18.6
2.24	204.1
24.07	235.7
keV/decay	68.7

$_{41}^{95}$Nb   (35.02 d)

Principal Beta Particles

Freq. (%)	E_{av} (keV)	E_{max} (keV)
99.95	43.3	159.7
keV/decay	43.5	

Conversion/Auger Electrons
keV/decay   1.09 *

Principal Gamma and X Rays

Freq. (%)	E (keV)
99.79	765.8
keV/decay	764.3

$_{42}^{99}$Mo   (65.94 h)

Radioactive Daughter   Half-Life
^{99m}Tc   (88.6%)   6.01 h

Principal Beta Particles

Freq. (%)	E_{av} (keV)	E_{max} (keV)
16.39	133.0	436.3
1.13	289.6	847.9
81.86	442.7	1214.3
keV/decay	389.2	

Conversion/Auger Electrons

Freq. (%)	E (keV)
3.44	19.5
keV/decay	3.51 *

Principal Gamma and X Rays

Freq. (%)	E (keV)
1.06	18.3
2.01	18.4
1.05	40.6
4.52	140.5
6.08	181.1
1.15	366.4
12.13	739.6
4.34	778.0
keV/decay	149.2

$_{43}^{99m}$Tc   (6.01 h)

Radioactive Daughter   Half-Life
^{99}Tc   (100.0%) 2.11×10^5 y

Principal Beta Particles
keV/decay   0.004 *

Conversion/Auger Electrons

Freq. (%)	E (keV)
8.84	119.5
keV/decay	16.1 *

more ...

$^{99m}_{43}\text{Tc}$ (continued)

Principal Gamma and X Rays

Freq. (%)	E (keV)
2.10	18.3
3.99	18.4
89.06	140.5
keV/decay	126.5

$^{103}_{44}\text{Ru}$ (39.26 d)

Radioactive Daughter   Half-Life
^{103}Rh (99.7%)   56.11 min

Principal Beta Particles

Freq. (%)	E_{av} (keV)	E_{max} (keV)
6.61	30.8	116.1
92.18	64.2	229.4
keV/decay	63.8	

Conversion/Auger Electrons
keV/decay   2.77 *

Principal Gamma and X Rays

Freq. (%)	E (keV)
90.90	497.1
5.73	610.3
keV/decay	494.7

$^{106}_{44}\text{Ru}$ (368.2 d)

Radioactive Daughter   Half-Life
^{106}Rh (100.0%)   29.9 s

Principal Beta Particles

Freq. (%)	E_{av} (keV)	E_{max} (keV)
100.00	10.0	39.4
keV/decay	10.0	

$^{103m}_{45}\text{Rh}$ (56.114 min)

Conversion/Auger Electrons

Freq. (%)	E (keV)
9.54	16.5
29.34	36.6
41.40	36.8
14.39	39.3
2.48	39.8
keV/decay	37.1 *

Principal Gamma and X Rays

Freq. (%)	E (keV)
2.20	20.1
4.17	20.2
keV/decay	1.71 *

$^{106}_{45}\text{Rh}$ (29.9 s)

Principal Beta Particles

Freq. (%)	E_{av} (keV)	E_{max} (keV)
1.70	779.2	1978.1
9.69	976.1	2406.4
8.39	1266.5	3028.2
78.84	1508.0	3540.0
keV/decay	1412.3	

Conversion/Auger Electrons
keV/decay   0.815 *

Principal Gamma and X Rays

Freq. (%)	E (keV)
20.60	511.8
9.81	621.8
1.46	1050.1
keV/decay	204.9 *

$^{110m}_{47}\text{Ag}$ (249.9 d)

Radioactive Daughter   Half-Life
^{110}Ag (1.3%)   24.6 s

Principal Beta Particles

Freq. (%)	E_{av} (keV)	E_{max} (keV)
67.49	21.8	83.9
30.59	165.5	530.7
keV/decay	65.7	

Conversion/Auger Electrons
keV/decay   6.54 *

Principal Gamma and X Rays

Freq. (%)	E (keV)
3.66	446.8
2.78	620.3
94.74	657.7
10.72	677.6
6.49	687.0
16.74	706.7
4.66	744.3
22.36	763.9
7.32	818.0
72.86	884.7
34.32	937.5
24.35	1384.3
3.99	1475.8
13.11	1505.0
1.18	1562.3
keV/decay	2750.5

$^{110}_{47}$Ag   (24.6 s)

Principal Beta Particles

Freq. (%)	E_{av} (keV)	E_{max} (keV)
4.40	893.9	2235.1
95.21	1199.1	2892.8
keV/decay	1181.4	

Conversion/Auger Electrons

keV/decay	0.110 *

Principal Gamma and X Rays

Freq. (%)	E (keV)
4.50	657.7
keV/decay	30.6

$^{109}_{48}$Cd   (462.9 d)

Conversion/Auger Electrons

Freq. (%)	E (keV)
1.29	17.8
2.04	18.1
1.72	18.3
5.89	18.5
2.40	18.7
1.87	21.1
1.46	21.4
2.52	21.6
41.77	62.5
3.26	84.2
18.86	84.5
21.93	84.7
8.95	87.5
1.61	88.0
keV/decay	82.6 *

Principal Gamma and X Rays

Freq. (%)	E (keV)
28.47	22.0
53.66	22.2
4.95	24.9
9.66	24.9
2.71	25.5
3.61	88.0
keV/decay	26.0

$^{111}_{49}$In   (2.83 d)

Conversion/Auger Electrons

Freq. (%)	E (keV)
1.01	18.6
1.57	18.9
1.34	19.0
4.48	19.3
1.85	19.5
1.46	22.0
1.13	22.3
1.96	22.5
8.41	144.6
5.04	218.6
keV/decay	34.7 *

Principal Gamma and X Rays

Freq. (%)	E (keV)
23.61	23.0
44.38	23.2
4.15	26.1
8.08	26.1
2.35	26.6
90.24	171.3
94.00	245.4
keV/decay	404.9

$^{113m}_{49}$In   (1.658 h)

Conversion/Auger Electrons

Freq. (%)	E (keV)
1.21	20.1
28.80	363.7
4.09	387.5
1.13	391.0
keV/decay	134.0 *

Principal Gamma and X Rays

Freq. (%)	E (keV)
6.99	24.0
13.12	24.2
1.24	27.2
2.40	27.3
64.23	391.7
keV/decay	257.7

$^{115m}_{49}$In   (4.486 h)

Radioactive Daughter   Half-Life
^{115}In   (96.3%)   4.36 h

Principal Beta Particles

Freq. (%)	E_{av} (keV)	E_{max} (keV)
4.95	279.3	831.2
keV/decay	13.9	

Conversion/Auger Electrons

Freq. (%)	E (keV)
1.65	20.1
39.29	308.3
5.78	332.0
1.51	332.5
1.69	335.6
keV/decay	158.5

Principal Gamma and X Rays

Freq. (%)	E (keV)
9.54	24.0
17.90	24.2
1.69	27.2
3.28	27.3
45.37	336.2
keV/decay	161.1

$^{125}_{51}$Sb   (2.77 y)

Radioactive Daughter   Half-Life
^{125m}Te   (23.1%)   119.7 d

Principal Beta Particles

Freq. (%)	E_{av} (keV)	E_{max} (keV)
13.54	25.0	95.4
5.72	33.1	124.8
18.15	34.8	130.8
1.50	67.6	241.8
40.32	87.0	303.5
7.22	134.6	445.9
13.54	215.5	622.2
keV/decay	86.6	

Conversion/Auger Electrons

Freq. (%)	E (keV)
1.80	22.7
6.17	30.5
1.35	34.6
keV/decay	13.6 *

more ...

$^{125}_{51}$Sb   (continued)

Principal Gamma and X Rays

Freq. (%)	E (keV)
13.43	27.2
25.01	27.5
2.43	30.9
4.73	31.0
1.45	31.7
4.31	35.5
6.70	176.3
1.50	380.5
29.50	427.9
10.33	463.4
17.64	600.6
4.84	606.7
11.33	636.0
1.72	671.5
keV/decay	430.8

$^{123}_{53}$I   (13.2 h)

Radioactive Daughter   Half-Life
^{123}Te   (100.0%) 1.3×10^{13} y

Conversion/Auger Electrons

Freq. (%)	E (keV)
1.24	22.1
1.10	22.4
3.30	22.7
1.42	23.0
1.19	26.0
1.57	26.6
13.62	127.2
1.61	154.0
keV/decay	28.4 *

Principal Gamma and X Rays

Freq. (%)	E (keV)
24.65	27.2
45.90	27.5
4.46	30.9
8.68	31.0
2.66	31.7
83.30	159.0
1.39	529.0
keV/decay	172.7 *

$^{124}_{53}$I   (4.18 d)

Positrons

Freq. (%)	E_{av} (keV)	E_{max} (keV)
11.29	685.9	1532.3
11.29	973.6	2135.0
keV/decay	188.4	

Conversion/Auger Electrons

Freq. (%)	E (keV)
2.20	22.7
1.05	26.6
keV/decay	7.14 *

Principal Gamma and X Rays

Freq. (%)	E (keV)
16.47	27.2
30.67	27.5
2.98	30.9
5.80	31.0
1.78	31.7
45.73	511.0
60.50	602.7
9.98	722.8
1.43	1325.5
1.66	1376.0
2.99	1509.5
10.41	1691.0
keV/decay	1083.2 *

$^{125}_{53}$I   (60.14 d)

Conversion/Auger Electrons

Freq. (%)	E (keV)
1.37	21.8
2.00	22.1
1.78	22.4
5.31	22.7
2.28	23.0
1.92	26.0
1.50	26.3
2.54	26.6
9.49	30.6
2.13	34.7
keV/decay	19.5 *

Principal Gamma and X Rays

Freq. (%)	E (keV)
39.71	27.2
73.95	27.5
7.19	30.9
13.98	31.0
4.29	31.7
6.65	35.5
keV/decay	42.0

$^{129}_{53}$I   (1.57×10^7 y)

Principal Beta Particles

Freq. (%)	E_{av} (keV)	E_{max} (keV)
100.00	50.3	154.4
keV/decay	50.3	

Conversion/Auger Electrons

Freq. (%)	E (keV)
2.27	24.5
1.13	28.8
9.64	32.1
2.14	36.7
keV/decay	13.0 *

Principal Gamma and X Rays

Freq. (%)	E (keV)
19.96	29.5
37.02	29.8
3.65	33.6
7.11	33.6
2.37	34.4
7.50	37.6
keV/decay	24.5

$^{131}_{53}$I   (8.04 d)

Radioactive Daughter   Half-Life

^{131m}Xe   (1.1%)   11.84 d

Principal Beta Particles

Freq. (%)	E_{av} (keV)	E_{max} (keV)
2.13	69.4	247.9
7.36	96.6	333.8
89.41	191.5	606.3
keV/decay	181.7	

Conversion/Auger Electrons

Freq. (%)	E (keV)
3.63	45.6
1.55	329.9
keV/decay	10.1 *

Principal Gamma and X Rays

Freq. (%)	E (keV)
1.40	29.5
2.59	29.8
2.62	80.2
6.06	284.3
81.24	364.5
7.27	637.0
1.80	722.9
keV/decay	381.6

$^{133}_{53}$I   (20.8 h)

Radioactive Daughter   Half-Life
^{133}Xe   (97.1%)   5.25 d

Principal Beta Particles

Freq. (%)	E_{av} (keV)	E_{max} (keV)
1.24	109.8	373.8
3.74	139.7	461.8
3.12	161.5	523.6
4.15	298.5	884.7
1.81	351.6	1016.2
83.52	440.6	1230.1
1.07	573.2	1526.8
keV/decay	406.4	

Conversion/Auger Electrons
keV/decay   4.28 *

Principal Gamma and X Rays

Freq. (%)	E (keV)
1.81	510.5
86.31	529.9
1.49	706.6
1.23	856.3
4.47	875.3
1.49	1236.4
2.33	1298.2
keV/decay	607.2 *

$^{135}_{53}$I   (6.61 h)

Radioactive Daughter   Half-Life
^{135}Xe   (83.5%)   9.09 h

Principal Beta Particles

Freq. (%)	E_{av} (keV)	E_{max} (keV)
1.04	86.4	302.4
1.41	103.1	353.8
4.79	137.5	455.5
7.42	145.4	478.0
1.59	195.9	618.1
1.11	213.4	665.1
8.00	242.9	742.7
8.81	312.5	919.8
21.77	358.4	1032.9
8.10	405.1	1145.7
7.70	450.4	1253.4
24.10	534.9	1450.6
1.11	591.1	1579.5
keV/decay	363.9	

Conversion/Auger Electrons
keV/decay   3.19 *

more ...

$^{135}_{53}$I   (continued)

Principal Gamma and X Rays

Freq. (%)	E (keV)
1.75	220.5
3.09	288.5
3.52	417.6
7.13	546.6
6.67	836.8
1.20	972.6
7.93	1038.8
1.60	1101.6
3.61	1124.0
22.53	1131.5
28.63	1260.4
8.65	1457.6
1.07	1502.8
1.29	1566.4
9.53	1678.0
4.09	1706.5
7.70	1791.2
keV/decay	1576.3 *

$^{127}_{54}$Xe   (36.4 d)

Conversion/Auger Electrons

Freq. (%)	E (keV)
1.18	23.0
1.06	23.3
3.08	23.6
1.33	23.9
3.93	24.4
1.15	27.1
1.50	27.7
1.54	112.1
3.65	139.0
6.62	169.7
keV/decay	32.4 *

Principal Gamma and X Rays

Freq. (%)	E (keV)
24.98	28.3
46.43	28.6
4.55	32.2
8.82	32.3
2.83	33.1
1.23	57.6
4.29	145.3
25.54	172.1
68.30	202.9
17.21	375.0
keV/decay	280.1

$^{133m}_{54}$Xe (2.188 d)

Radioactive Daughter   Half-Life
^{133}Xe   (100.0%)   5.25 d

Conversion/Auger Electrons

Freq. (%)	E (keV)
1.83	24.5
63.51	198.6
11.87	227.7
2.56	228.1
6.26	228.4
4.57	232.2
1.23	233.2
keV/decay	192.4

Principal Gamma and X Rays

Freq. (%)	E (keV)
16.05	29.5
29.78	29.8
2.94	33.6
5.72	33.6
1.91	34.4
9.99	233.2
keV/decay	40.8

$^{133}_{54}$Xe (5.245 d)

Principal Beta Particles

Freq. (%)	E_{av} (keV)	E_{max} (keV)
99.33	100.5	346.0
keV/decay	100.3	

Conversion/Auger Electrons

Freq. (%)	E (keV)
1.44	25.5
53.54	45.0
6.52	75.3
1.45	80.0
keV/decay	35.4 *

Principal Gamma and X Rays

Freq. (%)	E (keV)
13.71	30.6
25.34	31.0
2.52	34.9
4.89	35.0
1.70	35.8
37.42	81.0
keV/decay	46.1

$^{135m}_{54}$Xe (15.29 min)

Radioactive Daughter   Half-Life
^{133}Xe   (100.0%)   9.09 h

Principal Beta Particles
keV/decay   0.011 *

more ...

$^{135m}_{54}$Xe (continued)

Conversion/Auger Electrons

Freq. (%)	E (keV)
15.59	492.0
2.26	521.1
keV/decay	97.6 *

Principal Gamma and X Rays

Freq. (%)	E (keV)
3.94	29.5
7.31	29.8
1.40	33.6
80.66	526.6
keV/decay	429.0

$^{135}_{54}$Xe (9.09 h)

Radioactive Daughter   Half-Life
^{135}Cs   (100.0%) 2.3×10^6 y

Principal Beta Particles

Freq. (%)	E_{av} (keV)	E_{max} (keV)
3.12	170.7	549.8
96.00	307.4	908.2
keV/decay	302.1	

Conversion/Auger Electrons

Freq. (%)	E (keV)
5.64	213.8
keV/decay	15.1 *

Principal Gamma and X Rays

Freq. (%)	E (keV)
1.46	30.6
2.70	31.0
90.13	249.8
2.90	608.2
keV/decay	248.6

$^{138}_{54}$Xe (14.17 min)

Radioactive Daughter   Half-Life
^{138}Cs   (100.0%)   32.20 min

Principal Beta Particles

Freq. (%)	E_{av} (keV)	E_{max} (keV)
3.06	150.2	492.4
9.46	176.9	566.9
32.55	266.0	803.3
20.01	948.9	2379.8
13.34	966.2	2417.9
5.57	1036.7	2571.6
4.98	1139.5	2819.2
8.96	1148.5	2814.3
keV/decay	652.3	

more ...

$^{138}_{54}$Xe   (continued)

Conversion/Auger Electrons

Freq. (%)	E (keV)
2.56	10.9
1.74	117.8
1.80	222.3
keV/decay	20.3 *

Principal Gamma and X Rays

Freq. (%)	E (keV)
1.15	30.6
2.12	31.0
5.95	153.7
3.50	242.6
31.50	258.3
6.30	396.4
2.17	401.4
20.32	434.5
1.47	1114.3
16.73	1768.3
1.42	1850.9
5.35	2004.7
12.25	2015.8
1.44	2079.2
2.29	2252.3
keV/decay	1125.0 *

$^{129}_{55}$Cs   (32.06 h)

Conversion/Auger Electrons

Freq. (%)	E (keV)
1.33	23.9
1.19	24.2
3.38	24.5
1.47	24.9
1.29	28.1
1.68	28.8
3.81	34.1
keV/decay	17.8 *

Principal Gamma and X Rays

Freq. (%)	E (keV)
29.74	29.5
55.18	29.8
5.45	33.6
10.60	33.6
3.53	34.4
2.99	39.6
1.33	278.6
2.46	318.2
30.80	371.9
22.45	411.5
3.42	548.9
keV/decay	281.2

$^{130}_{55}$Cs   (29.9 min)

Positrons

Freq. (%)	E_{av} (keV)	E_{max} (keV)
44.02	893.0	1988.0
keV/decay	397.2	

Conversion/Auger Electrons

Freq. (%)	E (keV)
1.36	24.5
keV/decay	3.89 *

Principal Gamma and X Rays

Freq. (%)	E (keV)
11.95	29.5
22.16	29.8
2.19	33.6
4.26	33.6
1.42	34.4
89.29	511.0
4.10	536.1
keV/decay	516.7

$^{131}_{55}$Cs   (9.69 d)

Conversion/Auger Electrons

Freq. (%)	E (keV)
2.39	24.5
1.04	24.9
1.19	28.8
keV/decay	6.60 *

Principal Gamma and X Rays

Freq. (%)	E (keV)
21.02	29.5
38.99	29.8
3.85	33.6
7.49	33.6
2.50	34.4
keV/decay	22.8

$^{134m}_{55}$Cs   (2.91 h)

Radioactive Daughter   Half-Life
^{134}Cs   (100.0%)   2.062 y

Conversion/Auger Electrons

Freq. (%)	E (keV)
16.03	10.2
3.92	11.2
35.18	91.5
3.01	121.8
19.60	122.1
18.33	122.5
9.13	126.5
2.30	127.5
keV/decay	112.4 *

more ...

$^{134m}_{55}$Cs (continued)

Principal Gamma and X Rays

Freq. (%)	E (keV)
9.03	30.6
16.68	31.0
1.66	34.9
3.22	35.0
1.12	35.8
12.70	127.5
keV/decay	27.1

$^{134}_{55}$Cs (2.062 y)

Principal Beta Particles

Freq. (%)	E_{av} (keV)	E_{max} (keV)
27.28	23.1	88.6
2.47	123.4	415.2
69.80	210.1	657.9
keV/decay	157.4	

Conversion/Auger Electrons

keV/decay   6.83 *

Principal Gamma and X Rays

Freq. (%)	E (keV)
1.46	475.4
8.38	563.2
15.43	569.3
97.56	604.7
85.44	795.8
8.73	801.9
1.80	1167.9
3.04	1365.2
keV/decay	1555.3

$^{137}_{55}$Cs (30.0 y)

Radioactive Daughter	Half-Life
^{137}Ba (94.6%)	2.552 min

Principal Beta Particles

Freq. (%)	E_{av} (keV)	E_{max} (keV)
94.43	173.4	511.5
5.57	424.6	1173.2
keV/decay	187.4	

$^{131}_{56}$Ba (11.8 d)

Radioactive Daughter	Half-Life
^{131}Cs (100.0%)	9.69 d

Positrons

keV/decay 6.76×10^{-4}

more ...

$^{131}_{56}$Ba (continued)

Conversion/Auger Electrons

Freq. (%)	E (keV)
1.17	24.8
1.05	25.1
2.92	25.5
1.27	25.8
1.14	29.2
1.48	29.9
1.17	42.8
18.13	87.8
1.69	118.1
2.08	118.4
2.23	118.8
1.28	122.8
1.83	180.1
keV/decay	45.8 *

Principal Gamma and X Rays

Freq. (%)	E (keV)
27.72	30.6
51.24	31.0
5.10	34.9
9.89	35.0
3.43	35.8
29.05	123.8
2.19	133.6
19.90	216.1
2.41	239.6
2.81	249.4
13.33	373.3
1.29	404.0
1.89	486.5
43.78	496.3
1.23	585.0
1.57	620.0
1.19	1047.6
keV/decay	458.9

$^{133m}_{56}$Ba (38.9 h)

Radioactive Daughter	Half-Life
^{133}Ba (100.0%)	10.74 y

Conversion/Auger Electrons

Freq. (%)	E (keV)
15.85	11.2
5.36	12.3
1.47	26.4
59.24	238.6
10.87	270.1
2.31	270.5
4.88	270.8
4.04	275.0
1.16	276.1
keV/decay	221.3 *

more ...

$^{133m}_{56}$Ba (continued)

Principal Gamma and X Rays

Freq. (%)	E (keV)
1.39	12.3
15.18	31.8
27.95	32.2
2.80	36.3
5.42	36.4
1.96	37.3
17.51	276.1
keV/decay	66.8

$^{135m}_{56}$Ba (28.7 h)

Conversion/Auger Electrons

Freq. (%)	E (keV)
1.50	26.4
60.16	230.8
11.17	262.2
2.39	262.6
5.22	263.0
4.21	267.2
1.21	268.2
keV/decay	208.1

Principal Gamma and X Rays

Freq. (%)	E (keV)
15.41	31.8
28.39	32.2
2.84	36.3
5.51	36.4
1.99	37.3
15.64	268.2
keV/decay	60.1

$^{137m}_{56}$Ba (2.552 min)

Conversion/Auger Electrons

Freq. (%)	E (keV)
8.32	624.2
1.19	655.7
keV/decay	65.1 *

Principal Gamma and X Rays

Freq. (%)	E (keV)
2.13	31.8
3.92	32.2
89.78	661.6
keV/decay	596.5

$^{140}_{56}$Ba (12.74 d)

Radioactive Daughter

	Half-Life
^{140}La (100.0%)	40.27 h

Principal Beta Particles

Freq. (%)	E_{av} (keV)	E_{max} (keV)
24.60	136.5	453.9
9.82	176.6	567.4
3.80	305.6	872.4
38.86	339.6	991.2
22.92	356.9	1005.0
keV/decay	276.3	

Conversion/Auger Electrons

Freq. (%)	E (keV)
10.82	12.7
2.57	13.9
53.82	23.7
5.62	24.1
2.45	24.5
12.80	28.8
3.45	30.0
1.48	123.7
keV/decay	36.5 *

Principal Gamma and X Rays

Freq. (%)	E (keV)
1.22	13.9
13.78	30.0
1.01	33.4
6.21	162.6
4.30	304.9
3.15	423.7
1.93	437.6
24.39	537.3
keV/decay	182.7

$^{140}_{57}$La (40.272 h)

Principal Beta Particles

Freq. (%)	E_{av} (keV)	E_{max} (keV)
11.12	440.8	1238.2
5.71	443.2	1243.9
5.61	465.0	1295.7
44.87	487.1	1347.7
5.11	514.4	1411.8
20.83	629.2	1676.5
4.81	845.8	2163.5
keV/decay	525.8	

Conversion/Auger Electrons

keV/decay	9.33 *

more ...

$^{140}_{57}$La   (continued)

Principal Gamma and X Rays

Freq. (%)	E (keV)
1.05	34.7
20.74	328.8
2.99	432.5
45.94	487.0
4.41	751.8
23.64	815.8
5.59	867.8
2.68	919.6
7.05	925.2
95.40	1596.5
3.43	2521.7
keV/decay	2315.9

$^{141}_{58}$Ce   (32.50 d)

Principal Beta Particles

Freq. (%)	E_{av} (keV)	E_{max} (keV)
70.20	130.0	435.9
29.80	181.1	581.3
keV/decay	145.2	

Conversion/Auger Electrons

Freq. (%)	E (keV)
18.71	103.5
2.35	138.6
keV/decay	25.5 *

Principal Gamma and X Rays

Freq. (%)	E (keV)
4.86	35.6
8.87	36.0
1.76	40.7
48.20	145.4
keV/decay	76.5

$^{144}_{58}$Ce   (284.3 d)

Radioactive Daughter   Half-Life
^{144}Pr   (98.6%)   17.28 min

Principal Beta Particles

Freq. (%)	E_{av} (keV)	E_{max} (keV)
19.58	49.4	181.9
4.60	65.3	235.3
75.82	90.2	315.4
keV/decay	81.1	

Conversion/Auger Electrons

Freq. (%)	E (keV)
3.48	38.1
5.33	91.5
keV/decay	11.1 *

more ...

$^{144}_{58}$Ce   (continued)

Principal Gamma and X Rays

Freq. (%)	E (keV)
2.96	35.6
5.40	36.0
1.07	40.7
1.64	80.1
10.80	133.5
keV/decay	20.7 *

$^{144}_{59}$Pr   (17.28 min)

Principal Beta Particles

Freq. (%)	E_{av} (keV)	E_{max} (keV)
1.08	266.9	811.3
1.17	894.7	2300.5
97.74	1221.7	2997.0
keV/decay	1207.5	

Conversion/Auger Electrons
keV/decay   0.056 *

Principal Gamma and X Rays

Freq. (%)	E (keV)
1.48	696.5
keV/decay	31.9 *

$^{147}_{61}$Pm   (2.6234 y)

Radioactive Daughter   Half-Life
^{147}Sm   (100.0%) 6.9×10^9 y

Principal Beta Particles

Freq. (%)	E_{av} (keV)	E_{max} (keV)
99.99	62.0	224.7
keV/decay	62.0	

Conversion/Auger Electrons
keV/decay   0.003 *

Principal Gamma and X Rays
keV/decay   0.004 *

$^{169}_{70}$Yb (32.022 d)

Conversion/Auger Electrons

Freq. (%)	E (keV)
7.14	10.6
1.89	18.9
8.38	34.2
1.26	39.5
2.09	40.9
1.12	47.3
1.35	48.8
34.85	50.4
3.90	53.0
1.43	53.5
1.79	54.5
1.33	58.8
1.58	61.2
6.07	71.1
1.15	83.5
4.78	99.7
1.26	107.9
10.30	117.8
2.44	120.9
2.14	121.9
1.25	128.6
12.79	138.6
1.39	167.1
1.74	187.8
keV/decay	123.2 *

Principal Gamma and X Rays

Freq. (%)	E (keV)
53.00	49.8
93.73	50.7
9.94	57.3
19.17	57.5
8.20	59.1
43.74	63.1
2.66	93.6
17.37	109.8
1.88	118.2
11.11	130.5
21.45	177.2
34.94	198.0
1.90	261.1
10.80	307.7
keV/decay	311.4

$^{198}_{79}$Au (2.696 d)

Principal Beta Particles

Freq. (%)	E_{av} (keV)	E_{max} (keV)
1.30	79.4	284.8
98.68	314.5	960.7
keV/decay	311.5	

Conversion/Auger Electrons

Freq. (%)	E (keV)
2.88	328.7
keV/decay	15.5 *

Principal Gamma and X Rays

Freq. (%)	E (keV)
1.38	70.8
95.51	411.8
1.06	675.9
keV/decay	404.4

$^{197}_{80}$Hg (64.1 h)

Conversion/Auger Electrons

Freq. (%)	E (keV)
8.88	11.1
3.57	11.2
2.93	11.8
1.42	13.0
31.98	63.0
14.91	63.6
10.91	65.4
14.03	74.6
4.39	77.3
keV/decay	66.5 *

Principal Gamma and X Rays

Freq. (%)	E (keV)
1.81	11.2
12.84	11.4
4.39	11.6
2.31	11.6
2.57	13.4
20.47	67.0
34.96	68.8
18.00	77.3
4.14	77.6
7.94	78.0
3.27	80.2
keV/decay	70.3 *

$^{203}_{80}$Hg (46.612 d)

Principal Beta Particles

Freq. (%)	E_{av} (keV)	E_{max} (keV)
100.00	57.8	212.6
keV/decay	57.8	

Conversion/Auger Electrons

Freq. (%)	E (keV)
13.80	193.7
2.07	263.8
1.05	264.5
keV/decay	41.2 *

Principal Gamma and X Rays

Freq. (%)	E (keV)
2.17	10.3
1.62	12.2
3.86	70.8
6.53	72.9
1.49	82.6
81.46	279.2
keV/decay	238.0

$^{201}_{81}$Tl (73.1 h)

Conversion/Auger Electrons

Freq. (%)	E (keV)
7.09	11.5
3.23	11.5
1.55	12.1
1.14	13.4
7.30	15.8
6.28	17.4
1.90	27.8
1.63	29.3
7.47	52.2
15.44	84.3
1.14	120.5
2.35	152.6
keV/decay	43.4 *

Principal Gamma and X Rays

Freq. (%)	E (keV)
1.21	11.6
10.55	11.8
4.19	11.9
1.51	12.0
2.15	13.8
27.15	68.9
46.18	70.8
5.48	79.8
10.53	80.3
4.43	82.6
2.65	135.3
10.00	167.4
keV/decay	93.5

$^{204}_{81}$Tl (3.779 y)

Principal Beta Particles

Freq. (%)	E_{av} (keV)	E_{max} (keV)
97.42	243.9	763.4
keV/decay	237.6	

Conversion/Auger Electrons

keV/decay	0.229 *

Principal Gamma and X Rays

keV/decay	1.13 *

$^{208}_{81}$Tl (3.053 min)

Principal Beta Particles

Freq. (%)	E_{av} (keV)	E_{max} (keV)
3.11	340.9	1033.0
24.66	439.4	1285.6
21.94	533.2	1518.9
49.01	647.2	1796.3
keV/decay	557.1	

Conversion/Auger Electrons

Freq. (%)	E (keV)
1.27	10.1
2.84	189.4
1.97	422.8
1.28	495.2
keV/decay	53.9 *

Principal Gamma and X Rays

Freq. (%)	E (keV)
1.23	10.6
2.23	72.8
3.76	75.0
6.31	277.4
22.61	510.8
84.48	583.2
1.81	763.1
12.42	860.6
99.16	2614.5
keV/decay	3360.2

$^{210}_{82}$Pb (22.3 y)

Radioactive Daughter   Half-Life

^{210}Bi (100.0%)	5.013 d

Alpha Particles

keV/decay	7.07×10^{-5}

Principal Beta Particles

Freq. (%)	E_{av} (keV)	E_{max} (keV)
80.00	4.14	16.5
20.00	16.1	63.0
keV/decay	6.54	

more ...

$^{210}_{82}$Pb (continued)

Conversion/Auger Electrons

Freq. (%)	E (keV)
8.25	10.3
2.94	10.6
1.42	12.5
2.39	12.6
2.14	13.3
51.90	30.1
5.43	30.8
13.60	43.3
4.60	46.5
keV/decay	31.4 *

Principal Gamma and X Rays

Freq. (%)	E (keV)
8.37	10.8
2.13	12.7
2.09	13.0
4.11	13.0
2.46	13.2
4.05	46.5
keV/decay	4.81 *

$^{212}_{82}$Pb (10.64 h)

Radioactive Daughter
	Half-Life
^{212}Bi (100.0%)	50.55 min

Principal Beta Particles

Freq. (%)	E_{av} (keV)	E_{max} (keV)
5.08	41.9	157.5
82.98	94.4	334.2
11.77	172.6	572.8
keV/decay	101.0	

Conversion/Auger Electrons

Freq. (%)	E (keV)
5.76	10.3
2.19	12.6
3.42	24.7
32.32	148.1
1.29	209.6
5.04	222.2
1.32	235.5
keV/decay	74.4 *

more ...

$^{212}_{82}$Pb (continued)

Principal Gamma and X Rays

Freq. (%)	E (keV)
5.84	10.8
1.46	13.0
3.75	13.0
10.42	74.8
17.51	77.1
2.08	86.8
3.99	87.3
1.84	89.9
43.65	238.6
3.34	300.1
keV/decay	145.2

$^{214}_{82}$Pb (26.8 min)

Radioactive Daughter
	Half-Life
^{214}Bi (100.0%)	19.9 min

Principal Beta Particles

Freq. (%)	E_{av} (keV)	E_{max} (keV)
2.52	49.7	184.9
48.10	207.2	672.1
42.12	227.4	728.8
6.29	336.7	1024.0
keV/decay	219.2	

Conversion/Auger Electrons

Freq. (%)	E (keV)
5.06	10.3
1.79	12.6
9.50	36.8
2.48	50.0
5.32	151.4
7.43	204.6
9.48	261.4
1.16	278.8
1.47	335.5
keV/decay	74.0 *

Principal Gamma and X Rays

Freq. (%)	E (keV)
5.13	10.8
1.28	13.0
3.08	13.0
1.10	53.2
6.39	74.8
10.73	77.1
1.28	86.8
2.45	87.3
1.13	89.9
7.46	241.9
19.17	295.2
37.06	351.9
1.09	785.9
keV/decay	249.8 *

$^{210}_{83}$Bi (5.013 d)

Radioactive Daughter   Half-Life
^{210}Po   (100.0%)   138.7 d

Alpha Particles
keV/decay   0.006 *

Principal Beta Particles

Freq. (%)	E_{av} (keV)	E_{max} (keV)
100.00	388.9	1161.4
keV/decay	388.9	

$^{212}_{83}$Bi (60.55 min)

Radioactive Daughter   Half-Life
^{212}Po   (64.1%) 2.98×10^{-7} s

Alpha Particles

Freq. (%)	E (keV)
25.24	6050.9
9.64	6090.0
keV/decay	2174.7

Principal Beta Particles

Freq. (%)	E_{av} (keV)	E_{max} (keV)
1.89	190.5	625.4
1.46	228.6	733.3
4.49	530.6	1518.8
55.30	831.5	2246.0
keV/decay	491.9	

Conversion/Auger Electrons

Freq. (%)	E (keV)
18.25	24.5
1.87	25.2
4.73	36.9
1.57	39.9
keV/decay	11.7 *

Principal Gamma and X Rays

Freq. (%)	E (keV)
2.82	10.3
1.50	12.2
1.09	39.9
6.65	727.2
1.11	785.4
1.51	1620.6
keV/decay	106.3 *

$^{214}_{83}$Bi (19.9 min)

Radioactive Daughter   Half-Life
^{214}Po   (100.0%) 1.643×10^{-4} s

Alpha Particles
keV/decay   1.15 *

Principal Beta Particles

Freq. (%)	E_{av} (keV)	E_{max} (keV)
1.04	248.1	787.5
2.78	260.8	822.3
5.48	352.0	1065.9
4.31	385.0	1151.4
2.48	424.5	1252.7
1.49	427.0	1259.2
1.18	433.4	1275.4
1.57	474.8	1379.7
8.28	491.9	1422.6
17.60	525.2	1505.5
17.90	539.3	1540.4
3.32	615.2	1726.6
1.01	667.9	1854.5
7.52	683.6	1892.3
17.70	1268.6	3270.0
keV/decay	637.6	

Conversion/Auger Electrons
keV/decay   21.7 *

Principal Gamma and X Rays

Freq. (%)	E (keV)
46.09	609.3
1.56	665.5
4.88	768.4
1.23	806.2
3.16	934.1
15.04	1120.3
1.69	1155.2
5.92	1238.1
1.47	1281.0
4.02	1377.6
1.39	1401.5
2.48	1408.0
2.19	1509.2
1.15	1661.3
3.05	1729.6
15.92	1764.5
2.12	1847.4
1.21	2118.5
4.99	2204.1
1.55	2447.7
keV/decay	1508.2 *

$^{210}_{84}\text{Po}$ (138.376 d)

Alpha Particles

Freq. (%)	E (keV)
100.00	5304.6
keV/decay	5304.5

Conversion/Auger Electrons
keV/decay 9.34×10^{-5} *

Principal Gamma and X Rays
keV/decay   0.010 *

$^{212}_{84}\text{Po}$ (2.98×10^{-7} s)

Alpha Particles

Freq. (%)	E (keV)
100.00	8784.3
keV/decay	8784.3

$^{214}_{84}\text{Po}$ (1.643×10^{-4} s)

Radioactive Daughter   Half-Life
^{210}Pb (100.0%)   22.3 y

Alpha Particles

Freq. (%)	E (keV)
99.99	7687.2
keV/decay	7687.1

Conversion/Auger Electrons
keV/decay 8.21×10^{-4} *

Principal Gamma and X Rays
keV/decay   0.083 *

$^{218}_{84}\text{Po}$ (3.05 min)

Radioactive Daughter   Half-Life
^{214}Pb (100.0%)   26.8 min

Alpha Particles

Freq. (%)	E (keV)
99.98	6002.6
keV/decay	6001.4

Principal Beta Particles
keV/decay   0.014 *

$^{220}_{86}\text{Rn}$ (55.6 s)

Radioactive Daughter   Half-Life
^{216}Po (100.0%)   0.145 s

Alpha Particles

Freq. (%)	E (keV)
99.93	6288.4
keV/decay	6288.0

Conversion/Auger Electrons
keV/decay   0.009 *

Principal Gamma and X Rays
keV/decay   0.386 *

$^{222}_{86}\text{Rn}$ (3.8235 d)

Radioactive Daughter   Half-Life
^{218}Po (100.0%)   3.05 min

Alpha Particles

Freq. (%)	E (keV)
99.92	5489.7
keV/decay	5489.3

Conversion/Auger Electrons
keV/decay   0.011 *

Principal Gamma and X Rays
keV/decay   0.399 *

$^{224}_{88}\text{Ra}$ (3.66 d)

Radioactive Daughter   Half-Life
^{220}Rn (100.0%)   55.6 s

Alpha Particles

Freq. (%)	E (keV)
4.90	5448.9
95.08	5685.6
keV/decay	5673.9

Conversion/Auger Electrons
keV/decay   2.22 *

Principal Gamma and X Rays

Freq. (%)	E (keV)
3.90	241.0
keV/decay	9.90 *

$^{226}_{88}\text{Ra}$ (1600 y)

Radioactive Daughter   Half-Life
^{222}Rn (100.0%)   3.824 d

Alpha Particles

Freq. (%)	E (keV)
5.55	4601.8
94.44	4784.5
keV/decay	4774.4

Conversion/Auger Electrons
keV/decay   3.59 *

Principal Gamma and X Rays

Freq. (%)	E (keV)
3.28	186.0
keV/decay	6.75 *

$^{241}_{95}$Am   (432.2 y)

Alpha Particles

Freq. (%)	E (keV)
1.40	5388.4
12.80	5443.1
85.19	5485.8
keV/decay	5479.1

Conversion/Auger Electrons

Freq. (%)	E (keV)
22.30	10.6
11.59	10.8
3.64	11.6
16.04	13.1
7.94	13.4
2.29	14.2
20.61	14.6
2.19	15.6
2.71	16.0
11.87	16.1
1.38	16.9
5.49	17.1
1.72	17.9
2.74	21.0
3.50	21.8
3.68	21.9
2.79	25.8
1.44	26.3
12.01	27.7
4.38	28.8
4.76	32.2
1.73	33.2
3.24	34.1
11.06	37.1
10.06	37.9
1.24	38.5
2.40	39.0
10.10	41.9
7.72	55.1
2.71	59.5
keV/decay	58.6 *

$^{241}_{95}$Am   (continued)

Principal Gamma and X Rays

Freq. (%)	E (keV)
1.38	11.9
2.75	13.8
24.59	13.9
6.53	16.8
3.19	17.1
1.41	17.5
20.02	17.8
2.82	18.0
4.86	20.8
1.33	21.1
1.22	21.3
1.03	21.5
2.41	26.3
35.90	59.5
keV/decay	34.1

more ...

Index